LEARNING RESOURCES CTR/NEW ENGLAND TECH.
S0-CJE-388

MODERN PARTICLEBOARD & DRY-PROCESS FIBERBOARD MANUFACTURING

MODERN PARTICLEBOARD & DRY-PROCESS FIBERBOARD MANUFACTURING

THOMAS M. MALONEY

UPDATED EDITION

Miller Freeman Inc.
San Francisco

Library of Congress Catalog Card Number: 76-47094
International Standard Book Number: 0-87930-288-7

Printed in the United States of America
93 94 95 96 97 5 4 3 2 1

CONTENTS

3. Modern Processing Systems 87

4. Composition Board Materials: Properties and Testing 129

5. Parameters Affecting Board Properties 158

6. Raw Materials and Particle Geometry: Effects on Board Properties 178

7. Particle Generation, Conveying, and Storage 217

8. Drying Principles and Practices 282

9. Moisture Measurement and Control 306

10. Particle Separation: Principles and Equipment 333

11. Fire and Explosions: Prevention and Detection 352

12. Resins and Other Additives 367

13. Resin/Wax Applications and Blenders 413

14. Caul and Caulless Systems 458

15. Mat Forming and Formers 478

16. Prepressing 507

17. Hot Pressing and Presses 513

18. Finishing Board 573

19. New Developments 626

PREFACE

This book on composite manufacture was made possible only by the many professionals who so generously contributed their time in preparing presentations for 26 symposia on particleboard/composite materials held at Washington State University from 1967 through 1992. They are the true authors of this book and it is dedicated to them.

One of the original objectives of the symposia was the consolidation of the information presented in the series into one general book on the subject. To help meet this objective an outline was prepared for each symposium. The earlier outlines, in particular, were prepared in great detail in order to develop comprehensive information on the basics of board manufacture. Another reason for holding the symposia on particleboard was to "research" the experiences of many people in the field. Competent people with appropriate practical experiences were sought out and persuaded to talk and write about these experiences to complement the research and development papers presented. Such factual information is extremely difficult to obtain and these contributions have been most important to the success of the symposium series.

For detailed study, therefore, the several volumes of proceedings of the Washington State University symposia on particleboard/composite materials are recommended. In an attempt to tell the full story I have also included information from other publications. My apologies to the many authors if I have misconstrued any of their papers in preparing the text.

There is so much information available that it is difficult to include all of it in a single book. This effort attempts to highlight the important facets of the industry. Some may disagree with the author's selection of equipment and processes to be discussed in detail and those to be mentioned briefly or omitted entirely. However, for anyone wanting to enter the industry, this volume should provide basic information and background for asking detailed, intelligent questions before embarking on such a new venture.

The emphasis is on United States practices; however, I have included information from other parts of the world to try to provide different viewpoints on how to produce and handle composites.

Since 1977, when this book was originally published, great advances have taken place. The basic principles of manufacturing have remained the same, but enormous amounts of research have been done, new products and processes have been developed, and the true golden age of what we now call "composites" has started. Most of these developments are highlighted in Chapter 19 of this book. The future of the forest products industry, I truly believe, is in composites.

Thomas M. Maloney
Pullman, Washington
October, 1992

ACKNOWLEDGMENTS

A number of people at Washington State University have been of great assistance over the years in making the research efforts in the University's College of Engineering fruitful and in their contributions to the success of the particleboard symposia. Leadership has been exercised by Dean Carl W. Hall; Dean John P. Spielman (retired); the Assistant Dean of Research, George G. Marra, who for many years was Section Head of Wood Technology; the Director of the Division of Industrial Research, Dr. Eugene W. Greenfield (retired); and the Assistant Dean of Engineering Extension, William H. Knight.

Researchers on the staff who have made valuable contributions include Robert J. Hoyle, Jr., Roy F. Pellerin, M. D. Strickler, and John W. Talbott. Others on the staff who have contributed are Martin T. Lentz, Dayton A. Brewer, Gilmore A. Anderson, Nils A. Nilson, Bruce L. Shelton, Florence Jamison, Allan J. Ruddy, Herbert D. Howard, William R. Hawkins, and Glenn E. Sprouse. Many secretaries have provided invaluable help over the years. They include Esther Caprez, Patricia K. H. Baga, and Sue Vickerman. In particular, I wish to thank Kathleen N. Gill and Leslie Perri for their assistance in preparing this book.

Many book authors acknowledge their families, and this author will not be an exception. His family put up with the loss of one of its members for many evenings, weekends, and holidays while he worked on the book. The author's family includes a great wife, Donna, and (from a father's viewpoint) three marvelous children: William, Carol, and Joseph.

A number of individuals reviewed parts of this book for technical content and accuracy. Their kind assistance is deeply appreciated, and I wish to acknowledge their contributions. Reviewers included:

Laurel J. Allen, Plant Manager, Georgia-Pacific Corporation; Arnie Anson, Sales Manager, Clarke's Sheet Metal, Inc.; Howard B. Berrong, Product Manager, Mobil Oil Corporation; Michael K. Best, General Manager, Treated Fiber Products Division, Reichhold Chemicals, Inc.; H. H. Burkitt, The Rust Engineering Company; Craig C. Campbell, Sales Engineer, Littleford Bros., Inc.; Tony C. Carlson, Manager, Western Regional Sales, M-E-C Company; Bengt J. Carlsson, Chief Engineer, AB Motala Verkstad; Robert Carter, Production Superintendent, Boise Cascade Corporation; Elmer Christensen, Senior Project Engineer, Black Clawson, Inc.

Joseph B. Dede, Laboratory Manager, Chemical Division-Resins, Georgia-Pacific Corporation; Michael L. Dewey, Chemist, Koppers Company, Inc.; Roland Etzold, Engineer, Elmendorf Research Inc.; Karl Fischer, Prokurist, Pallmann KG; David C. Geisler, Specialist, Composite Panels, Weyerhaeuser Company; Walter H. Holt, Manager, Air Cleaning Division, Buffalo Forge Company; Julius S. Impellizzeri, President, Elmendorf Research Inc.; E. A. Jakaitis, Supervising Engineer, Technical Service Laboratory, Mobil Oil.

William F. Lehmann, Project Leader, Structural Particle Products, U.S. Forest Products Laboratory; Robert T. Leitner, Manager, Process Development, U.S. Plywood; Steve Mays, General Manager, Clarke's Sheet Metal, Inc.; Earl T. McCarthy, Owner, McCarthy Products Company; William M. McNeil, Consultant; George E. Mirous, Group Leader, Product Applications Specialist, Pacific Resins & Chemicals, Inc.; Charles R. Morschauser, Technical Director, National Particleboard Association; Robert G. Murdock, Western Manager, Mereen Johnson Machine Company.

Roy F. Pellerin, Associate Mechanical Engineer, Washington State University; John A. Phelps, Engineer, Northwest District Sales, Anacon Inc.; Irving L. Plough, Project Manager, Black Clawson, Inc.; Allan H. Quinby, Senior Engineer, Air Cleaning Division, Buffalo Forge Company; John Schuh, Manager, Particleboard, Boise Cascade Corporation; Robert D. Shumate, Chief Engineer, Clarke's Sheet Metal, Inc.; B. H. Shunk, Group Leader, Koppers Company, Inc.; Robert E. Silvis, Jr., Executive Vice President, Columbia Engineering International, Inc.; J. A. Stephenson, Account Executive, Forest Products, The Sherwin-Williams Company; Hans Sybertz, Jr., Technical Manager, Hombak Maschinenfabrik; Harry A. Raddin, President, Miller Hofft, Inc.

J. W. Talbott, Wood Technologist, Washington State University; Terry L. Twedt, Technical Director, Particleboard, Boise Cascade Corporation; Gerard P. Urling, Section Manager, Composite Panel and Molded Products Section, Weyerhaeuser Company; Thomas W. Vaughan, Secretary of the Corporation, Elmendorf Research Inc.; George D. Waters, Group Leader, The Borden Company; Irv Wentworth, Representative, Bison-Werke; Peter H. Wiecke, Vice President, Columbia Engineering International Ltd.; T. L. Wilson, Manager, Electronic Engineering, Votator Division, Chemetron.

A number of companies, laboratories, and individuals provided photographs of their equipment, and credit is given in photo captions. They are:

Allgaier-Werke GmbH; Alpine American Corp.; American Sheet Metal, Inc.; Atlas Systems Corporation; C-E Bauer; Becker & van Hüllen; Bezner Maschinenfabrik; Bison-Werke; Black Clawson, Inc.; Buffalo Forge Company; Büttner-Schilde-Haas AG; CAE Machinery Ltd.; CEA•Carter Day Company; Clark's Sheet Metal, Inc.; The Coe Manufacturing Company; Cothrell Machinery Sales, Ltd.; Draiswerke GmbH; Eastern Forest Products Laboratory, Canadian Forestry Service.

Fahrni Institute Ltd.; The Heil Company; Hombak Maschinenfabrik; I.T.P. Corporation; Keller-Peukert GmbH; Keystone Machine Works, Inc.; Littleford Brothers, Inc.; Otto Kreibaum Maschinenbaugesellschaft; Maschinenfabrik Rehau GmbH; McCarthy Products Corporation; M-E-C Company; Mereen Johnson Machine Company; Michigan State University; Miller Hofft, Inc.; Mitts

& Merrill; AB Motala Verkstad; North American Products Corp.; Pallmann KG; Ponndorf Maschinenfabrik KG; Potlatch Corporation; Rando Machine Corporation; Raute, Inc.; Redler Conveyors Ltd.; Edw. Renneburg & Sons Co.; Ribco Manufacturing Co., Inc.; Rotex Inc.; Saxlund A/S; Siempelkamp Corporation; Tallman Machinery Company; Fred W. Ter Louw; C-E Tyler Industrial Products; Underwriters Laboratories Inc.; Paul R. Walsh; Washington Iron Works; Weyroc Ltd.; Williams Patent Crusher & Pulverizer Company.

Technical journals, standards associations, industrial associations, trade publications, governmental organizations also assisted with providing information. These sources are identified in the text. Permission to use pertinent information was given by: American Board Products Association, *Brown Boveri Review*, Canadian Standards Association, the Food and Agriculture Organization of the United Nations, *Forest Products Journal*, *Holz als Roh-und Werkstoff*, National Particleboard Association, and *Wood & Wood Products*.

I would be remiss if I did not thank my publishers, Miller Freeman, Inc., and their very competent staff, in particular Peggy Boyer, the book editor assigned to this project. Over a period of several months, Peggy worked closely with me on the editing of this volume and her hard work and dedication to the project were invaluable to me and to the publishers.

Finally, I wish to express my thanks to Washington State University for permission to use the wealth of information contained in the proceedings of the many symposia held on particleboard and composite materials. The tremendous help of my good friend and colleague, Lucille Leonhardy, in editing the new material for this revised edition of the book is deeply appreciated.

T. M. M.

Following page: *press loader and hot press at the Post Falls, Idaho, particleboard plant. (Potlatch photo by Sue Reynolds, courtesy* Forest Industries.*)*

1. INTRODUCTION

The composition board industry, including hardboard, insulation board, particleboard (extruded and platen-pressed), medium-density fiberboard, cement-bonded board, and molded products, is of very recent origin. No one has been able to coin a good, descriptive term acceptable to everyone for this relatively new branch of the forest products industry. These products were conceived in the laboratory and brought to fruition through laboratory research endeavors as well as pilot plant development programs. Thus, these industries are tied very closely to science and technology.

Products made from comminuted woody materials in the shape of fiber, shavings, and the many other possible types of particles have special appeal, as they can be and mostly are made from woodworking waste, noncommercial or low-value tree species, and agricultural wastes. Future considerations indicate that bark, forest slash, industrial and municipal refuse, and other agriculture wastes will go into similar building products. This development is also ecologically sound, as previously wasted material is now and will be going into useful products, thereby conserving our natural resources. Relatively small amounts of energy and petrochemical derivatives are needed for this industry, further demonstrating its value to society in this day when people are becoming aware of the limitations of Earth's natural resources. Full use of nature's renewable resources such as wood places the forest products industry in a powerful position for maintaining and improving mankind's standard of living. No other building products industry has a resource base which can be maintained indefinitely.

Composition board products do, at present, require small amounts of other raw materials such as synthetic resin adhesives, but these are conserved because minor amounts are needed in a board or product consisting primarily of wood. Development of adhesive systems from paper-industry wastes or other plant sources now seems likely. It does not make ecological sense to produce buildings entirely of nonrenewable resources such as plastic or metal when wood is available. Our nonrenewable resources must be allotted carefully to provide the most for the common good; we must not use them if the renewable resource can provide the proper function.

Present commercial products of comminuted wood have some very desirable characteristics: availability in large sheets, smooth surfaces, uniformity in properties from sheet to sheet, and freedom from localized defects. However, such products have been largely excluded from primary structural uses because they have been unable to approach the longitudinal stiffness, dimensional stability, and long-term load-carrying ability of sawn lumber or plywood. Changes in construction design can accommodate many of the present products but this has not occurred to any extent yet.

The future situation seems likely to be dramatically changed. Combinations of recently developed methods of particle alignment, with the use of particles deliberately manufactured with optimum particle geometry, can yield materials equaling or surpassing the structural capability and reliability of sawn lumber and plywood. Another contributing factor is the ease with which properties of comminuted-wood products can be modified by treatment of the particles with fire retardants, preservatives, and stabilizing impregnations. Coupled with composition board's much higher efficiency of resource utilization, these factors must bring about extensive rethinking, reorganization, and realignment in the entire wood products industry.

Thus, while the quality of lumber and plywood trends downward as we enter the era of small-log, short-rotation forestry, the opportunity is opened for particle-based products of greatly improved structural utility to serve in much broader markets. This is in line with the continuing trend from labor-intensive processes to mechanized, automated, capital-intensive processes involving greater technological sophistication.

A particular advantage that platen-pressed, dry-process boards can have over other types of boards is unlimited size within practical consideration. Most plywood is 4x8 ft (1.22x2.44 m) in size. However, most particleboard and medium-density fiberboard is made in larger sheets for reasons which will be brought out later. The largest board made in the United States at present is 8x48 ft (2.44x14.63 m); however, a medium-density fiberboard plant will soon be producing board 8x65 ft (2.44x19.81 m). This size of press is already in use in Europe. Thus, board size can usually be determined by the end use rather than by the limitations of the board-producing equipment or the raw material, as with plywood.

This book will be about products made with comminuted wood. It will concentrate on the dry-process segment of the industry, as will be explained later. The dry-process board or products industry started its worldwide rapid expansion about 1949, as far as commercial production is concerned. The development of this industry will be traced later. The emphasis in this book will be on dry-process methods and technology for producing platen-pressed boards. Some discussion will cover extruded particleboard and molded products, but the main focus will be on those production processes used and products made in greatest volume in the United States and Canada.

Before discussing the various products made with comminuted wood, let us consider the wood elements that can be combined into a multitude of products. Dr. George G. Marra (1969) has eloquently spoken and written about these. His table of nonperiodic wood elements is presented in Figure 1.1. He shows 14 basic wood elements, of which 10 can be produced from residue wood or wood unsuitable for plywood and lumber. Seven or eight of these are of direct concern to the manufacture of composition board. The following discussion, which is of major importance to the forest products industry, is quoted from his treatise.

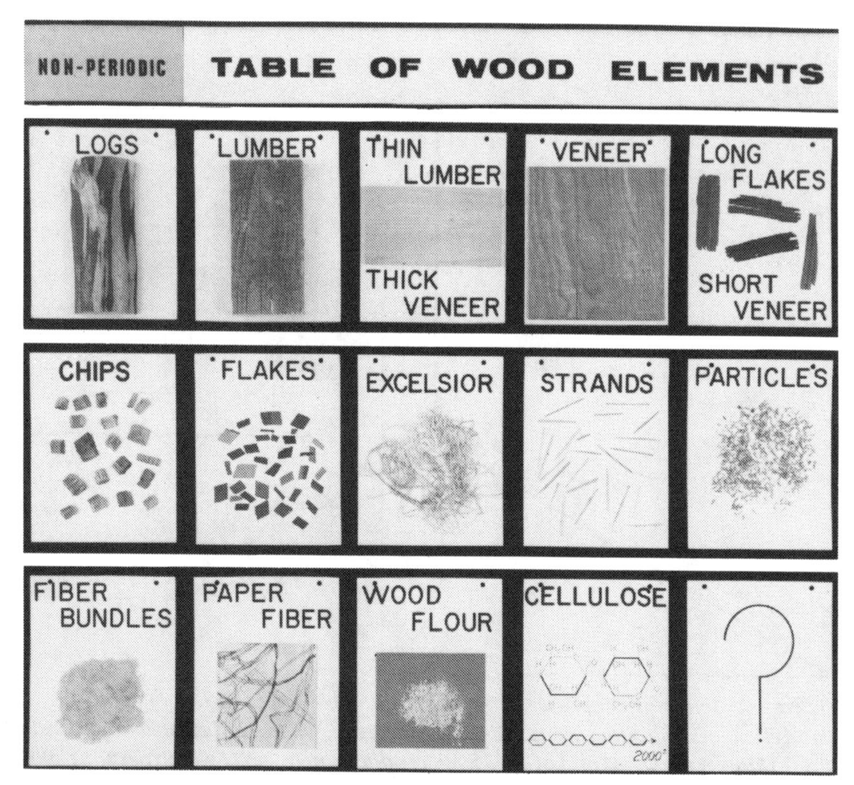

Figure 1.1. *Basic wood elements arranged from largest to smallest. Of the 14 elements shown, 10 can be produced from residue wood, or material unsuitable for lumber and plywood, and all can contribute to the development of new product concepts (Marra 1969).*

Hundreds of useful products can be produced from these elements individually in homogeneous fashion, and thousands can be produced from combinations of elements. Such factors as density, size of element, and wood species contribute greatly to the proliferation of products.

While a group of products may display differences in varying degrees, often the basic concept is the same. Therefore, these concepts, from which many products may spring for marketing pruposes, bear a special aura of importance and their genesis marks a significant innovative event. In order to analyze the output of past innovative effort and future potential, it is useful to consider each combination of elements as a distinct product concept. On the basis of this definition and the element shown in Figure 1 [see Figure 1.1], the number of possible concepts is seen to be mathematically limited. Taken in combinations of twos, for example, these elements would provide a maximum of 196 possible concepts, including homogeneous and seemingly redundant combinations. These redundant combinations are somewhat indeterminate in regard to concept status because of the wide difference in product properties resulting from varying proportions of each element, and because of the varying degree of deliberate orientation of which many elements are capable. Nevertheless, some real redundancy exists in this analytical scheme, but the main point I wish to make is not thereby invalidated.

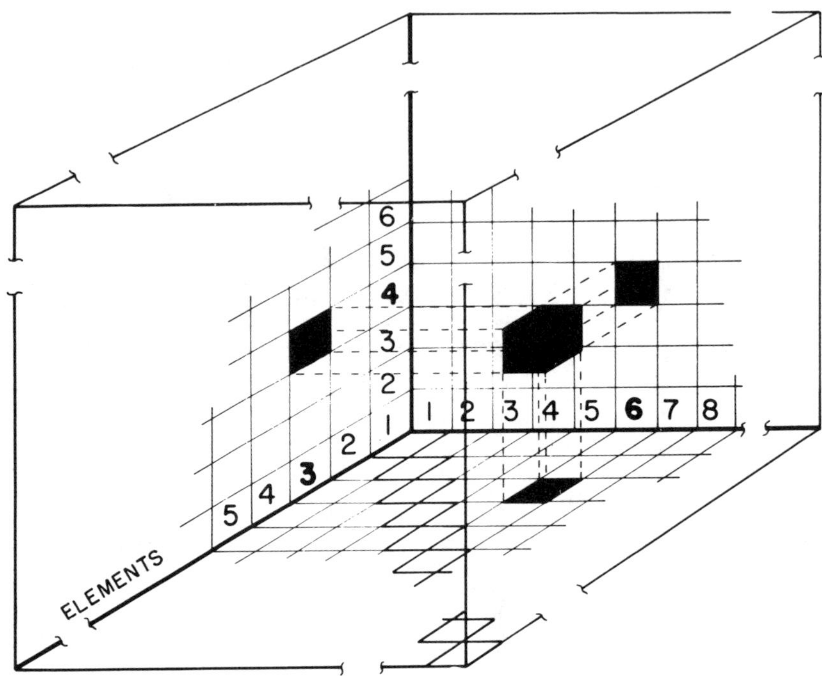

Figure 1.2. *Schematic representation of a three-element product concept. The 14 wood elements and 3 non-wood elements arranged in a three-dimensional array in this fashion generate a total of 4913 such product concepts, of which perhaps 3000 may have technical validity. The product concept shown, for example, contains element numbers 3, 4, and 6 (Marra 1969).*

A significant expansion in the number of possible product concepts occurs when three logical non-wood elements are included, namely, plastics, metals, and minerals. A very large increase in concepts occurs when both wood and non-wood elements are considered in combinations of threes. This is shown schematically in Figure 2 [see Figure 1.2]. On this basis, there is a theoretical maximum of 4913 concepts. However, with some discounting for redundancy and foolhardy combinations (i.e., logs and homogeneous non-wood), the theoretical maximum is probably nearer to 3000. Since each of these basic concepts can spawn a veritable galaxy of products, as some already have, it becomes clear that each such development is indeed an important event. Of course, other wood elements are likely to be discovered in the future, as indicated by the question mark in Figure 1, and these will further increase the concept possibilities. Bark, when it comes into its own, may develop another array of elements in similar fashion, and another array of products.

Analysis of current wood products reveals that most of them emerge from the application of relatively few basic concepts. For example, according to statistical summaries of the Department of Commerce, only three concepts account for practically the entire forest output of the State of Washington, and one of these, lumber, falls short of the status of a concept as I have defined it. . . This suggests, then, that if my premises are true, the State of Washington [for example] is operating on only 0.1% of the total possibilities. Actually, some other concepts such as laminated lumber, fiberboard, particleboard, and overlaid plywood are being utilized, which indicates a higher performance level, technically speaking. On a world basis, the operating level is probably about 20 time greater than this over-simplified example. According to my analysis, about 64 of the total of 3000 concepts are presently being used or being developed, or roughly 2%. The rest are waiting to be discovered and exploited by enterprising individuals.

Perhaps I have made a 100% error in tabulating existing concepts in use. If so, the percentage is about 4% instead of 2%, and one might well wonder why it isn't higher considering the pace of technological advances in other fields. Therefore at this moment, approximately 2872 concepts remain tantalizingly at large.

Without depreciating the value of single-element products, nor inflating the value of three-element products, it seems obvious that much progress lies beyond the horizon. In addition, we need to consider the tremendous potential inherent in particle products by virtue of the concept of molding to simple and complex shapes, and the concept of orientation of elements for directional strength. Furthermore, the unique susceptibility of all particle-type elements to special treatments for resistance to fire, insects, fungi, and moisture, adds a further dimension to product possibilities.

. . . people in the composition board industry are potentially in command of half of the basic wood elements known to man. It is unthinkable that [there has ever been] cause to complain about lack of opportunity for technical progress. . . .

It is rather tempting to cogitate on the 2872 product concepts awaiting discovery and to try artfully to select one for possible development. However, I believe that this would not be the most effective approach. Leading market analysts have long advocated that a thorough knowledge of the performance requirements of a proposed new product is necessary first in order to optimize the development effort and assure future market acceptance of the product. The sequence of action presumes that performance requirements can be translated into measurable properties, and these then scientifically designed into the product.

Wood, in general, has been slow to adjust to this principle, and understandably so because it is essentially a product of nature, and the initial processors have in the past been able to pass it on still largely in the form of a raw material with no essential change in its basic properties. However, in the hands of the particleboard technologist, wood can be made to absorb considerable amounts of science during processing—all in the interest of improved properties and better performance during ultimate use.

Here, then, lies one of the important keys to future industrial progress with wood. Given the service requirements of a product, add a knowledge of what elements contribute what properties, together with accessory knowledge of processing parameters, and it should be possible to zero in on an ideal product for the intended use.

The new fighting stance this kind of capability gives to wood may well be taken in stride and may be little noticed by our busy technologists and engineers. However, I believe that it is more than a pleasing academic platitude: It will be a whole new way of life for wood and it will embrace management attitudes as well as technical skill.

We are, of course, nowhere near [having] an adequate knowledge of wood elements to be able to make such optimistic advance guarantees of success. However, I believe that the trend of research effort is in that direction, or should be, and it is only a matter of time that such occurrences will be commonplace. The development of fine surfaces for certain types of particleboard in recent years is an excellent example of the operation of this principle.

DEFINITIONS

Composition boards have many different names and definitions, and they vary somewhat throughout the world. To start, let us define some of the products based on definitions set forth in the American Society for Testing and Materials (ASTM) Standard D 1554, "Standard Definitions of Terms Relating to Wood-Base Fiber and Particle Panel Materials," which covers U.S. practice. Terminology for worldwide use will then be reviewed.

General Definitions

Wood-base fiber and particle panel materials: A generic term applied to a group of board materials manufactured from wood or other lignocellulosic fibers or particles to which binding agents and other materials may be added during manufacture to obtain or improve certain properties. Composed of two broad types, fibrous-felted and particleboards.

Fibrous-felted board: A felted wood-base panel material manufactured of refined or partly refined lignocellulosic fibers characterized by an integral bond produced by an interfelting of fibers and, in the case of certain densities and control of conditions of manufacture, by ligneous bond, and to which other materials may have been added during manufacture to improve certain of its properties.

Particleboard: A generic term for a panel manufactured from lignocellulosic materials (usually wood), primarily in the form of discrete pieces or particles, as distinguished from fibers, combined with a synthetic resin or other suitable binder and bonded together under heat and pressure in a hot press by a process in which the entire interparticle bond is created by the added binder, and to which other materials may have been added during manufacture to improve certain properties. Particleboards are further defined by the method of pressing. When the pressure is applied in the direction perpendicular to the faces, as in a conventional multi-platen hot press, they are defined as flat-platen-pressed; and when the applied pressure is parallel to the faces, they are defined as extruded.

Wood-cement board: A panel material in which wood, usually in the form of excelsior, is bonded with inorganic cement.

Classification of Fibrous-Felted Boards

Structural insulating board: A generic term for a homogeneous panel made from lignocellulosic fibers (usually wood or cane) characterized by an integral bond produced by interfelting of the fibers, to which other materials may have been added during manufacture to improve certain properties, but which has not been consolidated under heat and pressure as a separate stage in manufacture, said board having a density of less than 31 lbs/ft^3 (specific gravity [sp gr] 0.50) but having a density of more than 10 lbs/ft^3 (sp gr 0.16).

Hardboard: A generic term for a panel manufactured primarily from interfelted lignocellulosic fibers (usually wood), consolidated under heat and pressure in a hot press to a density of 31 lbs/ft^3 (sp gr 0.50) or greater and to which other materials may have been added during manufacture to improve certain board properties.

Medium-density hardboard: A hardboard, as previously defined, in a density between 31 and 50 lbs/ft^3 (sp gr between 0.50 and 0.80).

High-density hardboard: A hardboard, as previously defined, with a density greater than 50 lbs/ft^3 (sp gr 0.80).

Classification of Particleboards

Low-density particleboard: A particleboard, as previously defined, with a density of less than 37 lbs/ft^3 (sp gr 0.59).

Medium-density particleboard: A particleboard, as previously defined, with a density between 37 and 50 lbs/ft^3 (sp gr between 0.59 and 0.80).

High-density particleboard: A particleboard, as previously defined, with a density greater than 50 lbs/ft^3 (sp gr 0.80).

By carefully studying these definitions, it can be seen that the terms are clearly defined. However, one of the most significant trends in the board industry is the elimination of the previously distinct differences between particleboard and hardboard, making it difficult to employ such clear definitions.

As noted above, hardboard is characterized as a product made of fibers. Most hardboards initially were of high density. In wet-process hardboard the lignocellulosic bonds hold the boards together, and it was not until the late 1940s or early 1950s that synthetic resins were used extensively in bonding fibers in the new dry-process hardboard industry. Particleboard, a dry-process product using a number of different types of particles, was at first usually made at low and medium densities. However, as both of these industries evolved, a wide overlapping has developed in their respective density ranges and particle types. Much fiber is now going into particleboard. Some very coarse fiber, which approaches a particle rather than a fiber, is employed in certain hardboards. In the previously mentioned ASTM D 1554, fibers are one type of element included in both the wood-base fiber and particle panel materials.

Certain hardboards such as the thick, medium-density board, are made primarily for the furniture market and are used interchangeably with particleboard. While usually classified as a hardboard, this medium-density board is marketed on the basis of the previous particleboard specifications for physical properties. In 1973 this board was referenced for the first time in PS58–73, which is a voluntary products standard for the Acoustical and Board Products Association (a recent merger of the American Hardboard Association and the Acoustical and Insulating Materials Association), covering the basic hardboards classed as tempered, standard, service-tempered, service, and industrialite. Industrialite is the new classification covering medium-density hardboard.

Also in 1973, the National Particleboard Association brought out the "Standard for Medium Density Fiberboard," NPA 4–73. The following definition is based on this standard:

Medium-density fiberboard: A dry formed panel product manufactured from lignocellulosic fibers combined with a synthetic resin or other suitable binder. The panels are compressed to a density of 31 to 50 lbs/ft^3 in a hot press by a process in which substantially the entire interfiber bond is created by the added binder. Other materials may have been added during manufacture to improve certain properties.

Presently, thin particleboard is marketed along with hardboard in the prefinished paneling market. Causing further confusion in definitions of board type is the product Cladwood, which is produced with a dry-process core and wet-process faces which are formulated as a fibrous slurry made from old newsprint. To complete the confusion, there is a Canadian process for making wet-felted particleboard by which the board can be composed of 85% particles and 15% fiber.

Consequently, the early distinct differences between these hardboards and particleboards are becoming more and more blurred, although both are still represented by separate and distinct associations.

It is not unreasonable to assume that the two groups of products at some time in the future may be jointly included under some new descriptive term such as *particle and fiber products, reconstituted products, composition products, engineered board,* or *wood-based panels*. Perhaps this latter classification is the most fruitful approach, as a family of related products will result which will be differentiated on the basis of properties rather than on the type of fiber or particle used.

Another recent development is the previously mentioned invention of a method for orienting fibers or other particles parallel to each other so that a board can be produced having far greater strength in one direction, compared with its other direction. Such a board can be comparable to lumber or plywood in properties. As such, it will be possible to use this board of oriented fibers for the face sheets on structural panels, in combination with a core of other types of particles, for a high-quality structural composition board. The distinction between types of boards therefore will be blurred further as composition board material moves more into the structural area.

Turning now to the world situation, the United Nations Economic Commission for Europe (ECE) and the Food and Agriculture Organization of the United Nations (FAO), in their "Classification and Definitions of Forest Products," present a number of definitions relative to wood-based panels which show the difficulty in pinpointing product descriptions. Wood-based panels (including similar panels from other lignocellulosic materials) cover plywood, particleboard, and fiberboard as a family. It is pointed out that fiberboard refers to fiber building board.

The product may be manufactured from wood in the form of solid wood, veneer, strands, particles, or fibers. Bonding agents and other materials may be added during manufacture to improve certain properties. The bonding agent can be an organic binder as in plywood and particleboard or may be inherent as in some fiberboards. The document goes on to state that the term *wood-based panel* is also applied to composite structures of the panel type such as cellular board (plywood of cellular construction) and those in which materials other than wood, such as foam plastic cores and plastic or metal faces or cores, constitute a small part of the whole material content, as well as to panels with an inorganic binder but in which wood or other lignocellulosic material constitutes the most important part of the product by volume, e.g., panel, mineral bonded, wood particle base. The interesting point with the panels made with inorganic binder is that volume is used in determining that the product is to be called *wood-based*. On a weight basis, the inorganic binder, which is usually cement, weighs more than the wood.

This definition goes on to state that non-wood lignocellulosic materials, including agricultural residues such as bagasse, flax shives, jute sticks, straw and hemp in the form of stalks, particles, and fibers, are also used to manufacture particleboard and fiberboard. In these the bond may be inherent or provided by synthetic resins. Additives may be applied during manufacture to improve certain properties. Veneer sheets are included in this product grouping.

The breakdown of wood-based panels shows veneer sheets, plywood, veneer plywood, core plywood other plywood, particleboard of wood (flat press), particleboard of other lignocellulosic materials (flat press), particleboard (extruded), fiberboard (fiber building board), insulating board, medium hardboard, medium hardboard (regular), hardboard, hardboard (regular), mineral-bonded panels (wood particle base), straw panel board, and other panels, which at the present time include combinations of two or more wood-based panels.

Blockboard, which is made of narrow widths of board overlaid with veneer, is a very popular product in Europe and elsewhere in the world. This is classified as a core plywood. It also appears that the composite panel of particles and veneer apparently will be called plywood for statistical purposes.

OBJECTIVE

It was decided to include in this book all those composition products manufactured by dry processing. Particleboard is made by dry processes while, as will be discussed, fiberboard can be made by wet or dry processes. The difference between dry processing and wet processing is basically in how the particles or fibers are laid up into the mat before it is consolidated into a board. Wet-process board can be considered an offshoot of the paper industry and involves some different technology from the dry-process board industry. The basic techniques for producing both dry-process hardboard, medium-density fiberboard, and particleboard (U.S. terminology) are quite similar and, as mentioned, will tend to overlap in the two industries as now represented by trade associations. As a case in point, the largest plant in the United States, if not in the world, is at the present time using a combination of planer shavings, other particles, and fiber for producing its particleboard product.

The objective of this book is to consider the dry-process board industry according to the rationale already described. The intended audience consists of those involved in the production of these types of boards and those who are interested in becoming acquainted with this industry. The presentation is meant to be practically oriented. Arguments as to whether a certain product should be a *hardboard, medium-density fiberboard,* or *particleboard* will be left to those involved in the trade associations.

Extruded board, molded products, and mineral-bonded products are minor parts of this industry and will be dealt with briefly. The book is organized to present historical, production, and market data first. This will be followed by an examination of the processes and product properties. For those unfamiliar with property terminology, it might be best to read Chapter 4 first in order to be familiar with these terms. Next, the scientific and technological parameters governing processing will be discussed, and the various process steps will be examined.

Both English and metric measurements will be used, as the United States is slowly moving toward the metric system. The metric or SI units will be those specified in ISO 1000 (International Organization for Standardization). The basic unit for pressure and stress is Pascal (Pa) or Newtons per square meter (N/m^2). The multiples include megapascal (MPa) and kilopascal (kPa). The choice of the appropriate multiple (decimal multiple or sub-multiple) of an SI unit is governed by convenience, the multiple chosen for a particular application being the one which will lead to numerical values within a practical range. The multiple usually chosen is between the numerical values 0.1 and 1000. A table of conversion values is in the Appendix.

HISTORY

Boards and building products of small particles or fiber are of very recent historical origin, although timber is man's oldest building material with the possible exception of stone. The earliest recorded history notes the use of timber and wood products for building.

Outside of building material, paper is the other major product of wood. The early Egyptians initiated its development when they used papyrus for the raw material. In A.D. 105 the Chinese developed the first paper out of wood by pounding the inner bark of the mulberry tree into pulp, adding water, and then drying the mixture in flat, fibrous sheets. It was not until 1799, however, that a Frenchman patented the first machine for making paper in a continuous sheet. In 1844 the Germans patented the first practical method of producing wood fiber by inventing a wood grinding machine, and in 1851 the English developed a system for breaking down the wood into fibers by chemical action. The manufacture of low-cost paper was then achieved by producing paper from wood, which was both cheap and plentiful, but the original concept was centuries old.

Plywood, a term that did not come into common usage until the late 1920s, was known in early times under the term *veneers*. The oldest portrayal of the art of veneering apparently was the mural, Sculpture of Thebes, which has been dated around 1450 B.C. The first commercial veneer firm in North America was established in New York City in 1824. In 1905, in Portland, Oregon, the idea of using veneer for a structural building panel was used to create the first structural plywood as we now know it. About 1934 waterproof adhesives were perfected for plywood manufacture, a development which enabled the use of plywood in exterior applications.

Timber, as mentioned previously, has always been known as a building material. The concept of plywood is centuries old, with the commercial exploitation taking place over the last fifty years. The concept of producing paper is also centuries old, with relatively recent commercial exploitation. The point of this brief discussion is that most of the useful materials made from wood have been known or the concepts have been established for centuries, and recent developments have been refinements of old concepts.

This brings us to boards or products of comminuted wood—the newest members of the forest products family. As is well known, a tremendous amount of residual material results from the production of lumber and plywood. Perhaps the

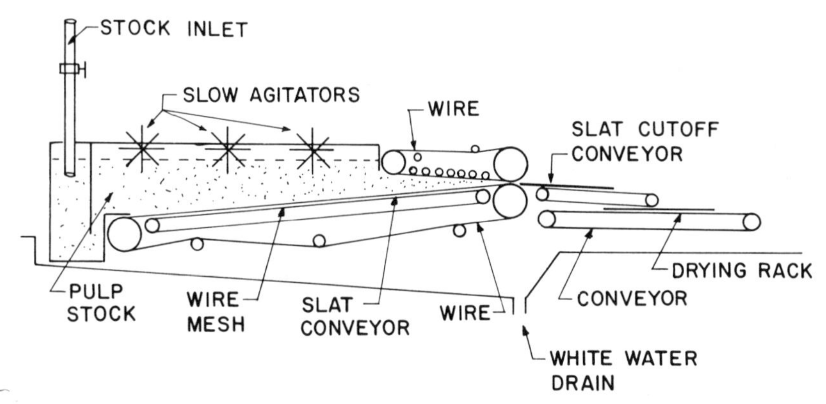

Figure 1.3. *Side view of the first machine for producing insulation board, taken from a drawing by Carl Meunch. Wood fibers carried in water flowed past rollers and onto a screen mesh where water was removed from the mat. Continuous mat was cut off at the end of the machine, then sheets went into a dryer. (Courtesy American Board Products Assn.)*

most widespread use of this material over the centuries has been for fuel. No doubt, innumerable people over the centuries have visualized taking the waste material and producing some type of man-made board. However, records show that such developments are of very recent origin.

Wet-Process Fiberboard

Lyman obtained a United States patent in 1858 on fiberboard. Fleury obtained a United States patent in 1866 on an artificial nonflammable wood, which was a wet-formed fibrous product which was to be compressed in a press or by heated rolls. Other early experiments are reported in the 1880s for producing a wet-process hardboard or insulation board type of product.

In 1914 Carl G. Muench had an idea for producing an insulating board out of wood fibers. He was hired by a paper company to develop this new product, and he was able to build a machine and make the world's first insulation board on it within ten weeks. This was a partial adaptation of the process used in making paper and consisted of a huge "head box" for holding wood fibers in water from which the fiber slurry was metered out onto a continuously moving screen. The water was drained away through the screen, leaving a damp mat of interlaced fibers as shown in Figure 1.3. This material was then oven dried, yielding the sheets of board. Later this process was adapted for making insulation board from bagasse, which is the residual sugar cane stalk after the sugar has been extracted. This process has been refined over the years, but the key point is that it is a wet-process system based on papermaking technology.

This type of board ranges in density from 10 to 25 lbs/ft^3 (0.16–0.40 sp gr), which is lower than the wood and other raw materials from which it is usually made. It is widely used for sheathing, interior paneling, rigid roof insulation, and

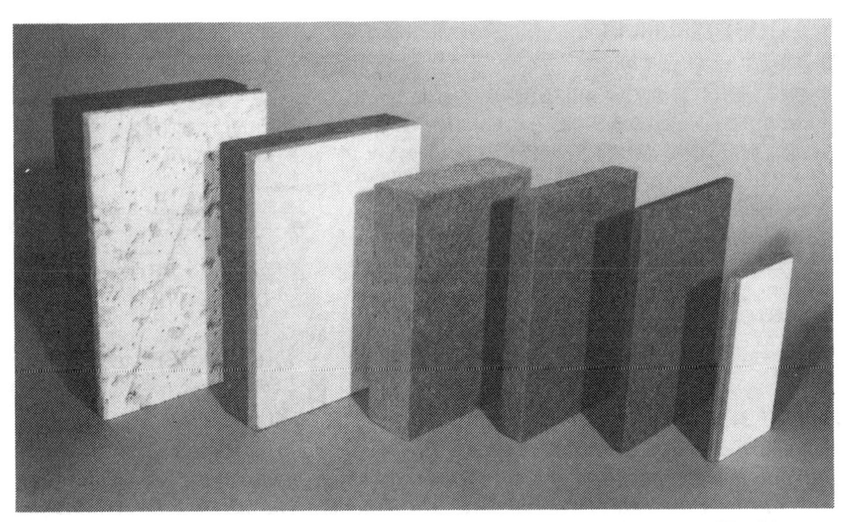

Figure 1.4. *Examples of typical insulation board products. (Courtesy Washington State Univ.)*

as siding. It is a relatively low-cost material which requires little, if any, binding agent. However, it should be mentioned that for higher-quality boards, materials such as asphalt or vinsol (a thermoplastic binder) are added to improve board properties, particularly those associated with the pickup of water or moisture from the atmosphere. Figure 1.4 is a photograph of typical boards.

It is possible also to densify the insulation board after drying by consolidation in a heated press. This results in a high-density hardboard product.

Wet-process hardboard, another refinement on paper-producing technology, was the next type of composition board material to appear on the scene. William H. Mason discovered the process in 1924, and his discovery established the Masonite process for producing hardboard. To prepare the fiber, chips are subjected to high-temperature steam in a pressure vessel called a *gun*. The chips are subjected to approximately 600-psi (4.1 MPa) steam pressure for one minute, and then the pressure is quickly raised to more than 1000 psi (6.9 MPa) for five seconds. The pressure is suddenly released and the chips are exhausted from the gun. The pressure built up within the chips literally explodes them into a coarse mass of fibers. This mass can be further reduced into individual fibers or fiber bundles by means of an attrition mill. Lignin, in general terms, is nature's glue for binding the wood fibers together. In the steaming operation the lignin bonds between the cellulose fibers are softened, which permits a more improved disintegration of the chips into the fiber bundles as compared to unsteamed chips ground to fiber in an attrition mill.

After the fluffy masses of fiber are refined, the fiber is washed and the fines and hemicellulose solubles are removed. The hemicelluloses in the wood are hydrolyzed in the gun so they can be removed by washing. Removal of this material improves the dimensional stability, as sites for combining with water molecules in the finished hardboard are eliminated. The remaining stock then goes to a former, somewhat similar to the one for producing insulation board, and is felted into a mat for the production of the hardboard. The mat is dewatered by draining most of the water through a fourdrinier-type screen which carries the formed mat. This type of mat is pressed into its final density in a hot press. A screen is used on the bottom of the mat to allow the moisture to escape during the pressing operation. If a smooth-two-sides (S2S) board is desired, the damp mat must first be oven dried before it is put into the hot press. Otherwise, the high moisture content of the mat will cause an excessive amount of steam to be generated, which will blow the board apart during or immediately after pressing. Board strength is obtained from the interlacing of the fibers developed during felting and by the rebonding of the lignin to the fibers, which occurs during the hot pressing of damp mats. Additives can be used to improve properties. The main point here is that this is again a wet-process method of producing a composition board.

It is interesting to note how Mason discovered the final steps of producing hardboard. Like many scientific breakthroughs, it occurred by accident. He had formed a wood fiber mat similar to the ones used in the insulation board industry. Just before his lunch hour he had placed this wet mat inside a crude hot press. He was using this method for drying the mat. As a precaution before he left for lunch, he turned off the steam so that excessive drying would not occur. However, the steam valve was faulty and steam kept on heating the press during his absence. When he returned he found the fiber mat had been pressed into a high-density hardboard. So from this initial discovery he proceeded to develop the modern wet-process hardboard industry.

This building material is well known throughout the world, as many plants have been constructed to produce this relatively new product. In comparison to insulation board, hardboard must have a density of at least 31 lbs/ft^3 (0.50 sp gr). For most production, the specific gravity is approximately 1.00. Figure 1.5 is a photograph of typical products.

Platen-Pressed Particleboard

For the platen-pressed particleboard industry as it is now known, an early reference for producing this type of board occurred when Ernst Hubbard in 1887, in a publication, "Die Derwertung Der Holzadfalle" ("Utilization of Wood Waste"), proposed to manufacture artificial wood from sawdust and blood albumin under application of pressure and heat, which illustrates the early conceptualization of the particleboard process.

Krammer in 1889 obtained a German patent for a method for gluing planer shavings onto linen cloth and then laying up the cloth layers in a cross-lap construction much like that for plywood.

Watson, an American, in 1905 referenced making thin wood particles for pressing into boards. This patent shows clearly a flakeboard very similar to some types of board made today. Watson could be called the inventor of the flake type of particleboard, but it is very difficult to state this as an absolute truth without a complete search of the world's scientific literature.

Figure 1.5. *Examples of typical hardboard products. (Courtesy Washington State Univ.)*

Beckman, a German, in 1918 suggested making a board with chips or wood dust in the center and surface veneers on the outside. This particular formulation is now coming onto the market as a "new" type of structural building panel. Another German, Freudenberg, in 1926 talked about utilizing planer shavings with the adhesives available at that time for making a board. He noted that the adhesive level should be between three and ten percent, which, interestingly enough, is about the range for the present-day particleboard.

Nevin, an American, in 1933 recommended the mixing of coarse sawdust and waste wood shavings with an adhesive and then forming and compressing them under the application of heat. A Frenchman, Antoni, in 1933 discussed boards of a mixture of wood fibers and particles and large elements such as excelsior or even metal netting in a board that was to be bonded with phenolic or urea glues. This, of course, was about the time of the development of these two new synthetic resins.

Samsonow, another Frenchman, in 1935 recommended using lengthy strips made from veneers in a board. These were to be arranged in a cross-lap manner, again much in the same fashion as a conventional plywood. This is a forerunner of the development of oriented flakeboard, which is going into production in the United States at this time. A Japanese, Satow, in 1935 obtained an American patent for making board with 75-mm-long (3 in.) chips arranged randomly within the board to prevent warpage. The German, Roher, in 1935 discussed pressing particles onto the surface of a plywood core in a single operation.

Carson, an American, was awarded a patent in 1936, which he applied for initially in 1932, for establishing a regular production line for producing particleboard. He proposed using a splintery type of sawdust with a moisture content of about 12%, which was to be first sized, impregnated against fungi growth, and applied with a fire retardant. A binding agent, which was to be a urea-formaldehyde-condensation product dilutable in water, was to be sprayed onto the wood particles in a rotary-drum blender. Before hot pressing, a prepressing operation was to take place, and he proposed covering the final board with a thermoplastic coating of synthetic resin. Much of what he discussed here will be found in many particleboard plants.

Another American, E. C. Loetscher, in a 1936 patent, provided interesting data on how to produce a particleboard in an automated system. In 1937 he discussed the production of a board made of sawdust with unconnected individual flakes on the surfaces to provide a decorative effect. These patents were the result of research initiated in 1933. Of great interest is the Farley & Loetscher Manufacturing Co., a millwork firm, which started pilot plant production in 1935 in Dubuque, Iowa, based on this research. Hammermilled particles were blended with liquid phenolic resin in an adapted concrete mixer. Four ⅛-in.-thick (3.2 mm) mats were formed for each press opening, prepressed, and then assembled into a package using metal caul plates between the mats. An 11-opening press was used; thus 44 boards were pressed at a time. Board specific gravity ranged from 1.2 to 1.3. Finished boards, trade named Loetex, were trimmed, sanded, and then used for core material to which a high-pressure thermosetting plastic was applied. The trade name Farloex was used for the board surfaced with a decorative laminate. This pilot plant ran until 1942, when it was discontinued because of difficulties in producing plastic sheet material near a dusty particleboard plant and because the company did not have enough raw material for developing a full-scale plant. Perhaps this could be called the first operational particleboard plant. The firm performed research on boards ranging from 0.70 to 1.8 in specific gravity with wood particles varying in size from coarse "hog chips" to fine wood flour. Many different species were investigated. Research also covered a variety of water-resistant additives and extenders. Figure 1.6 shows the Baldwin-Southwark press used in this plant.

In a Swiss patent, Chappuis in 1937 described producing board with dry particles with dry binder in the form of a powder, which, in this case, was to be Bakelite.

F. Pfhol obtained a Swiss patent in 1936 and described the use of long strips between 50 and 200 mm (2–7.9 in.) long, 4 to 8 mm (0.16–0.31 in.) wide, and 0.5 to 2 mm (0.020–0.079 in.) thick, which were to be arranged in a criss-cross fashion resulting in a board with high stability characteristics. He recommended covering the surfaces with finer strips 25 mm (1 in.) long, 3 mm (0.12 in.) wide, and 0.2 mm (0.008 in.) thick. This is the best-known European patent in the particleboard industry. Pfhol, however, also obtained a patent in Czechoslovakia in 1936 on "Boards for Cabinetmakers and Carpenters." The description of his process follows:

> The boards consist generally of wooden particles in the form of flakes cut from veneers, chips, etc., which for the coarser core layer are for instance about 50-100 mm (2-4 in.) long, about 4-8 mm (0.16–0.31 in.) wide and about 0.5–1 mm (0.020-0.039 in.) thick. For the finer outer layers on both sides, these wooden particles are about 25-50 mm (1–2 in.) long, about 2–4 mm (0.079–0.16 in.) wide, and 0.1–0.5 mm (0.004–0.020 in.) thick. These coarser and finer particles are separately soaked in a bonding agent or glue and the surplus of glue is thereafter removed by centrifugal force. Then it is required to

Figure 1.6. *Views of Baldwin-Southwark press used for producing particleboard in the 1930s (Ter Louw 1976).*

dry the particles so far as to reach the optimal gluing properties of the glue sticking to them. Thereafter a layer of fine particles is poured into a special press frame, followed by a layer of coarser particles and then again a layer of fine particles, all this in proper height to achieve the required board thickness.

This pouring has to be done in such a way so that the particles will lay in horizontal layers, i.e., in the plane of the board, but criss-cross over each other, which gives the board proper strength. The total height of the poured mat is governed by the required board thickness, and is about six times the final board thickness.

This mat is then pressed into a board in a heated press and after removal from the press can be used for overlaying with expensive veneers without any further preparation.

The board size depends only on the press size.

The Dyas Wood Products Industry, Ltd., bought the rights to this patent and, after one and a half years of experimental work, designed and built a crude production line and went into commercial production. World War II started and the production was shifted to aircraft plywood. After the war, production did not resume as other more refined processes were then available.

In 1938 and 1940, Torfitwerke G. A. Haseke obtained patents on methods of producing particleboard. The first one covered the use of liquid adhesives with a post-drying step, after application of the adhesive, to reduce the moisture content. The second patent covered gluing in the press before removing the board to prevent the blows in a high-density board.

This company in 1941 erected a commercial wood particleboard plant in Bremen, Germany, and this has been noted as the first operational plant. It could be argued, however, that the American plant of Farley & Loetscher Mfg. Co. was first, followed by the Dyas plant. It thus appears that universal agreement on who was the first manufacturer may not be possible. There probably are some other early producers not noted here. The Bremen plant produced ten tons of board a day from spruce chips using phenolic resin as the binder. In the years 1941–1943, two plywood plants in Germany began to produce similar boards with waste wood from their plywood operations, using urea resin as the binder.

In 1943 Fred Fahrni obtained a French patent covering the most favorable moisture content for pressing. Fahrni continued on to become one of the first great pioneers of this industry, developing the worldwide system of Novopan particleboard plants.

The first commercial particleboard in the United States after World War II was supposedly produced by the Southern Box and Lumber Company in Wilmington, South Carolina, although the Hu-Wood plant in North Sacramento, California, supposedly started operations about 1944. In 1947 the Plaswood Coporation in Wilton, New Hampshire, began to make itself known with a product called Plaswood. Other early plants in the Midwest were Swain Industries, Curtis Company, and Rock Island Millwork Company. In 1951 the Long-Bell Lumber Company commenced production with a small plant in Longview, Washington. At the same time, U.S. Plywood Corporation in Anderson, California, started producing Novoply under franchise from the Fahrni Institute in Switzerland. Shortly thereafter, the Pack River Lumber Company started to develop its Tenex plant in Dover, Idaho, which was designed to produce board with large flakes, now commonly called *wafers*, that yielded a board suitable for both structural applications and decorative effects.

Figure 1.7. *Examples of typical platen-pressed boards. (Courtesy Washington State Univ.)*

About 1960 the massive expansion of the United States industry commenced. Techniques were developed to produce boards with smooth surfaces, and resins were refined to speed curing times in the press. Breakthroughs into the furniture core and floor underlayment markets opened up huge markets that were exploited successfully. Plant sizes increased from about 100 tons (91 metric tons [mt]) per day to the massive Roseburg Lumber Company plant in Oregon, which can produce 2000 tons (1814 mt) of board per day. Some of the pioneer plants would probably fit into the largest forming line of this monstrous plant. Figure 1.7 shows examples of typical products.

Significant efforts are being made to bring particleboard and fiberboard further into the structural building panel market in direct competition with plywood. Some large breakthroughs have already occurred, most notably in the mobile home decking market, where urea-bonded particleboard has supplanted plywood for the floor membranes. These boards, 4 ft (1.22 m) wide or wider by 12 or 14 ft (3.66 or 4.27 m) long, are lower in cost, smooth-surfaced, and produced in lengths suitable for full-width spanning of the mobile home floors. Structural flakeboard is approved for use in Canada. Other particleboards are used structurally throughout the world.

The development of this particular segment of the board industry has been phenomenal since World War II. Many different types of board plants have been built around the world, based not only on wood waste and roundwood cut especially for particleboard but also from other lignocellulosic materials such as bagasse and flax. The ability to use the heretofore wasted raw materials has been a boon to all of society.

Extruded Particleboard

The extrusion method of producing particleboard was initially developed in Germany by Otto Kreibaum in the years 1947–1949. This production process has been expanded over the years in Europe to the extent that at least one manufacturer is producing both boards and factory-built houses in a single manufacturing complex. In the United States, several plants were built in the 1950s based upon the original concept from Europe as well as on developments by United States firms. The plants producing this type of board in the United States are relatively small and are captive in the sense that they produce board for use by other parts of their firms' operations, notably in furniture factories. No new plants of this type have been built in the United States for many years, probably because of their low production rates and because of inherent problems with the board, which will be mentioned later. Several plant expansions, however, have been made. Figure 1.8 illustrates typical products. A newer system for making a ribbed product has recently been developed in Texas and reportedly will be in use soon.

Dry-Process Hardboard

The next development in the composition board area was that of manufacturing hardboard by means of a dry process. In 1945 the Plywood Research Foundation began experimenting on and developing this process. The Plywood Research

Figure 1.8. *Typical extruded board. (Courtesy Washington State Univ.)*

Foundation is the research arm of the American Plywood Association, which was known as the Douglas Fir Plywood Association in 1945. The thrust of this development was to eliminate the waterforming of the mat found in the wet-process industry.

Normally the chips or other residual materials are conveyed mechanically through cooking and grinding units where the fiber is prepared. It is also now normal to apply the resin and wax to the raw material prior to grinding so that the attrition mill simultaneously prepares fiber and blends the additives with the fiber. In initial experiments and in the early plants, the resin was added after the fiber was prepared. The mats were then formed using air and mechanical means in a continuous system. This mat, however, did not have the fiber interlocking that is possible when forming mats by the wet process. Because of this lack of interlocking and the very low, if any, adhesion that can be obtained by lignin flow and hydrogen bonding in the hot press, the dry-process hardboard is dependent upon the binder for the development of physical properties. If the mat has a moisture content below 10%, it normally is pressed into a smooth-two-sides board which has a very light color in contrast to boards prepared by the wet-process system. At moisture contents over 10%, it is necessary to use a screen on one side of a mat in order to press the board, much in the same way as pressing a wet-process mat.

The boards manufactured by this method enter the same market as the wet-process board. Some people hold that the wet-process board is somewhat superior in properties because of the self-bonding of the fibers, the intermeshing of the fibers as developed in the wet-forming operation, and the removal of the hemicellulose. The elimination of the massive amounts of water necessary for forming the mat is an advantage of making this type of board, particularly now that tough constraints have been implemented governing the release of effluents from any industrial operation. However, synthetic resins, which are becoming more expensive, must be used for binders in the dry-process board.

Medium-Density Fiberboard

In the mid-1960s the most recent development in the board industry occurred. This breakthrough has been called medium-density fiberboard (MDF). Most of the plants producing this board use the pressurized refiner for generating finer fibers, which have far greater bulk than those usually produced by atmospheric refiners. These details will be developed later in this book. This particular board is usually much thicker than the traditional hardboards, which range from about 1/10 in. (2.5 mm) to about ⅜ in. (9.5 mm) in thickness. Most of the board produced in the MDF process is over ⅜ in. in thickness and goes to the furniture industry for use as core stock. Much of the recent interest in building plants has centered on the MDF process. Figure 1.9 shows typical products. This is a further refinement in the development of dry-process boards produced essentially with fiber.

Mineral-Bonded Products

Another type of panel building product that merits mention is mineral-bonded building products usually made of excelsior (wood wool, as it is known in Europe)

Figure 1.9. *Examples of medium-density fiberboard. (Courtesy Washington State Univ.)*

and cement. The first boards of this type had magnesite as a binder and were developed in 1914 in Radenthein, Austria. Over 600 million square meters of this material were manufactured by the Heraklith method in Austria through 1963. At that time the company was reporting that over 50% of the world's consumption was with the use of its product. Portland cement was used in this type of product in 1928, and later gypsum was used as a binder. The plants now manufacturing this product in the United States started operation in the late 1940s, although the product was manufactured much earlier.

Recently, greater interest has been shown in this product, as evidenced by the building of plants in both the underdeveloped and developed nations. It has particular appeal in the underdeveloped countries since it can be produced with crude methods using available cement for the binder. Many developing countries do not have the synthetic adhesives that are available for producing the fiberboard and particleboard as known in much of the world. However, good acoustic, thermal, fungi-resistant, and fire-resistant properties of this product are also appealing to the industrialized countries.

The use of the excelsior provides panels with a bending strength suitable for use as spanning members in construction. Other structural panels have been developed using wood flakes in the mix. These particular panels also have bending strengths that allow their use in standard building construction. Figure 1.10 is a photograph of typical panel products.

In another application, relatively lightweight building blocks approximately 8 in. square by 16 in. long (203x406 mm) or larger can be produced by using planer shavings or wood particles and the mineral binder. Walls are fabricated using the same techniques as for cinder block construction. Also, this wood aggregate

Figure 1.10. *Typical wood-based panel products using cement as the binder. (Courtesy Washington State Univ.)*

concrete can either be made into tilt-up walls or poured in place for wall sections. It is also used as lightweight concrete in apartment construction, particularly for its value in sound deadening.

Some feel that the further development of this part of the board industry will be accelerated because of the shortage of synthetic resins. This shortage is associated with the worldwide energy problems, as the conventional synthetic resins are petroleum-based derivatives. However, the fact that the production of cement is also closely associated with energy should not be overlooked, as a great amount of energy is necessary for cement manufacture.

Molded Products

Another type of comminuted particle product that is different from the conventional flat panel familiar to most people is molded particleboard (examples shown in Figure 3.9). This group of products has not received extensive attention in the United States because production of molded parts out of plastics has been more economical. With the advent of the energy and petroleum shortage, it is quite possible that the economics of producing this type of product out of wood particles in comparison to plastics will change dramatically. As a consequence, there is the likely possibility that this industry, which is fairly well developed in Europe, will find expanded applications in the United States. Details of this system will be covered later.

SELECTED REFERENCES

American Forest Products Industries, Inc. (American Forest Institute). 1960. *The Story of Pulp and Paper*. Washington, D.C.

American Hardboard Association (American Board Products Association), American Forest Products Industries, Inc. 1961. *The Story of Hardboard*.

American Society for Testing and Materials. 1974. *Metric Practice Guide*. Philadelphia, Pennsylvania: ASTM E 380.

———. 1975. Definitions of Terms Relating to Wood-Base Fiber and Particle Panel Materials. Philadelphia, Pennsylvania: ASTM D1554.

Baldwin, R. F. 1975. *Plywood Manufacturing Practices*. San Francisco: Miller Freeman Publications, Inc.

Blais, J. F. 1959. *Amino Resins*. New York: Reinhold Publishing Corporation.

Burrell, J. 1971. The 3,500-Year History of Hardwood Plywood. *Plywood and Panel,* Vol. 12, No. 1.

Carroll, M. N. 1966. Composition Board. *Encyclopedia of Polymer Science and Technology,* Vol. 4. New York: Interscience Publishers.

Chopra, N. N. 1975. International Standardization in the Field of Wood-Based Panels. Background Paper No. 12, third World Consultation on Wood-Based Panels, FAO, New Delhi, India, February 1975.

De Viley, J. T. 1968. The Earliest Hard Board? *Board Practice,* Vol. 11, Nos. 11 & 12.

Douglas Fir Plywood Association (American Plywood Association). 1955. *Facts about Fir Plywood*. Tacoma, Washington.

Fahrni, F. 1957*a*. A Documented Review of the Development of the Wood Particle Board and Novopan's Contribution to the Manufacturing Process. *Holz als Roh-und Werkstoff,* Vol. 15, No. 1.

———. 1957*b*. Essential Features of Batch Formation. Technical Paper No. 5.31. *Fibre-Board and Particle Board Technical Papers*. International Consultation on Insulation Board, Hardboard and Particle Board, FAO, Geneva, Switzerland, January-February 1957.

FAO. 1957. *Fibreboard and Particle Board*. Report of an International Consultation on Insulation Board, Hardboard and Particle Board, Food and Agriculture Organization, Geneva, Switzerland, January-February 1957.

———. 1966. Plywood and Other Wood-Based Panels. Report of an International Consultation on Plywood and Other Wood-Based Panel Products, FAO, Rome, Italy, July 1963.

———. 1976. *Proceedings of the World Consultation on Wood-Based Panels,* held in New Delhi, India, February 1975. Brussels: Published in agreement with the Food and Agriculture Organization of the United Nations by Miller Freeman Publications.

International Organization for Standardization. 1973. International Standards ISO 1000. Geneva, Switzerland.

Insulation Board Institute. 1964. *Insulation Board . . . A 50 Year Start on Tomorrow*. Chicago, Illinois.

Lewis, W. C. 1957. The Insulation Board and Hardboard Industry in the U.S.A. Technical Paper No. 3.2b. *Fibreboard and Particle Board Technical Papers*. International Consultation on Insulation Board, Hardboard and Particle Board, FAO, Geneva, Switzerland, January-February 1957.

Maloney, T. M. 1975. Composition Board and Composite Panels—Present and Future. *Proceedings of the Northwest Wood Products Clinic*. Pullman, Washington: Washington State University (WSU).

Maloney, T. M., and A. L. Mottet. 1970. Particleboard. *Modern Materials,* Vol. 7. New York: Academic Press Inc.

Marra, G. G. 1969. A Vision of the Future for Particleboard. *Proceedings of the Washington State University Particleboard Symposium, No. 3*. Pullman, Washington: WSU.

———. 1972. Particleboard and Materials Engineering. *Proceedings of the Washington State University Particleboard Symposium, No. 6.* Pullman Washington: WSU.

———. 1975. The Age of Engineered Wood Materials. Feature Address at the third World Consultation on Wood-Based Panels, FAO, New Delhi, India, February 1975.

Pfohl, F. 1936. Boards for Cabinetmakers and Carpenters. Patent No. 56350. Patent Office, The Republic of Czechoslovakia (November).

Pollard, N. E. 1973. Particleboard as a Structural Component in the Building Industry in New Zealand. *Proceedings of the Washington State University Particleboard Symposium, No. 7.* Pullman, Washington: WSU.

Roberts, R. J. 1958. A Brief History of "Loetex" High Density Wood-Particle Board Production at Farley & Loetscher. Dubuque, Iowa: Farley & Loetscher Mfg. Company.

Roubicek, T. T. 1974. Private correspondence on an early Czechoslovakian particleboard plant.

Roubicek, T. T., and A. Pantelejevs. 1964. Wet Felted Particle Board. Canadian Patent No. 683,842. Patent Office, Department of the Secretary of State, Ottawa, Canada (April).

Sandermann, W., and O. Künnemeyer. 1956. Historical Development of the Fiber Board Industry. *Holz als Ruh-und Werkstoff* (July), Vol. 13/14.

Schmierer, R. E. 1971. Australian Particleboard Practices. *Proceedings of the Washington State University Particleboard Symposium, No. 5.* Pullman, Washington: WSU.

Snodgrass, J. D.; R. J. Saunders; and A. D. Syska. 1973. Particleboard of Aligned Wood Strands. *Proceedings of the Washington State University Particleboard Symposium, No. 7.* Pullman, Washington: WSU.

Talbott, J. W. 1974. Electrically Aligned Particleboard and Fiberboard. *Proceedings of the Washington State University Particleboard Symposium, No. 8.* Pullman, Washington: WSU.

Ter Louw, F. W. 1976. An Early U.S. Particleboard Plant Operating in the 1930's. *Proceedings of the Washington State University Particleboard Symposium, No. 10.* Pullman, Washington: WSU.

Vajda, P. 1974. Structural Composition Boards in the Wood Products Picture. *Proceedings of the Washington State University Particleboard Symposium, No. 8.* Pullman Washington: WSU.

2. PRODUCTION AND MARKETS

DEVELOPMENT

High-density hardboard, as mentioned previously, can be made by either the wet or dry process. The boards produced by both processes compete in the same markets. The boards usually are relatively thin (about ⅛ in. or 3.2 mm) unless they are specifically laminated into much thicker products for specialty markets. A significant increase in the dry-process production over the last few years has been in siding, usually in thickness about 9/32 to 7/16 in. (7.1–11.1 mm). Several plants are now concentrating on this product.

At first, both the wet- and dry-process segments of the industry preferred using pulp chips for the raw material because a superior fiber could be generated from this source. However, as the demand for the prices of pulp chips increased, the hardboard industry moved into lower grades of residual materials such as sawdust and shavings for its raw material supply. Technological improvements in the processing had to be made to utilize this lower-grade raw material properly so that quality of the resultant board did not diminish, although chips are usually the preferred raw material. This trend of moving from a higher-grade residual material to a lower-grade residual material has been typical of the entire composition board industry because of the pressure applied by pulp mills on the raw material resources. Advances in technology in the pulp and paper industry have made possible the use of more and more shavings and sawdust, which puts further pressure on the amount of residual material available for the board industry.

For particleboard, the very first plants in Europe were designed to produce a product suitable for wall paneling and for some furniture. These initial plants were based on wood waste from manufacturing facilities. In the late 1940s good three-layer boards were developed in Europe from forest waste consisting of thinnings, material that was used for fuel wood, and branches. This material was usually cut into thin flakes for the faces of the boards and thicker flakes for the cores of the boards. This production was aimed at the furniture industry. Until this time, hardwoods such as beech, birch, and oak were usually used for fuel wood. It was found that these species, while not particularly suitable for normal production of lumber, were excellent for use in particleboard. The particleboard industry, therefore, moved to incorporate much of this material into its raw material supply. By 1956 there were 106 European mills producing 724,000 cubic meters of particleboard, according to the Food and Agriculture Organization (FAO) of the United Nations. This was a tremendous expansion in an industry that did not truly start its expansion until after World War II.

At the 1957 International Consultation on Fibreboard and Particleboard, fears were expressed about too rapid a growth of the composition board industry in Europe because of the inability of consumers to use such a tremendous amount of material. These fears have been shown to be unfounded, and the industry is still rapidly developing with about 27 times more production in 1973, compared with 1956.

In the United States, outside of a few plants based on shavings or hammermilled particles, most of the early board plants were based on flakes produced from either roundwood or very low-grade lumber considered waste material by sawmills. U.S. Plywood's early plant at Anderson, California, was built as part of the worldwide system of Novopan plants under franchise from the Fahrni Institute in Switzerland. They used thin flakes on the face of their panel and thicker flakes at the core. Long-Bell Lumber Company and Pack River Lumber Company also built early plants using large flakes. Much of the production, particularly that of the wafer-type plants such as Pack River's, was aimed at the structural board market. These efforts were a bit ahead of their time since the great increase in production of sheathing grade plywood was then taking place and there was an adequate veneer supply for this construction material.

The early board plants successful in using planer shavings for a panel product directed their efforts toward using waste material for making a building product that would realize a profit. However, much of the market as we now know it was yet to be developed, and a great amount of effort had to be expended to move such products into the furniture industry in place of solid lumber core in hardwood plywood. At about this same time it was found that shavings-type particleboard was suitable for use as floor underlayment, which is the panel material placed over rough subfloors providing a smooth, even surface for the application of sheet goods, tile, or carpet.

About 1960, breakthroughs in technology developed superior layered-type particleboards using planer shavings. These types of boards had fine particles on the surface and coarse particles in the core, which yielded panels with very smooth surfaces that could be easily overlaid with plastics or with veneers. Willamette Industries in 1960 built the first relatively large plant in the United States at Albany, Oregon, to produce this type of particleboard under the name of Duraflake. Other companies shortly followed them into the marketplace with similar board products, and the huge furniture market in the United States was opened up for widespread use of particleboard. Particleboard, since that time, has taken over virtually all of the furniture core market and also is widely used for other parts of furniture in which the particleboard is not covered with veneers or laminates.

A particular advantage that particleboard has had over plywood is the relative ease with which it can be made into larger panels in the production plant. For example, a common size for industrial grade (furniture board) is 5x18 ft (1.52x5.49 m). Large amounts of the particleboard furniture material is precut or cut-to-size, as it is known in the trade, into specified parts ordered by the furniture companies.

A 5-ft-wide panel can often be cut into furniture parts with less loss of material than a 4-ft-wide (1.22 m) piece of material. A case in point is the example of 30-in.-wide (762 mm) countertops of which two can be cut from a 5-ft-wide piece of particleboard. However, only one piece this width with a leftover 18-in.-wide (457 mm) piece can be cut from 4-ft-wide material. The narrow, 18-in.-wide piece is difficult to use efficiently.

Most of the initial plants making this type of board were built on the U.S. West Coast. Douglas fir had been found to be an ideal species for use in producing particleboard. The acid level in Douglas fir is excellent for speeding the cure of urea resins used for bonding, and the particle geometry is such that high-quality boards could be produced. However, this lucrative furniture market was apparent to others, and other plants were built in eastern Oregon, Montana, California, Arizona, and in the southern pine region. Problems with species were overcome during this expansion period. This marked the dramatic expansion of the particleboard industry in the United States when particleboard became a commonplace item of construction, both in the building and furniture fields.

In the mid-1960s particleboard penetrated the mobile home field, where it supplanted plywood for use as the floor panels. Until this time, particleboard in construction mainly was used for the previously described floor underlayment which was laid down over a subfloor, usually of plywood. In the mobile home application, the particleboard is nail-glued onto the floor joist system and serves as a structural member as well as providing the smooth surface for putting down the final floor covering.

Particleboard penetrated this market because of its lower cost, its smooth surface, and because it could be obtained in 10-, 12-, and 14-ft (3.04, 3.66, and 4.27 m) lengths which eliminated the need for end joints when laying the panels on the mobile home floor joist system.

Particleboard then moved into the interior wall paneling business where thin sheets somewhat similar to the thickness of hardboard were covered with vinyl overlays which were printed, usually with wood grain designs. This market is being further exploited by the development of continuous board presses specifically designed to produce the thin board products.

Most particleboard has been made with urea resin, which is a water-resistant but not waterproof adhesive. Thus, virtually all of the applications of particleboard in the United States have been for interior uses, although urea is used extensively for exterior applications in other parts of the world. Small amounts have always been made with phenolic resin, which is considered a waterproof adhesive. Some of this has been used for sheathing or made into special products for exterior siding or combination sheathing-siding materials. Some of the phenolic-bonded board has gone to applications in the furniture industry where laminates are applied at relatively high temperatures and long press cycles. Urea resins have a tendency to break down under such extreme secondary manufacturing conditions, and, therefore, phenolic resins are used instead.

The highest percentage of particleboard production goes into the industrial market for use in furniture. Board thicknesses range from ⅛ in. to 2 in. (3.2–50.8 mm). However, most production is in the range of ⅜ to ¾ in. (9.5–19 mm) thick. Board plants on the West Coast have competed well in this furniture market, even with extra freight costs for shipping to the east in many cases, because of better raw material for producing the boards. In the southern pine region, some difficulties have been encountered in using southern pine planer shavings for the industrial type of particleboard, although this has not precluded their use in this market. The southern pine species are higher in density than those used by the West Coast plants, which results in particleboards of slightly higher density. This, along with what is apparently silica in some of the southern pine, yields a board which at times is more abrasive to cutting tools. The southern pine boards, however, have found widespread use as underlayment and as mobile home decking, particularly in the latter case, since so many mobile homes produced in the United States are constructed in the southern pine growing region.

In the early 1970s more particleboard was manufactured with phenolic resin for use as structural material in direct competition with plywood. Earlier efforts had met with marginal success. Most of the production of this type of board has been using the familiar planer shavings or mixed planer shavings and flake furnishes common to the furniture and underlayment type of boards. Special approval has been made for boards from certain plants in order that they may sell in this market. Penetration of this market was made again because of lower price and because of the shortage of plywood for the construction industry.

It appears that further penetration will be made of the construction market by various types of particleboard. As will be discussed later, higher-quality particleboards quite similar in strength properties to plywood, particularly in bending and stiffness, should be on the market shortly, which will assist further in the penetration of this market.

The particleboard industry in Canada has not developed into as large an industry as in the United States. However, this is due to Canada's much smaller population rather than to lack of raw material or capital for building plants. A number of plants have been built to produce conventional boards of shavings or similar raw material.

Of particular interest, however, has been the rapid development of the waferboard part of the particleboard industry in Canada, where the plants are quite similar to the early Tenex plant of the Pack River Lumber Company in Dover, Idaho. This particular product is fully approved in Canada for structural applications such as roof and wall sheathing and for subfloors. It is a phenolic-bonded product, which permits its use under severe weather conditions, and as far as it is known, it is experiencing great success in the field. At the present time a number of new plants have recently opened or are under construction in Canada. A large plant for producing a similar board has been built in Grand Rapids, Minnesota, by Blandin Wood Products Company. The raw material for most of these plants is aspen or poplar, which is usually considered a low-value species for most applications; however, it is abundantly available in much of Canada and the lake states.

In the late 1960s the medium-density fiberboard segment of the industry was developed. The fiber for most of this industry is generated by grinding the chips or other residual material into fiber while under steam pressure. Boards made in this dry process are usually thicker ones, for example ¾ in. (19 mm), and are primarily marketed in the furniture market. These boards have smooth surfaces because of the fiber used, and they also have what is known as a "tight edge." Shavings-type particleboard is somewhat porous in the core due to the chunky shape of the particles or shavings making up the core layer and because of the press cycle, to be discussed later. In the furniture industry, it is usually necessary to edge band particleboard for profiling edges or developing smooth, square edges on the furniture parts. Usually it is possible to eliminate edge bandings or filling on the edges of the medium-density fiberboard. Consequently, at the present time the medium-density fiberboard is bringing a premium price in the marketplace.

The medium-density fiberboard segment of the industry is expected to have a strong impact on the board industry and markets. A number of plants have been built and others are in the process of being constructed. It is difficult to determine whether the premium price paid for this type of product will continue to be the case. Added competition from new (medium-density fiberboard) plants plus the competitive upgrading of conventional particleboards to meet the challenge of the medium-density boards may affect the price structure.

As mentioned previously, molding of products from particles is a fairly well-developed industry in Europe. In the United States, the Weyerhaeuser Company, Caradco, Inc., and Jeld-wen, Inc. have been producing molded door skins out of fiber. Weyerhaeuser has also produced a product called Pres-tock, from which parts can be molded for various industries. There is also the firm of Plex-O-Wood, in Bradford, Pennsylvania, that has been molding furniture parts out of particles with either the rough surface over which padding is applied or molded with overlays applied in a "one-shot" process.

The mineral-bonded product using cement and excelsior is also produced in the United States. This is a minor part of the industry and does not fit into the major thrust of this book, which covers products made basically with synthetic resins. The National Gypsum Co., for example, is marketing such a product called Tectum for roof decking and sound control panels in the USA, and a flakeboard is being produced in Europe and Japan.

The early extrusion plants built in the 1950s make up about all of the production capacity of this type, which is now less than one percent of particleboard production. Because of the low volume and problems with physical properties to be discussed later, this is a very minor part of the industry, and it is not expected to grow in relative terms unless major breakthroughs in technology occur.

PRODUCTION AND CONSUMPTION

Tables 2.1–2.7 from Basic Paper No. 1, "Production, Consumption and Trade in Wood-Based Panels—Present Situation and Alternative Outlooks for the Future," (presented to the World Consultation on Wood-Based Panels in New Delhi, India, in February 1975) show up-to-date information on production of wood-based panels and sawn wood. This information shows a substantial growth in production of all three major categories of wood-based panels. It also shows that panel production has expanded more rapidly than sawn wood. Particleboard growth has been spectacular, and it is obtaining a growing share. Production is fast approaching the same level of panel production as plywood. Fiberboard, on the other hand, is rapidly declining as far as its share is concerned, despite its continuing healthy growth.

Table 2.1 *World Production of Wood-Based Panels and Sawn Wood*

Product	1950	1960	1970	1971	1972	1973	Average annual growth 1950-60	1960-70	1970-73	Share of world total 1950	1960	1970	1973
			(million m^3)				(% per annum)			(%)			
Veneer	–	1.2	3.4	3.6	3.8	3.8	*	*	*	–	4	5	4
Plywood	6.1	15.3	32.7	36.1	39.8	42.5	9.6	7.9	9.1	53	53	47	44
Particleboard	0.02	3.1	19.3	22.9	27.2	31.5	*	20.1	17.7	–	11	28	33
Fiberboard	5.4	9.5	14.4	16.1	17.1	18.1	5.8	4.2	7.9	47	33	21	19
Total wood-based panels	11.5	29.1	69.8	78.7	87.9	96.0	9.7	9.1	11.2	100	100	100	100
Sawn wood	265.4	343.7	412.6	427.3	438.3	444.2	2.6	1.8	2.5				

Source: FAO, *Proceedings of the World Consultation on Wood-Based Panels*, February 1975, Basic Paper No. 1.

Note: Totals do not always agree with the addition of subtotals, due to rounding off.
*Not estimated because of low reliability of data.

Table 2.2 *Regional Production of Wood-Based Panels*

Region	1950	1960 (million m^3)	1970	1973	*Average annual growth* 1950-60	1960-70 (% per annum)	1970-73	*Share of world total* 1950	1960 (%)	1970	1973
North America	7.4	14.8	26.3	36.8	7.2	5.9	11.8	65	51	38	38
Europe	2.6	8.8	22.0	30.5	13.0	9.6	11.5	23	30	32	32
Japan	0.16	1.6	8.5	10.6	25.9	18.2	7.6	1	5	12	11
USSR	0.7	2.0	6.0	7.4	11.1	11.6	7.2	6	7	9	8
Asia (less Japan)	0.09	0.8	3.3	6.3	24.4	15.2	24.1	1	3	5	7
Latin America	0.15	0.5	2.1	2.5	12.8	15.4	6.0	1	2	3	3
Africa	0.04	0.3	0.8	1.0	22.3	10.3	7.7	–	1	1	1
Oceania	0.2	0.4	0.8	0.8	7.2	7.2	0.0	2	1	1	1
WORLD	11.5	29.1	69.8	96.0	9.7	9.1	11.2	100	100	100	100
Developed market economies	10.2	24.1	53.8	73.7	9.0	8.4	11.1	88	83	77	77
Developing market economies	0.3	1.4	5.5	8.9	16.7	14.7	17.4	3	5	8	9
Centrally planned economies	1.0	3.6	10.4	13.4	13.7	11.2	8.8	9	12	15	14

Source: FAO, *Proceedings of the World Consultation on Wood-Based Panels*, February 1975, Basic Paper No. 1.

Note: Totals do not always agree with the addition of subtotals, due to rounding off.

Table 2.3 *Regional Production of Sawn Wood*

Region	*1950*	*1960*	*1970*	*1973*	*Average annual growth* 1950-60	*1960-70*	*1970-73*
		(million m^3)				*(% per annum)*	
North America	107.1	97.0	108.1	125.6	−1.0	1.1	5.1
Europe	59.3	70.7	81.5	89.1	1.8	1.4	3.2
Japan	13.5	27.6	42.9	43.5	7.4	4.5	0.5
USSR	60.9	110.1	120.5	123.0	5.7	1.3	0.7
Asia (less Japan)	8.7	22.9	33.0	35.8	10.1	3.7	2.8
Latin America	9.6	12.1	16.4	16.8	2.3	3.1	0.8
Africa	1.5	2.5	4.6	4.8	2.4	6.3	1.4
Oceania	4.7	5.4	5.6	5.7	1.4	0.4	0.6
WORLD	265.4	348.2	412.6	444.2	2.6	1.8	2.5
Developed market economies	166.6	180.8	218.1	242.4	0.8	1.6	3.6
Developing market economies	19.4	24.6	37.9	39.6	2.2	4.4	1.6
Centrally planned economies	79.4	142.8	156.6	162.3	5.7	1.2	1.2

Source: FAO, *Proceedings of the World Consultation on Wood-Based Panels*, February 1975, Basic Paper No. 1.

Note: Totals do not always agree with the addition of subtotals, due to rounding off.

Table 2.4 *Regional Production of Plywood*

Region	*1950*	*1960*	*1970*	*1973*	*1950-60*	*1960-70*	*1970-73*	*1950*	*1960*	*1970*	*1973*
	(million m^3)				*Average annual growth (% per annum)*			*Share of world total (%)*			
North America	3.6	8.9	16.0	20.4	9.4	5.9	8.6	60	58	49	48
Europe	1.3	2.7	3.9	4.8	7.2	4.0	6.6	22	18	12	11
Japan	0.15	1.3	7.1	8.9	24.0	18.6	7.9	2	8	22	21
USSR	0.7	1.4	2.2	2.2	7.5	4.8	–	11	9	7	5
Asia (less Japan)	0.06	0.5	2.5	4.9	23.6	17.5	25.2	1	3	8	12
Latin America	0.13	0.3	0.7	0.8	8.1	9.0	5.2	2	2	2	2
Africa	0.02	0.1	0.3	0.4	15.1	10.3	7.2	–	1	1	1
Oceania	0.08	0.2	0.2	0.2	7.0	0.5	0.7	1	1	1	–
WORLD	6.1	15.3	32.7	42.5	9.7	7.9	9.1	100	100	100	100
Developed market economies	4.9	12.4	26.4	33.5	9.7	7.6	8.1	81	81	81	79
Developing market economies	0.3	0.8	3.2	5.7	11.0	14.9	21.2	4	5	10	13
Centrally planned economies	0.9	2.1	3.1	3.3	8.7	4.0	2.1	15	14	9	8

Source: FAO, *Proceedings of the World Consultation on Wood-Based Panels*, February 1975, Basic Paper No. 1.

Note: Totals do not always agree with the addition of subtotals, due to rounding off.

Table 2.5 *Regional Production of Particleboard*

Region	1950	1960 (thousand m^3)	1970	1973	Average annual growth 1960-70 (% per annum)	1970-73	Share of world total 1950 (%)	1960	1970	1973
North America	10	521	3,405	6,869	20.6	26.4	50	17	18	22
Europe	10	2,211	12,308	19,329	18.7	16.2	50	72	64	61
Japan	–	77	350	645	16.3	22.6		3	2	2
USSR	–	157	1,994	3,120	28.9	16.1		5	10	10
Asia (less Japan)	–	25	227	332	24.7	13.5		1	1	1
Latin America	–	41	558	716	29.8	8.7		1	3	2
Africa	–	30	113	143	14.2	8.2		1	1	1
Oceania	–	9	318	394	42.8	7.4		–	1	1
WORLD	20	3,072	19,272	31,548	20.2	17.9	100	100	100	100
Developed market economies	20	2,559	14,635	24,798	19.1	19.2	100	83	76	79
Developing market economies	–	77	824	1,113	26.7	10.5		3	4	3
Centrally planned economies	–	436	3,813	5,637	24.2	13.9		14	20	18

Source: FAO, *Proceedings of the World Consultation on Wood-Based Panels*, February 1975, Basic Paper No. 1.

Note: Totals do not always agree with the addition of subtotals, due to rounding off.

Table 2.6 *Regional Production of Hardboard*

Region	*1950*	*1960* (thousand mt)	*1970*	*1973*	Average annual growth *1950-60*	*1960-70* (% per annum)	*1970-73*	Share of world total *1950*	*1960* (%)	*1970*	*1973*
North America	383	709	1,521	2,174	6.4	7.9	12.6	40	26	26	29
Europe	498	1,423	2,551	2,952	11.1	6.0	5.0	52	52	44	39
Japan	9	98	398	432	27.0	15.0	2.8	0.5	4	7	6
USSR	15	128	542	1,030	23.9	15.5	23.9	2	5	9	14
Asia (less Japan)	5	85	338	414	32.8	10.4	7.0	0.5	3	6	5
Latin America	7	85	179	264	28.4	7.7	13.8	0.5	3	3	3
Africa	7	53	87	142	22.4	5.1	17.7	0.5	2	2	2
Oceania	42	128	185	168	11.8	3.8	3.2	4	5	3	2
WORLD	966	2,709	5,801	7,576	10.9	7.9	9.3	100	100	100	100
Developed market economies	938	2,252	4,064	5,033	9.2	6.1	7.4	97	83	70	67
Developing market economies	12	138	358	492	27.7	10.0	11.2	1	5	6	6
Centrally planned economies	16	319	1,379	2,051	34.9	15.8	14.1	2	12	24	27

Source: FAO, *Proceedings of the World Consultation on Wood-Based Panels*, February 1975, Basic Paper No. 1.

Note: Totals do not always agree with the addition of subtotals, due to rounding off.

Table 2.7 *Regional Production of Insulation Board*

Region	*1950*	*1960*	*1970*	*1973*	*Average annual growth* 1950-60	1960-70	1970-73	*Share of world total* 1950	1960	1970	1973
		(thousand mt)				*(% per annum)*			*(%)*		
North America	854	1,138	1,289	1,711	2.9	1.3	9.9	79	68	63	67
Europe	194	367	406	427	6.6	1.0	1.7	18	22	20	17
Japan	4	30	72	93	22.3	9.1	8.9	–	2	3	4
USSR	6	86	197	206	30.5	8.6	1.5	1	5	10	8
Asia (less Japan)	2	8	8	9	14.9	–	4.0	–	–	–	–
Latin America	3	10	57	62	12.8	19.0	2.8	–	–	3	2
Africa	3	13	12	12	15.8	0.8	–	–	1	–	1
Oceania	18	15	18	17	1.8	1.8	1.9	2	1	1	1
WORLD	1,084	1,667	2,059	2,537	4.4	2.1	7.2	100	100	100	100
Developed market economies	1,072	1,531	1,703	2,137	3.6	1.1	7.9	99	92	83	84
Developing market economies	5	11	63	68	8.2	19.1	2.6	–	1	3	3
Centrally planned economies	7	125	293	332	33.4	8.9	4.3	1	7	14	13

Source: FAO, *Proceedings of the World Consultation on Wood-Based Panels*, February 1975, Basic Paper No. 1.

Note: Totals do not always agree with the addition of subtotals, due to rounding off.

Tables 2.8–2.10 present data on apparent consumption of wood-based panels. In Europe the per capita consumption of particleboard is three times that of plywood. In the Federal Republic of Germany, 70 m^3 per thousand capita of particleboard are used. Another seven industrialized-type countries consume about 50 m^3 per thousand people. In North America 30 m^3 per thousand capita are consumed, which is approximately the rough European average. In contrast, developing countries have a consumption of 1 m^3 per thousand people. It is interesting to note, however, that consumption of particleboard exceeds that of plywood in the developing countries. Average per capita consumption is approaching that of plywood.

On a volume basis, fiberboard of all types is more important in North America and Japan at the present time. Only Sweden and Norway, of all the other countries in the world, have a similar situation. Hardboard consumption in North America is 10 mt per thousand capita. An important observation that has been brought forth in the study is that particleboard is best accepted and used in industrialized countries, although some early predictions were that the greatest production and use of this new product would be in the underdeveloped areas of the world.

It is impossible to break down the fiberboard information between wet processed and dry processed, without having the raw data carefully defined. Most of the dry-process production is in the United States. The fiberboard information as presented at the FAO meeting has been separated between hardboard and insulation board. North America and Europe are the leaders in the production of hardboard with North America far and away the predominant leader of production of insulation board. While North America has developed the particleboard industry rapidly over the last 20 years, Europe produces approximately three times more particleboard than North America. Some of this is of course due to the better-quality trees available for plywood production in North America contrasted to the smaller and more difficult-to-process trees presently found in Europe.

Moving to the United States alone, Figure 2.1 shows the production of hardboard, particleboard, and insulation board from the years 1961 through 1975. These curves have been developed from information supplied by the U.S. Department of Commerce/Bureau of the Census. Hardboard production is lumped between wet and dry processes and is reported under pulp, paper, and board by the government.

In examining this graph, it is apparent that the wet-process insulation board industry has grown at a relatively slow pace for the eleven-year period shown. The hardboard industry and the particleboard industry show about the same slope on the growth curves. However, it should be remembered that the particleboard is being reported on a ¾-in.-thick (19 mm) basis and the hardboard on a ⅛-in.-thick (3.2 mm) basis. Taking into account the normal higher density for most hardboard, there is about four times as much wood in a given square footage of particleboard as there is in the same footage of hardboard using the thickness basis reported. Thus, on a total usage basis, particleboard and hardboard account for about the same total weight of wood.

Much of the steep slope of the hardboard curve after 1970 can be accounted for by the rapidly developing medium-density fiberboard and siding products. Although these products are much thicker than basic hardboard, much of this

Table 2.8 *Apparent Consumption of Compressed Fiberboard (Hardboard)*

					Average annual growth						
Region	*1950*	*1960*	*1970*	*1973*	*1950-60*	*1960-70*	*1970-73*	*1950*	*1960*	*1970*	*1973*
		(thousand mt)				*(% per annum)*			*(mt per 1,000 capita)*		
North America	386	807	1,609	2,415	7.7	7.1	14.5	2.30	4.00	7.10	10.40
Europe	447	1,281	2,575	2,955	11.1	7.2	4.7	1.10	3.00	5.60	6.30
Japan	9	98	399	432	27.0	15.0	2.7	0.10	1.03	3.80	4.00
USSR	15	128	410	858	24.0	12.3	28.0	0.08	0.59	1.70	3.40
Asia (less Japan)	15	118	369	438	23.0	12.1	5.9	0.01	0.07	0.19	0.21
Latin America	21	96	150	176	16.0	4.6	5.5	0.12	0.45	0.53	0.47
Africa	11	42	87	100	14.0	7.6	4.8	0.05	0.15	0.25	0.26
Oceania	47	124	144	112	10.0	1.5	8.0	3.90	7.70	7.60	5.30
WORLD	951	2,694	5,743	7,486	11.0	7.9	9.2	0.38	0.90	1.58	1.95
Developed market economies	895	2,122	3,979	5,098	9.0	6.5	8.6	1.50	3.30	5.50	6.80
Developing market economies	40	213	419	494	18.0	7.0	5.6	0.03	0.15	0.24	0.26
Centrally planned economies	16	359	1,345	1,894	36.0	14.1	12.1	0.01	0.36	1.16	1.56

Source: FAO, *Proceedings of the World Consultation on Wood-Based Panels*, February 1975, Basic Paper No. 1.

Note: Totals do not always agree with the addition of subtotals, due to rounding off.

Table 2.9 *Apparent Consumption of Noncompressed Fiberboard (Insulation Board)*

Region	*1950*	*1960* (thousand mt)	*1970*	*1973*	*Average annual growth* 1950-60	1960-70 (% per annum)	1970-73	*1950*	*1960* (mt per 1,000 capita)	*1970*	*1973*
North America	861	1,124	1,299	1,733	4.2	4.4	10.1	5.20	5.60	5.70	7.40
Europe	174	308	413	437	5.9	3.0	1.9	0.44	0.72	0.89	0.92
Japan	4	30	72	93	22.0	9.1	8.9	0.04	0.31	0.69	0.86
USSR	6	86	197	206	30.0	8.6	1.5	0.03	0.40	0.81	0.82
Asia (less Japan)	5	10	13	17	7.2	2.7	9.4	–	–	–	–
Latin America	8	10	57	62	2.3	19.0	4.9	0.04	0.04	0.20	0.20
Africa	5	15	13	15	12.0	–1.4	4.9	0.02	0.05	0.03	0.04
Oceania	21	15	18	16	–3.3	1.8	–3.9	1.75	0.93	0.94	0.76
WORLD	1,084	1,598	2,082	2,579	4.0	2.7	7.4	0.43	0.53	0.57	0.67
Developed market economies	1.062	1,462	1,726	2,168	3.2	1.7	7.9	1.85	2.25	2.38	2.91
Developing market economies	15	16	71	78	0.6	16.0	3.2	0.01	0.01	0.04	0.04
Centrally planned economies	7	120	285	333	33.0	9.0	5.3	–	0.12	0.24	0.27

Source: FAO, *Proceedings of the World Consultation on Wood-Based Panels*, February 1975, Basic Paper No. 1.

Note: Totals do not always agree with the addition of subtotals, due to rounding off.

Table 2.10 *Apparent Consumption of Wood-Based Panels*

Region	1950	1960	1970	1973	Average annual growth 1950-60	1960-70	1970-73	1950	1960	1970	1973
	(million m^3)				(% per annum)			(m^3 per 1,000 capita)			
North America	7.40	15.20	28.2	39.2	7.5	6.3	12.0	45.0	77.0	125.0	168.0
Europe	2.60	8.50	23.1	32.4	13.0	10.5	12.0	6.7	20.0	50.0	69.0
Japan	0.22	1.22	8.3	11.3	18.0	21.0	11.0	2.7	13.0	80.0	105.0
USSR	0.80	1.90	5.5	7.1	8.9	11.0	9.1	4.5	8.9	23.0	29.0
Asia (less Japan)	0.14	0.70	1.5	3.0	19.0	8.4	24.0	0.1	0.4	0.8	1.4
Latin America	0.19	0.60	2.0	2.3	13.0	14.0	4.9	1.2	2.6	7.1	7.5
Africa	0.10	0.20	0.7	0.9	10.0	10.0	10.0	0.5	0.9	1.9	2.3
Oceania	0.20	0.40	0.8	0.8	6.4	7.1	1.7	18.9	25.0	41.0	39.0
WORLD	11.4	28.8	70.1	97.1	9.7	9.3	11.0	4.6	9.6	19.4	25.0
Developed market economies	10.10	24.0	56.5	78.8	9.0	8.9	12.0	17.6	37.0	78.0	106.0
Developing market economies	0.45	1.2	3.6	5.4	10.5	11.0	14.0	0.4	0.9	2.1	2.9
Centrally planned economies	1.00	3.5	10.1	12.9	13.0	11.0	8.6	1.1	3.6	8.7	10.6

Source: FAO, *Proceedings of the World Consultation on Wood-Based Panels*, February 1975, Basic Paper No. 1.

Notes: Includes veneer sheets except for 1950; consumption for: 1960 1.4 million m^3
1970 3.4 million m^3
1973 4.1 million m^3

Totals do not always agree with the addition of subtotals, due to rounding off.

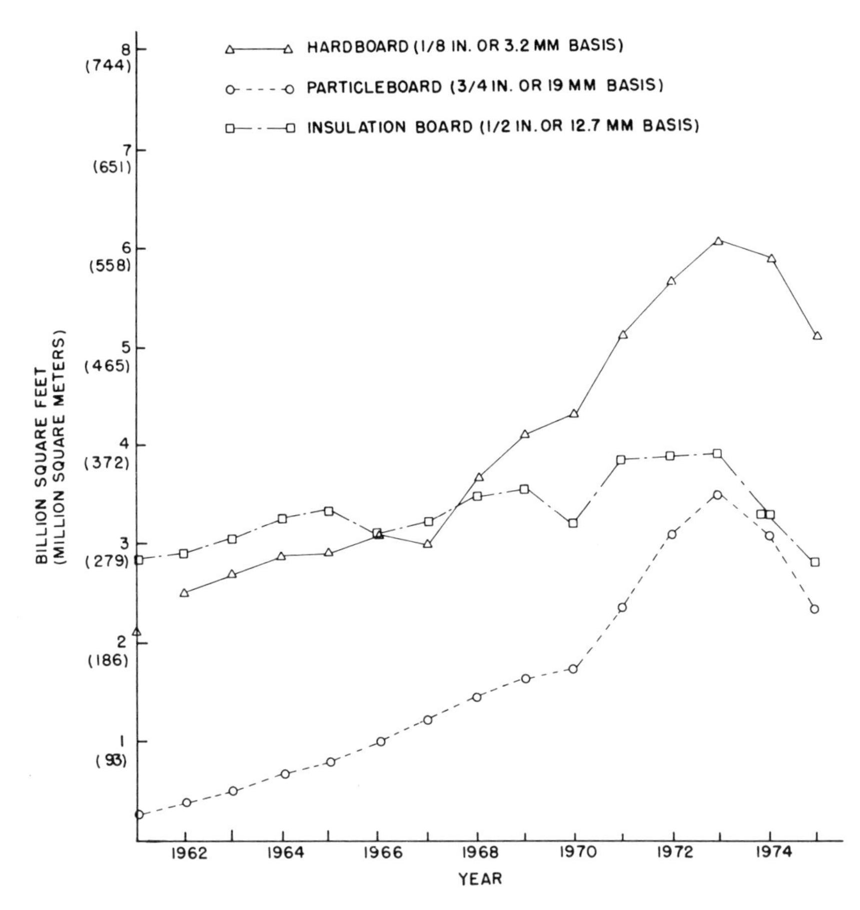

Figure 2.1. *Production of hardboard, particleboard, and insulation board in the USA. (U.S. Dept. of Commerce, Bureau of the Census.)*

production is reported as ⅛-in. (3.2 mm) hardboard. Consequently, as with much abbreviated statistical reporting, the built-in complications must be considered when interpreting the data. For an example in this case, converting ¾-in. (19 mm) production data to ⅛ in. (3.2 mm) increases production by six times because of the change in thickness.

Figures 2.2–2.5 show the consumption patterns for sawn wood and wood-based panels as a whole, plywood and veneer, particleboard, and fiberboard over the past 28 years with predictions through 1990. On these graphs is also the information developed for North America. This information was put together to consider the longer term outlook with regard to major forces shaping or changing the world economy and the position of wood-based panels within it.

Three different types of projections were made for future consumption: trend extension, balanced growth, and subjective judgment. Population and income

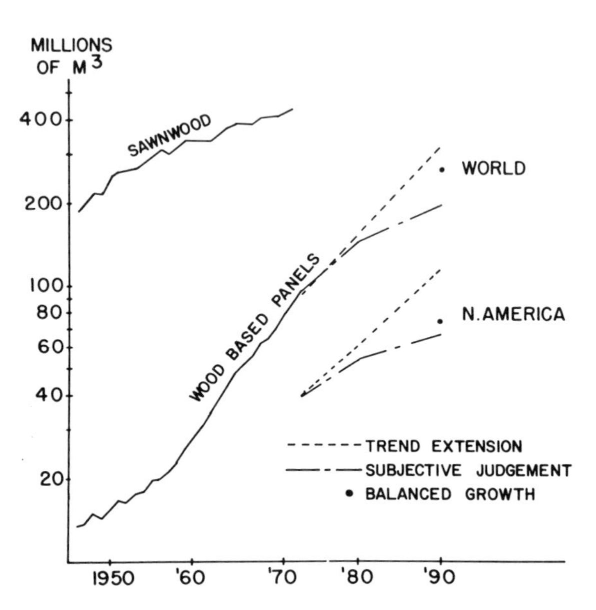

Figure 2.2. *Sawn wood and wood-based panel consumption. (Adapted from FAO 1975, Basic Paper No. 1.)*

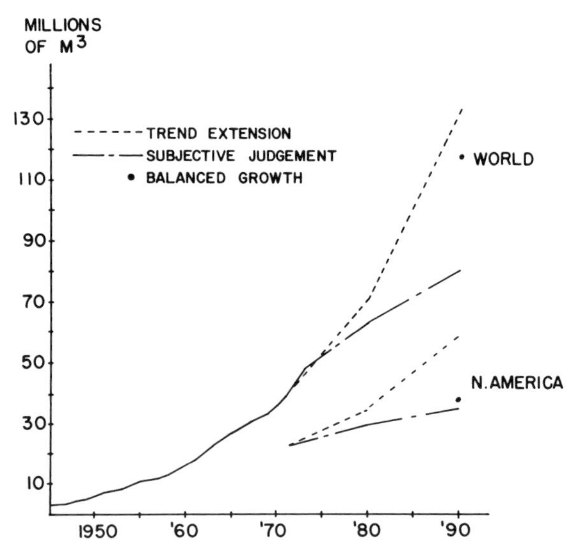

Figure 2.3. *Plywood and veneer consumption. Data on veneer included from 1960. (Adapted from FAO 1975, Basic Paper No. 1.)*

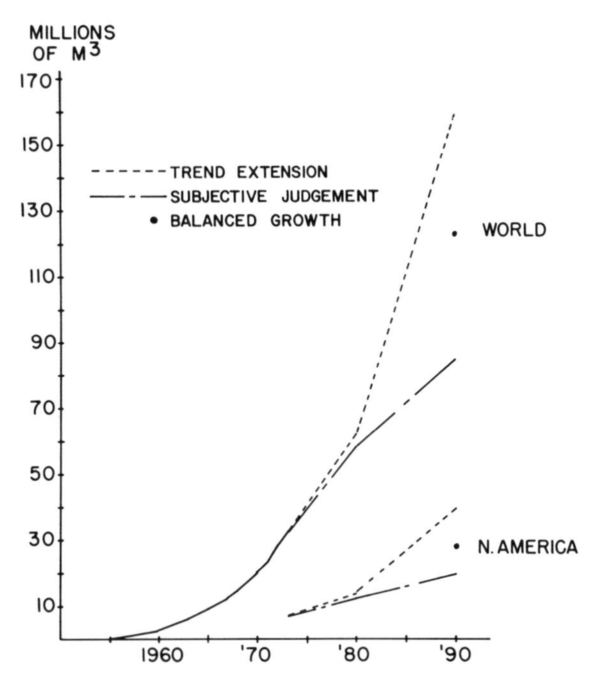

Figure 2.4. *Particleboard consumption. (Adapted from FAO 1975, Basic Paper No. 1.)*

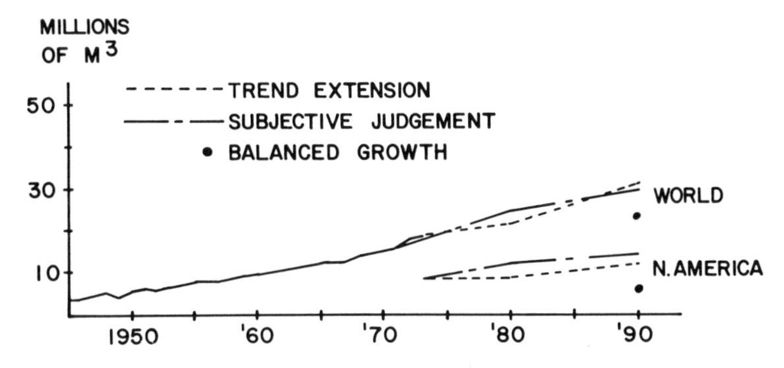

Figure 2.5. *Fiberboard consumption. (Adapted from FAO 1975, Basic Paper No. 1.)*

forecast for the trend extension estimate are those used for individual countries by FAO in its general projection work. The balanced growth is based on a set of population estimates, which assume the stabilization of population by 2001, through successive reductions in the expansion rate. This results in a 1991 population equivalent to that of 1981 in the trend extension alternative. The third projection is a conservative one, called subjective judgment developed on the considerations of market saturation in the developed regions, related to a decline in rates of substitution of wood-based panels for sawn wood and the continuing problems of investment funding of infrastructure and processing industries in developing countries, which will cause the rates of growth of consumption to fall rapidly on a world basis compared to recent levels and to a projected world income growth rate of 5%.

For this discussion the trend extension could be called the optimistic prediction and the subjective judgment a conservative prediction. The important point to observe, both on a world and North American basis is that all panel products are predicted to grow rapidly in the rate of consumption through 1990, according to all three predictions. Consequently, a two-fold problem of securing sufficient raw materials for production and the construction of many new plants needs to be solved.

BOARD GRADES

In the United States, extruded particleboard is exclusively used in captive markets where the board producer utilizes the product in secondary manufacturing. Mineral-bonded products are normally used in special applications such as ceiling tile, roof deck, sound control panels, and building blocks. Commercial standards similar to those of particleboard and hardboard have not been developed for these products as yet. The limited amount of molded product being produced out of comminuted products goes into specially produced products for the furniture industry and doors as mentioned earlier. The manufacturers have requirements they must follow to meet the various applications such as for chair seats, but no universally accepted standard has been developed.

In the case of the platen-pressed material, commercial or voluntary standards have been established to cover most of these products. As pointed out previously, there are overlaps between the product definitions. These standards will be discussed separately with some of the overlaps noted. Such standards are useful to establishing minimum properties and for providing useful data for specifiers, designers, and consumers.

Commercial Standard CS 251–63, "Hardboard," was established in 1963 for the production of hardboard and was an update of previous information. This standard has recently been supplanted by three voluntary product standards. The first, PS 58-73, "Basic Hardboard," was established to recognize dimensional and quality requirements for basic hardboard and to provide producers, distributors, and users with a basis for common understanding of the characteristics of this product. As shown in Table 2.11, basic hardboard is divided into five classes characterized by surface finish, thickness, and physical properties. (A number of these types of tables will be reproduced in this chapter. Pertinent properties will be shown; however, for complete information on how each property must be evaluated for meeting the standard or specification, the original document must be consulted.) The first four classes—tempered, standard, service-tempered, and service—are ones that have been around for many years. The physical properties cover water absorption and thickness swelling after a 24-hour water soak (which are maximum allowable values), minimum modulus of rupture or bending strength, and tensile strength parallel to and perpendicular to the surface. Many other physical properties are used in design for these products, but the ones just mentioned are those listed in the standard.

New in the standard is the fifth class, which is industrialite. This class was established (mainly) to cover the relatively new medium-density fiberboard. Maximum or minimum physical properties values have been established for this new product following the same format as for the other four classes of hardboard.

Tables 2.12 and 2.13 present the physical properties now required for hardboard siding as noted in the second voluntary product standard PS 60-73, "Hardboard Siding." Much of this material is produced by the dry process.

The third voluntary product standard is PS 59-73, "Prefinished Hardboard Paneling." This standard covers the four classes of hardboard: tempered, standard, service-tempered, and service. The modulus of rupture and tensile strength properties are the same as listed in the basic hardboard standard. Water resistance values are the same as the S2S standards listed in the basic hardboard specifications. Three thicknesses of board are covered: 1/8, 3/16, and 1/4 in. (3.2, 4.8 and 6.4 mm). Requirements are also set for abrasion resistance, adhesion, fade resistance, gloss, and so forth. These requirements can be found in this product standard but are not listed here since they will not be under discussion.

All of these standards also provide requirements for dimensions, thickness tolerances, squareness, edge straightness, and moisture content.

A standard was first produced for mat-formed wood particleboard in 1961. With more experience in the field and far more production developing in the 1960s this standard was rewritten in 1966 and is the present one in force. It is designated Commercial Standard CS 236-66, "Mat-formed Wood Particleboard." Table 2.14 presents the property requirements for particleboard as specified in this standard.

Two types have been established with Type 1 generally made with urea-formaldehyde resin binders for interior applications and Type 2 generally made with phenolic-type binders suitable for interior and certain exterior applications when so labeled. Products are further differentiated by density grade with an A, B, and C level for the interior applications and A and B density grade for exterior applications. At present, the C, or low-density grade, is not specified for exterior board. As a further differentiation, there are two classes in each of the density grades with each class having a separate set of physical properties established for modulus of rupture, modulus of elasticity, internal bond (tensile strength perpendicular to the surface), linear expansion, and face and edge screw holding.

In contrast to the hardboard products, no water absorption properties are specified for particleboard. Hardboard, on the other hand, does not have a linear expansion property specified. Hardboard is noted as having a minimum specific gravity of 0.50 (31.5 lbs/ft^3) with no differentiation established for specific gravity when discussing the five classes of hardboard. Particleboard, on the other hand,

Table 2.11 *Classification of Hardboard by Surface Finish, Thickness, and Physical Properties*

Surface	*Nominal thickness*		*Water resistance (max avg per panel)*				*Modulus of rupture (min avg per panel)*		*Tensile strength (min avg per panel)*			
			Water absorption based on weight		*Thickness swelling*				*Parallel to surface*		*Perpendicular to surface*	
			S1S	*S2S*	*S1S*	*S2S*						
	(in.)	*(mm)*	*(%)*	*(%)*	*(%)*	*(%)*	*(psi)*	*(MPa)*	*(psi)*	*(MPa)*	*(psi)*	*(MPa)*
Class 1: Tempered												
S1S	1/12	2.1	30	–	25	–						
S1S and S2S	1/10	2.5	20	25	16	20						
	1/8	3.2	15	20	11	16						
	3/16	4.8	12	18	10	15	7,000	48.3	3,500	24.1	150	1.03
	1/4	6.4	10	12	8	11						
	5/16	7.9	8	11	8	10						
	3/8	9.5	8	10	8	9						
Class 2: Standard												
S1S and S2S	1/12		40	40	30	30						
	1/10		25	30	22	25						
	1/8		20	25	16	18						
	3/16		18	25	14	18	5,000	34.5	2,500	17.2	100	0.69
	1/4		16	20	12	14						
	5/16		14	15	10	12						
	3/8		12	12	10	10						
Class 3: Service-tempered												
S1S and S2S	1/8		20	25	15	22						
	3/16		18	20	13	18	4,500	31.0	2,000	13.8	100	0.69
	1/4		15	20	13	14						
	3/8		14	18	11	14						

Class 4: Service											
	1/8		30	30	25	25					
S1S	3/16		25	27	15	22					
and	1/4		25	27	15	22					
S2S	3/8		25	27	15	22					
	7/16	11.1	25	27	15	22					
	1/2	12.7	25	18	15	14					
	5/8	15.9	–	15	–	12	3,000	20.7	1,500	10.3	75	0.52
	11/16	17.5	–	15	–	12					
	3/4	19.0	–	12	–	9					
S2S	13/16	20.6	–	12	–	9					
	7/8	22.2	–	12	–	9					
	1	25.4	–	12	–	9					
	1-1/8	28.6	–	12	–	9					
Class 5: Industrialite											
S1S	3/8		25	25	20	20					
and	7/16		25	25	20	20					
S2S	1/2		25	25	20	20					
	5/8		–	22	–	18					
	11/16		–	22	–	18	2,000	13.8	1,000	6.9	35	0.24
	3/4		–	20	–	16					
S2S	13/16		–	20	–	16					
	7/8		–	20	–	16					
	1		–	20	–	16					
	1-1/8		–	20	–	16					

Source: PS58-73 "Basic Hardboard." (See Table 19.1 for updated information.)

Table 2.12 *Physical Properties of Hardboard Siding*

Property		*Requirements*
Percent water absorption based on weight (max avg per panel)		Primed 15 Unprimed 20
Percent thickness swelling (max avg per panel)		Primed 10 Unprimed 15
Weatherability of substrate (max swell after 5 cycles)	(in.) (mm)	0.010 0.25 and no objectionable fiber raising
Sealing quality of primer coat		No visible flattening
Weatherability of primer coat		No checking, erosion, or flaking
Nail-head pull-through (min avg per panel)	(lbs) (kg)	150 68.2
Lateral nail resistance (min avg per panel)	(lbs) (kg)	150 68.2
Modulus of rupture (min avg per panel)	(psi) (MPa)	1,800 for 3/8- & 7/16-in.-thick siding 3,000 for 1/4-in.-thick siding 12.4 for 9.5- & 11.1-mm-thick siding 20.7 for 6.4-mm-thick siding
Hardness (min avg per panel)	(lbs) (kg)	450 204.5
Impact (min avg per panel)	(in.) (mm)	9.0 229
Moisture content	(%)	2.0–9.0 incl, and not more than 3% variance between any two boards in any one shipment or order.

Source: PS60-73, "Hardboard Siding." (See Table 19.2 for updated information.)

Table 2.13 *Maximum Linear Expansion*

Type of siding	*Thickness range*		*Maximum linear expansion*
	(in.)	*(mm)*	*(%)*
Lap	0.325-0.375	8.3-9.5	0.38
	over 0.376	over 9.5	0.40
Panel	0.220-0.265	5.6-6.7	0.36
	0.325-0.375	8.3-9.5	0.38
	over 0.376	over 9.5	0.40

Source: PS60-73 "Hardboard Siding."

has three definite density grades. Particleboard specifies minimum values for modulus of elasticity and face and edge screw holding, and these are not included in the hardboard standard. Besides the water resistance properties noted, tensile strength parallel to the surface is specified for hardboard but not for particleboard.

For the Type 2 or exterior-type particleboard, an accelerated aging test is required by the particleboard standard. This test is not required for hardboard used in exterior exposures. When applicable, the minimum average modulus of rupture after aging is not to be less than 50% of the minimum average modulus of rupture of samples not exposed to the aging tests. The calculation for determining modulus of rupture uses the specimen thickness before aging. During the aging tests a very significant increase in thickness can take place, and, if this thickness is used for the calculation, the modulus of rupture will be much lower than what is reported following the specifications of the standard. This is a somewhat confusing piece of terminology, and perhaps the term modulus of rupture should not have been used. What the test is designed to do is determine the amount of strength retention of the board after accelerated aging testing. The intent was to have an accelerated and replicable means of determining the durability characteristics of exterior particleboard and, in simple terms, is in actuality a quality control procedure. While termed *accelerated* this test still takes over two weeks to conduct.

When hardness is designated as an optional requirement for particleboard, the value is to be 800 lbs (364 kg) for A-density panels and 500 lbs (227 kg) for B-density panels.

As with hardboard, this standard also specifies dimensional tolerance for width, length and thickness, squareness and straightness, and specifies that moisture content of the panels at the time of shipment shall conform with one of the following requirements:

A. Moisture content consistent with a known end use of the panel.
B. Moisture content as specified by the purchaser.
C. Moisture content not in excess of 10%.

Table 2.14 *Property Requirements of Mat-Formed Wood Particleboard*

Density (grade) (min avg)	*Class‡*	*Modulus of rupture (min avg)*		*Modulus of elasticity (min avg)*		*Internal bond (min avg)*		*Linear expansion (max avg)*	*Screw holding*			
									Face (min avg)		*Edge (min avg)*	
		(psi)	*(MPa)*	*(psi)*	*(MPa)*	*(psi)*	*(MPa)*	*(%)*	*(lbs)*	*(kg)*	*(lbs)*	*(kg)*
*Type (use) 1.**												
A. High density, 50 lbs/ft^3 or 0.80 sp gr and over	1	2,400	16.5	350,000	2,413	200	1.38	0.55	450	204.5	–	–
	2	3,400	23.4	350,000	2,413	140	0.97	0.55	–	–	–	–
B. Medium density, 37-50 lbs/ft^3 or 0.59-0.80 sp gr	1	1,600	11.0	250,000	1,724	70	0.48	0.35	225	102.3	160	72.7
	2	2,400	16.5	400,000	2,758	60	0.41	0.30	225	102.3	200	90.9
C. Low density, 37 lbs/ft^3 or 0.59 sp gr and under	1	800	5.5	150,000	1,034	20	0.14	0.30	125	56.8	–	–
	2	1,400	9.7	250,000	1,724	30	0.21	0.30	175	79.5	–	–
Type (use) 2.†												
A. High density, 50 lbs/ft^3 or 0.80 sp gr and over	1	2,400	16.5	350,000	2,413	125	0.86	0.55	450	204.5	–	–
	2	3,400	23.4	500,000	3,447	400	2.76	0.55	500	227.3	350	159.1
B. Medium density, 37-50 lbs/ft^3 or 0.59-0.80 sp gr	1	1,800	12.4	250,000	1,724	65	0.45	0.35	225	102.3	160	72.7
	2	2,500	17.2	450,000	3,102	60	0.41	0.25	250	113.6	200	90.9

Source: CS236-66. (See Table 19.4 for updated information.)

* Type 1: Mat-formed particleboard (generally made with urea-formaldehyde-resin binders) suitable for interior applications.

† Type 2: Mat-formed particleboard made with durable and highly moisture- and heat-resistant binders (generally phenolic resins) suitable for interior and certain exterior applications when so labeled.

‡ Class: Strength classifications based on properties of panels currently produced.

It also takes into account that particleboard is a wood product and has a tendency to change moisture content with changes in atmospheric conditions; thus, no specific moisture content is guaranteed when the particleboard panels reach their destination.

With the rapid development of this industry, the market uses of the product have also changed as the particleboard has moved into new end uses in an effort to provide boards designed particularly for the more discrete end uses. The commercial standard is at present being rewritten. The basic change will apparently be the addition of a third level in the class category. These three classes will permit the specifiers to designate with more accuracy the particular board that they wish to use.

The Canadians established a standard for Mat-Formed Wood Particleboard under CSA Standard 0188–1968. This standard has been recently revised to CSA standard 0188–1975. Significant changes have been made in this new standard. Types of board based on the resin have been deleted as unnecessary. The distinction between *building* and *industrial* boards has been dropped. Grades P & Q refer only to waferboard which is an important product in Canada. New grades have also been introduced. The grades and property requirements are shown in Table 2.15. The description of end-use applications has been modified to provide an indication of the interchangeability of various grades of board (Table 2.16).

The physical properties for these boards are generally the same as those for the United States. Of particular interest are higher allowable linear expansion values for some grades of board and a lower internal bond in some grades as compared to United States commercial standards.

The methods for determining the various physical properties discussed are specified in the American Society for Testing and Materials (ASTM) D 1037, "Standard Methods of Evaluating the Properties of Wood-Base Fiber and Particle Panel Materials." It should be noted that some exceptions to the testing procedures are noted in the commercial standards. These test procedures will be dealt with later.

Other specifications are used when considering particleboard or hardboard for particular applications. These are usually Use of Materials Bulletins prepared under the auspices of the Department of Housing and Urban Development (HUD), Federal Housing Administration (FHA). These bulletins are used by the FHA to determine the eligibility of a property as security for an insured loan under the FHA program. The Use of Materials Bulletins are used preliminary to a material's being included in the FHA minimum property standards for one or two living units.

Particleboard floor underlayment has certain standards to meet as specified in "HUD Minimum Property Standards," 1973 edition. Underlayment shall comply with CS 236–66 type 1–B1 and, in addition shall not have a maximum thickness swelling of more than 0.063 in. (1.6 mm) when immersed in water for 24 hours in accordance with ASTM D–1037. For a nominal ⅝-in.-thick (15.9 mm) board, this is 10% swell.

The Use of Materials Bulletin No. UM-32 covers "Mat-Formed Particleboard for Exterior Use." Board properties listed must meet the standards established by the West Coast Particleboard Association specification for exterior-type wood

Table 2.15 *Test Requirements for Various Grades of Particleboard*

Property		*Grade P*	*Grade Q*	*Grade E*	*Grade F*	*Grade G*	*Grade H*	*Grade K*	*Grade L*	*Grade R*	*Grade S*	*Grade T*	*Test clause**	*Moisture content at test*
Thickness tolerance														
Maximum ± average deviation from nominal	(in.)	0.030	0.015	0.015	0.030	0.015	0.010	0.030	0.015	0.005	0.005	0.010		As shipped
	(mm)	0.76	0.38	0.38	0.76	0.38	0.25	0.76	0.38	0.13	0.13	0.25		
Maximum ± deviation													6.2	
from panel average	(in.)	0.030	0.010	0.010	0.030	0.010	0.005	0.030	0.010	0.005	0.005	0.005		As shipped
	(mm)	0.76	0.25	0.25	0.76	0.25	0.13	0.76	0.25	0.13	0.13	0.13		
Length and width tolerance		+0	+0	+0	+0	+0	+0	+0	+0	+0	+0	+0		
Maximum deviation from nominal	(in.)	−1/8	−1/8	−1/8	−1/8	−1/8	−1/8	−1/8	−1/8	−1/8	−1/8	−1/8	6.1	As shipped
	(mm)	3.2												
Squareness tolerance														
Maximum deviation from square	(in./ft)	1/64	1/64	1/64	1/64	1/64	1/64	1/64	1/64	1/64	1/64	1/64	6.3	As shipped
	(mm/m)	1.3												
Straightness tolerance														
Maximum deviation from straight	(in./edge)	1/16	1/16	1/16	1/16	1/16	1/16	1/16	1/16	1/16	1/16	1/16	6.4	As shipped
	(mm/edge)	1.6												
Density tolerance														
Maximum deviation from nominal	(%)													
Over 1/2 in. or 12.7 mm thickness		NS	NS	NS	NS	±10	±10	NS	NS	±8	±10	±10	6.5.3	As shipped
1/2 in. or 12.7 mm thickness and under		NS	NS	NS	NS	±15	±15	NS	NS	±10	±15	±15	6.5.3	As shipped
Minimum modulus of rupture	(psi)	2,000	2,000	1,500	1,300	1,300	1,700	2,400	2,400	2,000	1,700	1,700	6.6	EMC at 65% RH
	(MPa)	13.8	13.8	10.3	9.0	9.0	11.7	16.5	16.5	13.8	11.7	11.7		

Minimum modulus of elasticity	(psi) (MPa)	400,000 2,758	400,000 2,758	160,000 1,103	160,000 1,103	160,000 1,103	160,000 1,103	400,000 2,758	400,000 2,758	280,000 1,930	160,000 1,103	160,000 1,103	6.6	EMC at 65% RH
Minimum internal bond	(psi) (MPa)	40 0.28	40 0.28	40 0.28	NS	NS	50 0.34	NS	NS	65 0.45	50 0.34	40 0.28	6.7	EMC at 65% RH
Minimum hardness	(lbs) (kg)	500 227.2	500 227.2	500 227.2	NS	350 159.1	500 227.2	500 227.2	500 227.2	500 227.2	500 227.2	350 159.1	6.8	EMC at 65% RH
Minimum screw holding strength.														
Face	(lbs) (kg)	NS	NS	NS	NS	NS	250 113.6	NS	NS	250 113.6	200 90.9	NS	6.9	
Edge (for boards 5/8 in. or 15.9 mm and thicker)	(lbs) (kg)	NS	NS	NS	NS	NS	NS	NS	NS	150 68.2	100 45.5	NS	6.9	EMC at 65% RH
Maximum linear expansion	(%) (50 to 90% RH)	0.25	0.25	0.35	0.50	0.50	0.35	0.25	0.25	0.35	0.35	0.50	6.10	EMC at 50% RH EMC at 90% RH
Minimum direct nail withdrawal (for boards 3/8 in (9.5 mm) and thicker)	(lbs) (kg)	20 9.1	20 9.1	NS	NS	NS	NS	NS	NS	NS	NS	NS	6.11	EMC at 65% RH
Properties after accelerated aging														
Minimum modulus of rupture	(psi) (MPa)	1,000 6.9	1,000 6.9										6.12	Saturated

Source: CSA Standard 0188-1975

Note: The abbreviation NS (not specified) is used in the table whenever the grade has no requirement for the property.

* Noted in Standard.

Table 2.16 *Present-Day Uses of Wood Particleboard*

Uses	*Basic grade(s)*	*Alternate grades*										
		P	*Q*	*E*	*F*	*G*	*H*	*K*	*L*	*R*	*S*	*T*
High-quality furniture core	R											
Medium-quality furniture core	S									x		
Wall panel core	T						x			x	x	
Cabinets and vanities	H									x		
Interior wall sheathing	F	x	x	x		x	x	x	x	x	x	x
Mobile home decking "unsanded"	P or K		x						x			
Mobile home decking "sanded"	Q or L											
Underlayment	E		x				x			x	x	
Subflooring	P		x									
Exterior wall and roof sheathing	P		x									
Interior cladding	G						x			x	x	x
Exterior cladding, fences and screens	P		x									

Source: CSA Standard 0188–1975.

Note: This table provides a listing of the most common end-use applications of particleboard. For each application, a basic grade is shown. Also shown, by means of check marks, are other grades which meet (or exceed) the requirements of this standard for the basic grade, and are, in that sense, equally suited to the application associated with the basic grade. Other factors such as appearance, thickness, and fastening methods may also affect the suitability for the listed end uses.

particleboard as revised on May 1, 1961. The West Coast Particleboard Association is no longer in existence. It was one of the forerunners in the field and has been supplanted by the National Particleboard Association. The specified properties as developed at that time are shown in Table 2.17. This bulletin was being revised at the time this book was written.

Another type of federal specification includes the ones used by the General Services Administration of the United States Government for use in procuring materials by the various agencies of the federal government. For particleboard, LLL–B–800A dated May 15, 1965, "Federal Specification on Building Board (Wood Particleboard) Hard-Pressed, Vegetable Fiber," is used. This references the particleboard Commercial Standard CS 236–66 and covers other specifications, particularly in connection with packaging for shipment and procedures for inspection.

There are other government specifications regulating the use of particleboard in residential construction. An example is the Department of Defense "Guide Specification for Military Family Housing (DOD4270.21–SPEC)" covering floor underlayment.

Another type of standard includes those developed by the National Particleboard Association for particular end uses. NPA 1–73 is the standard for "Particleboard Mobile Home Decking." This specifies the modulus of rupture, modulus of elasticity, internal bond, and linear expansion—properties required for the two grades of decking. Also included in a concentrated load test. All boards must have a minimum hardness of 500 lbs (227 kg). Thickness swell cannot exceed 8%. NPA 2-72 is a "Standard for Particleboard Decking for Factory Built Housing." This references Commercial Standard 236-66 and notes that the board must be a Type 2-B-2 board which is bonded with phenolic or similar resin. A hardness requirement of 500 lbs (227 kg) is noted and the linear expansion is set at no more than 0.35% as against a level of 0.30% specified for mobile home decking. In addition, this type of board must pass a concentrated load test. In both standards a minimum board thickness of ⅝ in. (15.9 mm) is recommended.

In August 1973, the National Particleboard Association published NPA 4–73, "Standard for Medium Density Fiberboard." Table 2.18 presents the property requirements of this board, which are quite similar to medium-density particleboard. Of particular interest is the internal bond value of 100 psi (690 kPa). This is contrasted to the industrialite specification as noted in the hardboard voluntary standard PS58–73, where the internal bond value is a minimum of 35 psi (240 kPa). Also included in this NPA standard is a linear expansion of 0.30%, which is not specified at all in the hardboard requirements.

The National Particleboard Association has also developed specifications for particleboard floor underlayment coated with wax-polymer-type hot-melt coatings and for the wax-polymer-type hot-melt coatings.

Another approach for gaining approval for particleboard in certain construction markets is to apply to a building code group, such as the International Conference of Building Officials (ICBO). An independent testing agency provides ICBO with research data developed by conducting prescribed tests on the particleboard in question. ICBO then acts upon this information and publishes a research committee recommendation on the particleboard. An example of this

Table 2.17 *Required Properties of Particleboard for Exterior Use*

Property		*Value*
Thickness	(in.) (mm)	3/8, 1/2, 5/8, 3/4, 1 9.5, 12.7, 15.9, 19, 25.4 Tolerance ±0.010 in. or 0.25 mm
Size	(ft) (m)	4x2 to 4x12 1.22x0.61 to 1.22x3.66
Moisture Content (as shipped)	(%)	8 ±2
Density*	(lbs/ft^3) (sp gr)	41 min avg 0.66
Modulus of Rupture	(psi) (MPa)	1,800 min avg 12.4
Modulus of Elasticity	(psi) (MPa)	250,000 min avg 1,724
Internal Bond	(psi) (MPa)	65 min avg 0.45
Hardness	(lbs) (kg)	500 min 227.3
Dimensional stability† (% linear expansion)		0.35 max avg
Water absorption‡ (% weight increase – 24 hr)		18 max avg
Swelling§ (% thickness increase – 24 hr)		15 max avg

Nail holding (min avg)

Lateral nail resistance

Thickness	*1/4 in. or 6.4 mm from edge*		*1/2 in. or 12.7 mm from edge*	
	(lbs)	*(kg)*	*(lbs)*	*(kg)*
3/8 in. or 9.5 mm	70	31.8	150	68.2
1/2 in. or 12.7 mm	90	40.9	200	90.9
5/8 in. or 15.9 mm	130	59.1	250	113.6
3/4 in. or 19.0 mm	180	81.8	300	136.4

Direct withdrawal	*Board thickness*		*Load*	
	(in.)	*(mm)*	*(lbs)*	*(kg)*
	3/8	9.5	20	9.1
	1/2	12.7	30	13.6
	5/8	15.9	50	22.7
	3/4	19.0	80	36.4

Table 2.17 *Continued*

Nail holding (min avg) *(continued)*

Nail head pull-through	Board thickness		Load	
	(in.)	*(mm)*	*(lbs)*	*(kg)*
	3/8	9.5	175	79.5
	1/2	12.7	250	113.6
	5/8	15.9	350	159.1
	3/4	19.0	500	227.3

Accelerated aging

All aged specimens shall remain substantially intact and shall retain the following minimum properties:

Modulus of Rupture	(psi)	900 min avg
	(MPa)	6.2 (based on original thickness)

Nail holding (min avg)

Lateral nail resistance

Thickness	*1/4 in. or 6.4 mm from edge*		*1/2 in. or 12.7 mm from edge*	
	(lbs)	*(kg)*	*(lbs)*	*(kg)*
3/8 in. or 9.5 mm	15	6.8	75	34.1
1/2 in. or 12.7 mm	40	18.2	100	45.6
5/8 in. or 15.9 mm	60	27.3	125	56.8
3/4 in. or 19.0 mm	75	34.1	150	68.2

Direct withdrawal	*Board thickness*		*Load*	
	(in.)	*(mm)*	*(lbs)*	*(kg)*
	3/8	9.5	10	4.5
	1/2	12.7	20	9.1
	5/8	15.9	35	15.9
	3/4	19.0	50	22.7
Nail head pull-through	3/8	9.5	85	38.6
	1/2	12.7	125	56.8
	5/8	15.9	175	79.5
	3/4	19.0	250	113.6

Source: UM-32, "Mat-Formed Particleboards for Exterior Use."

* Board density based on volume and moisture content at time of shipment.

† Specimens shall be conditioned at 30% and 90% relative humidity. Conditioning shall continue at each humidity until practical equilibrium is reached.

‡ Specimens shall be submerged horizontally under 1 in. or 25.4 mm of water.

§ Caliper will be taken 1 in. in from the midpoint of each side.

Table 2.18 *Property Requirements for Medium-Density Fiberboard*

Density (lbs/ft^3)	Modulus of rupture (psi)	Modulus of elasticity (psi)	Internal bond (psi)	Linear expansion (%)	Screw holding	
					face (lbs)	edge (lbs)
31-50	3,500	300,000	100	0.30	350	300
(sp gr)	(MPa)	(MPa)	(MPa)		(kg)	(kg)
0.48-0.80	24.1	2,068	0.69		159	136

Source: NPA 4-73. (See Table 19.6 for updated information.)

approach is given in Figure 2.6, which reproduces one page of an ICBO recommendation covering the exterior particleboard produced by Humboldt Flakeboard of Sierra Pacific Industries (which is now part of the Louisiana-Pacific Corp.).

The National Particleboard Association has established a grading system covering underlayment-grade particleboard, mobile home decking, and factory-built decking. The minimum standards for board physical properties are those found in the applicable commercial or FHA specification. The prescribed number of samples are tested daily by mills and periodically by NPA; if the quality-control department of the plant producing the board is found to be doing its job and the boards are meeting these standards, the producer is allowed to grade stamp its product with the NPA stamp. If the quality-control department is found not to be effective, the grade stamp is removed.

Not all panel producing companies belong to the National Particleboard Association or the American Board Products Association (previously the American Hardboard Association and the Acoustical and Insulating Materials Association). Other plants or member plants may elect to use a private testing agency or do their own testing to establish the quality of their board. Similar techniques are used by these testing agencies for determining whether the board meets the specifications established. Implementation of these quality-surveillance and certification techniques will be discussed later under "quality control."

As would be expected, many different standards and specifications have been developed in other countries but discussion of these is not within the scope of this book.

TYPICAL PRODUCTS

A huge market for hardboard over the years has been in the furniture trade, where material can be used for cores over which are bonded fine veneers, decorative plastics, fabrics, or other materials. In this same marketplace hardboard is used in commercial displays, cabinetwork, as panels, shelves, doors, bulletin boards, and merchandise racks, in all sorts of shapes and patterns. In the

International Conference of Building Officials

RESEARCH COMMITTEE RECOMMENDATION

Report No. 2211 *March, 1976*

REDEX PARTICLEBOARD
LOUISIANA-PACIFIC CORPORATION
SAMOA DIVISION — HUMBOLDT FLAKEBOARD
POST OFFICE DRAWER "CC"
ARCATA, CALIFORNIA 95521

I. **Subject:** Redex Particleboard.

II. **Description: General:** The Redex particleboard is composed of wood particles bonded together under pressure by means of a ureaformaldehyde or a phenolic resin binding agent, varying in amount from 7 to 8 percent depending upon the board thickness. The panels have a density of 40 to 46 pounds per cubic foot, and are supplied in 4-foot wide units with lengths from 8 to 16 feet in 1-foot increments.

Smooth finish or sanded thicknesses vary from ⅜ inch to 1¼ inch in modules of ⅛ inch. Minimum nailing consists of bright or corrosion-resistant box, ring shank or common nails as specified for plywood in Table No. 25-P of the Uniform Building Code. Nails are placed with a minimum edge distance of ⅜ inch.

Roof and Floor Sheathing: Five-eighths and thicker panels bonded with phenolic resin may be used as roof sheathing, when protected by an approved weatherproof surfacing material. Not less than ¾-inch thick phenolic panels may be used as a combination subfloor and underlayment material. Allowable support spacings and loads are indicated in Table No. I.

All unsupported edges are blocked or have a tongue-and-groove configuration. The panels shall be laid normal to the span of the rafters or joists and shall be continuous over two or more spans. Tongue-and-groove panels shall be installed with the grade stamp up to ensure proper surface alignment of the joints. Cut-outs for items such as plumbing and electrical shall be oversized to avoid a forced fit. To avoid cracking of the panel edges, framing members shall be beveled to properly fit the panel surface.

Panels shall be installed with a 1/16-inch gap at all panel edges. A ¾-inch expansion joint shall be provided so that no continuous section of panels shall exceed 80 feet. A ½-inch gap shall be provided between the panels and concrete or masonry walls.

The Redex particleboard is capable of resisting roof and floor diaphragm shear loadings due to wind or seismic forces in accordance with Table No. II. All boundary members shall be proportioned and spliced where necessary to transmit all direct stresses. The nominal width of framing members and blocking shall be not less than 2 inches.

Underlayment: Three-eighths-inch and thicker panels may be used as an underlayment when required in Chapter 25 of the Uniform Building Code. Nailing shall be as set forth above.

Identification: All panels bear the stamp of Timber Engineering Company and the name of the manufacturer. The panels using ureaformaldehyde resin are identified as Type 1-B-2 and the panels using a phenolic resin are identified as Type 2-B-2 with a modulus of rupture of 3000 pounds per square inch and a modulus of elasticity of 500,000 pounds per square inch.

III. **Evidence Submitted:** Structural calculations, load tests, racking tests, nail tests and tests to establish the basic physical properties are submitted in conformance with U.B.C. Standard No. 25-25.

Recommendation

IV. Recommendation: That the Redex particleboard is an alternate type of material to that specified in the Uniform Building Code for the uses described in this report, subject to the following conditions:

1. **The panels are installed as set forth in this report.**
2. **The support spacings and loads do not exceed those indicated in Table No. I.**
3. **The allowable loads for the particleboard, used for roof or floor diaphragms which resist wind or seismic forces, do not exceed the values set forth in Table No. II.**
4. **The Timber Engineering Company, Eugene, Oregon, (AA-500) be maintained as the quality control agent for the fabrication of the material.**

This recommendation is subject to annual re-examination.

Figure 2.6. *A building group reports its recommendations for approval of a product in a particular construction market. This reproduction of an ICBO report does not include the tables. (Courtesy International Conference Building Officials.)*

construction field, concrete forms, ceiling material, door skins, and panels, are some examples of its use. In recent years the use of hardboard siding approximately 7/16 in. (11.1 mm) in thickness has become widespread. Also, 4x8-ft (1.22x2.44 mm) sheets of specially manufactured textured hardboard panels are used for exterior siding on houses. Hardboard is well known to most American families in the form of interior wood paneling which has photographic prints, usually of fine wood, applied. This is an easy-to-use product that the home handyman finds quite useful. This board can be used in many types of other items ranging from small applications in toys to paneling in highway truck trailers and railroad cars. Most of these markets have been well developed by the industry for many years.

Medium-density fiberboard has moved strongly into the furniture field as an industrial core material. It can be finished directly or overlaid with fine wood veneer or plastic laminates. It competes mostly with particleboard in this market.

As mentioned previously, extruded particleboard is used exclusively in the United States for core material in furniture applications and it is a relatively minor amount of the total production. Platen-pressed particleboard is used for four general categories of products. One is for floor underlayment which covers the rough subfloor in housing and over which sheet goods, tile, or carpet is laid. The use of a particleboard underlayment provides a smooth, even floor for this final building operation. Particleboard mobile home decking is another significant use developed since the mid-1960s. Most of the mobile homes now manufactured use particleboard for the floor system. The third major grouping is the furniture or industrial board, and the fourth category is that of ever-present miscellaneous uses which are open to the consumer's imagination as far as how the board is used. It is estimated that production can be divided up approximately as follows for particleboard:

- □ 50% industrial (mainly furniture board).
- □ 25% construction (mainly floor underlayment).
- □ 15% mobile home decking.
- □ 10% specialty or miscellaneous products (mainly industrial-type board).

The following listing will provide a sampling of a wide variety of uses for particleboard: countertops, shelves, sink tops, dinette sets, tabletops, cabinets, wall cases, tool benches, drafting tables, case goods, sewing tables, desks, pool tables, card tables, cabinet and drawer parts, show cases, chair components, table tennis tops, game boards, fence components, furniture partitions, benches, bookcases, chalkboards, kitchen cabinets, consoles, laboratory furniture, piano and organ parts, signs, door cores, bi-fold doors, church furniture, office furniture, computer cases and panels, children's furniture, garage doors, patio furniture, toys, pleasure-boat interior parts, tool chests, gun cases, casket components, metal work patterns, shuffleboard pucks, school furniture, mobile homes, massage tables, filing cabinets, engine crates, ironing boards, tote boxes, cutting boards, basketball backboards, floor underlayment, decking, sheathing, panels and wainscoting, soffits, wardrobes, interior window and door trim, baseboards, valances, shutters, face frames, stair risers, and treads.

TRENDS

The particleboard industry has been increasing its production at a rate of 15 to 20% every year for a period of 10 to 15 years. A much higher increase occurred in 1973. In 1974, because of the recession, the first decrease in production took place since the industry was established in the United States. Significantly, the decreased level of production would not have been sufficient to cover the market needs of 1970. It is anticipated that this phenomenal growth record will continue into the 1980s, perhaps with some slowing of the rate of expansion. During the past decade, fears have been periodically expressed that the particleboard industry in particular was overbuilt. A case in point was when the massive new production line at Roseburg Lumber came onstream in 1972. However, the consuming public absorbed this new production readily, and the particleboard producers in general during 1973 has their production sold out for three to four months ahead of time. Granted this was a particularly good year, but this has been typical of the industry as its production capacity has been enlarged.

The cost of producing board will rise as production costs will undoubtedly rise as in all types of industry. The low-cost raw material, planer shavings, is rapidly being used up throughout the country. Some areas, particularly in Oregon, have already used almost all of their planer shavings residue in areas where board plants are concentrated. Additionally, the cost of shavings will increase as its fuel value increases, as there will be competition for the residual material if the energy crisis becomes quite severe. It is well known that wood can be converted to power, and at this time sawmills, plywood plants, and composition board plants already have or are seriously considering using some of their material for their own in-plant power supply. In 1976, prices as high as $23 per ton for dry shavings were being quoted as equivalent value to oil. For years shavings sold for very low prices—about $3 per ton in some areas. The increased competition for shavings has already driven the price up to $10 to $15 in many areas. Such a demand for the residual material can cause some serious dislocations of the wood material supply for board plants which will cause a further increase in the cost of the wood.

Another cost problem associated with the energy crisis is the cost of the synthetic resins, which are derivatives of the petrochemical industry. It is anticipated that these higher costs for resin will continue throughout the 1970s. Costs for gas and oil will also increase, adding to the expense of manufacturing.

It is estimated that the demand for conventional products such as plywood and lumber will exceed the supply in the 1980s at the present growth rate in the United States. At that time, there will not be any additional volume from the commonly used construction species. However, there will still be great amounts of so-called residue materials available, particularly that left in the forest as slash. Also, material classified as noncommercial or low grade can be converted into excellent building products using particles, flakes, and fiber. Examples of these species are aspen, southern hardwoods, and alder; more agricultural wastes and even municipal refuse may be used for raw material.

More fiberboard and particleboard will be used in all types of building. Even now these products are supplanting lumber for casings, door jambs, and face frames on furniture and cabinets. The development of the market for overlayed panels with vinyl polyester, or low-pressure melamine should dramatically increase if these plastics are available for use. The problem in this case, again, may be a shortage of vinyl and melamine due to the oil shortage.

It is anticipated that composition board products will move strongly into the structural board field. One plant in 1973, due to shortages of sheathing grade plywood, marketed most of its production as phenolic-bonded sheathing grade board. Recent developments in producing structural panels with oriented or aligned particles or fibers, particularly on the faces, for developing high strengths will enable these board products to further penetrate higher-grade structural applications. However, it should be noted that waferboard is already approved for structural applications throughout Canada and this board is made without aligned particles on the faces. As far as it is known, this product is performing well in the field. As mentioned previously, several new plants are now being constructed in Canada for producing this particular product and great interest has been expressed in the United States by companies that are investigating the same type of production. A particleboard concrete form panel in sizes up to 8x24 ft (2.44x7.32 m) which is overlaid with a medium-density phenolic-impregnated paper has recently been placed on the market.

With all the factors contributing to the growth of the composition board industry, there is little chance that it cannot continue its phenomenal growth unless a severe economic recession occurs. In 1975, however, there were difficulties in justifying the capitalization costs of plants due to recession and the greater costs of construction, equipment, and operation. However, it appears that the industry is on the threshold of a new era of expansion. Composition boards can be produced using higher-cost material. Better flakes or fiber can usually be produced from this material, providing a better end product which can therefore command a higher price in the marketplace. Composite products which are combinations of particles and veneer have recently been introduced to the marketplace and appear to have a bright future.

This industry has been operating as a scavenger for the rest of the forest products industry; but the ability to produce structural materials, as well as those products already well established, should enable composition board products of comminuted wood to compete on a more equitable basis with lumber and plywood for the harvested forest. A far greater yield from the harvest when producing these new products is a further advantage. Sawmills and plywood plants usually have a yield of about 45%, which is about one-half that which can be expected from composition board plants.

The awesome prediction for growth in all parts of the forest products industry were summed up in Figures 2.2–2.5. It appears that conjecture on whether further growth in the industry can be tolerated has to be supplanted by plans on how the massive new demands for panel products can be met.

SELECTED REFERENCES

Blomquist, R. F. 1975. Com-Ply Studs—A Status Report. *Proceedings of the Washington State University Particleboard Symposium, No. 9.* Pullman, Washington: Washington State University (WSU).

Bonney, J. B. 1975. Marketing of Medium-Density Fibreboard in North America. Background Paper No. 125, third World Consultation on Wood-Based Panels, FAO, New Delhi, India, February 1975.

Brown, J. H., and S. C. Bean. 1974. Acceptance of StranwoodR Sheathing for Construction Uses. *Proceedings of the Washington State University Particleboard Symposium, No. 8*. Pullman, Washington: WSU.

Bruce, R. W. 1970. The Economics of Marketing Particleboard. *Proceedings of the Washington State University Particleboard Symposium, No. 4*. Pullman, Washington: WSU.

Canadian Standards Association. 1975. *Mat-Formed Wood Particleboard*. CSA Standard 0188-1975. Ottawa, Canada: Published by CSA and available from the association.

Carroll, M. N. 1975. Waferboard in Canada. Background Paper No. 86, third World Consultation on Wood-Based Panels, FAO, New Delhi, India, February 1975.

Chow, P. 1975. Dry-formed Composite Board from Selected Agricultural Fibre Residues. Background Paper No. 110, third World Consultation on Wood-Based Panels, FAO, New Delhi, India, February 1975.

Columbia Engineering International Ltd. (Consulting Engineers). 1975. An Assessment of the U.S. Particleboard Markets & the Industry, Vancouver, B.C., and Eugene, Oregon.

Countryman, D. R. 1974. Investigation of a Composite Veneer-Particleboard Structural Panel. *Proceedings of the Washington State University Particleboard Symposium, No. 8*. Pullman, Washington: WSU.

———. 1975*a*. Composite Panel Development. Background Paper No. 81, third World Consultation on Wood-Based Panels, FAO, New Delhi, India, February 1975.

———. 1975*b*. Development of Performance Specifications for Veneer/Particleboard Composite Panels. *Proceedings of the Washington State University Particleboard Symposium, No. 9*. Pullman, Washington: WSU.

Covington, K. H. 1975. The Use of Medium-Density Fibreboard in American Furniture and Woodworking Industries. Background Paper No. 87, third World Consultation on Wood-Based Panels, FAO, New Delhi, India, February 1975.

Dickerhoof, E. 1975. Particleboard Product and Raw Material Trends in the United States. Presented at the Annual Meeting of the Forest Products Research Society, Portland, Oregon, June 1975.

Duraflake Company. *101 Practical Ways to Use Duraflake*. Albany, Oregon.

Economic Commission for Europe and Food and Agriculture Organization of the United Nations. 1973. Classification and Definitions of Forest Products (Advance Version). Supplement 6 to Vol. 25 of the "Timber Bulletin for Europe." Geneva, Switzerland, August 1973.

Elmendorf, A. 1965. Oriented Strand Board. Patent No. 3,164,511. United States Patent Office (January).

Ericsson, P. 1976. Activities and Goals of the European Federation of Particle Board Manufacturers' Associations (FESYP). *Proceedings of the Washington State University Particleboard Symposium, No. 10*. Pullman, Washington: WSU.

FAO. 1970. FAO World Survey of Production Capacity for Plywood, Particle Board and Fibreboard 1970. Report for Food and Agriculture Organization Committee on Wood-Based Panel Products, Third Session, Rome, Italy, December 1970.

———. 1976*a*. Basic Paper I. Production, Consumption and Trade in Wood-Based Panels—Present Situation and Alternative Outlooks for the Future. *Proceedings of the World Consultation on Wood-Based Panels,* held in New Delhi, India, February

1975. Brussels: Published in agreement with the Food and Agriculture Organization of the United Nations by Miller Freeman Publications.

———. 1976*b*. Basic Paper IA. FAO World Survey of Production Capacity for Plywood, Particle Board and Fibreboard. *Proceedings of the World Consultation on Wood-Based Panels,* held in New Delhi, India, February 1975. Brussels: Published in agreement with the Food and Agriculture Organization of the United Nations by Miller Freeman Publications.

———. 1976*c*. Basic Paper III. Marketing of Wood-Based Panels. *Proceedings of the World Consultation on Wood-Based Panels,* held in New Delhi, India, February 1975. Brussels: Published in agreement with the Food and Agriculture Organization of the United Nations by Miller Freeman Publications.

———. 1976*d*. Basic Paper V. Economics of Investment and Production in the Wood-Based Panel Industries. *Proceedings of the World Consultation on Wood-Based Panels,* held in New Delhi, India, February 1975. Brussels: Published in agreement with the Food and Agriculture Organization of the United Nations by Miller Freeman Publications.

Fordyce, D. H. 1972. Particleboard from Forest Slash. *Proceedings of the Washington State University Particleboard Symposium, No. 6.* Pullman, Washington: WSU.

Gunn, J. M. 1972. New Developments in Waferboard. *Proceedings of the Washington State University Particleboard Symposium, No. 6.* Pullman, Washington: WSU.

Guss, L. M. 1975. Overall Implications of Panel Products in North America. *Proceedings or the Washington State University Particleboard Symposium, No. 9.* Pullman, Washington: WSU.

Hse, C-Y.; P. Koch; C. W. McMillin; and E. W. Price. 1975. Laboratory-Scale Development of a Structural Exterior Flakeboard from Hardboards Growing on Southern Pine Sites. *Proceedings of the Washington State University Particleboard Symposium, No. 9.* Pullman, Washington: WSU.

HUD, FHA. 1961. Mat-Formed Particle Board for Exterior Use. *Use of Materials Bulletin No. UM-32* (June).

Johnson, A. L., and R. T. Martin. 1975. The Washington Iron Works Continuous Thin-board System. *Proceedings of the Washington State University Particleboard Symposium, No. 9.* Pullman, Washington: WSU.

Jorgensen, R. N. 1975. U.S. Experience in Application of Structural Exterior Particleboard. *Forest Products Journal,* Vol. 25, No. 8.

Koenigshof, G. A. 1974. Forest Service COM-PLY Research Project. *Proceedings of the Washington State University Particleboard Symposium, No. 8.* Pullman, Washington: WSU.

Lahiri, K. L. 1975. Bright Future Predicted for Panel Producers. Closing address at the third World Consultation on Wood-Based Panels, FAO, New Delhi, India, February 1975. As published in *World Wood* (March 1975).

Lehmann, W. F. 1974. Properties of Structural particleboards. *Forest Products Journal,* Vol. 24, No. 1.

Lehmann, W. F., and R. L. Geimer. 1974. Properties of Structural Particleboards from Douglas-Fir Forest Residues. *Forest Products Journal,* Vol. 24, No. 10.

Lewis, G. D. 1969. Current and Future Particleboard Markets: Problems and Potential. *Proceedings of the Washington State University Particleboard Symposium, No. 3.* Pullman, Washington: WSU.

Lewis, W. C. 1975. Applications and Uses of Wood-Based Panel Products. Background Paper No. 139, third World Consultation on Wood-Based Panels, FAO, New Delhi, India, February, 1975.

Lin, R. T. 1975. Orienting Wood Fibers and Particles Electrically for Board Manufacturing. Background Paper No. 138, third World Consultation on Wood-Based Panels, FAO, New Delhi, India, February 1975.

Maloney, T. M. 1972. The Board Industry Today—A Status Report on Particleboard, Hardboard, and Fiberboard Manufacture. *Proceedings of Northern California Forest Products Research Society* (October).

———. 1975. Composition Board and Composite Panels—Present and Future. *Proceedings of the Northwest Wood Products Clinic* (May).

Maloney, T. M., and D. E. Lengel. 1973. Overview of the Board Industry Today. *Proceedings of the Northern California Forest Products Research Society* (October).

Maloney, T. M., and G. G. Marra. 1972. Trends and Developments in the Manufacture of Particleboard (North American Situation). *Proceedings, Seventh World Forestry Congress* (October). Buenos Aires, Argentina.

Maloney, T. M., and A. L. Mottet. 1970. Particleboard. *Modern Materials,* Vol. 7. New York: Academic Press Inc.

Marra, G. G. 1958. Particleboards . . . Their Classification and Composition. *Forest Products Journal,* Vol. 8, No. 12.

———. 1970. Small Steps for Particleboard, A Giant Stride for the Wood Industry. *Proceedings of the Washington State University Particleboard Symposium, No. 4.* Pullman, Washington: WSU.

Meakes, F. V. 1972. Acceptance in Canada of Waferboard. *Proceedings of the Washington State University Particleboard Symposium, No. 6.* Pullman, Washington: WSU.

National Particleboard Association. 1972. NPA 2-72. Standard for Particleboard Decking for Factory-Built Housing. Silver Spring, Maryland: NPA.

———. 1973*a*. NPA 3-73. Specification for Particleboard Floor Underlayment Coated with Wax-Polymer Type Hot Melt Coatings. Silver Spring, Maryland: NPA.

———. 1973*b*. NPA CO1-73. Specification for Wax-Polymer Type Hot Melt Coatings for Particleboard. Silver Spring, Maryland: NPA.

———. 1973*c*. NPA 1-73. Standard for Particleboard for Mobile Home Decking. Silver Spring, Maryland: NPA.

———. 1973*d*. NPA 4-73. Standard for Medium Density Fiberboard. Silver Spring, Maryland: NPA.

Office of Product Standards/National Bureau of Standards/U.S. Department of Commerce. 1963. *Hardboard.* CS 251-63.

———. 1966. *Mat-Formed Wood Particleboard.* CS 236-66.

Paulitsch, M. 1975. The Application of Wood Particle Board. Background Paper No. 44/3, third World Consultation on Wood-Based Panels, FAO, New Delhi, India, February 1975.

Pazina, R. V. 1969. Effective Marketing. *Proceedings of the Washington State University Particleboard Symposium, No. 3.* Pullman, Washington: WSU.

Phelps, R. B., and D. Hair. 1974. The Demand and Price Situation for Forest Products, 1973-1974. Miscellaneous Publication No. 1292. Forest Service/U.S. Department of Agriculture.

Pollard, N. E. 1973. Particleboard as a Structural Component in the Building Industry in New Zealand. *Proceedings of the Washington State University Particleboard Symposium, No. 7.* Pullman, Washington: WSU.

Quinney, D. N., and J. T. Micklewright. 1975. Markets and Marketing of Wood-Based Panel Products in North America. Background Paper No. 7, third World Consultation on Wood-Based Panels, FAO, New Delhi, India, February 1975.

Raddin, H. A. 1970. The Economics of the System Producing Dry Process Medium Density Fiberboard. *Proceedings of the Washington State University Particleboard Symposium, No. 4*. Pullman, Washington: WSU.

F. L. C. Reed & Associates Ltd. 1975. Markets and Marketing of Wood-Based Panels in and from Developing Countries. Background Paper No. 1, third World Consultation on Wood-Based Panels, FAO, New Delhi, India, February 1975.

Saunders, R. J.; J. D. Snodgrass; and H. B. McKean. 1975. Structural Panels from Aligned Wood Elements. Background Paper No. 75, third World Consultation on Wood-Based Panels, FAO, New Delhi, India, February 1975.

Schmierer, R. E. 1971. Australian Particleboard Practices. *Proceedings of the Washington State University Particleboard Symposium, No. 5*. Pullman, Washington: WSU.

Snodgrass, J. D.; R. J. Saunders; and A. D. Syska. 1973. Particleboard of Aligned Wood Strands. *Proceedings of the Washington State University Particleboard Symposium, No. 7*. Pullman, Washington: WSU.

Staepelaere, R. L. 1971. A Study of Particleboard in North America and Europe through 1980. Everett, Washington: Black Clawson, Inc. (April).

Talbott, J. W. 1971. High-Resin Particleboards for Structural Uses. Presented at Eighth Northwest Plastics Workshop, Washington State University, Pullman, Washington.

United Nations Industrial Development Organization. 1975. The Requirements and Use of Wood-Based Panels in Furniture Manufacture. Background Paper No. 130, third World Consultation on Wood-Based Panels, FAO, New Delhi, India, February 1975.

U.S. Department of Commerce/Bureau of the Census. 1970–1974. Particleboard. *Current Industrial Reports*. Series: MA-24L (for the years 1960-1975).

———. 1973. Pulp, Paper, and Board. *Current Industrial Reports*. Series: M26A (for the years 1961-1975).

U.S. Department of Commerce, National Bureau of Standards. 1973*a*. Voluntary Product Standard PS60-73. Hardboard Siding.

———. 1973*b*. Voluntary Product Standard PS59-73. Prefinished Hardboard Paneling.

———. 1973*c*. Voluntary Product Standard PS58-73. Basic Hardboard.

Vajda, P. 1969. Economics. *Proceedings of the Washington State University Particleboard Symposium, No. 3*. Pullman, Washington: WSU.

———. 1970. The Economics of Particleboard Manufacture Revisited or an Assessment of the Industry in 1970. *Proceedings of the Washington State University Particleboard Symposium, No. 4*. Pullman, Washington: WSU.

———. 1974. Structural Composition Boards in the Wood Products Picture. *Proceedings of the Washington State University Particleboard Symposium, No. 8*. Pullman, Washington: WSU.

———. 1975. A Comparative Evaluation of the Economics of Wood-Based Panel Industries. Background Paper No. 3, third World Consultation on Wood-Based Panels, FAO, New Delhi, India, February 1975.

Wentworth, I. 1975. Uses and Applications for Thin Boards. Background Paper No. 74, third World Consultation on Wood-Based Panels, FAO, New Delhi, India, February 1975.

3. MODERN PROCESSING SYSTEMS

For the uninitiated, the dry-process board industry is a confusing welter of terms. However, it is not unlike any other industry, and its distinct nomenclature must be learned to fully understand the process. In this chapter a general overview will be given to this industry so that the reader can understand more clearly the detailed discussions to be presented later. It must be realized, however, that with any generalized description there are always exceptions to the rule, and it will not be any different in this discussion.

This industry is highly mechanized, or, perhaps to be more correct, semiautomated. It is not a completely closed-loop system which can be controlled entirely by a computer program. Decisions by men are still necessary. However, manual input is held at a minimum, particularly as compared to other wood products processing operations.

In the early board plants, first established in Europe, many people were involved in production but, as mentioned, this has changed to the present semiautomated system. It is expected that advances in technology for controlling the process and greater understanding of the optimum methods for producing composition board products can eventually lead to fully automated production systems. As it is now, most plants run with very few men on the production line. However, behind the scenes are a number of highly qualified maintenance-type personnel such as electricians and mechanics. They are needed in this, as well as any, semiautomated or automated industry to keep the many controlling devices and the equipment functioning properly so that they can truly operate in an automated manner.

Preferably the plants run continuously, particularly the larger ones. As with any production facility, it takes a certain period of time to get the process line functioning properly. Once it is functioning, it is easier and more economical to keep the line running without having to shut it down unnecessarily. The dryers, some of them running with a combustion chamber temperature of 1600°F (871°C), must be heated up and cooled down if the plant is not running continuously. Once resin systems are flowing properly and correct amounts of resin are being applied to the wood furnish, it is best to keep them running. A particular process problem can be the mat former where slight changes, such as in ambient conditions or mat moisture content, can cause severe problems in properly forming a mat. Once the former is running properly, it is ideal to keep it running. It is convenient to keep hot presses operating continually once the "normal" temperatures ranging from 300° to 400°F (149°-204°C) are stabilized.

Most plants would like to run about 22 ½ hours per day with the remaining time allowed for normal downtime. An operating time of about 350 days during the year is desirable. The remaining time in the year is used for normal heavy maintenance. It is also normal for many plants to shut down for one eight-hour shift during the week for routine maintenance on the production line requiring more than a few minutes. Much of the cleanup needed in a board plant is performed during this time.

The major production lines operate with platen presses. Continuous press systems produce similar boards. The first part of the discussion will present general information on production factors for the platen system, followed by a general discussion on the different process lines. A more detailed discussion of the process steps will then be presented. Extruded board, molded processing, and mineral-bonded products will be presented next and not discussed further as they are relatively minor parts of the industry at this time.

PLATEN-PRESSED SYSTEMS

Dominant Production Factors

Some of the dominant factors governing the production of platen-pressed boards are type of raw material, wood species, particle size and geometry, resin level and type layering by particle size and density, mat moisture content and distribution, and the board density. Interactions between these and other factors result in many different variations, and these will be examined in detail later.

Ideally, it is best to keep each different type of raw material and species separate as it enters the plant, so that it can be metered in desired proportions into the board process. In the early days, with smaller plants, this was relatively easy to do. In fact, some plants could operate on exactly the raw material they found to be most desirable; however, this is no longer the case in most places. A wide range of residuals is brought into the board plant from many primary generators of a vast variety of residual material, making a complete separation very difficult.

As will be discussed later, particle geometry is of great importance in board manufacture. Particles with good slenderness ratios such as fibers and flakes provide products with high bending strength and stiffness, and for such products it is desirable to have such particles on the board faces. Short, chunky particles in the faces are detrimental to the development of these properties. However, such particles are excellent for improving bonding properties in the board core and, consequently, can be used there. Coarse fibers or thick flakes are also usually used in the core when they are part of the furnish.

Resin level is directly associated with the properties of a given board. Most particleboards are made with urea-formaldehyde resin at levels between 6 and 10%, based on the oven-dry weight of the wood. The finer particles used in faces need the higher resin levels because of their greater specific surface area. Phenol-formaldehyde resin, when used, is applied at slightly lower levels. However, boards made of large flakes such as waferboard, because of their lower specific surface area, require only 2 to 3% of phenolic resin. This is usually applied as a powder. Virtually all other resin is applied in the liquid form.

High-density hardboard is normally made with about 2 to 3% phenolic resin. The high-density product makes efficient use of the resin, as virtually none is lost in interparticle voids, such as found in low- and medium-density particleboards. Medium-density hardboard siding is made at resin levels up to about 6%. Normally, phenolic resin is used. Medium-density fiberboard is made usually with either urea-formaldehyde resin or urea-melamine concentrates applied at about 7 to 9%. Higher resin levels, when used, will result in better board properties. However, this expensive additive is kept at the minimum level required for producing products with acceptable properties. While many properties can indeed be improved dramatically by increasing the resin level, the interaction with other manufacturing conditions is of great importance. Bonding naturally will be improved, but particle geometry is dominant in determining ultimate bending strength, stiffness and dimensional stability.

Resin type determines the board's resistance to deterioration under severe environmental conditions, such as exterior exposure. Normally, phenolic resin is used to gain moisture resistance and improved weatherability. Urea resin is utilized when the board is used for interior applications.

Wax, usually in the form of an emulsion, is added to the furnish to improve dimensional characteristics, particularly those associated with the absorption of liquid water. Molten wax is also used in some of the processes. Levels in particleboard, medium-density fiberboard, and hardboard range between about 0.25% to 1.0%. Some hardboard siding has been reported to contain as much as 3% wax.

Other additives, such as preservatives and fire retardants, can be added for resistance to decay and fire.

All dry-process boards have a layered construction, which is usually of two types: particle size and density. The most obvious is particle layering, where fine particles are on the faces and coarser particles are in the core. The face-to-face density profile cannot always be seen visually and thus its effect on board properties can be overlooked by the untrained observer. This profile usually runs from high density in the faces to low density in the core.

There are three basic board configurations for conventional composition board as far as particle size is concerned. The first is homogeneous, where the mix of particles is randomly distributed throughout the board. A second formulation is the three-layer, where finer particles or thin flakes are in the face layers and the coarser particles are in the core. These boards, depending on the thickness, are made up of about 15 to 20% of fine particles on each face, with the remaining material being made up of coarse core particles. A third configuration is the graduated or multi-layer board. In this case, as illustrated in Figure 3.1, the finest particles are on the faces of the board, and there is a gradation of particle size ranging from the finest on the faces to the coarsest in the core. In boards of layered construction, the face layer thicknesses are usually kept constant and material is added or subtracted from the core to change board thickness.

All composition boards, if manufactured properly, are quite stable and resistant to warp. Those with fine particles on the faces have smooth surfaces which

Figure 3.1. *A graduated particleboard mat with fine face particles and coarse core particles (Haigh 1967).*

can be treated with standard finishes quite easily. Some of the finishing problems will be discussed later. This type of board is also an excellent substrate for overlaying with thin veneers or plastics.

It should be noted that, either by accident or by design, some boards are made with a two-layer construction. A case in point with dry-process hardboard is a siding product that has very fine fiber on the face that will be exposed to the weather, with coarser fiber making up the rest of the board. Such an unbalanced construction can provide a situation in which the board will tend to warp. Some processes have been refined so that the warp does not occur. In some of the older particleboard plants, two-layer construction occurred by accident because of equipment limitations. Some warp could be expected in these boards. However, by recognizing this fact, plants using the board in secondary manufacture for furniture developed design systems which minimized the warping in the final product.

As mentioned, some boards are made with a homogeneous construction, which means all particles are randomized throughout the board. However, the type of layering by density, mentioned previously as being less obvious than particle layering, also occurs in these types of board. Hot pressing, which will be discussed in detail later, develops a density layering throughout the thickness of the board, commonly called the *density profile*. As an illustration, the most common method of hot pressing ¾-in. (19 mm) board in the particleboard industry is to consolidate the mats in the press to final thickness in about one to two minutes. This yields a board with high-density faces and a low-density core. This method will be covered later under "hot pressing," but, in brief, the mat faces are consolidated first and the resin cured before the mat core is consolidated to final thickness, resulting in the density profile described.

The distribution of the moisture throughout the mat also affects the density profile. It is not uncommon to add more moisture to the faces to assist with the speed of heat penetration to the board core, which shortens the press time. However, this moisture in the faces, which turns to steam, plasticizes the face particles and enhances the densification of the board faces. In boards with the finer particles in the surface layers, more resin is required because of the greater specific surface area of these finer particles. This additional resin usually means a higher moisture content in the faces, which further helps with the face densification. It should be noted that it is also easier to consolidate these finer surface particles because they flow together easier than the coarser core particles. With such an environment, fast closing times in the press tend to accentuate the density profile. On the other hand, slower press closing times result in a decrease in face density and an increase in core. However, this is usually an undesirable situation in that the surface integrity and quality are diminished, which can result in boards with soft surfaces unsuitable for use.

It should go without saying that different densities of board will have different density profiles at different thicknesses and with different press closing times.

In general, the properties of dry-process board are dependent upon the specific gravity of the board in much the same way as properties of solid wood are related to the wood specific gravity. However, it must be remembered that high-strength board cannot be made out of a wood species that is approximately the same in specific gravity as that desired in the final board. An example would be trying to make medium-density particleboard at specific gravity of 0.64 out of white oak, which has a specific gravity at approximately this level. The pressing operation consolidating the particleboard and curing the resin is the same general type of function as gluing two pieces of wood together with a clamp. In normal gluing, a pressure of about 150 psi (1.0 MPa) is necessary for a good glue joint to be formed. The same type of pressure needs to be applied in the board-pressing process for obtaining good wood-resin-wood bonds throughout the final product. Using a relatively high-density wood for making medium-density particleboard would require very little pressure for consolidation into the final board. Thus, the surface of the adjacent wood particles would not come in intimate contact and poor glue bonds would occur. Consequently, for the dry process, the species going into a given specific gravity board must be lower in specific gravity, except for low-density boards which usually have low physical properties.

For hardboard, those boards made with fiber refined at atmospheric pressure must also be significantly higher in specific gravity as compared to the wood species used for generating the furnish. With the relatively new pressure-refined fiber, however, it is possible to produce medium-density boards using fiber from species having specific gravities much closer to that of the final board. This is a significant breakthrough, as species previously difficult to use for board products, because of their greater density, can now be used.

The fine pressure-refined fiber presents an anomaly when considering the specific surface area of the fibers, which is much greater than that of a comparable shavings- or flake-type material. It would be expected that the pressure-refined fiber would require several times as much resin as the larger particles, given the same composition board density. Surprisingly, this is not true, and the pressure-refined fiber requires about the same level of resin as the larger particles on a weight basis. This is an unexpected development, but one that has assisted in the rapid growth of the medium-density fiberboard industry.

Moisture content is a significant factor, in addition to the aforementioned distribution throughout the mat during pressing. Most composition board mats enter the press somewhere between 7 and 11% in moisture content. This level can be somewhat higher, ranging up to about 16%, particularly in some plants outside the United States. Unless special provisions are taken, moisture content of the mats cannot be higher than about 11%. Otherwise, the board will tend to blow, blister, or delaminate as it comes out of the hot press. Much of the moisture in the mat turns to steam during the pressing operation. This steam develops a significant vapor pressure within the board. Consequently, much of the vapor must be vented before the press is opened; otherwise the aforementioned destruction of the board will take place. In addition higher amounts of moisture will prevent the proper curing of the resin, which will also contribute to the blowing, blistering, and delamination problems.

The Different Process Lines

Before getting into a detailed discussion of the platen-press system, it seems proper to first provide an overview of the process. Afterwards, each step will be

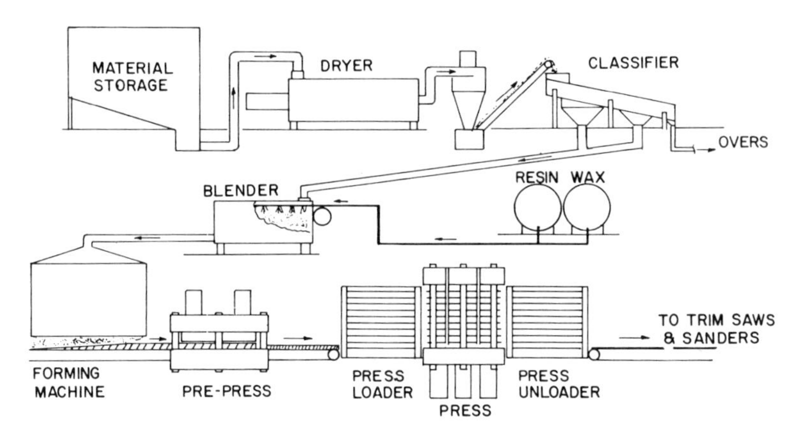

Figure 3.2. *Diagram of a typical composition board plant. (Courtesy Washington State Univ.)*

discussed in more detail as it relates to the fiber, particle, and the combination of particle and fiber types of board plants. Figure 3.2 is a drawing showing the general layout for a platen-pressed board plant. Not all of these pieces of equipment will be necessary in each plant because of the types of products being manufactured and the raw material being handled.

All-Fiber Production Lines: There are two general types of all-fiber production lines. The first produces relatively high-density hardboard products ranging in specific gravity from about 0.75 to above 1.0. These are the plants producing relatively thin board, which first came into operation in the late 1940s. The medium-density fiberboard lines are the newest member of the all-fiberboard process. Specific gravity of this product is normally in the range of about 0.72 to 0.93. It has particularly good properties for furniture, as the fiber provides boards with better bending strength and stiffness than can be achieved with shavings-type boards, while at the same time yielding boards with what is known as *tight edges*. The relatively fine fibers, when consolidated into the final board, provide board edges with little or no porosity as compared to shavings-type boards.

In the furniture industry, a medium-density particleboard usually will have to be edge banded with wood or a plastic material for many applications, particularly when the edges are to be profiled. The medium-density fiberboard, on the other hand, can be profiled and finished directly, eliminating the need for edge banding. Another advantage it has over some conventional particleboard is less tendency for developing "telegraph" or showthrough. In veneer or plastic overlayed particleboard, changes in moisture content can cause the outline of the thicker surface particles, and sometimes even the core particles, to show through the face of the thin overlay. This showthrough is caused by differential swelling and shrinking of the larger particles. Much of the telegraphing problem has been overcome in the particleboard industry by the use of finer particles throughout the board and only very fine particles on the faces. However, medium-density fiberboard is in an advantageous position because no large particles are in the board at all.

Particle and Flake Production Lines: In the United States and Canada, some of the plants producing what is known commonly as particleboard use an all-granular-type furnish such as planer shavings. The products produced go heavily into furniture applications, floor underlayment, and mobile home decking. Other plants are using a combination of ring-cut flakes, plus the small particulate material available from planer shavings and sawdust. The products go into the same market as the all-shavings-type boards. Another type of board is being made with all flakes and goes into the same market. Either ring or drum flakers are being used for generating the flake furnish from some type of solid wood material. This type of system is most applicable in those areas with green roundwood available, such as aspen, but other species are also used.

Another type of board product is the one made of all flakes for structural and decorative uses. These boards are made of relatively large flakes cut with huge disc flakers, although drum flakers are also used. This product, known as waferboard, is being used heavily for construction purposes in Canada at the present time, as stated previously.

Production Lines Using a Combination of Fiber, Flakes, and Granular Particles: The ideal composition board plant would appear to be one which could handle virtually any type of residual material, plus solid wood, which is available for board production. In this way, advantage could be taken of the particular characteristics of the raw material. Also it would give the plant the option of adjusting its process line to make the optimum products out of the raw material available.

Such a plant would include providing individual storage areas for each type of raw material according to geometry, species, and moisture content. It would include a variety of particle-generating equipment ranging from hammermills through flakers and atmospheric and pressurized refiners. It would also call for sophistication in the separation of particles so that the proper sizes with the right geometries could be used efficiently in the final board. Such a plant does exist at Dillard, Oregon. The Roseburg Lumber Company has built its Resin-Tite Fiberflake plant to operate on a wide variety of raw materials. It has two production lines—an older one producing approximately 700 tons (635 mt) of board per day, and a new one producing approximately 1300 tons (1179 mt) of board per day, depending on the board density and thickness. The old line is a caul system, and the new one is a caulless system, with a 20-opening hot press capable of producing 8x25-ft (2.44x7.62 m) boards. It is the ''Cadillac'' plant of the industry and is taking advantage of different ways in which particles, shavings, flakes, and fiber can be melded together to make superior board products.

Roseburg is using combinations of planer shavings, ring-cut flakes, and pressure-refined fiber in its production. The pressure-refined fiber is being produced from plywood panel edge and end trim, which has been a difficult material to use in particleboard. With the combination of particles going into this particular board, it can easily be seen that this is a hybrid type of composition board plant. It is using both hardboard and particleboard techniques. It is demonstrating the rational approach of using technical knowledge advantageously to produce a good product without being concerned particularly with the exact name of the resulting manufactured panel, but rather with its end-use characteristics and properties.

The board is being sold as a particleboard. However, this plant can also produce products made completely of pressure-refined fiber.

The Process in General

At first, the raw material is brought to the board plant and placed in some type of storage area. If it is relatively small particulate matter that can be used directly in the board process, some type of classification by size may be performed immediately and the material fed into the plant. Larger material must be reduced to flakes, particles, or fiber using the proper equipment, depending on the type of product being produced.

It has been pointed out that particle geometry is the most significant factor involved in most composition boards, as far as final physical properties are concerned. It is of particular importance in the conventional particleboard product. Some plants, because of the wood species of the planer shavings being used, find it difficult to obtain fibers or long, thin particles for use as face material. Chunky material does not have the right slenderness ratio (length/thickness) for use on the faces of particleboard. The chunky material, however, does work well in the core of a board, as there is good interparticle contact resulting in efficient applied resin use. This helps develop relatively good internal bond properties.

The furnish, or in other words the particles used for board manufacture, is handled as a single entity to a homogeneous board processing line. For layered boards, the furnish can be handled as a single entity until mat forming, where it is separated into coarse- and fine-fraction sizes. However, it is usually simpler to handle the face and core material separately throughout the process.

After the particles are prepared, the next general step is to reduce to 2 to 4% moisture content in a dryer. Plants operating with residual material generated from kiln-dried lumber and using powdered phenolic resin may be able to operate without a dryer, as the moisture content is low enough in the material as received.

Out of the dryer, the furnish may be classified into predetermined sizes as required for the production process. It is usually much easier to classify material after drying, particularly if the original raw material is green or wet. In the wet condition, the finer particles tend to cling to the larger ones and efficient classification is not obtained.

The furnish then proceeds to a blender, of which there are many versions. In the blender, resin and wax are applied to the furnish. Additional moisture can also be added if it is needed, as well as a catalyst if required. In some dry-process hardboard plants using liquid phenolic resin, the resin and wax can be added to the raw material before drying. The resin and wax are blended simultaneously with the furnish as it is being generated in the attrition mill. Thus, the attrition mill functions as both a particle generator and a blender. The phenolic resin has a higher curing temperature than the urea resin and can therefore pass through the dryer without being cured. The urea resin cannot be handled in this way, as it will cure out in the dryer. When drying this resin-treated furnish, it is not necessary to go to the very low moisture content of 2 to 4%. The final mat moisture content, ranging from about 8 to 11%, then becomes the specification for the material emerging from the dryer.

After blending, the furnish moves to the forming station, where it is laid down into mats. A number of different systems are used for performing this operation. Another term used besides forming is felting, which is a carryover from paper industry terminology. Depending on the board specifications of thickness, type of particle, and specific gravity, the mat thickness can range from less than 1 in. to over 12 in. (25.4 mm–305 mm).

Depending on the process, the formed mat can be partially consolidated to final thickness in a prepress. This is particularly important if the mat is very thick, as the thickness must be decreased sufficiently for it to fit into the hot press. As will be discussed shortly, the prepress is a necessary part of all but one type of caulless process line.

Multi-opening, single-opening, continuous, or stack presses can be used to produce a platen-pressed board. In the multi-opening system, the proper number of mats is accumulated in the press loader and then all are inserted simultaneously into the hot press for consolidation under heat and pressure into the final boards. In general terms, a ¾-in.-thick (19 mm) particleboard made with urea resin can be pressed in four to six minutes at temperatures ranging from 300° to 375°F (149–191°C). Boards produced with phenolic resin require slightly longer press times and higher press temperatures. Various kinds of catalyst systems can be used to speed up these press times. With single-opening presses, a heavily catalyzed resin system is used to speed up the press time because of the limitation of only one board being pressed at a time. It has been reported that press times of less than two minutes have been achieved, producing ⅝-in.-thick (15.9 mm) board on a single-opening press line. Naturally, thinner boards require less time and thicker boards more time in the press.

After the boards are pressed, they are removed by some type of unloading system and usually advanced through a cooler before being stacked. This cooler is important, particularly in lines using urea-formaldehyde resins, as the board temperature must be reduced before the board is stacked into a pile. The resin bonds of urea-bonded boards that are stacked without adequate cooling will deteriorate. In some cases, water absorption properties affected by the wax additive are enhanced by hot stacking. Phenolic boards can normally be improved in properties if they are hot stacked after pressing, but they are not always so stacked because of process line arrangements.

After either cooling or hot stacking, the boards are edge trimmed and moved through the sanding system. Some hardboards, as well as sheathing-type panels made of flakes, are not sanded, but in general most of the composition board products are sanded to dimension the boards to final thickness and to remove any surface imperfections before final processing. Layered boards with fine particles on the surfaces tend to have soft faces caused by resin precure about 0.020 in. (0.51 mm) thick, which need to be sanded off.

Many boards are shipped to market in a standard 4x8-ft (1.22x2.44 m) size. However, it is becoming more and more the practice to cut the final board to some predetermined size as specified by the customer. Hardboard lap siding is cut into the 4- to 12-in. (102–366 mm) specified widths. Particleboard, hardboard, and medium-density fiberboard are cut into a multitude of different sizes, depending upon the customers' needs. One of the more complicated parts of the plant process line is the cut-to-size operation which can be processing hundreds of different sizes during one operating shift. Some plants have gone to computers for programming the most efficient cutup of the boards into the final products. The cutup saws are very sophisticated, and some are controlled by punched cards or tapes. A number of plants also have finishing lines for applying filler, wet finishes, and overlays.

Process Details

Particle Preparation: First, the raw material must be assembled. It will usually come into a plant by truck or rail from other wood products plants in the form of residue such as planer shavings, pulp chips, plywood trim, sawdust, and in some cases solid or roundwood. Roundwood is especially harvested for board plants in some cases. This is very popular in countries other than the United States. Planer shavings are the dominant raw material used in the United States because of their low cost and availability. If a sawmill happens to be located nearby, the planer shavings and sawdust can be air-conveyed to the board plant, which provides an even lower cost raw material because of minimal transportation costs.

If at all possible, as the raw material comes in, the plants will attempt to segregate it by type of material, moisture content, and species. As will be developed later, all of these factors play important parts in board production. The finer types of raw materials are stored in huge buildings and protected from the elements. Large silos can also be used for storage. At one time, much of this type of raw material was stored outdoors. However, recently promulgated Environmental Protection Agency regulations have made it mandatory almost everywhere to store small particulate material indoors. This is to prevent the wind from blowing the small particles onto adjacent property. Larger materials such as logs, slabs, and pulp chips can be stored outdoors.

Particulate-type raw material can be conveyed into the board plant by means of screw augers operating from the bottom of a silo, belts, or flight conveyors. Material from a number of such silos can be mixed together to randomize species and material type variations. A particularly popular technique for handling this type of raw material is to use front-end loaders (as shown in Figure 3.3) for feeding the material into the plant. The raw material can be blended together much in the way one would prepare a large batch of food. The loader, for example might be used to intermix several different species. One scoop of Douglas fir shavings, one scoop of hemlock shavings, and one scoop of white fir shavings could be dumped into the hopper leading to the raw mateial conveyor. These scoop loads can also serve to mix materials of various moisture contents. This is a rather crude but effective means of controlling the raw material input. A more sophisticated procedure is to used constant-weight feeders.

Usually, some type of milling is required. For those plants using fiber, the raw material will be reduced by means of either atmospheric attrition mills or pressurized refiners. Green material such as chips, shavings, and sawdust can be milled into an acceptable fiber under atmospheric conditions, after first steaming the raw material. In at least one plant, however, green raw material is ground into fiber without steaming.

Figure 3.3. *A typical front-end loader. (Courtesy Washington State Univ.)*

For grinding under atmospheric conditions, both single- and double-revolving-disc types of mills are used. The previously mentioned steaming softens the lignin which is nature's glue holding the fibers together. Consequently, with presteaming a better fiber can be produced than with unsteamed raw material. This technique is well developed and used in many plants. If a very dry raw material is used, it is best to add moisture first to assist in the grinding of the fiber. The addition of water works best with small material such as shavings, as the water can easily penetrate into the particles. There is not sufficient time in a process line to allow water to penetrate into large particles such as chips.

The better-quality fiber is normally made from pulp chips. It is simply a case in which relatively long fibers can be generated out of this type of residual. It can easily be seen that long fibers cannot be generated out of very small particles such as dry planer shavings and sawdust. However, while some plants are still operating on chips, they have found it advantageous from the economic standpoint to substitute the lower-quality raw material, which is usually planer shavings and sawdust, for some of the pulp chips. This is a case of maximizing overall profits, although some of the fiber quality is sacrificed. Changes in the rest of the process line can be implemented to overcome some of the problems encountered with shorter-length fiber. This is particularly true in the case of high-density hardboard, where full use of the synthetic resins added as binder can be achieved.

Atmospheric attrition mills refine raw material into fiber under ambient conditions, whereas the pressurized refiners do so under steam pressure. The pressurized refiner is of particular value in producing good fiber out of dry raw material such as plywood trim or furniture waste. This material, of course, is first chipped or hogged into small pieces for feeding into the refiner.

The pressurized refiners operate in the same way as the atmospheric attrition mills do with one notable exception. Unlike grinding with the atmospheric attrition mill, the raw material passes from the steam-pressurized digester into the refining section while still under pressure, and the refining takes place under this same pressure. This particular type of system has been used in the board industry for only a few years. Much of the science or art of grinding this type of fiber is yet to be understood. However, the fiber is much finer and has different characteristics than that usually ground in the atmospheric attrition mill. The wood is broken down closer to individual fibers in the pressurized refiner than is possible under normal atmospheric attrition mill operating conditions, although fiber of similar bulk density can be generated in either mill. The material ground will tend to be bundles of several wood fibers rather than individual fibers.

Most of the newer plants producing fiber-type products use a pressurized refiner, and a number of the older ones have converted from atmospheric to the pressurized type. Using the pressurized refiner, as mentioned, is a particular advantage when utilizing residues of low quality. Specifically, these are dry hardwood residuals from furniture plants, low-quality hardwoods, and dry plywood trim. Just a few years ago, it was difficult to use any of these raw materials successfully in a board plant. Therefore, a great benefit of the pressure-refiner has been the generation of good fiber out of previously unusable or difficult-to-use raw material.

There are inherent variables operating in both of these refining systems that can cause a wide variation in the quality of the fiber generated. Besides moisture content of the raw material and variations in the steaming conditions, the variations in the raw material such as particle size and species, the variables inherent in the grinding process such as steaming time and pressure, disc diameter, plate pattern, loading rate, and the gap setting between the grinding plates can all influence the type and quality of fiber which can be generated.

As noted, most of the standard particleboard in the United States is produced using planer shavings as the basic raw material. While shavings are the dominant raw material, other residuals are included for use in the furnish in most plants. This is done to improve certain board properties and, in some cases, because of an inadequate supply of shavings. These other residues include solid wood materials such as planer trim ends, low-grade lumber, sawdust, and plywood trim. The solid wood material can be reduced by means of hogs or chippers into large chiplike material. The final breakdown into the desired particles can be achieved by using hammermills, ring refiners, ring flakers, or disc refiners or attrition mills. Some plants have combinations of these various machines.

For conventional particleboard, it is also sometimes necessary to reduce part of the granular material into fine fibrous material for use as part of the face furnish. To some extent, this depends upon the species being used. Some species normally will break up into very fine particles while being planed, and this will provide sufficient small material for the board surfaces. On the other hand, some species will tend to come off the planer as relatively large particles, and if fine material is required for the faces, part of these larger particles will have to be reduced into small ones. Ring refiners, hammermills, disc refiners, or attrition mills can be used for this type of milling operation. If at all possible, a fine fibrous-type material is generated because this provides a better particle for developing bending strength and stiffness while maintaining smooth faces on the boards.

If flake-type material is desired, several options are available. For producing higher-quality flakes, it is necessary to first start with green material either in slab or log form. This material can be chipped and then ring-flaked for the production of a quality flake. A very popular piece of equipment for the production of large flakes in North America, for use in structural flakeboard plants, is the disc flaker which has a flaking disc 93 in. (2.36 m) in diameter. Stock is fed edgewise against the disc, and the knives mounted in the disc slice off flakes to predetermined sizes.

Different types of drum flakers are available for reducing wood stock into flakes. One type takes roundwood, slabs, or rough lumber that is cut into predetermined lengths before it can be fed into the flaker. Another type takes random-length pieces of wood stock and reduces them to flakes. This flaker is widely used throughout the world but has only found minimal use on the North American continent to date, although it is an excellent machine. It is particularly appropriate for cutting flakes from small roundwood such as forest thinnings.

Drying: Several different types of dryers are available. These include single- and multiple-pass rotary-drum, stationary-drum and tube-flash dryers. Most of them are direct-fired and operate at very high temperatures using natural gas, propane, or oil for the fuel, although more sander dust is now being used to reduce fuel costs. These dryers can have very large combustion units where the heat is developed by open flames. It is possible to generate temperatures up to about 1600°F (871°C). When dryers of this type are used, it is imperative that the small particulate material pass rapidly through the dryer. If there is any hangup of material within the dryer, it will almost immediately catch fire and cause serious problems. Dryers are one of the more hazardous pieces of equipment in a plant because of the fire and explosion danger.

When there is a wide mixture of raw material as far as moisture content is concerned, it is advisable to treat the wet material separately from the dry. Normally the dryer handling the wet material will run at a higher temperature for fast removal of moisture. If a fairly dry material enters one of these dryers, it will almost immediately catch fire because the throughput rate is relatively slow for handling the wet material. Material such as planer shavings from kiln-dried lumber and plywood trim will be relatively low in moisture content as it enters the dryer. Most kiln-dried lumber is about 15% in moisture content; therefore, it only has to be dried very briefly to bring it down to the nominal 2 to 4% moisture content desired. Dryers handling the lower-moisture-content material will usually function at lower temperatures and higher throughput rates than those handling green material.

Particle Classification: Most classification is performed after drying, because wet or green particles tend to stick together. For particleboard, variations of screens and air-classifiers can be used to make the particle separation. The same type of separation can be made with flakes after they are cut. However, in many applications the cut flakes are treated as a single material for making homogeneous boards, and only very large and very fine material may be removed.

For hardboard and fiberboard, air-classifying of fiber to obtain fine material for faces can be used, but it is more common to prepare face and core material separately if layered board is to be made. Classifying of blended furnish at the mat former can also be done to select the finer fibers for faces. This is true for both the fiberboard and particleboard type of operations.

Blending: The next step is blending resin and wax with the furnish. If the board is to be homogeneous, all the material will pass through the same blender or set of blenders which function identically. In the case of multi-layer boards, separate surface and core blenders are generally used. This allows control of the amount of resin applied to each. Normally more resin is applied to the surface furnish than to the core furnish because of the higher specific area of the face particles, compared with the core particles. It is possible to apply resin to the face and core material simultaneously, but this type of blending is not as widely used as it once was. Catalysts can be added with the resin or immediately after blending. The wax, usually in the form of an emulsion, is also added at this time, either separately or mixed with the resin. Other specialized additives, such as fire retardants, need separate provisions for their addition. They may be added in the resin-wood blender or elsewhere in the process line.

Most resins are a liquid, but some plants use a powdered phenolic resin, and it is reported that at least one plant is now using a powdered urea resin instead, because of its lower cost. One way to apply the powdered resin is by tumbling the particles in a rotary drum with the powdered resin. In this type of operation, the molten wax or wax emulsion is applied ahead of time so that the powdered resin can adhere to the small, sticky drops of wax. Another method being used is to apply the powdered resin to the furnish as it is being reduced in size in an attrition mill. Thus, the attrition mill is simultaneously grinding the particles and blending the powdered resin, in much the same way as in the dry-process hardboard lines, where liquid phenolic is blended with the fiber in the attrition mill. A particular advantage in using powdered resin is that the moisture content of the furnish does not have to be lowered to the nominal 2 to 4% level. This has a twofold advantage: one of reducing the need for drying, and the other of reducing the fire and explosion hazard in the dryer.

Blenders applying liquid resin can be classified into two general types. One is the long-retention-time type which slowly blends the furnish with the additives, taking several minutes for completion. The second is the short-retention-time type which completes a blending in a matter of seconds. Both types have been used in the United States since the inception of the dry-process industry. Long-retention-time blenders gained the most popularity through the mid-1960s, but most plants now favor the short-retention-time type.

The additives are sprayed upon the churning or cascading furnish through air or airless atomizing nozzles, or they are applied by the use of centrifugal resin applicators. The intention is to get a fine dispersion of the resin over all of the furnish. A number of different approaches have been taken for applying the resin and wax to the furnish.

It was mentioned previously that, for hardboard fiber blending, liquid phenolic resin and wax can be added in the attrition mill simultaneously with the grinding of the fiber. It was also noted that ureas or melamines cannot be added as subsequent drying would cure the resin. Thus, such blending cannot be used in medium-density fiberboard plants. A partial exception is a process where part of the resin system is added in the attrition mill and the remainder after drying by means of a short-retention-time blender. The resin is then formed *in-situ* during hot pressing.

An early plant used specially developed urea-melamine concentrate for the binder system. This particular resin system has little to no tack. Tack simply is stickiness. A tack-free resin is a prime requirement for use in the fiber process because the slightest amount of tack will cause the fibers to adhere to one another, forming balls and causing many problems in handling and forming. Excess resin tends to accumulate on these fiber balls, resulting in undesirable "resin spots" on the surfaces of the pressed board. Such resin spots cause problems with appearance and with the adhesion of finishes and adhesives for applying overlays. The urea-melamine concentrate developed for this system has worked very successfully. However, competitive systems have been developed and are in successful use. It should be pointed out that all tack in the resin applied during the attrition milling of the fiber, such as in the hardboard production, is lost during the drying operation.

Short-retention-time blenders such as attrition mills and high-speed paddle types have been used successfully for applying urea resins to fiber.

Forming: In the next step, the blended furnish moves to the mat former, of which there are many different designs. Mat forming is extremely important as a misformed mat will yield a board with widely varying properties because of differential specific gravity throughout the board. A homogeneous board naturally is laid down with a random mix of the furnish. Multi-layer or graduated mats are laid down using appropriate equipment for keeping the particle sizes separate.

Particleboard formers meter the furnish by means of spreader rolls, belts, air, or combinations of these upon a caul plate, plastic sheet, or fabric belt. Fiber is usually formed onto a fine-mesh wire screen. It is handled by air or is forced through a screen to separate the fibers before they fall upon the screen. When forming fiber, a vacuum is pulled down through the fiber being formed. This vacuum compresses the fiber mat, and such compression is necessary because the bulky fiber yields a mat very thick in comparison to those found in the particulate and flake-type mats. Very thick mats cannot be inserted into the "daylight" of the hot press. As will be mentioned, prepressing is also necessary for further consolidation of the very fibrous mat.

Prepressing: The mats then move down the line, where in some cases they are prepressed. Prepressing consolidates a loosely formed mat into a relatively rigid "cake" which has a certain degree of cohesiveness. The mat is also reduced in thickness so that a press with minimum space (daylight) between the platens can be utilized. Prepressing can be accomplished by using either a platen, caterpillar tread, or belt prepress. Prepressing is not always necessary, particularly when using a caul system, but it is a must with all but one of the caulless systems. With low-bulk-density, pressure-refined fiber, it is usually necessary to reduce the mat thickness in order that the mat will fit into the press. Normally, a belt prepress is used with this system.

Pressing: After prepressing, the mats are then rough trimmed to length and, in most cases, width. The trimmed mats proceed to the hot press, which can be either a multi-opening, single-opening, stack, or continuous type. The mats can be carried on metal cauls, wire-screen belts, steel belts, or, for the caulless system, on plastic sheets or fabric belts.

The caul is a metal plate of steel or aluminum used as a support tray on which the mat is formed and conveyed. Special wire mesh can also serve as a caul. This support tray carries the mat through the prepress, if it is used, and on into the hot press. It stays beneath the mat until it is finally hot pressed into a board and is removed after pressing is completed. Cauls were used for all of the early plants in the United States, and it was not until the mid-1960s that the caulless lines became popular. Cauls are normally used with multi-opening and stack presses.

The caulless system uses multi-opening presses. Although a small single-opening press could also use the caulless system, its small size would not make the process economic. Briefly, with the caulless system, a forming sheet, which can either be a rigid or flexible material, is used for receiving the furnish dropped from the former. This forming sheet transports the formed mat into the prepress for consolidation. For successful prepressing, the resin used in the furnish must have a certain amount of tack which assists in holding the partially consolidated mat together. In addition, particles with high slenderness ratios (length/thickness) are of value in providing some of the mat integrity. The desired particle geometry of fiber provides sufficient mat integrity for operation in the caulless system.

After prepressing, in the most popular U.S. caulless sytem, the mat is removed from the forming sheet or belt on which it was formed and is transported into the press loader. When the press opens and the previously pressed boards are ejected, the press loader travels forward into the hot press carrying the prepressed mats. The loader goes to the far end of the hot press and then is slowly withdrawn. As it is withdrawn, the mats are carefully deposited directly onto the hot press platens or plates. As soon as the press loader finishes its operation, the press is closed and the mats are consolidated into the final boards. Other systems use different techniques to deposit the mat upon the press platens.

The press loaders are elevator structures that move up and down, receiving each new mat as it arrives. The process line is timed so that the press loader is ready with a full load of mats when the press is opened and the previous set of pressed boards is ejected. Once the newly pressed boards are removed from the hot press, the freshly formed mats are inserted; and the hot press consolidates them under heat and pressure into boards.

The boards are unloaded into an elevator structure, similar to the press loader, after pressing is completed. If the plant is using metal cauls for handling the mats and pressing the boards, the finished boards are then separated from the caul plates and usually advanced to a board cooler. The caul plates are then returned to the forming station where a new mat is laid down.

Single-opening presses function with either a steel belt or wire screen that carries the formed mat into the press. While the board is being pressed, another mat is being formed either by a traveling former, which moves back and forth over a stationary section of the belt or wire screen at the front of the press, or by a stationary continuous former which deposits the mat onto a moving belt conveyor. In the case of the continuous former, a speed-up conveyor is used to accelerate the finished mat forward, where it is then transferred to the metal belt carrying the mats into the press.

Most multi-opening presses employ simultaneous closing which permits faster press closing without blowing off the mat surfaces. The same density profile is developed in all boards because all boards are pressed alike.

At the completion of the press cycle, the press opens; the belt travels forward, carrying the finished board out and the freshly formed mat into the press. The single-opening presses are usually rather large, ranging up to 8 ft wide by 65 ft long (2.44x19.81 m). Even larger ones are possible. As in the case of the multi-opening press, the finished board usually passes to a board cooler for a short period of time so that board temperature can be reduced before stacking or further processing.

The final board thickness is controlled in the press by pressing to gauge bars or "stops," which are set between or outside the press platens, or by electrical means which hold the press platens at the proper spacing by what is known as *position control.*

With continuous presses, the formed mat is constantly being carried into and through the hot press. This is a platen-pressed type of board and should not be considered as extruded. Continuous pressing was a viable part of the industry in the 1950s. It then fell into disfavor for a number of years. However, it has recently made a comeback, particularly for pressing thin boards which can be classed either as hardboard or particleboard depending on the final board properties.

Another type of pressing is the stack press, which is not widely used. It will be discussed later, but basically it produces a platen-pressed board.

An exception to the pressing methods discussed previously occurs in the semi-dry hardboard plants. Mats are formed at moisture contents of about 25 to 30% (oven-dry basis). Screens similar to those used in the wet-process line are used to transport the mats. During hot pressing, these screens allow rapid venting of the moisture from the mat. The bottom surface of the pressed board has a screen impression, termed *screen-back,* like that found on wet-process hardboards. Only relatively thin boards can be made with this type of pressing.

Finishing: As mentioned, boards are usually cooled immediately out of the press or, at times, hot stacked. Skinner saws are used to trim the panels to final dimensions. Trimmed material can be recycled to the particle generation area or used as fuel. Huge belt sanders, which may have as many as eight heads (four above and four below) and are 8 ft (2.44 m) wide, dimension the board to final thickness. The sander smooths the surfaces and removes any soft material (precure) from the faces as well as surface blemishes caused by scratches or other blemishes on the caul plates or press platens. An exception to sanding may be found with products such as waferboard where surface smoothness is not a primary requisite or with hardboard products that have tight, hard surfaces created during the hot-pressing operation.

Heating mediums used in press platens include steam (the most popular in the USA), hot oil, or hot water. A combination of one of these plus radio frequency (RF) is being successfully used, primarily for pressing medium-density fiberboard.

After sanding, the panels are graded. They can then be shipped or cut to smaller sizes. Complicated loading procedures are required when shipping the cut-to-size pieces because of the many different sizes. As previously mentioned, some plants also prefinish panels with fillers, wet finishes, wood and plastic overlays, edge bands, and edge machining before shipping.

Post Treating: Several different post treatments can be performed. Boards can be impregnated with preservatives and fire retardants. Hardboard can be heat treated or oil tempered to improve properties, particularly those associated with dimensional stability and water absorption.

EXTRUDED PARTICLEBOARD

Kreibaum

It has been noted that the most popular method of producing particleboard is through the mat-formed system where the mats are compressed into the final board, usually by means of a platen press. Those boards being produced by continuous methods result in a board similar in properties to those manufactured by the platen press. Another method of producing particleboard is by means of the extruded pressing system. The first system was developed by Otto Kreibaum in the years 1947 to 1949 in his factory at Lauenstein, Federal Republic of Germany, and is called the Okal method. A few plants have been built in the United States to produce extruded particleboard either through this system or with similar ones developed by United States firms. However, since the first few plants were built approximately 20 to 25 years ago, no new ones of this type have been constructed. A new and different type will be mentioned later. Those in operation in the United States are captive plants which produce board used in another part of their business. These particular boards cannot meet the standards established for the platen-pressed particleboard. They must be overlaid with some other material such as thin particleboard, veneer, or a paper or plastic laminate.

The preparation of the furnish for this system is not much different from that used for the mat-formed process. The raw material must be first reduced to a small size similar to planer shavings. Many different species have been tried successfully in this process, as well as different forms of raw materials such as roundwood, wood with bark, forest thinnings and twigs, standard waste from sawmills and plywood plants, and agricultural residues.

The raw material is dried to somewhere between 4 and 6% in moisture content. Resin is then added in a blender at levels of 4 to 6%, based on the oven-dry weight of the wood. The quantity of resin, as in the mat-formed process, depends upon the desired properties of the final board. Most boards of this type are made with urea resin.

A former as used in the mat-formed process is not necessary. The blended particles are brought directly to the aperture of the extrusion press. This press is a heated die that is either vertically or horizontally oriented. The particles are spread out over the cross section of the heated die, and a reciprocating hydraulic ram then forces the particles through this die. The width of the die establishes the width of the board, and the distance between the die surfaces establishes the board thickness. This is a continuous process and is illustrated in Figure 3.4.

With a vertical die, an even feed of the particles into the die can easily be achieved. Also, a separation of the fine particles from the larger particles, which tends to happen when the die is oriented horizontally, is prevented. After each stroke of the hydraulic ram, more material is fed to the die and forced through by

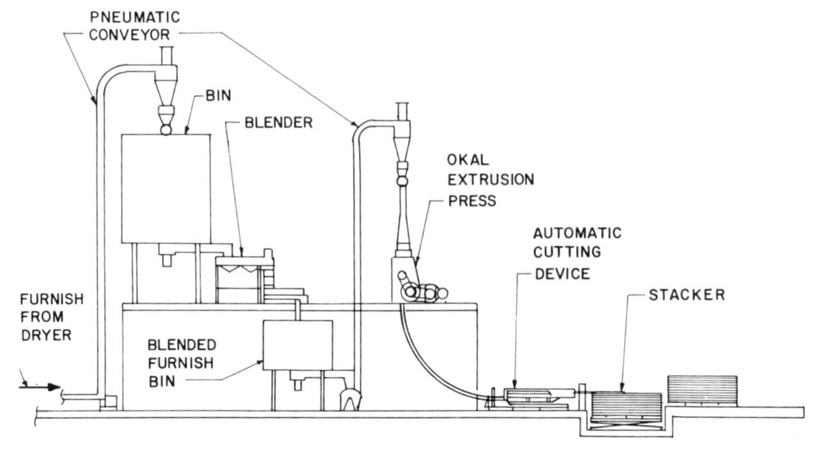

Figure 3.4. *Schematic outline for the preparation of extruded particleboard by the Kreibaum method (Kreibaum 1957).*

the next stroke of the ram. In this way, a continuous board is produced. A schematic drawing of this press is shown in Figure 3.5.

Normally, the board is in the hot die for only about 30 seconds and then it runs through a heating zone for another 40 to 60 seconds, depending upon the thickness of the board, for final resin curing. The density of the board depends upon the quantity of the particles being loaded into the press and the thickness is established by the cross section of the die. For producing thicker boards, tubular channels are inserted into the middle of the die. This can only be done with a vertical die. These channels are also heated and serve to heat the board being pressed from the inside. Such inner heating along with the heat from the hot die speeds the cure of the thicker boards. The final board thus contains tubular openings, which reduce the overall weight of the thicker boards. This type of thick board is used for the production of wall panels. It is possible to use these openings for installing the electrical wiring in a building. As the thinner boards leave the vertical press, they are brought to a cut-off saw by means of a gentle slope as shown in Figure 3.6. The cutoff saw then cuts the continuous panel to the desired lengths. The thicker boards cannot be bent in this way and are brought to a saw in the vertical direction.

The properties of these extruded particleboards are quite different from those manufactured by the platen system. In a platen system, the particles tend to be randomly oriented in a plane parallel to the surfaces of the board; thus, good bending strengths and stiffness are developed, taking advantage of the fiber length in the particle. In the extrusion system, the particles, as they are fed into the extruder, also lie flat. However, as can be visualized, they are randomly oriented in a plane perpendicular to the board surfaces. Thus, the extruded board has very low bending strength and stiffness as these properties depend on resin bonding between particles only. Fiber length in the particles is oriented in the wrong direction for developing good bending and stiffness properties. These boards have very little thickness swelling in moist conditions because of the orientation of these particles, since wood expands very little along the grain. Conversely, the linear expansion can be quite severe because what is thickness swell in platen-pressed board or solid wood becomes linear expansion in the extruded board.

To develop adequate bending strength and stiffness and to restrain the extreme linear expansion, extruded boards must be overlayed with some type of thin particleboard, veneer, or plastic laminate. Figure 3.7 shows a sample of this overlaid extruded board that was accidentally wetted. As can be seen, the high linear expansion of the extruded board core as related to the low expansion of wood veneer overlay resulted in the composite panel shearing apart in the center in a dramatic manner.

Another problem that occurs with this type of board, particularly those of thinner cross section that are curved from the extrusion die down to the cutoff saw as shown in Figure 3.5, is that some of this bow can be retained in the final board.

The relatively low production rates in these plants, plus the need for upgrading some of the physical properties of the final board, have prevented this extrusion board processing system from becoming a major factor in the United States or Canada. Even in other parts of the world where more extrusion plants are operating, the platen-type system for producing particleboard is by far the most popular.

Rib-Wood

Recently, a new extrusion system called Rib-Wood has been announced. It has a relatively low capacity but it is an important addition to the arsenal of composition board producing equipment and should be of great value to those wanting to

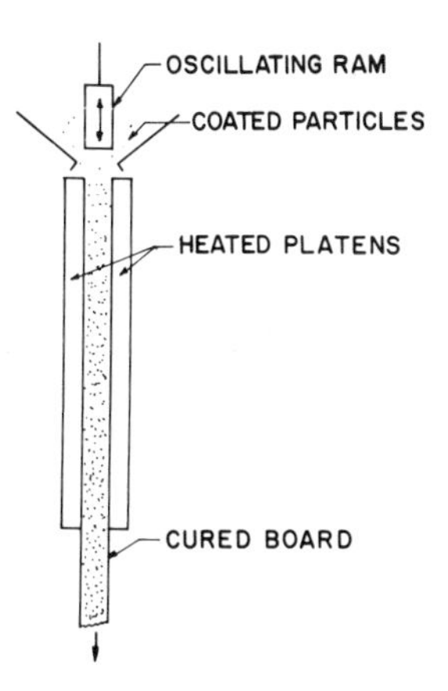

Figure 3.5. *Principle of particleboard manufacture by extrusion. Particles are forced by oscillating ram through die, consisting of two parallel heated plates. Distance between platens is fixed, and is equal to the thickness of the board (Suchsland 1972).*

Figure 3.6. *Extruded board leaving the die. (Courtesy Washington State Univ.)*

manufacture smaller amounts of product from their residuals. The first commercial machine will produce a continuous sheet 40 in. (1.02 m) wide with 13/16-in.-thick (20.6 mm) ribs every 2 in. (50.8 mm) across the width on one side (Figure 3.8). Relatively small particles are used to minimize detrimental particle orientation. The board density ranges from 40 to 65 lbs/ft^3 (0.64–1.04 sp gr). Gluing Rib-Wood panels together at rib ends or staggering ribs forms a hollow core panel 3/4 in. to 3 in. (19–76.2 mm) thick (Figure 3.8). It has a high-density surface that will take 1/28-in. (0.91 mm) veneer or 0.028-in. (0.71 mm) high-pressure laminate without crossbanding, which is an important feature. Overlaying is planned for this product.

Board width can be varied on different machines for various products, and mold variations will change with thickness, by length and distance between rib center.

Standard processing equipment is used except for the extruder. The extruder is heated to 400°F (204°C) and cures board at about 40 in. (1.02 m) per minute. A reciprocating ram is also used in this system for driving the furnish through the extruder.

MOLDED PRODUCTS

Molded composition board can take one of several forms. One type of molding consists of a product with over 25% of an adhesive binder mixed with wood flour. Such a product may be classified as a plastic. A familiar product of another, but lower-quality, type is a toilet seat which can be made with slightly larger particles and less resin. Another type of product is a panel which has embossed patterns imprinted on its face either during the pressing operation or by a post treatment. This is done with standard platen-pressed panels. A third type of molded product is that made with particles, fiber, or flakes for furniture parts, boxes, and countertops. Figure 3.9 shows examples of this type of product.

While the molded product segment of the industry is relatively minor, it is of interest for at least two reasons. One is that, while molding wood composition material has not been particularly economical up to now in the United States because of competition from all-plastic material, the great increase in prices for petrochemicals has possibly changed the economics of producing wood molded parts. Consequently, it may soon be economical for a significant expansion in this industry. Such molding has been taking place in Europe for quite a number of years. The second reason for interest in molded products is the future possibility of molding structural composite building panels that will have very high strengths. Such a molding has been visualized all along by those involved in the dry-process industry. Some initial work has been done along this line, but much is yet to be learned about molding such large pieces of material. Experience in molding the smaller pieces such as the furniture parts should be invaluable in assisting with the development of the molded structural board industry.

Embossed Faces

The imprinting of a wood-grain type or other pattern on faces of hardboard has been quite successful. (*Embossed* is the descriptive term used in this discussion

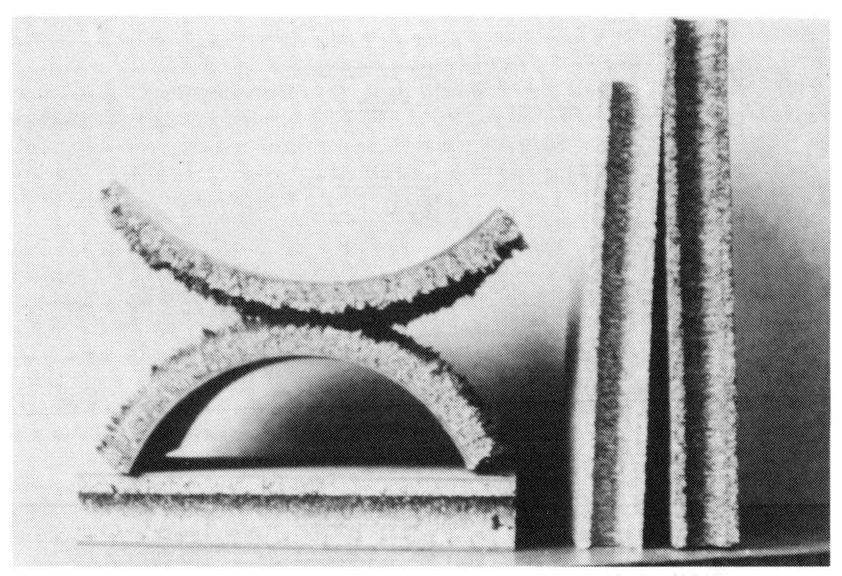

Figure 3.7. *This extruded board with veneer overlay was accidently water soaked and has sheared apart. (Courtesy Washington State Univ.)*

BASIC CONSTRUCTIONS

RIB-WOOD is formed in continuous ribbon.

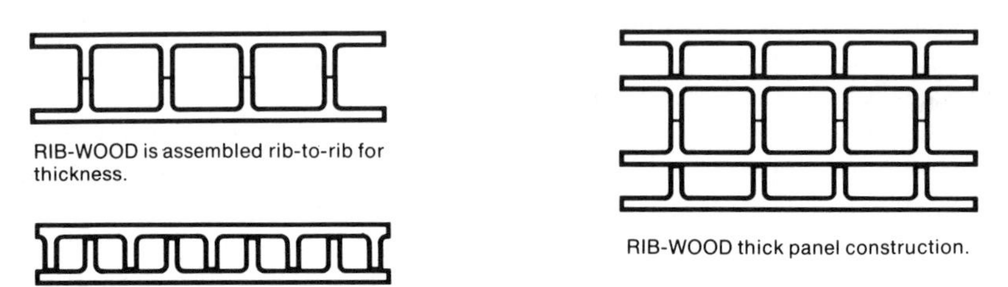

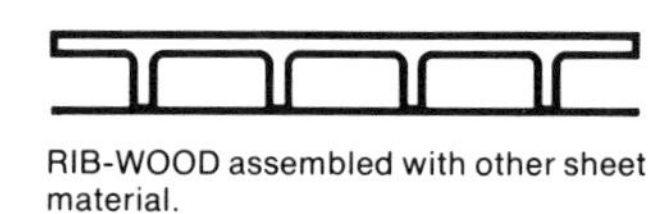

RIB-WOOD assembled with other sheet material.

EDGE CONSTRUCTIONS

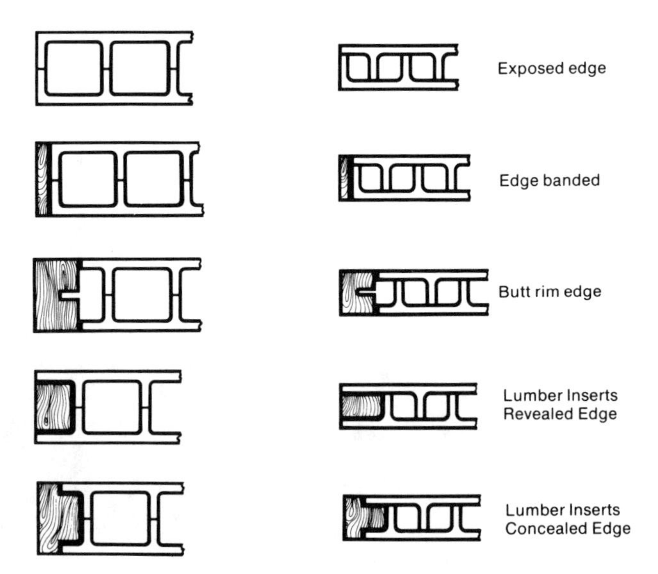

Figure 3.8. *Examples of Rib-Wood. (Courtesy ITP Corp.)*

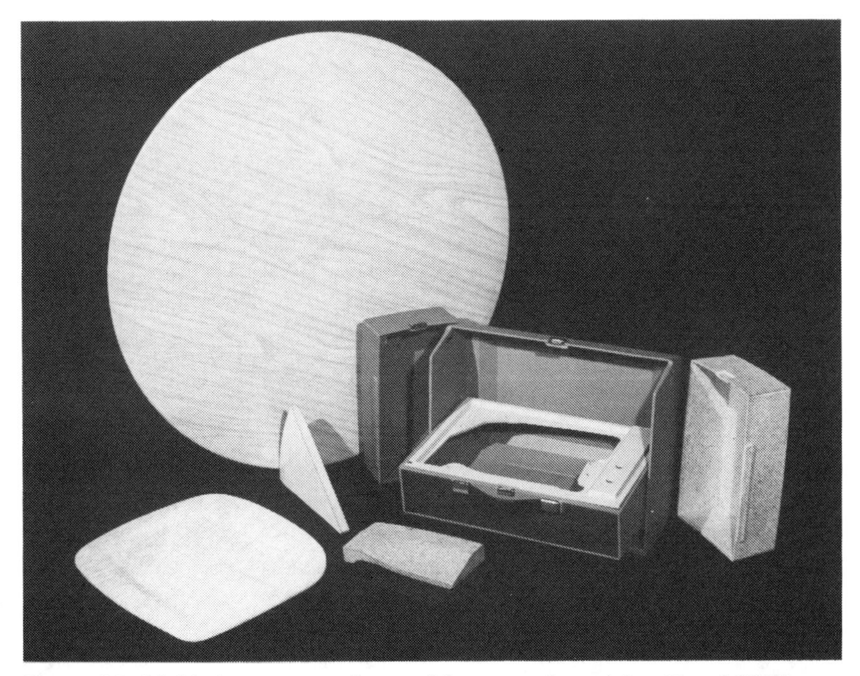

Figure 3.9. *Molded parts manufactured from wood particles (Pagel 1967).*

where the pattern sticks out from the plane of the board surface.) The pattern is imprinted by means of a false caul mounted in the press in which the desired pattern is etched or engraved.

In the case of particleboard, some of this imprinting or embossing has been performed successfully for many years on a small scale. It is done on a relatively high-density board with fine material in the faces. Figure 3.10 shows a sample of this particleboard. This same type of product is produced in Europe, as well as one where the pattern on the panel is impressed after the board leaves the hot press. *Repressed* board is the descriptive term used for this latter type of European board.

The U.S. board that has been produced as a flat panel with an embossed face has not had any paper-overlay type of material applied. For the repressed board, a white paper presaturated with resin is applied to both the top and bottom faces of the board. The pattern to be embossed into the board is etched into an aluminum false caul which is attached to the top of a hot press. A press temperature of 300°F (149°C) and a pressing pressure of 350 psi (2.41 MPa) is needed with a 40 lbs/ft^3 density (0.64 sp gr). The total cycle time is two minutes per panel. In general, the maximum depth or draw of the design is 1/16 in. (1.6 mm). This process is limited to quite shallow designs.

Aluminum cauls are used, rather than stainless steel, because they are easily etched and are relatively low in cost. In the U.S. plants, relatively high-cost dies have been used for embossing the panel.

Figure 3.10. *An embossed particleboard. (Courtesy Washington State Univ.)*

To obtain deeper designs in the panel, it is necessary to produce the embossed pattern as the board is being pressed. However, as with the repressed board, a low-density, resin-impregnated paper is used for surfacing both sides of the panel. The bottom sheet acts as a balance sheet, stabilizing the board and assisting in minimizing warp. However, the top paper does not have the resin applied until after it has been laid on top of the mat prepared for the hot press. A very fine spray of polyester resin is then applied to this top paper. The polyester resin then acts as a binding agent, gluing the paper to the upper surface of the particleboard mat, and it also gives the finished product a smooth, hard finish, which can be easily treated with paints or lacquer. It is important in this process to keep fine particles on the faces. Otherwise, the coarser particles usually found in the core would tend to puncture the surface paper.

As with the repressed embossing, this one-shot process also uses etched aluminum plates in the press. Press temperatures are between 300° and 325°F (149-163°C). The pressure applied is around 350 psi (2.41 MPa) for a board of 40 lbs/ft^3 density (0.64 sp gr). The curing time is, in general, about the same as for a standard particleboard being made without surface treatment. Figure 3.11 is a photograph of several panels produced by this one-shot process.

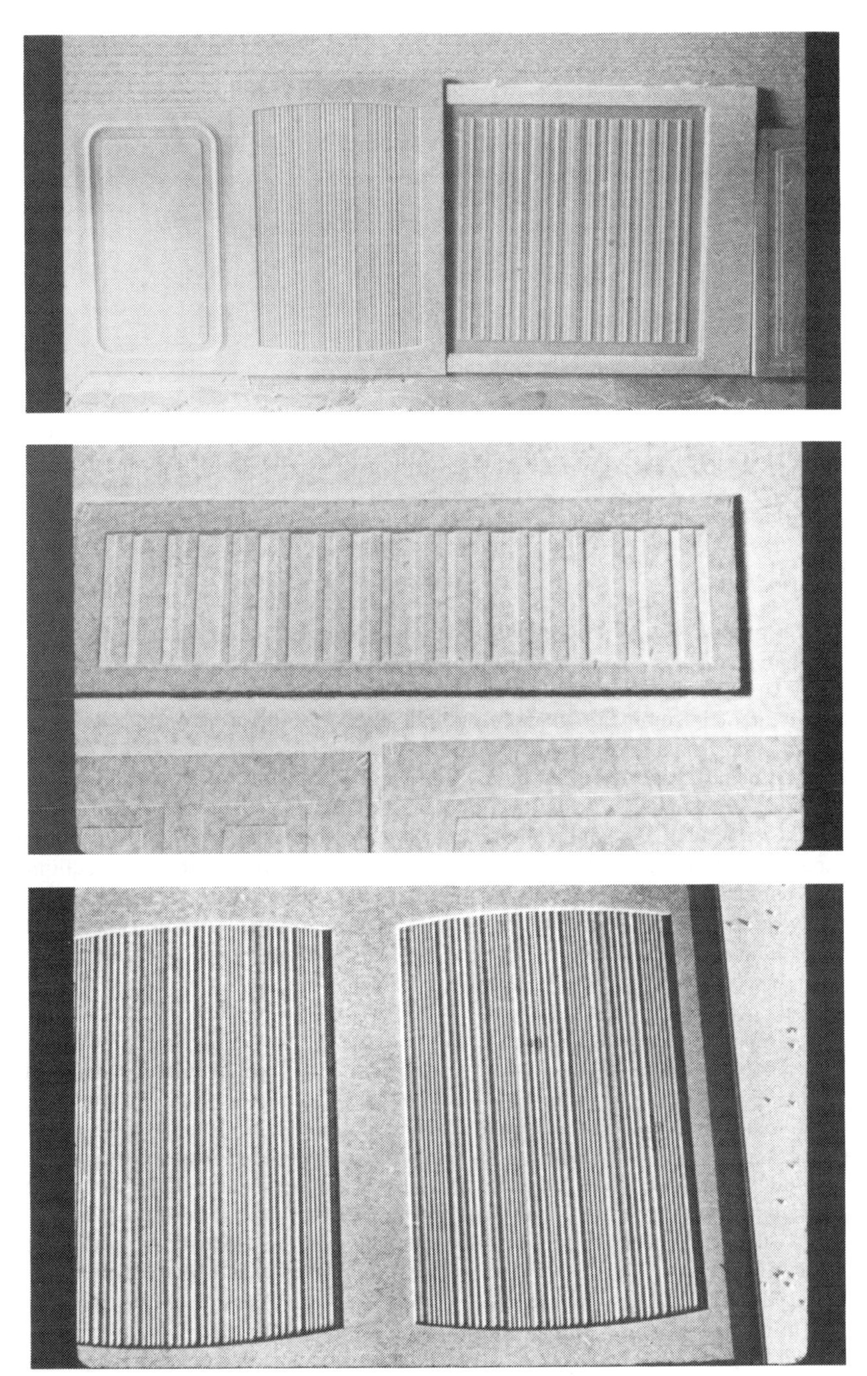

Figure 3.11. *Examples of embossed board (Wentworth 1969).*

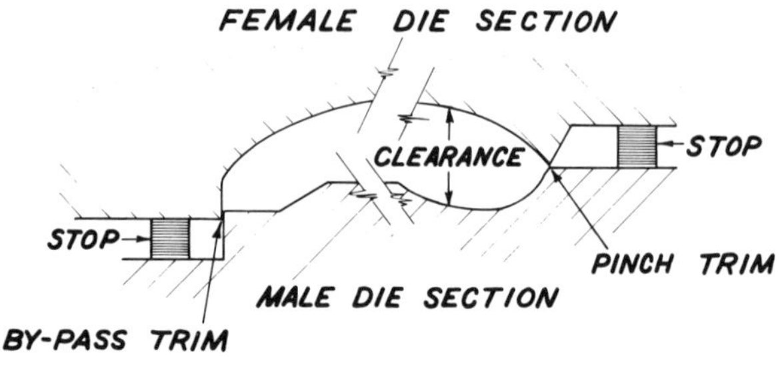

Figure 3.12. *Die-making terms (Pagel 1967).*

Many different patterns are embossed onto hardboards. It is relatively easy to emboss the higher-density fibrous products as the fibers will tend to flow into the design being embossed. The same is true, of course, with particleboard with fine fibrous material on the faces, which is necessary for getting good, clean designs with this type of operation.

Molded Parts

A standard flat-panel composition board could be characterized as being two-dimensional, whereas a molded part would be three-dimensional. The relatively simple molding is accomplished with the embossed type of board product. However, it becomes much more complicated when molding parts with deep draws such as tabletops with profiled edges, irregular boxes, curved drawer fronts, and beverage cases. Problems are encountered with mat formation, lack of particle and resin flow, expanding or contracting molds, and pressing conditions.

These parts are consolidated in matched metal dies or molds which are sometimes called *tools*. The female section is known as the *cavity* and the male section, the *punch*. The opening between the two parts of the die is predetermined and is fixed by stops. The distance between the male and female sections is called *clearance* or *closed opening*. The edges of the parts can be formed by pinch trimmers or bypass trimmers as shown in Figure 3.12.

This is a very sophisticated part of the particleboard industry. It took many years to overcome the problem encountered with the first crude attempts. However, there are some successful processes, particularly three that originated in Germany: Thermodyn, Collipress, and Werzalit. The Weyerhaeuser Company and the Plex-O-Wood Company in the United States have been using the Werzalit process, or parts of it. In addition, the Plex-O-Wood people have contributed their own developments to molded products. Newer developments, such as molding wood particles with a high level of isocyanate resin, may prove successful.

The general process conditions for producing molding products are not unlike the standard board system until the forming station is reached. Many different raw materials and particle types are used. Various amounts of resin are used depending upon the process and product. The commonly used resins are urea-formaldehyde and phenol-formaldehyde, although isocyanate resins have potential because of their lack of tack or stickiness which interferes with particle flow.

The prepared furnish is then formed and sometimes prepressed. In this mold, there may be a resin-impregnated paper, a veneer, or other type of sheet material added for decorative effect. This formed "mat" is then put into the hot press where the mat is consolidated into its final shape. Where the male and female parts of the metal dies meet, some material is normally squeezed out. This is called the *flash*. Some degree of deflashing is usually required. Other types of finishing, such as spray painting or vacuum-formed overlaid plastic sheets, may also be used to upgrade a raw part.

Figure 3.13 is a schematic drawing showing three different processes for making molded parts. A plant producing molded parts may have anywhere from

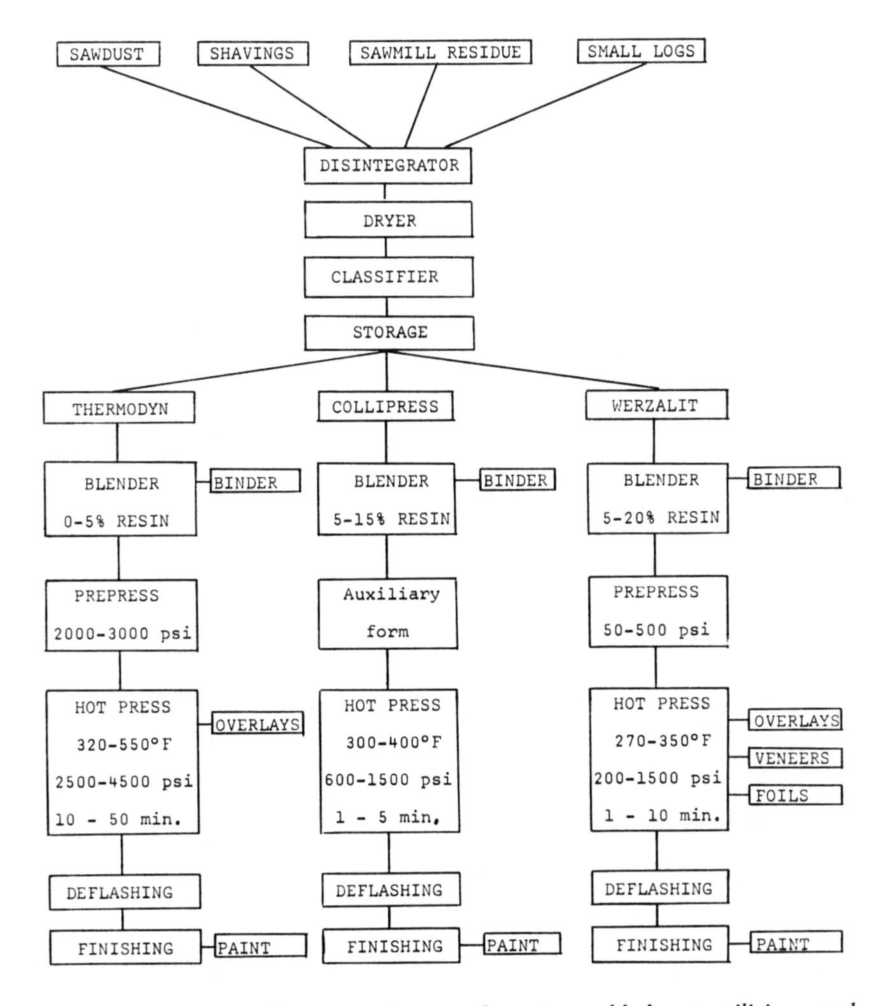

Figure 3.13. *Diagram of processes for manufacturing molded parts utilizing wood particles (Pagel 1967).*

Figure 3.14. *Press for Collipress process (Pagel 1967).*

1 to 40 or more hydraulic presses. It is in the blending and pressing that differences exist between the systems.

Thermodyn Process: This is a closed system which does not allow for any volatiles formed during pressing to escape. It is somewhat like pressing wet-process high-density hardboard where some of the wood components such as lignin are activated and also act as a binder. The wood particles also lose some of their features and are fused together. The resulting product is called Lignoplast, which is a high-density, high-strength product. Species with high contents of hemicelluloses, especially pentosan, promote the reaction process. Species having a high amount of pentosan, such as beech, birch, and poplar, are more desirable for use in this system as contrasted to low pentosan-containing species such as spruce, pine, and Douglas fir.

It is a characteristic of this process that certain parts of the wood itself are chemically altered. During the first phase of pressing, the hemicelluloses are hydrolyzed. Acetic and formic acids are formed and thereby cause further hydrolysis. These attack the lignin-cellulose bonds and form "active lignins." In the second phase of the pressing, the gases formed by the hydrolysis react with the altered wood substance and condense to a plastic. It is important that the die in this system is not open during pressing to let any gases escape. Because of the high vapor pressure being developed within the piece being molded, it is necessary to cool the die and the parts within below the boiling point of water before opening the press. This means a relatively long and expensive press time.

This material is very high in density—about 85 lbs/ft^3 (1.36 sp gr). A relatively fine particle is the most desirable for yielding a smooth-surfaced product. Thus, while it is good for some applications, its high density restricts its use in many others.

In this process, about 5% of phenolic resin is usually used. It is prepressed at about 2500 psi (17.2 MPa) before going into the hot dies. The dies can be heated between 320° and 550°F (160°-288°C) with pressures between 2500 and 4500 psi (17.2-31 MPa), depending on the product being produced. A conditioning time after being removed from the press is also required.

Collipress Process: This process was developed to produce heavy-duty containers for bottles, cans, and other packaged material requiring a box-shaped container. The preferred wood particle geometry is a flake with recommended flake size from 0.4 to 0.5 in. (10.2-12.7 mm) in length, 0.08 to 0.16 in. (2-4 mm) in width, and 0.008 to 0.016 in. (0.2-0.4 mm) in thickness. However, it has been found that any particle size common to the particleboard industry can be used, although final board properties may be changed. Fluffy fibers are not desirable because they are difficult to form into the die. Both urea and phenolic particleboard resins have been used at levels between 5 and 15%. The mat moisture content must be controlled between 8 and 18%, and thus in this respect is not unlike the regular dry-process system.

The resin-blended furnish is then formed into a preformer and then moved into the hot die. Figure 3,14 gives a general view of a typical Collipress which is equipped with one vertical and two lateral rams. The side rams compress two of the vertical sides while the other two side walls are compressed by wedge-shaped side members which are actuated by the lateral rams. The wall thickness of the

Figure 3.15. *Collipress containers (Pagel 1967).*

parts can be varied ±0.08 in. (2 mm) without major changes of the die. Figure 3.15 shows typical beverage cases which can be produced using this system.

Pressing conditions range from 700 to 1500 psi (4.8-10.3 MPa) pressure with temperatures of 320° to 400°F (160°-204°C). Pressing time can take from one to two minutes.

A volume of about 150,000 units has been found to be minimum for the economic success of a particular container. A press can turn out more than one million units per year on a three-shift basis. One press, however, is not economical to run because of all the other production equipment necessary. Thus, a minimum plant usually will have at least three presses. At least three million units per year could be produced by a single plant.

Finished parts for the furniture industry may be covered with wood veneer in a second operation in the same mold, or in a rubber bag press. Many of the parts, particularly the beverage containers, have some type of finish applied by spraying or dipping.

Werzalit Process: This process can use all raw material suitable for the manufacture of standard particleboard. However, a fine sliverlike particle is recommended. Standard board resins can be used, with urea-formaldehyde being the preferred one at the present time.

Furnish blended with resin normally is formed into preforms before going to the hot press, although in some cases the furnish can be fed directly into the heated die. The preformed dies are mounted in cold presses for prepressing. These preforms, when used, are then transferred into the hot dies where the temperatures can range between 270° and 350°F (132°-177°C). The pressure requirements can vary between 200 and 1500 psi (1.4-10.3 MPa), again depending on the density of the molded parts. In this operation, the molded parts can also be overlaid with fine wood veneers, polyester films, or melamine and phenol-formaldehyde resin-impregnated papers or any other desired sheet-type material. Press times may vary between 1 and 10 minutes, depending upon board thickness and the surface materials. The part density can be controlled between 35 and 70 lbs/ft^3 (0.56-1.12 sp gr).

The normal way of heating the dies is by conduction from the hot press platens. However, in the case of large dies such as those used for molded drawers, auxiliary heating can be placed into the dies.

The automated molding lines are very much like those of a flat-panel plant. The particles are generated, dried, passed through a continuous blender, and formed into a variable-thickness mat fitting the desired die. This process is quite flexible in that it is relatively easy to change the dies.

The wide variety of molded parts that can be made with this system is one of its strong points. Completely finished products such as countertops can leave the press with both the face and edges completed.

Design of Molded Parts

As has already been mentioned, there are many different types of molded products now being manufactured. Some of the dies being used at present are very expensive. However, unless large-volume sales are made with an expensive die, its costs cannot be recovered. Minimum volume requirements appear to be as low as 10,000 units up to a high of over 150,000 units per mold.

Some of the operating parameters of the three systems discussed here are mutually exclusive. Thermodyn, for example, must work with the absolutely closed system which eliminates the use of any side rams or other moving elements in the die. Also, a very high-density part is produced by this process. Collipress is designed for building containers. Werzalit is designed to produce complicated boxlike constructions as well as flat tabletops with molded edges.

Boxes or cases produced by molding have much better strength than those assembled out of flat pieces. The typical right-angle joint in wood products is very weak. This is eliminated in a molded product, and, by eliminating the weak joint, the molded box gains strength as well as an improved appearance. In the case of molded tabletops, the strength is not improved but the appearance of the tabletop is, as special resin-impregnated papers applied on the face and around the smoothly finished edge contours do not have any seams.

In any molded product, high-wear sections can be reinforced by increasing the density in that area, or reinforcing ribs can be added. The potential for improving the physical properties of the weak areas in a molded product is of great importance when considering building structural panels with a molded configuration. Molded products can be built with reinforced areas properly located to provide high working stresses. It should be possible to do this in a structural component production plant, opening up many new avenues of design.

Physical Properties

Because of the many different configurations of molded products, it is difficult to come up with absolute figures on their physical properties. They do not relate to the commercial standards established for the other dry-process panel products.

As with most of these products, density is a very important factor. Higher density, as with the other dry-process products, usually means better properties. It also means higher weight and the possibility of more warping, both of which are undesirable.

Because of the way they are produced, most overlaid molded parts will have an unbalanced structure, because different types of overlay are used on each face. Normally in an unbalanced structure, if more moisture penetrates through one face than through the other, the differential expansion will cause the board or product to warp. However, it has been found with the molded parts that, by using the proper overlay materials, the rate of moisture flow is impeded to such a degree that this warping does not take place.

Molded wood parts can be made to exhibit good acoustical properties, particularly for use in electronic cases, and they can be made to exceed the National Electrical Manufacturers Association (NEMA) specifications for the surface characteristics of high-pressure laminates. They have general mechanical properties similar to or better than the properties of the higher-quality particleboards. These improved properties are associated with the previously mentioned higher density and a usually higher level of resin in the molded products. Also, as in other board products, particle size, shape, and geometry have considerable influence on the properties.

MINERAL-BONDED PRODUCTS

There are three types of mineral-bonded products that are produced. The first type is made of excelsior or wood wool, as it is officially named, and is made in slabs ranging up to about 3 ½ in. (88.9 mm) in thickness. The second type consists of a product known as wood aggregate concrete. This product can be made as building blocks, full wall sections, or is poured in place as a slab for walls or floors. The third type is a building panel produced with thin flakes.

In comparison to concrete, these products are much lighter in weight. However, they are usually much heavier than the standard wooden building products that could serve the same end use. They are particularly good for acoustical and fire-resistant applications. The various mineral-bonded building products can range from about 18 lbs/ft^3 to about 78 lbs/ft^3 (0.20–1.20 sp gr).

Binders used for mineral-bonded products are magnesium oxysulfate, magnesite, gypsum, or Portland cement. The most important one is cement because of its high quality and availability in virtually all parts of the world. The cement should be rapid setting to enable a short working time in the process. However, the setting time of the cement is often strongly influenced by substances in the wood. These substances act as retarders to the cure of the cement and are said to "poison" the cement. The most important retarders appear to be wood sugars and/or the group of compounds known as *hemicelluloses*. They can sometimes be removed adequately by washing with cold or hot water. *Mineralizers*, a misleading term, can also be used to help with the bonding. Mineralizers are cement accelerators which speed the cure before the poisoning of the cement can occur. Three mineralizers are in common use: calcium chloride in Western Europe, sodium silicate in North America, and aluminum sulfate in the USSR.

In areas where many species are intermixed, this poisoning is a serious problem. Extensive work is going on in the United Kingdom and in the Federal Republic of Germany to overcome this problem. Recent reports indicate that pretreating by steaming or storage in hot water containing certain chemicals has increased the number of species suitable for use in cement-bonded products.

Excelsior Products

As noted in Chapter 1 the first excelsior boards named Heraklith were produced in Radenthein, Austria, in 1914. The binder used was magnesite. The first boards produced with cement came on the market in 1928. Before World War II, the binding agent used in Germany for the manufacture of this board was a combination of 39% cement, 35% magnesite, and 26% gypsum. However, after the war, the use of cement increased so that it became the dominant binder. Particularly in Europe, the processes and equipment have been highly developed and the dimensions, weight, and properties of excelsior boards are standardized.

This type of product has been made in the United States for many years. However, it has been difficult to determine the exact history of the early plants. It is known that such panels have been manufactured in currently operating plants since the 1940s but board was available much earlier from other sources, particularly one in Chewelah, Washington.

The Chewelah plant was the Thermax Division of the Northwest Magnesite Co. The process was based on the Heraklith process. A sawmill was built to cut large planks which were then cut into excelsior. Two products were produced: Thermax which was a "standard" excelsior panel and absorbex which was a higher-density panel made for acoustical and decorative purposes.

The superintendent of Northwest Magnesite Co. became interested in the manufacture of this type of board as a means of using the flue dust which was extracted from the kiln gases expelled during the burning of magnesite. For the manufacture of Thermax, which was made in 1-, 2-, and 3-in. (25.5, 50.8. and 76.2 mm) thicknesses, western larch and magnesium oxysulfate were used as the raw materials. To produce the cement, fresh flue dust was passed through a ball mill and then reacted with dilute sulfuric acid to form an emulsion. A small amount of caustic magnesite was added to improve the board strength. This emulsion was mixed with larch excelsior and then the material was spread onto a thin, steel belt and passed through a heated dryer or press. Steel belts were operated on the top and sides of the mat to hold to the desired dimensions. Heat was introduced into the shell of the dryer to speed curing. This was a continuous operation and the finished board was cut to the desired lengths as it emerged from the press.

Absorbex was made with cottonwood and caustic magnesite. It did not have the dark grey color of Thermax which was caused by the carbon content of the flue dust. It had a natural light tan color and was used extensively without further finishing. It was, however, a more costly product and this limited its use.

Production reached its peak in early World War II when much of the output went to the armed services throughout the world. However, limited production, and inadequate sales structure and difficulty in obtaining parts from Europe caused the plant to cease production in 1943.

In Germany, spruce has been found to be a popular species for use because it is relatively easy to prepare the excelsior and there are few extractives for inhibiting the cure of the cement.

Excelsior has been produced as a packing material since about 1900, with the first production occurring in the United States. These particles are long, thin wood strands about 19.7 in. (500 mm) in length with the width varying between 0.019 and 0.2 in. (0.5–5 mm). The thickness of excelsior ranges between about 0.010 and 0.025 in. (0.03–0.64 mm).

The prepared excelsior is mixed with the appropriate amounts of cement and water. Much less water is needed in this type of process than in a conventional concrete system because only a small amount of moisture is necessary for the cement to hydrolyze and cure.

In the early days, the excelsior board was produced in relatively small plants. The forming of the mats was done by hand. However, in the modern plants, the formation is now performed by mechanical processes.

For curing the mats into the final board, one system stacks the mats one on top of the other and then applies pressure by means of concrete weights. The mats are left under pressure for about 25 hours. Another system places the formed mats in a press for about 1 ½ minutes. The mats are then clamped together and removed from the press. After 15 to 20 hours, the boards are taken out of the forming mold and set aside for final curing. While it was mentioned that cement is now the favorite binder to be used, the Austrian system called the Heraklith procedure has apparently remained with magnesite as the binding agent (magnesium carbonate, $MgCO_3$). This board is continuously pressed by press rolls with a setting time in the order of about 30 minutes. It then takes one to two weeks, depending upon the thickness of the board, for final curing of the mortar to take place.

The first system developed by the National Gypsum Company, according to its patent information, encompasses the continuous forming system and a continuous pressing system that uses steam heating for rapid curing of the cement binder. A magnesium oxysulfate cement is used, which consists of a water-borne mixture of magnesium oxide in magnesium sulfate. These two ingredients are combined just before being sprayed upon the excelsior.

First the excelsior is deposited on a solid conveyor belt. After a smooth mat of excelsior is developed, it is then sprayed with the magnesium cement. A typical mixture will be 35% of the magnesium oxysulfate dissolved in water, to which is then added 65% of magnesium oxide. As soon as these two ingredients are mixed together, the cement starts to set. However, normal setting time will be over one hour. In this process, as soon as the cement is applied to the mat, the mat moves forward through a preheating oven. Heated, water-saturated air is blown down through the mat and through the perforated conveyor belt. The temperature of the air ranges from 110° to 140°F (43.3°-60°C). This particular temperature range will cause the magnesium oxysulfate cement to set in a time somewhere between four and eight minutes. The exact time is dependent upon the chemical activity of the magnesium oxide. The mat moves forward into a continuous press. The preheated mat must enter the press before the cement has set. The press itself consolidates the mat to final thickness, and this consolidation must take place before final curing of the cement. The press temperature is about 190°F (88°C). This press is composed of a solid rubber conveyor belt which is backed up by moving metal platens for forming flat surfaces. As the pressed board emerges from the press, the cement is set sufficiently to hold the board at its desired thickness. The boards then go through a series of drying ovens where heated air is blown through them, removing excess moisture.

This process has been improved so that curing can take place in a steam-heated hot press. These presses range up to 100 ft (30.5 m) long, with the first one-third to one-half of the length built to pass live steam through the mat. Experimentation showed that using live steam with more than 10% moisture reduced the setting time of the cement to as short a time as 30 seconds.

As in the original invention, the cement is sprayed on a thin, excelsior mat and subsequently distributed on the excelsior fibers by means of picker rolls. The picker rolls distribute the excelsior into a thick (12 to 20 in. or 305 to 508 mm) formed mat which is preheated by steam and then passed through the press, where it is compacted to the desired thickness. The press is composed of upper and lower conveyor belts. The bottom one is a perforated stainless steel belt. The belts are backed up by upper and lower platens, each composed of a number of hollow platen members extending crosswise in the press.

Live steam is introduced through the perforations in the bottom platen, and as the steam passes through the mat, the cure of the cement is completed. It takes from 30 seconds to 5 minutes to cure the boards, depending upon the board thickness and other production parameters. It is necessary to get the temperature in the board to 212°F (100°C) to effect the final cure.

The board density may range from 10 to 50 lbs/ft^3 (0.16–0.80 sp gr), although 15 to 26 lbs/ft^3 (0.24–0.42 sp gr) has been found ideal for most construction uses. The following table provides a typical materials balance for a board of 21 lbs/ft^3 density (0.34 sp gr) 12 in. square (30.5 cm) by 1 in. (25.4 mm) thick.

Material	Weight (g)	(%)
Excelsior at 20% moisture	448	33.9
MgO (magnesium oxide)	284	21.5
$MgSO_4\ 7H_2O$ (magnesium sulfate)	153	11.6
Water	435	33.0
	1320	100.0

This table shows that about one-third of the mix is wood. The water is mostly removed during the drying. After pressing, the wood consists of about 50% of the board on a dry-weight basis.

It takes about 8 psi (55.1 kPa) to compress the board of 15 lbs/ft^3 density (0.24 sp gr) and about 14 psi (96.5 kPa) to consolidate the board of 21 lbs/ft^3 density (0.34 sp gr).

The rapid cure of the cement in this process has reduced the importance of the relative activity of the cement ingredients and has provided a situation where a wider range of oxides can be used.

This product is now being manufactured and sold under the trade names of Tectum I and Tectum II. Tectum I is the product just described. Tectum II has a rigid polyurethane foam applied to the top, which penetrates from ⅛ to ¼ in. (3.2–6.4 mm) into the face of the board. The urethane foam has a density of 2 lbs/ft^3 (0.03 sp gr) and fuses in place after it is applied.

Tectum is made in thicknesses ranging from 1 in. to 3 ½ in. (25.4–88.9 mm). Depending upon the thickness and supplemental reinforcement, the roof panels can span up to 72 in. (1.83 m). Tectum is also used for ceiling and wall sound control, accent and decorative purposes, and in recent years as a complete side-wall and roof system in residential housing.

Building Blocks

Building blocks using cement for the binding agent have been used for many years. This type of product has been popular in Scandinavia. A number of buildings have been put together with this type of material in the United States.

One pilot plant has been built in Spokane, Washington, where blocks are fabricated by machine in a size 8 in. (203 mm) thick, 12 in. (305 mm) high, and 48 in. (1.21 m) long. This machine can also produce blocks 24 in. (610 mm) long and vary the thickness and height if desired. The blocks are also made with a tongue and grooved top and bottom for the do-it-yourself builder, if he does not want to use the standard procedure of laying up the block with mortar. The larger-size blocks described weigh approximately 100 lbs (45.5 kg) and compare favorably to the weight of a similar piece of wood of the same size. Because of its construction it is nonflammable and has excellent insulation properties. Its density is such that nails can easily be driven into it and it can be sawn, drilled, routed, sanded, and handled much like any piece of softwood.

Because of the ease in manufacturing this type of building product it is not only suitable for small operations in developed countries but it is an excellent product for production in developing countries that have wood residues and cement. As stated previously, cement may be a problem because of the high energy costs associated with its production.

Flakeboards

Elmendorf Research Incorporated has developed a flakeboard with Portland cement as the binder. Its name for this flakeboard is Century Board and at present it is being manufactured in Japan and, on a small basis, in Switzerland under the name Duripanel. The Elmendorf term for long, narrow wood flakes is *strands*. In its process, the strands are first treated to remove any extractive material that may cause problems with the cure of the cement. Portland cement is then applied to the strands, and the coated strands are formed into mats on moving metal caul plates. The cauls with their mats of cement and strands are stacked and subjected to pressures equivalent to those for making plywood, which are retained by clamps until the cement has set. When high-early strength cement is used, the pressing time is about 12 hours. With normal cement, the pressing time may be up to 16 hours. The green boards are solid piled for at least three days, after removal from pressure, and can then be forced dried in several hours. The bending strength of this board is determined by the strength of the wood strands and not by the full strength of the cement.

This particular board is made in specific gravities of 0.95 to 1.20 (59.4–75 lbs/ft^3 density). Three different types are produced, namely: Type A—boards in which the cement-coated strands are randomly distributed in one or three layers and then compacted; Type B—boards in which the cement-coated strands are mechanically deposited in three layers with the strands of the outer layers oriented into parallelism; and Type C—boards in which the outer surface is textured and is composed of cement and inorganic filler.

This type product can be used for floors, doors, load-bearing walls and partitions, and concrete forms. Unlike the excelsior-type product, it can also be used for exterior exposures, and exterior siding is the major utilization.

Recently a group of Swiss companies, together with a German equipment manufacturer (Bison-Werke), have been working with the Elmendorf concept for the production of cement-bonded wood particleboards. At present, an expanded pilot plant is operating producing Duripanel. The type of product being produced is similar to synthetic resin-bonded wood particleboard, as is the process system.

The raw materials are Portland cement, wood particles of the flake type, and any other necessary chemical additives. The wood must be carefully selected because wood extractives or hemicellulose have an unfavorable effect on the cement. The preferred wood species in Europe are spruce and fir.

The preparation of the particles, blending, and mat forming are quite similar to standard board production processes using synthetic resin. An air-classifying type of former is being used, which lays the finer particles on the mat faces. Thus, the board has relatively smooth faces. The mats are formed on metal cauls, then a package of cauls and formed mats is placed into a hydraulic press, where the mats are consolidated to final thickness. The package is rigidly clamped together and heated and humidified for about seven to nine hours to accelerate the curing. The

Figure 3.16. *Exterior-wall cladding of the Sandoz AG Laboratory in Basel, Switzerland, (photo of the Durisol AG Dietikon from Deppe 1974).*

panels are then strong enough to be handled. The clamps are released, solid piled for several days, then forced air dried. The finished panels are trimmed to size and sanded on one side, or both sides, for dimensioning.

Work is underway at the present time to reduce the pressing or curing time, but care must be exercised in adding substances to speed up the cure. Such substances cannot be used in quantities which impair the board properties, particularly their resistance to fire. Stiffness, bending strength, and tensile strength of resin-bonded particleboards surpass those of the cement-bonded boards. The compression strength of the cement-bonded boards, however, is superior to that of the resin-bonded particleboards.

In Germany, these panels are planned for use as ceiling linings, and interior and exterior paneling. Consequently, their fire behavior is particularly important. Tests so far show that in Europe this particualr product has outstanding characteristics in a fire compared with the normally inflammable wood-based materials and fire-retardant wood-based materials.

The development of this type of board in Europe is only in its infancy, although much of the product is in use. Technological problems still have to be solved; however, its excellent behavior in fire makes this a worthwhile endeavor. Figure 3.16 shows one of the applications.

SELECTED REFERENCES

American Society for Testing Materials. 1975. Definitions of Terms Relating to Wood-Base Fiber and Particle Panel Materials. ASTM D1554. Philadelphia, Pennsylvania.

Aspulund, A. 1972. Trends and Developments in the Manufacture of Fiberboard. *Proceedings, Seventh World Forestry Congress* (October). Buenos Aires, Argentina.

AVM Corporation, Plex-O-Wood, Inc. Pex-O-Wood. Bradford, Pennsylvania.

Brooks, S. H. W. 1972. Medium Density Fiberboard and Municipal Waste. *Proceedings of the Washington State University Particleboard Symposium, No. 6*. Pullman, Washington: Washington State University (WSU).

Caron, P. E. 1968. Molded from Wood Fibers . . . New Family of Products for Variety of Markets. *Wood & Wood Products* (September).

Carroll, M. N. 1966. Composition Board. *Encyclopedia of Polymer Science and Technology,* Vol. 4. New York: Interscience Publishers.

Chittenden, A. E. 1972. Wood and Cement—Past and Future. Proceedings, Seventh World Forestry Congress, Buenos Aires, Argentina.

Chittenden, A. E.; A. J. Hawkes; and H. R. Hamilton. 1975. Wood Cement Systems. Background Paper No. 99, third World Consultation on Wood-Based Panels,FAO, New Delhi, India, February 1975.

Collins, H. W. 1953. Method of Forming Wood Wool Panels.Patent No. 2,655,458. United States Patent Office (October).

de Jong, W. J., and D. E. Wood. 1976. The Cojafex Process for the Manufacture of Medium Density Fiberboard. *Proceedings of the Washington State University Particleboard Symposium, No. 10*. Pullman, Washington: WSU.

Deppe, H.-J. 1974. On the Production and Application of Cement-Bonded Wood Chipboards. *Proceedings of the Washington State University Particleboard Symposium, No. 8*. Pullman, Washington: WSU.

———. 1975. Manufacture and Utilization. Background Paper No. 44/6, third World Consultation on Wood-Based Panels, FAO, New Delhi, India, February 1975.

Elmendorf Research, Inc. 1970. Manufacture and Uses of Century Board. Report No. 199-G (April). Palo Alto, California.

European Federation of Particleboard Manufacturer's Association. 1975. The Present Situation and Trends in Development of the European Particleboard Industry. Background Paper No. 44, third World Consultation on Wood-Based Panels, FAO, New Delhi, India, February 1975.

FAO. 1976. Basic Paper IV. Technology and Techniques in the Manufacture of Wood-Based Panels. *Proceedings of the World Consultation on Wood-Based Panels,* held in New Delhi, India, February 1975. Brussels: Published in agreement with the Food and Agriculture Organization of the United Nations by Miller Freeman Publications.

Fordyce, D. H. 1972. Realism of Logging Residues—Utilization or Disposal? *Proceedings of the Washington State University Particleboard Symposium, No. 6.* Pullman, Washington: WSU.

Gold Bond Building Products. 1971. Tectum II. Buffalo, New York.

———. 1973*a*. Tectum Roof Deck and Form Systems. *Technical Bulletin No. 3-4175* (January). Buffalo, New York.

———. 1973*b*. Tectum Wall Panels. *Technical Bulletin No. 9-8008* (March). Buffalo, New York.

Goodman, W. 1975. Private Correspondence on Thermax and Absorbex, Chewelah, Washington.

Gran, G. 1975. New Techniques and Technologies in Fiberboard Production. Background Paper No. 6, third World Consultation on Wood-Based Panels, FAO, New Delhi, India, February 1975.

Gunn. J. M. 1972. New Developments in Waferboard. *Proceedings of the Washington State University Particleboard Symposium, No. 6.* Pullman, Washington: WSU.

Haigh, A. H., Jr. 1967. The Bison Blenders. *Proceedings of the Washington State University Particleboard Symposium, No. 1.* Pullman, Washington: WSU.

Harter. W. 1966. Wood Aggregate Concrete Building Blocks. *Proceedings of the 21st Annual Northwest Wood Products Clinic* held at Spokane, Washington. (April). Pullman, Washington: WSU.

Heebink. B. G., and F. V. Hefty. 1968. Steam Post Treatments to Reduce Thickness Swelling of Particleboard. *Proceedings of the Washington State University Particleboard Symposium, No. 2.* Pullman, Washington: WSU.

Huffaker, E. M. 1962. Use of Planer Mills Residues in Wood-Fiber Concrete. *Forest Products Journal,* Vol. 12, No. 7.

Impellizzeri, J. S.; T. W. Vaughan; and R. Etzold. 1976. Oriented Strand Board Goes to Western Massachusetts. *Proceedings of the Washington State University Particleboard Symposium, No. 10.* Pullman, Washington: WSU.

Johnson, R. E. 1974. A Development in Production, Sales, and Use of Structural Particleboard. *Proceedings of the Washington State University Particleboard Symposium, No. 8.* Pullman, Washington: WSU.

Keesey, L. H. 1970. Plant Experience Producing Medium Density Fiberboard. *Proceedings of the Washington State University Particleboard Symposium, No. 4.* Pullman, Washington: WSU.

Kreibaum, O. 1957. Extrusion Press Manufacture of Particleboard—Economics, Quality, Production. Technical Paper No. 5.34. *Fibreboard and Particle Board Technical Papers,* International Consultation on Insulation Board, Hardboard, and Particle Board, FAO, Geneva, Switzerland, January-February 1957.

Lagerquist, M. 1975. New Particleboard Extruder Stretches Wood Residue Profits. *Wood and Wood Products* (April).

Lehmann, W. F. 1968. Durability of Exterior Particleboard. *Proceedings of the Washington State University Particleboard Symposium, No. 2.* Pullman, Washington: WSU.

Maloney, T. M., and A. L. Mottet. 1970. Particleboard. *Modern Materials,* Vol. 7. New York: Academic Press Inc.

Pagel, H. F. 1967. Molding Wood Products. *Proceedings of the Washington State University Particleboard Symposium, No. 1.* Pullman, Washington: WSU.

Prior, W. L., and H. J. Snelson. 1960. Process for Steam Treating Magnesium Cement Fibrous Panels. Patent No. 2,944,291. United States Patent Office (July).

Raddin, H. A. 1970. The Economics of the System Producing Dry Process Medium Density Fiberboard. *Proceedings of the Washington State University Particleboard Symposium, No. 4.* Pullman, Washington: WSU.

Schmierer, R. E. 1971. Australian Particleboard Practices. *Proceedings of the Washington State University Particleboard Symposium, No. 5.* Pullman, Washington: WSU.

———. 1975. Particleboard Practices in Australia and New Zealand. Background Paper No. 22, third World Consultation on Wood-Based Panels, FAO, New Delhi, India, February 1975.

Snelson, H. J., and W. L. Prior. 1961. Apparatus for Steam Treating Fibrous Panels. Patent No. 2,975,470. United States Patent Office (March).

Suchsland, O. 1972. From Wood Waste to Particleboard. *Michigan State University Extension Bulletin E-735* (March).

Trutter, G., and M. Himmelheber. 1970. Die moderne Spanplattenfertigung: Stand der Technik bei den Fertigungsverfahren (Modern Particle Board Manufacture: State of the Art in Various Manufacturing Processes). *Holz als Roh-und Werkstoff,* Vol. 28, No. 3.

United Nations Industrial Development Organization. 1975. Review of Agricultural Residues Utilization for Production of Panels. Background Paper No. 127, third World Consultation on Wood-Based Panels, FAO, New Delhi, India, February 1975.

Vajda, P. 1969. Economics. *Proceedings of the Washington State University Particleboard Symposium, No. 3.* Pullman, Washington: WSU.

Vasishth, R. C., and P. Chandramouli. 1975. New Panel Boards from Rice Husks and Other Agricultural By-Products. Background Paper No. 30, third World Consultation on Wood-Based Panels, FAO, New Delhi, India, February 1975.

Verbestel, J. B. 1968. Some Experience with and Possibilities for the Manufacture of Particle Board from Non-Wood Fibrous Raw Materials. FAO Committee on Wood-Based Panel Products, Second Session, Rome, Italy, November 6-8, 1968.

Vermaas, C. H. 1975. Duripanel Cement-Bonded Particleboard—Economic Aspects of Production and Marketing. Background Paper No. 23, third World Consultation on Wood-Based Panels, FAO, New Delhi, India, February 1975.

Wentworth, I. 1969. Moulded Products and Laminates. *Proceedings of the Washington State University Particleboard Symposium, No. 3.* Pullman, Washington: WSU.

Wyss, R. 1975. Duripanel—The Cement-Bonded Particleboard. Background Paper No. 29, third World Consultation on Wood-Based Panels, FAO, New Delhi, India, February 1975.

4. COMPOSITION BOARD MATERIALS: PROPERTIES AND TESTING

Composition board properties include strength, fastener holding ability, responses to soaking in water, responses to changes in ambient humidity conditions, and changes caused by exposures to severe weathering conditions. All of these properties are important in design, for determining the quality of the product, and for acceptance by the consumer. This discussion will cover the properties now measured, their definition, and how they are determined according to the United States and Canadian procedures. It is recognized that different test and methods of test are found in other countries, but a full discussion of these is beyond the scope of this book. Also to be covered will be standards and their development and the properties required in the developing structural products. Actual minimum properties were reported in Chapter 2. Trying to report all the actual properties of various boards and comparing them to solid wood and plywood is beyond the scope of this book as this is a big subject deserving separate treatment. A simple comparison between types of material cannot be made as design factors must be considered and this complicates the comparison.

PROPERTIES

The strength properties of composition boards include static bending, tensile strength parallel to the surface, tensile strength perpendicular to the surface (internal bond), compression strength parallel to the surface, shear strength in the plane of the board, glue line shear, and impact interlaminar shear and edgewise shear. Important fastener holding properties are lateral nail resistance, nail withdrawal, nailhead pull-through, and direct screw withdrawal. Other properties include hardness and abrasion resistance. Properties associated with moisture include water absorption and thickness swelling, linear variation with change in moisture content, edge thickness swell, accelerated aging, cupping and twisting.

Not all of the properties listed above are important for all of the products manufactured today. Rather, each product has a particular set of properties which are important because of the intended end use. Therefore, the applicable properties are selected for particular end uses, and minimum or maximum allowable values are specified. In general, the physical properties usually evaluated for board products include static bending (which covers modulus of rupture and modulus of elasticity), internal bond, water absorption and thickness swell, and linear variation with change in moisture content. For those products going into structural applications where they will be nailed, lateral nail resistance, nail withdrawal, and nailhead pull-through properties are of importance. Panels going to the furniture industry must have good face and edge screw-holding properties. Products such as mobile home decking must have a certain resistance to impact. Those products being exposed to the weather normally will have to have a certain level of resistance to accelerated aging. Since properties are usually associated with the specific gravity of the product, specific gravity is usually determined along with the physical properties.

Other properties normally not measured, such as thermal conductivity, duration of load, and creep, will become important as composition board moves into new applications. Surface smoothness is of great importance in secondary finishing with paints and films, but tests have not been standardized for use in the United States. Profilometers are in use elsewhere for measuring this board characteristic.

DETERMINATION OF PROPERTIES

Measurement of the physical properties of composition boards is generally performed according to the American Society for Testing and Materials "Standard Methods of Evaluating the Properties of Wood-Base Fiber and Particle Panel Materials," ASTM D 1037. In this document there are two parts. Part A covers the general methods for evaluating the engineering and design properties of wood-base fiber and particle panel materials. It is used for obtaining the basic properties suitable for comparison studies with other engineering materials of construction or for engineering design. In Part B, "Acceptance and Specification Test Methods for Hardboard," the methods noted are employed generally for testing purposes in the industry. They are designed for specific use in specifications for procurement and acceptance testing of hardboard. These tests achieve adequate precision of results combined with economy and speed in testing and are useful for the development of specifications. They may differ to some degree from properties evaluated by using the more refined methods noted in Part A of this standard. In Part A, many of the tests are conducted on specimens in both the dry and the water-soaked conditions, which is not required in Part B.

For static bending, two properties are usually measured: modulus of rupture and modulus of elasticity. In technical terms, the modulus of rupture is the computed maximum fiber stress in the extreme upper and lower surface fibers of the specimen under test. It is an approximation of the true stress, as the formula for computing it makes assumptions that are valid only up to the proportional limit. In simple terms, this value is regarded as the breaking strength of the product under test and is reported in pounds per square inch, or pascals. Common terms used interchangeably with modulus of rupture are bending strength and flexural strength. Modulus of elasticity refers to the stiffness of the material and is reported in pounds per square inch, or pascals. This property is useful in calculating the deflection of the product under stress. In composition boards, both of these properties are determined parallel to the face of the panel.

An important point often overlooked is the marked improvement of bending and stiffness properties of furniture-type products in the final assembly when overlaid with some type of veneer or plastic sheet. The final strengths are the most important in such a composite and should be the ones considered. It is not uncommon for modulus of rupture to be doubled in such a construction.

Tensile strength perpendicular to the surface, more commonly called internal bond within the industry, provides a measure of how well the panel is glued together. A test specimen 2 in. (51 mm) square is pulled apart with the direction of the load as nearly perpendicular to the faces of the board as possible. This value is reported as the pounds per square inch or pascals required for causing failure. For measuring tensile strength parallel to the surface, specimens with a reduced cross section in the area under test are pulled apart to determine the value of the property. This test is not used extensively in the industry as yet but will become of more importance as the industry moves into the structural field.

A relatively complicated procedure is necessary for determining the compression strength parallel to the surface because of the large variation and the character of the wood-base fiber and particle panel products. It is one that is not used widely in the industry; however, as new products move into structural applications, the determination of this property and the development of design information on it should become more important.

As mentioned previously, lateral nail resistance is an important property of those panels being used in structural applications such as sheathing. In this test, 6-penny common nails (0.113 in. or 2.80 mm in diameter) are used and are located ¼, ½, or ¾ in. (6.4, 12.7, or 19 mm) in from the edge of the specimen under test. The nails are driven at right angles through the face of the board so that an approximate equal length of nail projects from each face. Hardened steel pins can be used in place of the nails as nails tend to bend in the test. These nails are then pulled laterally from the panel until the material between the nail and the edge ruptures. The property is reported as the pounds or kilograms required to cause failure.

For the nail withdrawal evaluation, 6-penny common nails are driven through the board at right angles to the face. The load in pounds or kilograms required to withdraw the nail from the product under test is reported. In the nailhead pull-through test, 6-penny common nails are again used. In this case the nail is driven through the board until the head sits flush with the top surface of the board. The shank of the nail exposed on the bottom side of the test specimen is then gripped and a load is applied until the nailhead pulls through the specimen. Again the load is reported as pounds or kilograms.

There are two direct screw withdrawal evaluations that are performed. One is withdrawal perpendicular to the plane of the board, or face screw holding, and the other is withdrawal from the edge of the board, or edge screw holding. The standard calls for the use of the No. 10, 1-in. (25 mm) wood screw to be used for this test. However, there is an exception where 1-in. No. 10 type A sheet metal screws can be used. This sheet metal screw is the one commonly used in the particleboard industry. The load required for pulling the screw from the specimen under test is recorded in pounds or kilograms.

For the hardness test a modified Janka ball test is used. This ball is 0.44 in. (11.2 mm) in diameter. The measure of hardness is recorded as the load at which the ball has penetrated to one-half its diameter. Hardness can also be measured by what is called the hardness modulus method of determining the equivalent Janka ball hardness. In this test, the hardness modulus in pounds per inch of penetration divided by 5.4 provides the equivalent Janka ball hardness in pounds. Normally, a specimen must be at least 1 in. (25 mm) thick or be made up of several thicknesses laminated together to provide a thickness of 1 in. to conduct the Janka ball hardness test. The modulus method of determining hardness can be used for specimens ranging from ⅛ in. to 1 in. (3.2–25.4 mm) in thickness.

The shear strength in the plane of the board is tested by forcing a shear failure within the board proper. For most applications this property is not measured; however, as mentioned before, as composition boards move more into the structural field this particular property will become quite important. Another shear test is the glue line shear test which evaluates the glued board constructions obtained when various thicknesses of board are laminated together to provide a product with greater thickness.

The falling ball impact test is used to measure the impact resistance of boards to the kind of damage that occurs in service when struck by moving objects. A 2-in.-diameter (51 mm) steel ball is dropped on a supported test specimen from increasing heights. Each drop is made at the same point at the center of the panel until the panel fails. The height of the drop in inches that produces a visible failure on the face opposite the one receiving the impact is recorded as the index of resistance to impact. Other tests not included in this test specification for impact include testing with 10- and 50-lb weights. Experimentation is going on at present in an effort to develop more meaningful impact tests.

There are several types of teting machines available for evaluating abrasion resistance. The one specified by ASTM is the U.S. Navy Wear Tester. An abrading disc rotates against the specimen under test. Abrasion is reported as loss in thickness in inches per 100 revolutions of wear up to 500 revolutions if the wear is uniform. Due to the density profile in most boards, the wear may not be uniform after the first 200 revolutions because of a change in the abrasion resistance as the test specimen is penetrated. In this case the amount of wear for each 100 revolutions is recorded.

The evaluations of water absorption and thickness swelling are made simultaneously on a test specimen. The standard size is 12x12 in. (305 mm square) in size; however, a 6x6-in. (152 mm square) specimen size can be used as well. The specimen is submerged horizontally under 1 in. (25.4 mm) of distilled water, and the increase in weight due to the absorption of water and the increase in thickness, also due to absorption of water, are measured after a 2-hour and 24-hour submersion. In most cases only the 24-hour test is conducted.

Linear variation with change in moisture content is a time-consuming test. Test specimens usually 3 in. (76 mm) in width and at least 12 in. (30.5 cm) in length are first conditioned at a relative humidity of 50% ±2% and a temperature of 20° ±3°C (68° ±6°F). After equilibrium is reached, the length of the specimen is measured and then it is placed in an atmosphere held at a relative humidity of 90% ±5% and a temperature of 20°C ±3°C (68° ±6°F). Again, after practical equilibrium is reached, the increase in length is measured and the increase is reported as a percentage. It is of value to also measure thickness swell although it is not required.

Some plants have been successful in correlating vacuum-pressure-soak (VPS) evaluations with linear expansion. In this test, specimens are placed in a pressure chamber. A vacuum is first applied, followed by a pressurized time period with the specimen in water for rapid impregnation of the test specimen with water. Such a severe test naturally forces a much greater linear expansion than would be found in the normal test. However, its speed is of value on a production line. Correlations can be developed versus the standard test. Efforts so far to develop a general correlation between the test and the standard linear expansion test for widespread use, however, have not been successful. Correlations must be developed with each type of board being evaluated.

Edge thickness swell is an important property in materials such as siding. If the edge under severe exposure conditions expands more than the rest of the board, the change in thickness is quite noticeable and detracts from the appearance of the structure to which the siding has been applied. A method for evaluating edge thickness swelling, called the *disk* method, has been developed. It is used to measure the maximum amount of edge thickness swelling caused by the 24-hour water soaking. A small disk 1 in. (25 mm) in diameter by the thickness of the board is used for this test. After proper conditioning, it is measured for thickness and then submerged horizontally under 1 in. (25 mm) of distilled water in the same way as the water absorption evaluation is conducted. After 24 hours it is measured again for thickness and the percentage of thickness is recorded. In addition, the amount of water absorbed is also recorded.

The next test to be discussed is the accelerated aging test, which is perhaps the most controversial test used in the industry. It is designed to obtain a measure of the inherent ability of a material to withstand severe exposure conditions. It is a very severe six-cycle test with each cycle consisting of the following: (1) immerse in water at 120° ±3°F (49° ±2°C) for one hour, (2) spray with steam and water vapor at 200° ±5°F (93° ±3°C) for three hours, (3) store at 10° ±5°F (-12° ±3°C) for 20 hours, (4) heat at 210° ±3°F (99° ±2°C) in dry air for three hours, (5) spray again with steam and water vapor at 200° ±5°F (93° ±3°C) for three hours, (6) heat in dry air at 210° ±3°F (99° ±2°C) for 18 hours. After the completion of the six cycles of exposure, the material is further conditioned before any other testing at a temperature of 20° ±3°C (68° ±6°F) and at a relative humidity of 65% ±1% for at least 48 hours.

At that time, static bending, lateral nail resistance, nail withdrawal, water absorption, or nailhead pull-through tests can be conducted. The most common test used after accelerated aging is modulus of rupture. The commercial standard for particleboard calls for a retention of 50% of the modulus of rupture property after accelerated aging; however, the thickness of the specimen before testing commenced is used for making the calculation. Since the panel under test usually has expanded considerably in thickness, this means the actual modulus of rupture property of the specimen under test would be much lower that that calculated using the original specimen thickness. Modulus of rupture is probably an incorrect term to be used in this case since the number that is developed is used more as a quality control index rather than as a bending strength indication.

As can be seen, this is a very extreme test. It is one that was first developed for use with low-density fiberboard and has been adopted for use with hardboard and particleboard as they developed. No board in service would be reasonably expected to have such a severe exposure. Its basic meaning appears to be to determine whether an interior-type resin such as urea-formaldehyde or whether an exterior-type resin such as phenol-formaldehyde was used. Usually, the urea-bonded boards will fall apart early in the test. A more reasonable and definitive test for evaluating composition board materials under exposure is needed. It should be noted that urea-bonded boards are used extensively for structural applications in other countries. Mobile home decking in the United States is urea-bonded and does not have to pass this test. Thus, to some this is an unreasonable test, as board serving successfully in the structural mode elsewhere would fail in the United States.

The Canadian standard calls for a different type of accelerated aging evaluation. In it, modulus or rupture specimens are prepared and then boiled for at least two hours. Immediately after boiling, the specimens are placed in cool water for at least one hour and then tested. As with the American standard, the calculation for modulus of rupture is made using the original thickness of the specimen. In Canada, only grades P & Q board must pass the accelerated aging test. A minimum value of 1000 psi is specified, which is 50% of the minimum dry value. This test is nowhere near as severe as the American test. Other countries also conduct such testing with the boil test.

Cupping and twisting of test specimens after accelerated aging can also be determined. This particular test is not used widely. Technically speaking, a properly manufactured composition board should be warpfree. Production of board with unbalanced construction, e.g., more fine particles on one side than on the other, usually results in warp. However, the subject is not completely understood. Warp will appear in board coming off a production line with all manufacturing variables apparently under control. Extensive investigation will not determine the cause. And then, just as mysteriously as it appears, warp will disappear. This has been a frustrating phenomenon to a number of plant managers.

It has been pointed out that specific gravity is usually determined in all specimens undergoing tests. The industry normally determines specific gravity by using the weight and dimension of the specimen at time of test. The standard, however, indicates that for precise measurement the oven-dry weight of the specimen under test should be used in this calculation.

Moisture content of the specimen under test can also be evaluated. Usually for most testing purposes it is not measured, although its determination can be useful in interpreting test results.

Interlaminar shear is of importance if the board will be used in glued structural assemblies such as sandwiches, webs in box and I-beams, and gusset plates in trusses. Data provided by the test for shear strength in the plane of the board or the glue line shear test are not comparable because of effect of friction in the two and because of the fact that failure in the block shear test can occur only in a ⅛-in.-thick (3.2 mm) area in the middle of the board. Shearing in the interlaminar shear test takes place in the weakest plane of the specimen under test. Again, this test, which provides a shearing modulus in psi or kg per cm squared, will become of more use as composition board moves into structural applications where shear properties will be of importance in engineering design.

Edgewise shear fits into this same category for structural applications. This test is designed in such a way that the specimen is loaded in compression in order to introduce shear forces that produce failures across the panel. This test produces the kind of shear stressing that occurs when the shear forces are introduced along the edges of the panel. This property is important for boards used where racking forces are involved as in sheathing. Racking is the type of load imposed by wind blowing against the edge of a wall section. Sheathing could therefore be visualized as the bracing in a wall resisting collapse induced by wind forces applied edgewise.

Other useful information can be determined using tests not included in ASTM D 1037. These include two tests developed by the Canadian Forest Products Laboratory in Ottawa: the torsion-shear and surface strength tests. The Japanese have a stripping resistance test in their standard and the British have a surface roughness test in their standard. The Federal Republic of Germany has many tests and standards beyond those presently used in the United States. Other important tests in the United States are the ones for thermal conductivity, fire retardancy, and resistance to decay.

The torsion-shear strength test is an important quality-control technique which accurately predicts the internal bond of the board under test. It is particularly useful because it eliminates the time-consuming conventional internal bond test, which requires a glued-up test assembly, and also eliminates the need for a testing machine. The measurement is made on 1-in.-square (25.4 mm) specimens using a standard torque wrench equipped with 1-in. sockets.

Figure 4.1 shows one of these tests being performed. The bottom socket is clamped in a vise and the top socket is fitted to a torque wrench. The 1-in.-square specimen under test is positioned with one-half its thickness inside both the top and bottom sockets. The torque wrench is then slowly turned until the specimen

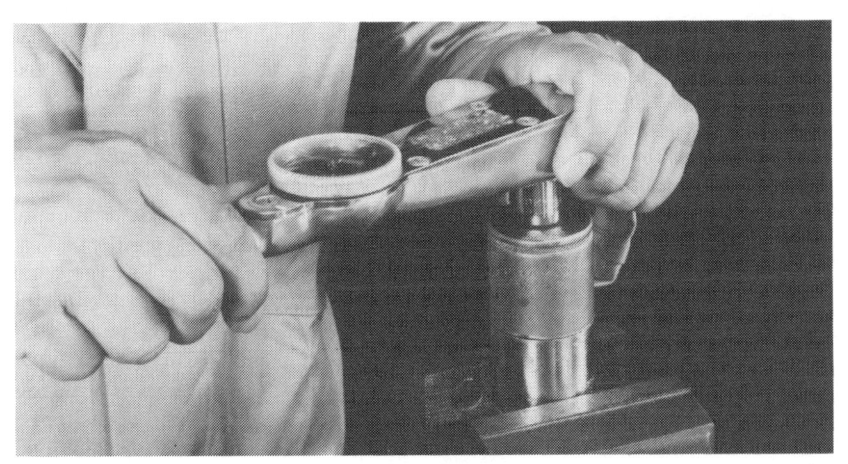

Figure 4.1. *Torque-shear test on a 1-in.-square specimen with the bottom socket clamped in the vise and the top fitted to a 12-ft-lb (16.3 J) torque wrench. A 1-in. (25.4 mm)-square specimen is positioned with half its thickness inside both top and bottom sockets (Shen and Carroll 1969).*

fails. The maximum torque in terms of foot-pounds is then read. This information can then be correlated with the internal bond properties determined according to the standard method. The test is performed at the center of the specimen under test because normally this is the weakest plane in a board. However, it is also possible to select other planes within the board for testing, showing the strength profile through the cross section of the board. It has been found that this strength profile is closely related to the density profile within a board; consequently, the density profile can be measured indirectly by this method.

A drawback of this test in terms of quality control is that the failure is being forced at a certain plane within the board. The core of the board is normally weakest in internal bond; however, because of process problems, the face of the board can, at times, be weaker in bonding than the core. Consequently, when using this technique for quality control, it is also wise to evaluate the surface strength of the particleboard. Using the standard internal bond testing technique, the failure occurs in the weakest plane. The use of the torsion-shear test with boards less than ¼ in. (6.4 mm) in thickness is not advised.

The Canadian surface strength testing method is listed as an appendix to their standard CSA Standard 0188-1975, ''Mat-formed Wood Particleboard.'' As an appendix, it is not a mandatory part of their test standard. In this test, the minimum thickness of the specimen required is ¾ in. (19 mm). For thinner boards it is necessary to laminate several together to obtain this minimum total thickness. In this test a metal cylinder with a 1-$in.^2$ (25.4 mm^2) cross section is glued onto the surface of the test specimen under evaluation using an epoxy or other suitable adhesive. This test assembly is then loaded into a testing machine and the glued-on cylinder is pulled from the surface of the board test specimen. The load is recorded in psi or pascals and provides an indication as to the strength of the faces of the board under test.

In the United States, attempts have been made to develop a surface roughness test procedure and these instruments have been used in laboratory work. However, it is not included in any adopted set of physical property test methods. Additional work in test method development and in understanding the information gathered will be necessary before such surface roughness tests are used widely; however, the adoption of such a standard will be important for those involved in finishing board. Other countries already employ such a test.

An important characteristic of composition board that has an effect on many board properties is *springback*. This characteristic is not measured, but it is important that it is recognized. Wood will set in a deformed shape due to a plastic flow at temperatures above 250°F (121°C) if the moisture content is relatively high. Considerably higher temperatures are apparently needed (over 330°F or 166°C) if the moisture content is around 7%. Thus, most composition board in the United States, which is pressed at moisture contents of about 7 to 10% and at temperatures throughout the board ranging from about 220°F (104°C) in the core to the press-platen temperature at the board surface of 300° to 400°F (149° to 204°C), do not set in a permanently deformed cross section. When moisture is absorbed by the board later, an increase in thickness takes place. If this absorption is high, the board does not return to its original thickness when the absorbed moisture is removed by drying. This increased thickness after swelling and shrinking of the board is called *irreversible swell*. In serious cases, this swelling can cause problems by changing board properties. In most uses, however, this has not been found to be a problem.

One technique developed to help alleviate this problem is post steaming of the panels, which relieves built-in stresses and reduces excessive springback. This also reduces surface roughening, which is called *fiber pop* by some. Effective use of this technique can apparently be made, particularly with boards that will face severe exterior exposure.

Some work has been conducted in determining the thermal conductivity of wood-base fiber and particle panel material. The test method followed is the one developed by the American Society for Testing Materials "Standard Method of Determining the Thermal Conductivity of Materials by Means of the Guarded Hotplate," ASTM C 177. This information is important when considering the heat flow through a composition board. Because of the many different formulations and densities, it is impractical to constantly conduct evaluations on the available boards used in building because of the long and tedious tests necessary for developing thermal conductivity information. The work performed so far has been designed to develop design values of thermal conductivity at 75°F (24°C) for all the common classes and kinds of insulation board, hardboard, and particleboard. Factors have also been developed to permit modification of the thermal conductivity factors when conditions require temperatures other than the mean of 75°F.

Due to the fact that there have been fires in which life and property were destroyed, efforts have been made through the years to reduce this loss. It is usually not the small fire that causes concern but a major tragedy which catches public attention and leads to the enactment of preventative codes. Such occurrences as hotel fires, school fires, and hospital fires are the catalysts which stimulate code revision in a series of increments rather than in an even progression. The solution in the prevention of such tragedies can proceed along three paths: (1) by eliminating the materials which furnish fuel for fire, (2) by rendering these materials noncombustible (a controversial subject), or (3) by early detection and suppression of the fire.

The first method means a loss of market or at best a reduced potential. The second and third methods allow the manufacturer to retain these markets. Eliminating products from a market is not an easy solution to the problem. The products in question are usually of widespread use and are of great economic importance. In addition, these products often have unique properties which are highly desirable and difficult to replace.

Fire retardancy of composition board and other wood products is not a property but a measure of its fire resistance or fire hazard. In terms of fire resistance, a rating is established such as found with a one-hour fire door. In terms of fire hazard, a classification is established and is designated as a certain class panel of plywood or particleboard depending upon the code authority involved. Fire problems, testing, rating, and classifying is in a considerable state of flux at present. Much effort is being directed into this field and significant changes may be in the offing.

Evaluation of the performance of wood building materials in fires involves two sets of considerations. The first relates to the combustion characteristics of the material itself—such as rapidity of flame spread, fuel contribution, and optical density and toxicity of smoke generated—which determine the degree of fire danger to occupants of a room containing exposed surfaces of the material in question. Test methods for measurement of these properties have been established to determine "fire hazard classifications" of building materials.

The second set of considerations relates to the ability of building assemblies, such as walls or doors, etc., to resist penetration of heat and flame while retaining structural integrity during fire exposure. Test methods have been developed to measure this ability of building assemblies to contain a fire to the room of origin to allow occupants of other parts of the building to escape, and to allow time for application of fire control measures. Such test are used to assign "fire resistance" ratings, measured in time, to building assemblies—such as a one-hour fire resistance rating for a fire door.

Fire-resistant products can be confusing as to definitions, and descriptive terminology is not universally agreed upon. Underwriters' Laboratories divides the classification of building material into two groups: fire hazard classification and fire retardant classification. The fire hazard group classifies materials as to surface burning characteristics, while the fire retardant group classifies structures fabricated from building materials. In the fire hazard classification the method of test provides a basis for comparing the burning characteristics of materials. It provides data in regard to flame spread, fuel contribution, and density of smoke developed during exposure to fire of the tested materials.

On the basis of these data the tested materials are compared to standard material in terms of applicable classification scales in which asbestos-cement board (rated 0) and red oak wood (rated 100) are the standard materials for flame spread, fuel contribution, and optical smoke density.

In the United States the method used to carry out these tests on panel products is the 25-ft (7.62 m) Steiner Tunnel Test as specified in ASTM E 84. Many fire codes base their classification of interior finish materials on this test. While Steiner tunnel values are required for wood building products for compliance with the fire hazard regulations in many building codes, other more convenient tests, using smaller apparatus, have been developed for product development and control purposes. These include the Standard Method of Test for Combustible Properties of Treated Wood by the Fire-Tube Apparatus (ASTM E 69), the National Bureau of Standards Radiant Panel Fire Test, and the Schlyter Test.

Figure 4.2 shows a schematic drawing of a Steiner 25-ft tunnel furnace, and Figure 4.3 is a photograph of the tunnel. This tunnel consists of a horizontal duct having an inside width of 17.5 in. (445 mm), a depth of 12 in. (304 mm), and a length of 25 ft (7.62 m). The sides and base of the furnace are lined with insulating masonry. One side is provided with draft-tight observation windows so that the entire length of the sample may be observed from outside the test chamber. The test specimen forms a roof over the furnace. An insulating cover, which is removable, is placed over the specimen and seals the main body of the tunnel. At the inlet end of the tunnel are two vertically directed gas burners which generate a flame impinging on the sample for a distance of 4.5 ft (1.37 m) down the length of the tunnel. Near the outlet end, the thermocouple sensing probe of a recording thermometer is located in the tunnel to measure the temperature increase during the test. In the exhaust duct at the tunnel outlet, a calibrated lamp and photocell combination measures the optical density of the smoke leaving the tunnel. The test is run for ten minutes and is standardized against select red oak flooring which is arbitrarily designated 100 and ¼-in. (6.4 mm) asbestos-cement board designated as zero. The test specimen is compared to these materials in flame spread rate, optical density of smoke developed, and fuel contribution.

There are several major code agencies; consequently, variations in classifications can be expected. In the Uniform Building Code, for example, interior wall and ceiling finish is classified I to III with respective flamespread in the tunnel test of 0–25, 26–75, and 76–225: the smoke density shall be no greater than 450 when the material is tested in the way intended for use. Thus, in the UBC one smoke density is used for all three classifications and fuel contributed is not considered. UBC also notes that the products of combustion shall be no more toxic than the burning of untreated wood under similar conditions.

Flame spread is a visual determination made by subtracting the distance of the ignition flame (4.5 ft or 1.37 m) from the total distance the flame traveled. The flame spread classification is determined by different formulas depending upon time of the distance of flame spread.

The optical density of the smoke, as measured by the decrease of the output of the lamp-photocell combination, is plotted as a function of time during the test. The smoke density rating is calculated from the measured area under the density-time curve for the sample expressed as a percentage of the area under the corresponding curve for red oak.

Fuel contributed is similarly determined from the time-temperature curves measured by a thermocouple placed 1 in. (25.4 mm) below the surface of the sample at mid-line and 12 in. (305 mm) from the end of the sample at the outlet of the tunnel. In this case the rating is calculated from the measurc area between the sample curve and the curve for the "0" rated asbestos-cement board expressed as a percentage of the corresponding area for red oak.

During the occurrence of a fire, the smoke developed is often considered to represent the major life danger to the building occupants. Although large quan-

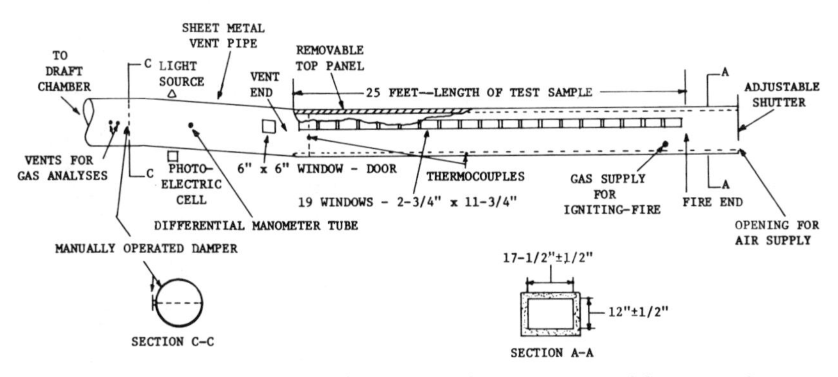

Figure 4.2. *Schematic drawing of Steiner 25-ft (7.62 m) tunnel furnace. (Courtesy Underwriters Laboratories Inc.)*

Figure 4.3. *A photograph of a Steiner tunnel furnace. (Courtesy Underwriters Laboratories Inc.)*

tities of toxic and flammable gases are present, the obscuration of vision by smoke, preventing occupants from escaping and hindering efforts of the firemen in the early stages of the fire, is held to be the greatest hazard. This occurs during the first minutes of the fire, before lethal conditions set in. Smoke is becoming of increasing concern to code authorities throughout the nation.

The Steiner test is large enough so that minor variations of a product tend to average out. It is, however, a large test which is expensive and time consuming to perform. Over the years it appears that this has been a realistic test and one of the most satisfactory methods available in measuring fire hazard flame-retardant-treated panel materials. Only fire hazard classifications are obtained in the tunnel. There is no pass or fail unless you are seeking a specific numerical or letter classification for a certain code.

The ASTM test for fire resistance is found in E 152, "Standard Methods of Fire Tests of Door Assemblies." In this test, door assemblies are evaluated for their ability to resist the passage of fire. The data developed by such tests enable regulatory bodies to determine the suitability of the particular door assembly in construction where fire doors are required.

The door under test is first placed in a test assembly. A furnace which is on one side of the door must generate sufficient heat to reach a temperature of 1000°F (538°C) within five minutes. The temperature at "0" is 68°F (20°C). The temperature is then increased slowly to 1700°F (926°C) at one hour. This can be further increased to 1925°F (1052°C) after three hours. Immediately after the fire endurance test, the assembly is subjected to a stream of water from a fire hose equipped with a nozzle having a 1 ⅛-in.-diameter (28.6 mm) discharge tip. The water pressure and the duration of application of the water stream vary according to the time the door is exposed to the fire. The tests are run for less than 1 hour, 1 to 1 ½ hours, 1 ½ to 3 hours, and 3 hours. As an example, a one-hour door would have to remain structurally intact after one hour of fire exposure and maintain contact with the door jamb so that no flame can be seen from the outside. Then the door would have to resist successfully a one-minute exposure to a 30 psi (207 kPa) stream of water from the firehose nozzle set 20 ft (6.1 m) away. The door is rated according to the length of time it is able to resist the fire and still stand up under the force applied by the water stream.

Some types of composition board are being used after being treated with preservatives. This is a very minor part of the market at the present time; however, because it is very easy to treat products made with comminuted wood when they are in the form of discrete particles or fiber, it is anticipated that this part of the market will grow immensely in the future. Some treated material is made for export to Hawaii and Puerto Rico.

At the present time, the test method used for evaluating the protection provided by the wood preservative is found in ASTM D 1758, "Standard Method of Evaluating Wood Preservatives by Field Tests with Stakes." In this evaluation, treated material in the form of stakes is placed in the ground at various locations throughout the country, if a wide-ranging test is desired. Perhaps the most severe testing is done in the southern states, where conditions are more amenable to decay. The material under test is measured periodically to determine the resistance to decay.

STANDARDS AND THEIR DEVELOPMENT

Standards for any product are needed by both the manufacturer and the consumer. The standard provides an estimate of the level of quality and provides a basis of comparison in judging that quality. When the manufacturer produces a product meeting the standards set forth for it, the consumer can then design and use the product with a high degree of confidence. In the early days of the board industry, when standards were not yet developed, it was difficult for the consumer to use the board products sensibly and with any degree of confidence. As would be expected in such a situation, many early board products were used incorrectly, the result being that the consumer was dissatisfied with the product or products. The standards and the minimum properties included therein were covered in Chapter 2.

Before any standard can be promulgated, it is necessary to first establish accepted test methods for determining the properties noted in the standard. In the United States, these are generally developed by the American Society for Testing and Materials. These standard test methods state how to perform a specific test. An ASTM standard deals with the characteristics of a material, product, system, or service. It has one of five purposes: a standard definition creating a common language for a given area of knowledge; a standard recommended practice suggesting accepted procedures for performing a given task; a standard method of test prescribing a way of making a given measurement; a standard classification setting up categories in which objects or concepts may be grouped; and standard specifications defining the boundaries or limits of the characteristics of a product or material. An ASTM standard is a voluntary standard. The word *voluntary* implies two things: (1) the standard is developed with the voluntary cooperation of all concerned and (2) the use of a standard by those affected is voluntary. An ASTM standard is full consensus standard. This means that everyone who is affected by the development or use of the standard has an opportunity to participate in the development of the standard in the ASTM committee involved.

Standards are developed in many different types of organizations such as companies, trade associations, governmental agencies, and technical and professional societies. Not all standards must be developed by the full-consensus system, as some are intended for use only within a company or an industry. If a full-consensus procedure is to be used for a standard dealing with a commodity in the marketplace or a standard that deals with a characteristic of the physical environment, all parties interested in either its development or use are fairly represented in the committee writing the standard. Such a standard enjoys a high credibility in the marketplace because the broad input into the standard should result in one that is technically valid.

In ASTM the committees dealing with materials, products, systems, or services having a commercial bearing have the following regulations governing their actions. The number of producer representatives shall not exceed the combination of consumer and general interest representatives. A producer representative is ineligible to serve as chairman of the committee or of its executive subcommittee. There are only two justifiable reasons for refusing voting membership on a technical committee to any ASTM member: (1) if his election would create an imbalance of the voting interest or (2) if he is not technically qualified or knowledgeable in the area of the committee's scope. A consumer is assumed to be knowledgeable. All negative votes on committee ballots must be considered by the originating subcommittee, and action taken in response to a negative ballot must be an affirmative vote of not less than two-thirds of those voting. All committee meetings considering technical matters relating to standards must be open to visitors. Actions must be evaluated by test data where applicable. This is an aspect of ASTM full consensus standards development often overlooked by many who are evaluating but not actually participating in the system.

It should be noted that ASTM has no certification program. Some manufacturers do label their products with an ASTM standard designation, thus implicitly certifying that the product conforms to that standard. However, ASTM plays no part in that certification. ASTM, however, is involved in developing criteria for evaluating agencies that test products for certification.

The commercial standards in the United States are developed by manufacturers, distributors, and user in cooperation with the Office of Product Standards of the National Bureau of Standards. Their purpose is to establish quality criteria, standard methods of testing, rating, certification, and labeling of manufactured commodities, and to provide a uniform basis for fair competition.

Adoption and use of a commercial standard are voluntary. However, when references to a commercial standard are made in contracts, labels, invoices, or advertising literature, the provisions of the standard are enforceable through usual legal channels as a part of the sales contract.

Commercial standards originate with the proponent industry. The sponsors may be manufacturers, distributors, or users of a specific product. One of these three elements of industry submits to the Office of Product Standards the necessary data to be used as the basis for developing a standard of practice. The Office, by means of assembled conferences or letter referenda or both, assists the sponsor group in arriving at a tentative standard of practice and thereafter refers it to the other elements of the same industry for approval or for constructive criticism that will be helpful in making any necessary adjustments. The regular procedure of the Office assures continuous servicing of each commercial standard through review and revision whenever, in the opinion of the industry, changing conditions warrant such action.

An important set of standards is published by the American National Standards Institute. In its literature, the Institute states:

> This Institute provides the machinery for creating voluntary standards. It serves to eliminate duplication of standards activities and to weld conflicting standards into single, nationally accepted standards under the designation "American National Standards."
>
> Each standard represents general agreeement among maker, seller, and user groups as to the best current practice with regard to some specific problem. Thus the completed standards cut across the whole fabric of production, distribution, and consumption of goods and services. American National Standards, by reason of Institute procedures, reflect a national consensus of manufacturers, consumers, and scientific, technical, and professional organizations, and governmental agencies. The completed standards are used widely by industry and commerce and often by municipal, state, and federal governments.
>
> The Standards Institute, under whose auspices this work is being done, is the United States clearinghouse and coordinating body for standards activity on the national level.

It is a federation of trade associations, technical societies, professional groups, and consumer organizations. Some 1,000 companies are affiliated with the Institute as company members.

The American National Standards Institute is the United States member of the International Organization for Standardization (ISO), the International Electrotechnical Commission (IEC), and the Pan American Standards Commission (COPANT). Through these channels American industry makes its position felt on the international level.

Other countries have developed their own standards following guidelines acceptable to their industries and governments. The International Standards Organization has a technical committee working on international standardization of derived wood products, It is looking at terminology, classification, test methods, and quality requirements. An international system would be of value in establishing uniform terminology and test methods throughout the world. However, on a practical base it would appear that such an international standard would be almost impossible to develop because of the many different opinions held on the subject of composition board testing and the physical properties of these products. Efforts are still continuing on establishing international standards.

There are two types of standards in use: end use and classification of characteristics. The end-use standards are basically consumer-oriented. Their establishment permits the pointing out of property requirements peculiar to any use. Where certain properties are very important to a successful use history, end-use standards can set proper requirements for these properties. The problem in using such standards comes from the complexity of requirements of some major markets. It has been suggested, for example, that an end-use standard would be suitable for particleboard core stock. However, the requirements for core stock are tremendously varied depending upon the overlay to be used, the size of the finished panel, the loads to be considered in the design, and the many ways this material can otherwise be used. Such a standard could have a series of narrowly based specifications or it could have very broad general requirements which would result in its having limited value.

The standards developed in the United States are of both types: end use and classification of characteristics. Examples are found in the FHA Use of Materials Bulletin No. UM-32, "Mat-formed Particleboard for Exterior Use," and Voluntary Product Standard PS60-73, "Hardboard Siding." As is obvious from their titles, these standards are aimed at a particular product for a predetermined use.

Examples of the classification of characteristics standard are Commercial Standard CS 236-66, "Mat-formed Wood Particleboard," and Voluntary Product Standard PS 58-73, "Basic Hardboard." Using the particleboard standard as an example, basic properties are listed for ten groups of boards in two types (interior and exterior) in a range of densities from below 37 lbs/ft^3 to over 50 lbs/ft^3 (0.59-0.80 sp gr) and in two classes of properties for each standard density range. The large number of board groups makes it possible for a consumer to accurately determine which board he needs for his use. Previously, the particleboard industry had Commercial Standard CS 236-61, "Mat-formed Wood Particleboard (Interior Use)," which specified the minimum values for particleboard. This was not a satisfactory standard because only the minimum values were noted and no allowance was provided for boards with superior properties. This was revised to the commercial standard now in use. This standard, CS 236-66, has received excellent acceptance by consumers because they have been able to specify the board with the properties needed for their end-use requirements.

In Canada, particleboard is given basic grades and these grades are assigned uses (see Table 2.16). Some grades can be used in more than one application, and this is shown in the table mentioned.

The listing of the board classifications, grades, and physical properties for hardboard and particleboard was presented in Chapter 2. These standards, of course, are revised periodically to cover new products coming into the marketplace and to meet the consumers' desires in being able to better specify board products for their use. A further new standard is expected to be developed by the particleboard industry covering the new structural board products.

Besides physical properties, standards cover fire resistance and adequacy of preservative treatment which were discussed earlier. Other important characteristics that may be included are surface finish, dimensions and allowable variations, squareness and straightness, decorative finishes, and weatherability.

STRUCTURAL MATERIALS

As mentioned previously, there are a number of important structural uses of composition board already established in the United States. A major one has been mobile home decking. Shelving, combination sheathing-siding, roof sheathing, and subfloor material are others that are in use, particularly in the siding market. Further developmental work is necessary before greater use of these products will be found in the marketplace, but it appears that the time span for this to occur will be relatively short. The important uses will be, as far as housing is concerned, panel products for sheathing and floors, and probably the Com-ply development, which is bringing the combination of particles and veneer or particles plus aligned particles into the marketplace not only as panels but as lumber. The driving force behind the move into the structural market is the coming shortage of plywood and wide-dimension lumber because of increased demand and shortages of raw material appropriate for plywood and the wider pieces of dimension lumber. These new developments are not designed to supplant plywood and lumber but to complement them. For example, the veneer supply can be expanded by using it in combination with particles for products using veneer on the faces and particles on the core of the panel.

Composition boards are being used successfully in structural applications in the USA and elsewhere in the world at present. To increase the use in the United States, the factors required for establishing structural composition board design properties are now being developed. Some thoughts along this line should be of value to those contemplating participating in this important development.

Every product has a set of critical strength properties associated with design. Regulatory groups have demanded that engineering design values be supplied. This is also of value to the producers as they can get broad use of their product in the market.

The composition boards have managed to find many markets where structural performance has been of secondary importance. The mobile home builders have been confronted with the need to use such material with defined structural properties. This has been supplied to them but no general structural design data have been developed and published in the engineering manuals. Structural design properties can be developed for the existing board products as soon as a number of mechanical and physical characteristics have been well established.

The first inroads are expected to be made with the panel-type products. Panel products will be compared with the existing plywood panel products which are well established throughout the marketplace. It would seem that it would only need a duplication of these products in order for structural particle products to move immediately into the market. In the case of panels, these properties can easily be found in the published technical literature of the American Plywood Association. The strength and elastic properties of these products are presented for a number of various grades and species groups in much the same way as they are available for lumber. These material properties are combined with the section properties by designers to engineer a structure. In the case of lumber, these section properties are related to the cross-sectional dimensions of the members and are easy to calculate. However, in plywood the section properties are related to relative amounts of material in a cross section with its grain parallel to the direction of the stress. For different plywood thicknesses, number of plys and mixtures of species allowed in the inner and outer plys, this becomes quite complex. Designers cannot be expected to have a thorough knowledge of manufacturing practices in order to work their way through this design procedure. The American Plywood Association has assisted designers by providing tables of standardized section properties for plywood sizes and grades.

By using the plywood strengths and elastic properties with the correct section properties, a designer can determine the calculated performance of plywood recommended for various spans and loads as approved by the building codes. There are differences between the calculated minimum properties of plywood and average or typical properties. It is normal for the building trade not to use the minimum sizes that appear acceptable on the basis of design calculations and published load deflection criteria. In many cases the builders prefer products which are thicker than calculated by engineering considerations. Consequently, a procedure that merely duplicates the published minimum performance of acceptable plywood will possibly underestimate the typical performance to which the user is accustomed. An immensely varied product such as plywood has a broad range of material quality within any category, averaging well above the lower limit that has been established. This is true for most wood products. However, it may not be true to a similar degree in composition board, since a closer control of the properties can be established in the manufacturing process.

Thus the direct copying of plywood properties by a manufacturer of structural composition board may not be the proper approach to take. As an example, a structural product that meets the deflection criteria may exceed the strength requirements. Because of the variations in the manufacturing process, probably no product can be manufactured to meet closely all of the design property requirements. Thus a different approach may be to determine the actual performance criteria that have been established from experience in building and to use these as a basis for establishing the properties that various types of panels should have to meet the design requirements. In this way, minimum property requirements can be developed; but it also should be noted, as in the case of plywood, the consumers and builders may prefer stronger materials and thus market requirements may override building code criteria.

The composition boards now available display a wide array of structural potential. Some of these may be usable for structural applications, while it is also probable that changes in board formulation will provide other products meeting the design criteria being developed. For structural board, properties can be published that are suited to the products now being manufactured. Another approach is to provide a means to indicate how composition board can meet the structural needs most economically. As an example, low-strength material can serve structural purposes if it is thicker. Such thicker material need not be higher in density as the *EI* product (elasticity x moment of inertia) governs design stiffness. Further improvements in strength and stiffness are expected to produce economically superior structural boards.

The usual approach to establishing allowable design properties of any material begins with the consideration of the shape of typical stress versus strain curves for each of the strength properties necessary. Also required is information about the variability of these properties within a given product. Since the stress-strain curves for wood are reasonably linear, it is only in the upper 1/3 of the range, from 0 stress to stress at failure, that strain ceases to be proportional to stress. The exact stress level at which a departure from proportionality starts is difficult to identify, whereas the stress at failure is a very discrete value. For this reason, allowable strength properties for wood are based on ultimate values rather than proportional limit or yield point values for the properties. An exception can be found with compression across the grain, because this ultimate strength is obscured by the nature of the compression failure; therefore, a proportional limit strength forms the basis of compression perpendicular to grain design strength.

Composition board properties have stress-strain relationships sufficiently like those of wood to suggest that their allowable properties for design can also be developed from information about their stress at failure, or their ultimate strength properties with the possible exception for compression across the grain.

Solid wood structural wood properties used for design are: elastic modulus (bending), allowable bending stress, allowable tension strength parallel to the grain, allowable compression strength parallel to the grain, allowable compression stress perpendicular to the grain, and allowable shear stress parallel to the grain. Because plywood has cross-oriented plys, it requires a pair of shear properties in lieu of shear parallel to the grain. These are rolling shear and shear through the thickness. Other properties may need to be examined to determine their importance and possible interaction with the properties published for design use.

In developing design properties, testing programs large enough to determine not only average value of a product but also its variance are required. With strength properties, the variance is a means of establishing a value low enough to be used confidently without risk of failure, but not too low to be viable. The transition from experimentally determined strength properties to usable design quantities gives consideration to the differences between test conditions in the laboratory and those existing in service. Because of economics, testing time for developing such information has to be short. However, wood strength is time dependent. Live loads are often applied gradually and dead loads must be sustained for long time periods so strength properties for these conditions will be the basis for design values. Thus, tests versus service load duration must be considered.

Another problem has existed with determining properties on clear straight-grained wood. In this case the weakening effect of growth effects and grain slope require consideration. This is accomplished for lumber and plywood by introducing a factor known as a strength ratio, which is a simple estimate of the weakening effect. Composition boards, however, do not have strength-reducing growth characteristics. An advantage can be taken of this characteristic. Lumber and plywood normally used for construction have a wide variability in strength properties. The potential for composition boards, therefore, is very favorably affected by their potentially low variance in properties. Consequently a composition board which has a lower average strength can serve a similar purpose to that of other products having much higher average strengths, but greater variability.

For any panel product serving a market where plywood is an accepted material, the low modulus of elasticity, for example, is a deceptive figure. The plywood design is based on the product of a wood modulus of elasticity and modified section property based on the parallel plys. Composition board stiffness, because the full depth is considered as active material, can have a lower modulus of elasticity, yet display equal stiffness. For example, for ½-in. (12.7 mm) standard sheathing uses, a modulus of elasticity for ½-in. composition board of about 800,000 psi (5500 MPa) is equivalent to plywood. For ¾-in. (19 mm) standard sheathing uses, a modulus of elasticity of ¾-in. composition board should be 575,000 psi (3900 MPa). Few existing boards have these properties; but by using particles with good length-thickness ratios such as found in flakes, aligned particles, or extra thick panels, boards can be made to meet these needs. The opportunity, therefore, for tailoring outstanding composition board structural products is quite possible. However, if some existing conventional products are introduced to the market for structural applications, there could, in some instances, be harmful implications to the overall reputation of the composition board material. A clear agreement on performance standards would be beneficial to the developing composition board market. Work is already well under way in developing the information for such applications.

QUALITY CONTROL

The subject of quality control is being dealt with at this point because of its association with the testing of composition boards. It is important, however, to consider that quality of a product is manufactured into it, not tested into it. Testing is used to verify the quality.

The control of the quality of a product has to be the responsibility of the top management. Policy must be established when the facility is being planned and it must be a prime consideration when contracting for wood raw material, binders, and other additive materials; and it also must be considered when selecting the equipment for the plant and when hiring the key plant personnel. The policy established must be firmly and continuously enforced to maintain a uniform product quality and to build the industry integrity.

Various standards and specifications have been developed for the composition board industry, covering many different types of products. Without these developments, it would have been quite difficult for the products to have penetrated the market as they have. To maintain the established position of the products, however, the quality of the products being manufactured must be maintained. Quality surveillance and certification can be handled by the plant itself, through a testing agency, or through an association. All three approaches are practiced in industry; however, the last two are by far the most popular.

The independent testing agency approach is to provide quality surveillance as accomplished by visual inspection to the production processes and by testing representative samples of the product on a predetermined sampling plan. The plant supervisory personnel use the quality reports to make process adjustments when necessary to conform to the requirments of the standard under the conditions stipulated in the agreement between the testing agency and the manufacturer. The testing agency provides appropriate stamps to mark the product and certificates to accompany shipment to the purchaser.

The manufacturer is authorized to continue marking the board with the agency stamps and to issue certificates as long as the product meets the minimum requirements of the specifications. The control of certification is delineated in the agreement between the agency and the manufacturer. The Federal Housing Administration (FHA) has made independent certification of softwood plywood mandatory for use in FHA insured housing. It has also specified such certification with particleboard manufacture according to the specifications found in UM 32, "Particleboard for Exterior Uses."

The arguments for the third party inspection center around the implication of non-bias. It is also felt that quality reports from such agencies reach top management and command more attention than those of a control program conducted within the plant. For the purchaser, a second party is provided to handle complaints if quality does not meet his approval. The agency is generally accepted as a referee on quality claims by performing reinspection and retests.

A type of third party agency program handled by an association is illustrated in the National Particleboard Association's Grademark Program. The rationale for this program is as follows:

> In order to improve the identification of mat-formed wood particleboard products conforming to standards developed by the industry, the National Particleboard Association has developed certain grade trade marks. Members of the Association are licensed to use these marks to identify products conforming to the standards. The right to use these marks is dependent upon maintaining production facilities capable of producing products conforming to the standards and upon the existence and performance of a Quality Control Department capable of reasonably determining conformance or nonconformance with the standard. In this manual [the Quality Control Manual] certain minimum requirements for a plant quality control department are outlined as well as methods by which the National Particleboard Association inspects Licensee, plants and products to determine their qualification to use the Association grade trade marks.

The minimum requirements for the plant quality program require daily testing by the plant of one sanded board per 100,000 ft² (¾-in. [19 mm] basis) for all physical tests required for whatever type board the manufacturer is certifying.

In addition to this physical property testing, five boards per 100,000 ft^2 (9291 m^2) must be examined for dimensional tests such as squareness, straightness, length, width, and thickness. Each day, plants are required to send copies of their daily test data to the Association.

Before plants may use the NPA grademark to identify their particleboard underlayment, they must pass an initial qualifying inspection. During this inspection, a qualified representative of the Association inspects the quality-control facilities of the plant to determine if they are adequate to conduct the tests required in the program and to determine if quality-control personnel are adequately trained to conduct the program. He reviews past quality-control data to determine the plant's ability to produce underlayment meeting program requirements. During this visit, samples are obtained for testing by the Association to determine conformance with program requirements.

These procedures are detailed in the license agreement entered into by participating member companies and in the quality manual made a part of that agreement. A separate agreement is used for each mill or plant, and each plant is regarded individually in respect to quality performance. After the initial qualification inspection, unannounced visits are made to each plant with the frequency based upon the amount of grademarked board produced, but at least once a month. During these visits samples are selected from stock in the finished goods inventory (or directly off the sander) for testing by the NPA inspector, using the plant's laboratory, or for shipment to the Association laboratory.

Failure to comply with program requirements can cause suspension of the rights to utilize the NPA grademarks. A plant is judged as failing to comply with the program requirements if the average property values for the last five panels tested in any particular thickness fail to meet the required values, or the sample panels chosen during plant visits fail to meet the property requirements.

Returning to the subject of quality control itself, it is generally agreed that quality-control procedures have evolved with the realization that there is need for some sort of inspection separated from the production's supervision. There has to be an effort to analyze the efforts of the various plant departments to point out errors and to provide data for corrective action. A complete quality-control program starts with the design for the product and does not end until that product is in the hands of a satisfied consumer. These concepts still embrace the inspection philosophy of acceptance or rejection applied to all functions that affect the product.

Quality control by statistics is a well-known subject, but it is only a technique to be used in a total quality-assurance program. A typical quality-control program in a plant will be somewhere between an extremely sophisticated system and one that operates from crisis to crisis.

The major emphasis, of course, should be an assurance that everything is being manufactured according to the specifications. All people in the plant, ranging from the company manager to the lowest production worker, are involved and each has his own quality function. Quality assurance, therefore, is the maintenance of quality, whereas the quality control itself is the function performed by a certain department within the plant.

Quality assurance starts, as mentioned earlier, with the responsibility of top management. The management should look at a quality-control program as another management tool that is vital to its success. It should be recognized that the plant quality control must operate at the same level as the production supervision. Management must actively participate by setting objectives and reviewing progress. Such recognition by the management broadens the scope, effectiveness, and acceptance of the program within the plant.

Once management has set its quality-control goals and assigned responsibility and authority, the entire organization can then work to obtain the objectives. The head of the quality-control section plans, organizes, and integrates the program functions. He in turn delegates authority, but cannot pass off any portion of his responsibility or accountability. He is judged by the degree to which the plant quality objectives are fulfilled.

A key to organizing a successful quality-control operation is to insure the fastest possible cycling of critical information. Closed-loop control is common in industrial instrumentation. These principles can also be applied to a quality-control organization, as illustrated in Figure 4.4. Two forms of plant structure are possible. These are decentralized and centralized. The decentralized organization, as shown in Figure 4.5, relies on part of the information to return from inspection not under the control of quality control. The graders or inspectors positioned after the sander frequently fall into this category. They have other plant functions which are better served by their being part of the production group. A centralized plant quality-control organization, as shown in Figure 4.6, differs by including the grade line inspection within the department. Regardless of the structure, the feedback loop must not be broken; because the faster the corrective action, the better the quality level of the product being manufactured.

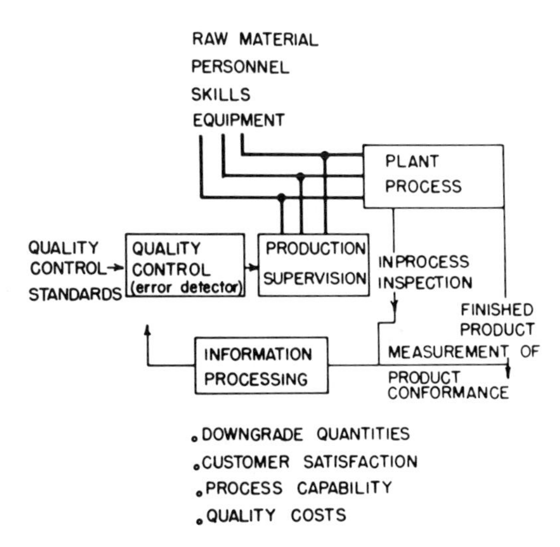

Figure 4.4. *Feedback control for quality aspects in a particleboard plant (Peters 1969).*

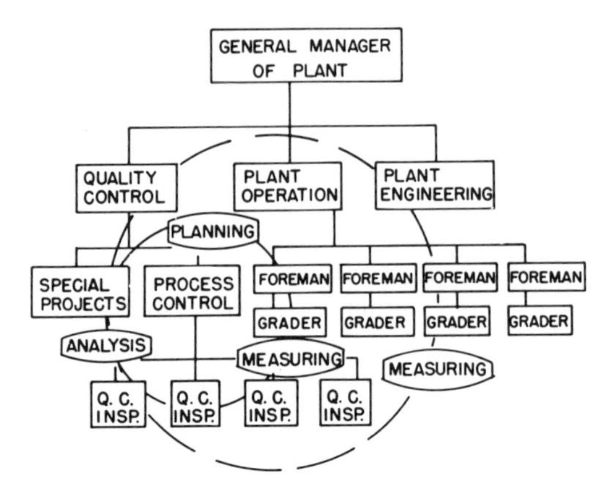

Figure 4.5. *An abbreviated diagram for decentralized plant quality-control organization (Peters 1969).*

The function of the quality-control group includes the establishing of product standards, evaluating conformance, initiating corrective action that must be taken, and studying and planning for improvements. In establishing product standards, the process and product specifications must be set, based upon the grading rules and other specifications adopted from commercial standards and Association specifications. The evaluation of conformance includes physical properties, grader inspection, process variables, process capability, quality costs, and some measure of the customers' satisfaction. The return and display of this information to the pertinent areas of the plant must be accomplished as quickly as possible.

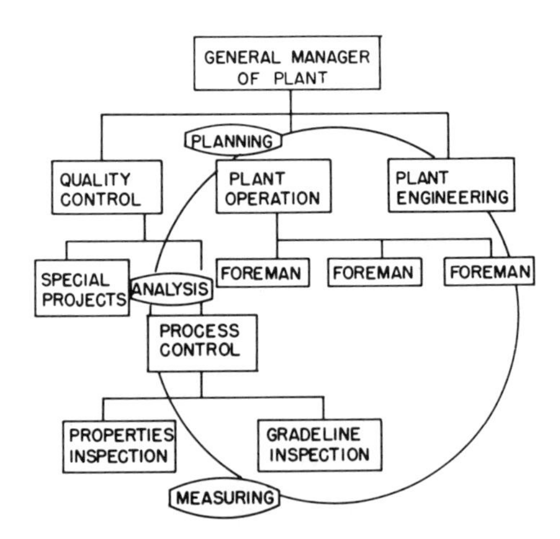

Figure 4.6. *An abbreviated diagram for centralized plant quality-control organization (Peters 1969).*

Whenever corrective action is needed, line supervision must be notified immediately about deviation from standards. Other action may include retesting and the isolation of suspect or downgrade production, and the assisting of the production department in taking corrective action. A quality-control program may have excellent data but the system will not function over a long period unless adequate time is spent analyzing the information. Thus the studying and planning for improvements are extremely necessary. Taking poor or meaningless measurements is just as bad as not taking a measurement at all.

Since management demands, or should demand, quality assurance, the quality-control department must contribute to the overall quality attititudes of the plant personnel. The personnel must feel quality conscious in that mistakes are avoidable. The quality-control program, however, must not relieve the production personnel of their responsibility for producing a quality board. An essential ingredient of the quality-control department is the involvement of a supervisory personnel in the details of the process. Good decisions are made by precise knowledge. Such knowledge is not obtained at a desk but rather is gained by experience in the actual operation of the plant.

There are other ways in which an in-plant quality-control program can contribute to quality improvement. Pressure can be exerted by measuring quality performance by shift and presenting the comparison between shifts and the expected standard. The extra effort and costs required of production personnel to sort out and regrade defective material emphasizes the importance of quality consciousness, and this can be made apparent to those working in the plant. The department can be utilized to provide exposure and training of potential supervisory personnel, as well as to provide quality consciousness for those that will work in many jobs within the plant.

It is extremely important that quality-control personnel work closely with the production supervisor. The needs of production are decisiveness, information on quality decisions, consistency on the reporting of information, understanding the need to produce the product in cooperation and help in trouble shooting. On the other hand, quality control needs assistance from the production people. This includes honesty in observing quality standards and procedures, providing reliable data from production measurements, allowing adequate time for testing, taking responsible corrective action when necessary, and establishing reasonable working relationships between line supervisors and inspectors.

Many different types of tools can be used in the quality-control effort. These can range from sophisticated instrumentation and computers to micrometers and hand rules. Most plants use statistical quality-control procedures which can provide data on process capability, assist in setting specifications, and help with control decisions. Further sophistication in the use of statistics is expected as the plants become larger and instrumentation becomes even more sophisticated.

When considering the key tests for moduli of rupture and elasticity and internal bond, an important development in nondestructive testing must be recognized. A method using longitudinal stress waves developed at Washington State University has been found to be suitable for measuring the board properties of modulus of elasticity, modulus of rupture, and in some cases internal bond. This method is quick and accurate and can be used to monitor board quality on the production line. It can evaluate the entire panel or parts of it. Data can be supplied to the plant control console by a readout device. Signaling equipment may be provided to identify substandard panels audibly. Information thus obtained might also be used in a feedback loop for automatic process control.

A block diagram of the basic principle is shown in Figure 4.7. Stress waves induced into any solid material will radiate from the point of impact in uniform patterns—quite similar to waves created by dropping a rock into a smooth pool of water. These waves move at a predictable speed but slow down when going through areas of low quality. Thus, the aforementioned board properties can be determined by measuring the time it takes stress waves to travel through a composition board. This enables the detection of problems such as insufficient resin, low press temperature or pressure, and insufficient curing time in the hot press, all of which affect modulus of rupture and modulus of elasticity which the stress wave transit time estimates.

The stress wave is introduced by means of a sharp impact at the end of the board. As the wave propagates through the material, time is measured between two accurately spaced sensors. The time interval thus measured is displayed in microseconds and reflects closely the properties of the material being tested.

In-line testing requires additional equipment, such as a temperature-measuring device to monitor board temperature as indicated board quality varies with temperature fluctuation. It has been found that testing after the board cooler is a good place to get fast and reliable data. The board is still warm but the temperature measurement can be used for calculating the correct physical properties of a panel when cool. Testing out of the board cooler means that the quality-control information is available within a very short time after pressing.

For testing, each board is momentarily stopped, two sensors are lifted against the bottom surface inside the rough edges and a stress wave is induced and travel time measured. The fast action of this test facility permits complete testing to be performed on the production line without the need for extensive equipment modifications.

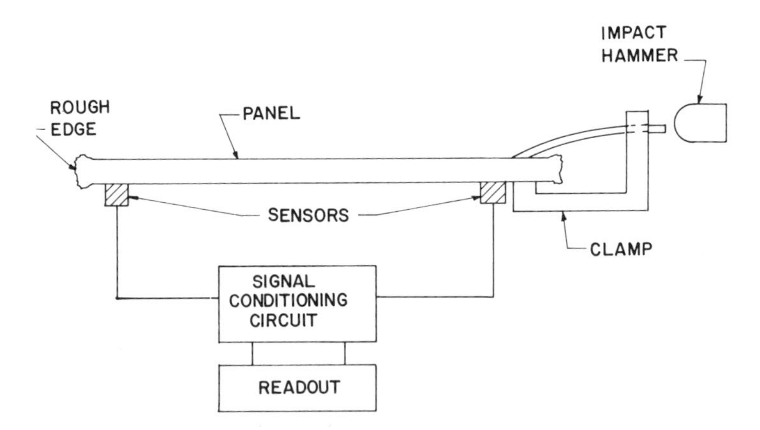

Figure 4.7. *Block diagram of stress-wave testing machine.*

Blows or delamination are defects in boards that are not always detected easily. Considerable adverse reaction from customers occurs if much of such low-quality product reaches the marketplace. Systems have been developed, however, that can evaluate each board quickly for blows. Large blows are easily seen by the eye as the board surface is ruptured or badly bulged. The troublesome type, which is difficult to observe, is "incipient," or blows that break the resin bonds within a board but do not bulge the surfaces.

The effectiveness of a quality-assurance program within a plant can best be measured by equating its results to dollars. The plant must know what the quality costs are. The costs of running the department are not the true costs. They must be balanced against other costs such as the dollar loss represented by downgraded material, the cost for inspection, the cost for reinspection and reprocessing downgraded board, and the cost of handling complaints.

Quality-control programs that do not meet the expectations of management generally fail because of organization, not because of technical orientation. The challenge in managing a quality-assurance program requires hard work and continual examination. Likewise, the challenges of producing quality board that performs satisfactorily in the market, and one that can be produced at economical costs, will be a greater challenge as the industry progresses. Product assurance is best served by elevating quality control from a service to a tool of management.

SELECTED REFERENCES

Albright, R. G., and E. T. McCarthy. 1975. Blow Detection + Continuous % Unbond Readout. *Proceedings of the Washington State University Particleboard Symposium, No. 9.* Pullman, Washington: Washington State University (WSU).

American Society for Testing and Materials. 1973 ASTM and thc Voluntary Standardization System. Philadelphia, Pennsylvania.

———. 1975*a*. Evaluating the Properties of Wood-Base Fiber and Particle Panel Materials. ASTM D1037. Philadelphia, Pennsylvania.

———. 1975*b*. Evaluating Wood Preservatives by Field Tests with Stakes. ASTM D1758. Philadelphia, Pennsylvania.

———. 1975*c*. Fire Tests of Door Assemblies. ASTM E152. Philadelphia, Pennsylvania.

———. 1975*d*. Surface Burning Characteristics of Building Materials. ASTM E84. Philadelphia, Pennsylvania.

Birks, A. S. 1969. Development and Application of the Ultrasonic Particleboard Blow Detector. *Proceedings of the Washington State University Particleboard Symposium, No. 3.* Pullman, Washington: WSU.

———. 1972. Particleboard Blow Detector. *Forest Products Journal,* Vol. 22, No. 6.

Bridges, R. R. 1971. Quantitative Study of Some Factors Affecting the Abrasiveness of Particleboard. *Forest Products Journal,* Vol. 21, No. 11.

Brown, J. H., and S. C. Bean. 1974. Acceptance of Stranwood Sheathing for Construction Uses. *Proceedings of the Washington State University Particleboard Symposium, No. 8.* Pullman, Washington: WSU.

Carruthers, J. F. S. 1975. Specifications for Wood-Based Board Materials. Background Paper No. 25, third World Consultation on Wood-Based Panels, FAO, New Delhi, India, February 1975.

Chopra, N. N. 1975. International Standardization in the Field of Wood-Based Panels. Background Paper No. 12, third World Consultation on Wood-Based Panels, FAO, New Delhi, India, February 1975.

Dewey, M. L. 1974. Private Correspondence on Fire Tunnel Testing, Koppers Co. (April).

FAO. 1976. Basic Paper IV. Technology and Techniques in the Manufacture of Wood-Based Panels. *Proceedings of the World Consultation on Wood-Based Panels,* held in New Delhi, India, February 1975. Brussels: Published in agreement with the Food and Agriculture Organization of the United Nations by Miller Freeman Publications.

Federal Specification: LLL-B-800a. 1965. Building Board (Wood Particleboard) Hard Pressed, Vegetable Fiber. Washington, D.C.: U.S. Government Printing Office.

Gertjejansen, R., and J. Haygreen. 1971. Torsion Shear Test for Particleboard Adapted to a Universal Testing Machine. *Forest Products Journal,* Vol. 21, No. 11.

Haygreen, J. G., and H. Hall. 1975. The Impact Strength of Particleboard and the Effect of Wetting. *Proceedings of the Washington State University Particleboard Symposium, No. 9.* Pullman, Washington: WSU.

Heebink, B. G. 1967. A Procedure for Quickly Evaluating Dimensional Stability of Particleboard. *Forest Products Journal,* Vol. 17, No. 9.

Heebink, B. G., and F. V. Hefty. 1968. Steam Post Treatments to Reduce Thickness Swelling of Particleboard. *Proceedings of the Washington State University Particleboard Symposium, No. 2.* Pullman, Washington: WSU.

Hoyle, R. J., Jr. 1973. Factors Required for Establishing Structural Particleboard Design Properties. *Proceedings of the Washington State University Particleboard Symposium, No. 7.* Pullman, Washington: WSU.

———. 1974. Property Requirements for Structural Roof and Floor Sheathing. *Proceedings of the Washington State University Particleboard Symposium, No. 8.* Pullman, Washington: WSU.

Hoyle, R. J., Jr., and R. D. Adams. 1975. Load-Duration Effects on Properties of Oriented Strand Structural Panels. *Proceedings of the Washington State University Particleboard Symposium, No. 9.* Pullman, Washington: WSU.

HUD, FHA. 1973. Combination Subfloor/Underlayment Particleboard for Factory-Built Modular Housing Units. Bulletin No. 57a.

Johnson, R. E. 1974. A Development in Production, Sales, and Use of Structural Particleboard. *Proceedings of the Washington State University Particleboard Symposium, No. 8.* Pullman, Washington: WSU.

Lehmann, W. F.; R. L. Geimer; and J. D. McNatt. 1974. The Engineering Properties of Structural Particleboards from Forest Residues. *Proceedings of the Washington State University Particleboard Symposium, No. 8.* Pullman, Washington: WSU.

Lehmann, W. F.; T. J. Ramaker; and F. V. Hefty. 1975. An Exploratory Study of Creep Characteristics of Structural Panels. *Proceedings of the Washington State University Particleboard Symposium, No. 9.* Pullman, Washington: WSU.

Lewis, W. C. 1967. Thermal Conductivity of Wood-base Fiber and Particle Panel Materials. U.S. Forest Service Research paper FPL 77 (June). Forest Products Laboratory/Forest Service/U.S. Dept. of Agriculture, Madison, Wisconsin.

McNatt, J. D. 1973. Basic Engineering Properties of Particleboard. *Proceedings of the Washington State University Particleboard Symposium, No. 7.* Pullman, Washington: WSU.

Morschauser, C. R. 1967. Standards. *Proceedings of the Washington State University Particleboard Symposium, No. 1.* Pullman, Washington: WSU.

———. 1969. Quality Control Program of the National Particleboard Association, A Third

Alternative. *Proceedings of the Washington State University Particleboard Symposium, No. 3.* Pullman, Washington: WSU.

———. 1975. Non-destructive Testing of Particleboard in the United States—A Status Report. Background Paper No. 84, third World Consultation on Wood-Based Panels, FAO, New Delhi, India, February 1975.

National Particleboard Association. 1967. Thermal Conductivity of Particleboard. Technical Bulletin No. 1 (September). Washington, D. C.

Patronsky, L. A. 1969. TECO Tested Particleboard. *Proceedings of the Washington State University Particleboard Symposium, No. 3.* Pullman, Washington: WSU.

Pellerin, R. F. 1969. Non-Destructive Testing of Timber at Washington State University. *Proceedings of Conference on Non-Destructive Testing of Concrete and Timber.* Institute of Civil Engineers. London, England: William Clowes & Sons Ltd.

———. 1974. New Concept Allows Non-Destructive In-Process Testing of Board Quality. *Forest Industries* (July).

Pellerin, R. F., and J. H. Kaiserlik. 1975. Grading of Wood Products by Stress Waves. Presented at IUFRO Division 5—Wood Engineering Group Meeting (April), Delft, Holland.

Pellerin, R. F., and C. R. Morschauser. 1973. Nondestructive Testing of Particleboard. *Proceedings of the Washington State University Particleboard Symposium, No. 7.* Pullman, Washington: WSU.

Peters, T. E. 1969. Quality Control in the Particleboard Plant. *Proceedings of the Washington State University Particleboard Symposium, No. 3.* Pullman, Washington: WSU.

Price, E. W. 1975. Properties of 4- x 8-foot Panels of Structural Exterior Flakeboard From Southern Hardwoods Flaked on a Shaping-Lathe Headrig—A Progress Report. *Proceedings of the Washington State University Particleboard Symposium, No. 9.* Pullman, Washington: WSU.

Shen, K. C., and M. N. Carroll. 1969. A New Method for Evaluation of Internal Strength of Particleboard. *Forest Products Journal,* Vol. 21, No. 6.

———. 1971*b*. Evaluating Particleboard Properties by Measuring Saw-Cutting Force. *Forest Products Journal,* Vol. 21, No. 10.

Shen, K. C., and M. N. Carroll. 1969. A new method for Evaluation of Internal Strength of Particleboard. *Forest Products Journal,* Vol. 19, No. 8.

———. 1970. Measurement of Layer-Strength Distribution in Particleboard. *Forest Products Journal,* Vol. 20, No. 6.

Shen, K. C., and B. Wrangham. 1971. Rapid Accelerated-Aging Test Procedure for Phenolic Particleboard. *Forest Products Journal,* Vol. 21, No. 5.

Snodgrass, J. D.; R. J. Saunders; and A. D. Syska. 1973. Particleboard of Aligned Wood Strands. *Proceedings of the Washington State University Particleboard Symposium, No. 7.* Pullman, Washington: WSU.

Suchsland, O. 1973. Hygroscopic Thickness Swelling and Related Properties of Selected Commercial Particleboards. *Forest Products Journal,* Vol. 23, No. 7.

Talbott, J. W. 1972. Random Thoughts (and Some Oriented Thoughts) on Structural Particleboard. *Proceedings of the Washington State University Particleboard Symposium, No. 6.* Pullman, Washington: WSU.

5.
PARAMETERS AFFECTING BOARD PROPERTIES

A number of parameters or factors affect the final board properties, whether the product is fiberboard or particleboard. Among the major factors are wood species, the type of raw material, the type of particle generated, the binder (resin) type, amount and distribution among layers, any special additives used, mat moisture levels and distributions, layering by particle size, layering by density or specific gravity, the level of the board specific gravity, and orientation of the particles. Almost all of these parameters interact with each other in one way or another.

Figure 5.1 is a drawing showing most of the interactions, although others most surely can be found. This drawing was put together in an effort to illustrate that the total interrelationship of these factors within the board process is a vast interrelated web. As shown in the figure, it is something like a spider's web. It should be immediately understood that a change in any one of these factors will result in a change in many of the other related factors in the board process. Consequently, each factor cannot be thought of as an individual entity which can be manipulated easily to control the board process as one sees fit. However, once it is recognized that there is an interrelationship between a number of factors, a more complete grasp of the process can be attained and actual manipulation can be successfully employed for controlling much of the process.

All of these factors will be discussed in this chapter except raw material type, particle type and geometry, which will be dealt with extensively in the next chapter.

SPECIES

Of all the variables present in the composition board process, species is one of the most significant. It interacts with virtually every other variable that can be imagined in the process. It determines how low in specific gravity the final board can be. It is reflected in the types of raw material available. To a great extent it governs the type of particles that can be generated economically. The formulation of the urea resin is determined by the species. Some species must have the mat moisture content more precisely controlled; otherwise the final board will blow or delaminate. When producing layered boards, particularly those in the particleboard class, some species provide better particles for producing boards with extremely smooth surfaces.

It has already been noted that the dry processes used for producing composition board employ some type of blender to distribute resin uniformly over the furnish. It has been pointed out that this should be properly referred to as a secondary blender. The primary and most important blender is the total manufacturing plant, which has to take into account a large number of variables, some of unknown and of great magnitude, and properly blend these variables to produce a product of sufficient quality and of uniformity to meet the customers' requirements. At the same time, this miracle has to be performed economically. Species is the most significant variable that the master blender has to contend with, particularly if there are a number of species entering the process line. If a plant is fortunate enough to have a single species or two similar ones to use, then the importance of species as far as that plant is concerned diminishes as long as the final boards are of the quality desired. However, this is rarely the occasion now, and the future trend is that mixtures of species from both hardwoods and softwoods of widely varying densities will become even more common than at present.

In the usual case, therefore, the process must be able to cope with wide variations in the species mix. It is recognized that a great number of these species variables are valuable and necessary to the end product, whereas some are of minor importance and can be tolerated or perhaps neglected. Others are of major

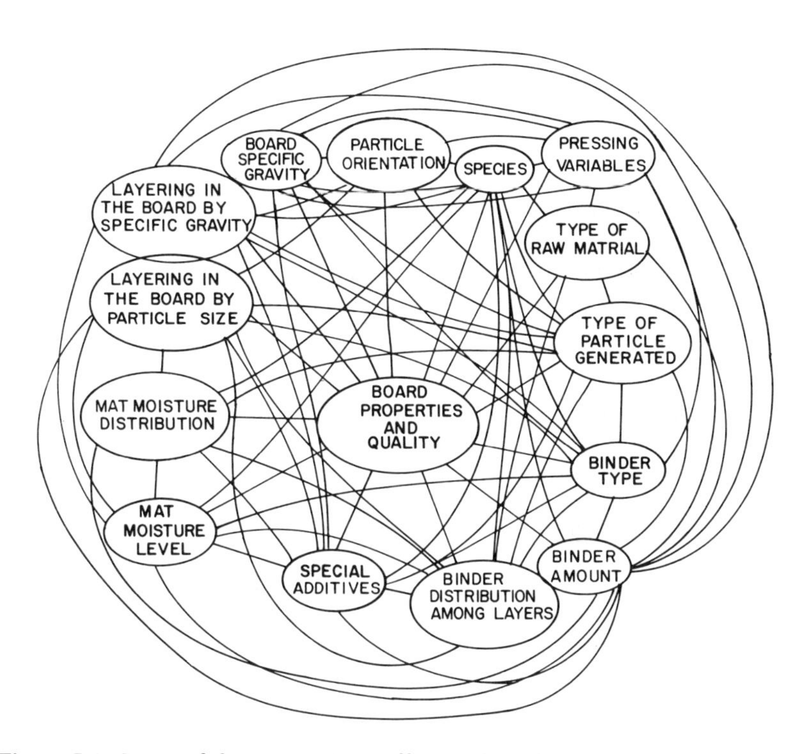

Figure 5.1. *Some of the interactions affecting board properties and quality.*

importance and at times create operating conditions that are chaotic, frustrating, and usually costly.

The species variation has been one factor that a number of people would like to ignore, at least at the plant level. Some feel that this variable is in the process and not much can be done about it. More research has been done along this line, in the laboratory and on the plant line, in the last few years because of the greater need to use a mixture of species. If a plant is fortunate enough to be able to keep the species separate until they can be metered into the production line, many of the problems associated with species can be overcome without too much additional effort or research. Australia and New Zealand are particularly fortunate, as Monterey pine makes up virtually the entire furnish. However, many plants are unable to determine which species are coming into the plant, let alone how much of each species. The problem is especially severe in those plants operating off comminuted wood such as planer shavings.

Such a plant may be purchasing shavings from as many as 100 sawmills distributed over a wide area. Perhaps six, seven, or more species will be involved. Many of these mills, if they were large enough and planed their lumber separately, could keep the planer shavings divided according to species if it was economical for them to do so. However, they normally sell the planer shavings for such a low price that such separation is an uncommon occurrence.

Planer shavings, which made up a majority of the raw material used in board processes in the United States at the present time, vary widely between species; and even within species they vary according to the moisture content of the lumber being planed. Some people do not recognize this wide variation which, depending on the planers being used, can range, for example, from very fine particles of ponderosa pine to quite large flakelike particles of Englemann spruce or white fir (all planed dry). Such wide particle geometry differences, which will be dealt with later, are very significant in determining the final board properties that can be achieved.

Returning to the species problem itself, the composition board industry has to some degree determined the extent that it will tolerate species variables. The industry, through research and operational trials, has determined which species are more suitable for the manufacture of board than other species. However, as the industry has been forced to use more and more species to secure sufficient material for operation, the number of variables introduced through species has dramatically increased.

Much of the problem of species variation can be handled by constantly subdividing the various types of raw material received from various sources and mixing them back together. Thus a randomization takes place, providing a more uniform furnish. As an example, raw material received from one source can be held in a storage silo with material from several other sources. The material is then sent to different pieces of particle-generating equipment such as refiners and flakers. Material can be combined out of these machines and dried together and then divided again, remixed, and randomized for blending and forming.

The producers using roundwood can be somewhat more selective, and are less vulnerable to species variations than those using waste material because, to some extent, they can separate the roundwood coming into their mill yards before it is reduced to particles. However, in some areas, such as in the southern pine country, where mixed hardwoods are used in some board plants, it becomes quite difficult to separate all species. In this case, if at all possible, particularly troublesome species are separated out so that they can be metered into the process in a known amount which can be handled properly. As mentioned, however, those plants operating on a wide variety of residuals without control of species can have an extremely difficult time.

The subject of species variation, therefore, takes a general twofold division: (1) the variations encountered from species to species, and (2) those variations within a species. Some of the major species variables are: density, acidity, moisture content, extractives, extraneous matter, growth location, seasonal changes, and supplier's practices.

Density

The most important species variable governing board properties is the density of the wood raw material itself. The density or specific gravity has been the important factor in determining which species are used for manufacture of composition board products. In general terms, the lower-density woods will produce panels within the present desired specific gravity ranges, usually with strength properties superior to the higher-density species. In the United States and Canada, this specific gravity range of presently used species is from about 0.30 to 0.50. Some of the higher-density species such as oak and hickory, which are about 0.65 specific gravity, are used in higher-density products, usually of the hardboard classification. A great amount of the higher-density species is available for use, particularly in the southeastern United States. An intensive research effort is under way to try to develop further ways to use these species successfully.

The reason for preferential use of the relatively light species is that they can be compressed into medium-density particleboards with the assurance that sufficient interparticle contact area is developed during the pressing operation to achieve good bonding. Heavier species simply cannot be compressed into medium-density particleboards that are well bonded. A certain amount of pressure is necessary while consolidating the mat into the final board to achieve maximum bonding. The appropriate pressure required is related to the ratio of board specific gravity to species specific gravity. A ratio of 1.3:1.0 provides a good guideline for determining if a species can be made into suitable medium-density products. (The species is given an index number of 1.0.) If high-density products are made from typical softwoods, this compression ratio increases dramatically, especially in terms of lower-density species such as redwood. For example, redwood converted into 1.0 specific gravity hardboard would have a compression ratio of about 2.9:1. Of course, this product would have much higher bending and stiffness properties than a medium-density product. Particles from higher-density species compressed into boards with acceptable properties for most uses would result in high-density panels (considering the 1:1.3 compression ratio). Such higher-density boards, of course, would be well bonded and would have suitable strengths. However, the greater weight plus difficulty in machining hinders the acceptance of such dense products in the conventional market at the present time.

A partial solution to the use of higher-density species is to mix them with low-density species to achieve a medium-density species blend.

Specialty products of high densities are made, but they normally are made with the lower-density species such as fir and pine. Higher-density species ranging up to about 0.65 in specific gravity are being used successfully for various types of fiberboard. When pressurized refiners are used with many of these species to generate the fine fibers suitable for medium-density fiberboard, it appears that a lower compression ratio can be considered as the good fiber quality apparently contributes significantly to the development of adequate board properties at lower board densities than otherwise possible.

The dominant species used at present include: the major ones of Douglas fir and southern pine, the associated other West Coast coniferous woods such as lodgepole pine, larch, pondersa pine, white fir, hemlock, spruce, and western red cedar. It should be noted that western red cedar is usually not used in combinations with other species. Its darker color is undesirable because it gives a mottled effect to the lighter-colored board made with the white woods. Redwood, also being a darker-colored wood, is normally used by itself in producing a product. Other significantly used species are aspen and the aforementioned mixture of southern hardwoods.

Thus the industry until now has concentrated on a number of lower-density softwoods and hardwoods. The lower density has been beneficial in these cases in establishing the desired physical properties in relatively low-density board products. Variations in the density of the material entering the plant do occur and cause severe operating problems. These include the milling operation, which provides particles with the desired geometry, the drying operation, the resin consumption, the bulking characteristics which affect metering and feeding equipment, the pressing operation and, as already implied, the desired physical properties of the final product.

Acidity

Another important species variable that requires attention is the acidity, as measured by pH and buffering capacity. Considerable research has been performed along this line by the resin companies. As will be mentioned later, to utilize adhesives that are both economical and suitable to the type of operation used in many plants, appropriate chemical conditions must be established in the board furnish for the proper cure of the resin binders. This is particularly important in plants using urea-formaldehyde resins. These conditions are dependent in part on maintaining a certain range of acidity in the curing mat. For a given species, the resin companies have developed binders that react properly within the normal range of acidity for that species. In some cases this can be achieved within the resin itself, whereas in others, it is necessary to add a separate catalyst to the resin before or after its application to the furnish. This separate addition of catalyst is required in urea resins for use with nonacid woods to avoid the very short storage life of highly catalyzed resins. A number of species with wide-ranging chemical characteristics can make it extremely difficult to provide the proper resin within a board plant.

Common species in use for composition board are on the acid side. Species such as Douglas fir that have pH levels of approximately 4.0 to 4.5 provide a chemical situation where urea resins cure properly; however, small amounts of catalyst can be used to speed the cure. Species with higher pH levels need a catalyst added to the resin in order that the resin will cure during hot pressing. Most phenolic resins, however, do not require acid conditions for curing to take place.

The other factor mentioned was that of buffering. While pH of the wood measures the specific level of acid activity under given conditions, the buffering capacity measures the resistance of the wood to change in pH level. A wood with high buffering capacity requires addition of a greater amount of acid catalyst to reduce the pH to the level required for optimum resin cure. This buffering factor can be important even with a single species if extreme raw material variations occur, but becomes a decisive factor when a mixture of species is involved, which presents a number of variables acting upon the buffering capacity of the furnish. Again, this notes the importance, if at all possible, of controlling the blend of species going into a board plant.

A method developed by one resin company for determining pH and buffering capacity is described below. The technique used is important as changes, particularly in particle size and soaking time, can give different results. Resin companies have learned to correlate their measuring techniques and results with their various resins.

First, in order to obtain a material of uniform size, the wood particle samples are screened to remove the coarse and fine particles. A Ro-Tap screen shaker is used with 35- and 48-mesh screens (the Ro-Tap is shown in Chapter 10). That fraction which passes through the 35 screen and is held on the 48-mesh screen is used for pH and buffer determination. The coarse fraction, which is retained on the 35-mesh screen, can be used for determining the moisture content of the material. The fines, those particles which pass through the 48-mesh screen, are simply discarded.

To maintain a uniform screening condition, the wood samples are screened in equal segments. Each segment is made up of 16 oz or 451 ml (by volume) of wood particles and is screened for 15 minutes.

Moisture contents of the samples are obtained by drying a 30-g sample of the coarse fraction in a 105°C (221°F) circulating oven for 24 hours.

Next, 30 g of oven-dry material is soaked in 400 g of distilled water at 20 ±1°C (68 ±2°F) for 30 minutes. The mixture is stirred after 0, 10, and 20 minutes. The liquid is then decanted into a number 2 Buchner (Coors) filter containing number 4 Whatman filter paper (7 cm). The liquid is drawn though the filter paper with the aid of vacuum.

A 150-g sample of the filtered liquid is then placed in a 400-ml beaker. A pH measurement is taken at 21°C (70°F) and this represents the pH of the wood being studied. Sulfuric acid (0.01N) is next added to the liquid in small increments (5 ml). The liquid is mixed and the pH is taken after each addition, until a pH of 3.5 is reached.

The pH and milliequivalents (N x ml) of acid needed to drop the pH to 3.5 are a measure of the buffer of the wood being tested.

Moisture Content

Moisture content of the raw material is important in planning any plant since it will determine the required dryer capacity. If high-moisture-content material is anticipated at some time in the future, this will have to be considered in the original design. Wide ranges in the moisture content of material entering a plant also causes production problems.

Extremely high-moisture-content species can be difficult to mill because they jam up in the equipment, are expensive to dry because of the greater energy required, and in some cases tend to yield particles with fuzzy surfaces which are more difficult to glue. Consequently, such particles can require more resin. Species with high moisture content normally give greater yields of desired particles, because few small particles (or "fines") are generated; and because of the higher moisture content, there is less breakage of the fiber when generating these particles. Extremely low-moisture-content species generally have the opposite characteristics of those noted above. All of these factors have to be taken into consideration in the following important areas: in the selection of the species to be used; in the selection of the plant location; in the design of the process; and, finally, in actual operation.

Once a plant is in operation, a change in moisture content of the raw material can create operational problems in almost all of the unit operations, including curing the resin in the press. A variation in moisture can occur with the reception of raw material from different locations, with a change in the sapwood to heartwood ratio, with the introduction of a different species, with a change in the ratio of a species mixture or, in the case of wastewood usage, with a change in storage and lumbering practices of the individual supplier or suppliers.

Extractives

Extractives are not part of the wood structure. They include tannins and other polyphenolics, coloring matters, essential oils, fats, resins, waxes, gums, starch, and simple metabolic intermediates. Extractives in wood may range between 5% and 30% in quantity. They can be removed by use of appropriate solvents. Extensive research has been conducted over the years on the variations and the types of extractives in species. Very little of this information has been related to the manufacture of particleboard, with the exception of troublesome species such as western red cedar, hemlock, and white fir. In cases where extractives and their variation play an important part in the process, problems can occur in the consumption of resin and its curing rate, poor water-resistant properties of the finished product, and problems with blows during the pressing. Such problems must be recognized and handled when using species that have extractives detrimental to the composition board process.

Some extractives in species cause problems such as the aforementioned blows. Western red cedar, for example, contains volatile material that turns to vapor during the hot-pressing operation, and this vapor can cause blowing and delamination problems at the end of the pressing period.

Extraneous Matter

Perhaps the terminology *extraneous matter* is not in its proper place when discussing species. It is associated with the raw materials received, which, as noted previously, are usually residue materials from other wood production plants. Some extraneous materials are associated with the wood, such as undesirable bark particles and mineral matter contained in the wood itself, which is a problem in some southern pine areas. Another problem is metallic particles that have found their way into the wood at some time either during its growth cycle or in previous production operations. Some of this material can cause discoloration of the board, affect resin consumption, increase the water absorption of the board, or adversely affect the finishing of the product with coatings and overlays. The mineral and metallic particles cause wear and damage to production equipment, cause fires in grinding and drying operations, and damage finishing saws both in the production plant and later on in secondary manufacturing plants.

Growth Location

Where a tree grows, even within a certain stand, materially affects it rate of growth, the chemical constituents within the wood structure, the cell structures, the sapwood-heartwood ratio and all other factors associated with the growth of trees. These differences can be shown between areas of high ridges and low valleys, areas of high and low soil nutrients, areas of heavy undergrowth, those with rocky soil, the amount of rainfall, and so forth. A board plant naturally has very little control over the raw material it receives from a wide variety of growing locations such as those noted above. However, such variables inherent in the wood received should not be glossed over or ignored. Important factors involved are how well the raw material can be milled, potential problems in drying and, perhaps most important of all, the previously mentioned differences in acidity which affect the curing of urea resin during hot pressing. Laboratory studies have shown that variations in the same species grown in different locations have enormously affected the cure rate of resin, but it should be recognized that what works in one location is not certain to be successful in another location as far as resin is concerned.

As a composition board producer selects material over a greater area, seasonal changes of suppliers can introduce variations from the different primary processing plants involved. Some of these are changes in logging practices dictated by the weather, changes in the method of log storage, and variations in sawing, drying, and planing the lumber manufactured. Because of the present dominance of planer shavings as raw material, changes in planing practice are important to the consumer in the composition board industry.

BINDERS

The predominant resins used in the composition board industry are urea-formaldehyde and phenol-formaldehyde. Urea resins make up by far the major portion of the binders used in the industry, not only in the United States but worldwide. Phenolic resins are favored for products that have some type of exterior exposure and indeed are, in effect, required for use with this type of product in the United States. This is not true elsewhere in the world, as products that have some exposure to the weather are made of urea resins. Ureas are favored because of low price, ease of handling and fast curing in the press. They are also colorless and do not lend any unfavorable color to the final board product. As mentioned, phenolic resins are the most popular throughout the world for those products that face severe weather exposures. Melamine resins or combination melamine-urea resins can be used for making some products that are fairly weather resistant, and a urea-melamine-phenolic is used in Europe in exterior-type boards.

In the last three years the worldwide increase in the cost of oil has caused the prices of synthetic resins for composition board or other products to increase tremendously. In some parts of the world the price increase has ranged as high as 1000%. This increase in price appears to be a permanent development because of its close association with oil. As will be described in a later chapter, these synthetic resins are products of the petrochemical industry. While they could also be produced from coal, facilities for doing so are not available at the present time and probably will not be for several years to come. With this increase in price, a greater interest has been focused upon adhesives that can be produced from a class of natural products called tannins. Work of this nature has gone on for many years, and these resins are successfully in use in some parts of the world.

Resin level is closely associated with board properties. However, because of the cost, minimum levels are normally used except for specialty products. Boards in the United States range between about 6 and 10% with urea resin and between 5 and 7% with liquid phenolic. Powdered phenolic resin is used at levels from about 1.5 to 5%. Higher resin levels, particularly with ureas, are found in Europe and in some other parts of the world.

Figures 5.2–5.4 show the effect of increased resin on the properties of homogeneous planer shavings boards made with three levels of resin. One percent of wax was used in these boards. The curves on the graphs are for boards at a specific gravity of 0.72 (45 lbs/ft^3 density). The Englemann spruce shavings were flaky and this, along with a lower wood density and thus better interparticle contact, contributed toward the higher bending properties. Ponderosa pine particles, however, were more blocky or cubical in nature and thus fiber length was poor for bending strengths. However, better intermeshing of the pine shavings and thus more edge-to-edge bonding was felt to be the reason for the higher internal bonds of the pine boards.

ADDITIVES

A number of special additives can be added to composition boards. The most commonly used is wax, which imparts resistance to absorption of liquid water by the final board product. Wax is normally applied as an emulsion at levels between 0.5% and 1%. It should be emphasized that wax resists the absorption of water and not the absorption of water vapor. It is particularly important in products such as exterior siding which are subjected to short periods of heavy wetting, as for instance during a rainstorm, and in underlayment, where accidental spills of water

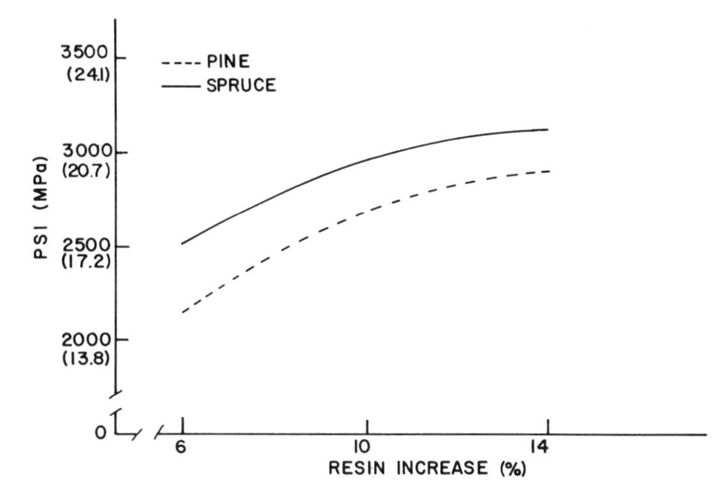

Figure 5.2. *Effect of increased resin on modulus of rupture of planer shavings board (Maloney 1958).*

can provide a situation where water is on the board for a short period of time. In the hardboard products, other oil-based additives can be used for "tempering," which hardens the board as well as giving it added resistance to water absorption and the attendant changes in dimension. This tempering is an oil usually of a synthetic base, although natural oils also are successfully used. Normally the oil is added to the board after it has been pressed and then oven cured. It should be

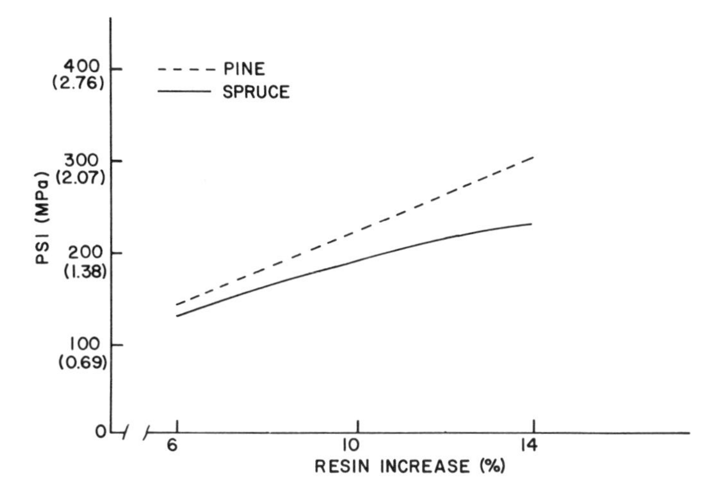

Figure 5.3. *Effect of increased resin on internal bond of planer shavings board (Maloney 1958).*

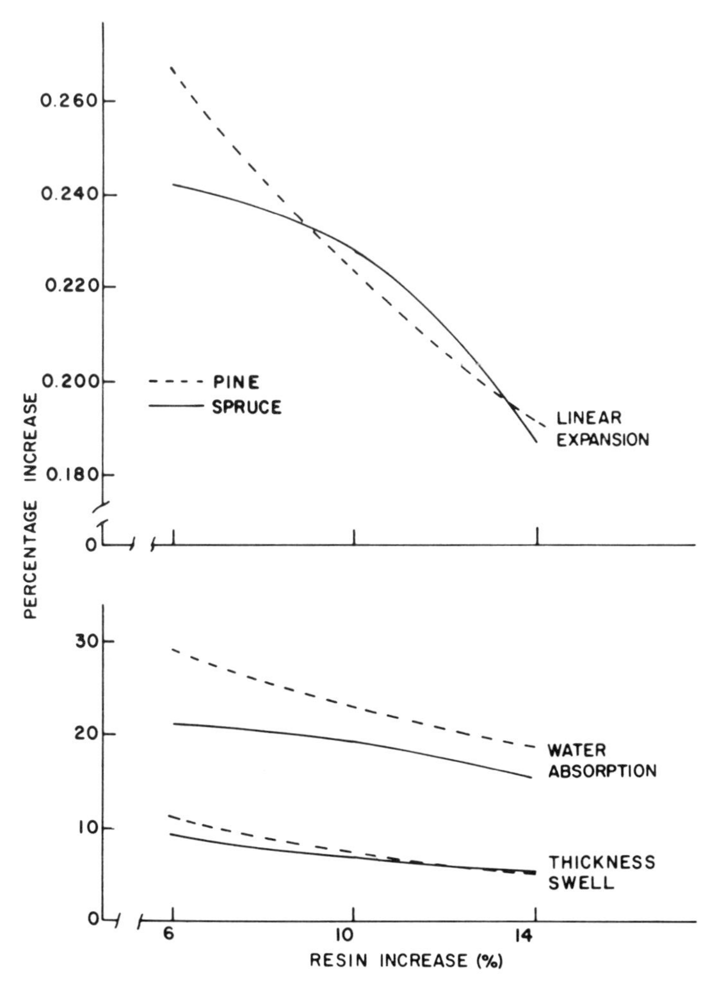

Figure 5.4. *Effect of increased resin on planer shavings board–24-hour water soak dimensional responses (Maloney 1958).*

mentioned, however, that it is also possible to add oil to the fibers either before they are formed into a mat or after the mat is formed. The oil is then polyermized in the hot press during the consolidation of the mat into the final board.

Other additives that are used in minor amounts at the present time are those for fire retardancy and preservative treating. Both of these markets are relatively minor at the present time in the United States and Canada. However, it is felt—particularly in the case of fire retardancy—that as the trend shifts more and more to mobile homes and apartment-type buildings, the building codes will demand the use of more fire-retardant panel products. Preservative treatments are important in the more humid tropical and subtropical regions of the world.

MOISTURE LEVEL AND DISTRIBUTION

The amount of moisture and its distribution through the thickness of the formed mat contribute significantly to the final board products. For example, if a high moisture level is in the faces of the mat and a low moisture in the core, the faces of the board will be highly densified in relation to the core; and thus bending strength and stiffness will be significantly greater than in a similarly pressed board with a uniform moisture content throughout the mat and having the same overall density. When the greater densification of the faces of this particular board takes place, however, the opposite effect takes place in the core, resulting in a core with a much lower specific gravity than in the comparison board made with a uniform moisture content. Thus, the board with a higher moisture content on the surface will have the aforementioned greater bending and stiffness strengths but a lower internal bond in the core of the board. As long as the total moisture content of the mat is kept within reasonable bounds so that the board does not blow apart when the press is opened, a considerable amount of manipulation of moisture content can be performed to alter board properties.

LAYERING BY PARTICLE SIZE

Layering by particle size is a conventional way of assembling a mat prior to pressing it into the final board both in fiberboard and particleboard products. The normal procedure is to place very fine particles or fibers on the faces, which results in a product with a very smooth surface. Such smooth surfaces are particularly needed by the secondary processing industries which laminate with thin films of vinyl, melamine, or polyester and by the firms that grain-print patterns upon the board substrate. Efforts are continually underway to provide panels with an extremely smooth and uniform face for this secondary processing. High-quality fibers of flakes placed on board faces yield panels with superior bending and stiffness properties. Blocky particles in the core result in panels with high internal bond.

DENSITY PROFILE

The density of successive layers in the board can be controlled though a considerable range. The previously mentioned distribution of moisture interacts significantly in this process. However, as pointed out later, it is normal for boards to have high face densities and low core densities with a few exceptions, probably most notably the very thin boards where the reverse type of density profile can be found.

At times references are seen to homogeneous particleboard or fiberboard. *Homogeneous* usually refers to the fact that the same mixture of particles is used throughout the board. This term is somewhat misleading because of the layering effect caused by the distribution of the furnish by particle size, resin, and moisture and by the way heat is applied during hot pressing. As will be dealt with in more detail later, the surfaces of the board are necessarily heated first as soon as the hot press contacts the mat. Since the hot press platens are normally well over 300°F (149°C), steam is immediately formed from the moisture in the surfaces of the mat.

This steaming plasticizes the wood, enabling the surfaces to be rapidly consolidated and densified. Quite a number of seconds go by, depending upon the board's thickness, before sufficient heat penetrates the board core to achieve the same type of plasticization. However, by the time this penetration to the core occurs, the faces have already been well densified and usually the face resin has been cured out. High-frequency curing of the mats as practiced in the medium-density fiberboard industry was specifically designed to alleviate this particular problem of a greatly unbalanced face-to-core density profile. However, it has been shown in commercial practice that in general with the thicker boards the same type of density profile exists. Density profiles will be discussed more comprehensively later on.

PARTICLE ALIGNMENT

Particles, particularly fibers and flakes, can be aligned to provide panel products with much greater bending strength and stiffness in the oriented or aligned direction. Such orientation again interacts with most of the major parameters involved in producing boards, and manipulation can change the level of strength possible through orientation.

Boards are usually referred to as randomly formed or aligned. Randomly formed boards, however, normally have their particles oriented. Only boards made with particles of uniform dimension in length, width, and thickness can have the particles randomly formed. In such an odd case, it would be like forming marbles into a mat in preparation for pressing into a board. However, virtually all particles used are not dimensionally uniform, such as a sphere or cube.

Let us use flakes to illustrate how a board is normally made of aligned particles even when called randomly formed. Flakes are aligned within the plane of the board. They will not stand on end or on edge if dropped from a mat former onto a caul plate. They are therefore highly aligned in the plane of the board. Consequently, oriented or aligned flakeboard means that the particles are lined up in one direction according to their length. The same general type of alignment exists with fiber and other types of particles formed into a mat randomly and which have a slenderness ratio over one. Such particles will line up according to length, but the more cylindrical type particles will have a more random alignment as far as width is concerned because the width may well be approximately the same as the thickness.

Two ratios should be understood when considering orientation. The aforementioned slenderness ratio is the particle length divided by the thickness. Any particle with a ratio over one will be longer than it is thick and thus will be amenable to orientation. The higher the ratio, the more slender the particle. The second ratio is the aspect ratio which is important when considering flakes. It is calculated by dividing the flake length by its width. A ratio of one means a square flake, which is one that cannot be oriented. A ratio of at least three appears to provide good orientation.

Particles with high slenderness ratios can be aligned to make composition boards strong enough to compete effectively with plywood. This potential may even extend into the structural lumber field. Mechanical means can be used to

Figure 5.5. *Aligned mat of Douglas fir fiber (Talbott 1974).*

align some particles, especially long, slender flakes. Panels made with oriented flakes aligned mechanically are coming into the marketplace as the core of veneer-faced composites. Panels made of thin oriented flakeboards cross-laminated into a panel, the same as plywood, have been produced and successfully test marketed. However, small shavings and fiber are quite difficult to align mechanically.

Figure 5.6. *Aligned mat of hammermilled Douglas fir particles (Talbott 1974).*

Figure 5.7. *Aligned mat of hammermilled red cedar mill waste including bark, planer shavings, and hogged solid residues (Talbott 1974).*

A new electrical method has been developed at Washington State University, using an electrical field for the orienting, which can orient small as well as large particles. Particles are dropped through an electrical field formed between two oppositely charged plates, and the electric field aligns them. This relatively simple device, using little energy, is incorporated as part of a former. It is now going into commercial application.

Figure 5.8. *Aligned mat of disc-cut redwood flakes (Talbott 1974).*

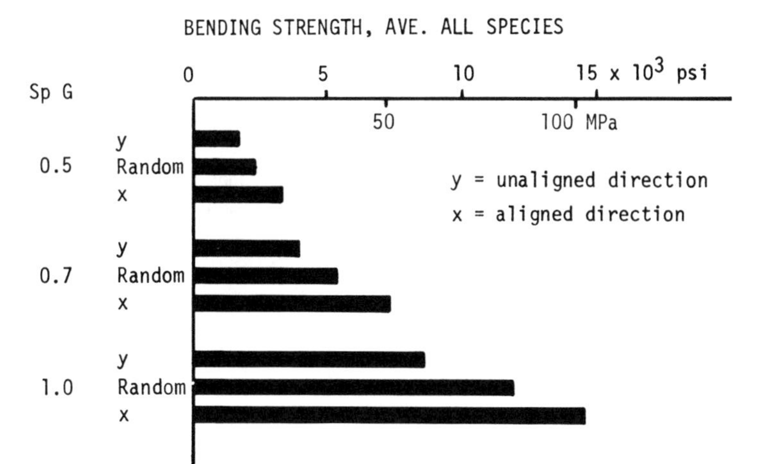

Figure 5.9. *Effect of alignment on the bending strength of fiberboard at three density levels averaged over the three species (Talbott 1974).*

The great advantage of this system is its ability to align many different types and sizes of particles even in admixture. Further, small particles and fibers can be aligned, which is quite difficult to do mechanically.

Figures 5.5–5.8 show several different types of furnish after being aligned in a mat. Figure 5.9 shows the improvement in bending strength (modulus of rupture) that can be achieved with fiber, and Figure 5.10 illustrates the comparatively high

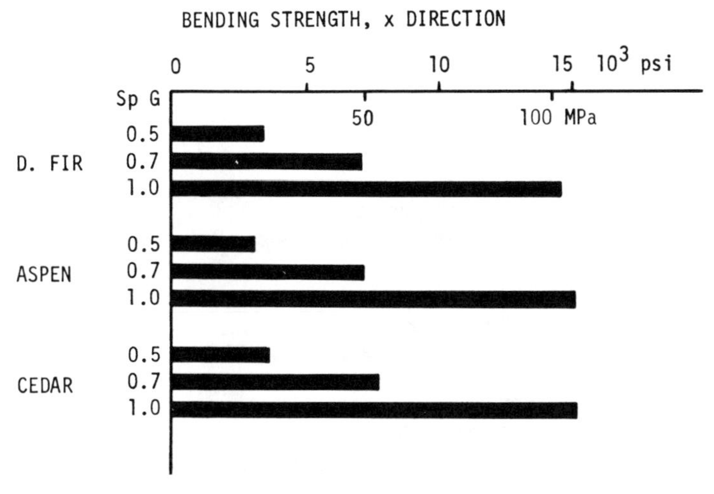

Figure 5.10. *Bending strength in the aligned direction for fiberboards of three different species and three density levels (Talbott 1974).*

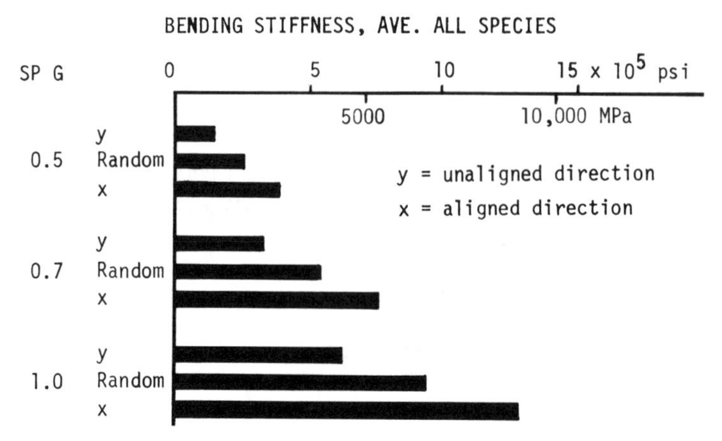

Figure 5.11. *Effect of alignment on the bending stiffness of fiberboard at three density levels averaged over the three species (Talbott 1974).*

bending strengths that can be developed in three different species. It is of interest that normally weaker species such as western red cedar and aspen approximately match the boards of the normally stronger species—Douglas fir. Similar comparisons were noted in bending stiffness (modulus of elasticity) as shown in Figures 5.11 and 5.12.

Of great import are the properties presented in Table 5.1 showing that boards of hammermilled particles have excellent bending strength and bending stiffness. Even western red cedar bark board had good properties. A combination panel of fiber faces and hammermilled particles in the core had a bending stiffness high

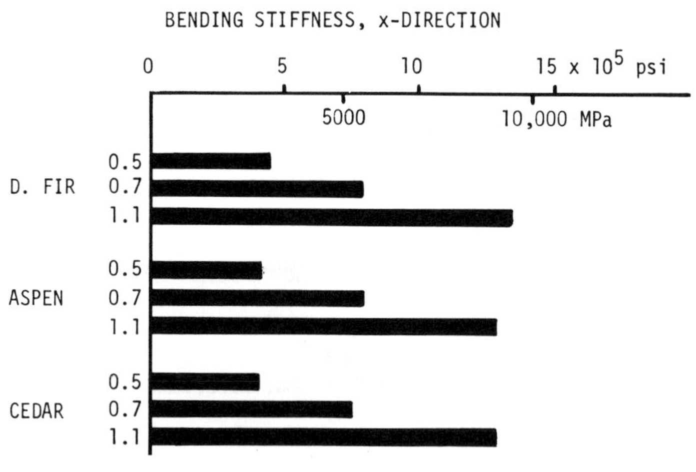

Figure 5.12. *Bending stiffness in the aligned direction for fiberboards of the different species at three density levels (Talbott 1974).*

Table 5.1 *Properties in the Aligned Direction of Several Types of Oriented Board Corrected to 0.7 Specific Gravity*

Formulation	*Modulus of rupture*		*Modulus of elasticity*		*Modulus of elasticity (degree of alignment*)*
	(psi)	*(MPa)*	*(psi)*	*(MPa)*	
1. Douglas fir fiber	8,590	59.2	750,000	5,171	2.4
2. Hammermilled cedar bark	3,350	23.1	510,000	3,516	2.6
3. Redwood flakes	7,020	48.4	1,120,000	7,722	3.3
4. Hammermilled Douglas fir particles	4,590	31.6	870,000	5,999	6.0
5. Three-layer boards (No. 1 formulation for faces, No. 4 formulation for the core)	8,500	58.6	1,000,000	6,895	3.8

Source: Talbott 1975.

* Degree of alignment = $\frac{\text{Modulus of elasticity in aligned direction}}{\text{Modulus of elasticity in nonaligned direction}}$

enough to justify design working stresses as great as those of high-quality sawn lumber. Boards of aligned redwood flakes fit into this same category. Thus, structural utilization of many raw materials previously unacceptable for use has become possible—a giant stride forward for the composition board industry.

BOARD DENSITY

The last, but certainly not least, factor to be mentioned is the level of the board's density or specific gravity. It was referred to earlier when discussing the specific gravity of the species used. Board specific gravity is a powerful factor affecting board properties. In most cases, an increase in board specific gravity results in a concomitant improvement in physical properties. This is a relatively easy way to improve properties. Many times no additional resin is needed because of the more efficient use of resin resulting from the increased board density.

In particleboard, it has been found that a board specific gravity of 0.15 to 0.20 above that of the whole wood is necessary to achieve minimum physical properties, unless low-density products are purposely being made. In all composition board, it is very noticeable that an increase in specific gravity improves the physical properties, with the exception of dimensional stability in water soak and exposure to high humidity. Because of more wood in the higher-density board, thickness swell and linear expansion can be greater after absorption of moisture.

In practice, the easiest way to improve most board properties is usually to increase the board specific gravity. The increased board specific gravity results in more intimate contact between the particles in the mat being compressed into the final board. As will be noted later, much resin is lost in interparticle voids in lower-density panels. If the board can be compressed to a higher density, more of this "dormant" resin is used effectively. Increasing specific gravity not only affords this increase in effectiveness of resin but, of course, causes more wood to be present to resist mechanical loads. This combination of effects results in board strengths being approximately proportional to the square of board specific gravity over the usual specific gravity range.

In places where wood costs are low, increasing board density can be the cheapest way to improve board properties. However, the increased board weight can result in a board harder to handle and in one which is more difficult to cut and machine. If freight rates are a factor, heavier boards can be more costly to ship.

SUMMARY

As perhaps can be seen more clearly now, the production of a composition board depends on the web of interrelated factors described plus many other minor but still important factors that also enter into this vast spider's web. In essence, the process of producing boards is simple. However, interactions occur in each of the process steps, which complicates the whole system. It does provide a situation, however, where numerous products can be made from the same general mix of raw material with proper understanding and control of the process. Man then becomes the master of a complicated situation and uses it to his advantage to produce a consistent and high-quality product, usually from waste materials.

Up until now we have discussed the subject in general terms. We will now turn to more of the basics involved in the process, as well as to discussion of the pieces of equipment necessary to put theory into practice. The entire process was described in a sequential form earlier to provide a general understanding of this exciting new industry. In the next several chapters, the fundamentals of the process will be the subject of discussion.

SELECTED REFERENCES

Brown, J. H., and S. C. Bean. 1974. Acceptance of Stranwood Sheathing for Construction Uses. *Proceedings of the Washington State University Particleboard Symposium, No. 8*. Pullman, Washington: Washington State University (WSU).

Carroll, M. N. 1963. Whole Wood and Mixed Species as Raw Material for Particleboard. Washington State University Bulletin 274. Pullman, Washington.

Crawford, R. J. 1968. Production Experiences in a Large Plant Using Several Species for Furnish. *Proceedings of the Washington State University Particleboard Symposium, No. 2*. Pullman, Washington: WSU.

Elmendorf, A. 1965. Oriented Strand Board. Patent No. 3,164,511. United States Patent Office (January).

Forest Service/U.S. Department of Agriculture. 1974. *Wood Handbook*. Agriculture Handbook No. 72. Forest Products Laboratory.

Foster, G. 1967. Species Variation. *Proceedings of the Washington State University Particleboard Symposium, No. 1*. Pullman, Washington: WSU.

Geimer, R. L.; H. M. Montrey; and W. F. Lehmann. 1975. Effects of Layer Characteristics on the Properties of 3-Layer Particleboards. *Forest Products Journal,* Vol. 25, No. 3.

Halligan, A. F. 1970. A Review of Thickness Swelling in Particleboard. *Wood Science and Technology,* Vol. 4.

Haygreen, J. G., and R. O. Gertjejansen. 1972. Influence of the Amount and Type of Phenolic Resin of the Properties of a Wafer-Type Particleboard. *Forest Products Journal,* Vol. 22, No. 12.

Hoyle, R. J., Jr. 1974. Property Requirements for Structural Roof and Floor Sheathing. *Proceedings of the Washington State University Particleboard Symposium, No. 8.* Pullman, Washington: WSU.

Hoyle, R. J., Jr., and R. D. Adams. 1975. Load Duration Effects on Properties of Clear Wood Plywood and Strandwood. *Proceedings of the Washington State University Particleboard Symposium, No. 9.* Pullman, Washington: WSU.

Klein, J. F. 1975. Problems in Using Mixed Species in Wood-Fiber Products. Background Paper No. 60, third World Consultation on Wood-Based Panels, FAO, New Delhi, India, February 1975.

Klauditz W. 1960. Manufacture and Properties of Particleboard with Oriented Strength. *Holz als Roh-und Werkstoff,* Vol. 18, No. 10.

Klauditz, W.; H. J. Ulbricht; W. Kratz; and A. Buro. 1960. *Production and Properties of Flakeboard with Oriented Strength.* Brunswick, Germany: Wood Research Institute, Brunswick Institute of Technology.

Maloney, T. M. 1958. Effect of Engelmann Spruce and Ponderosa Pine Dry Planer Shaving Particle Size and Shape on the Properties of Particleboard. Washington State University Work Order Number 897. Pullman, Washington.

———. 1960. Comparisons of Redwood and Douglas-fir for Particle Board Manufacture. Research Report 60/15-35. Washington State University, Pullman, Washington.

Maloney, T. M.; J. W. Talbott; and G. G. Marra. 1956. Effect of Species on the Physical Properties of Particleboard Composed of Dry Planer Shavings. Washington State University Work Order Number 865. Pullman, Washington.

Mottet, A. L. 1967. The Particle Geometry Factor in Particleboard Manufacturing. *Proceedings of the Washington State University Particleboard Symposium, No. 1.* Pullman, Washington: WSU.

———. 1975. Use of Mill Residues and Mixed Species in Particle Board Manufacture. Background Paper No. 59, third World Consultation on Wood-Based Panels, FAO, New Delhi, India, February 1975.

Rooney, J. E. 1970. Combining Species. *Proceedings of the Washington State University Particleboard Symposium, No. 4.* Pullman, Washington: WSU.

Snodgrass, J. D.; R. J. Saunders; and A. D. Syska. 1973. Particleboard of Aligned Wood Strands. *Proceedings of the Washington State University Particleboard Symposium, No. 7.* Pullman, Washington: WSU.

Snodgrass, J. D., and R. J. Saunders. 1974. Building Products from Low Quality Forest Residues. *American Society of Agriculture Engineers Paper No. 74-1549.*

Suchsland, O. 1962. The Density Distribution in Flake Boards. *Quarterly Bulletin* (August), Vol. 45, No. 11. (Michigan Agricultural Experiment Station, Michigan State University).

Talbott, J. W. 1974. Electrically Aligned Particleboard and Fiberboard. *Proceedings of the Washington State University Particleboard Symposium, No. 8.* Pullman, Washington: WSU.

———. 1975. Control of Directional Properties of Dry-Formed Lignocellulosic Composition Boards by Electrical Fibre Alignment. Background Paper No. 78, third World Consultation on Wood-Based Panels, FAO, New Delhi, India, February 1975.

Vital, B. R.; W. F. Lehmann; and R. S. Boone. 1975. Effects of Species and Board Densities on Properties of Exotic Hardwood Particleboards. Background Paper No. 58, third World Consultation on Wood-Based Panels, FAO, New Delhi, India, February 1975.

6.

RAW MATERIALS AND PARTICLE GEOMETRY: EFFECTS ON BOARD PROPERTIES

A number of different wood raw materials are used for composition board, ranging from logs to sander dust. Old packing cases are also used for raw material in other countries. Agriculture residues such as flax and bagasse are also of importance in various parts of the world. A wide range of particle types with varying geometries can be generated from solid wood or residues, either directly for the board process or as particles ready for use, such as planer shavings. The types of particles and the expense associated with generating them are significant factors in the cost of manufacturing board. Each type of particle, its inherent geometry, and combinations of various particles have profound effects on the board quality.

AVAILABLE RAW MATERIALS

The degree of freedom in controlling the particle geometry varies widely according to the kind of raw material available to the manufacturing operation. It has been noted that the maximum freedom exists when the available material is in the form of roundwood, preferably green, or in other forms of rather large solid wood. This permits the cutting or reduction of the wood in any way economically feasible within the plant. However, the large volume of raw material available as residue from other wood-processing operations, as noted, often constitutes the major raw material available for use in a composition board operation. Thus, depending upon the particle size of this residue as it enters the board plant, the degree of possible further reduction in size or shape is already somewhat limited. Significant improvement in particle geometry, however, can be accomplished by proper use of milling or flaking machines.

The principal types of raw material available in the United States at the present time, in descending order of freedom in further processing, are as follows:

- □ Green roundwood or logs.
- □ Veneer peeler cores.
- □ Cull lumber.
- □ Lumber slabs and edgings.
- □ Lumber mill end trim (green).
- □ Lumber planer end trims (mostly dry).
- □ Veneer log roundup (green).

- □ Veneer clippings (green).
- □ Pulp chips (green).
- □ Lumber planer shavings (dry).
- □ Sawmill sawdust (green).
- □ Miscellaneous dry woodworking residues from molding and furniture-type plants.
- □ Planer mill and similar dry sawdust.
- □ Plywood trim.
- □ Plywood sander dust.
- □ Miscellaneous woodworking sander dust.

Bark is not included in this list of material, although minor amounts are used throughout North America and elsewhere in the world. Most of the bark now being used ends up in the board furnish because the particles were generated from wood that was not debarked. Research has shown that certain board products can be successfully produced from bark. Additional research and development is under way at the present time, and a separate family of bark products should result. Other fibrous raw materials, such as bagasse and flax, are used elsewhere in the world and there is the possibility of using agricultural fibrous material here in the United States.

A 1974 survey of the major types of raw material used in North American particleboard plants is shown in Table 6.1 Softwood planer shavings make up a major share with 47.7% of the total. Softwoods account for 84.6%, with hardwoods being used for only 15.4% of production. It is notable that 3.7% of softwood chips and 2.1% of hardwood chips go into this type of board. Under present accounting systems they are too valuable for pulp; however, this situation may change as demand for high-value panel products increases.

When using any of the grass-type fibrous-type residues, it is important that shiny outside surfaces are eliminated by the splitting action of mills generating the final fiber. Preparation of the furnish is extremely important as the pith and dirt must be removed. In bagasse, for example, high levels of pith are found. Pith, which has no structural strength, is very absorbent and can rob the board of much of the applied resin. As demand for more building products increases worldwide, it is anticipated that more agricultural residues will be used for board.

PARTICLE TYPE AND GEOMETRY

Definitions of the various types of particles have been developed by the American Society for Testing and Materials as part of the "Standard Definition of Terms Relating to Wood-Based Fiber and Particle Panel Materials" ASTM designation: D 1554. The following definitions are based on the ASTM standard.

Chips: Small pieces of wood chopped off a block by axe-like cuts as in a chipper of the paper industry, or produced by mechanical hogs, hammermills, and so forth.

Curls: Long, flat flakes manufactured by the cutting action of a knife in such a way that they tend to be in the form of a helix.

Fibers: The slender threadlike elements or groups of wood fibers or similar cellulosic material resulting from chemical or mechanical fiberization, or both, and sometimes referred to as fiber bundles.

Flake: A small wood particle of predetermined dimensions specifically produced as a primary function of specialized equipment of various types with a cutting action across the grain (either radially, tangentially or at an angle between), the action being such as to produce a particle of uniform thickness,

Table 6.1 *Summary of the Major Types of Raw Materials Used in North American Particleboard Plants*

		Proportion of total particleboard raw material supply (%)	
Softwoods			
Roundwood		4.2	
Pulp chips		3.7	
Planer shavings		47.7	
Sawdust		11.8	
Hogged mill waste:			
Sawmill & planing mill	1.7		
Plywood plant	9.3		
Millwork plant	0.2		
Furniture plant	0.3		
Particleboard plant	0.9		
Unspecified	4.8		
Total hogged wood mill waste		17.2	
Total softwoods			**84.6%**
Hardwoods			
Roundwood		6.7	
Pulp chips		2.1	
Planer shavings		3.1	
Sawdust		0.4	
Hogged mill waste:			
Sawmill & planing mill	0.2		
Plywood plant	0.1		
Furniture plant	2.5		
Flooring plant	0.3		
Total hogged mill waste		3.1	
Total hardwoods			**15.4%**
			100.0%

Source: Mottet 1975, FAO Paper 59.

essentially flat and having the fiber direction essentially in the plane of the flakes, in overall character resembling a piece of veneer.

Shaving: A small wood particle of indefinite dimensions developed incidental to certain woodworking operations involving rotary cutterheads, usually turning in the direction of the grain and because of this cutting action, producing a thin chip of varying thickness, usually feathered along one edge and thick at the other, and usually curled.

Slivers: Particles of nearly square or rectangular cross section with a length parallel to the grain of the wood of at least four times the thickness.

Strand: A relatively long (with respect to thickness and width) shaving consisting of fat, long bundles of fibers having parallel surfaces.

Wood Wool (excelsior): Long, curly, slender strands of wood used as an aggregate component for some particleboards.

Present composition board processes utilize flakes, shavings, and fibers as the three most important particles. Many sizes of each are used in the industry. Hammermilled particles are generated from a number of raw materials such as chips and hogged waste. The only other particle specifically generated for panel products at the present time is wood wool. This is a relatively minor product line. Figure 6.1 provides photographs of the three particles mostly used at present in the United States. It must be borne in mind, however, that these are typical particles, and within each type there is a wide variation in size and geometry. Figure 6.2 shows other particles used in the process, ranging from large shavings and sawdust that require further milling through various flakes and wood wool.

There is a semantic problem with the particle nomenclature. The term *particle* is used generically, but it is also the most descriptive term in the English language for shavings and hammermilled furnish which have no outstanding geometric characteristics. If particles are slender in the grain direction, they are referred to as slivers and, if they are very slender, they are strands. If width is much greater than thickness, the particle is better described as a flake. Reduction in size results in either fibers or very small particles normally called fines. A *fiber* as a particle is actually a group of fibers in the botanical sense. *Particle,* however, is defined in ASTM D 1554 as the aggregate component of a particleboard manufactured by mechanical means from wood or other lignocellulosic material including all small subdivisions of wood.

Particle size and shape, or geometry, are essential characteristics of particle type. This can range from large wood chips or flakes with dimensions equaling or exceeding those of chips used in pulp and paper manufacturing down to fines such as sander dust. Particle geometry, as mentioned previously, is one of the basic factors determining the properties and characteristics of particleboard, along with wood species, type and amount of binder and other additives, and the board structure as determined by mat forming, layering, and pressing conditions. Particle geometry interacts intimately with virtually all these parameters in determining board properties.

Particle geometry has a tremendous influence in the selection and operation of production equipment. This influence is not only on the selection and operation of equipment involved in the particle generation and particle classification, which determines the actual particle geometry, but also on almost all the other major processing facilities. This includes the conveying equipment, dryers, blenders, mat formers, conveyor lines, and even the press itself through the effect on closing speed and on the "daylight" requirement of presses (the maximum opening between platens). It is not possible to plan a composition board operation, or operate it after it is constructed, without dealing with the factor of particle geometry.

The major factors governing the choice of geometrical characteristics of particles to be used in a particular manufacturing operation are: (1) mechanical properties and other characteristics required in the final board product, (2) nature of the available raw materials, and (3) the effect of particle geometry on the total manufacturing cost.

Particle geometry is extremely complex within itself, and, as already noted, has complicated interactions with the other board production parameters. It affects board characteristics in a wide variety of ways and, while one may completely understand how particle geometry affects board properties, he may not be able to completely control the effects in a given board process. A general guide can be developed on how to use particle geometry successfully in a board plant, but it must be remembered that much of the successful generation and use of suitable furnish is still part art and part technological skill.

EFFECT OF PARTICLE GEOMETRY ON MANUFACTURING COST

As will be discussed in detail later, the particle geometry determines to a great degree the final board properties, particularly in bending strength and stiffness. Panel products having high bending strength are not dominant in the part of the composition board industry using planer shavings for raw material. Achievement of superior bending strength and stiffness requires generation of higher-quality flakes or fiber at additional cost. Thus, in determining the answer to the practical question of how to use effectively particle geometry in composition board, the relationship of service characteristics to cost must be considered. This involves not only the cost of the raw material but also the influence that particle geometry has on the cost of preparing the final furnish for board production. For example, it is usually not economical to produce high-quality flakes or fiber for an end product that does not require the bending strength and stiffness afforded by such particles. These high-quality flakes and fibers are more expensive to generate and would end up, in the long run, in competition with particles that were much less costly to produce.

The various types of raw material listed above could have a wide range of cost when delivered to a board plant. The solid wood raw material usually has the highest cost, because, in many cases, it is harvested only for the composition board plant. Thus, all of the harvesting transportation and drying costs are borne by the composition board plant. On a dry-ton basis, the solid wood may cost $30

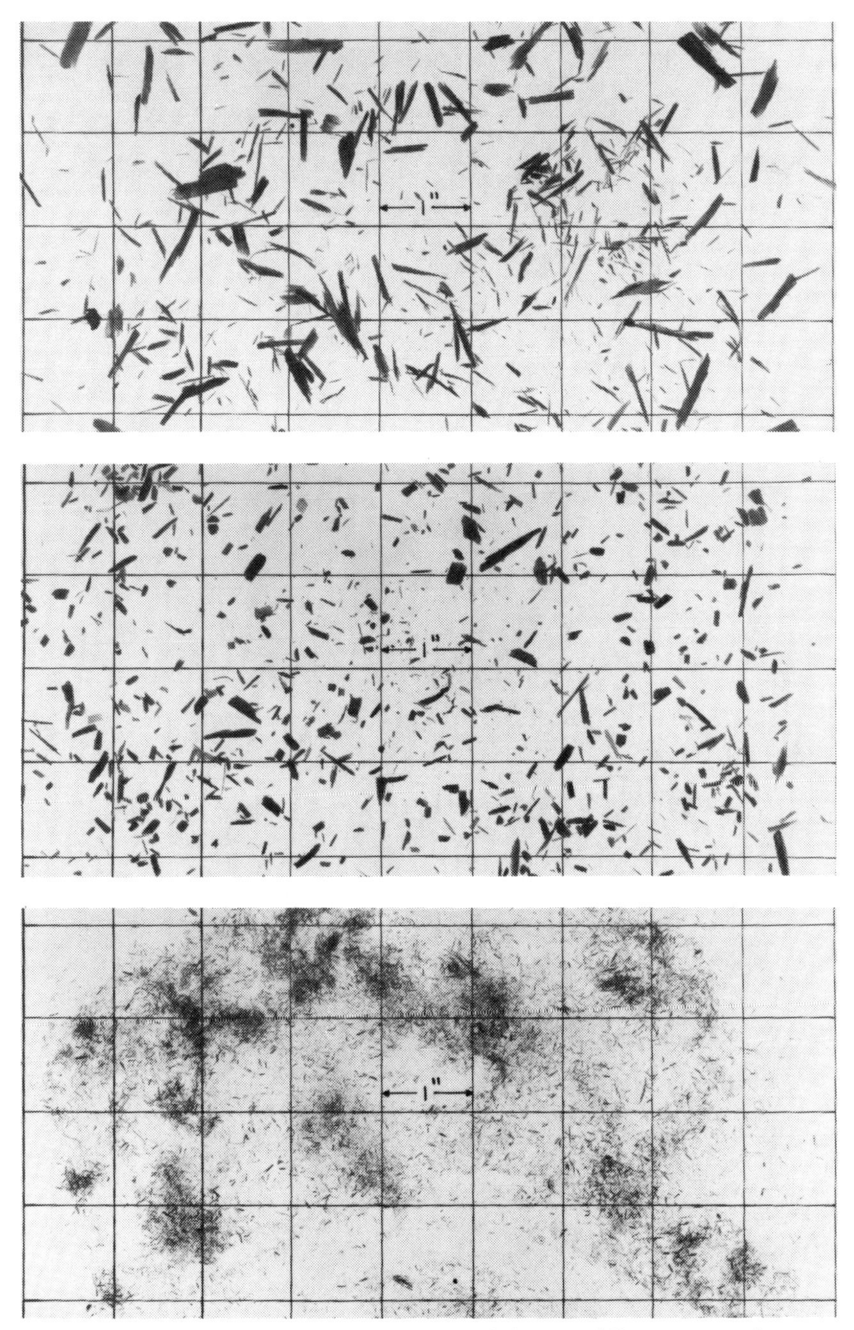

Figure 6.1. *Typical particles: top, ring-cut flakes; middle, ring-cut shavings; bottom, attrition-mill furnish from chips (Mottet 1967).*

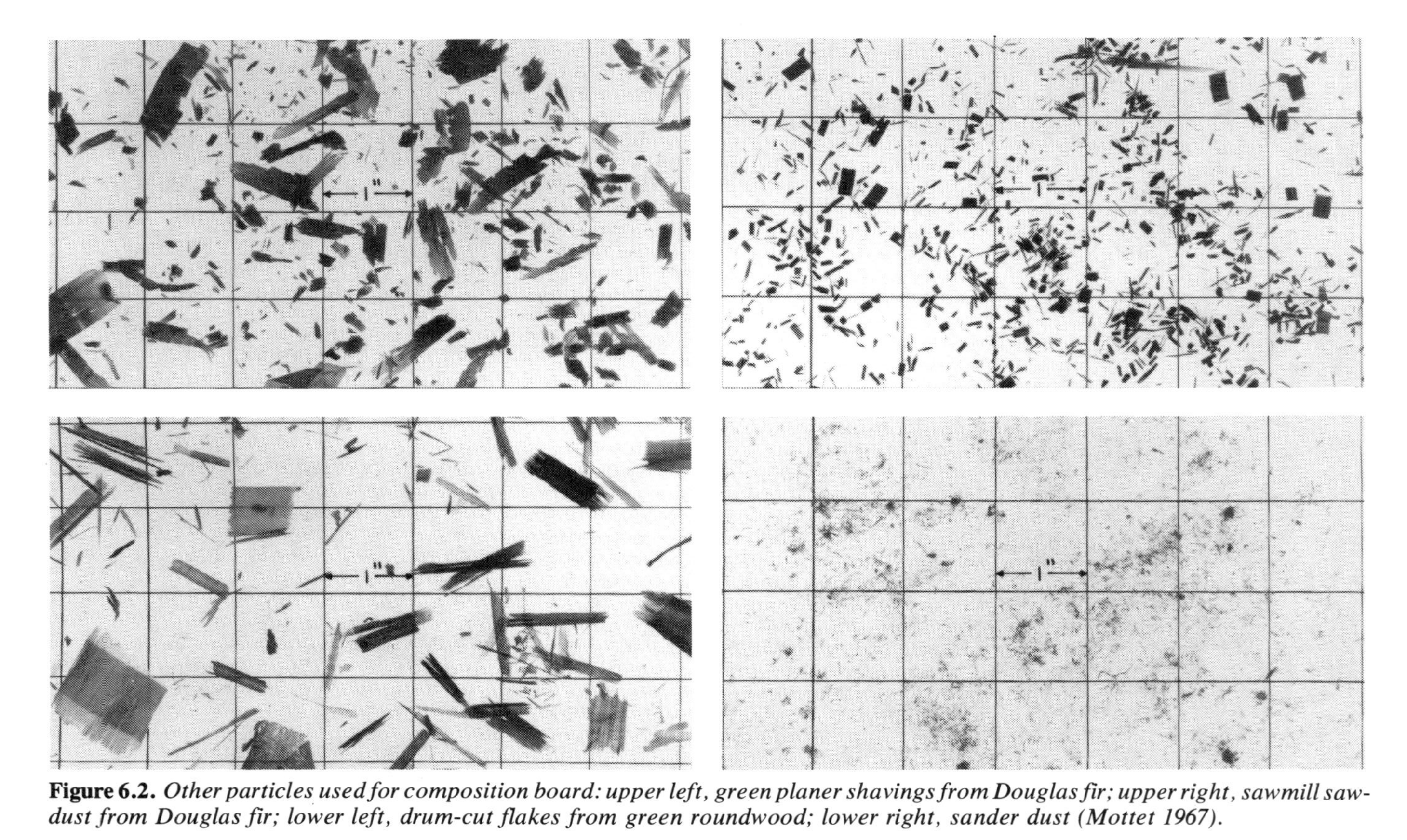

Figure 6.2. *Other particles used for composition board: upper left, green planer shavings from Douglas fir; upper right, sawmill sawdust from Douglas fir; lower left, drum-cut flakes from green roundwood; lower right, sander dust (Mottet 1967).*

per ton, while the planer shavings may cost $10 per ton or even as low as $3 per ton in some areas. This difference in cost has a powerful influence on the choice of the raw material as well as the ultimate success of the board plant in the marketplace. The cost situation, however, is rapidly changing because of increasing demand for comminuted wood as fuel.

The comminuted forms of raw material generally require less handling, fewer processing steps and consequently less labor in preparation than do the solid forms. Less machinery, therefore, is required for the comminuted forms, which results in less capital expenditure and a lower maintenance cost. In comparison, a sophisticated flaking operation may cost more than $1 million.

To meet special property requirements, the manufacturer may choose to prepare an "engineered" particle such as flake and fiber. However, considering the great difference in cost between using comminuted wood and roundwood, he may choose instead to change some other factor such as increasing the board density or the resin level of the board to gain the special properties. This needs an informed decision, balancing the higher cost of raw material and subsequent handling against the cost of additional chemicals or wood.

While considering flakes and fiber, it should be noted that there are variations in quality in both of these types of particles, as well as any others that are generated from primary wood processing plants. For example, relatively low-quality flakes that are still far superior to shavings in developing bending and stiffness properties can be generated through high-speed, low-labor chipping followed by mechanized flaking of the chips through a ring-type flaker. There are many different sizes, shapes, and thicknesses of these particular flakes, all generated at one time, but in the right situation they can be economically produced to provide a product superior to the one possible from finely comminuted wood.

PARTICLE GEOMETRY AND BOARD QUALITY

Several aspects of board performance are directly affected by particle geometry:

1. Mechanical properties such as bending strength, bending stiffness, tensile strength parallel to the surface, tensile strength perpendicular to the surface (internal bond), screw-holding strength, and nail-holding strength.
2. Board surface characteristics, particularly surface smoothness of faces and edges, which in turn affect finishing and secondary gluing characteristics.
3. Moisture responses such as moisture absorption from liquid or vapor phase and corresponding changes in dimensions, mechanical properties, and surface characteristics (these also involve exterior weatherability with cyclical moisture changes).
4. Behavior in machining operations such as sawing, boring, routing, shaping, planing, and sanding.

Particle geometry has important effects on all board properties and can interact with other basic process variables and with the equipment used to carry out the various steps of manufacture. It is a complex factor and its complexity is further increased by the fact that particles of varying geometric characteristics can be mixed in various proportions to make up the furnish for a board and can be distributed in different layers from the surface to the center of the board. Perhaps the greatest complication in this respect is that this mixture and distribution is not always under control. Changes in particle geometry in the particleboard process can be fairly obvious visually; however, it must be remembered that significant variation can occur in the fiber process as well. These "fibers" are usually a number of individual fibers making up a fiber bundle. However, the thickness, length, and surface quality can be vastly different between a fiber generated from one species under a given set of conditions and a fiber generated from a different species under a different set of conditions. Thus, particle geometry is as important when considering fibers as it is when considering flakes and comminuted particles such as planer shavings or sawdust.

Some of the important research on particle geometry is discussed below. A general summary is then presented on the effect of particle geometry on the most important board properties.

Flake Geometry

The cutting of flakes from solid wood offers the maximum freedom in controlling the size and shape of the particle. While this is not the dominant particle used

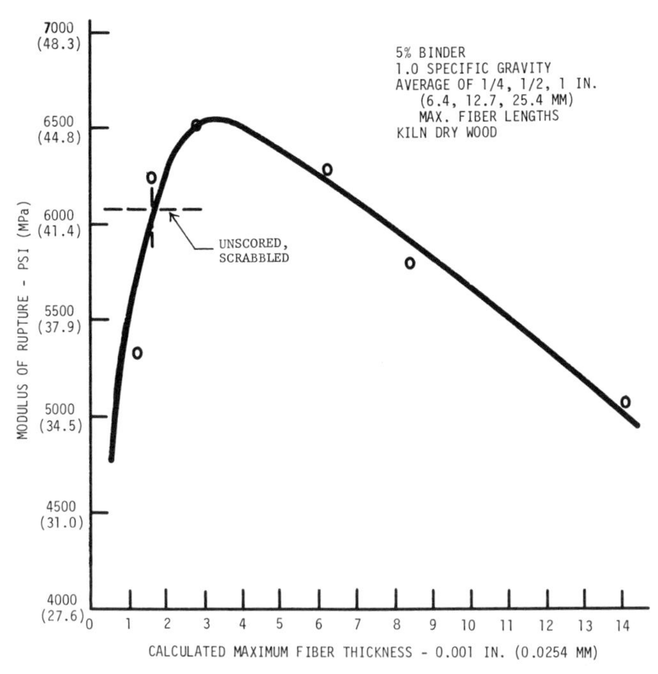

Figure 6.3. *Modulus of rupture vs the calculated maximum fiber thickness for board prepared from cross-cut fiber (Mottet 1967).*

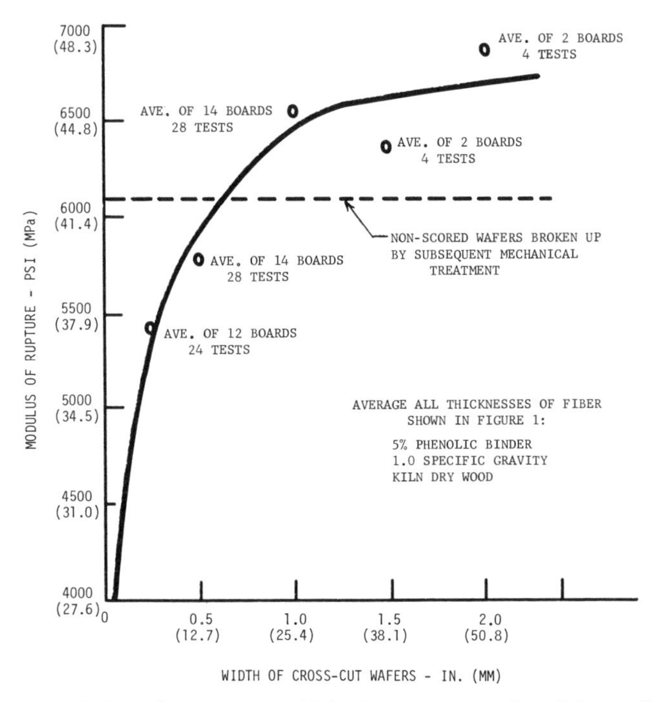

Figure 6.4. *Modulus of rupture vs width of cross-cut wafers (Mottet 1967).*

in the United States, flakes are predominantly used in other highly developed countries such as those in Western Europe, and Australia and New Zealand where the amount of planer shavings found in the United States is not available. It seems probable that if such low-cost raw material were available in those areas, they would also be the dominant source of raw material supply.

The first references to wood particle shape came early in the development of particleboard, particularly in the 1905 Watson patent. In later patents through the 1930s, flakes again were mentioned as a significant type of particle. The Long-Bell Lumber Company (now International Paper Co.) laboratory in 1949 conducted extensive work on particle dimensions by feeding blocks of wood at controlled rates into a rotating cylindrical cutterhead to yield various thicknesses of flakes. The width of the block was regulated to yield several different flake lengths. The results of this work are shown in Figures 6.3 and 6.4, which indicate that the best modulus of rupture values could be expected from relatively thin flakes 0.003 to 0.004 in. (0.076–0.102 mm) thick. Width of flake was also significant, indicating that flakes 1 ½ in. (38.1 mm) wide and over were superior. The particular boards referred to in Figures 6.3 and 6.4 were designed to be used as high-density, hardboard-type panels. Thus, the specific gravity of the boards was close to 1.0, which is much higher than the common particleboards now produced.

It should be mentioned here that the term *wafers* as used to describe the flakes in this research was a specially coined word describing flakes with beveled and

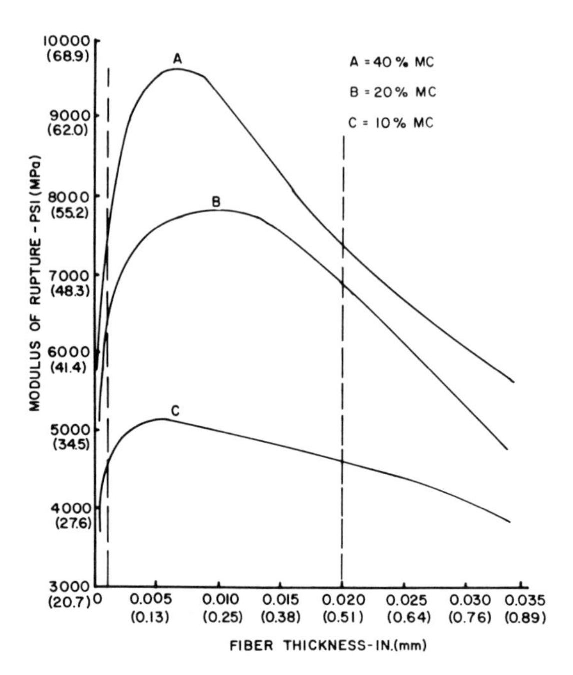

Figure 6.5. *Effect of fiber thickness on the strength of pressed fiberboards prepared from Douglas fir cross-cut fiber. (Mottet patent applications in 1953 and 1958.)*

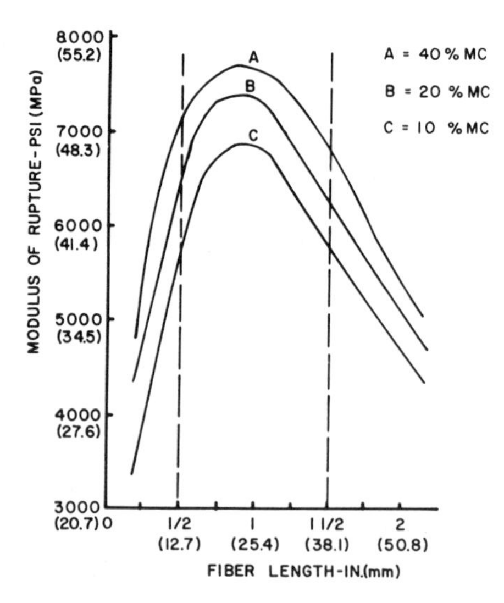

Figure 6.6. *Effect of fiber length on the strength of pressed fiberboards prepared from Douglas fir cross-cut fiber. (Mottet patent applications in 1953 and 1958.)*

jagged ends. While special machines were designed to make this type of flake, in reality the wafer, as it is now used in production, is a rather thick flake with relatively uniform thickness from one end to the other; and variations are mainly due to machining variables. Figures 6.5 and 6.6 show that flakes generated from material at about 40% moisture content were relatively high in quality. In this study, boards made with thin flakes about 1 in. (25.4 mm) long had the best modulus of rupture values. In this work, it was shown that, in cutting flakes from solid wood, it is important that the sharp bending of the particle at the knife tip be across the grain rather than parallel to the grain in order to prevent the development of numerous fine fractures of the flake, which cause weakening of the flake, and a reduction in the strength of the resulting board.

Figures 6.7–6.10 present some of the information developed by H. D. Turner in his classic research on the effect of particle size and shape on strength and dimensional stability of particleboard. In this work, precisely cut particles in the form of flat flakes, thin strands, helical ribbons and cubes were converted to resin-bonded panels ranging from 0.5 to 1.05 in specific gravity and having 2, 4, or 8% phenolic resin. He studied the effect of particle shape on modulus of rupture, thickness change or springback following two cycles of water soaking and oven drying, and by linear expansion during cyclic equilibrium exposure between 30

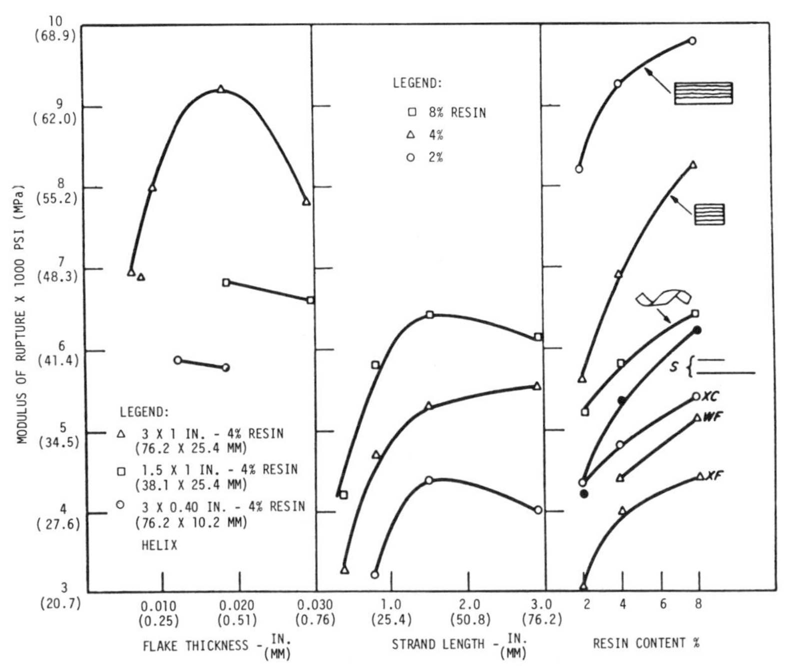

Figure 6.7. *Curves showing effects on strength at 0.80 sp gr of flake thickness, strand lengths, and resin content for several wood-particle shapes. S=straight strands; XC=coarse excelsior; WF=coarse fiber of aspen; XF=fine excelsior. (Turner 1954)*

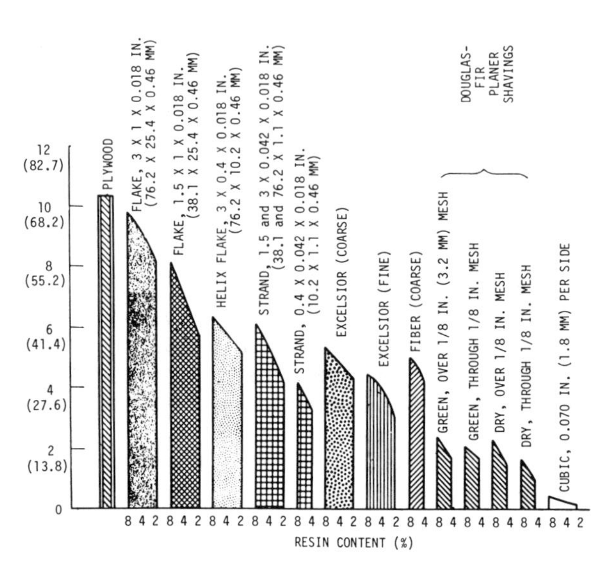

Figure 6.8. *Chart summarizing effects of resin content and particle shape upon strength, when referred to a common specific gravity of 0.80 (Turner 1954).*

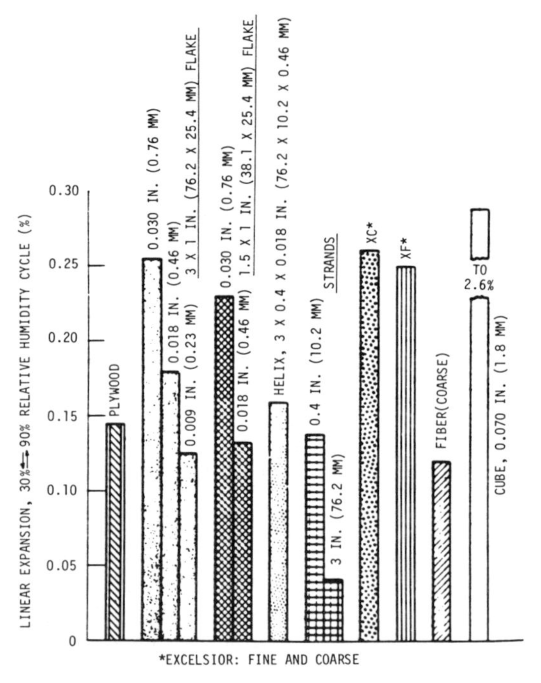

Figure 6.9. *Bar chart showing importance of effect of particle dimension and shape on lineal expansion during cyclic humidity exposure (Turner 1954).*

and 90% relative humidity. Particle sizes ranged from rectangular flat flakes varying from 0.006 in. to 0.030 in. (0.15–0.76 mm) thick, 45° pitch helical ribbons about ½ in. (12.7 mm) wide by 3 in. (76.2 cm) in fiber length and a series of straight strands 0.042 in. (1.06 mm) by 0.018 in. (0.46 mm) in cross section and having controlled lengths of about ⅜, ¾, 1 ½, and 3 inches (9.5, 19, 38.1, and 76.2 mm). Another precisely produced particle was a 1/16-in. (1.6 mm) cube. A series of fibrous materials were made from commercial grades of excelsior and a coarse mechanically made fiber. Douglas fir planer shavings were also included in this study.

Within the same particle class, it was determined that panel density or specific gravity had a direct and large effect upon the physical properties. Density was a primary influence and resin content, secondary, in control of strength properties.

The striking observation was that particle configuration and dimensions exerted the dominant control over both bending strength and dimensional stability. The relatively undamaged, long, flat flakes afforded boards nearly equal to plywood in bending strength, when using 8% resin. Boards made of helical ribbons and a 3-in. (76.2 mm) thin strand were inferior to the flat-flake panels in bending strength. In summation, the large flake showed the highest strength values; strands, excelsior and fiber, intermediate values; planer shavings, a lower value; and cubical particles, minimum strength values.

Particles that showed evidence of fiber damage or were seriously lacking in fiber quality displayed a comparatively high linear expansion. This was best

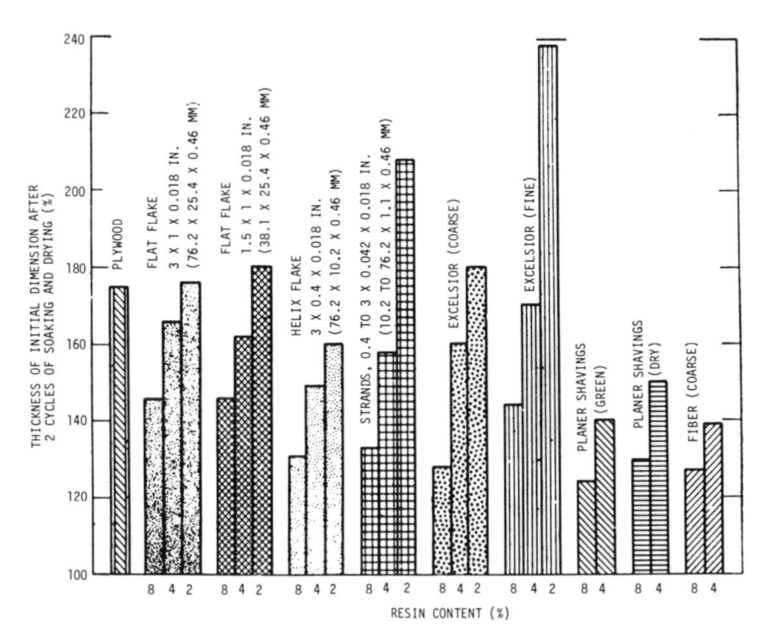

Figure 6.10. *Bar graph summarizing springback relationship to particle shape and resin content. These values correspond to test-panel sp gr 0.80 (Turner 1954).*

demonstrated by the cube-shaped material, whose expansion in boards was 10 to 20 times that of panels made from particles having sound fibers. Boards of thin flakes were better in linear expansion than those of thicker flakes. The least linear expansion, however, was observed in the boards of 3-in. (76.2 mm) thin strands. Thus, not only particle shape but dimensional variables within a particle class influence linear expansion.

The resin content was a governing factor within a given particle class in the springback behavior. This test employed six exposure cycles of 16-hour water immersion followed by 8 hours of oven drying, rather than equilibrium conditions. This may have been a case of retarding the rate of springback rather than an indication of a long-run performance under severe exposure. Other studies since, however, have shown that higher resin levels are of great importance when producing panels that must stand up to severe exposure.

One factor related to particle geometry that is important but difficult to control is the effect of surface characteristics of a particle on the efficiency of the utilization of the adhesive or binder. Rough surfaces make it difficult to bind two particles together with a number of very small droplets of adhesive. In such a situation, much of the adhesive is lost. While it can be categorically stated that surfaces of particles should be relatively smooth and undamaged and the cellular surface structure should be as perfectly intact as possible, the damaged surface quality of most planer shavings makes it almost impossible to follow the dictates of this scientific finding.

Flake thickness, flake length, and flake surface all influence the interrelationship of physical properties versus the resin spread. Figure 6.11 shows the relative

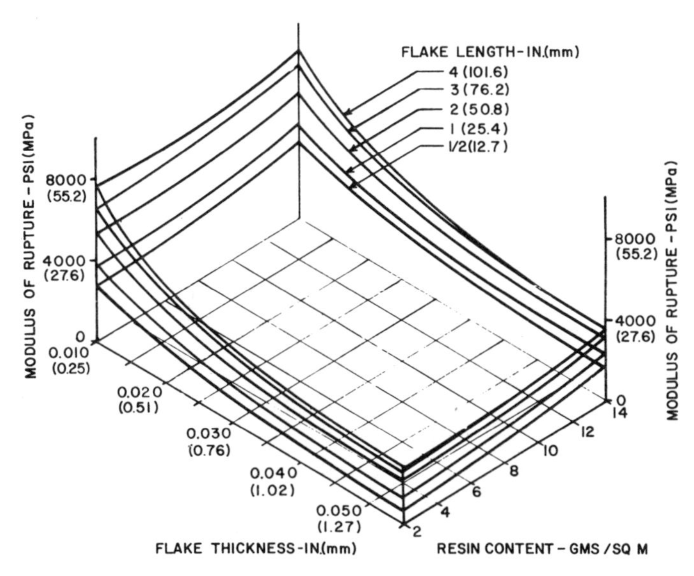

Figure 6.11. *Regression surfaces for various flake lengths of modulus of rupture in terms of flake thickness and resin spread (Post 1958).*

unimportance of resin content in terms of pounds per square foot or grams per square meter of flake surface as far as modulus of rupture is concerned. There is some influence, of course, but it is extremely difficult, from a commercial standpoint, to operate with a calculation of resin spread in terms of pound per square foot of particle surface. This is calculated in the industry on a percentage by weight and is an all-important factor of cost.

Other research along this line has shown that thickness stability is radically affected by flake thickness, particularly in connection with springback, which occurs when water vapor absorption proceeds beyond 15% equilibrium moisture content. Flake length and resin spread have a substantially less important effect on thickness stability.

Information on the effect of particle geometry on properties of northern oak particleboards showed that 1-in. (25.4 mm) flakes had the highest strength properties in stiffness, bending, and internal bond; boards from ¼-in. (6.4 mm) flakes ranked next in these properties, with planer shavings, fines, and sawdust ranking next in order. The dimensional stability properties of linear expansion and thickness swelling generally follow the same order. The results of this study are shown in Figures 6.12 and 6.13 and Table 6.2.

A great deal of work has been performed over the years on what is called the *length-thickness ratio*. It has been more or less determined that a ratio of length to thickness of 150 is the ideal; however, this must be considered in general terms because of the many variables involved. Also, contradictory observations have been made, as will be mentioned later. Returning to the 1:150 length-thickness ratio, a fiber 0.001 in. (0.025 mm) thick and 0.150 in. (3.81 mm) long has about this

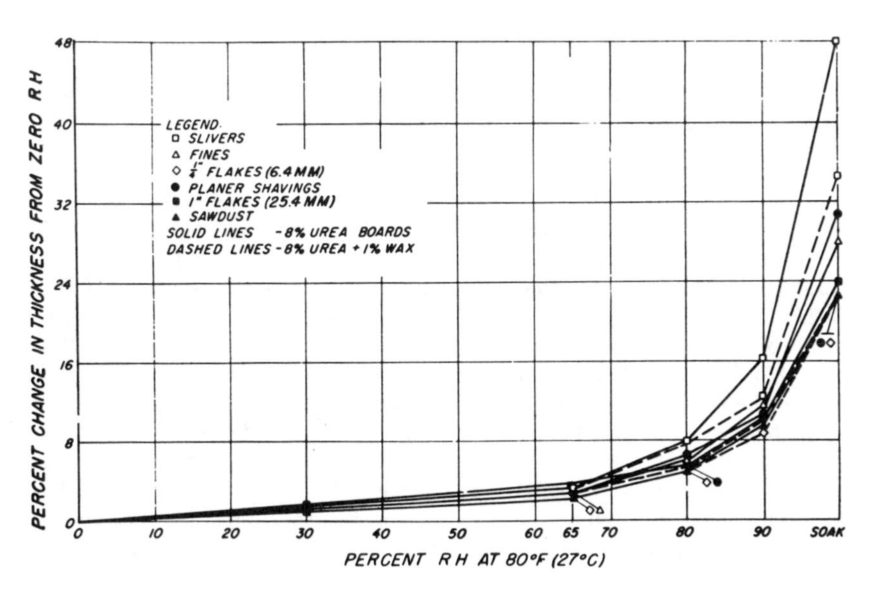

Figure 6.12. *The relation of change in thickness to percent relative humidity (Heebink and Hann 1959).*

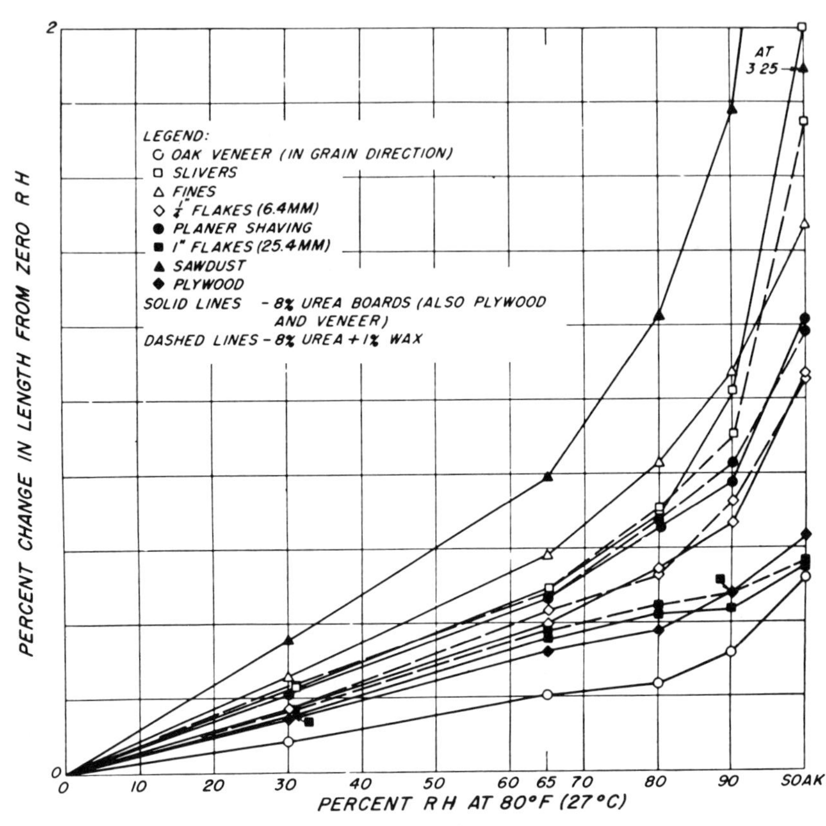

Figure 6.13. *Effects of particle shape on linear dimensional stability (Heebink and Hann 1959).*

ratio, as has a flake 0.030 in. (0.76 mm) thick and 4 ½ in. (11.4 cm) long. Figures 6.14–6.17 show some of the mechanical strength and dimensional stability results achieved in one experiment investigating four flake lengths (½ in. [12.7 mm], 1 in. [25.4 mm], 2 in. [50.8 mm] and 4 in. [101.6 mm]), four flake thicknesses (0.009 in. [0.23 mm], 0.012 in. [0.31 mm], 0.015 in. [0.38 mm] and 0.018 in. [0.46 mm]), two levels of resin content (3 and 5%) and three levels of specific gravity (0.65, 0.75, and 0.85). It was concluded that the longer the flakes, up to the 4-in. maximum, the greater the bending strength. Unlike most other findings, no significant effect of flake thickness was found in this work between 0.009 and 0.018 in. It was also concluded that internal bond is improved with shorter and thicker flakes. This may be because the shorter flakes intermesh well, and because thicker flakes, having less surface per pound, would have more resin per unit area at a given resin level. These data also showed that longer and thinner flakes improved thickness swelling and linear expansion properties after a 24-hour water soak.

When considering length-thickness ratios, particularly with flakes, a factor must be considered that can be easily overlooked. While a flake can be cut, for

Table 6.2 *Results of Particleboard Strength Tests*

Particle type	*Static bending*						*Tension perpendicular to the face*	
	Modulus of rupture		*Modulus of elasticity*		*Proportional limit*			
	(psi)	*(MPa)*	*(psi)*	*(MPa)*	*(psi)*	*(MPa)*	*(psi)*	*(MPa)*
Sawdust	1,615	11.1	250	1,724	909	6.3	307	2.12
Fines	2,402	16.6	273	1,882	1,085	7.5	345	2.38
Slivers	2,636	18.2	353	2,434	1,433	9.9	359	2.47
Slivers plus 1% wax	2,514	17.3	369	2,544	1,434	9.9	341	2.35
Planer shavings	2,891	19.9	417	2,875	1,699	11.7	286	1.97
Planer shavings plus 1% wax	2,905	20.0	466	3,213	1,784	12.3	261	1.80
¼-in.-long (12.7 mm) flake	3,425	23.6	483	3,330	1,991	13.7	377	2.60
¼-in.-long (12.7 mm) flake plus 1% wax	3,426	23.6	500	3,447	1,940	13.4	361	2.49
1-in.-long (25.4 mm) flake	6,318	43.6	744	5,129	3,908	26.9	423	2.92
1-in.-long (25.4 mm) flake plus 1% wax	6,303	43.5	691	4,764	3,389	23.4	436	3.01
Plywood made of quarter-sliced red oak:								
Parallel to face grain	11,493	79.2	1,444	9,954	5,090	35.1	319	2.20
Across face grain	7,352	50.7	724	4,991	5,090	35.1	–	–
Oak lumber (parallel to grain)	15,807	109.0	1,440	9,927	10,410	71.8	–	–

Source: Heebink and Hann 1959.

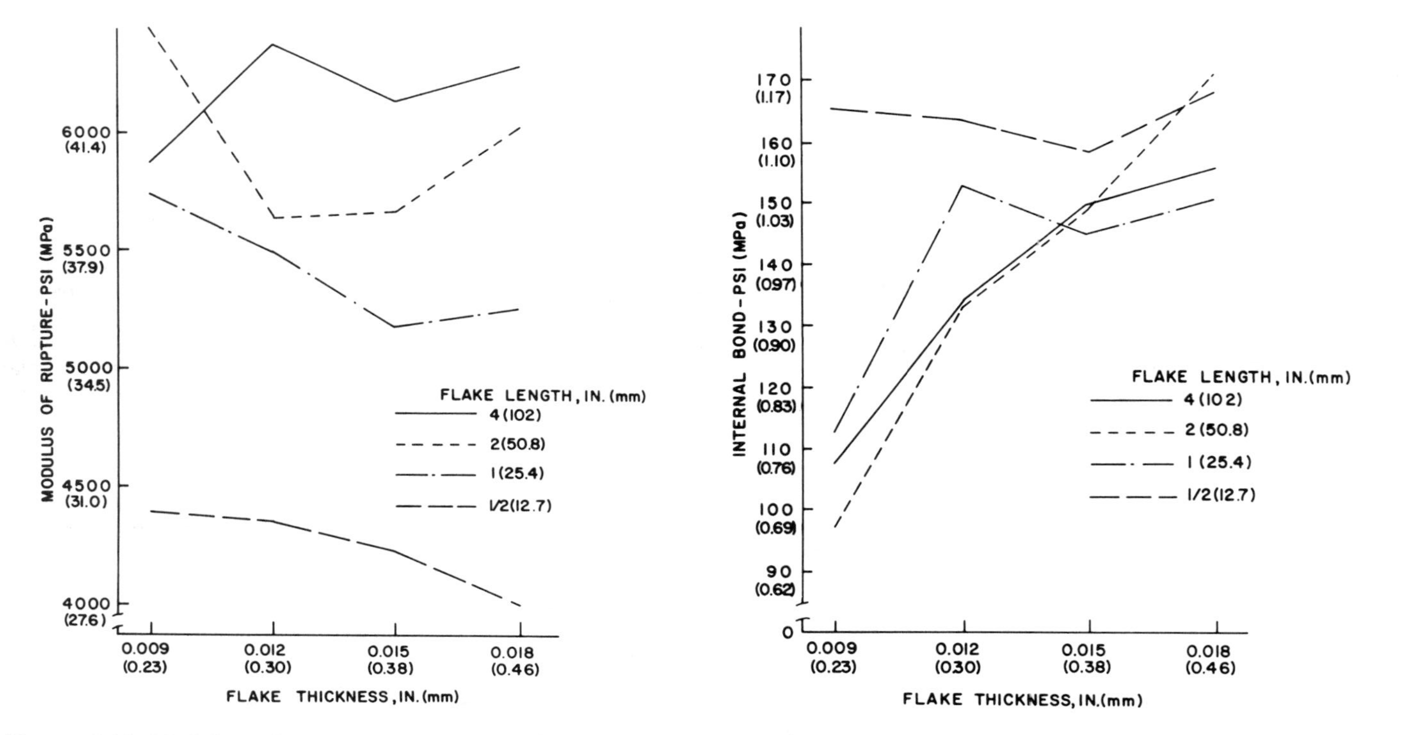

Figure 6.14. *Modulus of rupture of boards from four flake lengths and four flake thicknesses. Values averaged to 0.75 sp gr and 4% resin content (Brumbaugh 1960).*

Figure 6.15. *Internal bond strength of boards from four flake lengths and four flake thicknesses. Values averaged to 0.75 sp gr and 4% resin content (Brumbaugh 1960).*

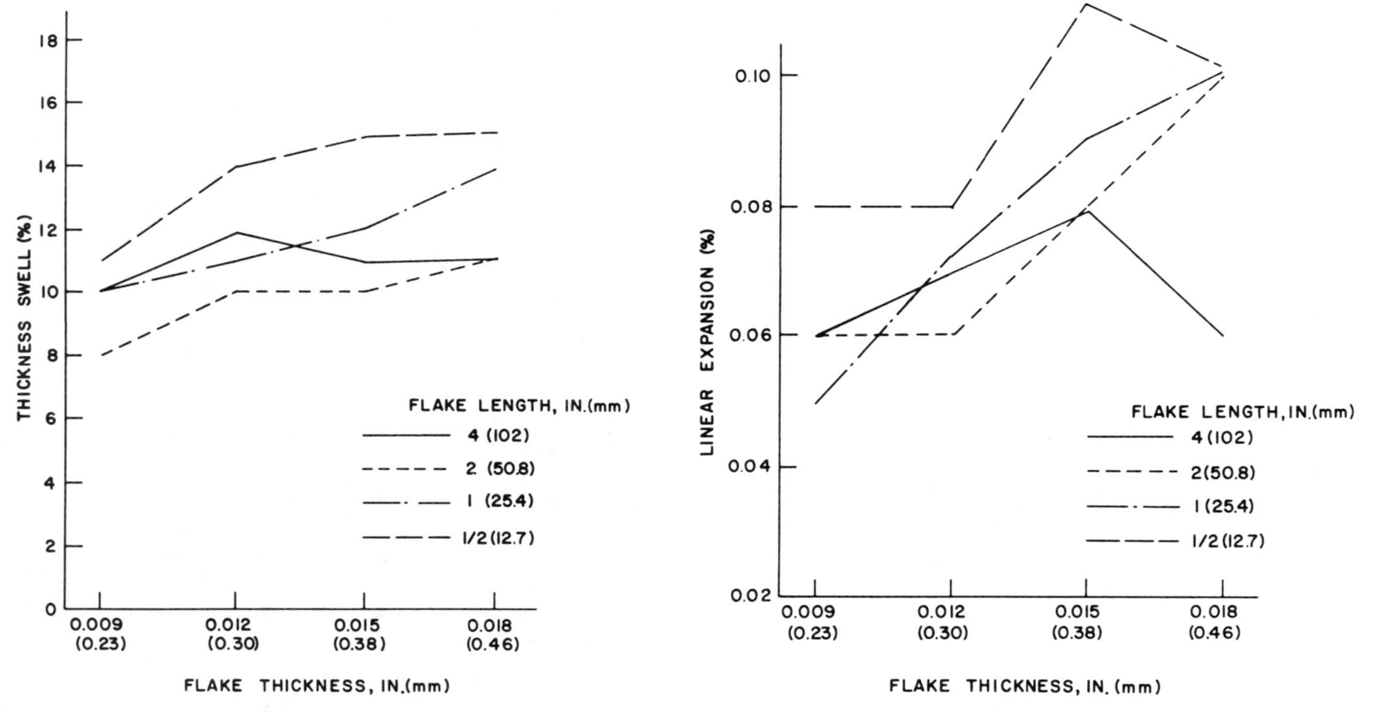

Figure 6.16. *Percent thickness swell after 24-hour water soak of boards from four flake lengths and four flake thicknesses. Values averaged to 0.75 sp gr and 4% resin content (Brumbaugh 1960).*

Figure 6.17. *Percent linear expansion after 24-hour water soak of boards from four flake lengths and four flake thicknesses. Values averaged as on Figure 6.16 (Brumbaugh 1960).*

example, 4 in. long, this does not necessarily mean it is truly 4 in. long in terms of well-aligned, strong fibers that can contribute to the bending strength and stiffness of a resultant flakeboard. A flake can contain damaged areas from the flaking operation, cross grain, compression failures from the tree itself, and other damaged fiber. In essence, such a flake will have much shorter effective length than the desired 4 in. for use in developing the best bending and stiffness properties. Thus, if a 1:150 ratio is desired, consideration must be given to developing a smaller particle that will not be so susceptible to a reduction in strength by some type of damage. All of this, of course, has to be considered in the framework of economics, but it should be apparent that it is extremely expensive to generate high-quality long flakes. A smaller particle such as a fiber may provide the same overall benefits in a finished board. However, all factors, such as increased resin cost for the fiber-type board, have to be considered.

In another study, it was shown that with increasing slenderness of flakes (length-thickness ratio), bending strength and compressive and tensile strength parallel to the plane of the board were found to increase to a certain point and then level off. The internal bond declined with the increase of slenderness within the entire range studied. Between flakes of 0.008 and 0.016 in. (0.2–0.4 mm) in thickness, the curves relating physical properties to length-thickness ratio were very similar. However, the absolute values differ considerably, indicating that there was no intrinsic significance to the ratio and that the correlation is essentially with flake length. This is in contrast with some research, mentioned previously, which considers length-thickness ratio to be of extreme importance. As the industry moves into more sophisticated properties, further work will have to be performed along this line to gather a greater understanding. At present, the wide range of particle sizes and geometries found in a given board have made precise determinations, at least from a commercial standpoint, difficult to achieve.

Figures 6.18–6.27 show the physical properties of ⅝-in.-thick (15.9 mm) boards prepared with 1-in.-long (25.4 mm) Douglas fir flakes, using 6% urea-formaldehyde resin and 2% wax size. From Figure 6.18 it is seen that, at lower specific gravities, bending strength increases steadily with increasing flake thickness and tends to level off at greater thicknesses. At the higher specific gravities, board strength goes through a maximum at around 0.012 in. (0.30 mm) flake thickness. Apparently as flake thickness increases, the decrease in specific surface and increase of resin spread per unit area results in increased strength. But at higher board densities, the factor of crushing and cross breaking of flakes comes increasingly into play as flake thickness increases, so that the benefit of increased resin spread is nullified. No doubt the optimum thickness for each board density will vary according to various factors such as wood species, wood density, moisture content of the furnish, and pressing conditions. This again illustrates that, while much is known about the composition board process, each case has to be considered individually when it comes to a particular board plant or a product to be manufactured.

With board stiffness, as shown in Figure 6.19, the tendency toward a maximum seems to exist at all levels of specific gravity, except it is more pronounced at higher specific gravities. With internal bond (Figure 6.20), the 0.012-in. (0.30 mm) flake thickness results in the highest values at all specific gravity levels. This is difficult

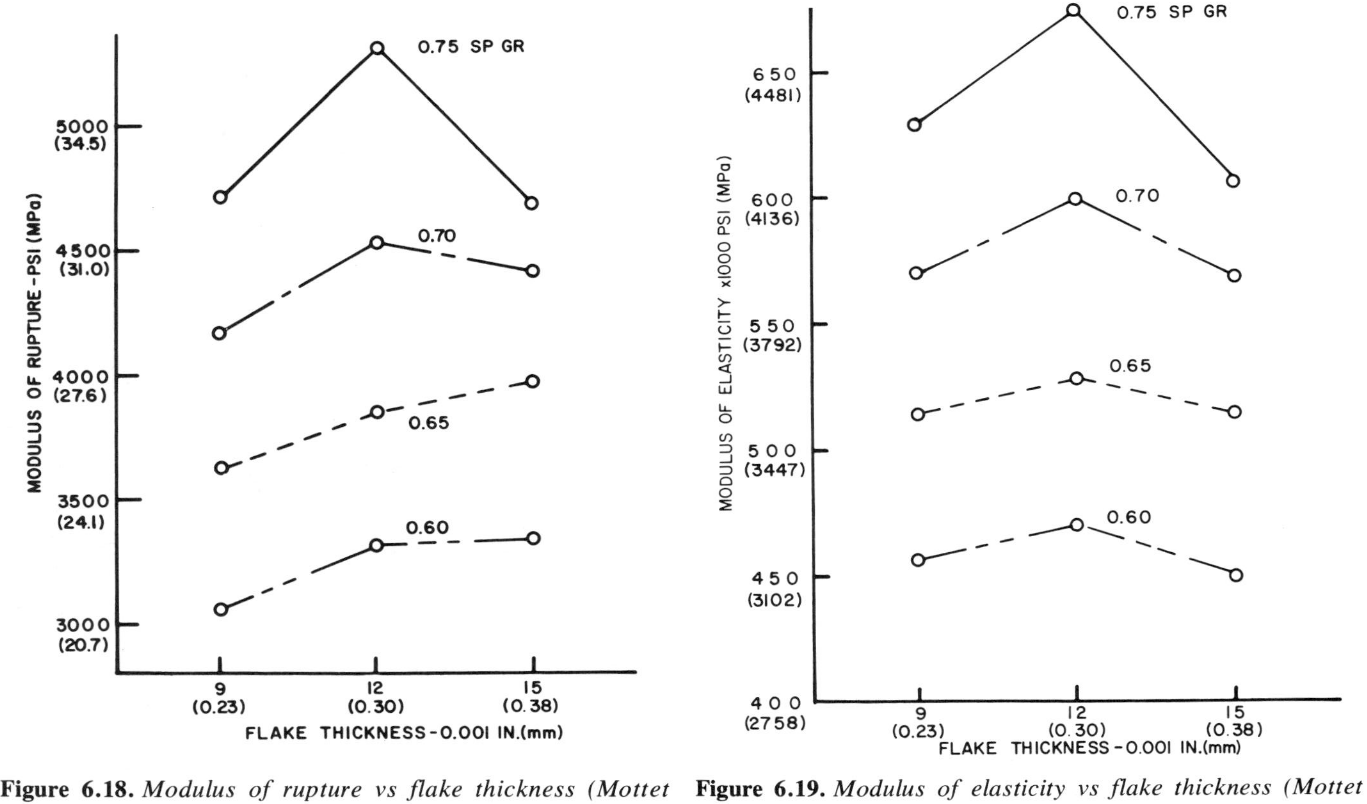

Figure 6.18. *Modulus of rupture vs flake thickness (Mottet 1967).*

Figure 6.19. *Modulus of elasticity vs flake thickness (Mottet 1967).*

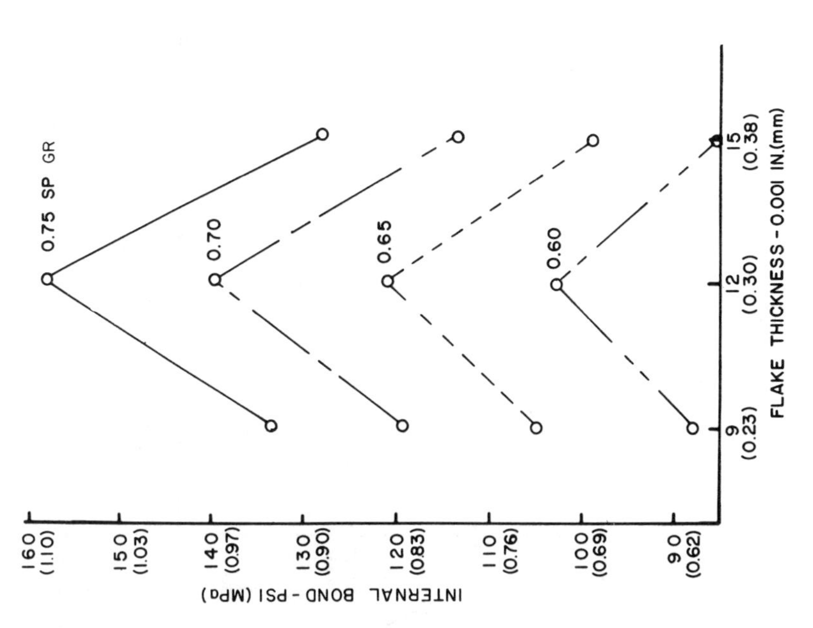

Figure 6.20. *Internal bond vs flake thickness (Mottet 1967).*

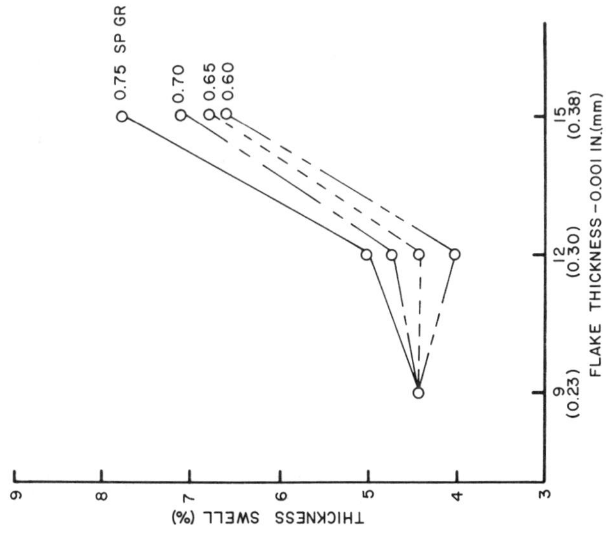
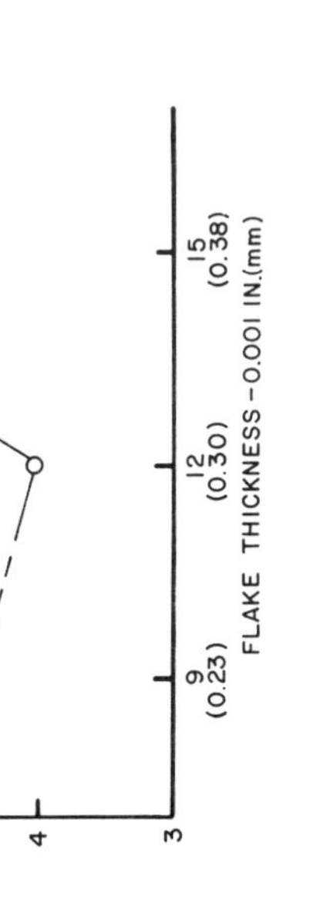

Figure 6.21. *Thickness swell vs flake thickness (Mottet 1967).*

to explain, as no effort was made to control flake width within the series, and there is a tendency for thick flakes to be wider. The increase in internal bond values from 0.009-in. (0.23 mm) to 0.012-in. (0.30 mm) flake thickness doubtless is due to the increased resin spread for unit flake surface area, while the drop-off in internal bond between 0.012 in. and 0.015 in. (0.38 mm) was due to decreased interfelting of particles perpendicular to the board surface as associated with increased flake width.

The increased resin spread per unit area illustrates another important particle ratio—that of surface area to weight. Theoretically, large particles with a low surface area to weight ratio, such as wafers, require much less resin for suitable binding than a particle such as a fiber with a high surface area to weight ratio. This is normally true; however, interactions with other board-producing parameters, different degrees of particle intermeshing, varying layer densities in the board, etc., can mask the true effect of the ratio. A case in point is pressure-refined fiber, which theoretically should require a very high level of resin; whereas in fact, it needs only the "normal" amount of about 9% in ¾-in.-thick (19 mm) board of medium density.

The relationship between flake thickness and thickness swell shown in Figure 6.21 also definitely changes with specific gravity of the board. Interestingly, at 0.009-in. (0.23 mm) flake thickness, all boards between 0.60 and 0.75 specific gravity have the same swelling value of approximately 4.5%. With thicker flakes, however, swelling of the lighter boards drops off before it starts to rise, while the swelling of the denser board rises steadily as flake thickness increases. A somewhat similar trend has been noted with water absorption. (This can be seen in Figure 6.22.

The effect of varying the proportion of fines was also determined in this experimental series. In the preparation of the flakes, all fines were removed with a minus-16-mesh screen and then varying amounts were added back to the furnish of different boards to give 5%, 10%, and 20% fines content in the furnish. Figures 6.23 and 6.24 show that both bending strength and stiffness decrease with increasing amounts of fines. Increasing fines content appears to increase internal bond a small amount in the case of 0.015-in. (0.38 mm) flakes, and has little or no effect on internal bond with 0.009-in. (0.23 mm) and 0.012-in. (0.30 mm) flakes, as shown in Figure 6.25. With thickness swell, 0.015-in. flakes which had the highest swelling tendency seem to benefit slightly by addition of fines, while no significant effect is noted with 0.009- and 0.012-in. flakes. Such a relationship also seems to prevail with water absorption, as shown in Figures 6.26 and 6.27.

A number of variables and their effect on the thickness swell of particleboard for exterior use are shown in Figure 6.28. Investigated were three flake thicknesses (0.007 in. [0.18mm], 0.015 in. [0.38 mm] and 0.030 in. [0.76 mm]), three flake lengths (½ in. [12.7 mm], 1 in. [25.4 mm] and 2 in. [50.8 mm]), one splinter board and one planer shavings board, all of Douglas fir. Other variables included resin content, wax content, board density, and mat moisture content. It was found that resin content far overshadows the particle geometry factors in permitting boards to withstand exterior exposure and accelerated aging without swelling in thickness. Planer shavings had good thickness swelling properties but were deficient in linear expansion.

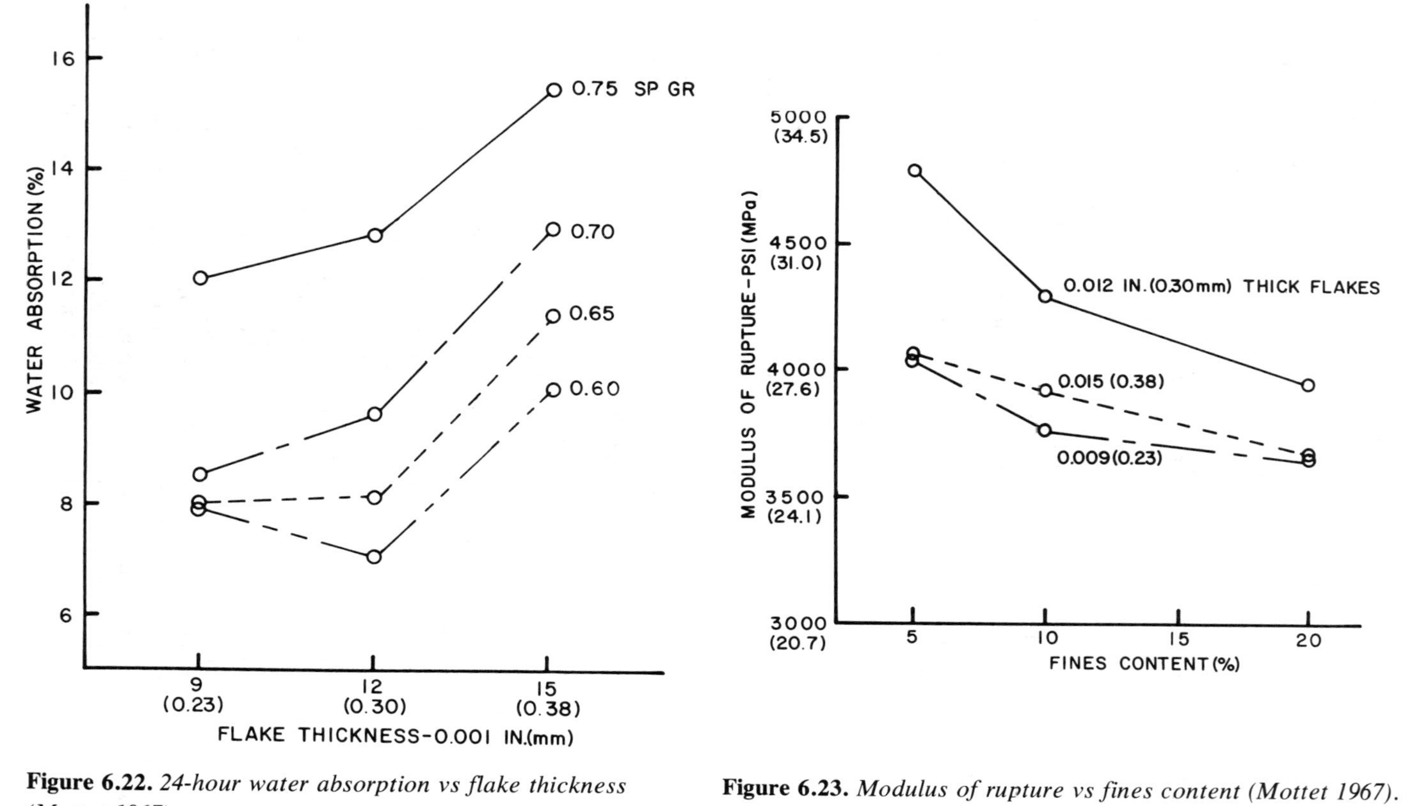

Figure 6.22. *24-hour water absorption vs flake thickness (Mottet 1967).*

Figure 6.23. *Modulus of rupture vs fines content (Mottet 1967).*

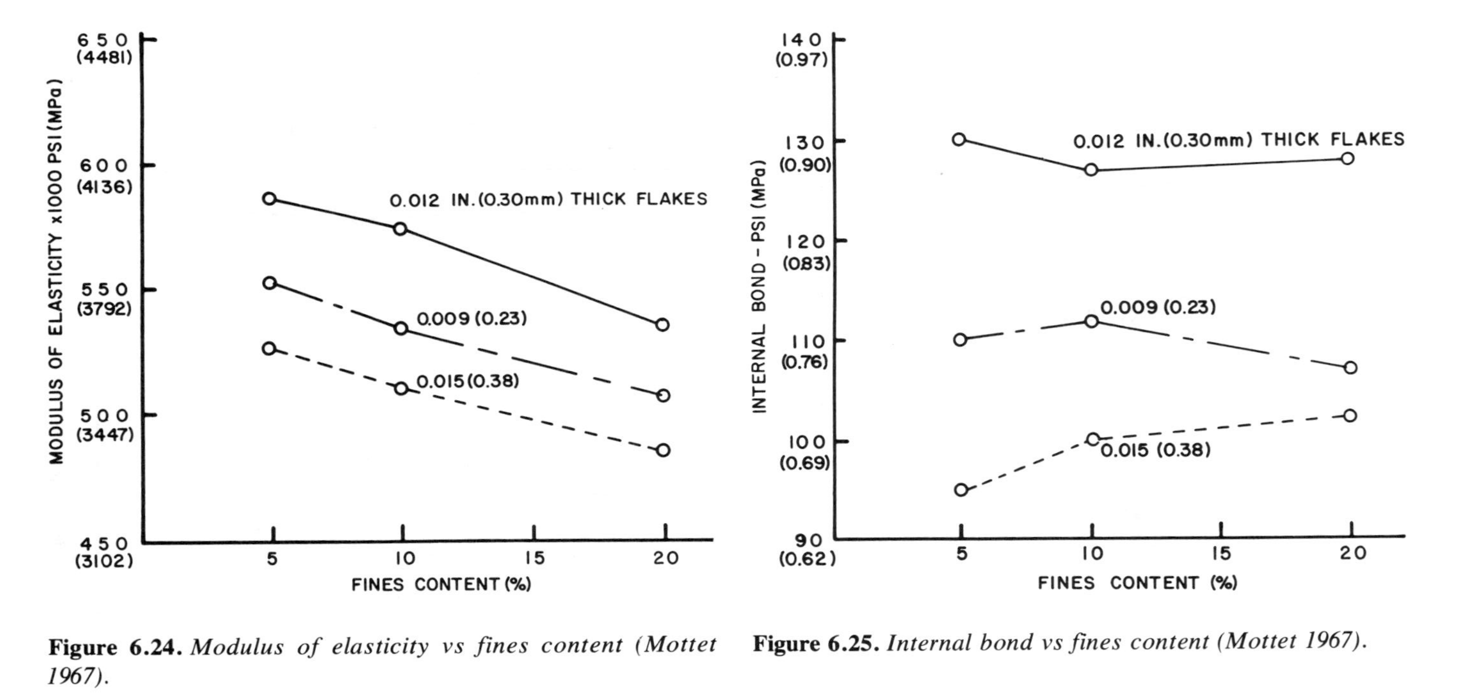

Figure 6.24. *Modulus of elasticity vs fines content (Mottet 1967).*

Figure 6.25. *Internal bond vs fines content (Mottet 1967).*

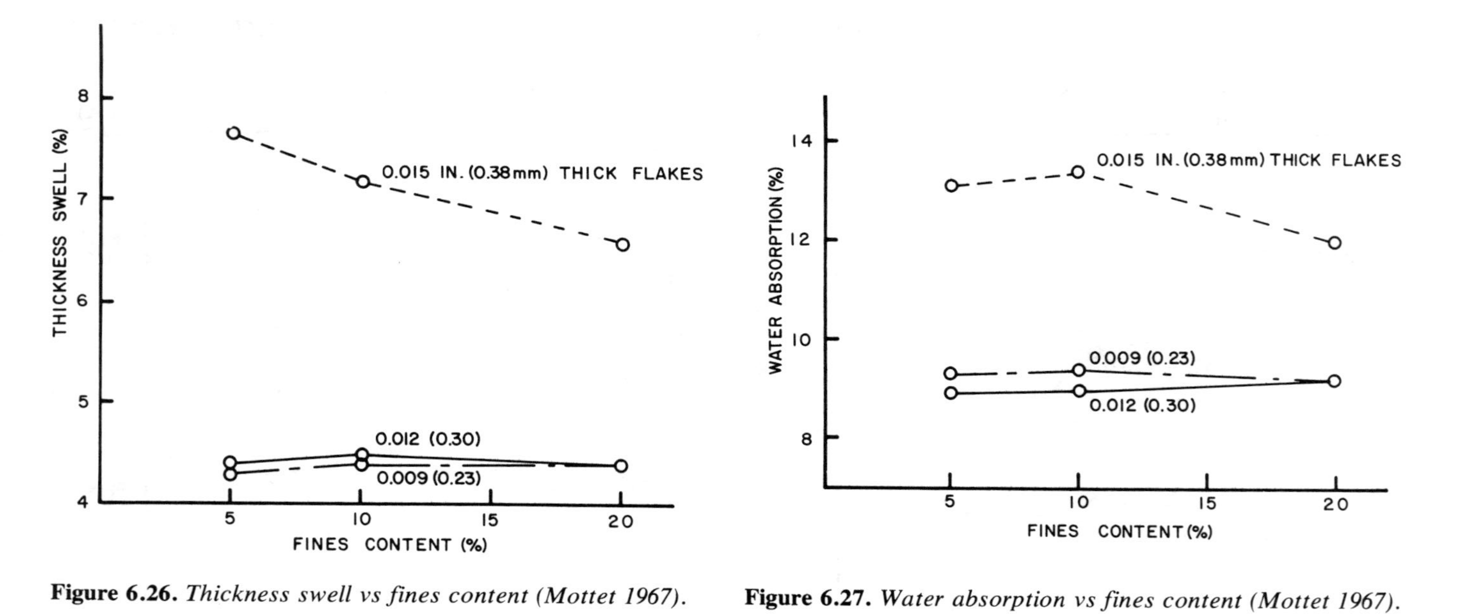

Figure 6.26. *Thickness swell vs fines content (Mottet 1967).*

Figure 6.27. *Water absorption vs fines content (Mottet 1967).*

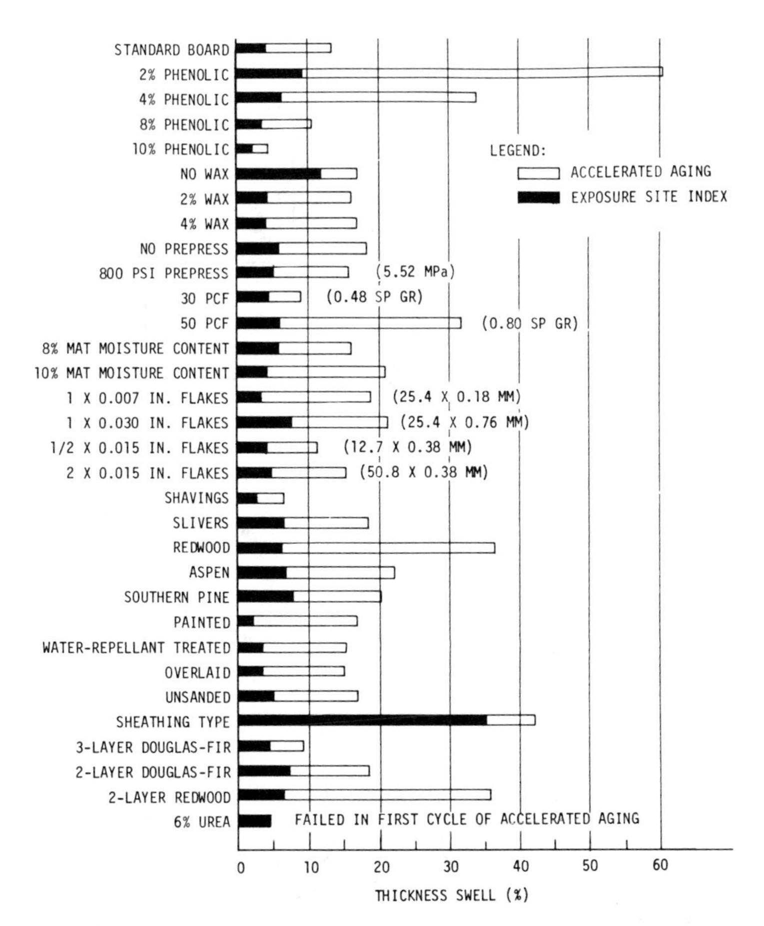

Figure 6.28. *Influence of accelerated aging and environmental exposure on thickness swell, based on measurements after equilibrating at 80°F and 65% relative humidity (Gatchell, Heebink, and Hefty 1966).*

Planer Shaving Geometry

As can be seen, there are some areas of dispute when considering the relation of flake length with thickness and their various ratios as they affect particleboard properties. Flakes are more subject to control in connection with shape and dimension, and thus it would seem that the wide variation of particle sizes and shapes found with planer shavings and their relationship with board properties would be even more complicated. Much research has been conducted on planer shavings, but only a minor part has been actually published. Much of the research has been in connection with the establishment of new board plants or through resin companies who were working on tailoring resins to the furnish of a particular particleboard plant.

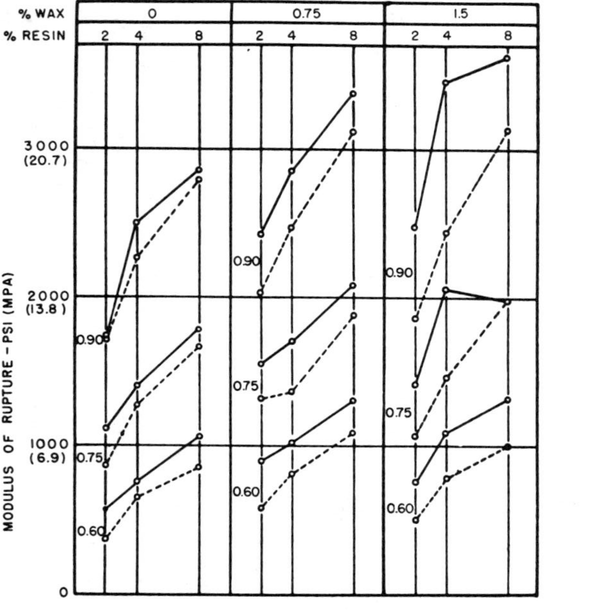

Figure 6.29. *Effect of wax and resin on modulus of rupture at specific gravity values of 0.60, 0.75, and 0.90. Full lines represent the +4 screen fraction, dotted lines the −4+16 screen fraction (Talbott and Maloney 1957).*

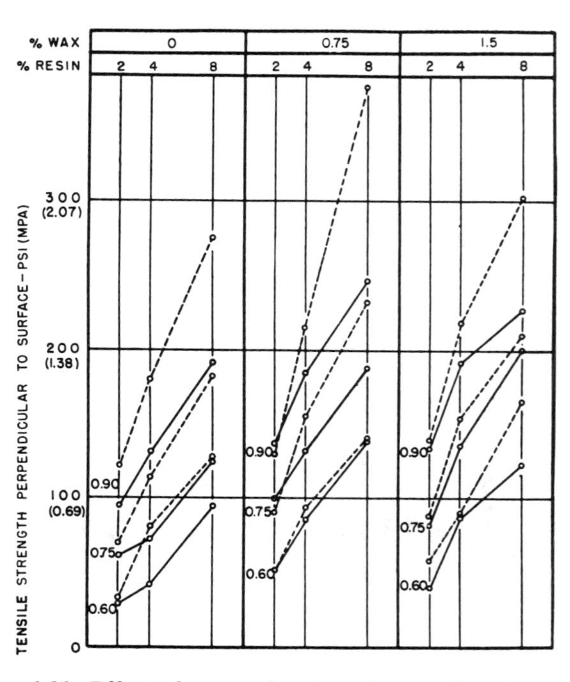

Figure 6.30. *Effect of wax and resin on internal bond at specific gravity of 0.60, 0.75, and 0.90. Full lines represent +4 screen fraction, dotted lines the −4+16 screen (Talbott and Maloney 1957).*

One research project noted that, with coarser fractions of planer shavings, higher bending strength values and lower internal bond values resulted through a wide range of specific gravities, resin contents, and wax contents, with the reverse being true for a finer fraction of planer shavings. This information is shown in Figures 6.29 and 6.30.

As mentioned previously, a newly designed planer capable of suitably planing lumber, while at the same time generating planer shavings of the proper size and geometry for particleboard, would be a welcome addition to the equipment in the wood industry. A research project was performed showing the advantages of such a development. In this research a wide variety of planer shavings was developed by varying the planing conditions, such as the effect of cutterhead diameters, cutterhead angle (between cutterhead axis and the longitudinal axis of the lumber), cutterknife rake angles, number of cuts per inch, depth of cut, knife jointing, knife face beveling, and moisture content of the wood being planed.

A very pronounced relationship was found between planer settings and conditions under which planer shavings were produced and the resulting properties of particleboards. Diameter of cutterhead, head angle, and rake angle were found particularly important in determining strength properties. The larger the head, the lower the head angle; and the greater the knife rake, the better the bending strength and linear expansion properties. Depth of cut, knife jointing, and knife face beveling were found to have a measurable, but slight, effect. Moisture content of the wood at time of planing was found the most potent variable, with green wood shavings yielding board with a quality index (a composite of bending strength and linear stability) about double that of board produced from shavings planed from wood in the 8 to 16% moisture-content range. This correlates with the industrial planer shavings used in the experiment mentioned immediately prior to this one.

Table 6.3 shows the results of an experiment using commercial planer shavings from five different species. This study evaluated white fir, Engelmann spruce, lodgepole pine, ponderosa pine, and a combination of Douglas fir and larch shavings as normally generated in the Inland Empire area of the United States. All planer shavings were $-4+16$ in screen size. Screen size means that a particle of minus dimension will pass through a screen with, in this particular case, four openings per inch and be retained on a screen with 16 openings per inch. It should be pointed out, however, that it is advisable to use a standard classification such as Tyler screens for making this determination. Even though a screen is purchased out of a hardware store with the right openings per inch, it can very well be the wrong one for making a good scientific comparison. Wires making up the screen vary in dimension tremendously. The key factor to remember is that the openings should all be of the prescribed dimension. A case in point is an order for a 10-opening screen that was placed without regard for properly specifying the screen openings, with the result that a screen was received that had openings that were closer to a 16-mesh screen than the 10-mesh screen specified.

Returning to the discussion on the comparison of these various commercial planer shavings, the Engelmann spruce and white fir had more of the larger size planer shavings and these had a flakelike geometry. As a consequence, the bending strength of these particular particleboards was superior to that of the other species made up of shavings that were more cubical in shape and with a higher

Table 6.3 *Average Properties of Homogeneous Planer Shavings Board*

Property		*Spruce*	*White fir*	*Ponderosa pine*	*Lodgepole pine*	*D. fir-larch mix*	*L.S.D.* at 95%*
Modulus of rupture of wood†	(psi)	8,700	9,300	9,200	9,400	13,000	
	(MPa)	60	64.1	63.4	64.8	89.6	
Sp gr of wood†		0.34	0.40	0.40	0.41	0.52	
Avg sp gr of boards		0.64	0.65	0.66	0.62	0.65	
Modulus of rupture	(psi)	2,030	2,180	1,750	1,830	1,680	215
	(MPa)	14.0	15	12.1	12.6	11.6	
Internal bond	(psi)	191	175	159	172	187	21
	(MPa)	1.32	1.21	1.10	1.19	1.29	
Thickness swell 24-hr water soak responses	(%)	7.7	7.6	6.1	7.1	4.7	1.25
Linear expansion 24-hr water soak responses	(%)	0.30	0.33	0.28	0.29	0.29	
Water absorption 24-hr soak responses	(%)	18.1	29.8	16.7	15.6	11.9	4.09

Source: Maloney 1956.

* Least significant difference.

† Average values at 12% moisture content, from "Wood Handbook."

percentage of smaller particles. Ponderosa pine shavings, in particular, had a very high percentage of very small particles and very few large particles, which was the opposite of the Engelmann spruce and white fir planer shavings. This is generally true in practice, with the ponderosa pine planer shavings tending to be small particles. As such, they are ideal for surface particles on particleboards that are being produced with very fine, smooth surfaces. However, because of this fine nature, and possibly resins in the wood, they tend to compress tightly together on the surfaces, producing what is known as a *tight face*, which makes is difficult to breathe vapor through the face of the board during the pressing operation. As will be discussed later, such particleboards have a tendency to blow or delaminate, and special care must be taken to prevent this. Figure 6.31 is a set of curves comparing the shavings by size in the −4+16 screen size range showing the variation between species within a relatively narrow spectrum.

In this particular study, Engelmann spruce and ponderosa pine were also evaluated for use in particleboard manufacture by separating the −4+16 screen fraction into five divisions. Boards were then produced at various resin levels and specific gravities and the results compared. The particle geometry, as mentioned, was shown to have dramatic effect on the board properties, indicating that if these species which were being considered for a board plant were indeed to be used, extreme care would have to be taken in the way each species was metered into the process. The flaky spruce particles assisted in developing better bending strength, while the chunky pine particles provided higher internal bond since they flowed together better and probably developed edge as well as face bonds. The low values in bending and bonding observed with −4+6 pine shavings were attributed to the fragile nature of the relatively thin particles. It appeared that they were damaged in planing and did not possess much inherent strength. These data are shown in Figures 6.32 and 6.33.

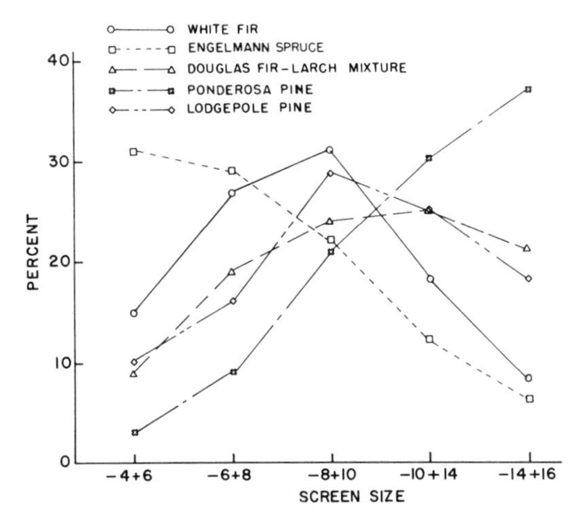

Figure 6.31. *Distribution of particle sizes found in commercial planer shavings, −4+16 screen size of five species (Maloney et al 1956).*

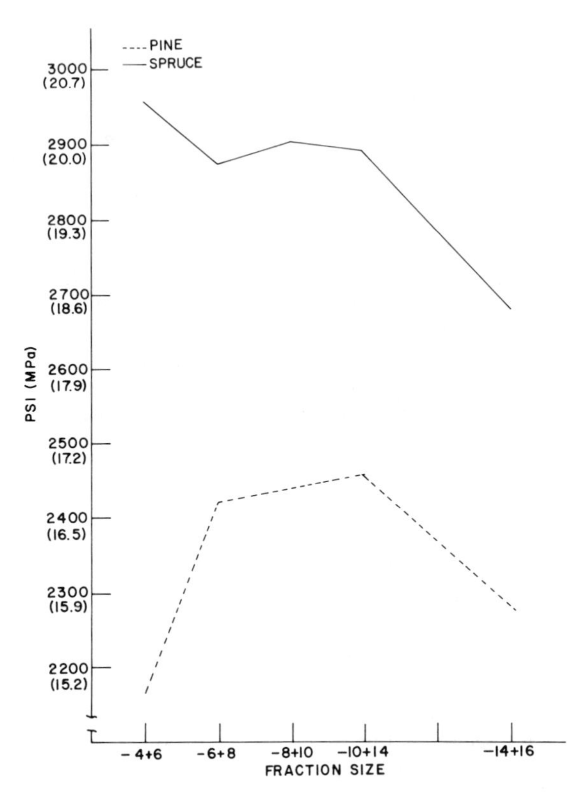

Figure 6.32. *Modulus of rupture vs screen fraction size for ponderosa pine and Engelmann spruce (Maloney 1958).*

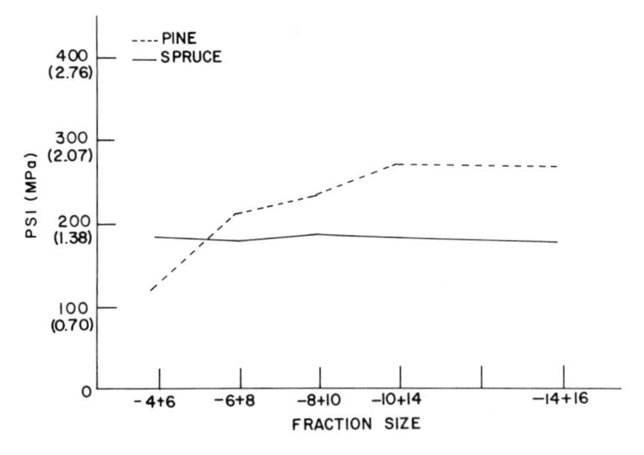

Figure 6.33. *Internal bond vs screen fractions size for ponderosa pine and Engelmann spruce (Maloney 1958).*

Sawdust Geometry

A significant amount of sawdust, particularly green sawdust, is used in board production. Green sawdust is usually much larger in size than dry sawdust. For fiberboards, this particular type of raw material, if it is ground by atmospheric refiners, provides a suitable fiber. The larger particles placed into the particleboard process normally must be refined to some extent, so that incipient fractures within the sawdust particles caused by the sawing action are eliminated. In essence, a large sawdust particle usually is composed of a number of smaller particles. Once these particles are broken apart, they can usually be effectively used in particleboard manufacture.

Figure 6.34 shows the general results obtained in a study using gang sawdust in the manufacture of particleboard. The material is described by means of a screen analysis, and board was produced with sawdust and flakes alone and with various mixtures of flakes or with an overlay of flakes. It is seen that the modulus of rupture of sawdust boards is increased in a regular way as flakes are added. However, in another series it was shown, as might be expected, that putting the flakes on the board surfaces contributed much more to bending strength and stiffness. A board with a straight sawdust furnish had a bending strength of 1330 psi (9.2 MPa) at 0.60 sp gr, one containing 30% flakes in admixture had a modulus of rupture of 2390 psi (16.5 MPa) and one with 30% flakes as an overlay had a modulus of rupture of 3360 psi (23.2 MPa). In comparison, the modulus of rupture of a board with 100% flakes was 4500 psi (31 MPa). It has been mentioned that long, thin particles, such as flakes, provide better resistance to the compressive and tensile stresses developed in a board in use than other particles, such as shaving or sawdust. When the flakes are used as an overlay in boards, excellent use is made of their particle geometry in developing superior bending strength and stiffness. The flakes are where fiber stress is at the maximum, which is the farthest from the board's neutral axis.

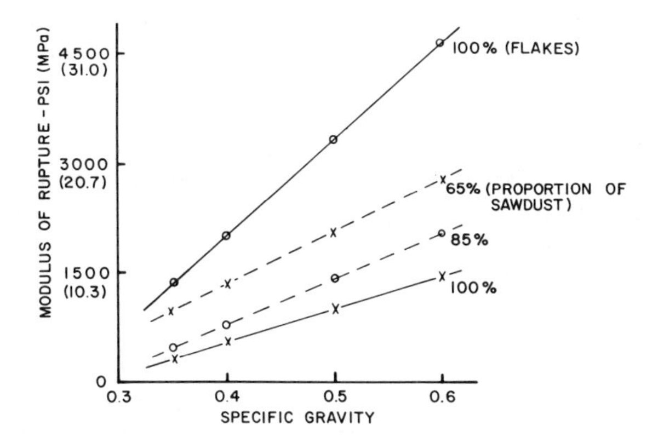

Figure 6.34. *Bending strength (modulus of rupture) of particleboards made from a mixture of sawdust and flakes as dependent on the specific gravity of the boards (Klauditz and Buro 1962).*

This also illustrates the problem of preparing a furnish with an admixture of a high-quality product for bending purposes, such as flakes. In order to be completely effective, the flakes have to be in contact with each other to develop the superior bending strength desired. If the flakes end up being isolated and surrounded by the other material, which in this case was sawdust, the flake cannot provide the ultimate in assistance to the bending strength. This same thought should be borne in mind when considering adding low-strength particles such as those found from bark of species like Douglas fir. With the addition of too much of such a low-strength particle, the bending strength, and for that matter, most other properties of the board take a tremendous drop.

Fiber Geometry

Fibers are always generated from some other particle before use in board. This type of particle is usually produced in attrition mills or refiners. The nature of the particle obtained is greatly influenced by the nature and condition of the raw material itself, its moisture content, the type of the grinding plates, plate clearance, size of the attrition mill, whether or not the raw materials is steamed before grinding, and many other operating variables which will be discussed later. The material is usually much finer than can be obtained by other means and, when used in particleboards, is usually used as overlay particles in multi-layer or three-layer boards. However, fiber is now being used extensively throughout some types of particleboard. In the fiberboard portion of the composition board industry, the fiber, of course, is the dominant particle with variations in size ranging from a relatively large fiber down to a very fine, almost dustlike, particle used for faces.

It is very difficult to measure the geometry of fiber. Flake-type particles can be measured with micrometers and rules. Planer shavings can be measured in a similar way and also can be classified with the screening systems available. Screening systems can also be used with fiber, but it is very difficult to use such information except as a quality-control tool within a particular plant. The popular method, at the present time, for measuring fiber is to fill loosely a one-cubic-foot container with the fiber and weigh it to obtain the bulk density, which for a high-quality fine fiber may be as low as 1 lb/ft^3. A lower-quality and heavier fiber would be in the range of 4 lbs/ft^3. A qualification should be made here as to the description of the fiber as far as quality is concerned. In some cases 4 lbs/ft^3 fiber may be the quality of fiber necessary to produce the product desired, whereas a 1 lb/ft^3 fiber may be undesirable. This method is, however, a relatively easy and quick way to measure fiber quality, and experienced individuals can relate this information to the process line under consideration.

Summary

This brief discussion has shown that the factor of particle geometry is an extremely complex one, affecting board characteristics in a wide variety of ways and interacting with almost every other important factor in board manufacturing. Such complexity makes it difficult, if not impossible, to "engineer" an ideal board furnish from most of the available raw material supply to fit specified requirements. Preparation of board furnish, therefore, is still largely an art in the United States because of the wide variety of raw material. Plants such as those located in New Zealand and Australia, which operate on a single species and cut flakes and other particles from solid wood, have the best opportunities to "engineer" optimum particles.

It is possible from the information at hand to state some broad generalities that can serve as a guide to plant operators and researchers.

There are three major particle types important at present in the field of board manufacturing: (1) specially cut flakes, (2) mildly refined planer shavings and similar comminuted forms of wood residue, and (3) attrition-milled fiber. These differ radically in their geometric character and effect on board properties. Some sawdust and sander dust is also used.

The bending strength and stiffness of flakeboard tend to increase with increasing flake thickness throughout the practicable flake thickness range for board at low and medium densities, but tend to pass through a maximum at some intermediate flake thickness for high-density boards when all other factors, including resin binder percentage by weight, are held constant. The exact values can vary depending on other production factors.

The bending strength and stiffness of flakeboard tend generally to increase with increasing flake length until a certain value is reached, and then to level off. However, in some instances, there is a tendency to reach a maximum and to drop off with further increases in flake length, due possible to faulty mat formation with longer flake lengths, faulty blending, or defects in the flakes that negate the potential value of their longer length.

The relationship between internal bond and flake thickness is somewhat ambiguous. In some cases, there is an indication that internal bond increases with increasing flake thickness until a maximum is reached, and then drops off with further thickness increase. In other cases, the internal bond curve levels off with further flake thickness increase. It should be pointed out that, with flakes, resin spread per unit surface area is directly proportional to flake thickness if resin percentage is constant. Internal bond would thus be expected to go up with flake thickness but can go no higher than the internal bond of the flakes themselves. The flake thickness where board internal bond levels off should thus be function of resin percentage.

The internal bond of flakeboard tends to decrease with increasing flake length throughout the practicable flake-length range.

The linear expansion of flakeboard tends to increase with increasing flake thickness throughout the practicable flake-thickness range and to decrease with increasing flake length.

The thickness swell of flakeboard tends to increase with increasing flake thickness except with low board densities. At low densities (below 0.64 sp gr or 40 lbs/ft^3) it appears there is an optimum at some intermediate flake thickness.

The conditions under which flakes are cut, i.e., disc versus drum cutterheads knife speed, knife angle, knife sharpness, etc., have significant effect on board properties.

With mixtures of particles having widely differing geometrical characteristics, particularly flakes and shavings, the strength properties tend in general to be intermediate between the properties obtained with the different kinds of particles separately. The moisture response properties follow this same pattern to a lesser degree.

Flakes tend to produce board with higher bending strength and stiffness, higher linear stability, lower internal bond, and lower thickness stability as compared with boards made from planer shavings. Screw-holding characteristics at a given board density are approximately the same for the two types of particles.

Particles having a length-thickness ratio of about 150 appear to be best for developing boards with the best bending and stiffness properties.

Incorporation of 5 to 20% of fines (below 30 mesh in screen size) with coarser flakes or shavings tends to reduce water absorption and thickness swell, but also reduces bending strength and stiffness. Such boards have smaller interparticle voids or passageways, and thus water absorption is slowed in a 24-hour test. Long-term testing could show little difference between these two types of board.

Conditions of production, i.e., planing conditions and sawing conditions, in connection with planer shavings and sawdust have pronounced effect on properties of board. Moisture content of the wood is of special importance in this regard.

The effect of planer shaving and sawdust geometry on particleboard properties is complex. From empirical field experience, plant personnel have been able to use planer shavings advantageously or have been able to overcome, to some extent, deleterious effects on board properties caused by poor planer shaving geometry. Shavings with higher length-thickness ratios yield boards with the better flexural properties. Blocky or cubical particles intermesh better and, with more edge-to-edge bonding, yield boards with better internal bonds.

The information on the effect of fibrous furnish geometry on board properties is somewhat limited. While it is anticipated that the length-thickness ratio of fiber would have about the same effect on board properties as does that of particulate and flake elements, further research is needed to understand thoroughly fiber geometry and its effect on properties.

SELECTED REFERENCES

Brumbaugh, J. 1960. Effect of Flake Dimensions on Properties of Particleboards. *Forest Products Journal,* Vol. 10, No. 5.

Currier, R. A., and W. F. Lehmann. 1975. Particleboard from Logging Residues, Bark and Other Sources. *Proceedings of the Washington State University Particleboard Symposium, No. 9.* Pullman, Washington: Washington State University (WSU).

Dost, W. A. 1971. Redwood Bark Fiber in Particleboard. *Forest Products Journal,* Vol. 21, No. 10.

FAO. 1976. Basic Paper II. Raw Materials for Wood-Based Panels. *Proceedings of the World Consultation on Wood-Based Panels,* held in New Delhi, India, February 1975. Brussels: Published in agreement with the Food and Agriculture Organization of the United Nations by Miller Freeman Publications.

Gatchell, C. J.; B. G. Heebink; and F. V. Hefty. 1966. Influence of Component Variables on Properties of Particleboard for Exterior Use. *Forest Products Journal,* Vol. 16, No. 4.

Heebink, B. G. 1972. Irreversible Dimensional Changes in Panel Materials. *Forest Products Journal,* Vol. 22, No. 5.

Heebink, B. G., and R. A. Hann. 1959. How Wax and Particle Shape Affect Stability and Strength of Oak Particleboards. *Forest Products Journal,* Vol 9, No. 7.

Heebink, B. G.; R. A. Hann; and H. H. Haskell. 1964. Particleboard Quality as Affected by Planer Shaving Geometry. *Forest Products Journal,* Vol. 15, No. 10.

Himmelheber, M., and W. Kull. 1969. Die Moderne Spanplattenfertigung: Rohstoffeinsatz und Plattenqualität (Modern Particle Board Manufacture: Raw Materials and Board Quality). *Holz als Roh-und Werkstoff,* Vol. 27, No. 11.

Iwashita, M. 1976. Tropical Hardwood from Papua, New Guinea, as Raw Materials for Particleboard. *Proceedings of the Washington State University Particleboard Symposium, No. 10.* Pullman, Washington: WSU.

Klauditz, W., and A. Buro. 1962. The Suitability of Sawdust for Particleboard Manufacture. *Holz als Roh-und Werkstoff,* Vol. 20, No. 1.

Koplow, R. A. 1976. Assuring the Fiber Supply for Composition Board Plants. *Proceedings of the Washington State University Particleboard Symposium, No. 10.* Pullman, Washington: WSU.

Lehmann, W. F. 1972. Moisture-Stability Relationships in Wood-Base Composition Boards. *Forest Products Journal,* Vol. 22, No. 7.

Lengel, D. E. 1962 (revised 1970). A Report on the Results of Investigations to Determine Optimum Methods of Producing Bagasse Fiber Boards in the Softboard, Particle Board, and Hardboard Density Ranges. Healdsburg, California: Rivenco. (Originally published in the *Proceedings of the 11th Congress of the International Society of Sugar Cane Technologists,* Mauritius. New York: Elsevier Publishing Co.)

———. 1969. Some New Technical and Economic Aspects of Bagasse Fiber Utilization. Healdsburg, California: Rivenco (January).

Lutz, J. F.; B. G. Heebink; H. R. Panzer; F. V. Hefty; and A. F. Mergen. 1969. Surfacing Softwood Dimension Lumber to Produce Good Surfaces and High Value Flakes. *Forest Products Journal,* Vol. 19, No. 2.

Maloney, T. M. 1958. Effect of Englemann Spruce and Ponderosa Pine Dry Planer Shaving Particle Size and Shape on the Properties of Particle Board. Washington State University Work Order No. 897. Pullman, Washington.

———. 1973. Bark Boards from Four West Coast Softwood Species. *Forest Products Journal,* Vol. 23, No. 8.

———. 1974. The Effect of Changing Raw Materials on Composition Board Production. Presented at the Joint Meeting of the Forest Products Research Society and the Society of American Foresters. (May), Vancouver, Washington.

Maloney, T. M.; J. W. Talbott; and G. G. Marra. 1956. Effect of Species on the Physical Properties of Particleboard Composed of Dry Planer Shavings. Washington State University Work Order No. 865. Pullman, Washington.

Maloney, T. M., and A. L. Mottet. 1970. Particleboard. *Modern Materials.* Vol. 7, New York: Academic Press Inc.

Maloney, T. M.; J. W. Talbott; M. D. Strickler; and M. T. Lentz. 1976. Composition Board from Standing Dead White Pine and Dead Lodgepole Pine. *Proceedings of the Washington State University Particleboard Symposium, No. 10.* Pullman, Washington: WSU.

Mottet, A. L. 1953. Crosscut Fiberboard and Method for Its Production. U.S. Patent Application No. 361,660 (abandoned before issue).

———. 1958. Crosscut Fiberboard and Method for Its Production. Canada Patent No. 565, 840.

———. 1967. The Particle Geometry Factor in Particleboard Manufacturing. *Proceedings of the Washington State University Particleboard Symposium, No. 1*. Pullman, Washington: WSU.

———. 1975. Use of Mill Residues and Mixed Species in Particle Board Manufacture. Background paper No. 59, third World Consultation on Wood-Based Panels, FAO, New Delhi, India, February 1975.

Post, P. W. 1958. Effect of Particle Geometry and Resin Content on Bending Strength of Oak Flake Board. *Forest Products Journal,* Vol. 8, No. 10.

Schulz, H. 1975. Raw Materials for Wood-Based Panels in Europe. Background Paper No. 46, third World Consultation on Wood-Based Panels, FAO, New Delhi, India, February 1975.

Snellman, E. J. 1973. Experiences Using Bark in Particleboard Manufacturing. *Proceedings of the Washington State University Particleboard Symposium, No. 7*. Pullman, Washington: WSU.

Steiner, K., and H. Sybertz. 1975. Technical and Economic Aspects of Use of Thin Flakes in Structural Flakeboard. Background Paper No. 56, third World Consultation on Wood-Based Panels, FAO, New Delhi, India, February 1975.

Steward, H. A., and W. F. Lehmann. 1973. High-Quality Particleboard from Cross-Grain, Knife-Planed Hardwood Flakes. *Forest Products Journal,* Vol. 23, No. 8.

Suchsland, O. 1972. Linear Hygroscopic Expansion of Selected Commercial Particleboards. *Forest Products Journal,* Vol. 22, No. 11.

Talbott, J. W., and T. M. Maloney. 1957. Effect of Several Production Variables on Modulus of Rupture and Internal Bond Strength of Board Made from Green Douglas-Fir Planer Shavings. *Forest Products Journal,* Vol. 7, No. 10.

Turner, H. D. 1954. Effect of Particle Size and Shape on Strength and Dimensional Stability of Resin-Bonded Wood-Particle Panels. *Journal of the Forest Products Research Society,* Vol. 4, No. 5.

Van Sickle, C. C. 1975. Raw Materials for Wood-Based Panels Produced in North America. Background Paper No. 76, third World Consultation on Wood-Based Panels, FAO, New Delhi, India, February 1975.

Woodson, G. E. 1976. Fiberboard from Southern Hardwoods. *Proceedings of the Washington State University Particleboard Symposium, No. 10*. Pullman, Washington: WSU.

7. PARTICLE GENERATION, CONVEYING, AND STORAGE

As we have seen, some particles are produced elsewhere, particularly in the case of planer shavings, where much of the material as received can be used in the composition board plant. The same is true with some of the sawdust and some of the particles from other millwork waste. Very small particles must be accepted in the shape received as little milling can be performed. However, other previously comminuted particles can be cut and refined to the advantage of the board properties. Costs are low in handling particle generation in this particular situation, but the particle geometry has for the most part already been dictated by other woodworking processes.

On the other hand, a number of particles can be generated from the larger shavings and other material listed under "wood raw materials." The opportunity for producing "engineered" particles, then, is at hand. The best type of particle or particles needed for a particular product can then be prepared. Several different machines are available for this operation, ranging from chippers and hogs for reducing larger pieces to attrition mills for generating fiber. Hammermills, ring flakers, attrition mills and ring mills can be used for producing smaller particles or fibers out of shavings and chips for use in particleboard. Ring flakers, disc flakers, and drum flakers produce flakes. Excelsior machines are used for generating the strands necessary for excelsior or wood wool products.

Each particle has its place in the scheme of production. Fiberboard plants will exclusively use attrition mills. Flakeboard plants will concentrate on one of the various types of flakers. Standard particleboard plants based on granular material will use combinations of hogs, chippers, hammermills, ring flakers, ring mills, and attrition mills; and if they wish to move into higher-strength products, some type of sophisticated flaker or pressurized refiner may be part of the equipment.

Particle generation used to be a relatively simple concept as far as types of machines were concerned. Flaking, hammermilling, and attrition milling covered the types of generation being performed. Nowadays, these same general types of machines are used, but there is a wide variation within each grouping as well as newer developments. Machines range from chippers, hogs, and cuttermills for rough particle generation through hammermills for either rough or final particle preparation to drum flakers, disc flakers, knife-ring flakers, wing-beater mills, turbomills, counter-rotating mills, cross-stream impact mills, double-stream ring refiners, counter-rotating double-stream ring refiners, atmospheric attrition mills, and pressurized attrition mills. Such a variety of machines leaves the mind reeling over which one or ones should be employed in a board plant.

The raw material for the board industry, as already noted, can be roundwood, slabs, edgings, end trim, shavings, sawdust, annual plants, etc. The generating conditions vary widely according to size, wood species and moisture content. To obtain particleboards with good strength, smooth surfaces, and equal swelling, manufacturers ideally want a homogeneous material with a high degree of slenderness (long, thin particles), no oversize particles, no splinters and no dust. Depending on the manufacturing process, the specifications for the ideal particle size are different. For a graduated board, wider tolerances are acceptable; for a three-layer board, the core particles should be longer and surface particles shorter, thinner, and smaller; for a five-layer or multi-layer board, the furnish for the intermediate layer between surface and core should have long and thin particles for building a good carrier for the fine surface and to give the boards high bending strength and stiffness. This is particularly important as the fine surface particles on the top of the mat will sift through towards the core before pressing is completed, leaving a rough texture rather than smooth surface.

Fibrous material for hardboard or medium-density products is usually generated as a homogeneous blend and is used this way, or the finer material can be separated or ground specifically for use in the board faces.

In considering the various machines available, care must be exercised to choose the one or the combination that will economically produce the desired particles. Energy requirements, maintenance, ruggedness, flexibility, capacity, capital cost, and operation cost all must be considered.

The machine or combination of machines must do the best possible job of particle generation. Few oversize or undersize particles should be tolerated. Conditions change in our rapidly changing world, including the type of raw material available. Thus, flexibility is needed in machines to handle these changes as well as those necessary on the production line for shifting to a different product.

Capacity depends on the type of infeed material, its moisture content, and the type of particle desired. The capacity must be sufficient to take into account normal wear in the machine. As parts become worn in the machine, the capacity as well as the quality of particle may decrease, because no machine remains the same as it was new without a major overhaul.

Uniform and controlled feeding systems are needed with, perhaps, the exception of chippers and hogs. Many do not realize the importance of a uniform feed as related to the quality of the final product in particle-generating machines. Vibrators, double-screws, and high-speed single-screws with a short pitch are used to provide uniform feeds. Flight conveyors, bucket conveyors, and slow-turning single-screw conveyors provide pulsating feeds which are anathema to quality particle generation.

The machine price and the investment cost for a complete particle-generating installation should be considered. This includes the machine itself with all accessories, such as drives, safety devices, motors with slide rails, switch gear, cables, feeding equipment, foundations, supports, discharge transport, erection, and freight. The total investment should be in reference to the capacity as this will give a specific figure in dollars per ton of hourly capacity.

Besides the depreciation, and most such equipment is relatively short-lived, the operating costs are mainly for labor, electric power, and wear. With good control equipment the installation should work without an operator, although some means of observing the complete installation is necessary. Machines having knives, however, need men for knife changes, knife setting, and knife sharpening. When calculating power cost, the installed motor size is less important. The actual metered power consumption for a complete installation should be taken into account, however, including all necessary auxiliary equipment and transport systems. An accurate figure, however, is often not obtainable before starting the installation.

Larger motors are preferable, in North America in particular, not only because they have more spare capacity for overcoming short-time overloads, but because the plants are usually operating well over the designed capacity.

In the following discussion, various types of particle-generation equipment will be examined with special emphasis on the more important pieces of equipment.

HOGS

Hogs are relatively crude machines for breaking larger pieces of residue material, such as plywood trim, down into smaller ones for further refining. Figures 7.1 and 7.2 show the cutting action of two types of hog. These types of machines have been around the forest products industry for years for use in preparing residue material for fuel, and the term *hog fuel* is well known.

CHIPPERS

Chippers, which are sometimes called *hackers,* particularly in Europe, prepare the familiar pulp chip from roundwood or other wood residues. The rotating-disc

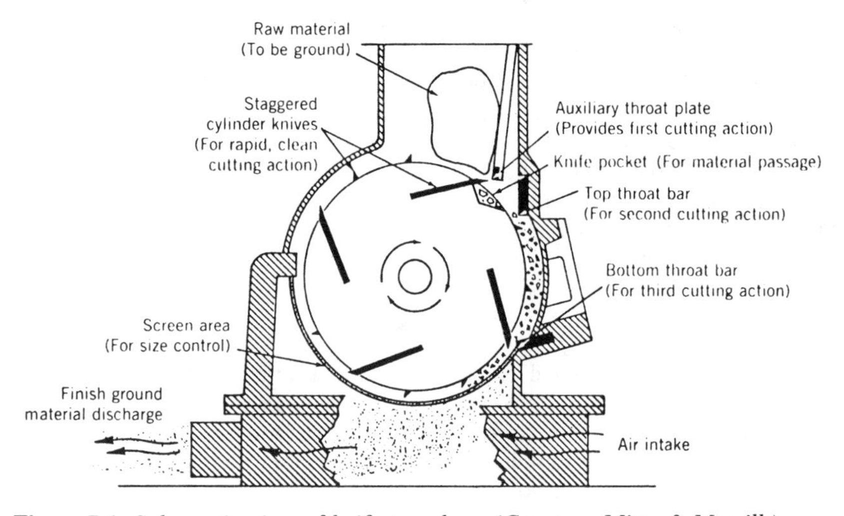

Figure 7.1. *Schematic view of knife-type hog. (Courtesy Mitts & Merrill.)*

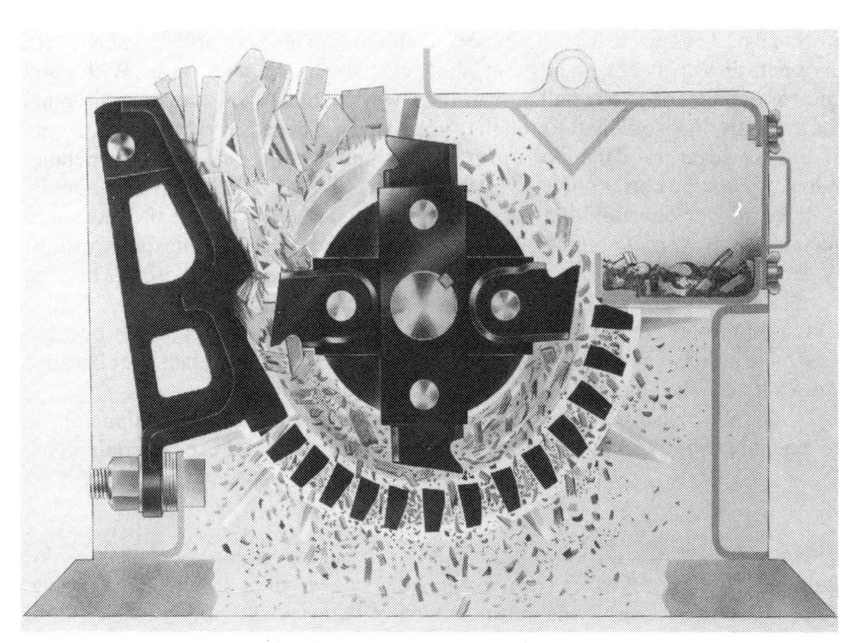

Figure 7.2. *Sectional view of the No-Nife hog, showing how the fast-revolving hammers cut the wood by impact. (Courtesy Williams Patent Crusher and Pulverizor Co.)*

type, as shown in Figure 7.3, is popular for use with larger-diameter raw material. Another type uses two conical discs and an infeed set at 90 degrees to the shaft. Drum-type chippers (Figure 7.4) are used for chipping veneer, slabs, edgings, and small-diameter wood. Screens can be built into drum chippers to reduce oversize particles as illustrated in Figure 7.4. In the composition board industry, chippers are used to reduce roundwood or other large material into a predetermined chip size for further refining. In operation, the knives cut across the fibers at an angle. The wood splits in the fiber direction because of a shearing action caused by the cutting and the hard impact of the knife striking the wood. A new chipper is now in operation to make very large *maxichips* for further cutting into long, slender flakes for use in oriented flakeboard. These chips are cut approximately 3 in. (76 mm) long.

CUTTER MILLS

Short wood offcuts and mill ends can be chipped better in a cutter mill (Figure 7.5) because the feeding system of a drum-type chipper can cause trouble with such short material. Cutter mills for boards (Figure 7.6) are specially constructed to chip all kinds of boards (offcuts from cut-to-size operations, rejects, laminated boards, furniture parts, and so forth). These cutter mills usually have a screen that has small-size perforations.

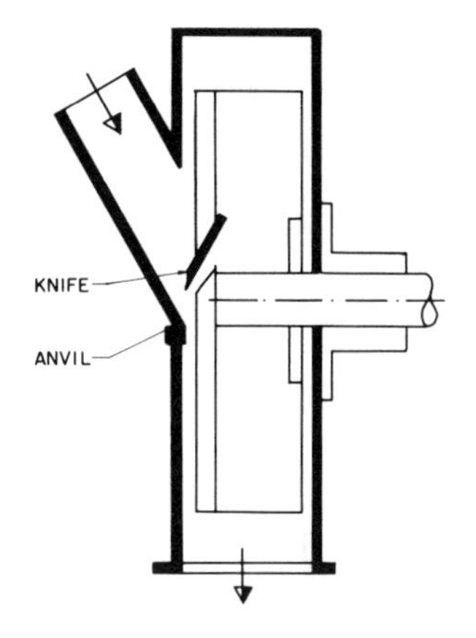

Figure 7.3. *A disc chipper (Fischer 1972).*

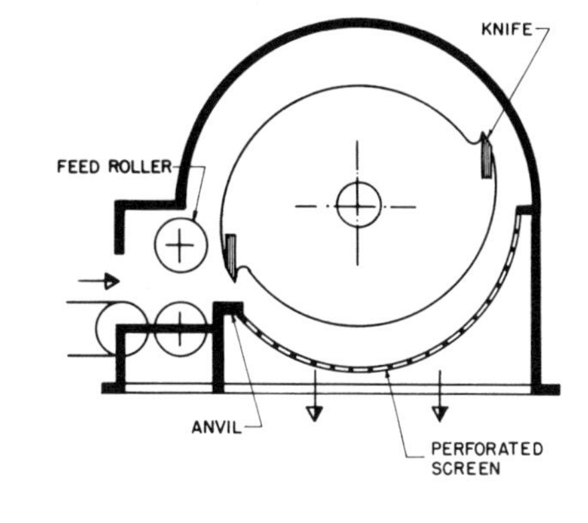

Figure 7.4. *A rotary- or drum-type chipper with screen (Fischer 1972).*

HAMMERMILLS

Hammermills (Figure 7.7), wing-beater mills, and similar-type machines use a beating action to reduce material. Particles which are too large after the first impact are pushed along the screen and are further reduced by friction and by squeezing between the edges of the screen holes and the beaters. The size of the screen perforations determines the size of the ground material. Beaters in hammermills are free-swinging hammers or steel strips. Wing-beater mills have an impeller with stiff arms, armoured with wear plates (Figure 7.8).

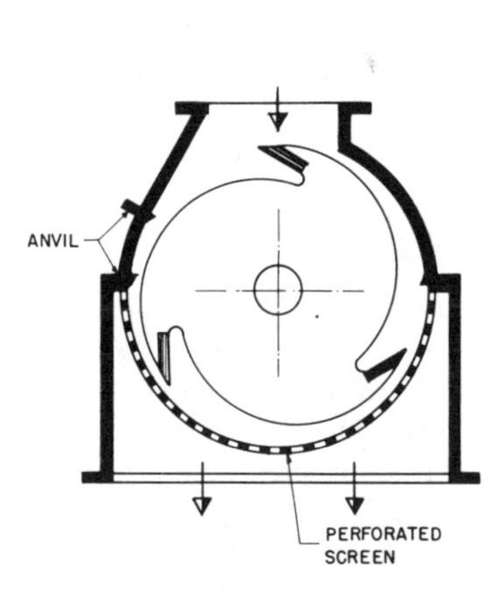

Figure 7.5. *A cutter mill or chipper for mill ends (Fischer 1972).*

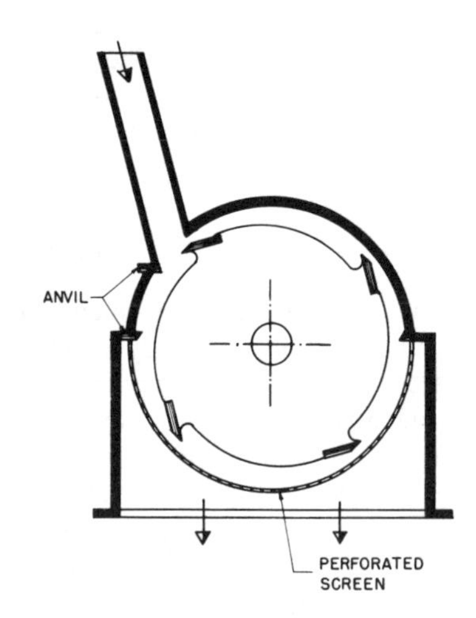

Figure 7.6. *A cutter mill or chipper for boards (Fischer 1972).*

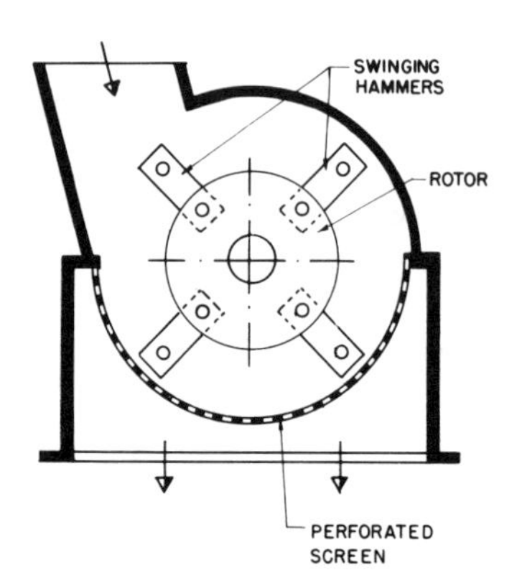

Figure 7.7. *A hammermill (Fischer 1972).*

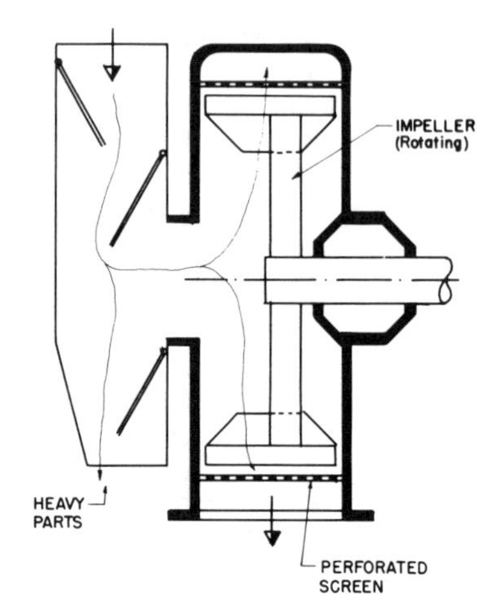

Figure 7.8. *A wing-beater mill (Fischer 1972).*

Hammermills and wing-beater mills are simple machines with low-cost wear parts. However, since size reduction is accomplished by splitting and breaking the infeed material into smaller pieces, little control of particle geometry is possible. When using a small-size screen, a high percentage of unwanted dust will also be produced. Because of their simplicity and ease of operation, these mills have been used extensively in the United States. Different screen sizes and screen openings enable varying sizes and geometries of particles to be generated.

When handling material with a higher moisture content, machines with a stationary screen-ring tend to plug. In this case, machines with a counter-rotating screen-ring are preferred. They assure a steady self-cleaning of the machine housing behind the screen (Figure 7.9). An additional advantage of the counter-rotation is a higher throughput capacity, and with a given screen perforation, a finer grinding is achieved in comparison to a stationary screen-ring.

IMPACT MILLS

Impact mills use as their principle the fact that wood particles thrown against a grinding path will split mainly along their fiber length (Figure 7.10). Impact mills operate with very high air speed (more than 300 ft—91.4 m—per second) to get the wanted hard impact. High air speed, however, does not necessarily produce large amounts of air. The air intake cross section is designed to limit the air throughput.

In the sophisticated ring refiner (Figures 7.11 and 7.12), heavy particles with a relatively small surface (in particular cubical ones) will hit harder against the grinding path than lighter particles of dust. As a consequence, particles already fine in size do not become much finer. A further sifting effect in the machine is caused by the airstream, because it will take the fine material away from the grinding path and lcad it through the screen. The few larger-size particles, which

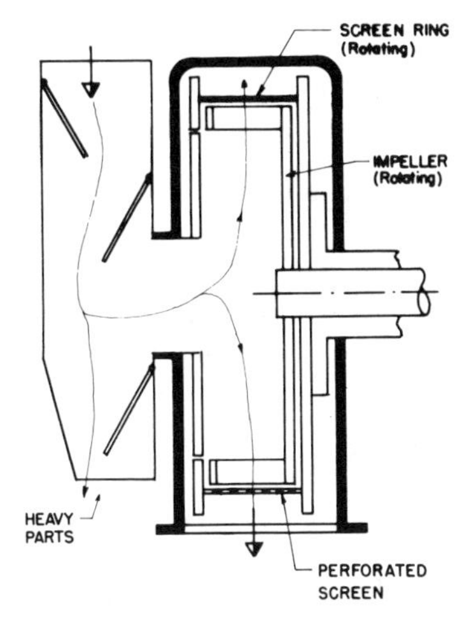

Figure 7.9. *A counter-rotating mill (Fischer 1972).*

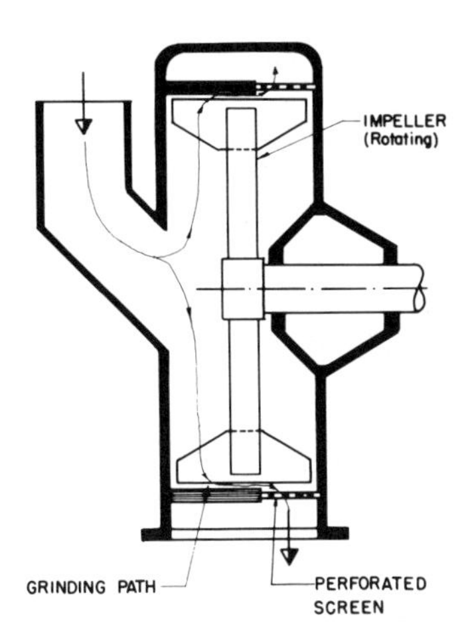

Figure 7.10. *A cross-stream impact mill (Fischer 1972).*

were not ground fine enough on the grinding path, are crushed between the impeller blades and screen.

For material with higher moisture content, counter-rotational machines are in favor, as illustrated in Figure 7.13, which is the same principle as discussed previously under "hammermills."

Impact mills have little wear, because there is little friction and no sharp edges are needed. The specific power consumption is comparatively low, because there is little braking effect. Flake material can be handled well; thus, little unwanted dust is generated. The amount of air going through the machine affects cooling and

Figure 7.11. *A double-stream ring refiner (Fischer 1972).*

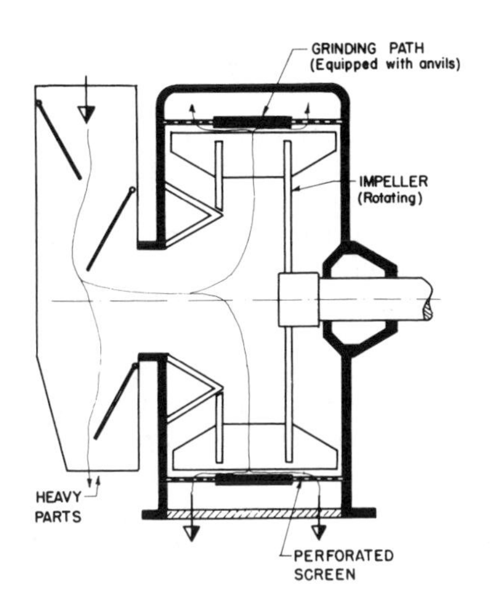

Figure 7.12. *View of a ring refiner. The easy-to-change grinding path is sitting in front of the refiner (Fischer 1972).*

cleaning. Anvils for the grinding path are available in different patterns, and screens can also be used with different perforations and shapes. This makes the machine very flexible and useful for different materials, such as splinters, sawdust, and planer shavings. Many plants throughout the world are now using impact mills to process fine surface furnish, and their use is increasing in the United States.

ATTRITION MILLS

Attrition mills are also known as *fiberizers* or *disc refiners*, and in the board industry the term is broadly interpreted to cover machines which can convert wood chips or other small waste material into board furnish. The impact mill (ring refiner) just discussed can accomplish somewhat the same result. The fiber generated for composition board, either in the particleboard or fiberboard area, is really a bundle of fibers and not a particle made up of a single fiber. The pulp and paper industry requires much smaller fibers than those normally used in composition boards. To obtain such fine fibers, some chemical cooking may be used to assist in breaking down the lignin in the wood. This technique can also be used in the composition board area and in some cases, such as in wet-process hardboard, it can be of value. The refining of raw or streamed wood is extensively used to produce fiber for mat-formed products such as insulation board and hardboard. As the industry moves more and more towards using high-density species, the possibility exists that some type of chemical cooking or pressurized refining may also be used for preparing fibers out of difficult-to-mill species for the board processes.

The actual particle shapes usually take a variety of forms such as splinters, coarse and fine chunks, flour, or individual fibers and fiber bundles. The end results depend upon a variety of factors, such as the type of raw material, moisture content, density of the raw material, the machine operating parameters, and the types of desired product.

Attrition mills or disc refiners mainly use attrition and shear as the particle generating principles (Figure 7.14). In this way, a fine defibration can be achieved, and fibers with a high degree of slenderness can be obtained. Steaming the feed material and grinding under high-temperature pressure increases fiber length. The raw material is fed into the center or "eye" of the mill and is thrown towards the outer periphery of the plates by centrifugal force as it is converted into fiber between the plates.

It should be mentioned that there is a combination impact and attrition mill called a *turbo mill*. This is a counter-rotating machine without screens (Figure 7.15). The ground material leaves the grinding chamber through an adjustable gap. Turbo mills are not sensitive units and thus need little maintenance. For coarser reductionizing operations, they are still in use in the board industry. For fine grinding, impact mills work more economically.

Attrition-type mills have been used for many years in the paper, insulation, and hardboard industries. Two general types are available: the single-revolving-disc and the double-revolving-disc designs. Each is operated at atmospheric pressure on raw, presteamed, or chemically treated wood feed material. These machines are also available for operation at elevated temperature and pressure. The choice of types will depend upon the types of raw material and the desired product result.

The type of fiber refined by means of these two types of attrition mills is usually quite different. In addition, the variations in the raw material particle size, species, moisture content, and the variables possible in the grinding process, such as steaming time, pressure, refiner-disc diameter, plate pattern, loading rate, and

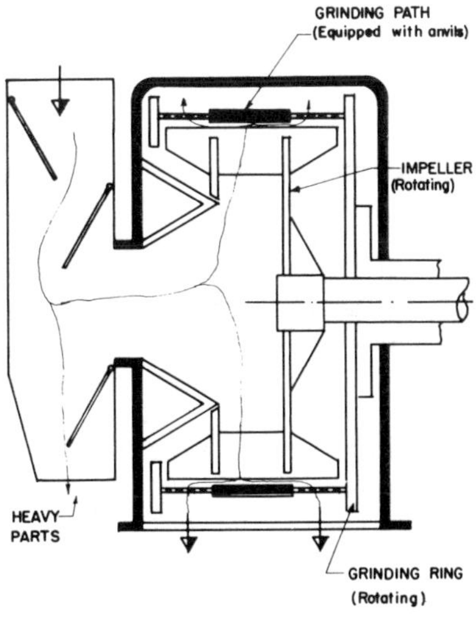

Figure 7.13. *A counter-rotating, double-stream ring refiner (Fischer 1972).*

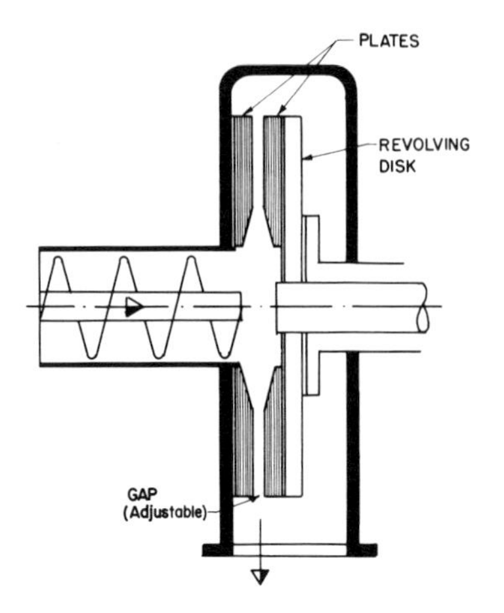

Figure 7.14. *A disc refiner (Fischer 1972).*

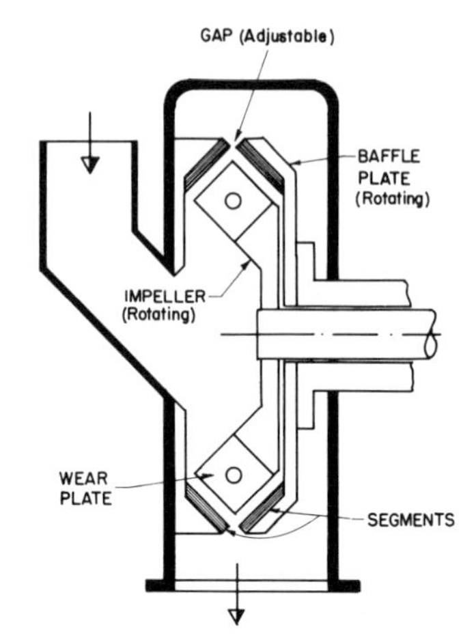

Figure 7.15. *A turbo mill (Fischer 1972).*

the gap setting between the attrition mill plates, all influence the type and quality of fiber that can be generated. In general terms, the pressure-refined fiber is much finer in character than the atmospherically ground fiber. The wood is being broken down closer to individual fibers in the pressurized refiner than is possible in the atmosphere refiner. However, even with the pressurized refiner, the fiber generated in board practice is usually bundles of several wood fibers rather than individual fibers. Figures 7.16–7.18 are photographs of various fibers ground by atmospheric attrition milling and pressurized refiner milling. This illustrates to some degree the differences that can be expected between these two types of fibers. It is very noticeable that the raw refined material did not yield a very fibrous furnish. Better results are attainable on low-density species.

When using atmospheric refining, the ideal raw material appears to be green wood chips, particularly of lower-density species such as softwoods or low-density hardwoods such as aspen or cottonwood. Even without presteaming, a fairly acceptable fiber for certain products can be ground in atmospheric attrition mills when using green raw material of the low-density type. However, drier material such as kiln-dried planer shavings cannot be ground into any type of acceptable fiber without at least presteaming. To produce the better fiber, moisture must also be added to this very dry material to assist in the grinding operation. The words shavings and sawdust can mean many things as to size, condition, and shape. In general, the smaller the size, the poorer the fiber quality will be.

Most plants using refiners to produce fiber for high-density hardboard were originally built to use green chip material. Some plants still are operating on such raw material; however, it has been found more profitable in the overall economic picture for most companies to sell their green chips for pulp and paper production and to bring in lower-quality material for the board plant, consisting mainly of planer shavings and sawdust. Sometimes such use of residuals is limited to 10 to

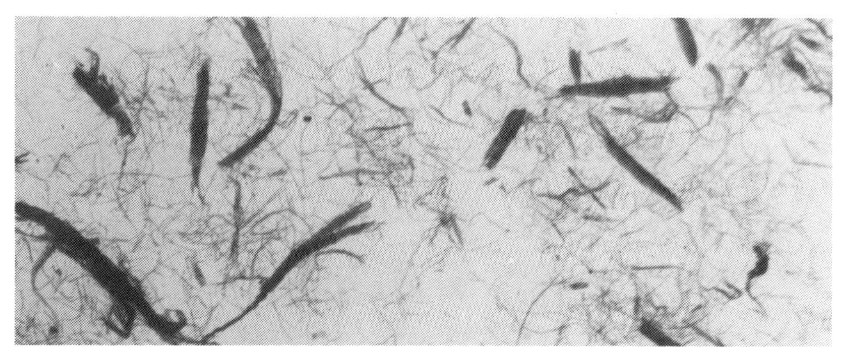

Figure 7.16. *Dry-process fiber from steamed chips ground at atmospheric pressure; magnification 6 X (Rauch and Cheney 1970).*

20% of total needs. The newer medium-density fiberboard (dry formed) plants are also using residuals other than chips for raw materials. Chips, of course, are still used for a raw material, particularly in cases where the chips are not especially suitable for pulp and paper and when companies' profits are maximized better by producing boards out of these chips rather than selling them for pulp.

Better-quality fiber can normally be made from pulp chips. It is simply a case where relatively long fibers can be generated out of pieces that are about ¾ in. (19 mm) long. Fibers are not as long as the chip since attrition milling reduces their length. Extremely long fibers also can present problems in handling through the remainder of the process line. Fiber length will also be influenced by the attrition mill type.

The two types of attrition mills or refiners will be discussed separately. The general operating principle of the two basic refiner types is the same; however, they do produce somewhat different products. Results will also differ with the processing of raw, presteamed, chemically treated wood or the use of pressure and elevated temperature with refining.

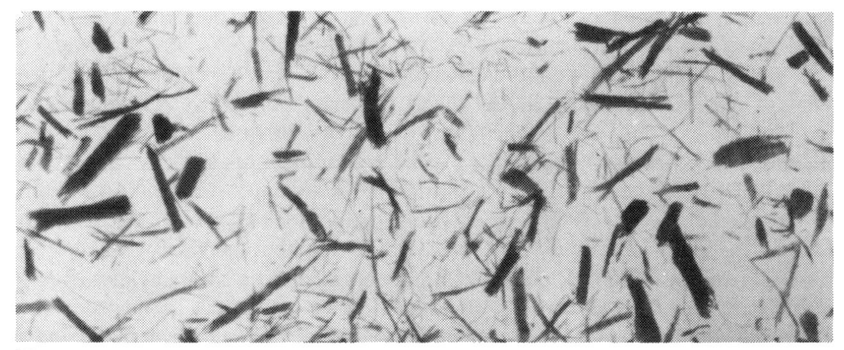

Figure 7.17. *Dry-process fiber from cold refined shavings and sawdust: magnification 6 X (Rauch and Cheney 1970).*

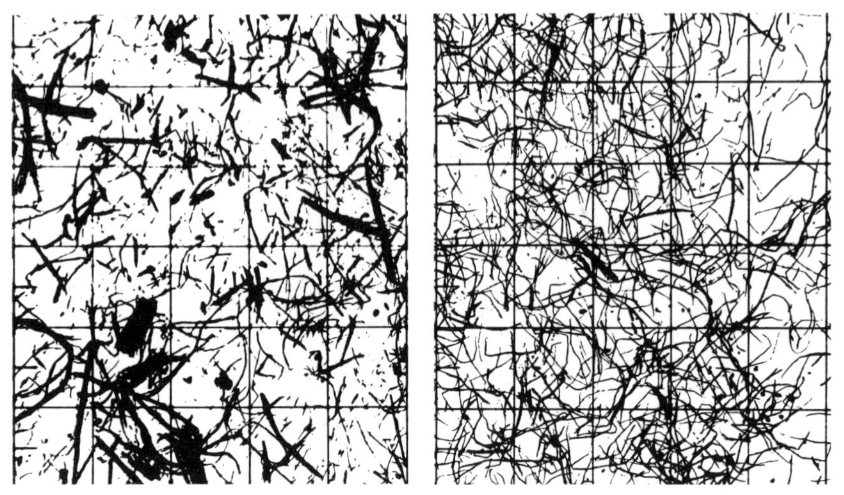

Figure 7.18. *Comparison of atmospherically refined, left, and pressure-refined, right, fiber of white pine (both types of fiber about 1 lb bulk density). Grid spacing = 0.10 in. or 2.54 mm (Maloney, Talbott, Strickler, and Lentz 1976).*

Atmospheric Refiners

There are two types of machines: (1) a single-revolving disc, and (2) a double-revolving disc. Figure 7.19 shows a cross section of a single-revolving-disc mill which has a built-in motor. The shaft motor rotor and disc at the right-hand side can be moved horizontally by means of a hand wheel on the right-hand end. This adjusts the space between the face of the rotating disc and the stationary surface, which is part of the mill door. Feed material passes into a spout at the left and is fed in between the plates or surfaces of the rotating and stationary discs. The feed material is thrown by centrifugal force towards the periphery of the machine and is ground into the fiber enroute. The resulting product is discharged at the bottom of the case which surrounds the discs. These machines may have screw-type feeders instead of a spout and they are available with a variety of disc diameters and motor sizes.

Figure 7.20 illustrates a double-revolving-disc mill which is equipped with two motors, one for each disc. The left-hand disc has spokes through which feed material passes from a spout which may or may not be fitted with a screw-type feeder. The disc on the right-hand side is driven in the opposite direction by its separate motor. This latter disc motor rotor and shaft may be adjusted to or from the left-hand disc by means of a hand wheel, as shown. This gap can range down to 0.001 in., which is impractical for operation, up to about 0.060 in. (1.52 mm) or more. As can be seen in the figure, the opening between the plates is usually widest near the center of the discs and tapers down toward the periphery of the disc. This sets the stage for a gradual reduction of the larger material to the finest fiber. Double-revolving-disc mills are also available with a variety of disc diameters and a variety of motor sizes.

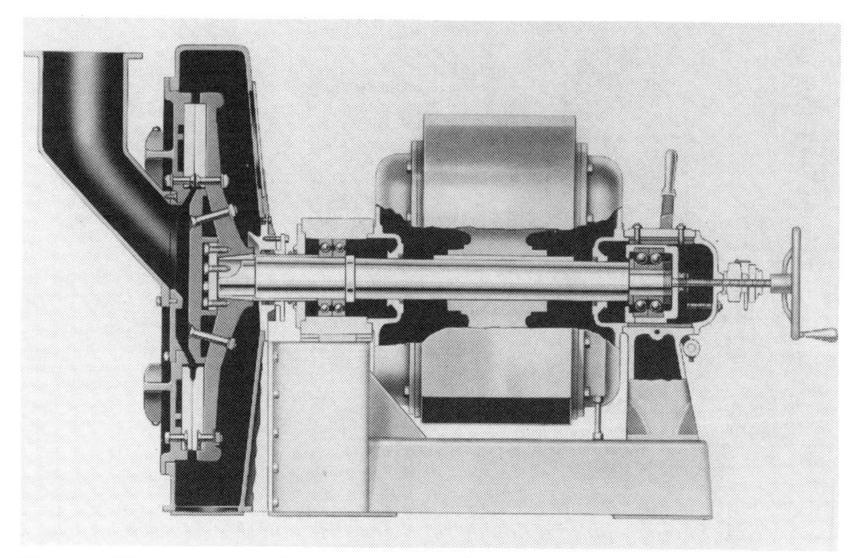

Figure 7.19. *A cutaway view of single-revolving-disc mill (McNeil 1967).*

In comparison to the single-revolving-disc mill, using a given disc speed, the double-disc mill will have twice the relative rim speed as the single-disc mill because of the counter-rotating operation. The double-disc mill is extensively used for generating fiber for the fiberboard operations.

The previously mentioned blending of phenolic resin and wax simultaneously with the conversion of the raw material into fiber is extensively used in dry-process hardboard operations. The feed end or left-hand shaft is extended through the outer bearing and is rifle-bored throughout its length. The outer end is fitted with a rotary joint which is connected to a metering pump. Resin, usually phenolic at about 15 to 20% in solids, is fed to the center of the mill, where it meets the

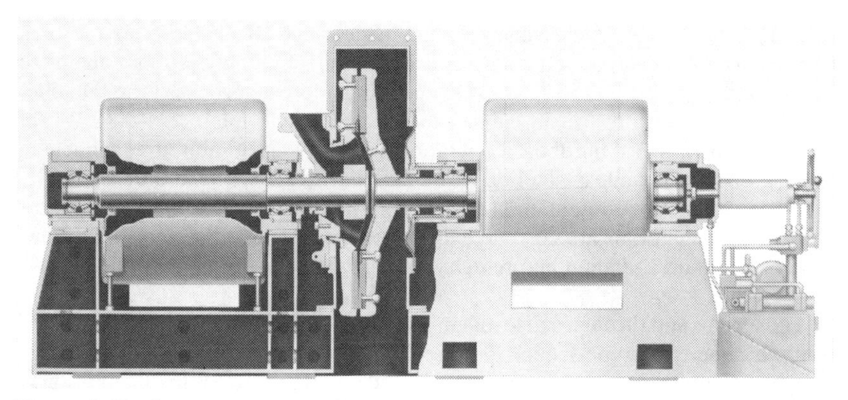

Figure 7.20. *A cutaway view of double-revolving-disc mill (McNeil 1967).*

Figure 7.21. *Photograph of a double-revolving-disc mill (McNeil 1967).*

incoming chips or wood waste. The resin distribution is excellent, and this system has replaced most of the blenders previously used. Attrition mills can also be used as blenders with previously refined fiber or particles, as will be discussed later. However, the binder cannot be tacky or viscous or heat sensitive.

Figure 7.21 illustrates an atmospheric double-revolving-disc mill which is equipped with a twin-screw feeder, circulating oil lubrication, hydraulic means for holding a manually adjusted head or disc position, and openings for forced ventilation of the main motors. The latter feature is quite important in dry-process operations, and outside clean air can be supplied to prevent an accumulation of wood dust in the motor, which could cause a fire. This same idea is used for single-revolving disc-mill motors.

In general, disc mills of the atmospheric type are available with most standard motor speeds, but 1200 and 1800 rpm are the most common types for 60-cycle current. Although larger single-disc mills are made, a 250-HP (187 kW) size is about the largest practical type for most particle operations. Fiber operations can utilize from 300 to 3000 HP (224 to 2238 kW) depending upon the process type. Double-revolving-disc mills are usually limited to 300 to 600 HP, total, (224-448 kW) for particle or dry-formed hardboard operations. Both types of mills are available in either carbon or stainless steel construction. Particleboard feed stocks usually have a low moisture content and are therefore not corrosive, while many wet or steamed feed materials are corrosive to carbon steel. Thus, these considerations must be taken into account before determining which type of construction to use for a mill.

The feed to and through a disc mill in a dry or semi-dry operation is improved if a suitable metering-type feeder is utilized. The stock quality does not change greatly with minor feed volume variations, but a variable motor load could cause electrical trouble due to overloading.

Three other items are desirable for disc mill applications on dry-process refining:

1. Balanced case wipers which are two balanced metal lugs. They are attached in diametrically opposite positions to one rotating disc. They serve to sweep or fan away any material which may tend to remain in the mill case and constitute a fire hazard.
2. Case vents in the refining cases of double-disc mills will permit ventilating air to circulate through the case and assist in maintaining a low temperature.
3. An air suction pickup means to receive stock as it is discharged from the mill and convey it to the next process stage. It is desirable to arrange this means so that air is pulled through the mill. Such action increases capacity in many instances, cools the mill interior, and aids in preventing an accumulation of material.

It has been pointed out that single-revolving-disc mills are available with up to 250 HP (187 kW) at 1200 or 1800 rpm. They appear to be most suitable when a coarse type of product is required. Such mills usually have a 36-in. (914 mm) disc diameter, although smaller and larger ones are produced. They may be fed through a spout or screw feeder, as shown previously, depending upon whether the feed is free-flowing or requires pushing or driving. Mills with much greater power are made; but in most cases they cannot be used on particle stock, because the required amount of feed to load them will not be accepted by the mill. For example, chips with a moisture content of about 50% will have power rates which vary by species and product type from 6 to 12 lbs (2.73–5.45 kg), oven-dry basis, per horsepower hour (HPH) (0.746 kWh). Feed rates of 18 to 40 tons (16.3–36.3 mt) per day (oven dry) would be required to load the machine. It must be recognized that the maximum feed which a mill will accept is not an exact science and will vary with the plate style, plate condition (wear), wood species, moisture content, and plate setting. Coarse particle stock power rates and output will change considerably from the finer stock.

Although there can be some overlap in the performance of single- and double-revolving-disc mills, the double type is generally regarded as the best unit for generating slender or fibrous material. The opposing rotation of the disc provides a relative speed of 2400 to 3600 rpm. In addition, the predominant particle orientation between the discs seems to be radial which tends to reduce the fine or short portions of a given stock.

The size of double-revolving-disc mills in the particle or dry-process fiber field is limited by practical considerations. Power rates tend to be low with as much as 14 horsepower days (HPD) per dry ton (251 kWh) on fiber of the type used for dry-formed hardboard or as little as 2 to 3 HPD/ODT (36–54 kWh) on particle stock. Practice in the fiber field with chips at 50% moisture seems to indicate that a 500 to 600 HP (373–448 kW) unit is large enough, assuming atmospheric refining. With both double- and single-revolving-disc mills, the determination of maximum size is not an exact science.

There are many styles of grinding plates available in several alloys. Selection of the right type in any given case is sometimes a trial-and-error procedure. They can be found in either a flat or tapered design. Since it is impossible to see between the operating plates, we can only guess as to what goes on. Actually, usage has provided useful background, and the fact that many producers find a given style to be satisfactory cannot be ignored. A great number of plate styles are available among the manufacturers of mills and plates. Following is information on a few of the most popular types used for grinding.

Figures 7.22–7.24 illustrate the two opposing discs in a set. There are usually six segments on each disc. Notice that plate 36301 has no dam at the outer edge and that the outer edge of plate 36302 has a dam (Figure 7.22). Actually, the dams on opposite sides are staggered to form a wave-type path for the material being ground to pass through. Other versions of this same design involve differences in the length of the large breaker ribs, which extend to the inner edge, and dam arrangements at the inner side. This style, or some variation of it, is frequently used to make a low-bulk-density fibrous stock from damp or steamed wood for medium- or high-density boards. It has also been popular in pressurized refiners.

Figure 7.23 illustrates plate designs which are mirror images of each other. Again, one set of six segments each is mounted on each opposing disc. This style is used without the outer dam when a coarser product is required. It can also be used without any dams for maximum coarseness. These plates with the outer dam have

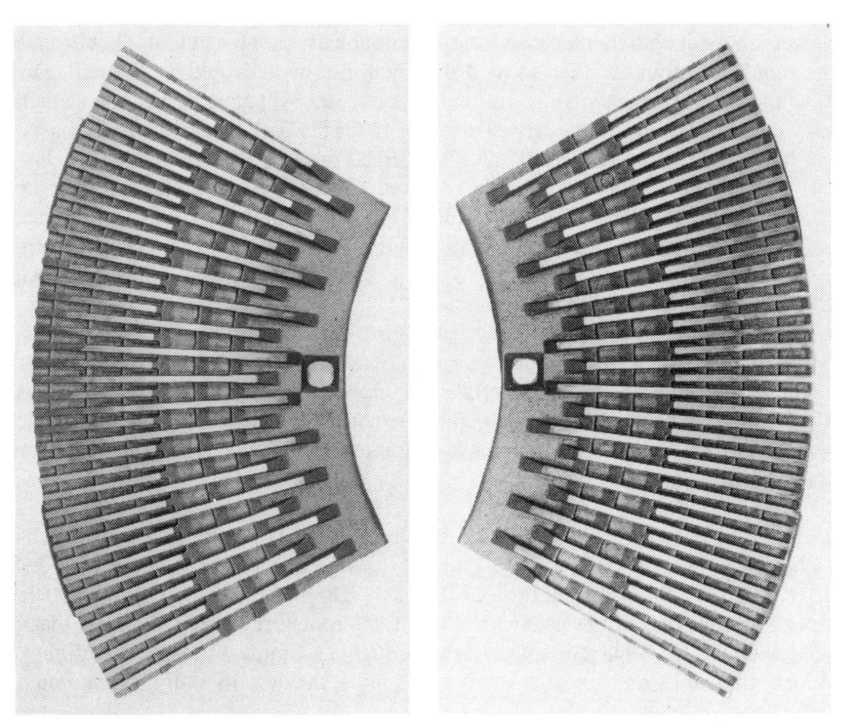

Figure 7.22. *Opposing segments of discs: left, No. 36301; right, No. 36302 (McNeil 1967).*

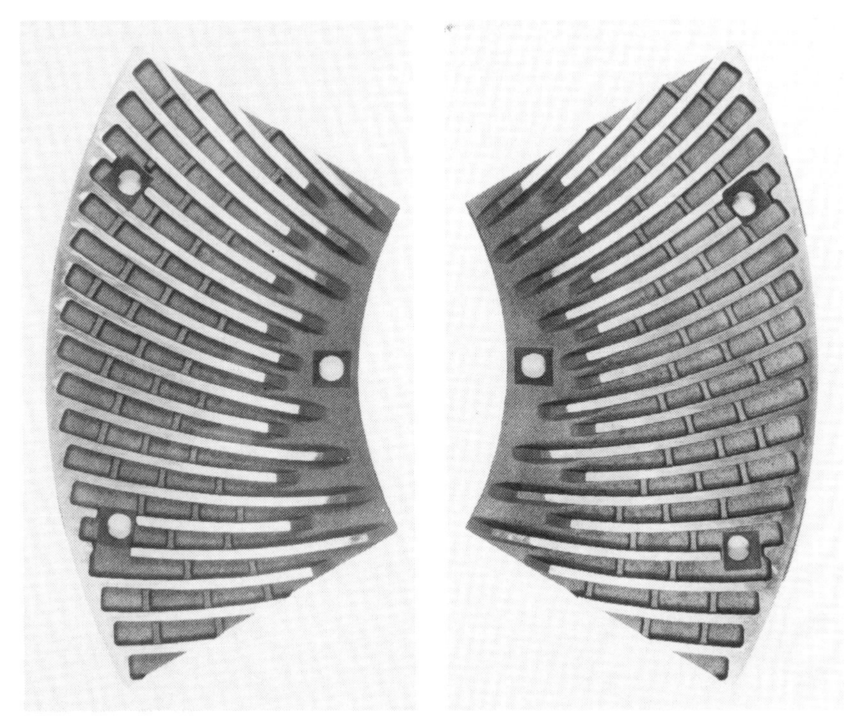

Figure 7.23. *Opposing segments of discs: left, No. 36504; right, No. 36505 (McNeil 1969).*

been very popular in the board industry. If used on fibrous stock for medium-density products, the bulk density could be somewhat more than that obtained through the use of the plates shown in Figure 7.22.

Figure 7.24 shows the waveline design which is useful for making coarse stock from various raw materials. They are not usually suitable for fibrous products.

Now that the general operation of an atmospheric disc refiner has been related, the critical factors governing actual refining will be presented. One of the first considerations is the type of raw material and final product. If it is a mixture of sawdust and shavings, perhaps a large percentage of the total requires no milling or only a light treatment for particleboard. Separation of such material by screens, or other means, and bypassing the refiner could reduce refining equipment investment. Also, feeding fines through a mill will only make them finer. The separate milling of the fractions could be advantageous if a size-graduated formation is employed for the final board. If waste material contains large pieces, they should be reduced in a hog or chipper to a maximum size of about 1 to 1¼ in. (25.4–31.8 mm).

The moisture content of feed material has a considerable effect upon particle shape. If a large percentage of fines or wood flour is wanted, a moisture content of 10% or less is desirable. Dry wood makes a chunky rather than a fibrous particle,

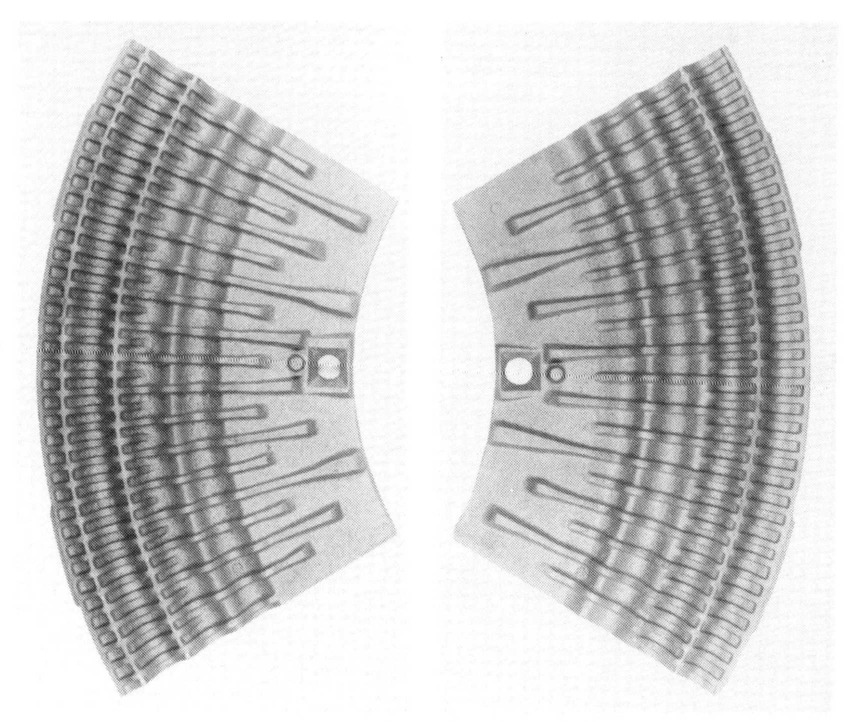

Figure 7.24. *Opposing segments of waveline discs: left, No. 36708; right, No. 36709 (McNeil 1967).*

which is large or small depending upon mill treatment. The particle becomes more fibrous in any mill as the moisture content of the raw material reaches the 40 to 60% level. If a fibrous type of particle is desired from a dry material, hot or cold water soaking may be used in some suitable and controllable manner. If the raw material is large, the water addition will be largely a surface type. A more costly and sometimes more satisfactory procedure involves steaming for five to ten minutes at a pressure of 80 to 120 psi (552–827 kPa). Usually water must be added to raise the moisture content.

If small dry particles, such as planer shavings, are to be refined, water can be sprayed upon them in a blender or water can be added in the feeding spout to the refiner. Experience has shown that the 50 to 60% moisture-content level can be appropriate for obtaining a good fiber from such dry material. Fiber length, of course, is determined for the main part in the length of the shavings used. Long fiber, naturally, cannot be made from extremely short particles.

High-density species are usually more difficult to refine because of their nature, and greater amounts of energy may be required. Rocks and large pieces of metal, or course, cause severe problems when passed through a refiner. If such problems are regularly encountered, the installation of a chip washer, or a separator of the air or magnetic types, may be warranted.

To obtain better fibers or, in more correct terms, *fiber bundles*, presteaming of the raw material is preferred and is a common practice in the hardboard industry. It has also been used in the particleboard industry when a fibrous face material has been needed to improve the bending strength. This can be accomplished in the more popular continuous steam digesters as shown in Figure 7.25 or in batch digesters. Figure 7.26 illustrates one of the sizes of the auger which can be as much as 52 in. (1.32 m) in diameter. Grinding hot, steamed raw material, while improving the fiber quality, also requires less energy in the refiner. Steaming pressure ranges from about 10 to 100 psig (68.9–689 kPa). Specific treatment depends upon material size, density, moisture content, and product needs. It might be desirable to also add some water to assist in plasticizing the wood. Heated wood has the lignin (nature's glue holding the cellulose together) softened which enables the subsequent refining to separate the raw material into fibers. Rotary valves, or an equivalent type, are necessary to seal this type of digester and to make it work. The Bauer-Grenco type is a pocket valve (Figure 7.27). One valve is used to deliver the raw material into the digester and another one removes the steamed chips at the other end. Screw-type valves, which form a plug from feed material, force feed a digester while forming a pressure seal. They are extensively used as inlet units.

Feeding refiners must be carefully controlled. If at all possible, the feed material should be of a relatively uniform blend of sizes to keep the loading of the refiner constant and for making a uniform product. Actual feeding should be at a constant feed rate. In practice, the gap between the plates in the refiner should be set for producing the fiber desired. Next, the best approach is to feed as much material as possible into the refiner to build up the motor load to the rated level. The actual governing factor then becomes the electrical load on the motors. Some minor adjustment of plate setting may be required. In other words, the refiner is run by establishing the plate setting and then adjusting the feed rate to load the motors. It should be remembered that there is a limit to the feed quantity for a

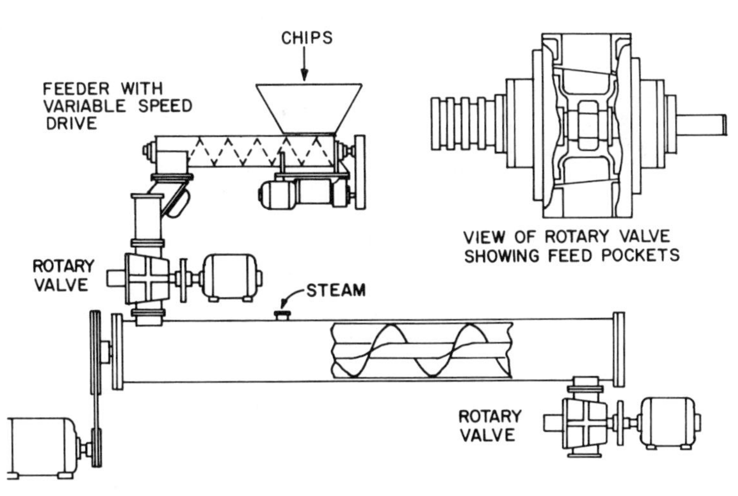

Figure 7.25. *A Grenco continuous digester (Textor 1957).*

given refiner under required process conditions. It may not be possible to load a large motor and still get the desired product.

A problem can arise when feeding low-bulk-density materials such as waste paper, or raw material as typified by bagasse. These materials are so bulky that a relatively small amount of stock by weight can be ground in a given time. Systems for pelletizing, compressing, or a different refining concept could open the door for greater use of these raw materials in the board industry. Single-revolving-disc refiners can handle the high-bulk-density stock better, as the screw feed can compress the stock while the stock must be fed between the spokes of one of the discs in the double-revolving type. It is important to watch for start-up problems with a cold machine.

Fortunately, disc mills can be readily adjusted to make a variety of particle sizes. As indicated before, feed material, mill speed, plate style, and plate setting are determining factors. Particle size and shape, together with fines content, are important considerations. Surface or overlay stock for particleboard may have a top screen size limit of 12 to 30 or 40 mesh (openings per inch or 25.4 mm). Core stock may have a top limit of 4 mesh. Fibrous furnish is more difficult to measure, as noted, and each plant has established some type of measurement system, most notably bulk density (lbs per ft^3), appearance, screen analysis, or drainage rate in a freeness test.

It seems that word or number descriptions of stocks from various raw materials are usually meaningless except for local comparison. By comparison, a description of a flake stock is easy. Most disc mill manufacturers are in a position to demonstrate their equipment and this service should be utilized by any board producer who is considering the use of attrition mills.

Pressurized Refiners

The pressurized refiner is the newest type of refiner used in the board industry. Refiners of this type have been used previously in the wet-process hardboard and

Figure 7.26. *Auger for continuous steam digester. (Courtesy C-E Bauer.)*

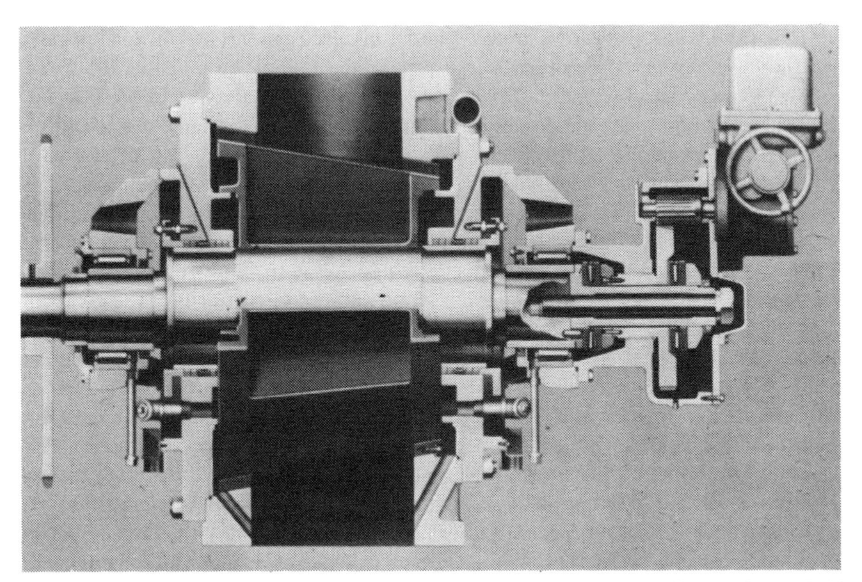

Figure 7.27. *Infeed valve to a digester, Bauer-Grenco type, 8 pockets, 10 HP–7.46 kW (Rauch and Cheney 1970).*

paper industries. However, in the mid-1960s such units were built for use in medium-density fiberboard plants, and their use has expanded into dry-formed hardboard and particleboard plants.

Pressurized refiners have been a major breakthrough in the wood composition board industry. Hitherto difficult-to-use raw materials, such as dry hardwood residues, dry shavings and plywood trim, can now be refined into acceptable fiber. Greater mixtures of species, particularly those with widely differing specific gravities, appear possible. Dry material can be refined into acceptable fiber, and amazingly, to a number of people, the moisture content of the refined material is not significantly increased. This refiner and more advanced and sophisticated versions are expected to play an important role in the development of the composition board industry.

As a salient point, these refiners are starting to be used heavily in the pulp and paper industry as the quality of their raw material is declining much as it is in the composition board industry. As noted before, composition board is for the most part being forced lower and lower on the scale of raw material. Interestingly, pulp and paper is also moving lower on the scale to secure its raw material. Some of the deterioration of raw material quality has come about by improvements in sawing and planing in sawmills. While they have improved, the resultant residues are of poorer quality for pulp and composition boards.

Pressurized refiners have been promoted by some as a panacea for the problems found when refining low-quality raw materials. They are not a cure-all solution. Significant improvements in particle generation are possible. However, the factors threading their way through all composition board processes—such as species; raw material type and quality, and moisture content; mixture of species; and interspecies specific gravity variations—must still all be considered.

Two types of pressurized refiners are being used and they shall be differentiated here as the single- and double-revolving disc which is the same classification as used to identify the atmospheric refiner.

The known fact that heat has a plasticizing effect on wood, as mentioned, is used to assist in refining with some of the atmospheric refiners. Wood becomes more plastic with increasing temperature, especially in the presence of moisture. The wood constituents that are the most affected are hemicellulose and lignin. The hemicellulose and lignin components of wood and the microfibrils in the cell wall are softened before the cellulose skeleton of the fiber walls. Since the lignin binds the cellulose fibers together, it is easier to fiberize wood at higher temperatures as contrasted to ambient conditions because the binder (lignin) has been plasticized. This results in horsepower savings because it is easier to grind the fiber, and the quality of the fiber is better when ground under the heated conditions. The pressurized refiners were developed upon the principle not only of grinding steamed wood but grinding while the steam pressure was still applied.

Field results to date indicate that it takes from 0.25 to 1.0 lb (0.11–0.45 kg) of steam to grind 1 lb (0.45 kg) of oven-dry fiber. Interestingly, material fed into such a refiner can have about the same moisture content after refining as before. When using dry raw material, minimum drying after refining is needed for dry forming lines and operating costs are reduced.

Bulk density of pressure-refined fiber ranges from about 1.0 lb/ft^3 for the finest fiber to 3.5 lbs (1.59 kg) for coarser fiber. Atmospheric ground fiber will be near the top of the density range although lower density fiber can be ground.

Figure 7.28. *General view of refiners in a plant (Runckel 1973).*

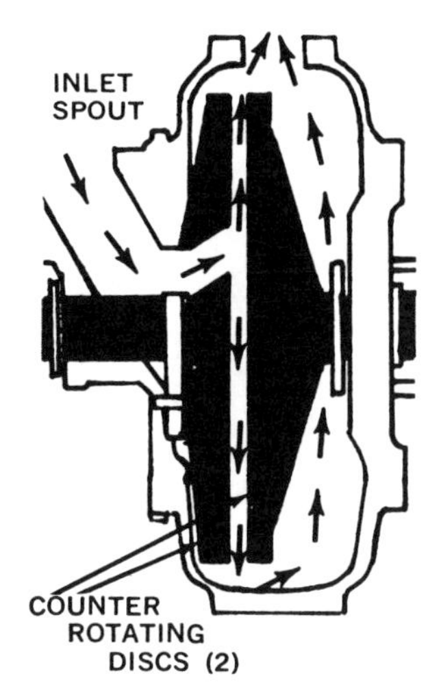

Figure 7.29. *Cutaway view of double-revolving-disc grinding chamber (Runckel 1973).*

Double-Revolving-Disc Mills: Raw material for a double-revolving-disc system is fed into a horizontally oriented, continuous steam digester through a rotary valve of the type used on digesters for atmospheric refiners. A typical system is shown in Figure 7.28. Figure 7.29 is a cutaway view of a refiner's double-revolving-disc grinding chamber. Figure 7.30 shows the major operating parts. Pressures in the digester up to 150 psig (1.03 MPa) reportedly have been used. The pressure has to be determined by experimentation for each application. Although more and more knowledge is being gained about this relatively new piece of equipment, each installation has sufficient inherent differences to require some experimentation to establish the operating parameters. After steaming is completed, the raw material enters a transfer feed screw which presents the material to the refiner. After refining, the stock and steam go to the blow cyclone, where the steam and fiber are separated.

In a double-disc mill, feed material must pass between the spokes in one disc into the refining zone. It is necessary to have a seal at this point to prevent infeed material from passing around the back of the disc without going through the refining area. A shive problem would result without a seal. Shives are larger sticklike particles which are undesirable in a fibrous board. Figure 7.31 is a view showing the inside of such a refiner. Note the initial taper between the plates where the raw material is first reduced in size before final fiberizing.

The double-revolving-disc refiner has two motors driving the discs in the opposite directions to each other for the grinding operation and, since the grinding is performed under pressure, the shafts must be sealed to prevent steam loss and wear in the bearings. Stellite sleeves and lantern rings are among the sealing methods used. Some problem was found by some plants at first with seals but, reportedly, research and development have solved these problems.

Single-Revolving-Disc Mills: The other type of pressurized refiner is the single-revolving disc. The general operating principle of refining under pressure is the same as for the double-revolving-disc refiner. However, the techniques of

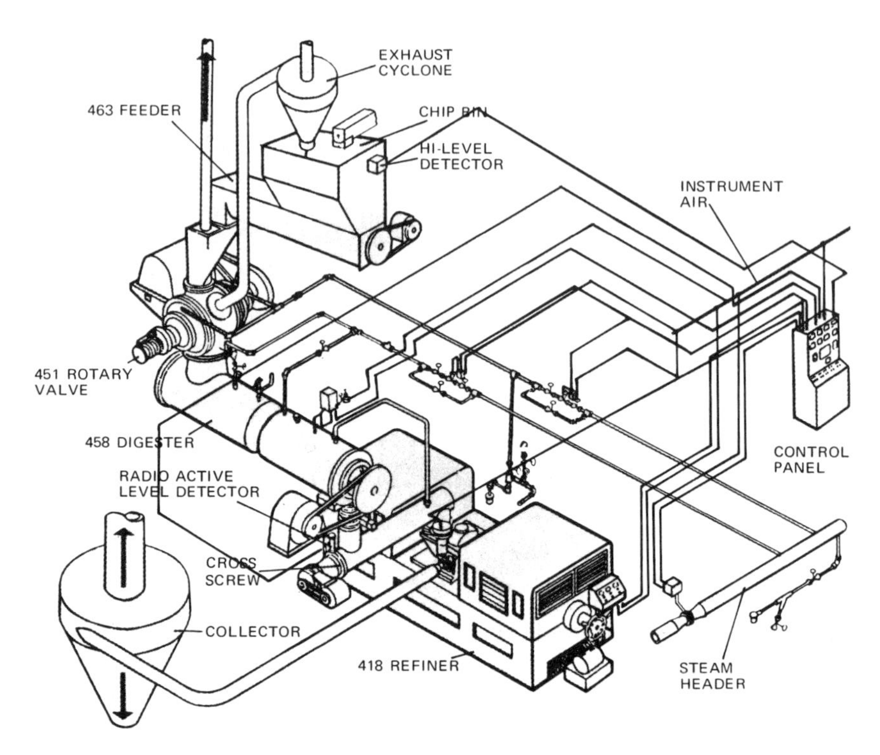

Figure 7.30. *Schematic of C-E Bauer 418 refiner system (Johnson 1973).*

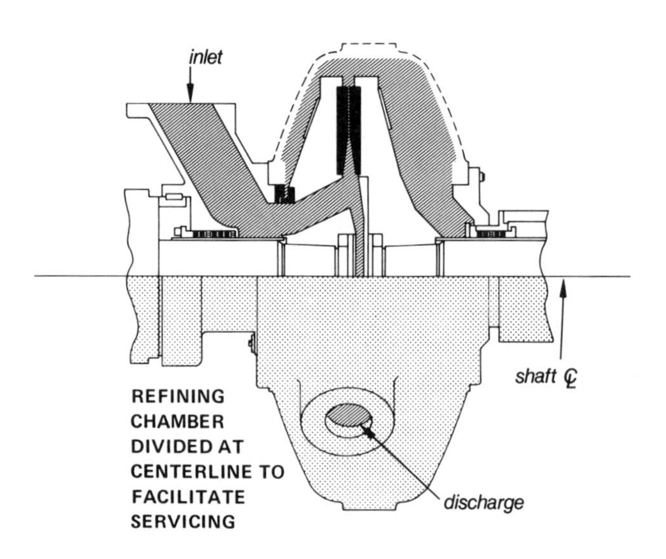

Figure 7.31. *Cutaway view of split case (Runckel 1973).*

accomplishing refining are different. Figures 7.32 and 7.33 are schematic drawings of a typical installation. The typical installation under discussion has a refiner with 42-in.-diameter (1.07 m) discs. The steaming vessel for this refiner is vertical, contrasted with the horizontal feeder of the previously discussed double-revolving-disc refiner. The discussion here will be generally covering the Defibrator L-42 refiner; however, it should be noted that single-disc pressurized refiners are also made by Sprout Waldron.

The raw material for the refiner, after passing under a magnet and going through a rock separator, goes to a storage bin. From this bin the material is fed by screws to the feed chute which in turn feeds the screw feeder. The screw feeder conveys the wood into the vertical preheater and in so doing must overcome the steam pressure in the preheater. At the same time, it provides a seal which retains the steam in the preheater.

The seal is made by forming a wood plug at the throat of the screw feeder and in the plug pipe, which feeds directly into the preheater. The screw feeder is the most common feeder used to introduce the raw material into the steaming vessel. It is a rugged, dependable unit. The only parts that are subject to wear are the screw and the throat, which are manufactured of special wear-resistant materials. They are easily replaceable and can be rebuilt for further use. Extensive investigation has shown that there is no damage to the fibers caused by the compression in the screw feeder under the conditions of the thermomechanical process. On the other hand,

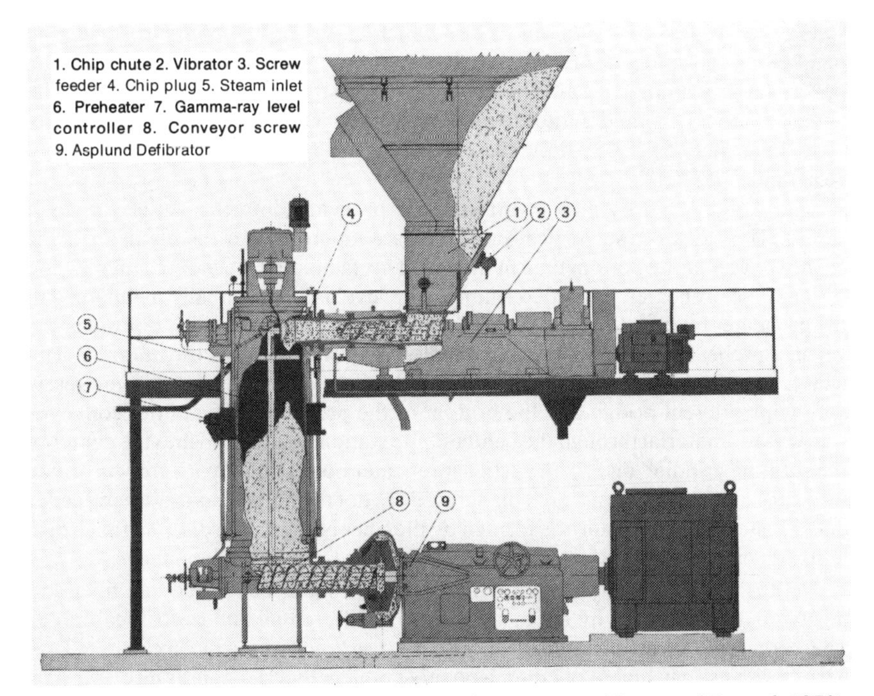

Figure 7.32. *Modern Asplund Defibrator pulping system (Gran and Bystedt 1973).*

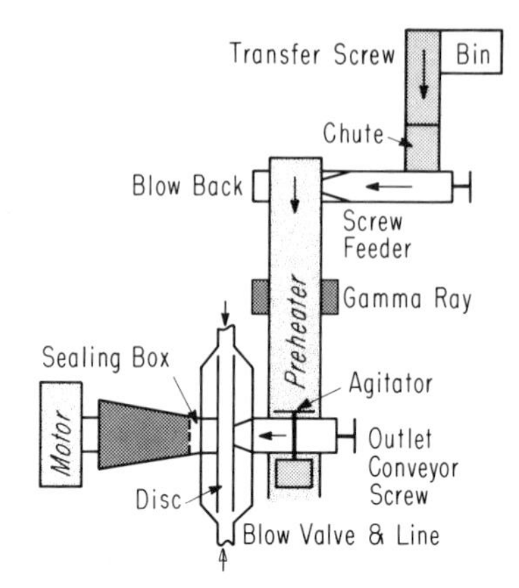

Figure 7.33. *Schematic drawing of L-42 Defibrator installation (McVey 1973).*

it has been found that when the wood is cooked to pulp yields below approximately 70% in the presence of strong acid or alkaline liquors, the compression may have an adverse effect upon the pulp quality. For these and other applications, where the use of the screw feeder may not be desired, rotary-valve type feeders are available.

The refiner is equipped with a blowback control system which prevents steam from blowing back through the screw feeder when a chip plug is not formed. This is done by an air-activated conical disc which closes off the plug pipe when the plug is lost. This plug of compressed chips emerges from the screw feeder into the pressurized steaming vessel. The steaming causes the plug to fall apart and the chips are quickly heated to the steam temperature. The steam pressure is usually 80 to 150 psig (0.55–1.05 MPa), corresponding to a temperature range of 320° to 365°F (160°-185°C).

The wood is fed by gravity through the preheater into the outlet conveyor screw. The sides of the vertical preheater are tapered, with the diameter being slightly larger at the bottom than at the top. This facilitates the movement of wood through the preheater. A predetermined chip level is automatically maintained in the preheater by the use of a gamma-ray source and detector. At the bottom of the vertical preheater is a slow-moving paddle-type scraper called the *agitator*. The agitator facilitates the movement of wood down into the outlet conveyor screw and helps prevent bridging at the bottom of the preheater. The outlet conveyor screw feeds material through the center of the stationary disc and into the center of the rotating grinding disc. The vertical preheater outlet conveyor screw is driven by a variable-speed DC motor. This screw does not form a plug as the screw feeder does. Instead, a plug of sorts is formed by the fiber between the discs so that a disc housing and preheater steam pressure differential can be maintained.

In the grinding chamber, the steam-softened wood is converted into fiber and thrown out of the discs by the steam pressure differential and centrifugal force. There is a separate steam control for both the disc housing and the preheater. This permits the establishment of either a positive or negative steam pressure differential between the preheater and the housing. Steam is introduced at the top of the steaming vessel, at the infeed tube of the refiner and at the disc housing. The pressure differential between the inlet and outlet of the refining discs can be controlled in order to obtain the proper flow of steam and fibers. The disc housing is equipped with a drain valve and a product blow valve and line. The product blow valve has an adjustable orifice which controls the rate at which fiber and steam are vented from the housing. The set pressure in the disc housing is maintained automatically by adjusting the steam pressure controls and blow valve opening. At the disc housing end of the main frame bearing bed is located the sealing box for the machine shaft. Water is used to cool and lubricate the packing and to prevent fiber penetration into the packing. Seal water is always maintained at a pressure greater than the working pressure in the disc housing. This means there is continuous discharge of water, internally into the disc housing and externally to the atmosphere. This discharge of water also directly cools the stellite wear sleeve which protects the machine shaft.

Between the front and rear bearings there is an integral hydraulic cylinder for controlling the gap setting between the discs. The piston of this cylinder is fitted to the front bearing housing. When oil is pumped to the rear cylinder chamber, the whole shaft assembly with the rotating disc is moved forwards, i.e., the disc gap is closed. To open the discs, oil is pumped to the other cylinder chamber.

At a set disc gap, the closing and opening hydraulic pressures are automatically controlled to maintain the disc gap with an accuracy of 0.0004 in. (0.01 mm). The hydraulic servo system increases or decreases the disc clearance and, thus, the grinding pressure to balance the grinding loads imposed by the variation of infeed material. Manual disc adjustments can also be made.

The L-42 refiner usually has the rotating disc turning at 1800 rpm and is driven by a 2000 HP (1.5 MW) motor. A contant-speed AC motor can be used to drive the screw feeder or a variable-speed DC drive.

The screw feeder on the Defibrator is run at a constant speed; the amount of wood going to the preheater is controlled by the variable-speed screw feed out of the chip bin. This "starved screw" principle has worked very well in the field. There are some obvious limitations, as for instance if a certain minimum wood feed rate related to the speed of the screw feeder drops too low, there is insufficient wood to form a plug. If a certain maximum wood feed rate related to the speed of the screw feeder is exceeded, the wood will not be carried away fast enough and the incoming wood feed system will plug. The wood feed rate range over which these two extremes will occur is quite wide (3 to 9 tons/hr or 2.72 to 8.2 mt). The wood feed rate range can be changed by varying the size of the screw feeder drive pulley.

The type of wood raw material that is being processed through the refiner makes a big difference in the selection of the feeder screw. For example, the screw for running green material may not work with dry material. A plug must be formed and driven into the preheater as shown in Figure 7.34. In one example, the compression ratio of the screw was not high enough when running dry material to form enough pressure on the grooved walls of the throat or the smooth walls of the plug pipe to prevent the plug from turning. The turning of the plug in the throat and plug pipe prevented it from being extruded into the preheater. This rotating plug would stay in the plug pipe, preventing the passage of additional

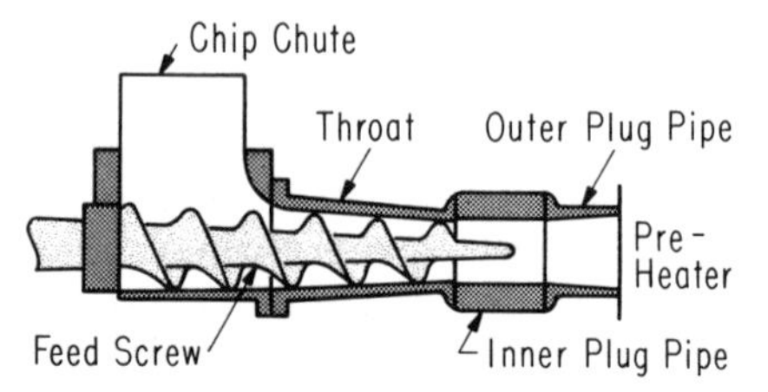

Figure 7.34. *Schematic view of plug forming operation (McVey 1973).*

material into the preheater. This problem was corrected by replacing the feeder screw with one which had a higher compression ratio and by also replacing the smooth bore plug pipe with one machined with grooves. Such refinements may have to be made in the field once a plant commences operation. These refiners can have steam blowback problems, where steam goes back through the screw feeder and chip chute.

It is possible for small chips or particles to foul the blowback control. Manual operation may correct the problem or it may be necessary to reduce the steam pressure in the preheater to allow the chip plug to form again. A modification by one plant to help minimize the problem of plugging the vertical preheater was to weld a couple of inches on to each end of the agitator blade (Figure 7.35). This prevents an inner shell of wood from forming on the inside of the preheater walls and holding up the pile of chips. It also facilitates the removal of wood from the preheater if it should become plugged. The speed of the agitator is important for uniform feed to the refining discs. One plant uses an agitator rpm from 4 to 8 to achieve this uniformity.

Serious consideration must be given to the location of the discharge from the waste blow line used to exhaust the steam, water, and coarse fiber from the housing when starting a refiner. Each time the machine is started, it must be vented until near normal operating conditions are reached; otherwise serious pipe and dryer build-up will occur from the excessive amounts of water being vented with the fiber. This material presents a potential pollution problem as it is vented to either a drain or to the atmosphere. The best way of handling this material appears to be to mix it back in with the incoming raw material. However, this idea involves handling problems in storage, metering, and blending.

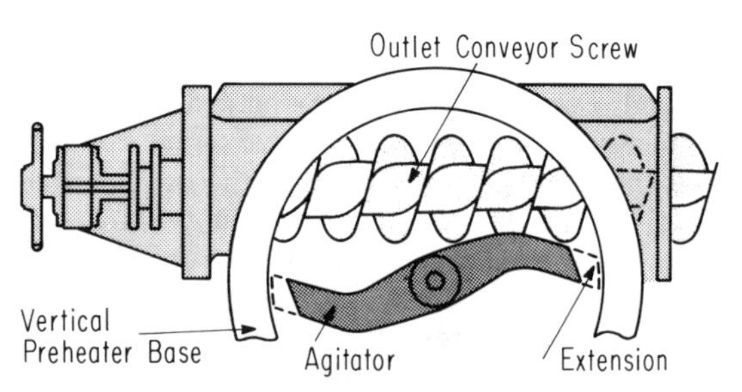

Figure 7.35. *View of agitator extension in vertical preheater (McVey 1973).*

In small machines, a reciprocating-valve-type discharger is used because it is easier to control the steam consumption with the reciprocating valves. The blow valve is used in larger machines, but it has to be adjusted very carefully because, with a small opening, there is always the hazard of plugging the valve; whereas, a large opening caused excessive steam consumption. At a high production rate, however, where most of the machines run, the opening has to be larger to allow for the increased flow of fibers and steam. The steam consumption per ton of fiber is not increased. Also the larger opening is not so easily clogged. The flow through the blow valve is continuous and less restricted than by a reciprocating valve. Sampling the fiber shortly after it leaves the housing is a problem with pressurized refiners. The product blow line has an above atmospheric pressure which makes sampling somewhat difficult.

Two potential problems, among others, may be present with pressurized refiners. First, the entire machine is hot, and under this condition the proper plate gap setting is more or less fixed. However, if the steam pressure is lost, the entire machine starts to contract which can result in the discs being ground together if the initial gap setting was small. Some type of interlock may be necessary to open the gap, thus preventing the plates from grinding together and causing damage to each other as well as posing a potential fire hazard. The second problem is the necessity for stopping the machine because of malfunctions on the production lines. As noted before, pressure-refined fiber has a relatively low bulk density. Consequently, on a weight basis, a relatively small amount of storage space for freshly refined fiber will be all that can be normally expected. Electrical and mechanical interlocks are therefore necessary to protect the equipment and to make it possible for an orderly shutdown when stops of more than short duration are necessary.

Acidity and Moisture Content

Acidity and moisture content of the raw material can be a problem in grinding good fiber. Under conditions of steaming, pressure, and temperature, various amounts of weak and strong organic acids are produced. The amount and type of acids have been found to vary considerably, depending on species and, to a lesser extent, within a species. It has been observed that the pH of condensate from steamed chips has been on the acid side, and at equilibrium seems to buffer between pH 4 and 5. Research on digestion of various species has shown that great variations may be expected between species at the same oven-dry weight of wood with regard to the amount of acid titratable at a given steam pressure. As pressure increases from 50 to 125 psi (0.34–0.86 MPa) in a species, the total acid produced may be doubled. In digestion and refining, the range of acid content can produce many problems. As the acid content increases, chemical acid hydrolysis of lignin and cellulose increases; and when combined with the already high temperature and pressure in the digester-refiner, the action on the fiber could be either troublesome or desirable.

Investigations of the chemical nature and quantity of acids have shown that different species of wood will be attacked at different rates and when mixtures of wood species are made, one must be careful not to mix highly acidic species with wood species that are subject to acid attack. Some woods, such as gums, produce small quantities of acids even under severe conditions (150 psi [1.03 MPa] for five minutes) while others, such as red oak, produce four to eight times the titratable acid under the same conditions. Acid attack on gum is vigorous and a 50–50 mixture of gum and oak can lead to a fiber that is short in length, difficult and dangerous to dry, and one that will cause excessive sticking in the press. This type of fiber will possibly require an excessively high resin level in order to produce boards of the desired quality. Water absorption will be high and water soluble material will be higher in these boards than in the corresponding 100% gum or 100% oak boards.

There are other variables that can cause problems, such as treatment of wood, storage, washing and/or drying. When there is a mixture of wood species, control of the mixture of species is necessary or the mixture must be treated for the most troublesome member. Penalities may consequently be imposed by the restrictions on pressure and time. The acidity of digestion conditions has one more effect, which is corrosion. Special alloy plates probably are necessary in the refiner to achieve a reasonably long plate life when using highly acidic woods. Abrasion, plus high acidic conditions, wears plates and drops fiber quality while increasing power requirements for a given fiber.

FLAKERS

Three types of flaking are available at present: drum or cylindrical, ring, and disc. The first two are the most popular throughout the world. Disc flaking produces the highest quality flake and, while used extensively in the early days of particleboard, it has been mostly supplemented by drum and ring flakers because of cost and capacity considerations. A sophisticated flaker is now being developed for cutting flakes from roundwood being cut into timber and lumber products, working much like a chip-and-saw headrig.

All cylindrical-type flakers (Figure 7.36), knife-ring-type flakers (Figure 7.37), and disc flakers use the same cutting principle. Adjustable, sharp knives penetrate the wood and cut flakes of a predetermined thickness. Closer control is usually found with disc and drum flakers. Planers and molders also use the cutting principle; however, the feed direction of the wood is perpendicular to the rotor shaft. Therefore, the planer shavings usually are curled with one end thicker than the other. As a reminder, however, it should be noted that each species planes differently and the many factors involved in planing can result in large flaky particles through tiny cubical ones.

Roundwood and split wood of large dimensions are usually flaked with drum flakers, although the disc flaker is popular in Canada. Wood of small diameter and sawmill waste are usually converted into chips first and, in a second operation, the chips are flaked in a knife-ring-type flaker.

A knife cuts into the wood relatively easily, particularly if the wood moisture content is above fiber saturation. That means the power consumption in kWh/ton is low. A second advantage of the cutting is that the flake thickness can be set to certain tolerances. Cut flakes have a relatively smooth surface which assists with resin bonding. Resin is not lost in rough surfaces. Very green wood, however, can also result in flakes with rough surfaces as the green wood can crush under the

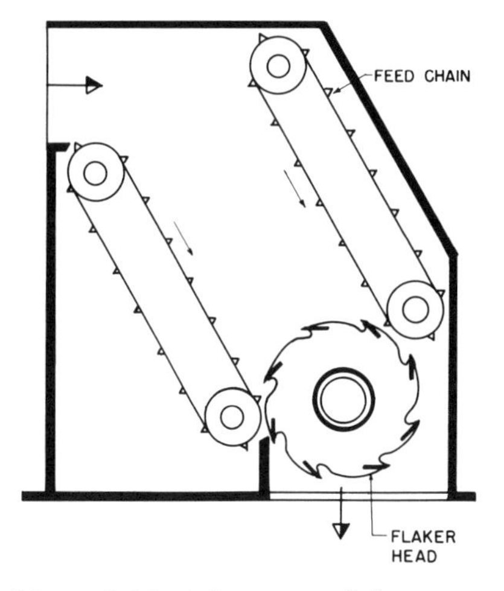

Figure 7.36. *A drum-type flaker (Fischer 1972).*

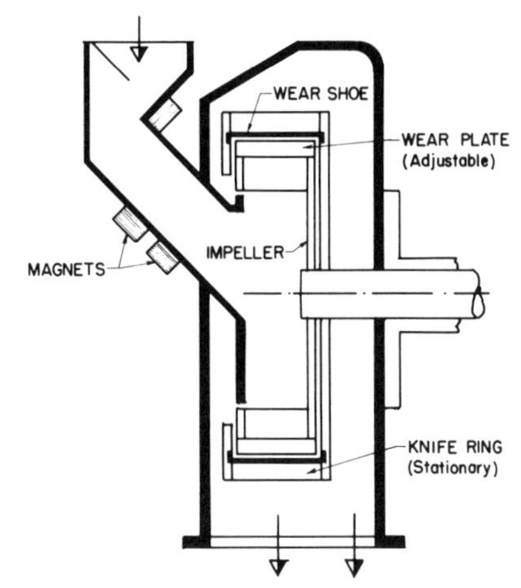

Figure 7.37. *A knife-ring-type flaker (Fischer 1972).*

flaking pressure. Unfortunately knives become blunt and dull after a short time of operation. Knife changing, knife sharpening, knife setting and the knife itself are expensive items. Flake quality often depends on the maintenance by operators. For the cylindrical-type flaker an operator is needed.

Disc flakers are usually relatively slow operating. Flake quality can be high, however, and for certain types of plants this type of flaker is competitive. Usually the cost of flaking is too expensive for products that can be made with shavings or ring-cut flakes.

Ring Flakers

The most widely used flaker, particularly in North America, is the ring flaker. Previously described in general, the machine is capable of reducing chips, hogged material, and large shavings into acceptable flakelike furnish. The resultant furnish usually has a wide variation in size, as shown in Figures 7.38 and 7.39. However, a certain amount of size variation can be of advantage as particles can intermesh better, usually resulting in boards with better internal bonds.

Two general types of ring flakers have been developed: one which has a stationary knife ring and the other that has a revolving knife ring which is either counter-rotating to the impeller or rotating in the same direction as the impeller.

Figure 7.40 shows the general cutting action of the counter-rotating flaker. Figure 7.41 is a close-up of a section of the knife ring, and Figure 7.42 is a schematic drawing illustrating the cutting action.

Feed material enters through a feed chute into a flaker proper where the fast-rotating impeller catches it and forces it against the counter-rotating knife ring. The flakes are cut off according to the knife projection. Cut flakes are discharged through the gap between the knife and pressure lips.

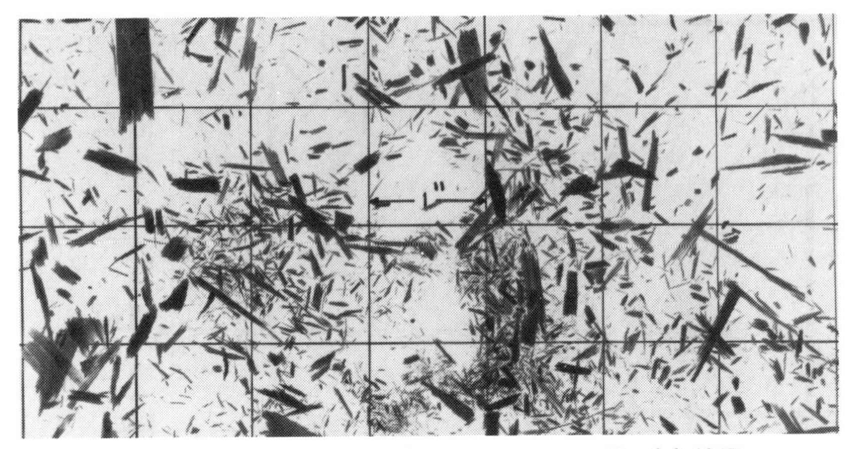

Figure 7.38. *Chips ring-flaked at 70% moisture content (Surdyk 1967).*

Figure 7.43 shows a sophisticated new model of this machine. The hydraulic release system and spring system are part of a new development which permits extremely fast changing of the knives. Previously, the whole knife ring had to be removed for knife sharpening. Now with the new system, knife changing time has been sharply reduced.

Hundreds of these flakers are in use throughout the world, handling a wide range of wood species of varying densities. They can reduce a variety of sizes and forms of wood into a broad range of flake sizes at a fast rate.

Three important steps are interlocked for successful operation: proper feeding of material into the machine, actual flaking, and proper discharge of the newly cut flakes. Failure in any one of the steps destroys optimum flaking action and flake quality.

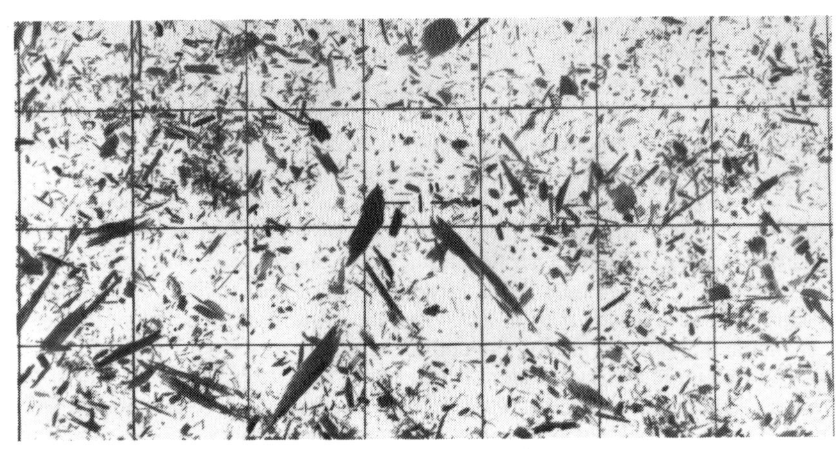

Figure 7.39. *Flakes from planer shavings (Surdyk 1967).*

Figure 7.40. *Cutting action of a knife-ring flaker with a counter-rotating impeller (Surdyk 1967).*

A number of factors affect the performance of any ring flaker and control of these are necessary for achieving optimum performance by the machine. Factors related to the machine are as follows:

1. Sharpness of the knives.
2. Amount of wear on the other interior parts.
3. The gap between knife edge and impeller counterknives.
4. The projection of the knife into the ring.
5. The gap width between the knife edge and leading wear plates (pressure lip).
6. The angles of the knife edges.
7. The angle of the knives in the machine.
8. Impeller rpm and horsepower.
9. Knife-ring rpm and horsepower (if the ring is rotating).

Factors related to the wood are:

1. Amount of metal, stones, and other debris in raw material.
2. Chip shape and size and their range.
3. Moisture content.
4. Chip feed rate and uniformity of feed rate.
5. Species.
6. Chip temperature (whether frozen, warm or steamed).
7. Species density.

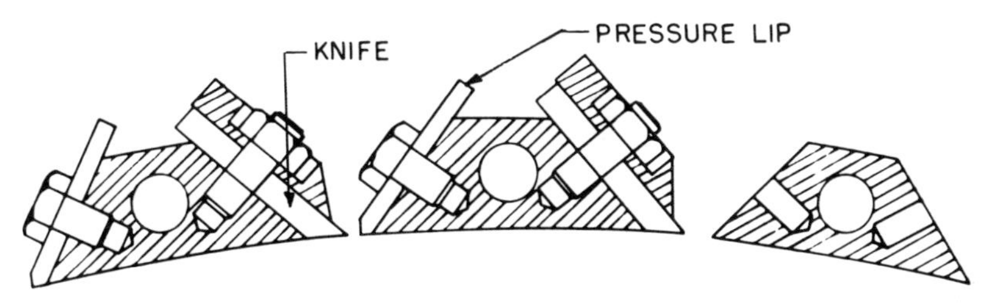

Figure 7.41. *A section of the knife ring (Surdyk 1967).*

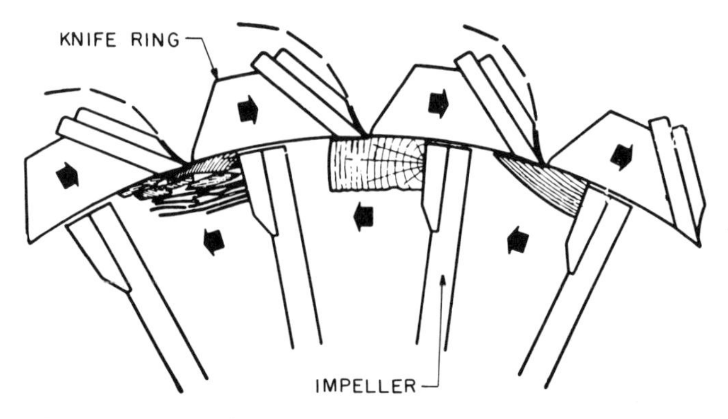

Figure 7.42. *A section of the knife ring and impeller showing flaking taking place (Surdyk 1967).*

Since there are so many factors, it is a problem to determine the right combination for producing the desired flake. Controlled factors specified by the machine manufacturer are horsepower, impeller and knife-ring rpm, gap between knife edge and leading wear plate, knife angle, and feed rate range. Wood species and density will depend on the plant location. Some experimentation may be needed to find proper angles of the knife edges and the projection of the knives into the ring.

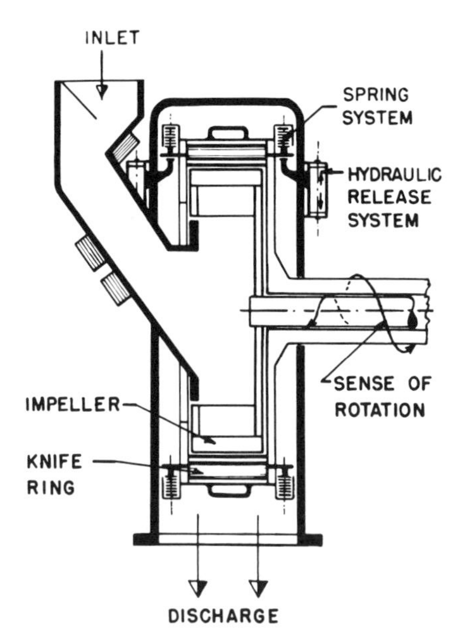

Figure 7.43. *New design of a Pallmann ring flaker with fast knife-changing system (Fischer 1974).*

Further experimentation can establish limits for knife sharpness, machine wear, moisture content, and range of chip sizes.

There remains the establishment of an adequate quality-control program to insure that the machine is operating within the established limits for producing the desired flakes. One system is to run frequent tests on the flakes, such as thickness measurements and screen analysis, and to watch the impeller-motor ammeter (a high load indicates dull knives). Experience in a given installation will establish the optimum time interval for changing the knives and other wearing parts to maintain flake quality. However, the ammeter on the drive motor must constantly be checked in case the knives are dulled rapidly from foreign material such as rocks or metal.

Eliminating rocks and metal from the material going into the flaker is critical. Metal detection equipment and magnets are absolutely necessary. Suitable rock drops are effective for separating large rocks and heavier metal pieces. A chip-washing unit is especially effective because it will remove sand, small rocks, and metal fragments as well as large rocks and heavy metal objects. The increase in knife life, reduction of interior parts wear, and increased flake quality due to cutting wet material probably will more than offset the added cost of drying a higher-moisture-content furnish.

Depending on plant location in the United States, one could flake wood at temperatures of 60°F (16°C) or more, or frozen wood. Better-quality flakes are cut from the warmer wood especially if the moisture content is above the fiber saturation point. Far more finer particles can be generated from frozen wood. This reaction to frozen wood is particularly true if the rotor counterknife to knife clearance is 0.040 in. (1 mm), or more, or if the knives are dull. With a carefully maintained machine and well-adjusted clearances, good flakes can be produced from frozen wood. Many fines can also be generated from very wet chips (200% moisture content chips), evidently because this very wet material tends to compress and perhaps disintegrate somewhat as it does in an attrition mill.

One of the virtues of this type of machine, however, is its ability to produce acceptable particles from a wide variety of raw material. Figure 7.44 shows a normal distribution of flake thickness from a typical flaker setup. The thicker particles can be reground either in an attrition or an impact mill.

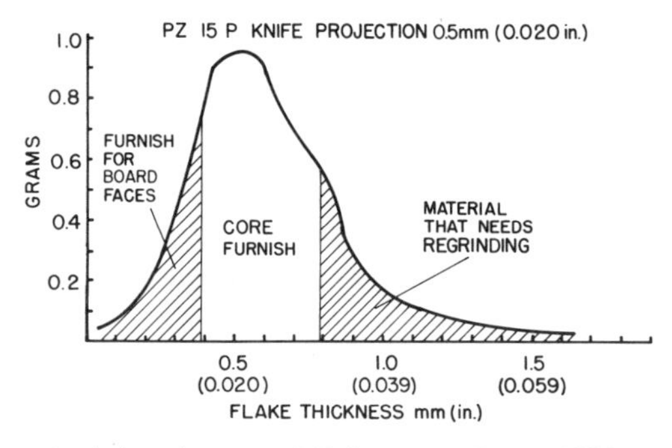

Figure 7.44. *A typical distribution of flake sizes (Jager 1975).*

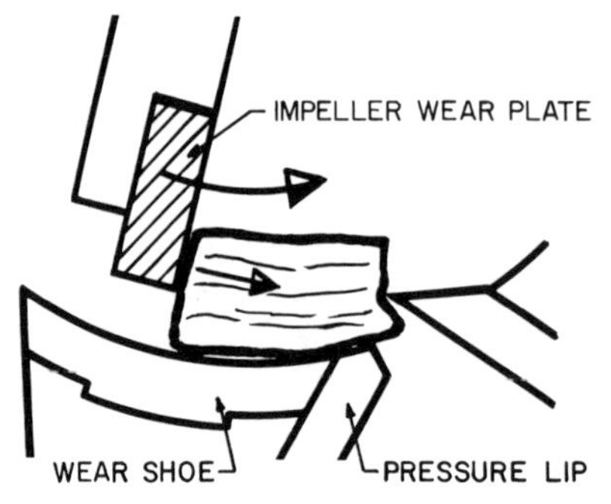

Figure 7.45. *Flaking action in a well-maintained machine (Fischer 1974).*

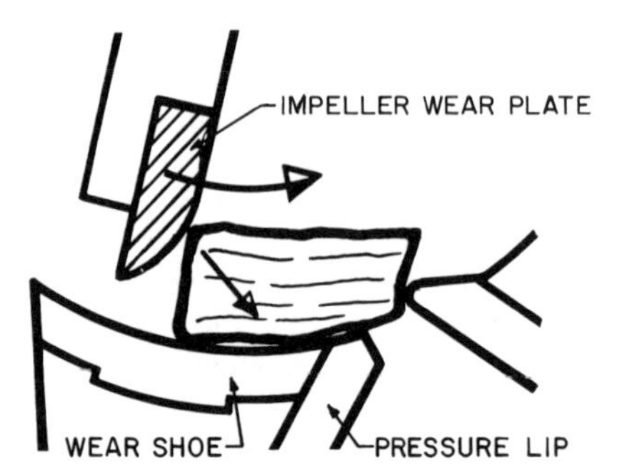

Figure 7.46. *Flaking action in a machine with worn knives and impeller wear plates (Fischer 1974).*

Naturally the range of chip or other raw material sizes has a great effect on flake size and shape as well as on the range of sizes. Additionally, the knife sharpness and maintenance of the machine are of importance. Figure 7.45 shows the flaking action in a well-maintained machine. The impeller wear plate drives the chip parallel to the inner face of the knife ring. However, this ideal condition changes with time. Blunt knives cut poorly and more power is required for operation. The rounded edge of the impeller wear plate also causes significant problems. Figure 7.46 illustrates what happens as the impeller imparts the driving force into, rather than parallel to, the knife ring. This causes even more wear on the wear shoes and impeller wear plates. The gap between the two widens into a hopper-like shape. Small chips can be squeezed into this gap, causing further wear, and they also function as a brake within the machine thus requiring more power for operation. Flake quality, of course, also diminishes. A similar condition exists when the counterknife to knife clearance becomes excessive

The other type of ring flaker has a stationary knife ring and both the impeller and knife ring are in the shape of a truncated cone (Figure 7.47).

Figure 7.48 is a schematic diagram of this flaker showing the key parts.

The material to be flaked is fed into the machine through the hopper shown in Figure 7.48. This material is uniformly distributed by the rotor over the length of the flaking knives. Counterknives placed on the ends of the rotor blades then force the material to be flaked across the knives. The rotor blades and the counterknives are set diagonally to the cutting knives to counteract the tendency of the chips to move toward the large diameter of the rotor.

The projection of the flaker knives within the inner surface of the knife ring determines the flake thickness to be cut. However, splinters over minimum thickness, resembling broken match sticks or toothpicks, will occur in increasing percentages as the clearance between the knives and counterknives increases beyond 0.040 in. (1 mm). Flakers with cylindrical knife rings and rotors have a running clearance between the knives and counterknives which varies inversely with the flake thickness for which the knives are set. At a given knife setting, the running clearance is essentially fixed.

With a conical knife ring and rotor, the running clearance can be adjusted independently of flake thickness by moving the rotor axially toward or away from the knife ring. This is done by rotating the adjustment nut (Item 8, Figure 7.48) to move the rotor toward the knife ring until contact is made between the knives and counterknives. The calibrated adjusting nut is then rotated in the opposite direction until the desired running clearance of about 0.010 in. (0.25 mm) is obtained. The adjusting nut is then locked to hold the rotor in the set position. Readjusting the running clearance is not normally required unless the flake thickness or rotor counterknives are changed.

The material is fed through an infeed chute where an adjustable vane splits the material flow into the flaker just above the center of the rotor.

The flaker infeed is designed to accept material either wet or dry having a maximum piece size of 2 ⅜ in. (60 mm) in the greatest dimension. Occasional splinters 5 to 6 in. (127–152 mm) long have been found to feed without difficulty.

Permanent magnets are mounted in the infeed hopper and should be cleaned at each knife-ring change. Additional magnets or metal detectors are recommended in the feeder to the flaker. The vertical drop of material going into the flaker should not drop more than 2 ft (610 mm) before entering the infeed hopper. If material is allowed to drop more than 2 ft, tramp metal may attain enough velocity to bypass the magnets and damage the knives. As with all flakers the feed rate to the machine should be as uniform as possible.

Guide plates within the housing divide the infeeding material into two parts. As viewed through the casing door, the material falling down to the right of the shaft is directed towards the door to the small diameter of the rotor. Material to the left of

Figure 7.47. *A conical flaker with the knife ring removed (Schumann 1970).*

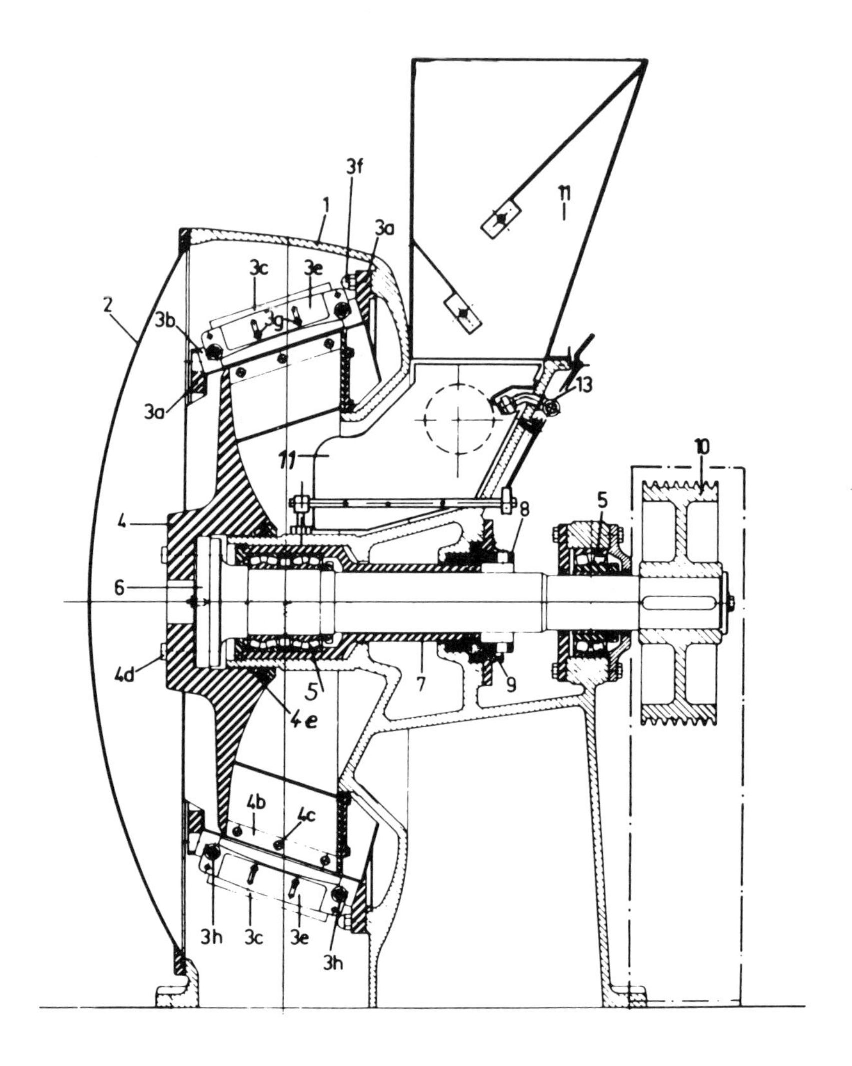

1. Machine housing
2. Housing door
3. Knife ring
3a. Knife ring flanges
3b. Knife supports
3c. Flaking knives
3e. Knife holding plates
3f. Bolts holding knife ring in flaker
3g. Bolts securing knife holding plate to knife
3h. Bolts holding knife assembly in place in the knife ring

4. Rotor
4b. Counterknives
4c. Counterknife bolts
4d. Bolts for holding rotor to shaft
4e. Dust seal
5. Bearing
6. Rotor shaft end plate
7. Rotor shaft housing
8. Clearance adjusting nut
9. Locking piece
10. Drive pulley
11. Material infeed chute
13. Water spray arrangement

Figure 7.48. *A cross-sectional view of a conical flaker (Schumann 1970).*

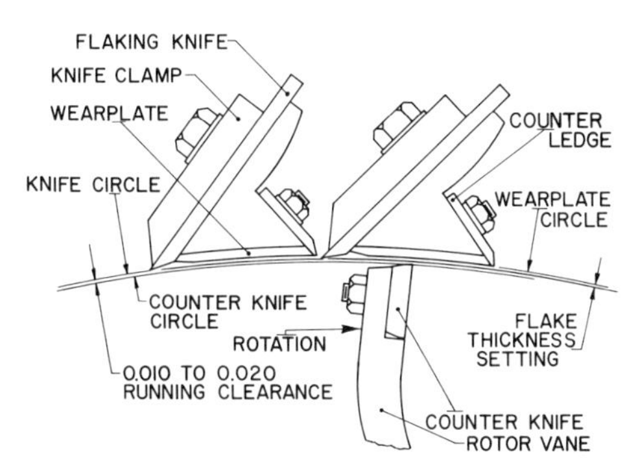

Figure 7.49. *The arrangement of knives and rotor counterknives for the cutting of flakes (Schumann 1970).*

the shaft falls relatively straight down to the large diameter of the rotor. By this means, the material feeding into the central area of the rotor is distributed uniformly over the full length of the flaking knives.

In Figure 7.48, a counterknife is shown on the tip of one of the rotor vanes, and its exact placement in relation to the knife ring is shown in Figure 7.49. This counterknife is made of abrasion-resistant steel with the leading edges protected by a hard-surface material. Both edges of the counterknife are designed for use; when one edge becomes worn, the knife is inverted and the other edge is used until regrinding is required.

Besides the knives themselves, critical parts of the ring are the knife clamps, the counterledges, and the wear plates. Each knife assembly, consisting of a knife and knife clamp, is bolted with three bolts to the knife support. The knife supports are an integral part of the knife ring. Below each knife support is a wear plate. The wear plates are exposed to heavy abrasive action because of the constant rubbing of wood material and, at times, of foreign material such as sand or small rocks.

The counterledges serve two functions:

1. They provide a renewable wear surface above which the flaking knife projects, establishing the projection of the flaking knife to determine flake thickness.
2. They establish the width of the opening between the counterledge and the leading edge of the flaking knife. The size of this opening has a major effect on the uniformity of the flakes produced.

Figure 7.49 shows schematically the cutting action of the flaker (cross section of knife ring and rotor). The rotor of this flaker functions as a pump impeller that conveys wood material entrained in air. As the rotor revolves within the flaker, flow conditions develop which are similar to a pump, and the rotor distributes the material evenly across the cutting knives. A row of fan blades on the rotor provides air in addition to that generated by the rotor vanes to assure complete scavenging of the cut flakes from the machine.

This flaker has the advantage of two adjustments which control flake thickness. First, a predetermined knife setting is fixed when the knife assembly is put

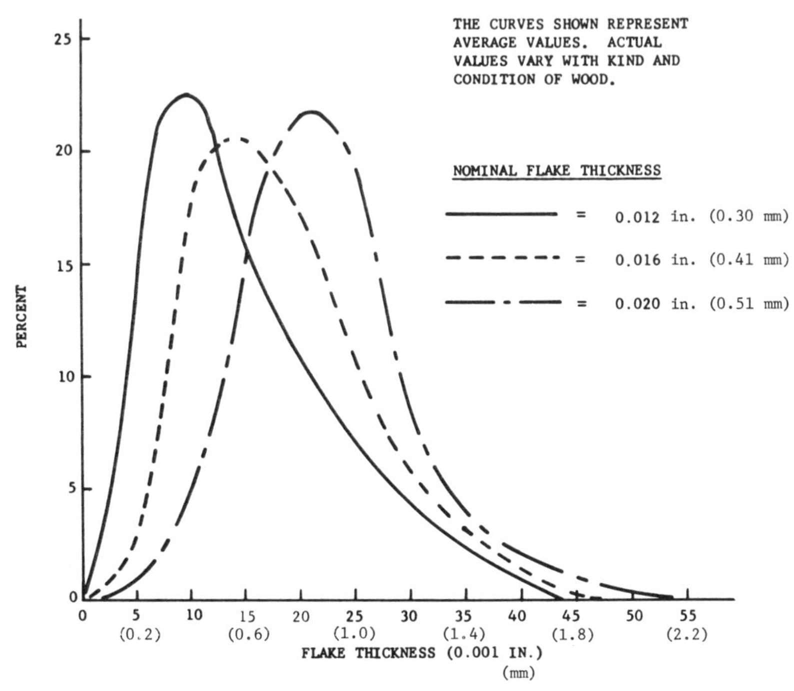

Figure 7.50. *Flake thickness distribution when cutting three different nominal thicknesses of flakes (Schumann 1970).*

together. Second, the axial adjustment of the rotor within the truncated-cone knife ring allows for precise final control over the flake thickness.

A water spray is available to assist with the flaking of very dry chips and particles (Figure 7.48).

At least two knife rings generally should be provided for each machine. As soon as the flaking knives of one ring become dull, the complete ring can be exchanged for the second one, which normally requires a downtime of 5 to 10 minutes.

After the knife ring is taken out of the machine, it is placed on a ring carrier equipped with rollers. Thereafter, no additional handling of the ring is required for the knife changes. By simple rotation of the knife ring on the carriage, dull knife assemblies can be removed quickly and easily and new sharp ones inserted. Bolt pressure is regulated by using a torque wrench. Any conventional, standard knife grinder can sharpen the flaker knives, and no special tooling is required.

The stationary truncated-cone-shaped knife ring is of simple construction. Problems of unbalanced knife rings due to different knife weights are eliminated because the ring does not rotate. Since only the rotor revolves, there is just one drive motor and belt drive required.

Figure 7.50 presents summarized data on flake thickness distribution in an experiment cutting three different thicknesses of flakes.

Drum or Cylindrical Flakers

The other popular flaker, particularly in Europe and other parts of the world outside of North America, is the drum or cylindrical flaker. Its use, however, is growing in North America. It works particularly well when reducing roundwood into high-quality flakes either for furniture-type panels or structural particleboards. Several designs are available but the basic difference between the two is that one style has to have the stock cut to predetermined lengths while the other style can cut flakes from random-length material. These flakers are in common use where roundwood is used for board. Figure 7.51 shows a typical wood yard when such flakers are used, which is uncommon in the United States. Figure 7.52 illustrates a large drum flaker for precut logs.

For cutting flakes, the long axis of the stock is fed parallel to the flaking drum. The cylindrical flaking drum then slices off the flakes. It is obvious that the rate of feed of the wood into the cutterhead must be regulated in accordance with the rate at which wood is cut away, which in turn is determined by the knife projection. (The cutterhead speed, of course, also affects this, but speed is constant for the machine.) If the wood is fed too rapidly, the cut surface will press against the cylinder, tending to brake the head and overload the cutterhead motor. If the wood is fed too slowly, flakes will be thinner than intended and more variable and uncontrolled in thickness. For each knife projection setting, there is therefore a proper wood feed against the cutterhead. Consequently, the output of the flakers

Figure 7.51. *Random stacking of 1-m round logs and billets in a round timber loader with following distribution. (Courtesy Hombak GmbH.)*

depends on two main factors: (1) the thickness of the flakes being cut and (2) the steadiness with which wood is fed against the cutterhead.

Flake length is established by using scoring knives or by setting the knives in staggered rows, in which case the flakes are torn from the stock leaving ragged ends. The use of scoring knives results in neatly cut ends on the flakes and fewer fines.

Flakers are available to produce flakes from stock lengths ranging from about 3 ft 7 in. to 7 ft 2 in. (1.09–2.18 m). Figure 7.53 shows a schematic drawing of one of these machines and Figure 7.54 shows a cutterhead.

In another smaller-type flaker, shown in Figure 7.55, the wood need not be reduced to a specified length. The wood is fed by gravity in a steeply sloping chute, which in reciprocating up and down alternately feeds the wood against the cutterhead in the up stroke and allows the wood to slide forward into place for the next cut after the down stroke. The reciprocation of the chute is effected by means of a hydraulic cylinder. This type of flaker has been mostly replaced by the larger types.

With the larger types, a horizontal conveyor carries the wood forward to a reciprocating cutterhead. This type also handles wood in any length and therefore eliminates the necessity of precutting to specified lengths. When ready to operate, the stock to be flaked is first clamped to prevent movement during flaking. The cutterhead, in the largest machine which is 4 ft 10 in. long by 3 ft 3 ½ in. in diameter (1.47x1 m), is then driven by hydraulic cylinders across the stock, reducing it to flakes. The cutterhead then returns to its original position, the stock is unclamped and indexed forward and reclamped ready for the pass by the cutterhead. Flake thickness usually ranges from 0.008 in. (0.2 mm) minimum to a maximum of 0.030 in. (0.76 mm) depending upon the thickness desired.

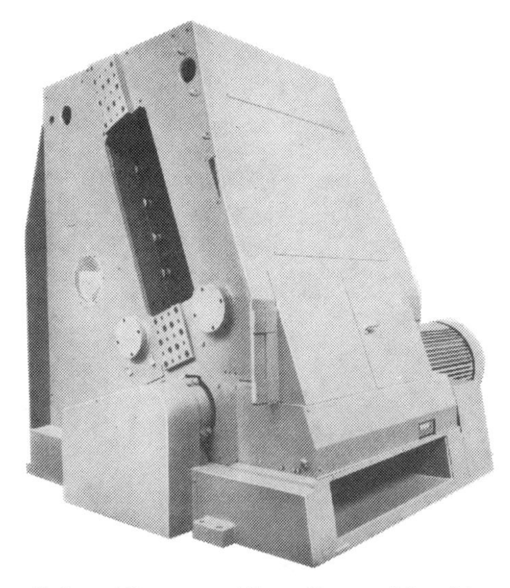

Figure 7.52. *Drum flaker. (Courtesy Albert Bezner Maschinenfabrik.)*

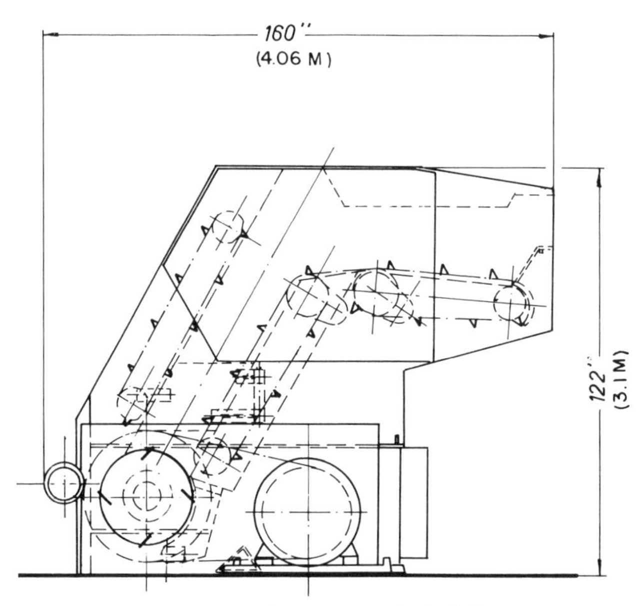

Figure 7.53. *Schematic drawing of PRZ Hombak flaker. (Courtesy Hombak GmbH.)*

Drum flakers are available which can handle tree-length logs up to 30 in. (762 mm) in diameter. Figure 7.56 shows logs being fed into the plant. The cutterhead and fixed-position backstop for one of these are shown in Figure 7.57.

When tree-length flakers are used, operations such as cutting, splitting, transporting, and feeding precut logs are eliminated. In addition, the tree-length logs lose less moisture in short-time storage than short logs, which makes it easier to maintain flake quality.

In addition to their use on tree-length roundwood, these drum flakers have been used successfully to flake green slabs and edgings which are fed into the machine in bundles.

Skidding logs in the woods results in having large quantities of dirt and stones adhering to the bark. These impurities will cause severe knife damage if the logs are not debarked or cleaned; thus it is recommended that logs be cleaned before flaking. By bouncing the logs several times on a heavy steel-beam structure before they are put into the feeding unit of the flaker, a large amount of sand, stones, and loose bark particles can be removed, and much longer duty cycles of the knives can be achieved. It should be noted that in many cases logs are flaked in Europe without debarking.

A commonly used method of fastening the knives in the cutterhead is shown in Figure 7.58. The knife (*A*) is fastened onto a knife carrier (*B*) by means of screws. The knife and knife carrier are held in place by a spring-loaded gib (*C*).

When the knives need resharpening, the cutterhead is turned to bring the gib of the first knife to be removed directly under a lever mounted on the cutterhead housing. The gib is forced down by means of this lever, compressing the spring (*E*), and as a consequence the wedging force of the gib against the knife carrier and knife is removed. The knife and carrier are slid out of the slot, and a fresh knife and carrier are slid back in place. The cutterhead is then moved to bring the next knife slot into position. The process is repeated until all of the knives have been changed.

The dull knives are sharpened by conventional means and then repositioned on the knife carriers by means of a jig. The knife is positioned on the carrier with the assistance of a micrometer dial indicator mounted on the jig.

Figure 7.54. *Hombak drum flaker, Type Z. (Courtesy Hombak GmbH.)*

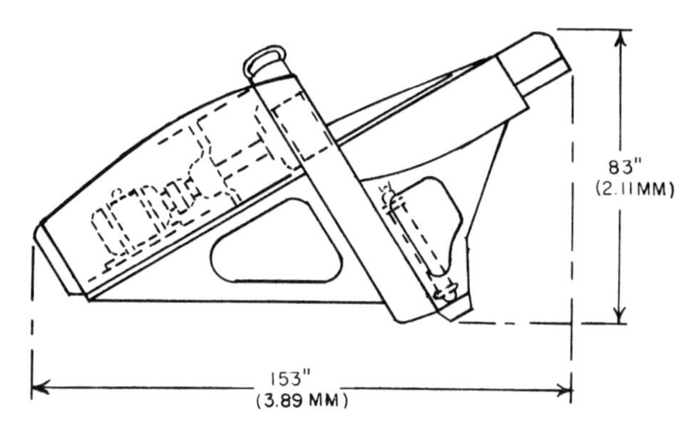

Figure 7.55. *Schematic drawing of the Type PRZ Hombak flaker (Mottet 1967).*

The position of the knives on the carriers determines their projection beyond the surface of the cylinder which in turn determines the thickness of the flakes cut. It can be seen in the diagram that the position of the knife carrier in the cutterhead always remains the same because of a tongue-and-groove arrangement (*D*) for positioning the knife carrier in the cutterhead.

This knife-changing arrangement provides for a very rapid method of changing knives. With a 10-knife cutterhead, the knives can be changed in as little as three to four minutes. On very large machines, knife-changing time is only 20 to 30 minutes. The ease and speed of knife changing encourages operating with sharper

Figure 7.56. *Logs being fed into plant for drum flaking. (Courtesy Hombak GmbH.)*

Figure 7.57. *Cutting area in a large drum flaker. (Courtesy Hombak GmbH.)*

knives on the average and therefore contributes to the quality of flakes. Interestingly, an even faster system of changing knives is being developed and should be available soon.

A wide range of quality flakes can be generated with drum flakers. Outputs can be from about 2600 lbs (1182 kg) per hour with a small unit to 30,000 lbs (13,636 kg) per hour with the largest units available. It should be noted that a greater production of thicker flakes is possible as compared to thinner flakes. When cutting 0.020-in.-thick (0.51 mm) flakes, for example, only half as many cuts are required as would be needed if 0.010-in.-thick (0.25 mm) flakes were being cut. With the rapidly developing structural particleboard product, greater use of the large flakers is expected.

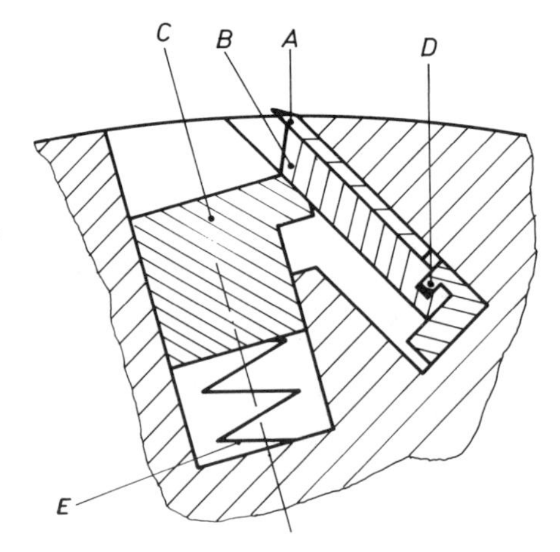

Figure 7.58. *A method of assembling knives in the cutterhead. (Courtesy Hombak GmbH.)*

Disc Flakers

Disc flakers can provide the highest quality flakes because of the precision with which stock is presented to the knives and the way the knives can be oriented to perform the cutting. Such precision, however, is not economical for most of the industry and some quality must be sacrificed for speed. It is interesting to note, however, that in some cases, thin and large flakes, particularly wide ones, can cause problems in board. These flakes when used as surface material can block the venting of vapor through the board faces during hot pressing, which can lead to blows. Such is the case especially with species that contain volatile extractives as for instance cedars and redwood. In these cases, it may be necessary to use a hammermill without a screen, wing-beater mill, or some flailing device to break the wide flakes lengthwise, thereby providing more surface vents.

Material to be flaked is fed edgewise against the cutting disc, whereas the same material would be fed endwise into a disc chipper for the cutting of chips. Otherwise, both types of machines operate on the same general principles.

Disc flakers can operate with the stock being fed against a horizontal, angled, or vertical disc. The most popular in North America uses a vertical disc. The flaker shown in Figure 7.59 is a 93-in.-diameter (2.36 m) disc oriented vertically. The wood length fed to the flaker is 24 to 24 ¼ in. (610–616 mm). Minimum and maximum log diameters are 4 and 20 in. (102 and 508 mm) respectively. Flake thicknesses can be adjusted between 0.010 and 0.060 in. (0.025–1.52 mm). The disc requires about a 200-HP (149 kW) motor and the disc revolves at about 500 rpm. Twenty knives are mounted in the disc. A sophisticated feeding system consisting of two sets of spiked chains on each side is needed to hold the stock in place until it is completely flaked as the last small piece of a log. Unless the stock is held

Figure 7.59. *A large disc flaker. (Courtesy CAE Machinery Ltd.)*

Figure 7.60. *A machine for producing wood wool. (Courtesy Maschinenfabrik Rehau.)*

properly, large chunks or highly irregular flakes may be generated. This flaker is popular in the rapidly expanding Canadian waferboard industry although drum flakers are also used. Most flakes being generated are about 1 ½ in. to 2 ⅛ in. (38.1–54 mm) long with a great variation in widths since the material being flaked is roundwood and it is impossible to predetermine widths with roundwood. For establishing flake length, scoring knives are set in the disc face. They cut tiny grooves across the face of the stock prior to the slicing off of the flakes by the flaker knives. If boards are being flaked, the basic flake width is determined by the board thickness.

In the waferboard industry, to improve the flaking operation, logs are placed in hot water ponds (temperature about 195°F [90°C]) for periods up to about four hours before flaking. Shorter conditioning times are used in the warmer summer months. This type of conditioning is used with both drum and disc flakers.

EXCELSIOR OR WOOD WOOL PREPARATION

Excelsior or wood wool products are relatively minor at present. Shortages of synthetic resins, which encourage the use of Portland cement as a binding agent (the normal for wood wool products), and the fire resistance of such products may lead to a significant expansion of this part of the industry. It should be mentioned, however, that resin can be used for wood wool panels. Figure 7.60 shows a typical machine for producing wood wool. In most machines, the log is held parallel to a reciprocating slide on which are mounted alternate slicing and scoring knives set for the desired thickness and width of excelsior. The log is fed downward against the reciprocating slide during the cutting operation.

MATERIAL HANDLING AND STORING

Conveyors and bins are used throughout the process. For convenience, they are being discussed here.

Storing and handling chips and particles, as with any part of the composition board process, result in a number of interactions between the various particles and their characteristics and the conveyor or bin configuration. For example, there are differences between wet and dry material. Flakes, being small wedges, tend to accumulate and jam into some pieces of conveying equipment, and this has to be taken into account in the design. Fiber, because of its low bulk density, can cause all kinds of problems in conveying and handling. All factors must be considered before choosing the final equipment. Many pitfalls are waiting to trap the unwary if they rush this part of the design phase of a new plant. There is no universal conveyor or bin that works equally well on all types of particles.

Conveyors

A number of different types of conveyors can be used, including screws or augers, belts, chains, vibration, and air or pneumatic systems. Air systems now have the problem of pollution to contend with as air exhausted to atmosphere must be cleaned of small particulate matter.

Simple belt-type conveyors are popular in the industry for fast movement of material such as shavings, chips, and blended furnish (Figure 7.61). Screw or auger conveyors are used for more precise operations such as feeding attrition mills where they can also serve as metering devices. Screws can be set in either the horizontal, inclined, or vertical position. In the vertical position, the simultaneous lifting-turning motion conveys granular material uphill along the helical trough.

Chain conveyors (Figure 7.62) can serve as a simple conveyor in many variations or as a metering conveyor as shown in Figure 7.63. Such a conveyor also can act as an elevator (Figure 7.64) in either the vertical or inclined mode.

Vibrating conveyors can move fragile particles without damage. If desired, other processing steps could be employed, such as screening, drying, and scalping. However, these are uncommon in the industry at present. Vibration can also be used effectively on material feeders for granular material, but they cannot be used too effectively with fiber.

Figure 7.61. *Belt conveyors carrying furnish from face and core blenders to the former. (Courtesy Washington State Univ.)*

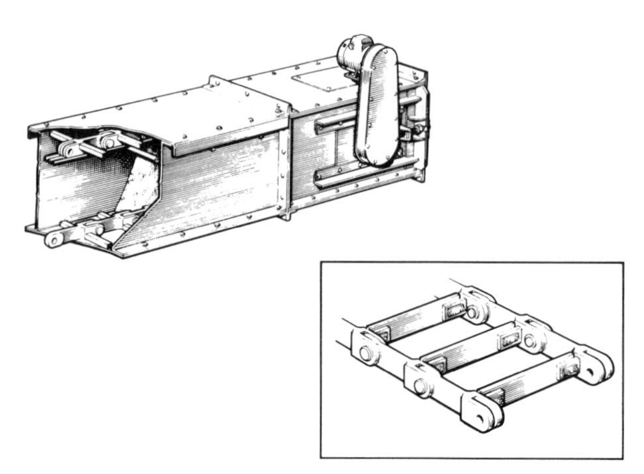

Figure 7.62. *Two types of conveyor chains. (Courtesy Redler Conveyors, Ltd.)*

Basic Conveyor	*Horizontal or inclined – controlled or flood feed – multiple inlets and outlets.* 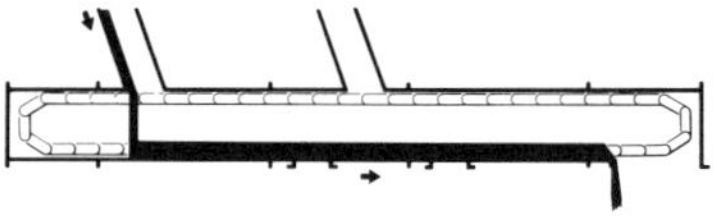
Basic Conveyor with Hoppered inlet	*For materials which must be fed directly into the conveying run – controlled or flood feed.* 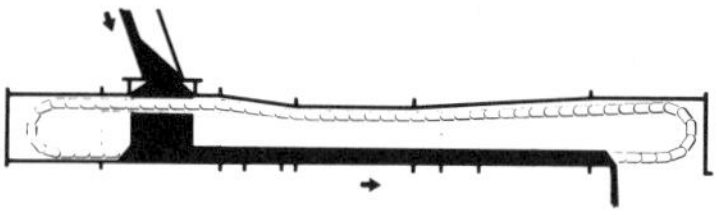
Split-leg Conveyor with Hoppered inlet	*Multiple inlets feeding directly into the conveying run – controlled or flood feed.* 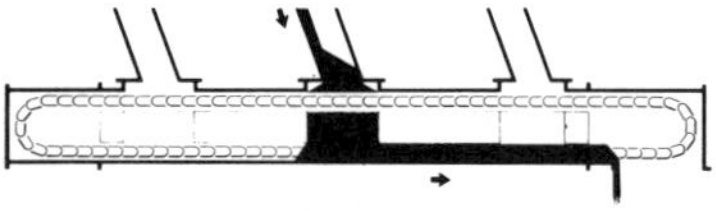
Two-way Conveyor	*Conveying in both directions at the same time – multiple inlets and outlets – controlled or flood feed.* 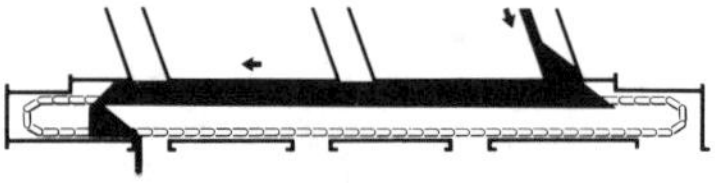
Metering Conveyor	*Discharges from a flood feed at a rate determined by an integral metering control gate. Can also be fitted with variable speed drive for variable rate feeding, or with totalising counter for batch measurement.* 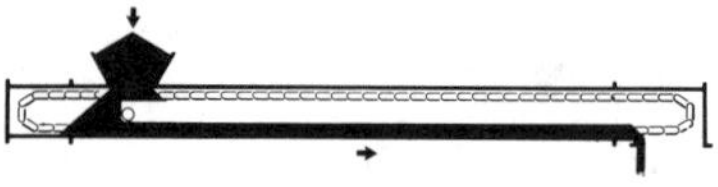

Figure 7.63. *Typical conveyor arrangements. (Courtesy Redler Conveyors, Ltd.)*

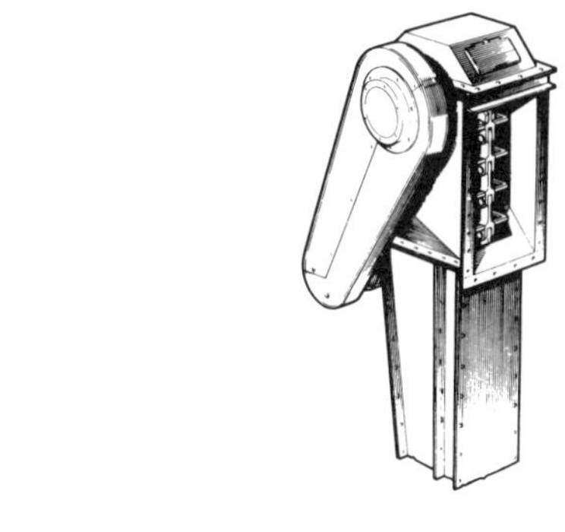

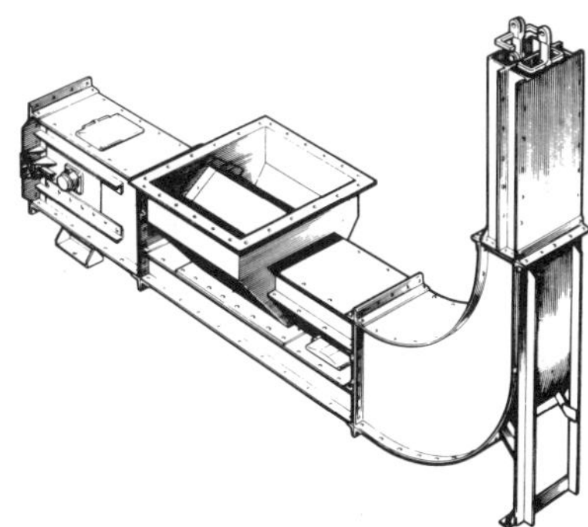

Figure 7.64. *A chain elevator. (Courtesy Redler Conveyors, Ltd.)*

Bins

To eliminate binning problems with the raw material as received, in many U.S. plants the particles are stored outside, as shown in Figure 7.65, or in large storage sheds. Material usually comes in by truck and is dumped (Figure 7.66) and then blown by air conveyors to the storage area. Conveyor belts mounted overhead deposit the particles in the desired pile (Figure 7.67). The dump operator controls the conveyor belt or belts, which can be adjusted over the length of the storage building. The operator can check on the storage area by using a television setup. Thus the large building becomes a large "bin." Usually such material must be stored in a building rather than outside as the finer particles can be blown about, causing air pollution. Front-end loaders (one shown in Figure 7.65) are used to meter the raw material into the plant by the scoopful.

Characteristics which affect bin design and operation are: particle size, moisture content, bulk density, angle of repose, angle of slide, coefficient of friction, abrasiveness, corrosiveness, compressibility, fibrosity or bulk density, fluidity, and the tendency to arch or bridge. It would be desirable never to stop furnish once it starts through a plant. This is very difficult to achieve as plants have failures in operating. Most of these are minor, but provisions must be built in to accumulate material and hold it so that the overall production can proceed as continuously as possible. Thus *surge* bins are necessary. They serve as a buffer in the flow of material between the particle-generation equipment and the dryers and

Figure 7.65. *A raw material storage yard and a front-end loader. (Courtesy Washington State Univ.)*

between the dryers and the blenders. Consequently, they are multifunctional as they store, meter, and proportionately deliver material.

Ideally, they are built low and long so that material cannot pack into them to a great degree. However, such bins take up large areas and so bins that are more compact—that is to say, tall and relatively small in area—are used in many cases. These types have more trouble with packing and bridging because of their configuration. All types of bins or silos should operate on the "first in—first out" principle so that no material is left unused for long periods of time.

Bins are built as silos or with a rectangular-shaped configuration. Some would define silos as *round storage structures,* and bins as the *rectangular shaped structures*. Bins are also known as *bunkers*. Wood particles, as is well known, tend to arch or bridge when confined in a bin. A number of approaches are taken to overcome this tendency of the particles to arch or bridge, ranging from live bottom screws to arch breakers. Size of the bin is also important, particularly if material is allowed to sit in them for several days. For example, particles left to sit in a full, large bin can become so well packed that extreme measures are necessary for breaking them loose.

The particle geometry is also quite important when considering storage. Pulp chips are relatively easy to handle, followed by sawdust, shavings, flakes, wafers, and fiber.

Silos: Silos for large bulk storage use several methods for unloading, including a rotating screw revolving on the horizontal axis, a screw operating on a V pattern, a single rotating shaft or arch breaker set on an inclined angle up into the bin operating in conjunction with a plate feeder, by using chains of sweep buckets, and a hydraulic rotor.

Figure 7.68 shows the first type. The single-rotating-screw unloader operates on the bottom of a suspended, circular storage bin. Screw unloaders of this type are particularly effective with wet material. The screw rotates about its own axis forcing the binned material out through an opening in the centrally located rotating thimble. The whole mechanism travels in a circular path about the vertical binned axis. A conveyor carries the discharged material away. The carriage travel speed and screw speed are adjustable. The pitch volume of the screw flights from end to end is varied so that the same amount of material is picked up along the entire length of the screw. As the screw moves forward, all material in its pathway on the bottom of the bin is moved out. The cavity left behind is filled by the material above it breaking and falling of its own weight. The agitation of the screw helps

Figure 7.66. *A truckload of wood raw material being dumped (Johnson 1973).*

with this filling action. Picker bars welded onto the screw flighting can help this filling action.

The V unloader is shown in Figure 7.69. It was designed as an inexpensive positive -discharge unloader adaptable to a wide variety of bin sizes and types of construction. It consists of a single operating screw at the bottom of a suspended circular storage bin. However, the screw does not rotate around its own axis, as with the previous type, but traverses back and forth in an arc. The discharge end is mounted on a pivot outside of the bin. The free end of the screw traverses an arc of about 64 degrees.

The type of discharge shown in Figure 7.70 is a combination of an arch breaker and circular bin discharger of a plate design. The arch breaker rotates inside the bin preventing bridging and allowing the rotary plate feeder to provide a positive discharge of the particles. Variable-speed drives can be used to control the discharge rate. A similar machine developed by Ehrsam has worked well in the United States.

The system using chains of sweep buckets is illustrated in Figure 7.71. This system can work either with bins or with open storage. From three to six sweep chains are used, depending on bin diameter and volume of flow required. Each sweep chain is fixed at one end to a powered rotating "pull ring" encircling the storage area. The other end is free or trailing. As the pull ring rotates around the periphery of the bin, the sweep chains automatically trail towards the center. The sweep buckets contact the stored material outside of the pile and as the pull ring continues to rotate, the buckets fill and the material is swept through the openings between "grizzly" bars into an outfeed conveyor recessed in the floor.

Figure 7.67. *Overhead conveyor in storage building. (Courtesy Washington State Univ.)*

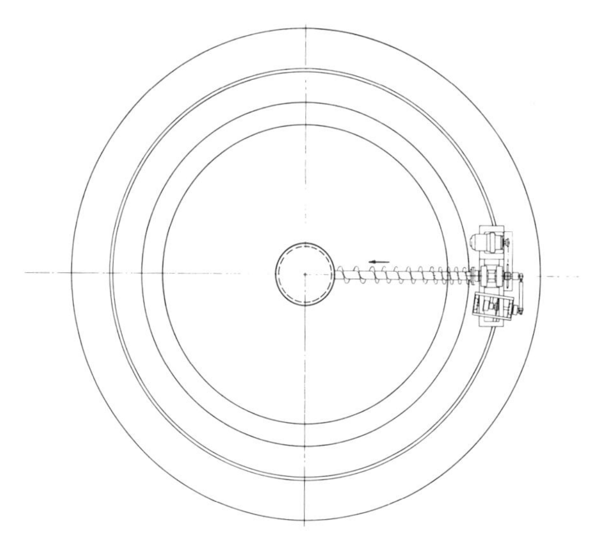

Figure 7.68. *Plan view of a Roto-Centric unloader. (Courtesy Miller Hofft, Inc.)*

The buckets feed from the outside of the pile under all conditions. Bridging of compacted materials thus does not interfere with the discharging process.

The outfeed is uniform and has been found as an effective surge and feeding system in volumes as small as 300 ft^3 (8.4 m^3). Very large installations have been made with this system.

In many parts of the world, large storage silos are in use that have a hydraulic rotor at the bottom for removing particles from the silo. As far as is known, all particles except fiber can be successfully handled by this system. Standard sizes range up to 600 m^3 (875 yd^3) in capacity.

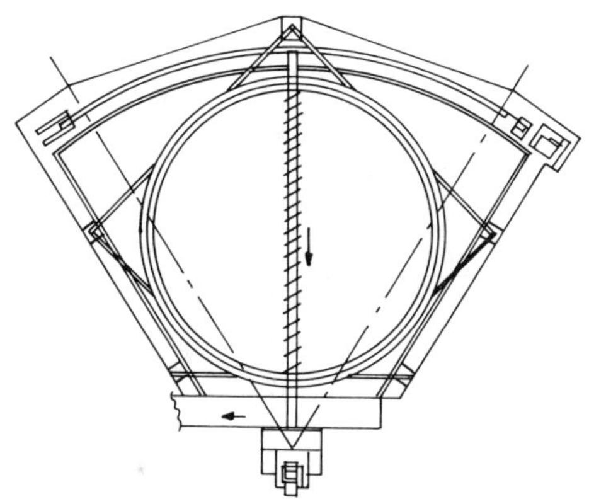

Figure 7.69. *Typical plan view of a type V unloader with suspended tank. (Courtesy Miller Hofft, Inc.)*

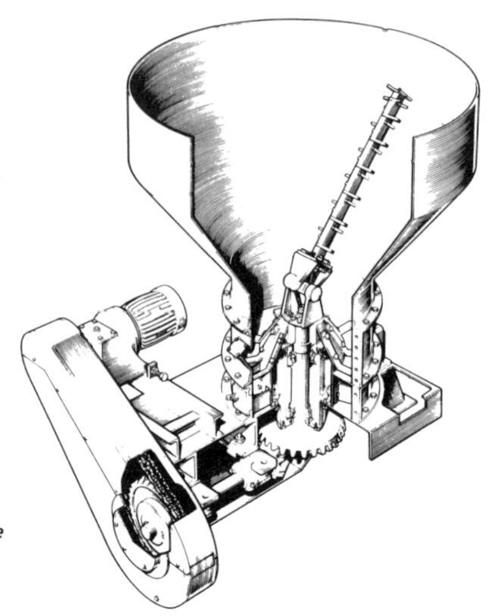

Figure 7.70. *An arch breaker system. (Courtesy Redler Conveyors, Ltd.)*

Figure 7.72 illustrates the rotor system which is mounted on the bottom of the silo. Material is fed into the silo and leveled by means of a distributing rotor. The hydraulic rotor system then removes the material at a predetermined rate by raking the material over openings in the silo floor. Both arms are mounted under the rotor, or one can be mounted above the rotor so that any bridging of the material is broken up. The set working pressure of the hydraulic cylinder is adjustable from the outside. (Another company's design uses spring-loaded arms, Figure 7.73.) The rotor arms are very rigid and work with full force right up to the silo wall, which helps prevent bridging. If a high resistance is met, however, the

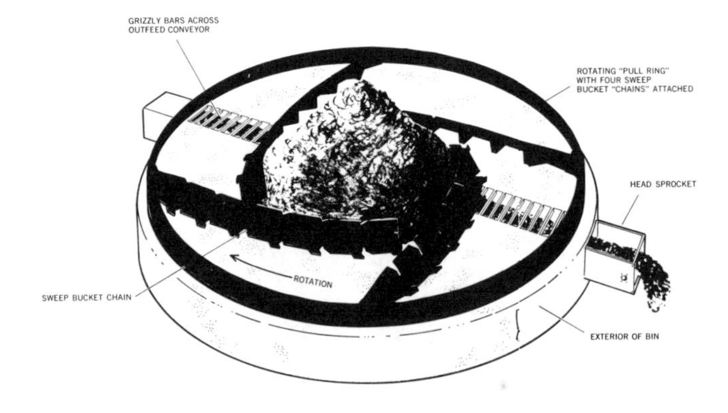

Figure 7.71. *System using chains of sweep buckets. (Courtesy Atlas Systems Corp.)*

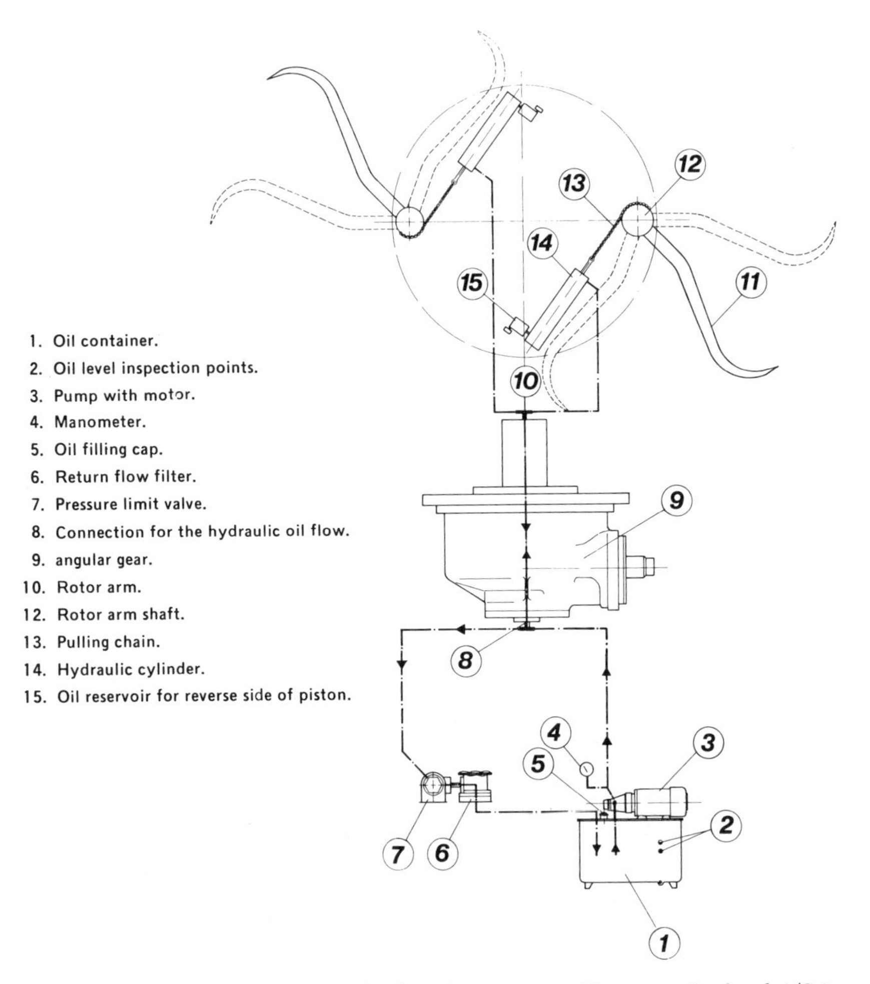

Figure 7.72. *Plan view of rotor discharging system. (Courtesy Saxlund A/S.)*

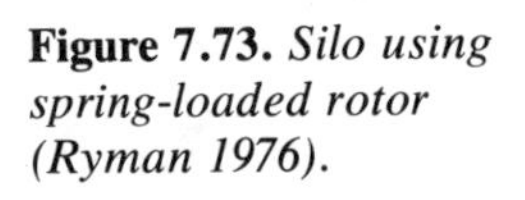
Figure 7.73. *Silo using spring-loaded rotor (Ryman 1976).*

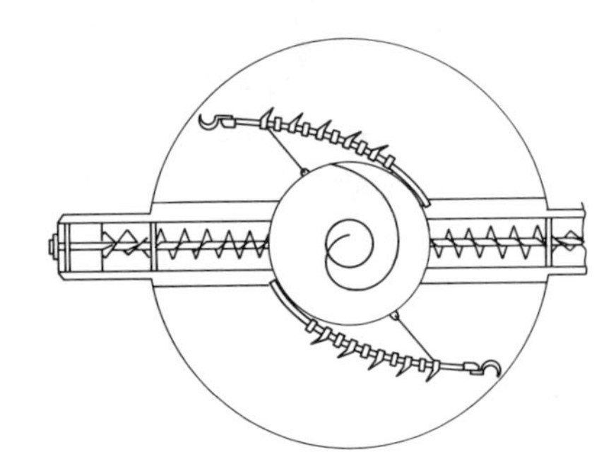

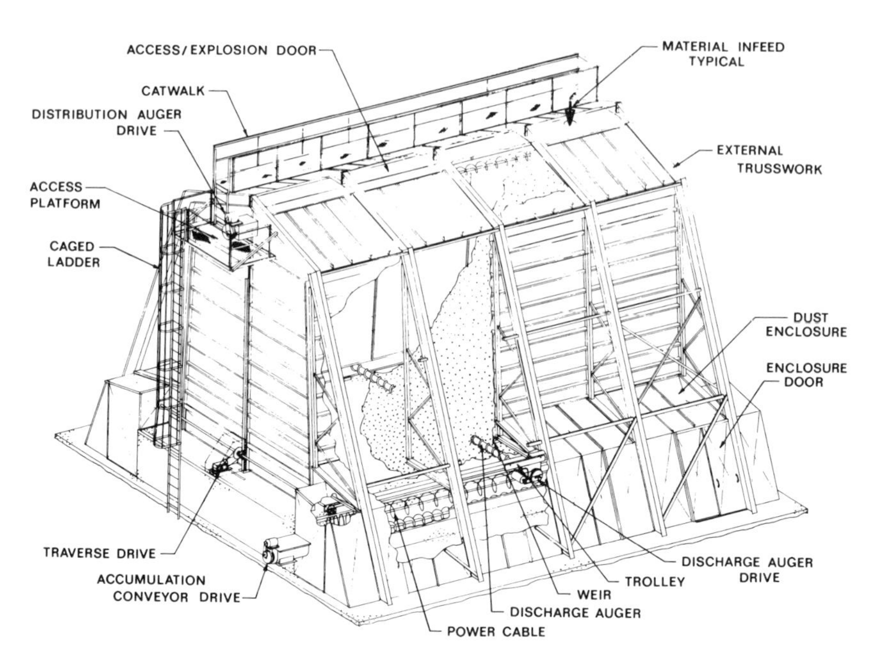

Figure 7.74. *A traversing screw unloader. (Courtesy Clarke's Sheet Metal, Inc.)*

Figure 7.75. *Discharge end of a traversing screw unloader. (Courtesy Miller Hofft, Inc.)*

rotor arms give way (shown in Figure 7.72) as a pressure limit valve is included as part of the hydraulic package.

Material is carried away by screw conveyors or bucket conveyors mounted beneath the silo in various ways. These conveyors can be installed to help randomize and blend the material.

Traversing Screw Unloaders: This type of unloader operates beneath a suspended wall rectangular bin, a below grade-level pit or an A-frame-type structure. The bin sides are normally inward sloping to help prevent bridging of the material. This is an all-purpose unloader operating with single or twin counter-rotating feed screws as shown in Figure 7.74. The screws traverse back and forth along the entire length of the bin discharging particles. Figure 7.75 shows one in operation.

Live Bottom Bins: Live bottom bins in use utilize screws, drum-rolls, chains, and belts as the active bottom of the bin. They are rectangularly shaped and normally have inward sloping sides so that there is a larger area at the bottom to minimize bridging or arching.

Figure 7.76 shows a typical bin configuration under which screws or drum-rolls are mounted. Constant- or variable-speed drives can be used. The screws drive the material across to one side of the bin to the discharge end. Screws are excellent with chips, shavings, and sawdust. The drum-roll-feeder type has a high discharge rate and passes the material between the rolls as shown in Figure 7.77. The surfaces of the rolls are interspersed with welded lengths of bar stock or other similar material for a positive feeding action.

Belt bottom feeders are low-volume, high-discharge metering bins effective with low-bulk-density materials. The material is advanced ahead by the belt bottom while being scalped across the top to keep a constant head. The material is raked off the front and deposited into a conveyor, as shown in Figure 7.78. This type of bin works well with flakes and wafers.

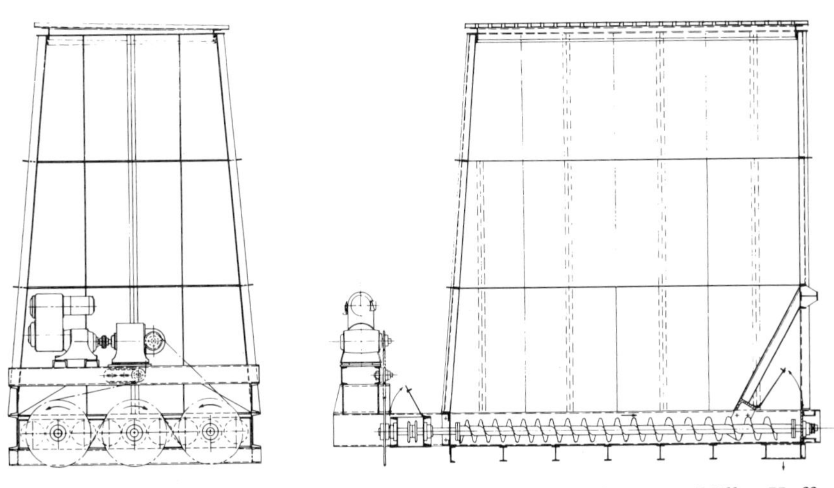

Figure 7.76. *Typical arrangement of a live bottom bin. (Courtesy Miller Hofft, Inc.)*

Figure 7.77. *Drum rolls turning, showing positive, controlled feeding action. (Courtesy Miller Hofft, Inc.)*

Chains are effective means of discharging fine, dry material such as sander dust. They also serve as the live bottom of the fiber bin that has performed excellently in the field (Figure 7.79). Wing-type drag chains pull the fiber into "doffing" rolls (Figures 7.80 and 7.81). The fiber is metered between these doffing rolls (14 in. or 356 mm in diameter turning at about 175 rpm) where it is aerated so that it can be handled by air systems or belt conveyors without any clumping.

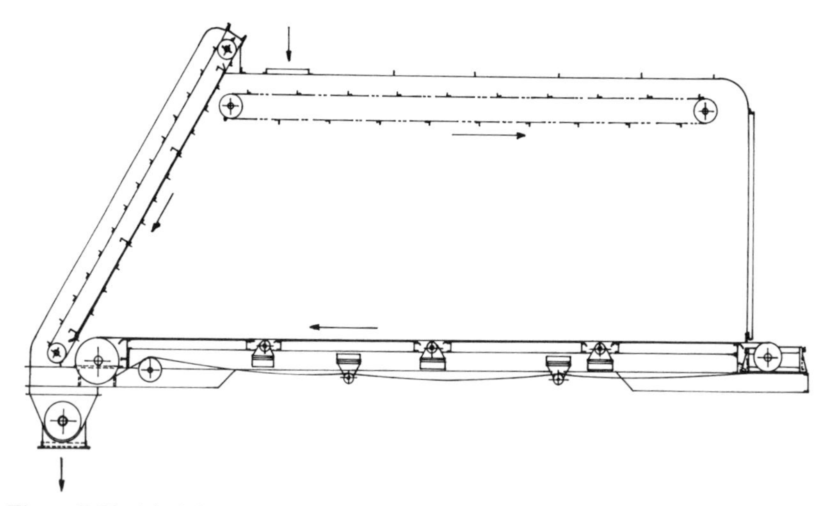

Figure 7.78. *A belt bottom feeder. Arrows indicate material flow. (Courtesy Miller Hofft, Inc.)*

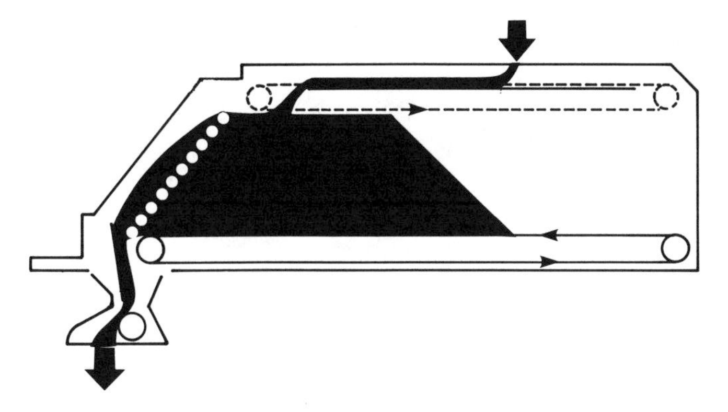

Figure 7.79. *Live bottom bin with doffing rolls (Ryman 1976).*

Figure 7.80. *Inside of doffing roll bin, showing drag chains on bottom, doffing rolls at the far end, and the overhead leveling screws (Patten 1976).*

Figure 7.81. *Close-up of doffing rolls (Patten 1976).*

Overhead screws serve to level the incoming material and move it to the rear of the bin, keeping a constant head of material in the front of the bin near the doffing rolls. The fiber should be fed into the bin immediately behind the doffing rolls to maintain a full bin in the metering area.

Explosions are always a problem with fiber. From field reports, it is advised that at least 1 ft^2 of venting to each 35 ft^3 (0.99 m^3) of capacity be used. Covers on explosion vent doors should be solidly chained to the bin to prevent injury to operating personnel. Venting should be directed away from the operators. The door at the end of the bin can be opened and the material dumped by reversing the chains on the bin bottom in case of fire. This has been a particularly effective system when using pressure-refined fiber.

When dealing with fiber, only those bins absolutely necessary should be used to avoid the plugging and variable-feed type of problems associated with fiber.

SELECTED REFERENCES

Albert Bezner Maschinenfabrik. 1975. Cylinder Type Chipping Machine MW. Ravensburg, Federal Republic of Germany.

Allgaier-Werke GmbH. 1975. Allgaier Sifter. Uhingen, Federal Republic of Germany.

American Sheet Metal, Inc. 1975. *American Sheet Metal's Doffing Roll Metering Bins.* Bulletin DRB-75. Tualatin, Oregon.

Atlas Systems Corp. 1975. *The Atlas Storage and Reclaim System.* Spokane, Washington.

Barrington, B. C. 1967. The Ortmann Flaker. *Proceedings of the Washington State University Particleboard Symposium, No. 1*. Pullman, Washington: Washington State University (WSU).

C-E Bauer. 1970. Specific Gravity Separators. Bulletin G-18A. Springfield, Ohio.

C-E Ehrsam. 1975. Rotary Plate Feeders and Bin Dischargers. Abilene, Kansas.

C-E Tyler Industrial Products. 1976. *Testing Sieves and Their Uses*. Handbook 53. Mentor, Ohio.

Chittenden, A. E. 1972. Wood and Cement—Past and Future. *Proceedings, Seventh World Forestry Congress*. Buenos Aires, Argentina.

C. J. Wenneberg Inc. 1974. Wenneberg Reclaiming Systems. Glen Ellyn, Illinois, subsidiary of C. J. Wenneberg AB, Karlstad, Sweden.

Clarke's Sheet Metal, Inc. 1975. *Clarke's Flo-matic Bins*. Eugene, Oregon.

FAO. 1976. Basic Paper IV. Technology and Techniques in the Manufacture of Wood-Based Panels. *Proceedings of the World Consultation on Wood-Based Panels,* held in New Delhi, India, February 1975. Brussels: Published in agreement with the Food and Agriculture Organization of the United Nations by Miller Freeman Publications.

Fischer, K. 1972. Modern Flaking and Particle Reductionizing. *Proceedings of the Washington State University Particleboard Symposium, No. 6*. Pullman, Washington: WSU.

———. 1974. Progress in Chip Flaking. *Proceedings of the Washington State University Particleboard Symposium, No. 8*. Pullman, Washington: WSU.

Gran, G., and I. Bystedt. 1973. Latest Developments in Pressurized-Refining with Defibrator Equipment. *Proceedings of the Washington State University Particleboard Symposium, No. 7*. Pullman, Washington: WSU.

Gunn, John M. 1972. New Developments in Waferboard. *Proceedings of the Washington State University Particleboard Symposium, No. 6*. Pullman, Washington: WSU.

Jager, R. 1975. Preparation of Quality Particles from Low-Quality Raw Material. *Proceedings of the Washington State University Particleboard Symposium, No. 9*. Pullman, Washington: WSU.

Johnson, D. R. 1973. Panel Discussion on Experiences and Problems with Pressure-Refined Fiber. *Proceedings of the Washington State University Particleboard Symposium, No. 7*. Pullman, Washington: WSU.

Johnson, R. L. 1973. Panel Discussion on Experiences and Problems with Pressure-Refined Fiber. *Proceedings of the Washington State University Particleboard Symposium, No. 7*. Pullman, Washington: WSU.

Koch, P. 1964. Square Cants from Round Bolts Without Slabs or Sawdust. *Forest Products Journal,* Vol. 14, No. 8.

———. 1975. Shaping-Lathe Headrig Will Convert Small Hardwoods into Pallet Cants Plus Flakes for Structural Exterior Flakeboard. *Proceedings of the Washington State University Particleboard Symposium, No. 9*. Pullman, Washington: WSU.

McNeil, W. M. 1967. Defiberizors for Particleboard. *Proceedings of the Washington State University Particleboard Symposium, No. 1*. Pullman, Washington: WSU.

McVey, D. T. 1973. Plant Experiences with a Defibrator Pressure-Refiner. *Proceedings of the Washington State University Particleboard Symposium, No. 7*. Pullman, Washington: WSU.

Maloney, T. M.; J. W. Talbott; M. D. Strickler; and M. T. Lentz. 1976. Composition Board from Standing Dead White Pine and Dead Lodge Pole Pine. *Proceedings of the Washington State University Particleboard Symposium, No. 10*. Pullman, Washington: WSU.

Maschinenfabrik Rehau. 1975. Wood-Wool Machines. Rehau (Bavaria), Federal Republic of Germany.

Miller Hofft, Inc. 1968. *Handling of Bulk Materials*. Bulletin No. 681. Richmond, Virginia.

Mottet, A. L. 1967. The Hombak Flaker. *Proceedings of the Washington State University Particleboard Symposium, No. 1*. Pullman, Washington: WSU.

Patten, J. M. 1976. Experiences with Doffing Roll Bins. *Proceedings of the Washington State University Particleboard Symposium, No. 10*. Pullman, Washington: WSU.

Plough, I. L. 1974. The Effect of Running Clearance on Flake Quality and Power Requirements of Conical Knife-Ring Flakers. *Proceedings of the Washington State University Particleboard Symposium, No. 8*. Pullman, Washington: WSU.

Rauch, A. H., and R. A. Cheney. 1970. Pressurized Refining of a Variety of Residuals. *Proceedings of the Washington State University Particleboard Symposium, No. 4*. Pullman, Washington: WSU.

Redler Conveyors Ltd. 1974. *Bulk Materials Handling*. Publication No. 74/14. Gloucester, England.

Rooney, J. E. 1970. Combining Species. *Proceedings of the Washington State University Particleboard Symposium, No. 4*. Pullman, Washington: WSU.

Rotex Inc. 1975. Rotex Chip Screeners. Catalog No. 606. Cincinnati, Ohio.

Runckel, W. J. 1973. C-E Bauer Pressurized Double-Disc Refining Systems—Application and Development in the Board Field. *Proceedings of the Washington State University Particleboard Symposium, No. 7*. Pullman, Washington: WSU.

Ryman, U. 1976. Comparison of Metering Bins Used in Europe and North America. *Proceedings of the Washington State University Particleboard Symposium, No. 10*. Pullman, Washington: WSU.

Saxlund A/S. 1975. Saxlund Rotor-System. Risør, Norway.

Schmierer, R. E. 1976. The Bunkering of Raw Material Furnish—Some South Australian Experiences. *Proceedings of the Washington State University Particleboard Symposium, No. 10*. Pullman, Washington: WSU.

Schumann, E. W. 1970. B. Maier—Black Clawson MKZ Flaker. *Proceedings of the Washington State University Particleboard Symposium, No. 4*. Pullman, Washington: WSU.

Simon, J. A. 1973. Panel Discussion on Experiences and Problems with Pressure-Refined Fiber. *Proceedings of the Washington State University Particleboard Symposium, No. 7*. Pullman, Washington: WSU.

Stewart, H. A. 1971. Chips Produced with a Helical Cutter. *Forest Products Journal,* Vol. 21, No. 5.

Stewart, H. A., and W. F. Lehmann. 1973. High-Quality Particleboard from Cross-Grain, Knife-Planed Hardwood Flakes. *Forest Products Journal,* Vol. 23, No. 8.

Surdyk, L. V. 1967. The Pallmann Flaker. *Proceedings of the Washington State University Particleboard Symposium, No. 1*. Pullman, Washington: WSU.

Sybertz, H. 1974. Energy Requirements for the Production of Cut Flakes Used for Flakeboard. *Proceedings of the Washington State University Particleboard Symposium, No. 8*. Pullman, Washington: WSU.

Sybertz, H., and K. Sander. 1972. A New Flaking Machine for Structural Board. *Proceedings of the Washington State University Particleboard Symposium, No. 6*. Pullman, Washington: WSU.

———. 1975. Special Equipment to Split Flakes into Strands. *Proceedings of the Washington State University Particleboard Symposium, No. 9*. Pullman, Washington: WSU.

Textor, C. K. 1957. The Bauer Method of Preparing Furnishes for the Manufacture of Insulation Board and Hardboard. Technical Paper No. 5.3. *Fiberboard and Particle Board Technical Papers*. International Consultation on Insulation Board, Hardboard and Particleboard, FAO, Geneva, Switzerland.

Vajda, P. 1973. Panel Discussion on Experiences and Problems with Pressure-Refined Fiber. *Proceedings of the Washington State University Particleboard Symposium, No. 7*. Pullman, Washington: WSU.

———. 1976. Introduction to Problems with Storing and Handling Various Types of Particleboard and Fiberboard Furnish. *Proceedings of the Washington State University Particleboard Symposium, No. 10*. Pullman, Washington: WSU.

Veuger, F. 1975. Disc Flaker Forms Heart of Structural Waferboard Plants. Information from CAE Machinery, Ltd. (May), Vancouver, B.C.

Wiecke, P. H. 1973. Panel Discussion on Experiences and Problems with Pressure-Refined Fiber. *Proceedings of the Washington State University Particleboard Symposium, No. 7*. Pullman, Washington: WSU.

———. 1975. Private conference on problems with conveying and storing wood particles. Columbia Engineering International, Ltd.

Young, D. R. and H. G. Moeltner. 1976. Storing and Metering Green and Dry Wafers. *Proceedings of the Washington State University Particleboard Symposium, No. 10*. Pullman, Washington: WSU.

8. DRYING PRINCIPLES AND PRACTICES

Drying is a critical step in the processing of composition board. Rarely does the furnish arrive at the plant low enough in moisture content for immediate use. Exceptions to this rule will be found in the few plants using powdered phenolic resin and dry shavings or flakes generated from kiln-dried lumber. In these cases drying may be omitted. But these plants produce a minute amount of the total production. Furnish can range from about 10% in moisture content to over 200%. For use with liquid resin, the major binder in use, the furnish must be reduced to about 2 to 7% in moisture content depending upon process parameters.

Composition board furnish is done in a much different way than conventional wood drying. In a dry kiln, close attention is paid to wet-bulb and dry-bulb temperatures, and sophisticated drying schedules have been developed. If done improperly, the lumber is severely damaged. Drying can take days. In contrast, drying composition board furnish takes place in minutes, or seconds at relatively high temperatures.

Research on particle drying has not been as extensive as on other parts of the process. Some work has been done, as in 1958 when experimental boards were made comparing flash drying with more severe oven drying, where the material reaches much higher temperatures. There was a definite indication of significant differences in the modulus of rupture values for these boards—ones made from oven-dried furnish were lower. Similar differences might be assumed for the modulus of elasticity. Factors that contribute to the drop in strength values might be case-hardening, a problem that occurs to some extent in any lumber or veneer drying operation; changing of the chemical environment on the particle surfaces causing problems with resin bonding; or a flow of natural wood resins and chemicals to the surface if the material temperature becomes too high, which could cause further resin bonding problems. An example of the wood-resin problem is the buildup on the surfaces and ends of lumber that has been dried in a kiln. Such an accumulation could easily interfere with the adhesive bonding properties that are so crucial in board production. These are just some examples of problems faced when drying composition board furnish.

Research is now taking place on particle drying and the application of fundamental knowledge gained should help improve practices and equipment. For now, however, present principles and practices are those to be discussed. Much of the drying equipment used is modified from the agricultural industry, where dryers are used extensively with alfalfa, sugar, corn, etc.

When drying any type of furnish that is to be used in producing composition board furnish, the primary objective should be to achieve uniform moisture content, while minimizing cost and the fire/explosion hazard. The different types of drying equipment, now used in the industry almost without exception, involve the principle of moving the furnish rapidly through a drying medium by means of comparatively high-velocity, high-temperature air. A few dryers still are used that mechanically propel furnish through relatively slowly. Drying is accomplished by heat transfer (convection, conduction, and radiation) and conveying. These dryers are efficient and those that use them are satisfied. However, the large high-volume production plants use the high-speed type of dryer.

In the modern drum and tube dryers the furnish is introduced into a high-temperature, high-velocity airstream to expose the maximum furnish surface area for a minimum period of time. It has been shown that this basic principle does, in fact, provide efficient drying, as indicated by uniformity of final moisture content and drying cost.

Two classifications of dryers are: (1) the horizontal rotating and (2) the horizontal fixed type. Variations of each type are found. Later discussion of various dryers will provide more detailed information.

Three types of fuel are used in the present dryers: (1) gases—natural or liquid; (2) oils of various grades; and (3) fine wood fuel such as sander dust, sawdust, and finely ground bark and larger wood material such as board trim. It is common to find combinations of the three types of fuel. More of the wood fuel is being used because of increased costs and less availability of the oil and gas types. Another type of "fuel" is to recycle the stack gases to reduce the total fuel consumption.

Of interest is the trend reversal involved with composition board trim. After many years of experimentation, it was possible to refine the trim into particles and introduce them along with sander dust into the furnish. Ten years ago this was rarely done; but it is quite common now, although there is some question about detrimental effects on board properties. However, it appears now that board trim and sander dust will have more value as fuel than as furnish. More and more plants are moving towards energy independence, especially for dryer and boiler fuel.

Most drying equipment manufacturers have several sizes available, depending upon the drying requirements. The determination of dryer size is based primarily on the amount of water to be evaporated, the quantity of furnish to be handled, and the particle size or geometry effects on the length of the drying period. Multiple-stage drying seems to be increasing in popularity. This concept uses two or more dryers in series, removing part of the moisture in one stage and the remainder in a second stage. This approach also makes it possible to combine materials of widely varying initial moisture contents by drying the higher-moisture-content material first in one dryer and combining the two materials in a second dryer for final moisture removal. This can also reduce the fire hazard found when drying high- and low-moisture content material simultaneously. Another benefit of multiple-stage drying is that it produces a furnish with more uniform moisture content.

The economics of the drying operation, like most cost factors, is extremely difficult to determine, as reliable information from operating plants is usually not available. The final drying cost depends on many variables such as: (1) fuel cost; (2) inlet temperature, which determines fuel quantity; (3) initial moisture content;

(4) final moisture content; (5) size and shape of furnish; (6) efficiency in the use of the heating medium (which depends on all the previous factors plus equipment design); and (7) ability of the operating personnel to get maximum heat from a given fuel (this depends on good control of inlet versus outlet dryer temperatures).

A number of operating factors are associated with drying, such as initial moisture content of the furnish and its variations, particle geometry of the furnish, variable ambient conditions, the feeding system to the dryer, contamination and discoloration, and last but not least, fire and explosion problems. Each of these will be dealt with in general before discussing individual dryer designs.

INITIAL MOISTURE CONTENT OF FURNISH

Souce of Raw Material

The intitial moisture content of the furnish entering the dryer varies considerably depending primarily on the type of raw material being used. If the raw material originates from green wood, the initial moisture content will be much higher than in a raw material from residue materials which have already been partially or completely dried.

Examples of green material are:

1. Green planer shavings which result from surfacing of green lumber.
2. Green residue material (normally chips) from a plywood plant, which result from the roundup of veneer bolts and veneer clippings.
3. Green residue materials (normally chips) from sawmill slabs, edgings, and trimmings.
4. Green roundwood which has been processed into furnish.
5. Green sawdust.

Examples of dry raw material are:

1. Planer shavings from surfacing kiln-dried lumber of any species.
2. Residues from plywood plants (mostly plywood trim).
3. Mill ends from kiln-dried boards from lumber, and millwork plants.
4. Sawdust.
5. Sander dust.

Mixtures of relatively dry and green material, particularly if uncontrolled, can be disastrous. Dry material can be overdried and ignite. It may also be difficult to glue this material. Green material will be too wet and probably will cause blows in the press.

Species

The species of wood being used has a very significant effect on the green moisture content. For example, in the coniferous softwoods, the pines in general have much higher green moisture content (80 to 200%, oven-dried basis) in comparison with Douglas fir (about 40%). Even though the range in green moisture content among hardwood species is not so great, there are significant differences in the order of two to one among the common species being used. There are major differences between the sapwood and heartwood of a given species, particularly in the pine group; but since there are no practical means of segregating sapwood and heartwood materials, this usually is not an important factor from an operating standpoint. However, such problems may be handled to a degree if it is possible to segregate raw material by large and small trees and top versus butts.

Average Level of Moisture Content

The average initial moisture content is very important. The amount of moisture in the raw material affects the economics of drying since, regardless of type of equipment, it requires somewhere between 1400 and 2000 Btus (1.48 and 2.11 megajoules) to evaporate a pound of water. The range may be even wider in some plants, depending on the efficiency of the dryer operations. If the initial moisture content is high relative to outlet moisture content (i.e., if the evaporation rate is high), comparatively large drying equipment both in terms of the burner or combustion chamber, drying drum, and conveyor will be required to produce a given quantity of dry furnish. Three basic cost factors are involved: (1) the initial capital investment of the drying equipment; (2) the cost of the fuel to generate the drying medium; and (3) the efficiency of the drying equipment (included in the cost of fuel), which is also influenced by the average moisture content, with higher efficiency expected as the initial moisture content increases.

Variation in Initial Moisture Content

The variation in the initial moisture content of the raw material is more important in determining the final range in moisture content than is the average level of the moisture content. If a wide range exists in the material entering the dryer, the drying equipment must make rapid responses in the burning of the fuel to adjust for the changing heat requirements. Two examples will illustrate this more effectively. Assume in working with green planer shavings that the average moisture content is 60% (oven-dry basis). Also, assume that after the drying equipment has reached an equilibrium temperature state (meaning a fairly stable inlet temperature) and therefore fuel level burning rate has been achieved, the average moisture content of the green shavings suddenly drops to 40%. The types of drying equipment used in many cases will immediately respond to this change by reducing the fuel consumption rate. For a comparatively short period of time, even after the fuel rate has been reduced to correspond to the lower drying load, there will be a tendency to over-dry the furnish because of the residual heat which is contained in the drying drum, resulting from operating at the previous higher inlet temperature level. On the other hand, if the initial moisture content should suddenly increase to a level of 100%, the dryer would immediately respond by increasing the inlet temperature resulting from the higher fuel consumption rate to handle the added drying load. In this particular case, part of this added fuel will be used to increase the equilibrium-state-temperature of the drying drum rather than for drying the furnish. For a short time there will be a tendency for furnish to be discharged at a higher than desired moisture-content level.

Some of the drying equipment makes adjustments in retention time in the dryer rather than by adjusting fuel consumption rate. Regardless of the method used to correct for the sudden change in initial moisture content, the effect on uniformity in final moisture content will be the same. If the initial moisture content changes gradually from high to low or low to high, the effect on the final moisture variation will be indiscernible. Therefore, if it is possible to have prior knowledge of the variation in the moisture content of the incoming material and if it is practical to uniformly mix high- and low-moisture-content material, the problems previously described will not be serious. The most important factor in drying furnish to a uniform moisture content is to have furnish with a relatively uniform moisture content entering the dryer.

PARTICLE GEOMETRY

The size and shape of the individual particles in the furnish also are significant factors affecting the drying operation. Variation in size and shape during the drying operation can be extremely troublesome.

Size and Shape

Assuming the objective is to supply the plant with furnish at a given moisture content, increasing the size of a particle of a given shape will tend to increase the cost of drying. The larger-size particle will require more heat to remove a given percentage of moisture because of the exposure of less surface area per unit of volume and time.

Unfortunately, it is difficult, and often undesirable, in most of the raw material processing operations to achieve uniformity in particle configuration. For example, less variance in particle size would be expected in starting with roundwood that is reduced to flakes by a disc or drum flaker, since the thickness and the length of the flakes are more closely controlled as compared with raw material such as planer shavings, hogged fuel, or furniture millblocks that are hammermilled. If the range in size is large, there will be a tendency to over-dry the small-size and under-dry the large-size material. The over-dried small particles may present a fire and explosion hazard. From a practical operating standpoint, as far as over-dried small particles are concerned, this may not be a serious problem if the dried material is allowed to remain in an in-process storage bin for a sufficient length of time to allow some equalization of moisture content, namely, a transfer of moisture from the large- to the small-sized particles. However, fire and explosion problems in the dryer still exist.

VARIABLE AMBIENT CONDITIONS

Changing ambient temperature and humidity conditions often make adjustments in the dryer inlet temperature necessary in order to maintain a constant desired final moisture content. For example, if the relative humidity increases significantly, dry furnish from the discharge of the dryer can be expected to pick up one or two percent in moisture content as it is conveyed through the forming and pressing operations. The opposite is true if relative humidity decreases. This can be a serious problem in a plant, particularly in the southeastern United States, where hourly or daily fluctuations in relative humidity are common.

In the winter months in the northern states, the climatic conditions involve many days of subfreezing weather which affects the capacity of the drying equipment for three reasons. First, if the wood material is frozen, additional heat is required for evaporating the water or ice. Second, the air used to support combustion of the fuel and as the drying medium must be heated from a much lower temperature level. Third, the furnish itself must also be heated more. Daily and hourly savings, however, are much more important than seasonal ones according to the experience of dryer manufacturers.

FEED SYSTEM TO THE DRYER

The method of introducing the furnish to the dryer inlet can also affect the uniformity in moisture content of the dried furnish. In virtually every process step, uniform feeding is extremely important. By far the majority of the feeding systems use a volumetric rather than a weight principle. For example, a screw conveyor operating at a given speed will deliver a given volume of material. This would also be true if a belt conveyor were used and even if the dryer is being fed directly from the tailspout of a cyclone, assuming the material is being fed to the cyclone by a screw or belt conveying system.

The major problem with the volumetric approach in the feeding system is the influence of changing bulk density in the furnish and short-term volume changes. The bulk density (weight per unit volume) is influenced by uncontrollable factors such as species (where more than one species is being used), change in sharpness of flaking knives over the life of the knives, or influence of particle configuration resulting from variation in moisture content entering the initial processing equipment. As the bulk density of the furnish changes, and even assuming the initial moisture content has remained fairly constant, the drying load on the dryer will change. For example, if the bulk density of the incoming material has been at a level of 10 lbs/ft^3 (161 kg/m^3) and gradually changes to 15 lbs/ft^3 (241 kg/m^3), other factors being equal, the evaporable load on the dryer has increased by 50%. If this change in drying load is slow and gradual, no particular problem will result because dryer controls can respond.

Practical operating experience with different types of feeding equipment indicates that comparatively small-diameter, high-speed screw conveyors are preferred over large-diameter slow-speed screw conveyors. Minimum variations in the material load coming to the screw conveyors can be leveled out better with the first type of feed system. This same principle applies if the material is being conveyed by a belt to the dryer inlet. It is much better to operate the belt at a comparatively high speed, delivering a small quantity of material to the belt, rather than delivering a comparatively large quantity on a slow-moving belt.

Feeding pressure-refined fiber into a dryer presents a different set of problems. Tube-type dryers are commonly used with fiber. One method exhausts fiber and steam from the refiner directly to the dryer inlet. At the dryer outlet, the cyclone separates the fiber from the air and steam. The cyclone is equipped with a rotary valve for discharging the fiber. The rotary valve keeps the steam and conveying air from passing with the fiber to a storage bin. Another method has a cyclone before the dryer for separating the steam from the fiber. The fiber is fed through a rotary valve into a tube dryer. Slugs of excessively wet material can cause problems in this arrangement as it is too heavy to be conveyed by air. Thus, it cakes in the dryer entrance and eventually drys out and ignites. A solution to this problem is to use a drop box after the rotary valve cyclone discharge. The heavy material thus drops out because it cannot be conveyed by the airstream and is not introduced into the dryer.

Fiber from atmospheric refiners can also be fed to the dryer by means of a cyclone discharge through a rotary valve.

CONTAMINATION AND DISCOLORATION

Contamination

A problem with certain types of fuel, particularly oil, is the possible contamination of the furnish. The importance of getting complete combustion of the fuel must be emphasized. Complete combustion is more difficult to achieve with oil than natural gas. If complete combustion is not achieved, contamination can be expected in the form of unburned fuel particles which, in the most critical state, could affect bonding in the final pressing operation by interference with the resins that have been added as bonding agents. This condition is not likely to happen if reasonable control of fuel combustion is exercised.

Discoloration

Another problem encountered in high-temperature dryers is the discoloration of the furnish resulting from using excessively high inlet temperatures. In many products this is not critical unless actual charring of the finely divided particles takes place, which again could affect the particle-to-particle resin bond. On the other hand, some manufacturers have found that it is important to maintain color uniformity in the furnish, since it affects the appearance of the finished products. It is possible at high-inlet-temperature levels to scorch the finely divided particles and thus give a grayish cast to the final product.

FIRE PROBLEMS

Instrumentation

A few points on fire problems will be mentioned here, with more details in Chapter 11. With high-temperature dryers, regardless of the particular type of equipment, one must expect to have occasional fires. All of the major manufacturers of this type of drying equipment incorporate in their control systems good instrumentation to indicate either the presence or probability of fires; suppression systems are also built into the dryers. In all cases there is some sort of high-limit-temperature control which automatically keeps the outlet temperature below a preset level. There are also provisions in the equipment to take care of "flame-out" in the combustion chambers. Despite these precautions, the very nature of this type of drying presents fire and danger, particularly when there are sudden changes in the infeed of material to the dryer. This can result from change in the condition of the material, discussed previously, or from such unexpected conditions as partially plugged cyclones, screw conveyors, etc., that affect the regular feed to the dryer.

Fuel

The type of fuel also affects the probability of a fire. Experience indicates that it is more difficult to achieve complete combustion of oil than gas. More problems can also be expected when a combination of fuels involving oil and/or gas and residual wood materials such as sander dust are used in combination. One of the major problems in the use of the lower grades of oil is achieving complete combustion and minimizing the chances of an oil-film buildup near the discharge end of the combustion chamber. When this condition reaches a certain point, after a certain inlet temperature level has been reached, the oil will ignite and cause a fire.

Control Methods

Certainly the chances of having a fire are greatly reduced if a heat exchanger is used between the combustion chamber and the dryer, because the exposure of the furnish to open flame is eliminated. On the other hand, this is not the most efficient and economical method of drying the materials normally used in board manufacturing.

Most operating plants use the high-temperature drying principle mixing hot gases from a burner with the furnish. Thus, they accept the fact that it is virtually impossible to eliminate fires completely. Newer developments are lowering the oxygen level in the dryers, thus reducing the risk of fire. One system on a tube dryer shown in Figure 8.1 recycles part of the dryer discharge gas directly to the dryer inlet. Some of the discharge gas also goes to the incinerator. Thus, the oxygen content in the dryer is reduced, heat is also conserved, and the particulate emission to atmosphere is either reduced or eliminated. Three gas flows are mixed going into the dryer furnace, depending on the desired conditions: dryer discharge gas, hot gas from the solid wood incinerator or burner, and makeup air from atmosphere.

The main concern is extinguishing the fire once it starts and, in particular, keeping the burned material from getting into storage bins and other confined areas where the danger of explosion is high. In some systems, it is desirable to keep the fans running for cooling and to remove the material from the dryer. Two methods are popular for handling a fire. One is to contain the fire in the drum and to suppress it. The other is to get the burning material out of the drum and to dump

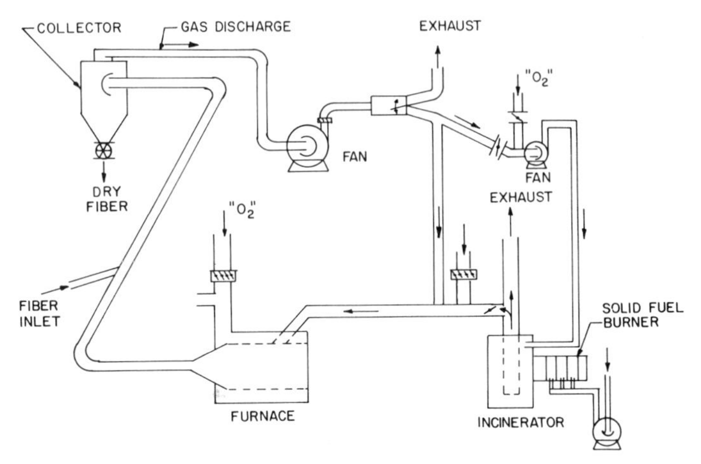

Figure 8.1. *A discharge-gas recycle system for a dryer. (Courtesy M-E-C Co.)*

it before it reaches the storage bin. With the first method, it is not easy to contain the material in the dryer drum because most of the furnish is usually being conveyed by air. Even if the fan systems are interlocked with the fire control instrumentation, there are still several seconds before the fans can be completely stopped. Due to the high velocity of air (perhaps 4000 to 6000 ft/minute or 1220 to 1830 m/minute) in these air-conveying systems, it is almost impossible to keep some burning material from getting into the storage bins. One of the more practical methods of hedging against this problem is to discharge from the dryer onto an open conveying system, such as a screw of belt conveyor, which can be reversed or stopped quickly when the proper signal is received. Beyond this approach, there remains only to provide adequate suppression equipment, and the discharge conveying systems, including cyclone and conveyor belts and screws.

In the system using a fire dump, the fans keep running to cool the dryer and to remove the burning material. The infeed of material is stopped when a fire occurs, as it is with the previously discussed system. The fire dump can be anywhere between the dryer discharge and the conveyor belt leading to the storage bin. An extinguishing system is still required in the dryer in case the fire cannot be handled by the fire dump approach. With bins storing dried fiber, which usually have belt bottoms, the belt direction can be reversed to dump the stored fiber in case the fire reaches the bin.

A development pioneered by the industry itself in certain dryers has been to increase the distance between the combustion chamber and the feed entrance for the furnish. Small sparks thus have more time to extinguish before reaching the furnish. Another approach has been to mount the combustion chamber at 90 or 180 degrees to the dryer itself to provide more time for any sparks to extinguish before they reach the furnish.

Often fires exist in the furnish before it is fed into the dryer. A case in point would be when rocks or metal get into a flaker or refiner, spark, and set a fire in the furnish.

OTHER PROBLEM AREAS

Greater sophistication in manufacturing procedures, all designed to produce a broader range of products and/or more versatile products have resulted in other drying problems, For example, in systems producing multi-layer or graduated boards, it has been found desirable in many cases not only to vary the type of furnish with respect to particle size and configuration in each layer, but also to vary the moisture content in the different layers. It is generally desirable to have a higher moisture content in the surface layers than in the core layers. The reasons for high moisture in the surface layers is discussed later under "pressing." Usually, however, different dryers are needed for face and core furnish when the moisture contents are different.

In a few of the manufacturing processes it is desirable to introduce the synthetic resins ahead of the drying operation. This occurs, for example, with the blending of resin with fiber as it is being ground in an attrition mill. At the present time, the choice of the resins (usually phenolic) which can stand the heat treatment involved in the high-temperature dryers is limited. Recently, however, several plants have been adding the urea part of urea-formaldehyde resin in the blow line leading to the dryer or raw material being ground into fiber in the attrition mill.

ROTARY DRYERS

Rotary dryers can be broken down into three types: three-pass, double-pass, and single-pass. A number of designs of this third type have been popular in the past but their use has been superseded by the three-pass and the fixed horizontal dryer. It appears, however, that this trend is reversing with more single-pass dryers going into use.

Three-Pass Dryers

Three-pass dryers have been built for over 40 years for use in the agricultural industry, commonly for drying alfalfa. In addition to drying many different forms of wood products such as paper pulp, flakes, fibers and particles, these highly efficient, high-temperature dryers have also been used successfully in the processing of various by-products, industrial wastes, and chemical residues. A representative sample of these other products includes antibiotic residues, beet pulp, canning company by-products, kelp, peat, cattle manure, citrus waste, cooked blood, crab scrap, feathers, fish meal, and packing house by-products. Figure 8.2 shows a bank of these dryers. Figure 8.3 is a schematic drawing showing the operation of such a dryer. Figure 8.4 is a cutaway view of a typical three-pass installation. Several companies supply this type of dryer. These dryers can range in size to over 100 ft (30.48 m) in length and over 14 ft (4.27 m) in diameter.

The three-pass dryer can best be described as a concurrent, multiple-cylinder rotary-drum, direct-fired-type dryer which utilizes heated air as the drying medium. The product to be dryed is transported through the system in a pneumatic fashion by means of an induced-draft fan on the discharge side of the drum, with a retention time commensurate with particle size, geometry, and moisture content.

Figure 8.2. *A bank of three-pass dryers. (Courtesy The Heil Co.)*

The major components of the three-pass dryer are illustrated in Figure 8.4. The first is the combustion chamber which is designed to be in complete balance with the attached drying drum and air-conveying equipment to permit efficient and economical drying for a given drying load. The burner of the combustion chamber can be designed to permit the burning of natural gas, propane, fuel oil, sander dust or combinations of fuels. The second major component is the rotating drying cylindrical drum, which basically consists of three concentric cylinders tied together as illustrated in Figure 8.5. The third major component, known as the exhaust unit, consists usually of one or two fans. The air system can be either positive or negative as shown in Figure 8.6. With a positive air system, the fan is located between the dryer and the cyclone. With the negative air system, the fan is located beyond the cyclone. This system has the advantages of less wear on the fan impeller and no generation of fines by passing the furnish through the fan.

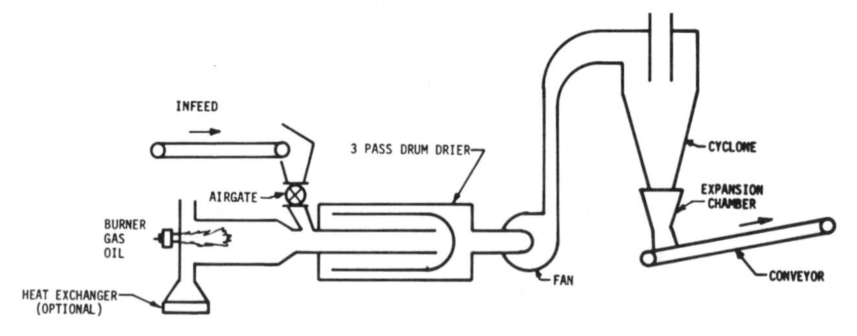

Figure 8.3. *A conventional three-pass drum dryer (Buikat 1971).*

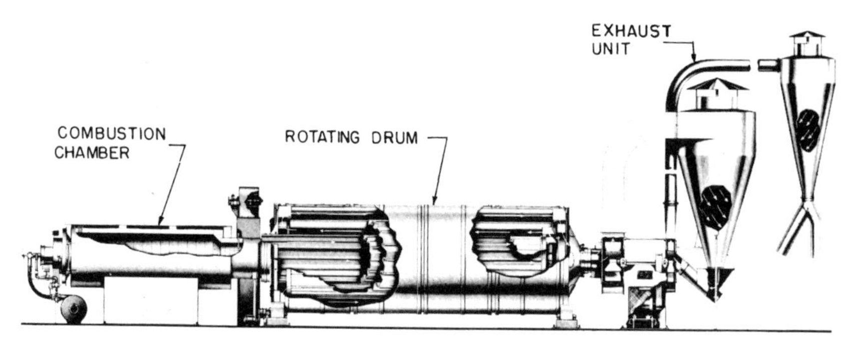

Figure 8.4. *Major components in a typical three-pass dryer installation (Stillinger 1967).*

One cyclone is used if the material load is relatively light. A second cyclone is used with heavier loads. The first one takes out the heavy particles and the second one, the fine particles. The use of two cyclones also makes it easier to control particulate emissions.

The major components of the electrical control system can involve a temperature controller with thermocouple located between the drum and fan, a high limit recorder device with thermocouple located between the drum and the fan, and an indicating pyrometer with thermocouple located between the furnace and the drum. The temperature controller with thermocouple measures the temperature of the gases at the outlet end of the dryer drum and is the primary control instrument. A given outlet temperature corresponds to a specified outlet moisture content for a given size and configuration of furnish. The high limit system

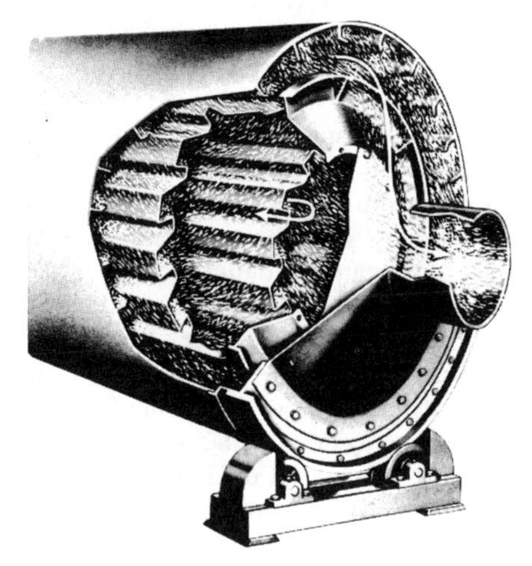

Figure 8.5. *Cutaway section illustrating the "showering action" in a three-pass dryer (Stillinger 1967).*

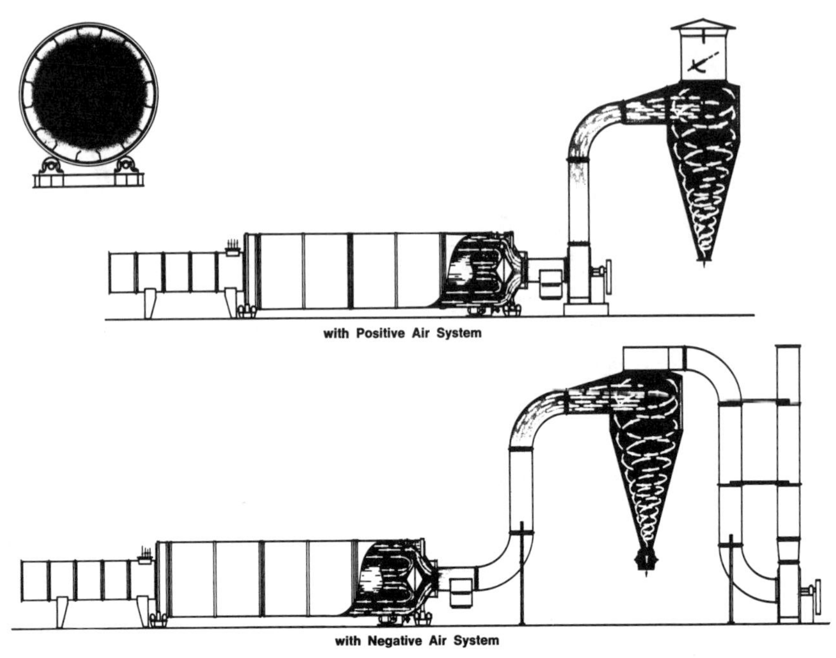

Figure 8.6. *Positive and negative air systems for three-pass dryers. (Courtesy M-E-C Co.)*

automatically stops the entire drying operation when the outlet temperature exceeds a certain preset level. Inlet temperature can range as high as 1500° to 1600°F (816°-871°C) with wet furnish. This is far above the burning point of wood (about 450°F or 232°C). The "blue haze" problem with dryers occurs at temperatures above 750°F (399°C). However, the drying taking place keeps the wood cool and as long as moisture is present in the wood, its temperature stays below the combustion level. Once all the moisture is removed, fires can be started in a matter of seconds. For drying relatively dry furnish such as kiln-dried planer shavings, inlet temperatures are reduced to about 500 to 600°F (260°-316°C) or lower.

The drying load for a given-size dryer is dependent upon the initial moisture content, inlet and outlet temperatures, outlet moisture content, ambient conditions, air velocity, speed of the drum rotation, etc., and the geometric configuration of the material being dryed. For example, the higher the initial moisture content, the higher inlet temperature to which the material can be exposed; and, therefore, a greater quantity of water can be evaporated per unit of time. The configuration of the material being dryed influences the evaporable load from two standpoints: in some materials there is a definite upper limit in inlet temperature which the material can stand without burning or discoloration; and, in a furnish consisting of various-sized particles, the evaporable load will normally be controlled by the smaller-size material because it will be the first to dry, although it passes through the dryer faster.

The concept of three-pass drying is as follows. When heat and air turbulence are applied to the furnish, surface moisture evaporates (being absorbed by the heated air), during the constant-rate-evaporation period. This takes place predominantly in the first cylinder where the highest air temperature and velocity can be employed. As the vapor pressure increases within the particle, moisture diffuses to the surface. The larger particles are allowed to slow up in the intermediate cylinders where they can warm up in a moderate temperature zone; thus, the absorbed moisture is removed during the diminishing-rate-evaporation period in the second and third cylinders.

The three-pass cylindrical drum uses comparatively higher air velocities and heat in the center cylinder (first pass) because of its smaller diameter where the products are the heaviest and the wettest. Lower velocities and heats are used in the intermediate drum, and even lower velocities and heats are used in the outer drum. The actual velocity of the heated air varies with temperature and drum diameter as it moves through the three drums. With green planer shavings being dried in any of the larger size dryers and assuming a maximum inlet temperature of 1200 to 1400°F (649°-760°C), the entrance velocities for each drum would be in the order of 1600, 640, and 320 ft (498, 195, and 98 m) per minute. Fine dust emerges in seconds while large particles may take more than five minutes to travel through the drying drum. The three-pass drum provides gentle handling of the particles as they fall a few inches from flight to flight. This is particularly true in the intermediate and outer drums, although there are more "bounces" of flights as compared to single-pass dryers.

Heat is transferred to the product primarily by convection air currents and secondarily by radiation and conduction. The latter plays an almost negligible part when processing refined wood particles. To obtain maximum and intimate contact of the heated air with the product, the interior and exterior of the various cylinders are equipped with properly designated lifting flights which provide the so-called showering action. Air turbulence surrounding each particle must be sufficiently high to prevent or destroy the cool film of vapor enveloping each particle which otherwise would tend to insulate against further evaporation. It follows that if a good differential between air and particle velocity can be achieved, a maximum evaporation rate can be achieved. Care has been exercised in designing and proportioning the drum cylinders, so that velocity of combustion gases and heated air is not sufficiently great to adversely influence the movement of the product through the drying drum.

Electronic temperature control systems usually installed have a built-in anticipating device which minimizes over and under shooting of the desired temperature control point, resulting in accurate control of the product moisture content. Sensitive electrical controls adjust to changes in moisture and changes in product volume while holding the outlet air temperature within close tolerance. Accurate temperature control is further assisted from the nominal amount of refractory used in the combustion chamber; therefore, only a nominal amount of heat is stored. The development of quick and accurate instrumentation for determining moisture content nowadays makes it possible to operate the controls based on the moisture content of the furnish as it enters and leaves the dryer. Naturally, wild swings in incoming moisture levels will make it almost impossible for the controls to work properly, even if their measurements are precise. The temperature within the dryer just cannot be adjusted quickly. Therefore, as mentioned, reasonably even feed of furnish within predetermined moisture content limits is still of paramount importance even with the sophisticated new moisture measuring and controlling instrumentation.

Two-Pass Dryers

Two-pass dryers have also been developed but their use has not caught on in the United States. The drying action is the same as in the three-pass dryers. An advantage reported is easier access to the interior for maintenance because of more room between the concentric shells.

Single-Pass Dryers

Figure 8.7 shows a parallel-flow direct-heat single-pass dryer used for years successfully in the USA. It is 10 ft (3.05 m) in diameter by 100 ft (30.4 m) long. The rotation is 3 rpm. The gases and dry flakes are pulled from the dryer by two fans turning at a calculated rate of 720 rpm through an opening that splits into a Y configuration. Hot combustion gases are mixed with diluent gases and fed into the dryer drum. Exposure of the wet furnish to the hot gases is accomplished by means of peripheral lifters on the drum wall.

Single-pass dryers, as mentioned, are being used more, although their use is not widespread yet. Hard data comparing these two types are hard to find. Some differences found with the single-pass dryer apparently are less initial expense for

Figure 8.7. *Feed-end view of single-pass particle dryer. A conveyor belt feeds the wet flakes to the dryer through the circular chute shown at top center (Surdyk 1971).*

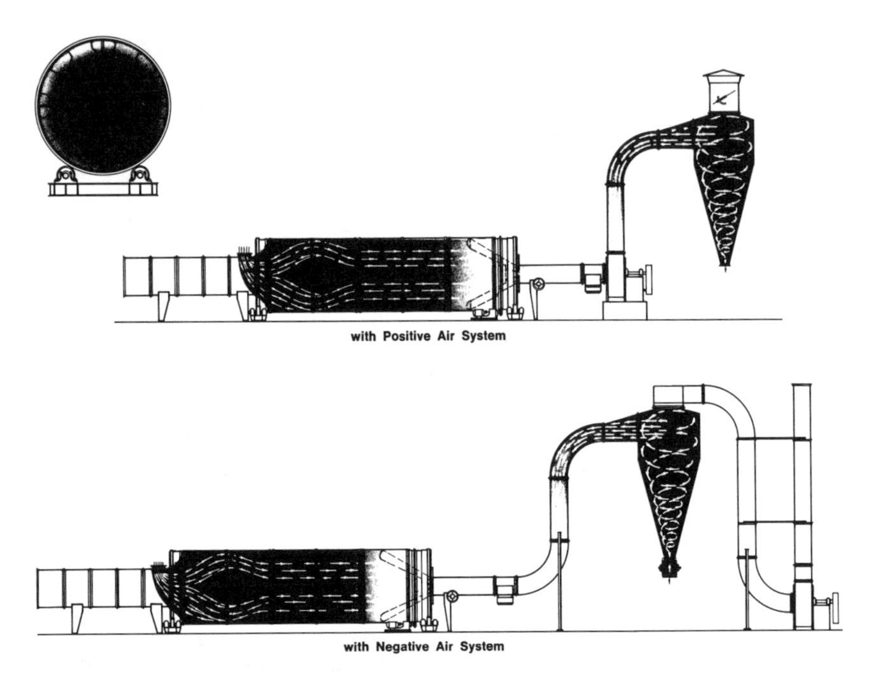

Figure 8.8. *Positive and negative air systems for single-pass dryer. (Courtesy M-E-C Co.)*

the same capacity, less maintenance because of simpler construction, a requirement for a slightly higher outlet temperature, and slightly more attention to operation (automatic moisture controls alleviate this problem). The single-pass type takes up more space than the three-pass. The air drawn through the dryer moves the fines through rapidly, preventing over-drying. Controls are the same as with the three-pass dryers.

Figure 8.8 shows this type of dryer with positive and negative air systems. A "target" is shown about one-fourth the way down the dryer. This is simply a metal disc which causes turbulence in the airstream. This turbulence provides for better use of the hot air in the dryer.

Another design from Europe has been used extensively there and has found applications in Canada. This dryer works on a different principle than the dryers described so far, where hot air is mixed with the furnish to be dried at extremely high temperatures. It must be recalled that wood particles at temperatures over 400°F (204°C) will burn before 500°F (260°C) is reached. With dryers operating at inlet temperatures that may reach 1600°F (871°C), particles must pass through the dryer quickly before their temperatures are raised to the charring or burning point.

In this single-pass dryer, also known as a *bundle-tube* dryer, the wet furnish is tumbled in a rotary drum over steam, hot water, or oil-heated tubes similar to the tubes in a boiler. Figure 8.9 is a schematic drawing showing the general arrangement of such a dryer. Figure 8.10 is a photograph of the heated tubes.

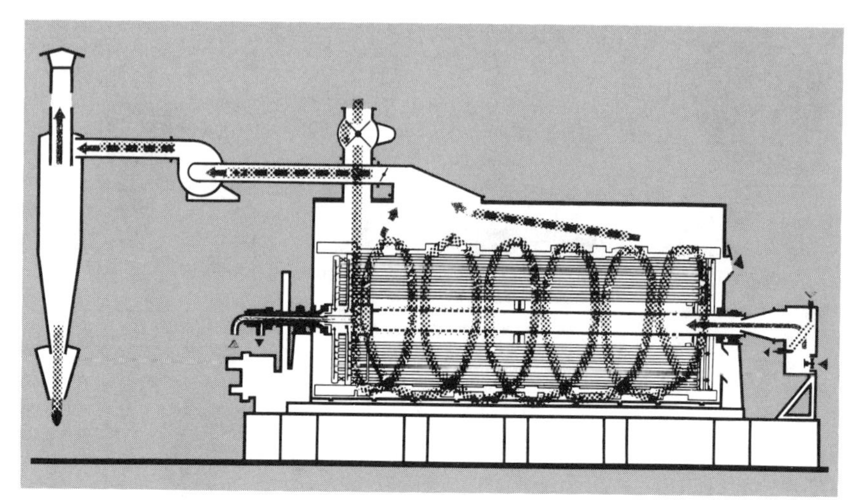

Figure 8.9. *Schematic drawing of a typical dryer equipped with heated tubes. (Courtesy Ponndorf Maschinenfabrik KG.)*

Figure 8.10. *Photograph of the heated tubes (Young 1967).*

The amount of evaporation which takes place is controlled by the temperature of the water or steam and by the speed at which the furnish is passed through the dryer. Because of the direct contact of the furnish with the hot tubes, the temperature in the tubes is normally held below 400°F (204°C).

Uniformity of feed into the dryer and uniformity of the incoming furnish moisture content is extremely important. An air-classification such as the one found in the three-pass dryer is not taking place. A preheating drum has been used to help control the incoming moisture content.

Efficiency is rated as high and the electrical energy required is low because no fans are required for moving large volumes of air. Some air must be exhausted, of course, to carry the evaporated water out of the dryer.

This dryer is relatively safe from fires and explosions but they have happened. The dryers are supplied with automatic sprinkler systems. A dangerous time for fire is when the dryer is shut down. Furnish then sits on top of the hot tubes and can catch fire. This is a particular problem when power failures occur, but safeguards are built in to cool the dryer and the automatic sprinkler system is available in the event fires do start.

HORIZONTAL FIXED DRYERS

Two types of horizontal fixed dryer have been developed, one commonly known as a *jet dryer* and the other as a *tube dryer*. The jet dryer was first manufactured about 20 years ago. Tube dryers are of relatively recent origin in the industry as they are specifically used in the plants using fiber for the furnish. Some of the drying tubes are in S shapes but the burners are horizontally oriented.

Jet Dryers

The jet dryer was developed in the Federal Republic of Germany and many have been used throughout the world, although only a few are in use in the United States. Its use has been mostly in the particleboard industry, but some installations have recently been made in fiberboard plants. One of these dryers is shown in Figure 8.11.

The jet dryer operates basically on the fluidized-bed principle. The wet furnish is fed into one end of a stationary drum and heated air is blown into the drum through a slot at the bottom of the drum extending its entire length. The heated air blows through the bed of material; the fluffing and forward movement of the bed is assisted by rotating raker arms or paddles. Figure 8.12 diagrams the general arrangement of the jet dryer. The four major parts are: the heating unit, the jet dryer drum, the circulating fan, and the cyclone collector.

The wet material forms a bed along the length of the drum one-fourth to one-third of its depth. The fan forces the heating medium around the dryer circuit, from the heating unit, through the drum and cyclone collector, and back through the heating unit. The air-lock system is very efficient as heat for drying is mostly kept within the drying unit.

The slot through which the heated drying medium enters the drum is arranged so that the medium enters tangentially and thus causes a rotation or spinning of the material in the bed as it moves forward through the drum (Figures 8.13 and 8.14). The dried material is drawn out of the dryer drum, through the circulating fan, and is discharged from the drying system through a rotary air lock at the bottom of the cyclone collector.

Figure 8.11. *A jet dryer. (Courtesy Büttner-Schlide-Haas AG.)*

Part of the spent air is exhausted and the remainder is recirculated through the heating unit. For maximum heat economy only enough spent air is exhausted to equal the volume of moisture evaporated plus the volume of makeup hot combustion gases from the heating unit. Offgas will probably have to be utilized or cleaned to meet pollution standards in many countries.

Of paramount importance in the control of the dryer operation is the series of vanes situated in the air intake slot along the bottom of the drum. These determine the angle of spin of the heating medium in the drum and the speed of forward movement of the furnish, and, through this, the dwell time of the material in the drum. It is to be noted in the diagram that the vanes in the first third of the drum are disposed to hasten the movement through that zone, in the second third to be neutral toward forward movement, and in the last third to tend to hold back the forward flow.

It is characteristic of this system of moving furnish for the finer particles, which need less drying time, to move faster through the drum and for the coarser material, which needs longer drying time, to move more slowly and thus have a longer dwell time. The adjustment of the vanes therefore affords not only control

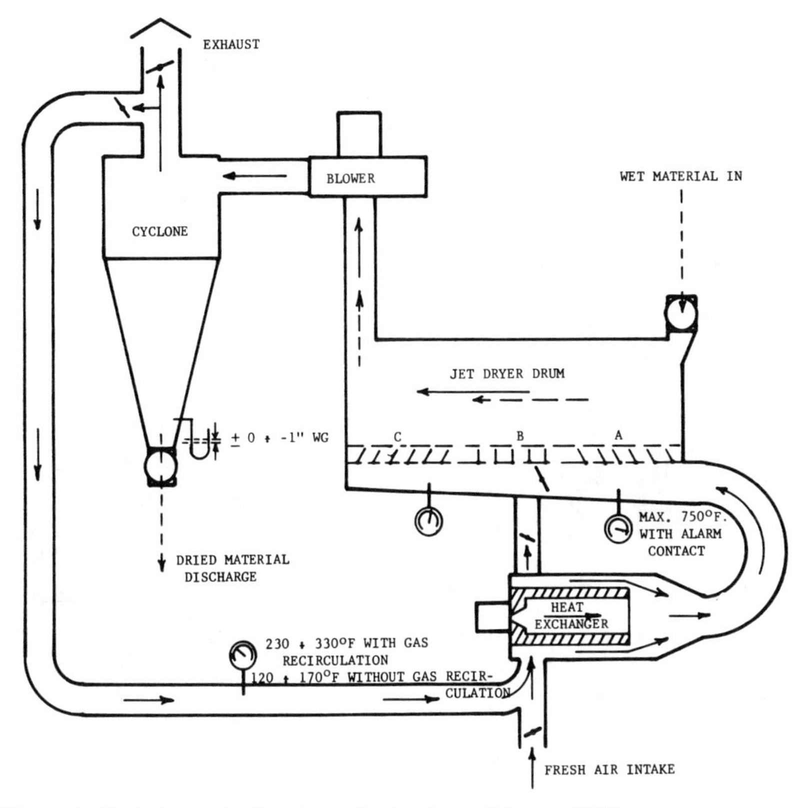

Figure 8.12. *Schematic drawing of a jet dryer (Mottet 1967).*

of the average dwell time of the material in the drum but also the relative dwell times for fines and coarse material. Thus, the material moves through the dryer at times corresponding to particle size even though the drum is stationary.

If the feed rate, furnish moisture content, and drying characteristics of the furnish are known, the factors under control in the jet dryer are: (1) outlet drying medium temperature—via thermostatic control of the firing rate of the heating unit, (2) material dwell time in jet drum—via jet vane control, and (3) amount of recirculation of heating medium and fresh air makeup.

The jet dryer shown in Figure 8.13 is a direct-heated type which constitutes the majority of installations, employing either gas or oil fuel, which can be supplemented with sander dust fuel from the board sanders. The other type (Figure 8.14) involves indirect heating by means of steam, hot water, or thermal oil. In this type the spent air cannot be recirculated.

The direct-heated dryers operate with inlet temperatures up to 700° to 750°F (371°-399°C) while the indirectly heated types, operating usually with the heating medium in the range of 360° to 400°F (182°-204°C), have dryer inlet temperatures of 320° to 370°F (160°-188°C).

Due to the fact that the direct-fired jet dryer operates predominantly with recirculated air comprised mainly of inert gases—nitrogen, carbon dioxide, and water vapor—fire danger is minimized. However, because a combustible material (wood) is dried in a system employing heating medium temperatures above the ignition point of wood, fire danger remains.

Some recent improvements on the jet dryer have been offered, all concerned with maximizing fire safety: First, speed governors for fan, rake shaft, and discharge rotary air lock so that, if one of these fails, both the material supply and furnace will be shut down. Second, automatic cleanout of the dryer cylinder. If the feed material contains large chunks of wood, these tend to accumulate in the dryer

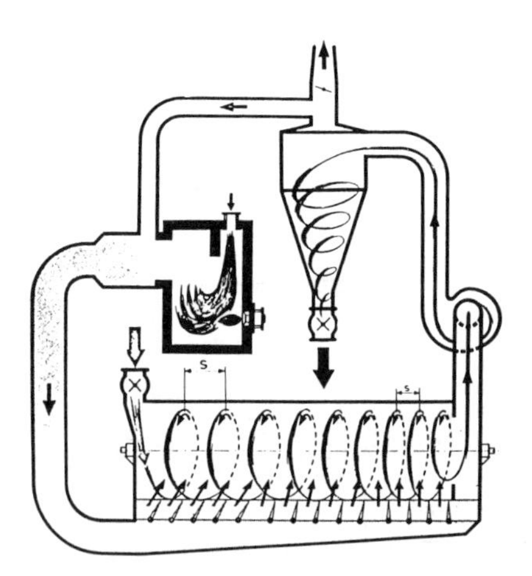

Figure 8.13. *Direct-fired dryer showing movement of furnish (Mottet 1967). This type is generally used.*

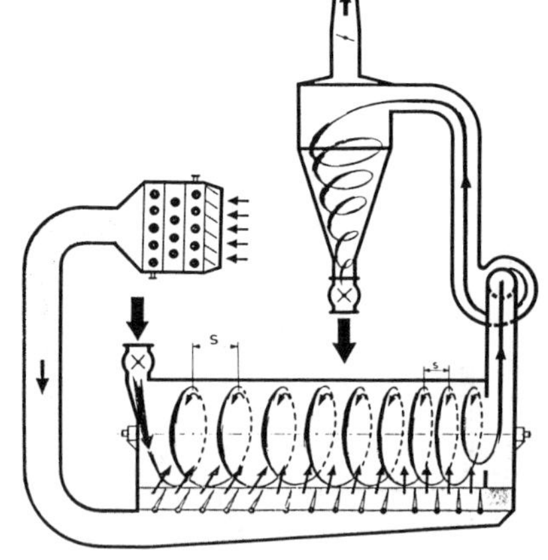

Figure 8.14. *This dryer is indirectly fired through means of a heat exchanger (Mottet 1967).*

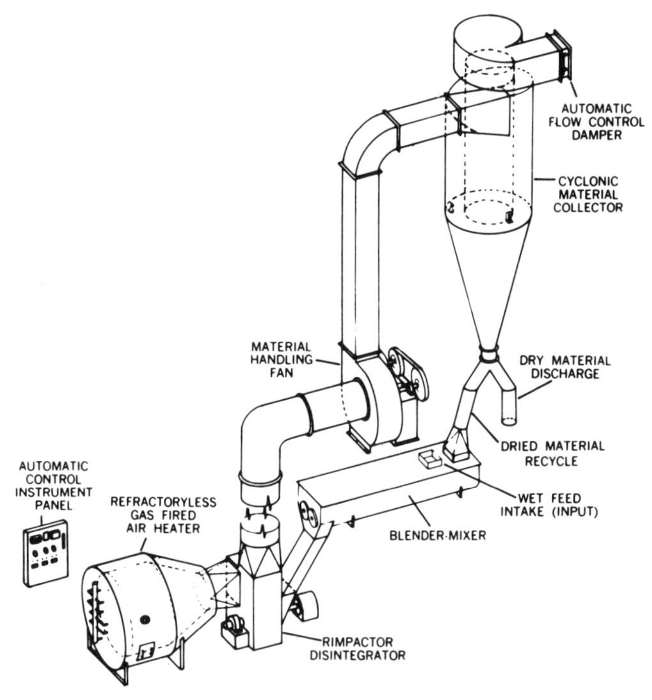

Figure 8.15. *A single-tube dryer. (Courtesy Edw. Renneburg & Sons Co.)*

drum, particularly if the control vanes are set for long dwell times. With automatic cleanout, the vanes are automatically switched to fast travel briefly to move out the heavy chunks. Third, a nylon thread connected to a limit switch so that, if the thread overheats, it melts and both material supply and furnace are switched off. Fourth, automatic water sprays in the dryer and in the cyclone collector if a specific temperature limit is exceeded. Fifth, hot gas damper control. When the dryer and furnace are shut off, an automatic damper discharges the air from the furnace to an outlet stack instead of into the dryer, where some remaining material may be overheated by the heat stored in the furnace brickwork, shell, etc. The aforementioned new moisture measuring and control systems are also being used to precisely control this dryer.

Tube-Type Dryers

The tube dryers (single-stage or double-stage) are shown in Figures 8.15 and 8.16 and are used mostly for drying pressure-refined fiber. The single-stage tube dryer consists of one long tube of about 225 ft (69 m) or longer with a pneumatic pull-through system (Figure 8.15). A tube dryer installed is shown in Figure 8.17. This particular dryer has tubes about 54 in. (1.37 m) in diameter. When using pressure-refined fiber, the heat can be generated from gas, oil, sander dust, or combinations of these, or a heat exchanger. The double-stage dryer (Figure 8.16)

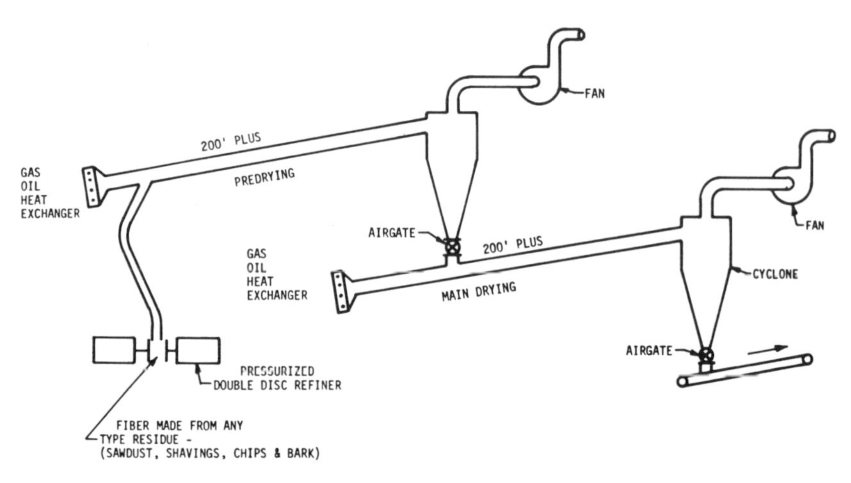

Figure 8.16. *A double-stage tube dryer (Buikat 1971).*

does the drying in two steps in a total tube length that can be ±400 ft (122 m) in length. In the first stage of drying, more heat is used, at about a range of 400° to 600°F (200° to 316°C). Temperatures in the second stage are 100° to 180°F (38° to 82°C). In both stages, the moisture is discharged through a pneumatic pull-through system. Moisture control is effected by controlling temperature and, of course, the amount of material entering the dryer. The drying is accomplished in a matter of seconds.

Dry fiber is extremely sensitive to fire and explosion and great care is exercised in handling these problems. One of the most important features is to feed the refined fiber evenly into the dryer, thus preventing a plastering of wet fiber onto the hot tube. If such plastering occurs, the fiber mat, when dried, catches fire and sends sparks through the dryer. In the extreme case, sparks could carry on through the cyclone and into dry fiber storage; should this occur, the sparks would cause wholesale fires and explosions.

Figure 8.17. *M-E-C flash tube dryer system. (Courtesy M-E-C Co.)*

SELECTED REFERENCES

Arnold Dryer Co. 1971. Heil Dehydration Systems. Bulletin No. ARD 71669-1171. Subsidiary of The Heil Co., Milwaukee, Wisconsin.

Buikat, E. R. 1971. Operating Problems with Dryers and Potential Solutions. *Proceedings of the Washington State University Particleboard Symposium, No. 5.* Pullman Washington: Washington State University (WSU).

Edw. Renneburg & Sons Co. 1975. Advanced Processing Equipment. Bulletin G871. Baltimore, Maryland.

The Heil Company. 1975. Heil Ardriers. Bulletin No. ARD-61406. Milwaukee, Wisconsin.

Klamroth, V. K., and J. Hackel. 1971. Dryers for the Particle Board Industry. *Holz als Roh-und Werkstoff,* Vol. 29, No. 12.

Lengel D. L. 1967. The Reitz Dryer. *Proceedings of the Washington State University Particleboard Symposium, No. 1.* Pullman, Washington: WSU.

McNeil, W. M. 1975. Private communication on drying of pressure-refined fiber. Springfield, Ohio.

Mottet, A. L. 1967. The Buettner Jet Dryer. *Proceedings of the Washington State University Particleboard Symposium, No. 1.* Pullman, Washington: WSU.

Ponndorf Maschinenfabrik KG. 1975. Heavy Duty Chip Driers. Kassel, Federal Republic of Germany.

Porter, S. M. 1971. Gas Recycling and Dryer Modifications to Reduce Smoke Emissions. *Proceedings of the Washington State University Particleboard Symposium, No. 5.* Pullman, Washington: WSU.

Stillinger, J. R. 1967*a*. Drying Principles and Problems. *Proceedings of the Washington State University Particleboard Symposium, No. 1.* Pullman, Washington: WSU.

———. 1967*b*. The Heil Dryer. *Proceedings of the Washington State University Particleboard Symposium, No. 1.* Pullman, Washington: WSU.

Surdyk, L. V. 1971. Dryer Modifications to Reduce Smoke Emissions and Improve Drying. *Proceedings of the Washington State University Particleboard Symposium, No. 5.* Pullman, Washington: WSU.

Westphal, J. L. 1974. Trends Toward Self-Sustained Dryer Energy Through Utilization of In-Process Sources. *Proceedings of the Washington State University Particleboard Symposium, No. 8.* Pullman, Washington: WSU.

Young, D. R. 1967. The Ponndorf Dryer. *Proceedings of the Washington State University Particleboard Symposium, No. 1.* Pullman, Washington: WSU.

9. MOISTURE MEASUREMENT AND CONTROL

The moisture content level of the raw material and furnish is of extreme importance in all stages of the dry-board process. Moisture content is an important factor in determining how particles will be generated and their resultant physical characteristics. As already discussed, drying of the furnish prior to blending is an important step in producing acceptable boards. Excessive moisture content extends hot pressing times and causes problems with blows and delaminations. Over-drying possibly will cause problems with resin bonding. The mat moisture content and its distribution is reflected in the board's density profile, and hence physical properties. Moisture contents that are out of control are normally reflected in process problems, and wide variation in board physical properties. The most noticeable problems with processing are the aforementioned blows or delaminations in the boards caused by excess moisture in the mat being consolidated in the hot press. Such problems are usually obvious because the blow is usually accompanied by a popping sound of escaping steam, or bulges on the board surfaces can easily be seen; however, small incipient blows can go undetected by plant personnel.

The board industry is known for its capital-intensive processes and sophisticated instrumentation. Thus, it is somewhat surprising that the critical step of quickly and accurately measuring moisture content is the major one to evolve slowly from a manual to an automated operation. Many plants still do not use continuously operating automated systems.

The laboratory method used for measuring wood's moisture content is the oven-dry method. This is literally taking a piece of wood, and weighing it before and after oven-drying and then calculating its moisture content based on the weight before and after drying. Instruments correlate their readings with the oven-dry method of moisture measurement. For the sake of clarity, percent moisture content, percent moisture, and moisture level, all terms which are used to indicate the amount of water present in wood, should be defined. In the lumber, plywood, and composition board sectors of the forest products industry, percent moisture content (oven-dry basis) is the common terminology used. It is defined as follows:

$$\text{Percent-Moisture Content (MC)} = \frac{\text{Weight of Water}}{\text{Weight of Dry Wood}} \times 100$$

In the pulp and paper segment of the industry, however, it is most common to speak of percent moisture, defined as:

$$\text{Percent Moisture} = \frac{\text{Weight of Water}}{\text{Weight of Green Wood}} \times 100$$

Moisture contents greater than 100% are common to lumbermen, while 100% moisture means *all water* to a pulp and paper mill operator. These differences in meaning are the result of the basis selected for the calculation—and this is the important point—these are calculated values. All instruments, however, measure moisture level, as a function of the amount of water in the sensing zone. This is defined as:

$$\text{Moisture Level} = \frac{\text{Weight of Water}}{\text{Volume of Wood}} \text{ (in sensing zone)}$$

REQUIREMENTS

Board plants have for the most part suffered along with measuring moisture by the old method of *grab* sampling. A handful of furnish is grabbed, put in a drying device, and the moisture content calculated. This requires time and other problems are attendant. The sample may have been small (10 g) and not representative. The technician may have carried the sample in his sweaty hand for 100 ft and changed the moisture content. The ambient atmospheric conditions may have changed the moisture level by the time the measurement was completed. During all of this time the process line keeps on running. Therefore, reliable moisture-measuring instruments with quick responses have been of imperative need. Such instruments ideally should be able to feed signals for adjusting process controls into the system. A number of these have been developed and will be discussed; however, as mentioned, not all plants use them.

Measuring and controlling moisture is of the highest importance out of the dryer. At this critical juncture the moisture has to be at the right level. As mentioned previously, moisture control is important elsewhere in the process but very little variation can be tolerated out of the dryer because of downstream effects.

A meter sensing the dryer input moisture content should have a range of about 0 to 200% moisture content (oven-dry basis). A meter sensing the output moisture content should have a range of 0 to 20% MC with an accuracy of about 0.2% so that dryer control can be maintained to 1%.

Specifications for a meter should include the following:

1. It must measure moisture to the desired accuracy regardless of variations in bulk density, particle size and shape, and ambient or particle temperature.
2. It must have a sufficiently fast response and desired accuracy to be able to control the dryer to the desired output range.
3. It must be simple to operate. The adjustments should not have to be changed

after the initial installation.
4. It must be able to survive the environmental extremes of dust, heat, vibration, and chemicals.
5. It must be as near failure-proof as possible and it must require minimal maintenance by plant personnel.

METHODS AND LIMITATIONS

A number of possibilities for continuously measuring moisture are: electrical resistance or conductance, capacitance, humidity sensing, microwave absorption, and optical infrared absorption, which is quite popular at present. The only direct measure of moisture is that obtained by oven-drying, and that is the method used during the calibration of all the various instruments for defining either moisture level or percent-moisture content.

Resistance

Pin-type meters using resistance are based on the fact that the moisture content of wood is a logarithmic function of resistance as illustrated in Figure 9.1. At moisture contents of 7% (oven-dry basis) and lower, the electrical resistance is very high (10^{10} ohms and higher). This type of meter has been in existence for a long time and is used heavily in sawmills. It requires manual application.

Electrical resistance as a continuous measuring method did not find much favor initially because the signal was too weak to measure without resorting to exotic instruments. Furthermore, the resistance is very temperature sensitive. However, in recent years an effective system based on resistance has been developed. Early attempts used roller electrodes that had to be fed a continuous, partial stream of the particle flow. This was not practical and resin also built up on the electrodes. Further attempts were made to develop this system because even small moisture content changes result in large resistance changes, according to a logarithmic function shown in Figures 9.2 and 9.3.

It may be clearly seen that even a small moisture content change results in a large resistance change. This is particularly true for the moisture-content range up to 100% based on oven-dry weight (all moisture contents in this discussion of

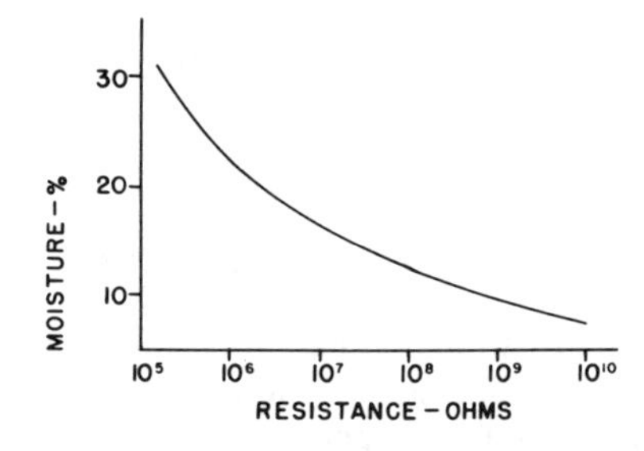

Figure 9.1. *Moisture content of solid wood as a function of electrical resistance of the wood as measured with a pin-type meter (Perry 1972).*

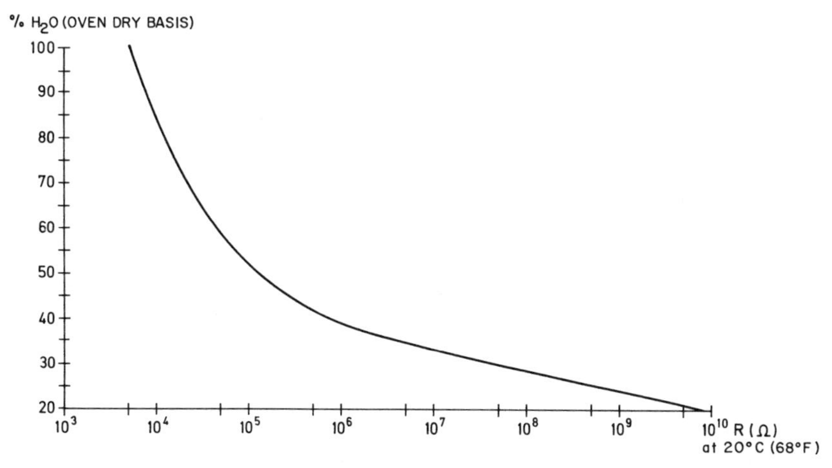

Figure 9.2. *Resistance of wood particles as a function of moisture content–high moisture content range (Klinkmueller 1972).*

resistance are based on oven-dry weight). Even beyond this is the resistance change per 1% moisture content change, which is large enough to obtain measurements with satisfactory accuracy. In the particularly important moisture-content range from 0 to 15%, the resistance change is considerably larger than in the range of moist particles. It is here that this method is most accurate.

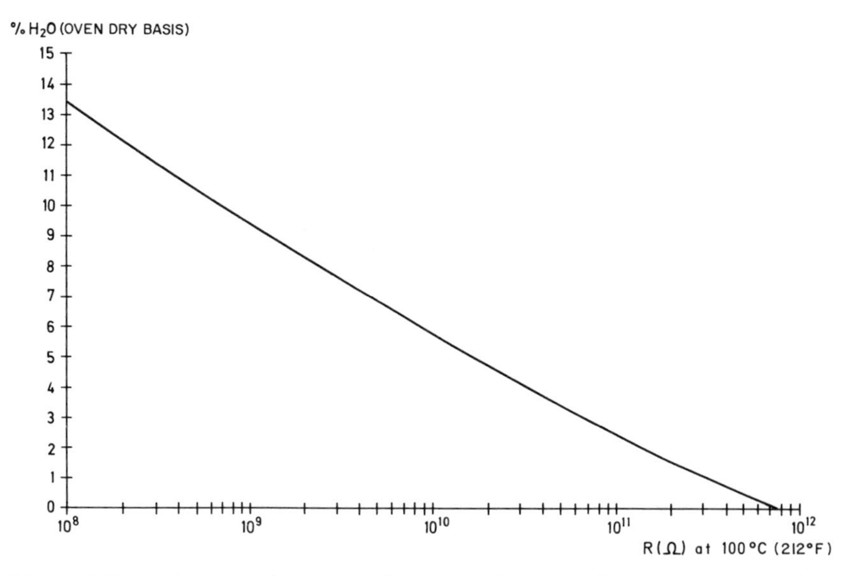

Figure 9.3. *Resistance of wood particles as a function of moisture content–low moisture range (Klinkmueller 1972).*

Figure 9.4 gives an indication of the severity of the temperature effects. According to this figure, a temperature change of t =104°F (40°C) in particles having a moisture content of 5% will cause an absolute measurement error of 2% and an error of 5.5% (based on oven-dry weight) in the case of particles having a moisture content of 20%. Therefore, temperature compensation is absolutely necessary if useful results are to be expected. Effects of density of the furnish are considerably smaller than temperature effects. They may not be disregarded, but they can be eliminated by appropriate transducer design.

Development of the resistance method resulted in a transducer which meets the practical requirements to a considerable degree. It consists of a steel measuring cylinder with four ceramic-insulated hemispherical electrodes arranged on the inside wall (Figure 9.5). The construction assures complete shielding of the measuring section against exterior AC stray currents. The measuring cylinder is placed inside of a chute containing a falling particle stream. An oversupply of particles is thus constantly falling on the entrance aperture of the measuring cylinder. The feeding screw picks up a certain fraction of the total particle flow and pushes these particles into the front measuring space. The exit aperture of the cylinder is constricted by four blade springs which cause crowding of the particles. Thus, a compact particle plug is being formed which constantly passes the four electrodes as shown in Figure 9.6. The springs support the particle plug in such a way that only the protruding part is allowed to break off and return automatically to the main particle flow.

Where the particle structure is fine and fiberlike, as for example in the manufacture of fiberboard, a conical nozzle can be installed in order to increase the crowding pressure.

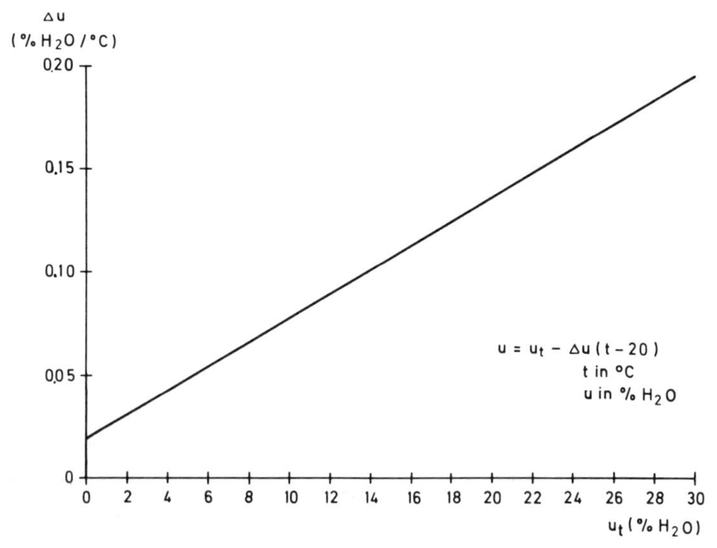

Figure 9.4. *Effect of temperature on moisture-content readings as a function of moisture content (Klinkmueller 1972).*

Figure 9.5. *The novel transducer (Klinkmueller 1972).*

The four electrodes form four measuring channels with the measuring current flowing from the top of each electrode to the cylinder wall. The measuring span extends therefore along the peripheral zone of the passing particle plug. A DC amplifier with added power amplifier averages the signals from the four channels. The average is then displayed by an integrating indicator. Erroneous measurements that could be caused by single moist or thick particles are thereby largely eliminated.

Figure 9.6. *Transducer swung out of the particle chute showing the particle plug (Klinkmueller 1972).*

The temperature compensation device is arranged in the front part of the measuring cylinder, in the area of the electrode location, and always senses the temperature of the particles being measured. For technical reasons it is designed for two temperature ranges, namely for the measurement of wet particles in the temperature range from 14° to 122°F (−10° to +50°C) and for the measurement after drying in the range from 104° to 248°F (40° to 120°C). The moisture content indication, therefore, occurs independently of the temperature of the particles in percent based on oven-dry weight.

The electronic test amplifiers developed for this method use field-effect transistors and operational amplifiers. They have such a great stability that provisions for subsequent calibration were found to be unnecessary.

The rate of material flow through the transducer is about 2.2 to 4.4 lbs/minute (1-2 kg/min) for dry particles and 6.6 to 8.8 lbs/minute (3-4 kg/min) for wet particles (at about 40% moisture content). The accuracy claimed with these transducers considering all influencing factors is ±0.5% moisture content for the range from 0 to 10% moisture content and ±5% of the indicated value in the range above 10% moisture content. Thus, the tolerance of the indicating instrument is ±1% at a moisture content of 20% and ±2.5% at a moisture content of 50%. These tolerances have been verified for 95% of all measurements by extensive series of measurements (Figure 9.7). Errors resulting from variations inherent in the production process, such as variations in species combinations and particle geometry, are included in these tolerances. This means that a measuring system has been created which is sufficiently accurate and reliable to be used as the basis for the control of the drying process.

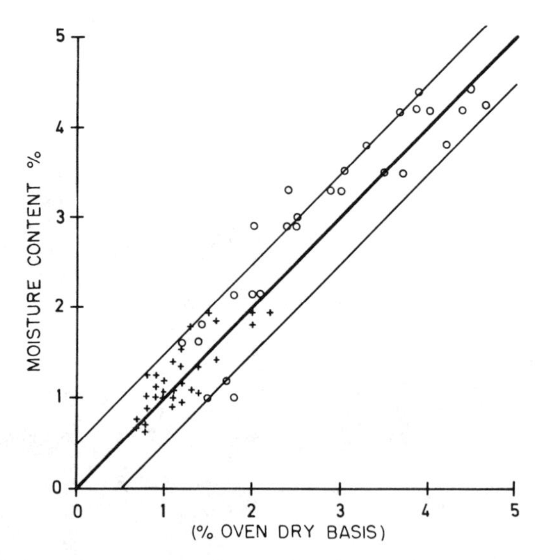

Figure 9.7. *Moisture-content readings related to moisture content by oven-dry test of face and core materials for three-layered board, obtained from two different jet dryers (Klinkmueller 1972).*

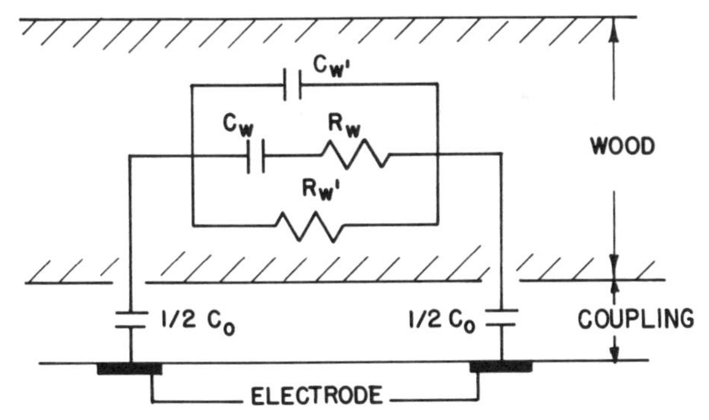

Figure 9.8. *A diagram of an equivalent circuit in wood for moisture measurement by capacitance method (McCarthy and Parkinson 1969).*

This measuring instrument, developed in Europe, has found use elsewhere but is not yet popular in the United States.

Capacitance

Capacitance instruments are based on the determination of the dielectric properties of wood particles. Capacitance of wood is directly proportional to the dielectric constant and depends on the wood specific gravity and moisture content. Such units have been used for years in plywood plants. The principle for measurement depends upon the detection of changes in the dielectric constant of the wood with changes in moisture. The dielectric constant for dry wood is about 2, while the dielectric constant for water is much higher at about 81. The addition of a small amount of water to the wood reflects a significant, easily measured change in the dielectric constant. The changes which occur are detected by changes in what may be considered the equivalent electrical circuit within the wood as approximated by the circuit model shown in Figure 9.8. The capacitor, C_o, represents the coupling capacitance created by the air gap between the electrode and the wood. There is also a resistance component in the coupling, but its effect is so small that it is neglected. In the wood, C_w' is the capacity of the oven-dry material and R_w' is the resistance of relatively wet material, especially when the moisture forms conductive paths. C_w in the series R_w-C_w circuit is the circuit component whose changes are measured. With sensing elements in a fixed location on either side of the sample, the effect of the coupling capacitance, C_o can become quite high because of the physical configuration that requires adequate space be allowed between the elements to permit the wood to be passed. However, when a fringe-field or spray-field electrode principle is applied, as shown in Figure 9.9, the wood to electrode distance can be minimized by bringing the wood into contact with the electrode. Most of the problems caused by the effect of coupling capacitance in early models of this type of moisture detector have been overcome by sophisticated developments in circuitry. The earlier models had problems where surface moisture or very wet material close to the sensor masked the reading of drier material farther away.

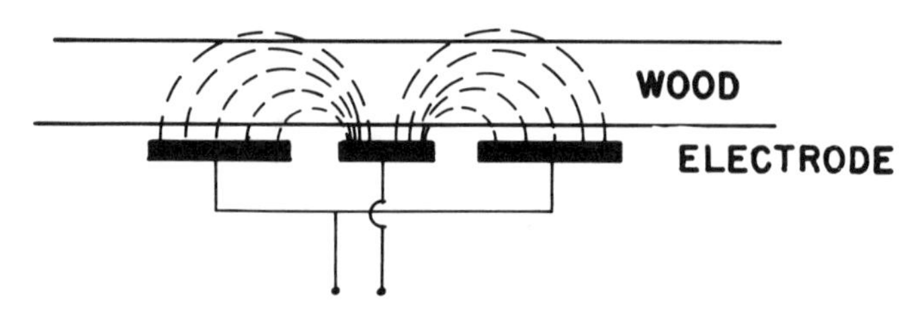

Figure 9.9. *A diagram of a fringe-field electrode (McCarthy and Parkinson 1969).*

Newer models have up to six sensors attached to a single control console for use in monitoring or controlling in either open- or closed-loop modes. No bridge balance adjustment is needed, and the amplifiers are permanently calibrated. Thus no adjustments or periodic recalibration after installation is needed. Each sensor is a modular self-contained unit containing an RF bridge and measuring capacitor assembly that produces a DC signal voltage proportional to total moisture content per unit volume of material being measured. The preferred method of application is on a screw conveyor, as shown in Figure 9.10.

This type of unit has had many years of dependable service in pulp mills and has been adapted for use on composition board plants. For operation, the full stream of material can be measured, or a partial stream of furnish is run through the instrument for moisture measurement. Extremely wide bulk density variations, which usually develop quickly, can cause problems with the dielectric measurement, but these problems have been minimized in the design of the newer detectors of this type.

Figure 9.10. *Capacitance sensor mounted on screw conveyor (McCarthy and Strom 1972).*

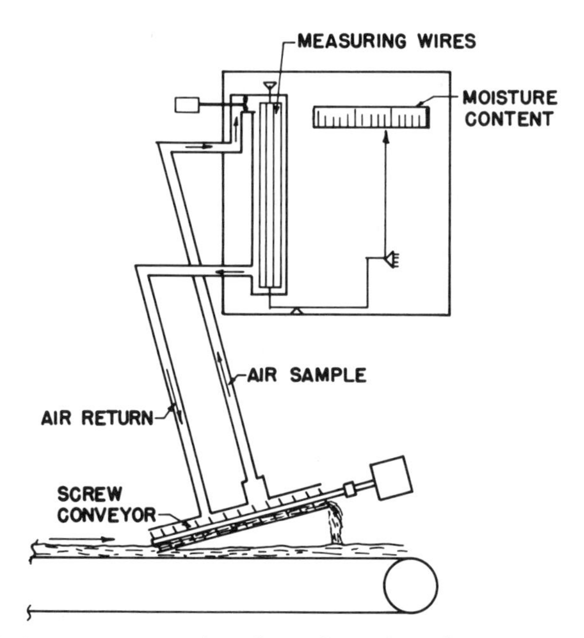

Figure 9.11. *A moisture meter based on relative humidity sensing operating in a particleboard production line (Brumbaugh 1969).*

Relative Humidity Sensing

Moisture content measurement by the determination of the equilibrium moisture content is based on the fact that a hygroscopic equilibrium is established between particles of a given moisture content and a sufficiently small surrounding air space. With the use of sorption isotherms, the moisture content of the particles can be established by measuring temperature and relative humidity of the air space. Considerable inertia and hysteresis, as well as the moisture content range being limited by the fiber-saturation point, apparently make this method unsuited for automatic control application.

A continuous moisture meter is based on the principle of drawing a continuous sample of air from the passing wood raw material, then measuring the relative humidity of the air sample, and recording this relative humidity as moisture content (Figure 9.11). The collecting of a continuous sample of raw material is accomplished by imbedding the lead end of a small screw conveyor in the raw material flow. As the screw draws the sample of raw material by the inlet end of the air sampling line, a continuous air sample is being pulled from the stock. A small motor driving an in-line fan draws the air through the sampling tube and blows it over the measuring wires. The expansion or contraction of the wires with changing relative humidity is transmitted by levers and is measured as the moisture content. The measured air is then blown through a return line to the raw material sampling tube which is a few inches ahead of the air intake tube. This process tends to dampen sudden small changes in relative humidity.

Microwave Moisture Measurement

Microwave energy is a form of electromagnetic radiation having frequencies between 0.1 and 100 billion cycles per second (0.1-100 gigahertz [GHz]). The wavelength of energy used in the system under discussion is approximately 3 cm (1.18 in.) in free space, corresponding to a frequency of 9.4 GHz. Microwaves lie between the television band and the infrared spectrum of light and combine the properties of radio waves and optical waves. Microwaves are commonly found in radar and communications systems.

Water molecules in the free state will absorb and also reflect microwave energy. It has been determined that most host materials, e.g., particleboard furnish and its constituents, are virtually transparent or invisible to microwave energy in the oven-dry condition. Thus, this technique offers excellent sensitivity for detection and measurement of the free water molecular quantity in a given volume of material.

Chemically or molecularly bound water molecules, such as in ice or hydrates, are not free to rotate and resonate in the microwave beam, and are therefore unable to absorb microwave energy. Also, water in the vapor form cannot be measured by the microwave absorption method, since the water molecules are too widely spaced. This is advantageous as changes in relative humidity or water vapor level in the environment will not affect the gauge readings.

The microwave detector senses the amount of energy received and converts this to a voltage signal. This signal is processed by the electronic circuits, in such a way that the amount of energy is displayed on the instrument outputs. Thus, the gauge can be calibrated in terms of actual water content.

The microwave absorption technique measures the total mass of water in the microwave beam. This is of particular value in drying operations. When percent-moisture measurements are to be made, control or measurement of density must be maintained. Several approaches are possible here. A filled test chamber can be weighed and a nomograph employed to provide accurate percent-moisture measurements when bulk density varies. A load cell providing a signal which can automatically correct for density variations can be employed under the test chamber. Another approach is to fill the test chamber with a known and constant weight of furnish, then compact it to a constant volume. A more practical approach has been to use a screw auger to maintain a constant head of material above the sensor area and to drive the furnish through at a constant rate. Another approach is to use a weigh belt feeder where "total mass" signals are fed into the microwave gauge and averaged with the microwave-moisture-mass signal over a short period of time, then electronically divided to give percent-moisture output signals.

Another variable which must be controlled is furnish temperature. It is preferable to maintain operating temperatures at a constant value. If this is not possible, then the material temperature can be measured manually and corrections can be made with a suitable nomograph. A second approach is automatic temperature measurement and correction, utilizing an internal computer board within the microwave gauge. In most materials, a temperature change of 10°F (−12.2°C) will usually appear as a moisture change of a fraction of a percent. The temperature coefficients can be either positive or negative, and the slopes may also vary as a

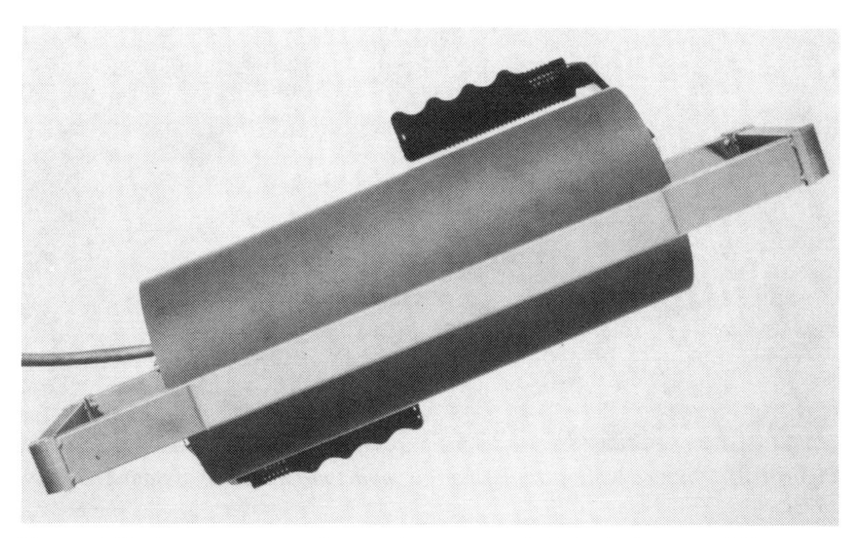

Figure 9.12. *Contact sensor for use in conveyor belt bed (Kenton 1972).*

function of the class of material, nominal temperature level, and percent-moisture range involved.

Figure 9.12 illustrates a relatively new development for conveyor-belt applications. This contact sensor measures moisture in a zone approximately ½ in. (12.7 mm) from the sensor surface. A single-sided measurement may be obtained, and it is unnecessary to measure total mass or density. One application under development at this time is the measurement of individual layers of particleboard furnish coming out of the blenders for resin control. Another possible application is the measurement of individual layers of furnish during particleboard buildup on the caul during forming. For measurement of the finished board, a pair of noncontact through-transmission sensors is most practical (Figure 9.13). These sensors are contained, rugged, aluminum housings, and the microwave window is made of a durable Teflon cap. Sensors of this type are designed specifically for hostile environments and are totally waterproof.

Infrared Absorption

This type of instrument is a near infrared spectrophotometer in which two active wavelengths of near infrared radiation, one that is sensitive to moisture and one that is not, are used to measure the change in the reflected energy caused by an increase or decrease in moisture in the product being measured.

A tungsten lamp which emits a large spectrum of radiation at a constant level is used as the infrared source. Light from the tungsten lamp is focused and directed to a rotating filter wheel which contains two precise optical filters to allow very narrow constant-amplitude wavelengths of energy to pass (Figure 9.14).

The resulting discrete wavelengths are directed toward the sample in question and the reflected energy focused on a lead sulfide detector for conversion into

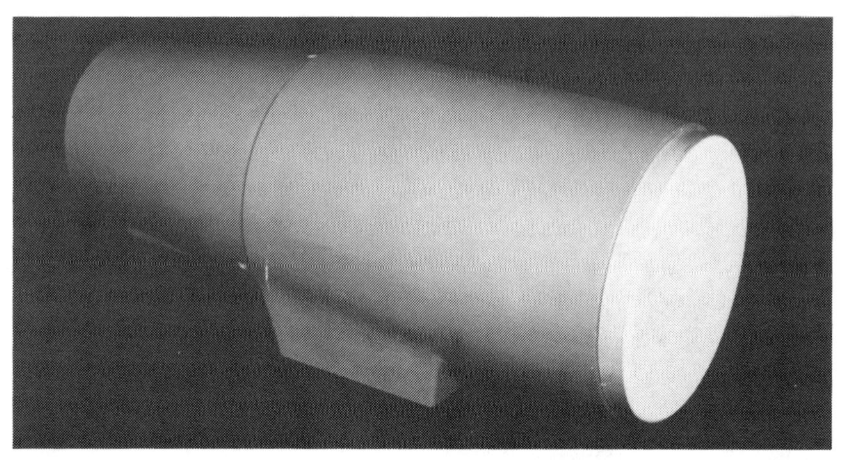

Figure 9.13. *Rugged waterproof sensor for noncontact testing of board (Kenton 1972).*

electrical signals. The conversion of light energy to electrical energy occurs at the lead sulfide detector and the resultant signal is processed by the preamplifier which shapes and conditions the pulses than have been created for easy processing in the main electronic unit. Figure 9.15 illustrates typical frame of reference and measure signals being generated at the detector stage of the analyzer. The change (△E MEA) in measure signal is a result of increasing or decreasing moisture. The measured wavelength energy is inversely proportional to the amount of moisture present; therefore, when the measured energy decreases, the moisture in the product has increased and vice versa. Completely independent controls for zero or intercept and sensitivity allow the calibration to be very narrow (0-1%) or very wide (0-90%) absolute moisture.

This type of moisture content instrument can be calibrated quickly using an optical standard. With previous moisture analyzers, in order to verify calibration, one usually had to go through the tedious process of taking dozens of samples in order to establish the calibration curve and to be certain that the calibration curve had not shifted. Figure 9.16 is an example of the accuracy of one of these meters as compared to oven-dry samples.

An interesting development has been an automatic solid material sampler which can be used on materials being conveyed in enclosed air or gravity systems. Samples are withdrawn automatically from these conveying systems at predetermined times as short as 15 seconds. These samples can then be quickly measured for moisture content by the infrared analyzer.

Nuclear Gauge Measurement

Nuclear bulk moisture gauges are used for continuous measurement of moisture in bulk solids on belt conveyors. The moisture gauge, consisting of a *neutron gauge* and a *gamma gauge*, continuously samples a cross section of the conveyed

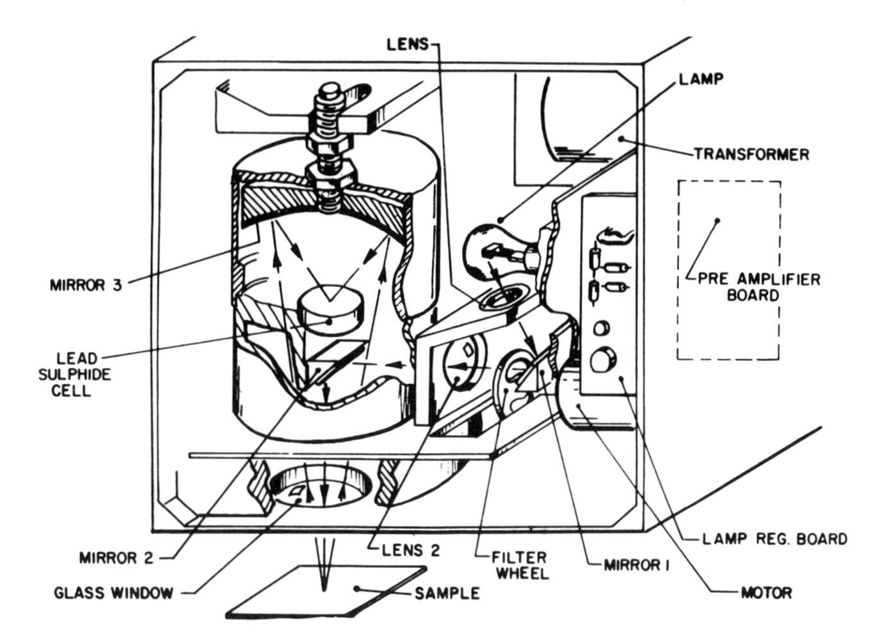

Figure 9.14. *Cutaway drawing of the Anacon analyzer in operation (Fordham and Buba 1974).*

material and provides a continuous measurement of water content and bulk mass content. The water content is measured by the neutron gauge, and the bulk mass content by the gamma gauge. The signal outputs of each gauge are then used in a ratio circuit to provide a signal corresponding to the percentage of moisture.

The gauge configuration is a C-frame (Figure 9.17). The combination neutron and gamma source is housed in one arm of the C-frame. The radiation detectors are housed in the other arm of the C-frame. The C-frame is mounted

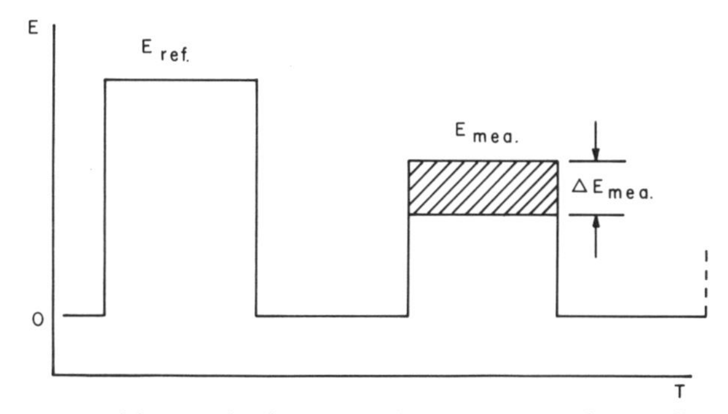

Figure 9.15. *Typical frame of reference and measure signals (Fordham and Buba 1974).*

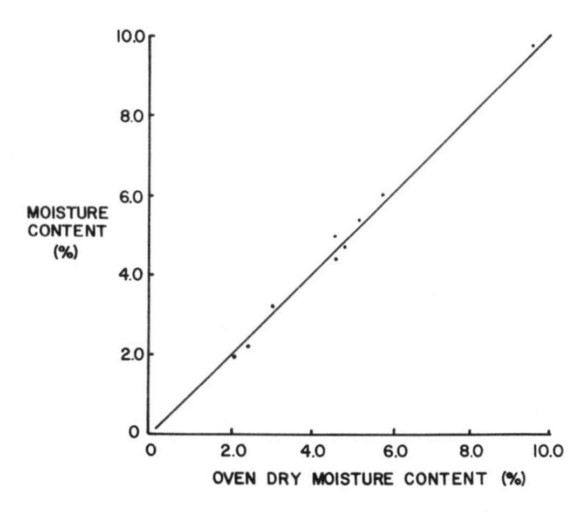

Figure 9.16. *Oven-dry moisture content vs content measured by an infrared moisture analyzer on furnish dried in an industrial dryer (Fordham and Buba 1974).*

about the material conveyor. The same C-frame houses both the neutron gauge and the gamma gauge.

Neutron Gauge: The neutron gauge consists of a source of fast neutrons (average energy about 4 to 5 MeV) and a slow neutron detector which in this case is a BF-3 gas proportional counter tube. The source material is Am-Be. The fast neutrons, in passing through the process material, are slowed down primarily by collisions with hydrogen nuclei. The neutrons lose a large fraction of their energy when they encounter hydrogen nuclei. The amount of energy loss depends upon the impact conditions of the neutron and the hydrogen nuclei (e.g., if the impact is head-on, the energy loss will be nearly total). If the fast neutron source and the

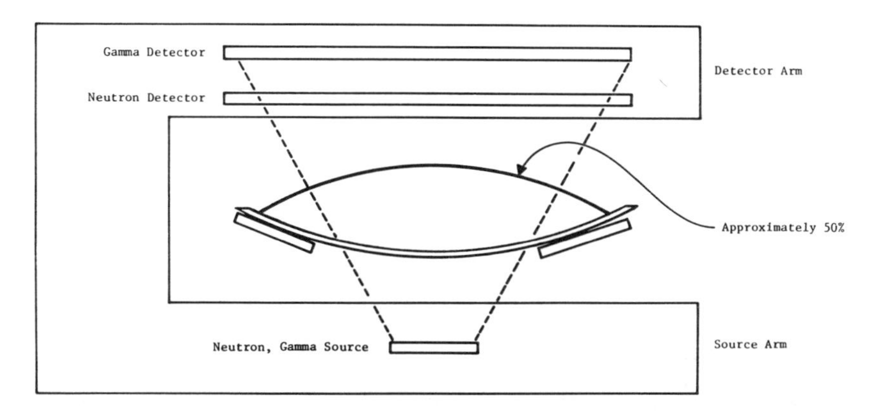

Figure 9.17. *Moisture gauge configuration (Chen 1972).*

slow neutron detector are arranged in a geometry such as in Figure 9.16 the number of slow neutrons arriving at the detector will be directly related to the mass of hydrogen in the process material. When the quantity of hydrogen becomes large enough to slow down all fast neutrons to thermal energies, then the slow neutrons detected attain saturated level. As long as the hydrogen loading is kept below the saturation level, the neutron technique is useful. Basically then, the neutron gauge measures hydrogen. Water contains 11.1% hydrogen by weight. From the measurement of hydrogen the amount of water present in the process can then be inferred.

If hydrogen is in the measured product or conveyor belt, this must be taken into account and compensated for in order to provide a correct moisture determination from the neutron gauge.

Gamma Gauge: The gamma gauge, like the neutron gauge, consists of a source and detector arranged in the transmission of the same C-frame as the neutron gauge. The source material is Cesium-137, which emits 0.66 MeV gamma rays. The detector is a gas-filled ionization cell.

As the radiation is continuously passed through the process material, the gamma radiation is absorbed by the process in relation to its total mass in the radiation beam. The amount of radiation arriving at the detector is an inverse function of the total mass in radiation beam. The gamma radiation absorption is dependent upon bulk mass (H_2O plus oven-dry material).

Moisture Gauge: It has been explained that the neutron and gamma gauge are arranged independently of each other. The gauges themselves are independent entities, but their geometries must be interrelated to provide a precise percent-moisture gauge. The unique source-detector configuration of the bulk moisture gauge provides equivalent cross sections for both measuring channels. The advantage of this configuration is that gauge measurement precision is unaffected by changes in process loading profile (slump angle) and load shifts.

The bulk moisture gauge consists of two channels of data, moisture weight and total weight, both being generated simultaneously as shown in Figure 9.18. This information is amplified and fed into the moisture gauge computer. The computer samples both channels of data for a given increment of time, which is determined by a variable synchronizing timer, common to both channels. The signal outputs of the moisture weight and total weight are properly converted into signals useful for feeding into a percent-moisture computer which provides an output for percent-moisture recording. The computer also provides a signal which can be fed into an optional oven-dry computer. The oven-dry computer accepts this signal and an external total weight signal from a conveyor scale to provide a totalized oven-dry digital counter readout. A further option of providing an oven-dry control signal is also available. Figure 9.19 shows an installation for evaluating the total oven-dry chip discharge to the digester on a batch digester system. In conjunction with the moisture gauge there was installed on the same conveyor a pair of belt weigh scales, whose configuration is also that of a C-frame. The belt weigh scale provides a total weight throughput signal, which is required for the computation of total oven-dry throughput.

This type of moisture measuring instrument has not been refined for widespread use on furnish with low moisture contents.

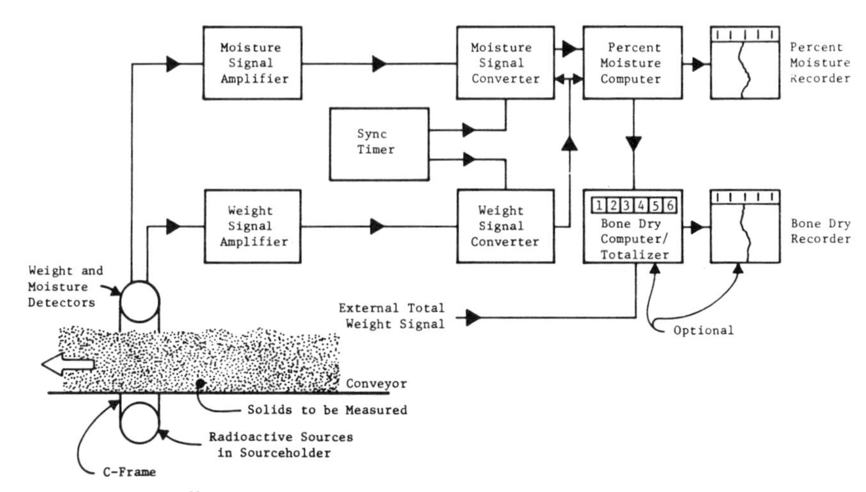

Figure 9.18. *Bulk moisture gauge system (Chen 1972).*

APPLICATION OF MOISTURE-MEASURING INSTRUMENTS

As mentioned in the introduction, the development of control instrumentation is of immense value to the industry, especially in conjunction with the dryer. The resistance, capacitance, and infrared devices are being used to measure moisture levels in and out of the dryer and to also operate the controls. Success is also being found in monitoring and controlling resin application in the blender, thus saving expensive binder.

Dryer Control

Successful applications on dryers result in uniform drying, savings of fuel, reduction of fires because over-drying is less of a factor, and reduction in the number of blown boards.

Figure 9.19. *A view of a moisture gauge mounted on a conveyor (Chen 1972).*

The drying of particles occurs in continuous dryers of widely varying constructions. The three most important types are the jet, rotating drum, and tube. As much as these three dryer types differ in construction, from the standpoint of automatic control technology, only one difference is of importance and that is the dwell time of the particles in the dryer. In the case of jet dryers, one can figure on very short dwell times, i.e., the particles pass through the dryer within one to two minutes at the most. In the case of rotating-drum dryers, one has to figure either on dwell times of from five to eight minutes or a very short time with three-pass dryers. Furnish passes rapidly through the tube dryers. Dryer types offer, even at present, three possibilities of manual control by the operator:

1. Control of furnish feed rate.
2. Control of temperature.
3. Control of dwell time.

A further control is maintaining a relatively constant moisture content of the furnish entering the dryer. This is a very popular method used in lieu of automatic dryer controls.

Opinions regarding the merit of quantity control are widely divergent. It affects the continuity of the production process to a considerable degree. In the view of the manufacturers of the resistance device, the only practical solution appears to be the control of the temperature and the variation of the dwell time. A limitation of this approach is the fact that the control point temperature of the return air or the exhaust gas temperature of the dryer should, for safety reasons, not exceed a certain level. In order for the output moisture content not to exceed the desired level, this limitation would require that the input quantity be reduced when the maximal evaporating capacity of the dryer has been reached and while the input moisture content is still rising.

While both the desire to keep the input at a constant level and the limitations due to safety considerations are justified, one can clearly discern efforts to base the control of the drying process on constant input, if at all possible. Next to these purely control technological considerations, it is necessary, of course, for an automatic control system, which is supposed to automate the entire drying process, to contain a number of safety provisions to meet practical requirements.

A number of resistance-type control systems are in operation today, particularly in Europe. Some of these have been operating for a number of years. The combination of quantity of furnish and dwell-time control, as well as temperature and dwell-time control, and a combination of both systems have proved their functional capabilities.

The capacitance type of detector is being used to control dryers by modulating two variables: feed rate and burner temperature. The operator dials in the desired feed rate and establishes a nominal burner operating temperature as well as the desired target moisture for dried material. With the system on automatic control, the water load is computed and, dependent upon the infeed and outfeed furnish moisture levels, the incoming feed rate is modulated. Burner temperature is also modulated a nominal amount as required so that the dried furnish is at a uniform moisture content. Figure 9.20 illustrates a typical installation.

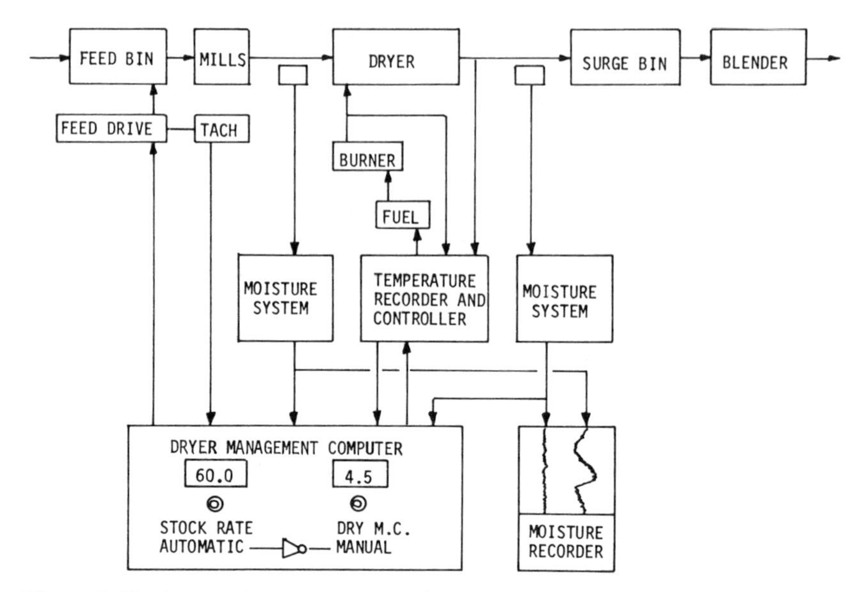

Figure 9.20. *A capacitance moisture detection and control system for dryers. (Courtesy McCarthy Products Corp.)*

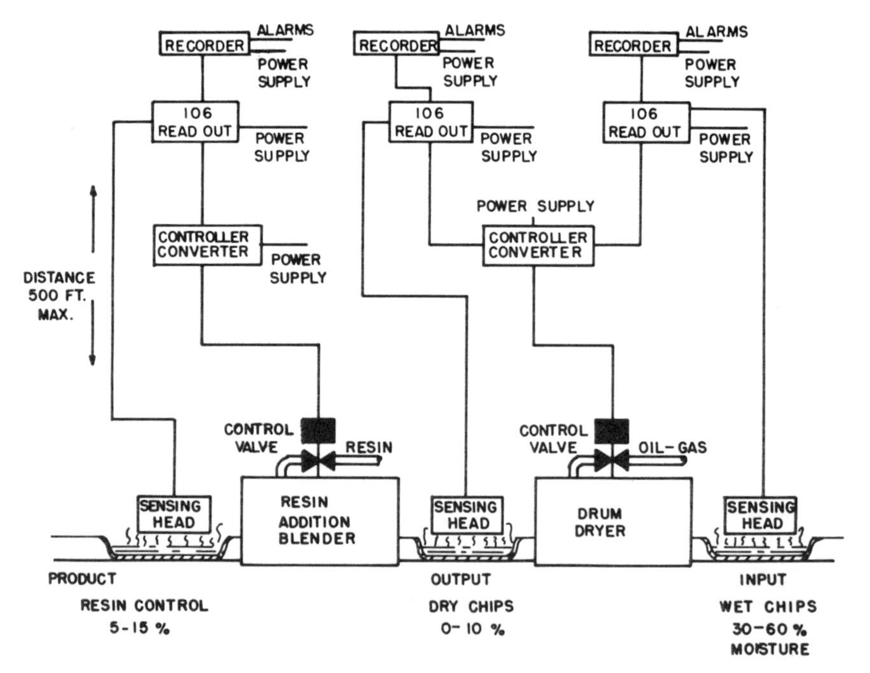

Figure 9.21. *Particleboard moisture and resin control (Fordham and Buba 1974).*

Figure 9.21 illustrates a feed-forward/feed-back control system as a closed-loop infrared system for a rotary-type drum dryer. The moisture analyzer is monitoring both the input to the dryer and the output. In the case of the input to the dryer, the signal from the moisture analyzer is fed to the recorder and the primary controller; in the case of the output, it is fed to a secondary controller which in turn directs a 4- to 20-milliamp signal to the primary controller. The output of the primary controller is then used to adjust the fuel input to the dryer and control is maintained strictly on the basis of moisture. Data available indicate that the system will be able to control dryer operation to extremely narrow margins and eliminate many of the problems that are now being encountered in plants due to manual operation of dryers through outlet temperature control.

Most of the dryer systems now in operation in the particleboard industry utilize an outlet-temperature-control system. The theory behind this control system assumes that changes in outlet temperature can be used to control the dryer operation well enough to produce constant moisture level in wood particles.

Now that the moisture can be measured continuously, better control can be established. Figure 9.22 shows inlet and outlet temperatures plotted versus moisture content for a typical five-hour period in a dryer. This dryer is operating on outlet control, yet the output moisture level is varying from 10 to 30%, even though the outlet temperature is relatively constant.

Figure 9.23 is a strip-chart record showing several hours of moisture profiles from two rotary-drum dryers typical of many plants throughout the southern United States. The objective of the system is to control core and face material at moisture levels of 4 and 5% respectively. It can be seen that wide variations from as low as 2% to as high as 8% occur. Plant personnel were infrequently monitoring outlet temperature and sampling the material on a half-hourly basis. The profiles show considerable variations in moisture content. In particular, at about 10:45 a.m. the graph shows excessively high moisture above 8%. Shortly after the tests on this dryer were completed, serious problems were encountered at the press section, which were directly related to the high moisture content measured.

Figure 9.24 shows the performance data on the dryers shown in Figure 9.23 when the operator manually corrected variations in moisture by adjusting the outlet temperature controller. When comparing the strip-chart recordings in these two samples, it can be seen that a significant improvement was made in dryer operation with this manual system; but further refinement would be of value.

The main objective in the control of a dryer can be to hold the inlet and outlet temperatures constant, e.g., 1000° and 150°F (538° and 67°C), respectively. However, when a feed-forward control system is installed to maintain moisture content at a fixed level within a 2% band, it is necessary to vary the inlet and outlet temperatures of the dryer significantly. In one run, inlet temperatures ranged from 300° to 1200°F (149° to 649°C) and outlet temperatures ranged from 100° to 210°F (37° to 99°C).

Figures 9.25 and 9.26 show the operation of a rotary-drum dryer before and after a control system was incorporated to adjust dryer fuel control and to hold regular moisture content between 5 and 6%. Previously, the wood particles varied from as high as 7% moisture to as low as 2% moisture when no control system was used. Significant improvement was made with a control system (Figure 9.26).

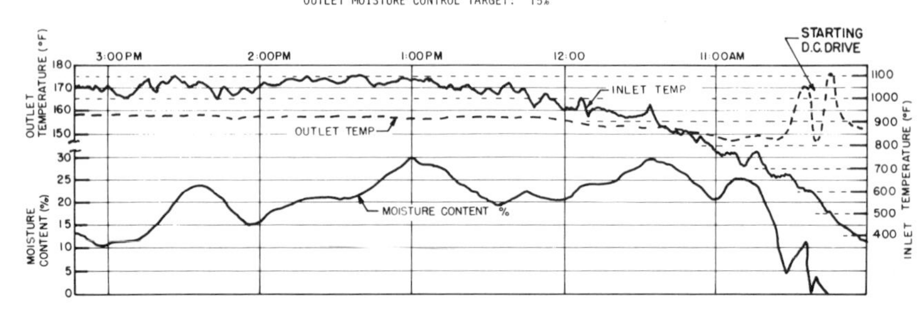

Figure 9.22. *Strip chart recording of dryer inlet and outlet temperatures in a dryer operating on outlet control (Fordham and Buba 1974).*

Blender Control

Precise control of moisture is of great importance also through the blending and pressing steps of the board production process. Out-of-control moisture levels of a few percentage points can result in serious production problems. Problems with moisture can occur with blending, such as adding too much or too little moisture or excess or insufficient resin. With the wider use of short-retention-time blenders, where the dwell time of the furnish in the blender is only of a few seconds duration, better moisture control is desirable. In the long-retention-time blenders where blending may take as long as 20 minutes, small variations in moisture can even out during the blending time.

The resistance, capacitance, and infrared absorption devices can be used to monitor and to control both moisture and resin level of the furnish being blended. If the furnish is too dry, water can be added to the furnish in the blender. If the furnish is too wet, the operator can be notified of this fact by various readout systems. Resin level (liquid types) can be measured by comparing the moisture content of the furnish as it enters and leaves the blender and calculating the amount of resin added based on the solids content of the resin. The sensitivity of such instruments is illustrated in Figure 9.27 where an infrared moisture analyzer is monitoring the output of a blender.

Mat Moisture Detection

The resistance system has been adapted for use in measuring the mat moisture content after forming (except on continuous pressing lines). A transducer has been developed for use in the separation zone where the continuously formed mat is separated by a saw and vacuum pickup system. A narrow width of mat is cut out in this zone and the cut-out furnish is picked up by the vacuum system and recycled to the former. A transducer is mounted on the saw carriage and is run through the mat ahead of the vacuum pickup. The results of the measurement across the width of the mat are integrated and displayed on a readout and are also recorded.

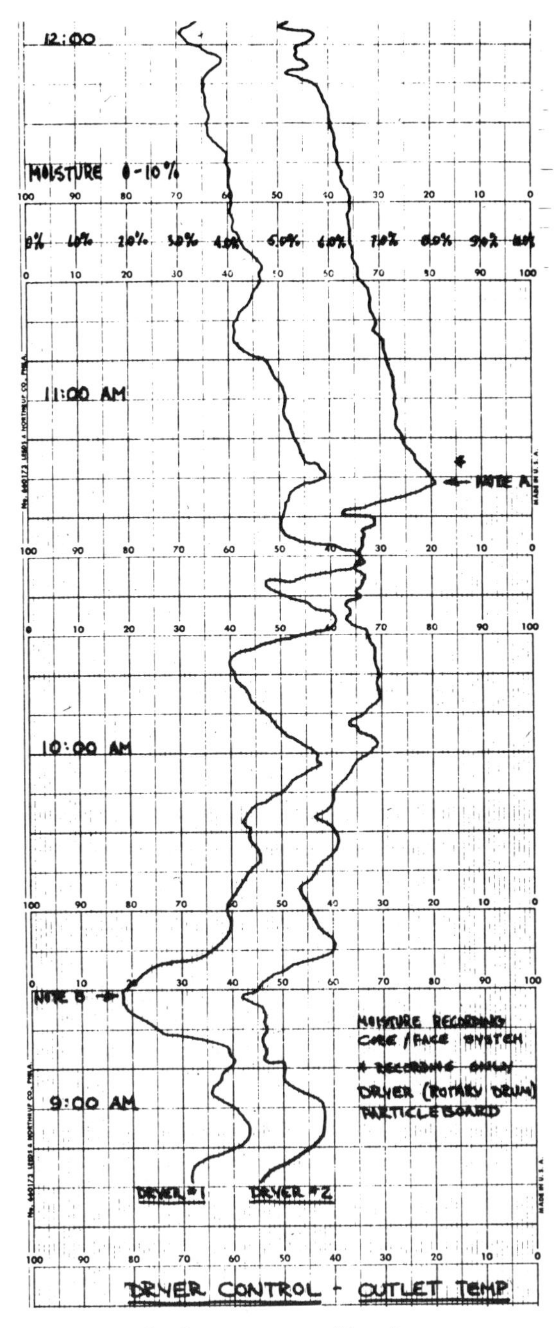

Figure 9.23. *Chart record of moisture profiles from two rotary-drum dryers (Fordham and Buba 1974).*

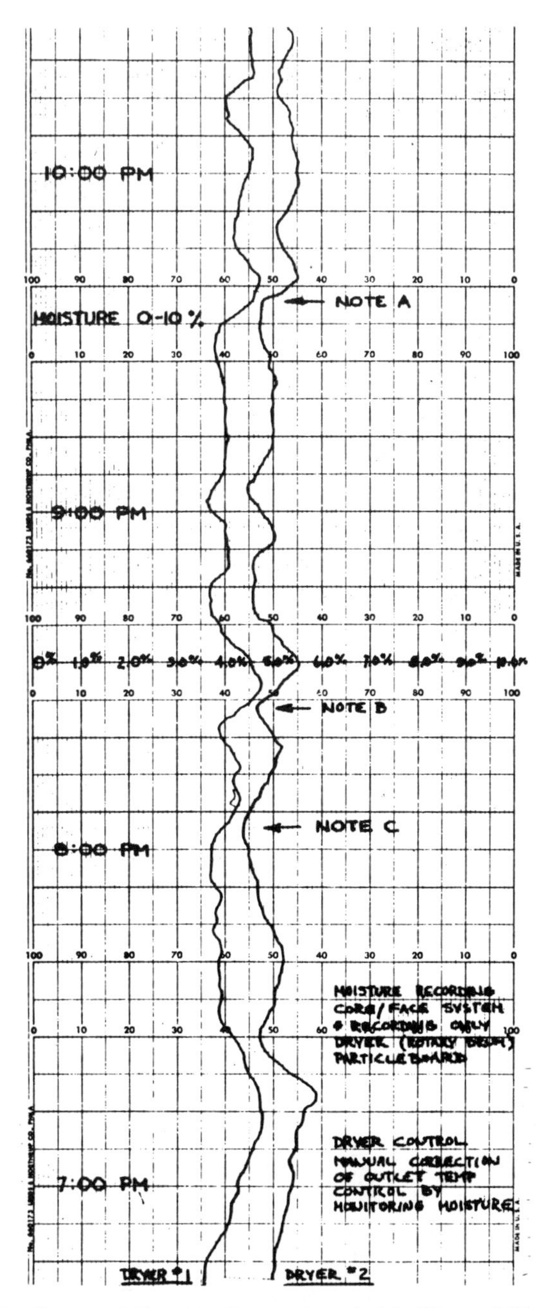

Figure 9.24. *Performance data on dryers recorded in Figure 9.23 when operator manually corrected for variations in moisture by adjusting the outlet temperature controller (Fordham and Buba 1974).*

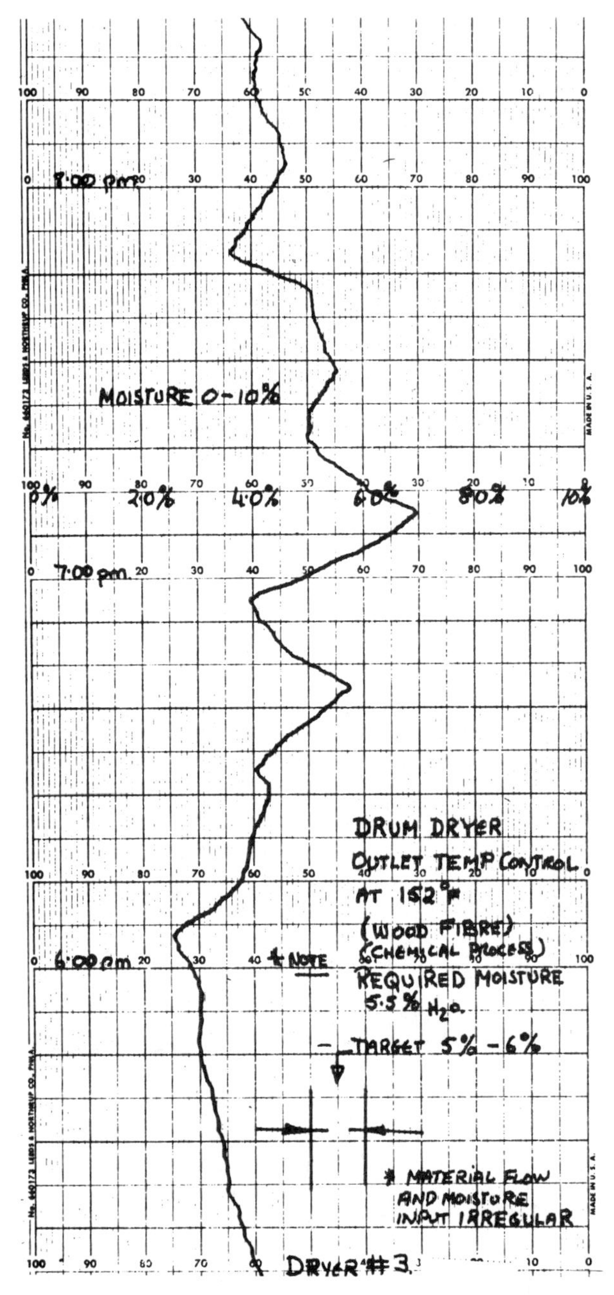

Figure 9.25. *Strip chart recording of the furnish moisture content out of a rotary-drum dryer before a moisture-control system was installed (Fordham and Buba 1974).*

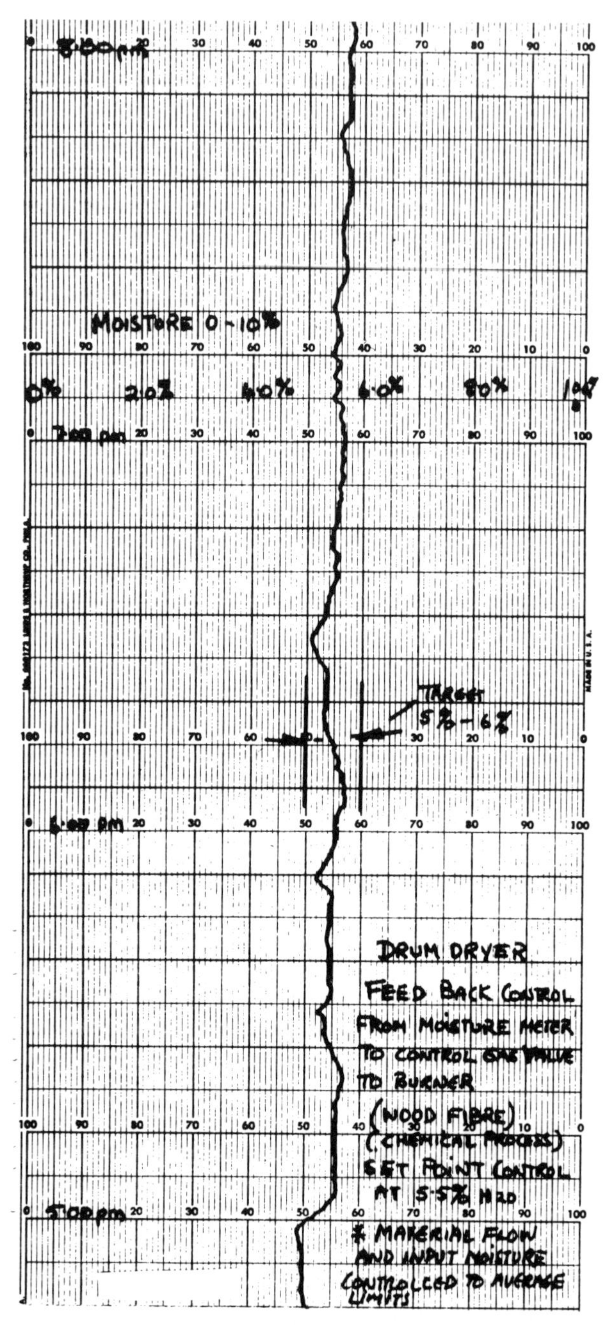

Figure 9.26. *Strip chart recording of the rotary-drum dryer considered in Figure 9.25 after a moisture-control system was installed (Fordham and Buba 1974).*

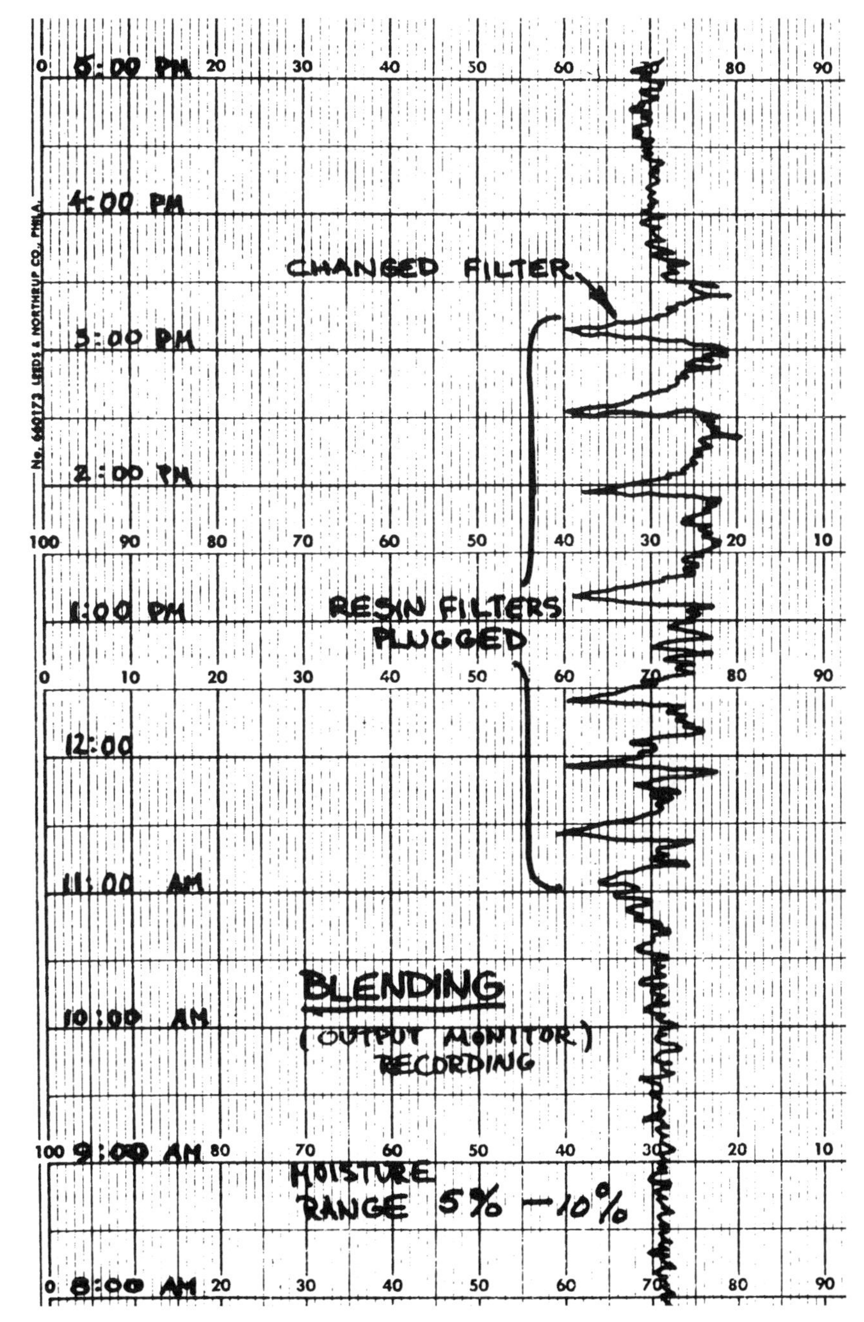

Figure 9.27. *Chart record of an infrared analyzer monitoring the output of a blender (Fordham and Buba 1974).*

SELECTED REFERENCES

Brumbaugh, J. I. 1969. Plant Experiences in Controlling Moisture. *Proceedings of the Washington State University Particleboard Symposium, No. 3*. Pullman, Washington: Washington State University (WSU).

Chen, Y. C. 1972. Nuclear Gage Measures Moisture in Wood Chips. *Proceedings of the Washington State University Particleboard Symposium, No. 6*. Pullman, Washington: WSU.

Donald, J. C. 1972. Monitoring Moisture in Particleboard Manufacture with an Infrared Moisture Analyzer. *Proceedings of the Washington State University Particleboard Symposium, No. 6*. Pullman, Washington: WSU.

Fordham, J. 1975. Use of a Continuous Non-contact Infrared Moisture Gauge in Particleboard Hardboard and Medium-Density Fibreboard Plants throughout USA and Europe. Background Paper No. 129, third World Consulation on Wood-Based Panels, FAO, New Delhi, India, February, 1975.

Fordham, J. and R. J. Buba. 1974. Use of a Continuous Infrared Moisture Gauge in Particleboard Plants. *Proceedings of the Washington State University Particleboard Symposium, No. 8*. Pullman, Washington: WSU.

Kenton, R. H. 1972. Microwave Moisture Measurement of Furnish and Particleboard. *Proceedings of the Washington State University Particleboard Symposium, No. 6*. Pullman, Washington: WSU.

Klinkmueller, H. 1972. Continuous Moisture Content Determination and Control in the Manufacture of Particleboard. *Proceedings of the Washington State University Particleboard Symposium, No. 6*. Pullman, Washington: WSU.

McCarthy, E. T., and J. R. Parkinson. 1969. Measuring Moisture for Particleboard Manufacturing. *Proceedings of the Washington State University Particleboard Symposium, No. 3*. Pullman, Washington: WSU.

McCarthy, E. T., and J. R. Strom. 1972. Moisture—Continuous Measurement and Information Application. *Proceedings of the Washington State University Particleboard Symposium, No. 6*. Pullman, Washington: WSU.

Perry, W. D. 1972. Moisture Detection and Control in the Particleboard Process. *Proceedings of the Washington State University Particleboard Symposium, No. 6*. Pullman, Washington: WSU.

Swearingen, G. V. 1972. Green and Dry Furnish Moisture Measurement. *Proceedings of the Washington State University Particleboard Symposium, No. 6*. Pullman, Washington: WSU.

10. PARTICLE SEPARATION: PRINCIPLES AND EQUIPMENT

Almost every particleboard preparation system requires the separation of particles. Some fiberboard operations also require separation, but it is not as common as in particleboard. The requirements vary, but fall generally into the categories of separation in terms of particle size by screening, separation in terms of particle surface-to-weight ratio by air-classification, and a third method which is a combination of screening and air-classifying. Input materials can range from dustlike material to chunks of wood. Smaller particles are often directly usable in the process, and their removal means less material needs to be run through particle-generating equipment.

The screens commonly used are vibrating and gyratory. Capacity and efficiency are the two principal criteria for judging screens. Capacity represents the quantity of material a screen will satisfactorily handle under a given set of circumstances. Efficiency describes the quality of the particle separation and can be expressed as a percentage by dividing the amount of a certain size material removed by the amount of such material in the particle mix.

Air-classifiers pass the wood particles through a moving airstream for stratification of the particles. Unlike screening, where size is the only variable which affects the separation, air-classification is affected by both particle weight and surface area. Part of the material air-classified can then be screened if desired. Also, some air-classifiers use screening simultaneously with air separation.

Probably no two board plants in existence have the same screening and classifying requirements. This relates to differences in raw materials, differences in end product, and differences in the plant design. Decisions about particle separation also come from within the process itself. These decisions also must consider the methods and equipment available. This discussion reviews the general principles involved and offers several examples of machines developed for particle separation or classification. Some have high capacities and some that operate more precisely have relatively low capacities. A number of companies produce such equipment using the approaches presented.

SCREENING

By definition, screening is accomplished by passing material over a surface with sized openings through which the undersized particles will pass. The differences between the various types of screens relate to the means of transporting and agitating the material bed.

The choice of the type of screen best suited to any given separation is not always easy. Screening is an empirical field lacking in precise mathematical theory. The best practical approach is to select the screen type or types that appear best suited to the desired separation and have laboratory tests performed to forecast the actual commercial screening operation.

A number of factors affect screen performance, such as material bulk density, particle shape, and moisture content as well as feed rate to the screen, retention time on the screen, and screen motion. Before discussing these factors, however, it will be well to consider how various screens function. None of them can economically perform as well as the various laboratory classifiers such as the Ro-Tap sieve shaker (Figure 10.1). Particles are retained long enough in these classifiers to completely and precisely separate the particles. The quality of particle separation required in an industrial screening operation can be evaluated by using a sieve analyzer.

The Ro-Tap testing seive shaker or any other similar analyzer, such as Syntron, Sweco, or Williams, is a valuable piece of equipment for determining the ultimate screening that can be achieved. For economical reasons, such precise screening cannot be done in the plant, but a correlation can be worked out for the plant screens that are used.

Figure 10.1. *Ro-Tap testing sieve shaker. (Courtesy C-E Tyler.)*

The testing sieves have woven wire screens with standardized openings. In the United States, Tyler Series and U.S.A. Sieve Series are used, and which is used should be noted as the sizes differ. For this discussion the Tyler Series will be considered.

In rough terms, the sieves are known as having so many openings per inch, e.g., a Number 4 sieve has 4 openings per inch (25.4 mm). However, care must be exercised in such a description, particularly with industrial screens, as the diameter of the wire can be different than the standard and thus change the screen opening. Consequently, it is desirable to specify the screens by both the size of the opening and the openings per inch or by the size of the openings and the wire diameter. In doing so, a closer correlation between the industrial screen and the Tyler Sieve can be made.

To run the tests, a set of sieves is nested together and placed in the shaker. For particle testing, a typical set of sieves could be 6, 9, 14, 20, 32, and 42 mesh. The bottom of the assembly is a pan. The assembly has the largest screen at the top and the smallest at the bottom. A 100-gram sample (usually dry) is randomly selected, making sure a representative sample is chosen. The sample is placed in the top sieve and the shaker is then turned on. A typical shaking time is five minutes. Afterwards the amount of material retained in each screen is weighed and reported as a percentage of the total. The normal way of reporting this information would be the percentages for +6, −6+9, −9+14, −14+20, −20+32, −32+42, and −42 (using the screens described in the above example). These sieve designations mean the material passed through the sieve shown as minus and was retained on the sieve shown as plus. For accuracy, at least three tests should be made, and the average reported.

Since severe shaking takes place, some long, slender particles will be upended and they will drop through the screen opening according to the width and thickness of the particle. Such particles passing over an industrial screen, however, will tend to slide over the small openings since they probably will not be upended as frequently because of less agitation in the short period of time they are traversing the screen. There are also fewer opportunities for these particles to be screened out. Additionally, the bed depth of the particles possibly will restrain such upending. With flake-type particles, therefore, it may be advisable to perform some type of air-classifying to better determine the sizes of the flaked particles. Laboratory devices for doing such a test are available, but no commercial ones are known.

Industrial screens can be regarded as falling into two basic categories: vibrating screens (high frequency, small amplitude) and shaking or gyrating screens (low speed, large amplitude). The latter type is the most popular in the industry.

Inclined Vibrating Screens

Inclined vibrating screens, with the exception of a few specialized magnetic or pneumatically vibrated units, are driven by a rotor whose axis lies parallel to the screen plane which develops a circular motion in the plane at right angles to the screen plane. Because this circular motion has little or no conveying property, the screen plane must be inclined in order that the material will move from feed to discharge (Figure 10.2).

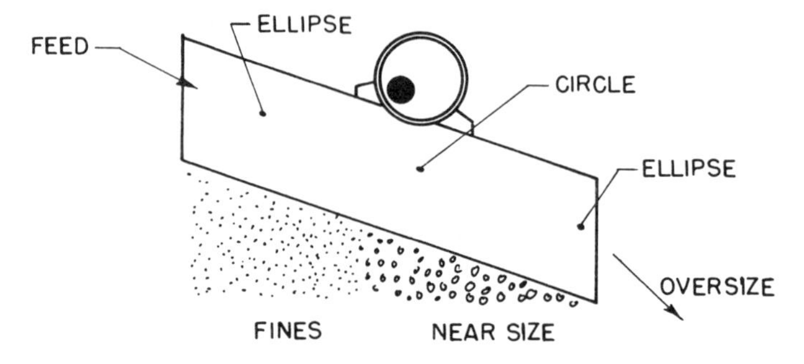

Figure 10.2. *The inclined vibrating screen (Ryan 1967).*

In the most common two-bearing design, an unbalanced rotor is mounted in the vibrating body, or *screen box,* which is supported on elastic mounts of low stiffness. The resulting motion is circular if the rotor is coaxial with the center of gravity of the vibrating body. If the rotor is displaced from the center of gravity, a rocking action around the center of gravity results in elliptical motion at the extremities of the screen as illustrated in Figure 10.2.

Vibration transmitted to the supports is determined by the stiffness of the springs, and is normally limited to 2% of the exciting or driving force. In addition to its extreme simplicity and low vibration transmission characteristics, the inclined vibrating screen has the important advantage of easy amplitude adjustment.

Horizontal Vibrating Screens

Horizontal vibrating screens operate with a reciprocating motion inclined at an angle to the horizontal. Material is conveyed across the screen in a series of hops, with virtually no friction with the screen surface (Figure 10.3). The direct-connected eccentric drive, commonly used with slow-speed shaking screens, is not used in the high-speed designs because the large inertia forces developed by the high-speed motion are very difficult to harness in any economical manner. Inertia drives and natural frequency drive systems are used instead.

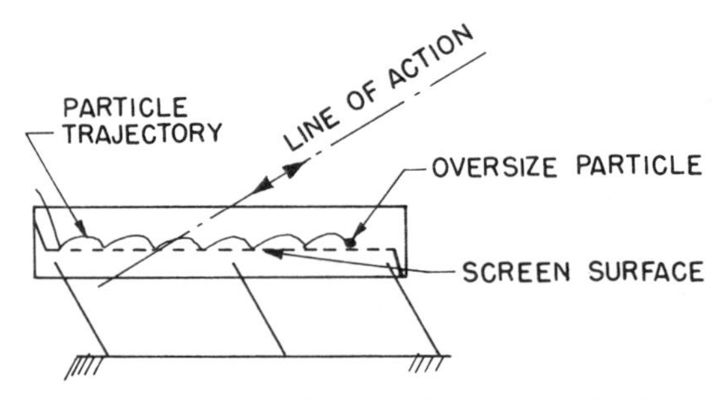

Figure 10.3. *Schematic diagram of material travel on the horizontal vibrating screen (Ryan 1967).*

Circular and Gyratory Screens

Under the category of circular and gyratory types falls a wide variety of screens which are distinguished from the inclined and horizontal vibrating screens in that some component of the motion is circular or elliptical in the plane of the screening surface. The screening surface may be horizontal or inclined, the latter relying at least partially on gravity to move the material from feed to discharge. The following descriptions of popular screen types within this category illustrate the operating principles of this type of screen.

A circular screen popular outside the United States is shown in Figure 10.4. This screen has a three-dimensional movement which combines horizontal and vertical screening actions, thereby achieving a tumbling effect as illustrated in Figure 10.5. The material is fed into the screen from above. As it spreads over the screen, the finer material falls through the screen and the oversize passes out through the spout on the side of the machine. Three screen panels are stacked in the machine shown in Figure 10.5. The entire unit is supported on soft springs which isolate vibration from the support structure. Air can also be used with these screens to assist with the particle separation. These screens are fully enclosed, thus reducing dust pollution.

A popular inclined rectangular screen is illustrated in Figure 10.6. The screen box is driven at the feed end by a vertical eccentric unit which imparts circular motion in the plane of the screening surface. The discharge end of the screen box

Figure 10.4. *A circular screen for separating four fractions. (Courtesy Allgaier-Werke GmbH.)*

Figure 10.5. *Three-dimensional movement of material on the screen surface. (Courtesy Allgaier-Werke GmbH.)*

rests on slide blocks which restrain the box so as to create linear motion. The motion at other points on the deck is elliptical, with the minor axis of the ellipse becoming smaller as the discharge end is approached. The manufacturer feels that this motion gives higher efficiency in the heavily loaded feed zone and greater conveyance in the more lightly loaded area of the deck.

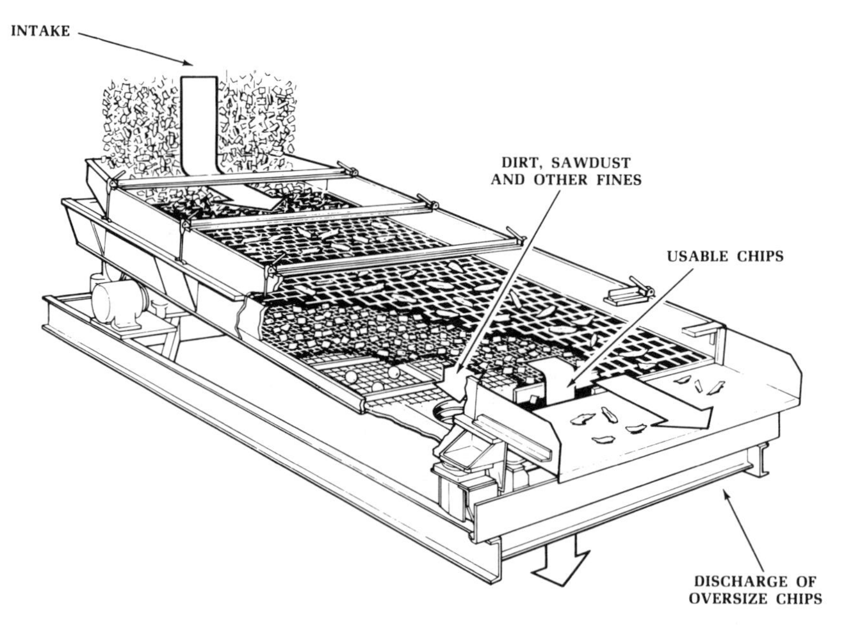

Figure 10.6. *A Rotex screen working with chips. (Courtesy Rotex Inc.)*

Figure 10.7. *A full rotary particle screen. (Courtesy Black-Clawson, Inc.)*

Another popular type of gyratory screen is shown in Figure 10.7. It is an inclined rectangular screen with a full circular motion, the plane of which is horizontal. This type of design allows a practically complete dynamic balance of the inertia forces. It is a versatile design concept adaptable to two-, three-, or four-sort screening operations and features the *bouncing balls* when screen blinding is a problem, as does the previous type of screen. These balls continually tap the bottom side of the screen, thus preventing screen blinding (Figure 10.8).

Figure 10.8. *Bouncing balls used to prevent screen blinding (Christensen 1975).*

Factors Affecting Screen Performance

Because a number of variables affect screen performance, some definition for the means of evaluating and measuring results is needed. Efficiency and capacity are the two principal measures of screen performance. Screen efficiency concerns the quality of the separation, and is not related to power consumption or operating costs. There are several ways of measuring efficiency, the most common and useful being the *percentage of recovery*. This is the ratio of the amount of undersize pieces passing through a certain size screen opening to the total amount of undersize in the initial feed.

Capacity is the maximum feed rate at which the desired results of size separation and efficiency can be produced. No statement of capacity, whether in terms of weight or volume, has any significance unless the corresponding efficiency is known. In an extreme example, an overloaded screen could transport a very high tonnage, acting more as a vibrating conveyor than a screen. This can also happen in the laboratory where an overeager technician, anxious to perform, can destroy an experiment by using the screen as a conveyor rather than as a separating device.

Another consideration deserving mention is the definition of an *undersize* particle. A wood particle, because of its fibrous structure, is inherently irregular in shape, usually with one of its principal dimensions (width, length, and thickness) far greater than the other two. Theoretically, the size of the particle is that of the smallest screen opening through which the particle could be forced to pass. However, it is known that this particle will pass through the opening only if it is brought into contact with the screen opening in that particular orientation. For long, slender particles such as flakes, the statistical probability of this occurring becomes more remote, even with the most violent screening action. The definition of size, then, is not very meaningful when screening such irregular particles.

A practical solution is to use the standard sieve analysis as a basis for establishing particle size and size distribution within a sample of material. One must concede, however, that the screening action of the sieve vibrator, acting for a long enough time on a shallow bed of material, will remove all of the undersize capable of passing any screen. A realistic ratio between screen analysis and production screening must be developed.

Along with the problem of defining particle size for efficiency determination is the question of acceptable shape in terms of the ultimate use. A common example in particleboard furnish preparation is the separation of small *fines* from long, slender particles or *slivers*. If both can pass through the screen, then high efficiency is not the objective. Here the nature of the screening action becomes critical and the results must be judged visually as well as statistically.

There are many variables which affect the capacity, efficiency, and sensitivity of a screen. These should be considered in terms of material factors and machine factors.

Material Factors: Most of the material factors, such as bulk density, particle shape, and moisture content are determined elsewhere and cannot be regarded as controllable as far as the screening operation is concerned. However, problems with screening may indicate that particle generation is out of control.

The most important characteristic of a particulate material is its tendency for natural size classification. A particle size mix classifies itself according to size, with a minimum amount of mechanical agitation. The smallest particles end up at the bottom and the largest at the top. Natural size classification helps bring the smaller particles to be extracted in contact with the screening surface first, thus promoting screening efficiency.

Factors such as increased moisture content and pitch content affect the natural classification tendency adversely. Moisture content above a certain limit is a deterrent to screening efficiency on many materials. For example, wet sawdust particles tend to cling together and it is impractical to screen them commercially. However, these same particles dry are very mobile and can be classified efficiently. The effect of moisture content is dramatically illustrated in Figure 10.9.

A screen is normally selected to achieve a certain size split in a given material as determined by a sieve analysis. The size of the screen opening, defining the desired size split relative to the frequency distribution of sizes in the mix, significantly affects the efficiency of the separation and thus the amount and size of equipment required to handle a given amount of material. The relative amount of particles that are "near size" with respect to the size of the screen openings is especially important, because these particles have low *drop-through* probabilities and tend to cause plugging or blinding problems.

High screening efficiency, therefore, is often difficult to achieve and 100% efficiency would usually be impractical. The particle shape must also be considered by visual examination in determining efficiency. Figure 10.10 compares cost versus efficiency in production screening in a hypothetical illustration. Very high efficiency is costly and the throughput is low in this example. Thus efficiency has to be balanced against cost and undesirable effects in the final product. Each operation will have to determine what level of efficiency is most appropriate for the cost.

Machine Factors: The machine factors to be considered include the size and shape of the screening surface, the amplitude and frequency of the vibration,

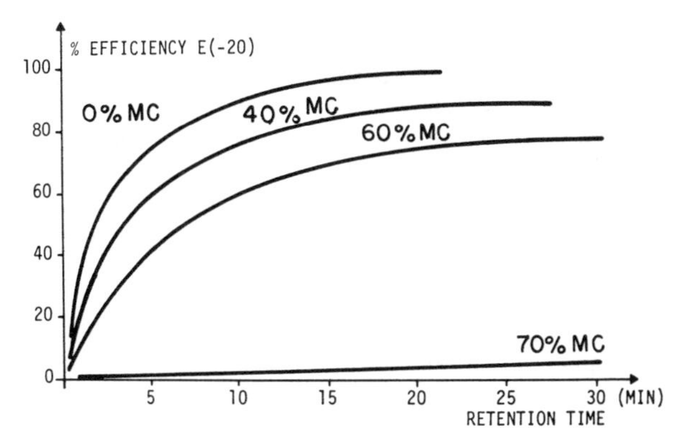

Figure 10.9. *Efficiency vs retention time for sawdust at four different moisture levels (Christensen 1975).*

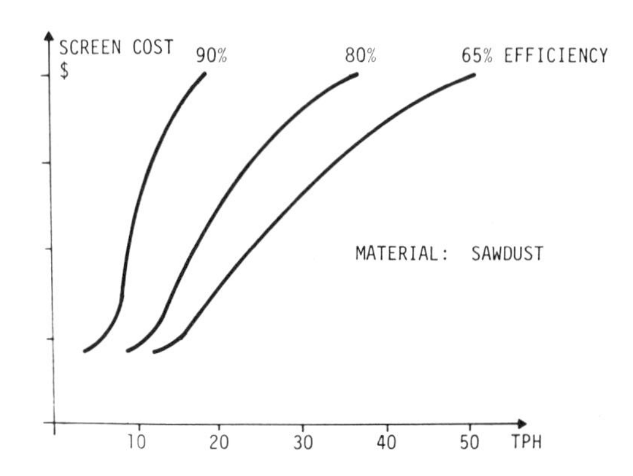

Figure 10.10. *Cost vs efficiency in production screening (Christensen 1975).*

inclination of the screening surface, effectiveness of cleaning devices (bouncing balls), and many others. It is often more significant to group these variables in terms of their net combined effect on the separation and then analyze each variable within that framework. These will be discussed under the divisions of retention time and feed rate. This discussion is based on the performance of the more popular gyratory, circular or "orbiting" type of screen motion. On this type of screen, all points of the screen surface move in a horizontal, circular path at a relatively low speed. This gentle type of motion is especially effective on materials with a high tendency for natural classification, such as dry wood particles. The circular motion of the inclined screen surface gives the particles three-dimensional motion that improves the mobility of the particles within the material blanket, thus aiding the natural classification tendency by letting the smaller particles filter through the material blanket to the screening surface, improving the particle drop-through probability and hence the efficiency.

Retention Time: The total time the material is on the screen depends on the screen. Assuming a constant rate of travel over the screen, the retention time equals the length of the screen divided by the rate of travel. The rate of travel, and thus the retention time, is a function of the amplitude and frequency of the screen motion, the inclination of screening surfaces, and, to some extent, the material characteristics and the screening medium. The retention time is a measure of the number of opportunities a given particle has to find an opening to go through. Even with long screens, however, wet material is not screened efficiently.

Feed Rate: The quantity of material handled by the screening surface is the factor to consider, not necessarily the gross feed rate to the screen. The screen may have several screening decks operating at different feed rates.

The number of particles of a given size increases proportionately with feed rate. A larger number of particles will be competing for the same number of openings, with the result that the probability of each particle finding an opening to go through diminishes, lowering the overall screening efficiency. With the increased competition for openings, the size of particles to be extracted shifts toward the smaller end of the size spectrum, because of the higher drop-through probability of smaller particles. Figure 10.11 shows the effect of feed rate on efficiency for screens of different lengths and for material with different moisture content. The effect of feed rate on efficiency and consequently (initial) machine cost was illustrated in Figure 10.10.

AIR-CLASSIFICATION

Quite often particle separation on the basis of size does not satisfy the board maker's requirements. The physical properties of board are dependent on particle shape and surface-to-weight ratio as well as size. Separations made on air-classification equipment help to meet these other requirements. It should be noted that some air-classification may be done further down the production line in the blender or former.

In order to appreciate and evaluate the performance of air-classification equipment, a clear concept of the scope and limitations of this equipment must be grasped. The common use of low-velocity air systems for dust control leads to an association of air-classification with only small particle separation. In actuality, neither the size of the particle nor its weight is the controlling factor. The basis of separation is the particle surface-to-weight ratio, sometimes referred to as specific surface. Any particle situated in a rising column of air is acted upon by two opposing force vectors. The airstream exerts an upward force on the particle called *drag*. The magnitude of the drag force is directly proportional to the area of the particle, the only other factors affecting the force equation being those related to the density, velocity, and viscosity of the fluid medium which in this case is air.

Resisting this vertical drag force is the gravitational force due to the weight (or, more correctly, the mass) of the particle. The vector resultant of these two forces determines whether the particle will rise, fall, or float in the airstream.

On this basis, the wood particle entrained by an airstream might actually be larger in one case, or heavier in another case, than the particle dropping out. Since

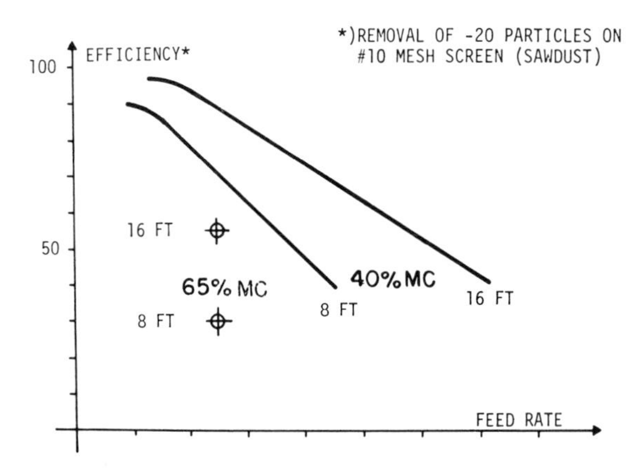

Figure 10.11. *Efficiency vs feed rate for different lengths of screens with sawdust at two moisture contents (Christensen 1975).*

more than one independent variable is affected in the separation, the evaluation of results requires judgment as well as standard quantitative procedures.

If the feed to a classifier can be limited to a uniform particle size in terms of width and length, then the classifier can perform a reasonably precise thickness separation due to the weight difference between particles of fairly constant surface area. For this reason, it is common practice to use air-classifiers in conjunction with screening equipment.

The differences between the various air-classifiers used in the industry lie mainly in the means of presenting the flow of material to the separating airstream. Some of these types of air-classifiers are the vertical air leg, suspension, and fluidized-bed classifiers.

Vertical Air Leg

The simplest, least expensive, and least sensitive air-classifier is one often referred to as an *air leg*. This device consists of a vertical pipe or chute through which air is lifted by an external fan. Material is fed into this column at or near the top and a two-way vertical separation is made. The liftings are released through a cyclone collector located in the exhaust air system. The material not lifted drops out the bottom into a collecting bin or conveyor.

From this basic concept a variety of machines has evolved, with refinements consisting in most instances of integral fan arrangements, rotary air locks, and feed regulators.

Suspension

The air-classifier shown in Figure 10.12 falls into the general classification of circular classifier. The material enters the top of a cylindrical chamber by gravity flow and is fed onto a horizontal rotating disc or distributor plate which throws the material outward towards the walls in a uniform curtain. Air is drawn upward through the material stream, lifting the lighter (high surface-to-weight ratio) fraction upward into an external fan and cyclone system. The heavy (low surface-to-weight) product falls downward onto a sloped perforated screen and into a discharge spout. Any material screening through this deck may be collected separately as a third product or recombined with the main heavy product discharge. A similar unit uses a closed-circuit recirculating air system with an expansion chamber for precipitation of the aspirated product from the airstream (Figure 10.13). This is a relatively simple, easy-to-adjust machine. Separations are made by particle size or shape or by specific gravity.

The advantage of this type of separator over a simple air leg is the broad curtain of material through which the airstream flows. The entrained particle can be easily lifted from the stream of heavy material without the resistance encountered in a vertical material flow.

There is a very popular suspension sifter in use in many parts of the world that is available in four basic designs: single- and two-stage sifters and fresh-air or air-recirculation sifters. In the single-stage sifter (Figure 10.14), the material to be screened is divided into two fractions, acceptable and oversize; while in the

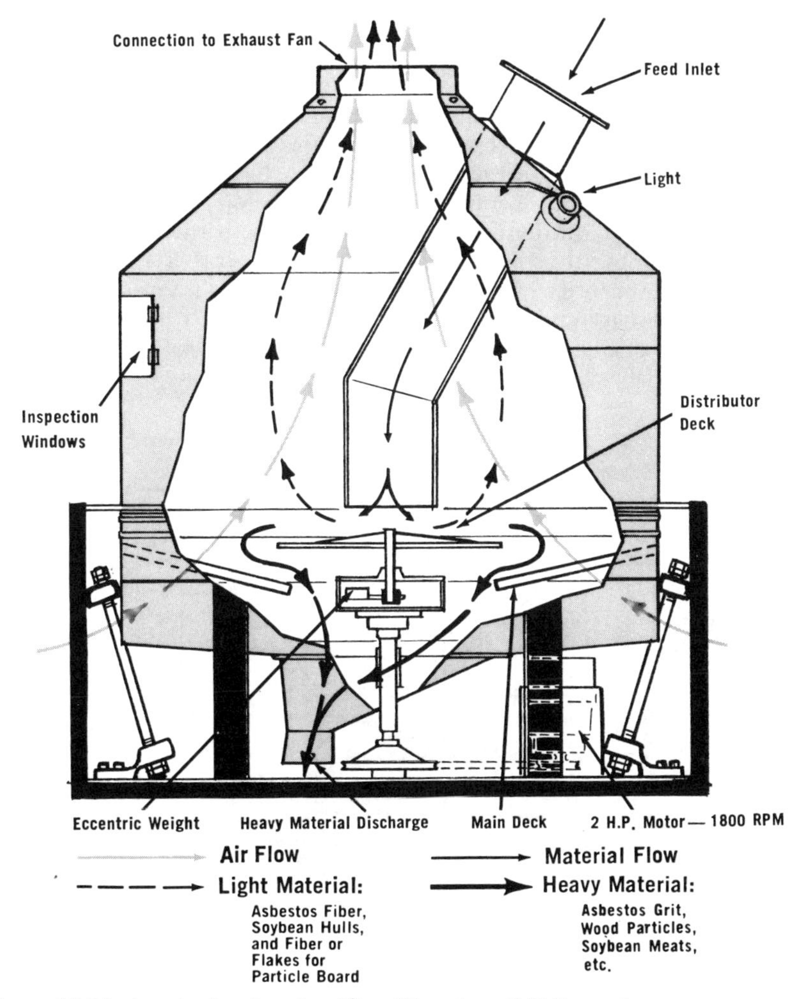

Figure 10.12. *An air circular classifier. (Courtesy C-E Bauer.)*

two-stage sifters, the furnish is divided into three fractions, for example, face material, core material, and oversize material. The so-called fresh-air sifter uses the air from the immediate surroundings of the sifter and discharges air directly to the atmosphere through a separator. Air-recirculation sifters, however, operate with approximately 95% recirculated air. Thus, only 5% of the air is discharged from the system, either through a separator or through a separate pneumatic system. These types of sifters have the advantage of reducing the amount of dust in the atmosphere around the machine. This has a further advantage in preventing the pickup of moisture by very dry furnish, because the air is constantly recirculated. This can be a significant problem when using fiber which will rapidly pick up or lose moisture depending upon the ambient conditions.

In special cases these sifters have also been used as dryers, where a heat source is connected to the air inlet; thus, drying of the furnish takes place simultaneoulsy with the sifting.

The furnish to be separated falls through a rotary valve and a central feed pipe into the suspension chamber. Rotating stirring arms above the screen floor distribute the material evenly over the area of the screen, through which the airstream enters the suspension chamber from below. With appropriate air velocities, any desired separation of the furnish can be obtained. The portion of the furnish in suspension is removed by the airstream and the oversized material which does not float in air is moved by the stirring arms to the periphery of the separating area and is discharged through rotary valves. Any fines left among the oversized material will be drawn back into the suspension chamber by the bypass air entering the oversize material rotary valves.

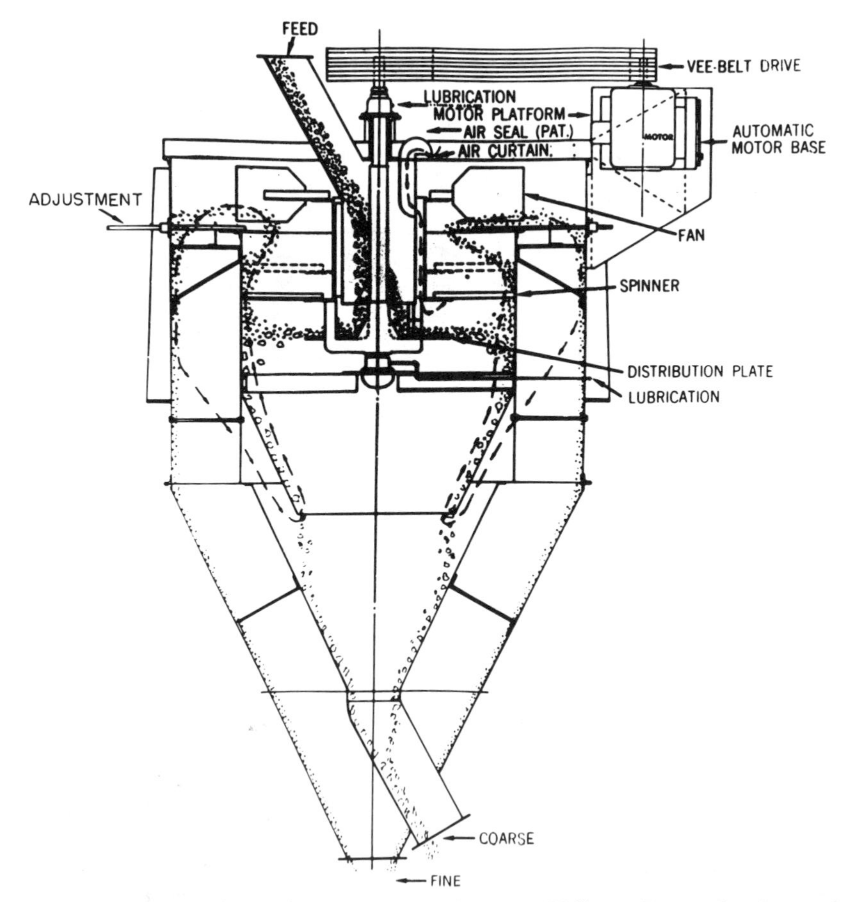

Figure 10.13. *A mechanical air separator. (Courtesy Williams Patent Crusher and Pulverizer Co.)*

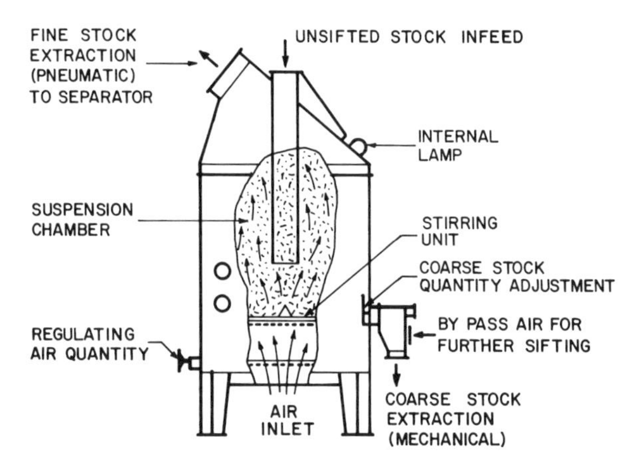

Figure 10.14. *A single-stage suspension sifter plant (Shumate et al 1976).*

In two-stage sifters, illustrated in Figure 10.15, this same process is carried out twice, first in the upper chamber and then in the lower chamber. Once the fine material is removed in the first stage, the medium and oversized material is fed to the second stage through two exterior rotary valves and attached ducts. The medium material is separated out in this second stage and oversized foreign bodies are removed.

It is reported that this type of sifter is excellent for removing foreign bodies such as sand, stones, metal, or bark from the furnish.

The zigzag type of classifier, which has no moving parts, is used elsewhere in the world. For classifying, it utilizes a number of vertically arranged zigzag channels as shown in Figure 10.16. As many as 12 of these channels have been used in a single classifier. This classifier can be used for separating flakes, fibers, and granular particles. Classification is made by particle thickness, not by the length or cross-sectional area. It is used extensively for separating oversize particles from core furnish, oversize particles from face furnish, and fractionation of particulate furnish into core and face fractions.

The feed material is normally introduced into the classifier (see Figure 10.16) by means of a closed-type feed screw (3). As it flows by gravity across an inclined perforated sheet (1) the furnish is fluidized by jets of high-velocity air blown up through the sheet by the blower (8). The air imparts a high-velocity turbulence to the flowing feed material, which lifts even larger-size particles off the perforated sheet. This effect has been termed the *flying bed,* and is much like the molecular movement in a gas.

The high fluidization of the feed material enables most of it to reach the zigzag channels by air. The feed material is presented to a series of channels, and is classified until what remains at bed level is an extremely clean, coarse material which drops out of the discharge (6). Each channel accepts only as much material as permitted by the classifier air adjustment. The finer particles ascend with air and pass through a cyclone (5) for the separation of the particles from the air. Since

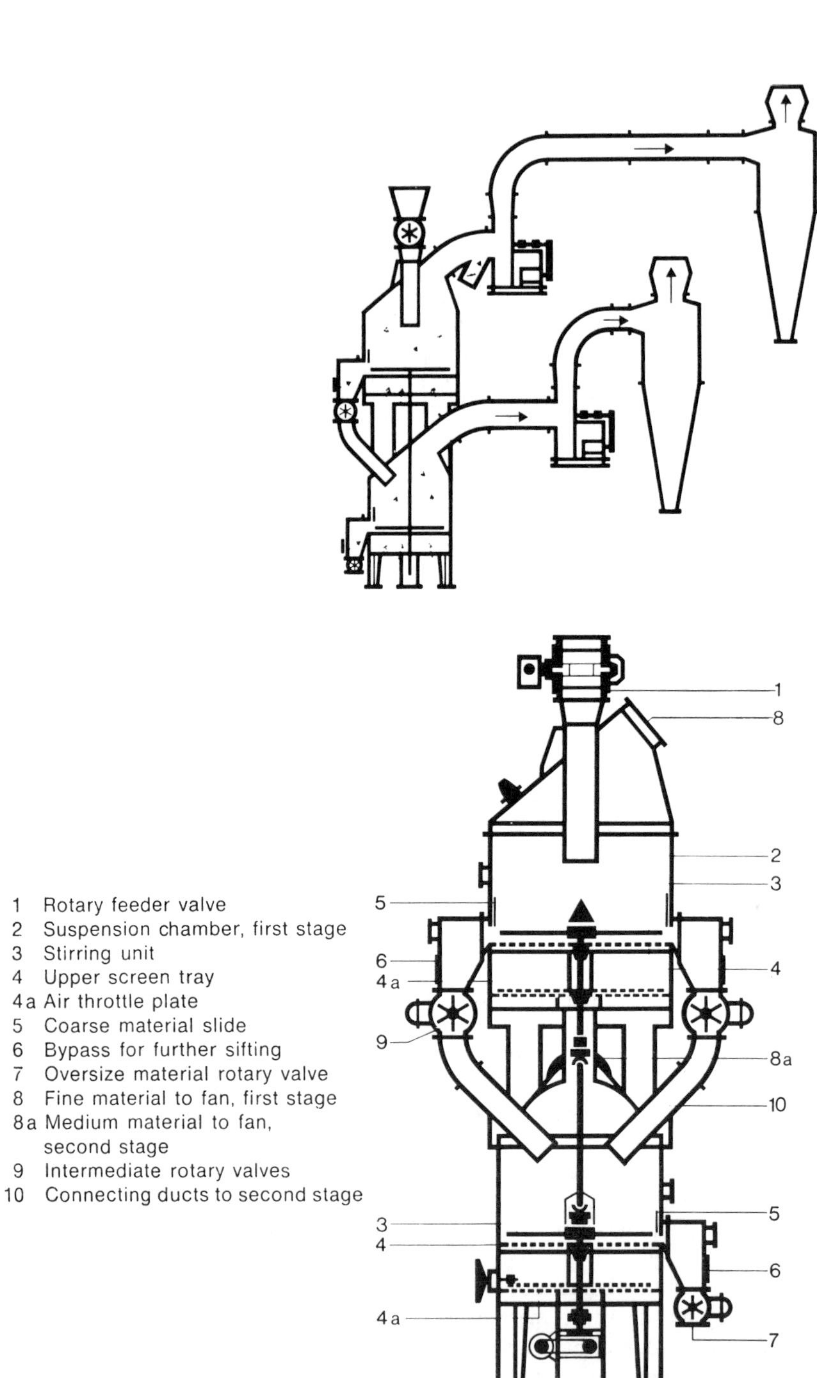

Figure 10.15. *Two-stage suspension sifter plant (Shumate et al 1976).*

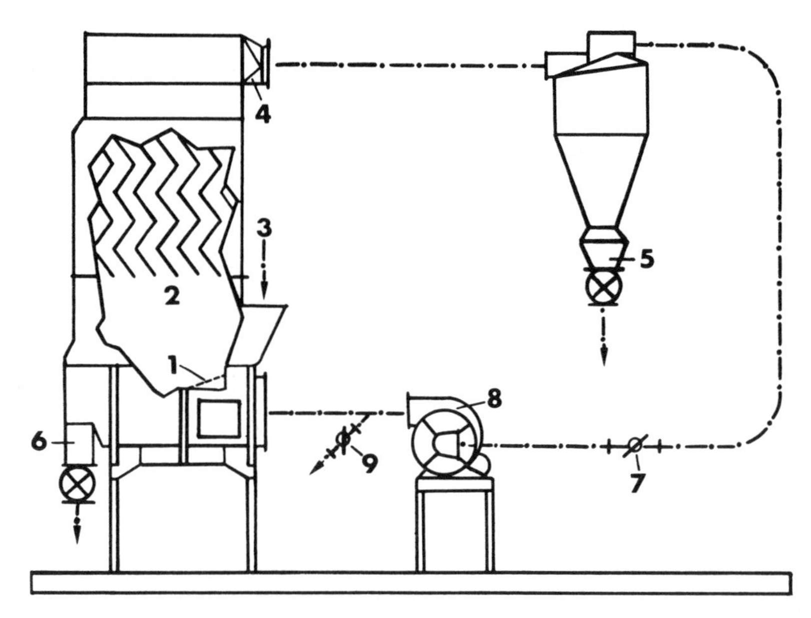

Figure 10.16. *A flying-bed zigzag classifier. (1) perforated sheet, (2) preclassifying chamber, (3) particle feed, (4) classifier head, (5) cyclone, (6) coarse material discharge, (7) throttle flap for classifier adjustment, (8) blower, (9) throttle for removal of leaking air. (Courtesy Alpine American Corp.)*

this is a closed-circuit machine, the air from the cyclone is returned through the blower to the flying bed.

The classification action in the zigzag channel is shown in Figure 10.17. The flow pattern as illustrated shows how the finer particles ascend the channel and the coarser ones descend as classification takes place. A horizontal vortex is developed in each limb of the channel (Figure 10.17), and classifying is effected at every joint of the channel. Imperfect separations at the lower end of the channel are corrected with each change of direction in the zigzag channel. Thus only the desired small particles are allowed to ascend with the air going to the cyclone.

The classification point (cut point) depends upon air velocity, which can be steplessly adjusted from 1.0 to 10 meters per second, and the feed rate of the material.

Fluidized Bed

In fluidized-bed classifiers, the material is presented to the airstream in the form of a fluidized bed instead of a falling curtain of material. A fluidized bed is achieved by distributing the material evenly over a porous vibrating deck through which air is blown at relatively low velocity. The material is stratified with particles of high surface-to-weight ratio floating to the top of the bed. The final products are derived from this preseparated bed in several ways.

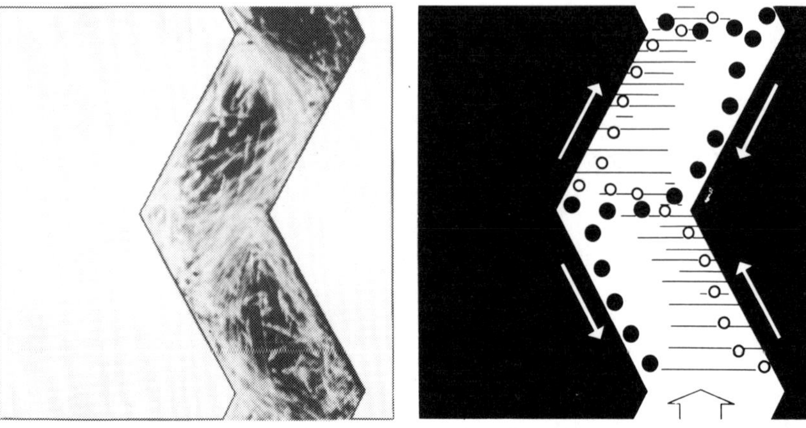

Figure 10.17. *A zigzag channel showing the classifying action. (Courtesy Alpine American Corp.)*

One type of this classifier performs a straight two-way separation using the fluidized-bed principle. In this unit the deck motion conveys the material down an inclined deck from a feed pan. The air flowing through the deck is controlled and accelerated to lift the desired proportion of the material. A controlled air bleed-in through the "heavy" discharge assists in entraining the more marginal particles. Similar classifiers are built as an integral part of a horizontal resonant screen.

A more sophisticated fluid-bed classifier makes a three-way separation by rejecting a small percentage of very low surface-to-weight particles. This is accomplished by vibrating the inclined deck with an uphill conveying motion which counteracts the gravity flow of the material. With the bed in a fluidized condition, the low surface-to-weight particle lying nearest the deck surface receives this conveying impulse and is transported up to the high end of the deck. There it is restrained and banked into a segregated zone of particles and bled off through a regulated gate.

The high surface-to-weight particles are floated in the material bed and flow by gravity down the inclined deck. The highest surface-to-weight particles are lifted through the hood outlet.

The most sophisticated fluid-bed classifier design is the type which uses an uphill conveying motion whereby the particles near the bottom of the fluidized bed are transported up the deck. However, instead of discharging material at each end of the deck, the entire flow discharges over one side in a continuous stream with a gradual decrease in surface-to-weight ratio from the lowest end of the discharge lip to the highest. The flow is divided into the desired products by means of adjustable splitters which are positioned to divert the material into a series of spouts. An exhaust hood is provided for lifting an additional product. A small unit of this type is shown in Figure 10.18. It is an excellent laboratory tool.

The advantages of fluid-bed classification lie in the ability to produce a multiplicity of products in a single operation, also in the sensitivity gained by stratifying and preseparating the material bed.

Figure 10.18. *A fluid-bed classifier in the Washington State University Laboratory. (Courtesy Washington State Univ.)*

SELECTED REFERENCES

Albert Bezner Maschinenfabrik. 1975. Roller Screen. Ravensburg, Federal Republic of Germany.

Alpine Aktiengesellschaft. Alpine Multi-Plex Zigzag Classifier. Folder No. 50/le. Federal Republic of Germany.

Christensen, E. 1975. Screening for Particleboard. *Proceedings of the Washington State University Particleboard Symposium, No. 9.* Pullman, Washington: Washington State University (WSU).

Keller-Peukert. Keller Suspension Sifters. Printed by Buch-und Offsetdruck Thyssen, Federal Republic of Germany.

Ryan, D. M. 1967. Equipment and Methods for Separating Wood Particles. *Proceedings of the Washington State University Particleboard Symposium, No. 1.* Pullman, Washington: WSU.

Shumate, R. D.; J. E. Sloop; and A. H. Schenkmann. 1976. Experiences with Air Suspension Classifiers in Particleboard Manufacture. *Proceedings of the Washington State University Particleboard Symposium, No. 10.* Pullman, Washington: WSU.

C-E Tyler Industrial Products. 1976. *Testing Sieves and Their Uses.* Handbook 53. Mentor, Ohio.

11.

FIRE AND EXPLOSIONS: PREVENTION AND DETECTION

The change in recent years to finer furnish in the particleboard industry has accentuated the fire and explosion problem. The development of the pressure-refiner and the resultant very fine fibrous furnish has created a situation where the fire and explosion potential is extremely critical. The fiberboard part of the industry has always had to fight these problems. Further, the implementation of pollution control devices such as baghouses has added new pieces of equipment with severe fire and explosion problems.

When looking at the subject, it divides neatly into fire and explosion prevention on one hand and fire and explosion detection on the other hand. Fires and explosions usually are not treated separately, however, as both are closely interrelated. Fires are more frequent and occur most times without explosive force. Explosions are the result of fires in which combustion occurs nearly instantaneously due to the presence of highly flammable fuel, such as wood fiber dust. Fire prevention and explosion prevention therefore go hand in hand.

CAUSES OF FIRE

The following are seven commonly found sources of fires and possible explosions in plants. They are not necessarily given in order of their frequency:

1. Foreign material such as metal or rocks in the furnish.
2. Fires caused by maintenance personnel welding electrically or cutting with acetylene torches.
3. Mechanical breakdown of equipment such as overheating of bearings, motors, non-turning idlers, rubbing belt or chain guards, dragging chain or screw conveyors.
4. Overloading or exceeding the design capacity of milling and drying equipment or uncontrolled circulation of dry furnish mixed with green furnish through dryers.
5. Electrical faults on short circuits.
6. Uncontrolled distribution of wood fiber over the plant facility, or, in other words, a dirty place.
7. Improper startup procedures.

This chapter discusses ways to eliminate or at least diminish the problems.

Foreign Material in the Furnish

Most plants receive their furnish such as shavings, sawdust, plywood trim, and chips from outside sources over which they have very little control or no control at all. It is therefore up to the plant to eliminate the metal or rocks in the furnish before it reaches the milling and drying equipment. The removal of metal from the furnish normally is not too much of a problem. Ferrous metals can be removed by permanent magnets or electromagnets. It is desirable, therefore, to install these magnets as close as possible to the mills. It has been recommended that plate magnets installed in chutes leading into attrition or flaker mills are not the whole answer, even when installed in dogleg-shaped chutes. These plate magnets are usually installed on only two sides of the chute, instead of all four. Even when all four sides are covered, these magnets are only about 60% efficient. Consequently, additional magnets should be installed on conveying equipment feeding into the metering bins of the mills. These should be electromagnets over belt conveyors, magnetic-head pulleys on conveyors running 200 ft (61 m) per minute or less, or permanent-plate magnets inserted in vibrating conveyors. Experience has shown that the magnets inserted in vibrating conveyors are the most effective way of removing tramp metal from furnish.

The elimination of nonferrous metals can be accomplished by metal detectors mounted on belt conveyors, which electrically control a diversion gate that dumps the material containing the nonferrous metal off the conveyor into a collection box. Nonferrous metals are a problem to some plants, but in general do not pose as much of a problem as ferrous metal since they are not as hard.

The removal of rocks and sand from the furnish presents the greatest problem by far. Large rocks can be eliminated by screening, if the overs are not taken to hammermills directly for reprocessing. Otherwise, rocks are presently being removed by either commercially available rock traps, which are usually of a small capacity, or by rock traps which are used in conjunction with pneumatic conveyor systems. A design of the latter type of rock trap, which is commonly used, has the following arrangement. A belt or vibrating conveyor discharges into the suction line of a pneumatic conveyor. The inlet opening of the suction line is horizontal and is located some distance below and away from the discharge of the conveyor. The suction line is telescoping so that the inlet can be moved closer to or away from the conveyor discharge. With this arrangement, the discharge is then adjusted so that the furnish is picked up and heavier particles, such as rocks, fall past the inlet into a collection box. In essence this rock trap works like an air separator and is efficient when the furnish is light. The possible use of X-ray, ultrasonic, or nuclear density-measuring devices for the detection of rocks has been suggested, but no system of this type apparently is being applied in the industry. Rocks can also be removed by chip washers; however, chip washers, which also remove sand very effectively, can only be used on chip-type material. The removal of sand and grit from incoming furnish requires screening and discarding out the fines portion of the incoming material, preferably after milling and drying. The preference for screening after milling and drying is based on in-plant experience which has shown that sand and grit are, in most cases, embedded in the furnish and will not come free until after milling and drying.

Welding and Acetylene Torch Fires

Most of the reports on serious fires or explosions in plants, which originated by maintenance personnel welding or cutting, had this in common. The fire broke out during the periods when the plant was operating or when equipment was only partially shut down for maintenance. In most cases, when serious fires or explosions resulted, pneumatic conveyors were left running and sparks or burning metal from the welding operation were picked up by these blow systems, which dumped the burning metal into bins or into conveyor belts or dryers where fiber residue was present. Recommendations have been made to exercise all precautions and, if possible, shut down all machinery surrounding the area of maintenance when welding or cutting has to be done on equipment. The area surrounding the welding operation should be completely covered by nonflammable or fire-retardant blankets to prevent any burning metal from entering or falling into places where fiber has accumulated.

Mechanical Breakdown of Equipment

Equipment breakdown as a hazard includes, of course, the overheating of bearings, gear boxes, motors, non-turning of idlers, rubbing belt or chain guards, and dragging flight or screw conveyors. The elimination of this source of fire is to a large extent a question of maintenance. Constant supervision and inspection of all equipment can prevent most fires caused by these sources.

Overloading or Exceeding the Design Capacity of Milling and Drying Equipment

Overfeeding of mills is a constant hazard. Large attrition mills and flakers have drive motors which in many applications exceed the actual demand required to mill the furnish. It is therefore very difficult to tell by ammeter readings whether the mill is being choked or not since all of the available horsepower of the drive is not being used. It is therefore important to establish the permissible feed rate for each mill and to correlate the feed rate with the sharpness of the knives or plates. This can best be determined by continuous inspection of the furnish coming out of the mill.

The overloading or exceeding of the design capacity of drying equipment is a problem, especially where the production of a plant has been increased considerably over what the plant was originally designed for, or where the furnish has changed from what was anticipated during the time of the design. In both cases, it is more than likely that the inlet temperature has been raised either to cope with the additional material which has to be handled or with furnish which is much wetter than the furnish previously used by the plant. High inlet temperatures become especially hazardous when the dryer is fed unevenly because of insufficient metering or when green and dry materials are fed simultaneously into the dryer. The latter should be avoided if at all possible. A further potential hazard in dryer operation is the possibility of an insufficient dryer exhaust fan capacity, which could allow the furnish to be retained in the dryer for a longer duration than acceptable. Such a situation may eventually result in fire. If preheaters are used on dryers for makeup air, it is important to check at all times that no unnecessary pressure drops occur over the preheating equipment. The dryer must be kept clean of the buildup of fiber, resin, rosin, wax, etc. A dirty dryer is sure to cause fires.

When starting up a system with a digester and refiners, provision must be made to prevent getting sloppy wet fiber into the dryer. This type of fiber plasters itself to the metal of the dryer; and as the fiber dries, it adheres to the metal and finally catches fire. Some dryers have been kept clean of buildup of resin, rosin, and wax sufficiently to avoid fires by washing them once a week with a hot alkaline cleaner.

If dryers are stopped and cleaning takes place, blowing fine furnish into the air, all open flames must be extinguished; since an explosive mixture of dust and air may develop, and this can be touched off by the gas or oil flame.

Electrical Faults

The majority of fires caused by electrical faults and short circuits in electrical equipment are the result of insulation failures or overheating. Insulation failures are caused either by faulty installation, vibration, or overheating. There is very little that can be done about faulty installation except a thorough inspection during installation. Any conduits containing feeder cables or terminal boxes which vibrate are always apt to cause insulation failure through the constant pounding on the insulation, and it is therefore very important that all conduits and panels be mounted free from vibration and be rigidly secured. Overheating, resulting in the eventual breakdown of insulation, occurs mostly in terminals on starters, relays, and circuit breakers, and on undersized wires and cables. It is therefore recommended that all terminals in starters, circuit breakers, and relay panels be retightened at quarterly intervals during periods of complete shutdown, when all equipment can be electrically de-energized.

To prevent fires caused by arcing contacts, electrical equipment such as power distribution centers, motor control centers, relay panels, SCR control panels and MG sets should be kept in pressurized dust- and dirt-free electrical rooms.

Uncontrolled Distribution of Wood Fiber Throughout the Plant

The last source of fires is the even distribution of wood fiber dust over the plant, a problem of great concern. It appears to be a serious problem in many board plants. The best way to eliminate this source of fire would be, of course, to prevent the dust from settling on pipes, trusses, joists, and purlins within the plant. It is therefore recommended that the areas generating dust in large proportions be screened off as far as possible, that all bins be closed up tightly to prevent the escape of dust from them, that all transfer points of conveyors be properly chuted, and that the plant be kept clean on a day-to-day basis rather than on a month-to-month basis. The practice of blowing dust off equipment with compressed air should be used only where absolutely necessary, such as on trusses, rafters, and joists. All equipment which can be reached or cleaned by other means should be cleaned without the use of compressed air. The conditions which exist immediately after or during the blowing of dust from trusses, and from equipment with compressed air, are highly explosive and should be avoided if at all possible.

In conclusion to this analysis of the main sources of fire, it should be mentioned that board plants using pressurized refiners for milling and the radio frequency (RF) presses for curing of their mats have additional problems to deal with. Since the material is so highly fiberized before entering the drying system, the control and supervision of the inlet and outlet temperatures of the dryers have to be very reliable and fast to prevent the ignition of the fiber during the drying process. With regard to the RF press, the presence of metal in the press during the RF curing cycle usually leads to fire immediately, and it is therefore very important to have reliable metal-detection equipment and reject facilities on the conveyor system leading to the press.

Improper Start-up Procedures

Plants are normally designed so that improper start-up cannot occur. However, instruments and machines do fail, and a failure, combined with operator error, can cause fires and possible explosions. Everyone involved should be thoroughly familiar with the proper start-up (and shutdown) procedures, so that malfunctions can be detected and possible accidents avoided.

FIRE PREVENTION

It is apparent from the previous description that a considerable number of precautions can be taken to prevent fires from occurring in plants. Nevertheless, it is also clear that in the present state of the art, and with equipment breakdowns and human errors, the possibility of fires and possible subsequent explosions cannot be completely eliminated. It is therefore necessary to give the operating personnel all tools economically available to detect fires as quickly as possible, to make the operation of the plant under these conditions as safe as possible, and to provide fire-fighting and explosion-suppression equipment where necessary.

Isolating Sources of Potential Fires

Looking at the total picture of fire and explosion prevention and detection, several additional points need to be considered. First of all, to prevent the spreading of fires, it is immediately apparent that it is very desirable to separate physically, as much as possible, the different functions of the board process. The raw material storage should be physically isolated from the rest of the plant facilities; and the milling and drying operation should be separated, or at least partitioned off, from the blending and forming operation. The dryers should never feed directly into bins or forming machines but should be separated by at least one conveying means, so that, in case of fire, the diversion of the material flow from the process is possible. The blending area in board plants is usually very dusty, and consideration should be given to enclosing the blending area completely to prevent the dust from this area from spreading into the rest of the plant. The forming and pressing line should, if at all possible, be separated from the warehousing area by an adequate aisle space to allow free access to both these areas from either side. Sander dust bins and dry fiber bins for pressurized-refined fiber should be located outside, remote from operating personnel, and be provided with sprinkler piping.

Sprinkler Systems

All board plants are provided with sprinkler systems, and it is, of course, important that these be adequately maintained at all times. It should be mentioned here that sprinklers have to be installed over all equipment which is covered by a roof, whether it is indoors or outdoors. Sprinklers also have to be installed in all buildings which have been or are being added to existing plant facilities. Often the necessity for installing sprinklers in plant additions is overlooked. Fuel and lubricants, if stored indoors, should be stored in buildings isolated from the rest of the plant.

Detection Tools

For high-speed fire detection, a number of tools are available. Among them are preset thermostats, which may be either adjustable or nonadjustable in temperature setting, that operate when a set temperature limit has been reached. These thermostats utilize either bimetallic sensors, thermocouple sensors, resistance thermometers, melting of metal fuses, or capillary action. These types of thermostats are generally reliable, but their operating speed is relatively slow. If they are made fast acting, nuisance trip-outs will result.

Rate-of-rise thermostats operate when the rate-of-temperature rise exceeds a preset limit. The sensors used by these thermostats are the same as described previously for the fixed-temperature thermostats.

Ultraviolet detectors work on the photoelectric principle and detect ultraviolet radiation. This radiation is present only in burning material and not in glowing or smoldering wood fiber. The usefulness and application of ultraviolet detectors to the industry are therefore limited.

Another important "detection device" available is the operator controlling the equipment. The operator senses fire by sight, noise, or smell, and he is ultimately the most reliable detector available to us. Unfortunately, due to the type and layout of the equipment used in plants, his reaction time can be quite slow.

The most important means of detecting fires is infrared detectors which sense radiation emitted in the infrared frequency band photoelectrically. They will detect glowing or smoldering fires which emit these frequencies, and, thanks to the development work done by Harold Ely of Pope & Talbot, Inc., and the American District Telegraph Company, there is now a reliable high-speed detection unit available for board plants. Pope & Talbot, Inc. has been a pioneer in solving fire and explosion problems. For more detailed information on this subject, the TELEMAC Ultrahigh-Speed Fire Detection System of American District Telegraph will be considered. This system was developed to provide split-second, spot-type protection needed for rocket fuel processing as well as for an increasing variety of industrial hazards.

Mounted to survey critical points in a process or operation, supersensitive photoelectric sensors "watch" for the sudden appearance of a telltale flame or burning ember developing within, or passing through, the detector's field of view.

Solid-state control circuitry, which responds instantaneously to changes in detector reaction, can be used to immediately signal an alarm, shut down machinery, fire the explosive primer of a specially designed deluge-type sprinkler system, or trigger other automatic fire-control action.

The system provides special detection and response capabilities not obtainable with more conventional automatic fire-alarm systems usually operated by heat, smoke, or other airborne products of combustion. For example:

1. It responds with lightning speed to flame or glow in the infrared range, initiating fire-control functions within two to three thousandths of a second.
2. It responds to even small quantities of infrared energy appearing within the detector's fields of view for only a split second.
3. It is virtually immune to atmospheric conditions—dead air, turbulent air, rapid temperature changes, etc.—which may adversely affect the operation or stability of detectors responding to convected heat or airborne products rather than to radiant energy.
4. It can operate reliably in duct applications where dust, lint, etc., may clog the sampling tubes of systems depending upon fire detection by sampling, which makes their use unfeasible.

Other features made possible by the sophisticated electronic technology employed in the system, include the following:

1. Compact detectors can be mounted unobtrusively in conveyor duct walls, with minimum interference with normal operations, with only the lens protruding into the duct itself, or on the carriage of a machining tool, or at a point several feet from the supervised operation.
2. The compact control unit includes facilities for testing system operation without operating any associated fire-control equipment.
3. The solid-state circuitry ensures long life and low power consumption.
4. The system can be operated either continuously or only when a hazard is present. The control unit can be interlocked with the power to the supervised operation, if desired, to ensure that the system is always turned on before commencing operations.
5. Although these detectors are not explosionproof, they operate on such low power that they may be considered intrinsically safe for use in hazardous locations. Full knowledge of this by the user and considered judgment are essential in every case. Light-conducting tubes containing optical fibers have been used successfully to achieve detection within a hazardous area, with the actual detector located outside of the area. Optical fibers have also been used in high-heat areas and in areas subject to RF heating to transmit infrared energy to detectors remotely located.

6. Where special circumstances demand even more stringent safeguards against false operation, "duality" can be provided. The control circuit then responds only to simultaneous alarm indications from two or more detectors viewing the same hazard from different angles.

Being designed for special fire hazards requiring a split-second response, this system is generally used to trigger automatic fire-control action of some kind, ranging from simple machinery shutdown to sophisticated extinguishing and/or airborne product diversion systems. This system is used to provide spot-type protection demanding an instantaneous response to fire in either of two ways:

1. To detect flame or embers at the point of ignition—point source protection—in operations involving highly flammable and hazardous materials, or machinery in which overheating must be promptly detected to avoid costly repairs and/or disruption of operations.
2. To detect burning materials rapidly passing a critical point in an air duct or conveyor belt transport system.

It is the latter method that is of primary concern to the wood processing industry.

Most automatic fire-detection systems are designed to detect the burning of cellulose materials such as wood, fiber, paper, and their products, coal and petroleum products, or cotton flyings and lint. Conventional heat- or smoke-actuated fire-detection systems are admirably suited for this purpose, providing the incipient fire remains reasonably stationary.

Where fire must be detected in the fine materials—such as used in the dry-process industry—which are moving rapidly on a conveyor belt or in an air duct, an infrared system may provide the only feasible method of detection since it reacts instantly to small quantities of infrared energy visible to the detectors for only a brief period. This is a particular problem where transport velocities reach 3000 ft (914 m) per minute or more.

The accidental ignition of these materials while moving within a duct is not likely to cause duct damage itself; however, such fire must be prevented from rapidly spreading and from being carried into bins, stacks, and hoppers. Therefore, extremely rapid detection of small fires within a duct is essential so that the material flow can be instantly and automatically stopped or diverted, and the fire extinguished or controlled with minimal damage and interruption of operations.

Some recent problems have occurred where baghouses are connected directly to sanders. Sparks from the sander, caused by broken belts or sanding ferrous metal in a panel, can be propelled into the baghouse so fast (possibly at 5500 ft [1676 m] per minute) that fire and immediate explosion take place before detection and suppression can take place. More on this will be mentioned later.

Locations of Detectors

Considering the detection devices available, it is desirable to apply these in the locations where fires are most likely to occur. From the available evidence, this area starts at the milling equipment and terminates at the dryer outfeed (except for trim hogs and sanders). The majority of fires in this area are either caused by rocks or tramp metal in the furnish (causing arcing in the mills which ignites the furnish before it reaches the dryers) or by dryer overtemperature which causes the furnish to ignite in the dryer.

The most suitable device for detecting fires originating in the mills is the infrared detector mounted in the discharge blower lines of the mills. It is recommended that the infrared detector be incorporated into the electrical control circuit and should initiate upon actuation the following procedure immediately:

1. Divert the fiber flow from the mill and/or from the dryer to a fire dump.
2. Initiate water spray in the mill discharge cyclones and into the diverter screws.
3. Shut down the dryer flame.
4. Stop the infeed into the mill.
5. Give audible alarm and visual indications to let the operator know where the fire has originated.

The application of thermostats or rate-of-rise thermostats for the detection of fires originating in the mills is not recommended, since they are substantially affected by ambient conditions and react too slowly. If the furnish being milled is very dry, the danger of explosion in the exhaust pipe exists. It is recommended that, under these circumstances, the furnish be wetted prior to entering the mill to give the added benefit of a better milling operation. In addition, pressure sensors with explosion-suppression equipment may have to be installed in the discharge piping and cyclone of the mills.

The most commonly used devices for detecting fires in dryers are thermostats. Most dryers are equipped with a fixed and an adjustable overtemperature shutdown thermostat, which sense the dryer outlet temperature. The fixed thermostat is intended for a backup protection and is usually set at a temperature slightly above the maximum operating temperature. The adjustable thermostat is intended to provide the actual fire protection and the operator should be instructed to adjust this setting at 10° to 20°F (6° to 12°C) above the setting of the dryer temperature controller. In rotating dryers, this overtemperature sensor has worked very effectively in detecting fires due to the long retention time of the furnish within the dryer. When dryer overtemperature occurs, the thermostats should initiate the following operations automatically:

1. Divert the dryer discharge from the process flow to a fire dump.
2. Initiate water spray injection into the dryer, the dryer discharge cyclone, and the fire dump.
3. Shut down the dryer flame.
4. Shut down the dryer exhaust blower.
5. Stop the infeed of furnish into the dryer.
6. Give an audible alarm and indicate where the fire has originated.

It is also recommended that, where there is more than one dryer in the milling and drying area, if one dryer goes into overtemperature, all other dryers be shut down simultaneously to free the operator from further equipment supervision.

The application of infrared detectors in the detection of fires in dryers is very desirable due to their high-speed characteristic. In tube dryers and in drum dryers drying pressurized-refined fiber, thermostatic detectors are generally too slow. With infrared detectors on the dryer outlet, the automatic shutdown sequence to be initiated is identical to the sequence described above, with the exception that, in tube dryers, the exhaust fan should not be shut down.

Experience has shown that, with the start of a fire, there is no flame, only a shower of glowing embers traveling down the duct. These glowing embers are emitting infrared rays which can be picked up by infrared detectors and actuate a fire-extinguishing system. Infrared technology is now highly developed and reliable equipment is readily available from several sources.

Figure 11.1 shows the general layout of a fiber preparation line and indicates the location of the infrared detectors. The infrared detectors are located immediately after the refiners to catch fires caused by them. Also infrared detectors are located just after the dryer and after the mixer. It is felt that infrared detectors should not be placed too close to the refiners since quite a few sparks come from these units and are extinguished after traveling through a short distance in the duct. The detectors are located about 10 to 12 ft (3.05 to 3.66 m) from the refiner. Many variations are possible and are designed to fit the plant.

Fighting Fires

The constant supervision and maintenance on fire hose stations, hydrants, and extinguishers is important. Unfortunately, the wrenches for opening hydrant valves make very handy hammers and have a tendency to disappear at the most awkward moments.

There have been instances with very sad results where operating personnel were not aware of the shutdown procedure of the equipment during fires or explosions. It is recommended that these shutdown procedures be reviewed from time to time with operating personnel. Some modifications may be required in the control circuitry so the operator can shut down equipment safely and rapidly.

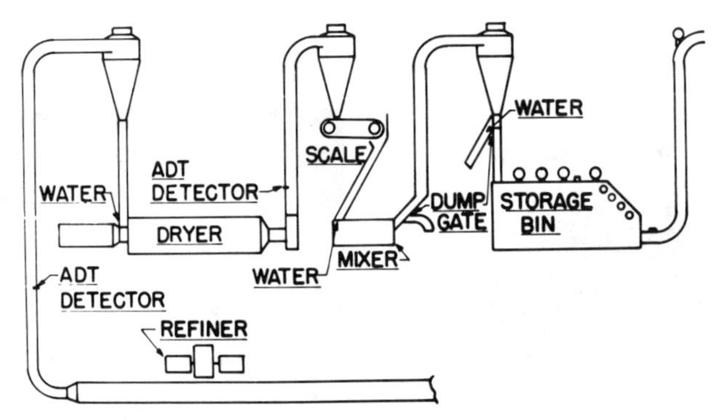

Figure 11.1. *Pope & Talbot fire and explosion detection and suppression system (Ely 1970).*

The training and familiarization of the personnel in fire fighting and the use of the fire-fighting equipment has to be stressed. Local fire departments are usually very helpful and interested in training plant personnel in fighting fires and handling emergency procedures. It is advisable to familiarize the local fire-fighting personnel with the plant facilities so that they know their way around in case they are called upon to help fight a fire.

Proper clothing should be worn by those in areas susceptible to fire and explosion. Clothing of polyester ignites and burns rapidly and melts into the flesh, causing terrible blisters that can be avoided by clothing of natural fibers. Severe fires, of course, will also ignite and burn the natural fibers, but time is gained in dousing the flames on an afflicted worker.

EXPLOSIONS

Sources

As an introduction to the basic techniques of explosion protection, explosion with reference to fire should be defined. Figure 11.2 illustrates these phenomena. *Fire* may be defined as the rapid oxidation of a fuel with an attendant evolution of light and heat. In an explosion or a deflagration, there is not only the evolution of heat and light, but also an increase in pressure and mass flow. In the initial stages of the combustion reaction, this flow is directed away from the point source of ignition; and, in the later stages, it reverses itself towards the point of ignition.

In the burning of most substances, the actual combustion takes place only after the solid fuel has been vaporized or decomposed by heat to produce a gas. The visible flame is the burning gas. If the flammable mixture is enclosed, the pressure front or shock wave will equalize at sonic velocity throughout the enclosure. The maximum pressure which can be expected will seldom exceed a ratio of 8 to 1, assuming ordinary temperatures; e.g., if an explosion begins at 14 psig (96.5 kPa), the maximum pressure will be 112 psig (772 kPa). The flame velocity during the reaction will generally move at approximately 8 to 10 ft (2.44–3.05 m) per second during the initial stages of the reaction as opposed to the 1100 ft (335.3 m) per second of the shock front.

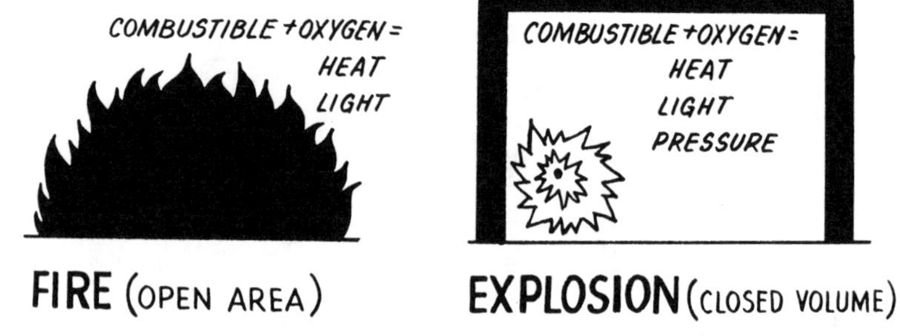

Figure 11.2. *A comparison of fire vs explosion (Charney 1970).*

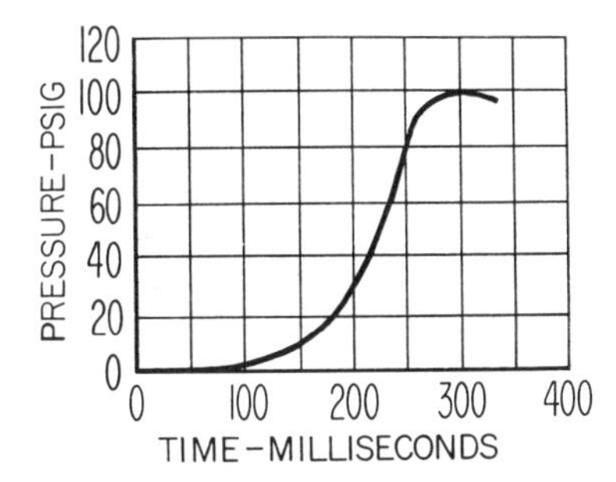

Figure 11.3. *A typical pressure-time curve for a wood dust explosion (Charney 1970).*

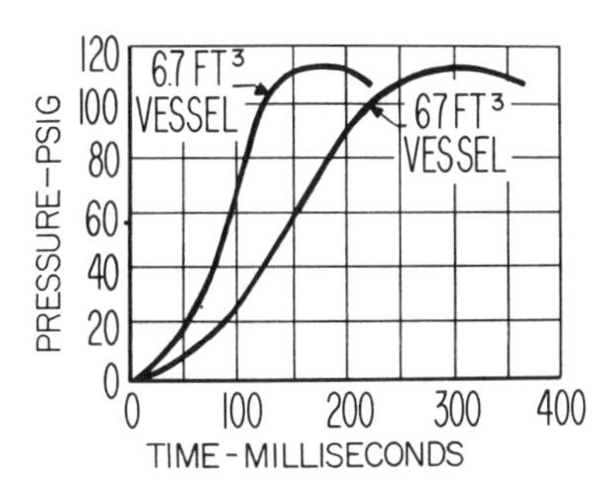

Figure 11.4. *A graph of the volume effect on explosions (Charney 1970).*

Figure 11.3 illustrates a typical pressure-time curve for a wood dust explosion allowed to proceed to completion. The curve is basically logarithmic in that, in the initial stages of the reaction, the rate of pressure rise is relatively slow; as the reaction continues, however, the rate of pressure rise reaches a maximum, indicating the reaction is complete.

There are a significant number of parameters which, when varied, can produce significant changes in the intensity of the explosion. For example, Figure 11.4 illustrates the volume effect. Two explosive reactions of the same material, under identical conditions except vessel size, are depicted. As noted, the difference in volume significantly affects the rate at which the pressure increases or the rate at which the explosion reaches completion. The time required to achieve completion of an explosion may be approximated by $t = 75 \sqrt[3]{V}$, with the volume being expressed in cubic feet and the time in milliseconds. This time is generally conservative for a dust explosion.

A second variable of interest is the moisture content of the material. As expected, Figure 11.5 indicates that, in general, the higher the moisture content, the lower both the rate of pressure increase and the maximum pressure. High moisture content will also significantly increase the amount of energy required to ignite the same flammable mixtures.

The amount of available oxidizer which could contribute to the reaction will also affect significantly the pressure-time history of the reaction. Figure 11.6 graphically illustrates a series of tests performed at the Fenwal Company Reaction Test Facility to determine the effects of incremental changes in the oxygen-nitrogen ratio on wood dust explosions. As would be expected, as one increases the diluent and decreases the percentage of oxygen, the pressure-time history is altered downward.

Other variables that can practically affect explosion reactions include particle size, ignition energy, initial pressure and temperatures, and flammable limits of material. Particle size in particular can be significant in that particle sizes larger than 40 mesh are generally considered nonexplosive; e.g., when suspended in air, normal ignition sources such as static discharge, hot metals, etc., will not produce a combustion reaction.

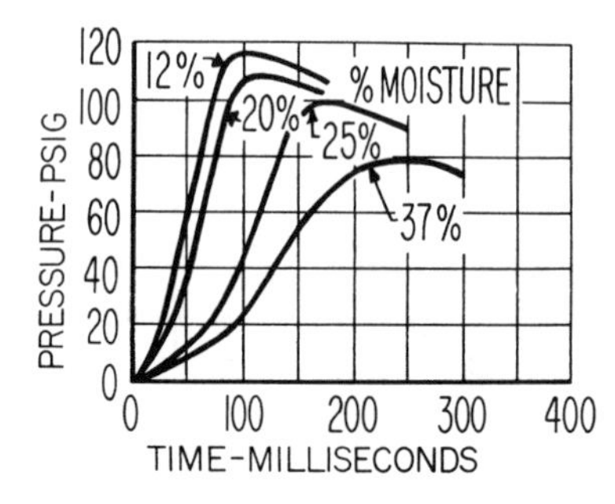

Figure 11.5. *The effect of moisture content on explosions (Charney 1970).*

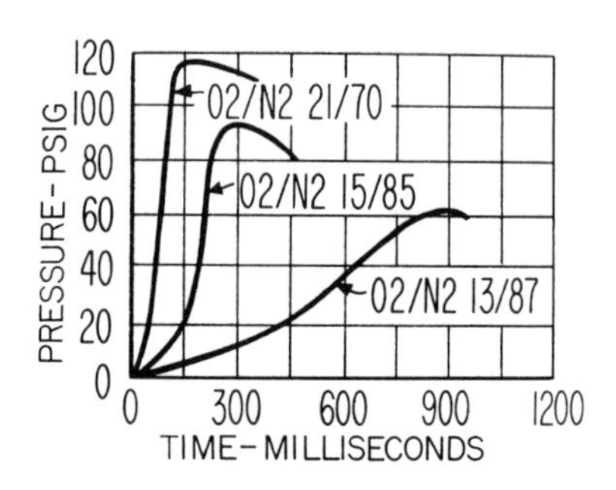

Figure 11.6. *The effect of a diluent on wood dust explosions (Charney 1970).*

In attempting to apply explosion data to practical processes, one must not only judge the hazardousness of the process by the end results of an incident, but one must also be concerned with the possibility of what might have happened if some of the parameters previously discussed had been altered.

Explosion Control

Three methods are available for preventing or controlling explosions. The first and the least expensive method, and the one most neglected with disastrous results, is plant housekeeping. Articles and books have been written describing housekeeping and its beneficial effects regarding safety. Some plants support housekeeping in practice, but the results are quite the opposite.

The importance of housekeeping, particularly in the board industry, can be illustrated quite easily since, in most dust explosions, there is in fact more than one explosion. Generally, the result of a disaster may be eight or even ten successive explosions. The cycle is as follows:

1. Ignition occurs within a piece of process equipment and the pressure is sufficient to rupture the equipment.
2. The shock wave, traveling at sonic velocity, exits the process equipment shaking loose dust from various places such as beams, stair railings, tops of equipment, etc.
3. The flame front traveling at 10 ft (3.05 m) per second then exits the piece of process equipment and to its delight finds another flammable mixture located within the room, produced by falling dust shaken loose by the shock front.
4. Steps 1 through 3 are repeated until either there is no longer a building or all of the dust has been consumed.

A second and less satisfactory method of protection, rather than prevention, is explosion venting. Unfortunately, venting is more of an art than a science and, therefore, recommendations regarding it are subject to a number of interpretations. Insurance estimates are generally and understandably significantly more conservative than others. In order to intelligently evaluate the explosion-vent ratios, not only must the explosion potential of the dust under consideration be evaluated, but the propensity to ignition must also be known.

For large storage bin volumes, ratios of 1 ft^2 of vent to 100 ft^3 of volume (1 m^2–30.4 m^3) may be considered. Unfortunately, the effectiveness of these vents depends to a large extent on the location of them relative to the ignition of the explosion, the inertia of the vent itself, etc.

The final method of protection to be discussed concerns the use of suppression of explosion as opposed to preventing or venting. An explosion-suppression system, like any other protection system, is made up of a device to sense a manifestation of the explosion, a control unit to translate the signal into action, and a device to prevent the reaction from continuing.

Figure 11.7 depicts the most basic of explosion-protection systems. The system is comprised of a sensing detector electrically coupled to an explosively actuated extinguisher. The explosion detector may be sensitive either to light or pressure, and ignition within the protected volume will be sensed by the detector before the flame can propagate more than a few inches. This action then completes the electrical circuit from the power supply to the suppressing device. This particular device consists of a frangible container filled with an inerting agent and having a small electro-explosive charge within the center. Because the electro-explosive device is fired nearly instantaneously and imparts extremely high distribution speeds to the inerting agent, the expanding flame is rapidly extinguished and the unburned portion of the mixture is chemically inerted.

The elapsed time from detection of an incipient explosion to suppression of the mixture may be only a few milliseconds. Typically, suppressed explosions in practical applications will peak at a pressure anywhere from 1 to 3 psi over a maximum time of 150 milliseconds.

The area which must be considered an explosion hazard in most fiberboard or particleboard plants usually starts after the dryer. Here the material itself is well

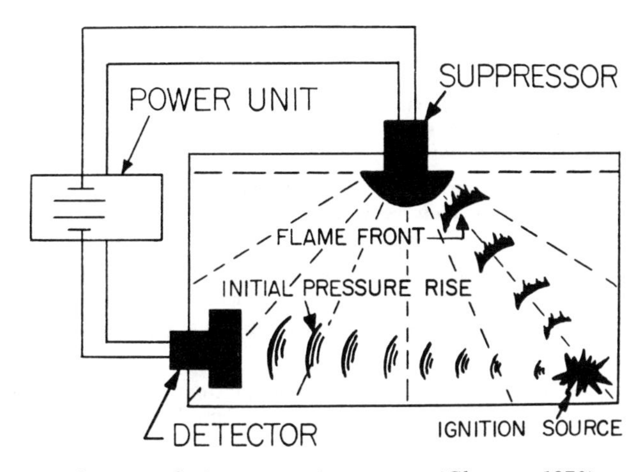

Figure 11.7. *A basic explosion protection system (Charney 1970).*

within the explosive range from the standpoint of particle size (if fine) and moisture content. From this point on cyclones, duct work, forming heads, transfer fans and collectors, and sanders are all potential explosion hazards. Sources of ignition within the process would include: burning embers, static discharge, foreign material, etc.

As mentioned, the problems from explosions are potentially much more severe than those from fires and explosions can result in catastrophic losses both in personnel and production. Plants must be protected against this catastrophe and a brief review of the factors which influence wood dust explosions is of importance.

Plant experiences with explosions indicate that they have started in the dryer and bins with fire as the ignitor which is carried over into the cyclones, ducts, and felter. Wood particles discharged from the bins, cyclones, etc., combined with dust shaken from beams and ducts, have caused secondary explosions. Such explosions in the equipment have resulted in little equipment damage; but the secondary explosions have injured workmen, and buildings have suffered severe blast and fire damage.

The fiber bins are a major source of explosions. One safeguard is to locate them outside to avoid secondary explosion problems. They can also be protected with a Fenwal explosion-suppression system.

The aforementioned problem with sparks from the sander entering baghouses is causing significant problems at present. Serious explosions have occurred as the rapid transit of the sparks into the baghouse is too fast for the explosion-suppression system to work. Explosion vent doors (blowout doors) can be hazardous to anyone nearby when an explosion occurs. Some baghouses vent to atmosphere, eliminating the need for explosion doors. Another approach is to use a primary cyclone before the baghouse, giving the explosion-suppression system time to react.

Fires and explosions are still occurring at an undesirable level. Continuing development and education are needed to eliminate these serious problems.

SELECTED REFERENCES

Charney, M. 1970. Protection Systems for Dust Explosions. *Proceedings of the Washington State University Particleboard Symposium, No. 4*. Pullman, Washington: Washington State University (WSU).

Ely, H. M. 1970. Fire and Explosion Prevention, Detection and Extinguishing. *Proceedings of the Washington State University Particleboard Symposium, No. 4*. Pullman, Washington: WSU.

Loock, G. E. 1970. Space Age Fire Detection. *Proceedings of the Washington State University Particleboard Symposium, No. 4*. Pullman, Washington: WSU.

Nestelle, F. H. 1975. Private communication on fire and explosions. Archer Blower & Pipe Co.

Wiecke, P. H. 1970. Introduction to Fire and Explosion Prevention and Detection. *Proceedings of the Washington State University Particleboard Symposium, No. 4*. Pullman, Washington: WSU.

12.
RESINS AND OTHER ADDITIVES

Synthetic resins were the fundamental factor in the successful development of the dry-process composition board industry. Glues common to woodworking and plywood manufacturers for a number of years, such as casein, soybean, bone and hide, and blood have not been found to be suitable for composition board. While synthetics have been the dominant resin used so far, it is also possible to manufacture suitable resins from tannins (natural phenols) and spent sulfite liquor. The amounts of these resins used are negligible at the present time. However, if shortages in the synthetic materials normally used continue, extra emphasis is expected on the development of these types of resins as well as the establishment of production facilities for manufacturing them.

Both the synthetic and natural adhesives will be covered in this chapter. This will be followed by a discussion of the fundamental parameters of resin. Other discussion will cover the other additives either necessary or desired for some products. Prominent machines in use for blending resin and other additives with the wood furnish will be considered in the next chapter.

SYNTHETIC RESINS

Three major synthetic resins types are used in the board industry. By far the most dominant is urea-formaldehyde (UF), followed by phenol-formaldehyde (PF) and melamine-formaldehyde (MF). Another interesting but little used synthetic binder is polyisocyanate.

Urea and melamine resins are also known as amino resins and are the reaction products of amino and amido groups with aldehydes, most commonly, formaldehyde. The reaction causes formaldehyde to condense with the amino/amido compound to form methylol derivatives which upon further heating form the cured, colorless resin. These resins and phenol-formaldehyde resin can be cured rapidly in the presence of catalysts with the application of heat. As noted, the resins are built up by condensation polymerization.

Soluble and fusible (thermoplastic) intermediate products can be isolated during the production of the particular resin derived and these intermediates can be used to fabricate the final products. This is of great importance to the technology of these resins. Most applications are based on the ability to use intermediate condensation products to generate adhesives suitable for manufacturing wood panels bonded together with high-molecular-weight condensation polymers. The intermediate products are fusible or thermoplastic low-molecular-weight polymers. To convert to infusible or thermosetting products, cross-links are built up between the low polymers giving a three-dimensional structure in the cured resin.

About 90% or more of the world's composition board production is made with UF resins. They are the lowest in cost and have the fastest reaction time in the hot press, and they are easy to use. Additionally, they are white or colorless. However, they do not produce boards with good exterior exposure resistance. Thus, UF boards are not suitable for exterior applications. In some parts of the world they are used in protected exterior exposures, where they are covered with suitable coatings such as paints. The PF resins have found the most favor to date when exterior-type products are desired. The liquid types are reddish in color and give a darker shade to boards, or possibly show as tiny red spots. Much of the deleterious effect can be removed in the sanding operation. MF resins are better than UF resins but cannot match the exterior performance of the PF resins. Melamines are used in some board processes, especially when blended with UF resins. With such a mixture, they enhance the moisture-resistance properties of the UF resin. A recent development in Europe for exterior resin is one made of urea, melamine, and phenol reacted with formaldehyde. It will reportedly cure as fast as urea-formaldehyde resin.

The resins commonly used in the board industry are products of the reaction between urea, melamine, or phenol and formaldehyde. Figure 12.1 shows the generalized diagram followed for the production of UF and PF resins. It should be seen immediately that the key raw materials for the two resin starting materials are natural gas and crude oil, both of which are in demand throughout the world at the present time for uses other than resins.

Only about 9% of the urea goes into the manufacture of UF resins, and these resins are used in other woodworking industries besides that of composition board products. About 58% of the phenol manufactured in the world ends up as phenol-formaldeyde resin. Again, this resin is used in other applications, so that the total amount is not reserved for composition board.

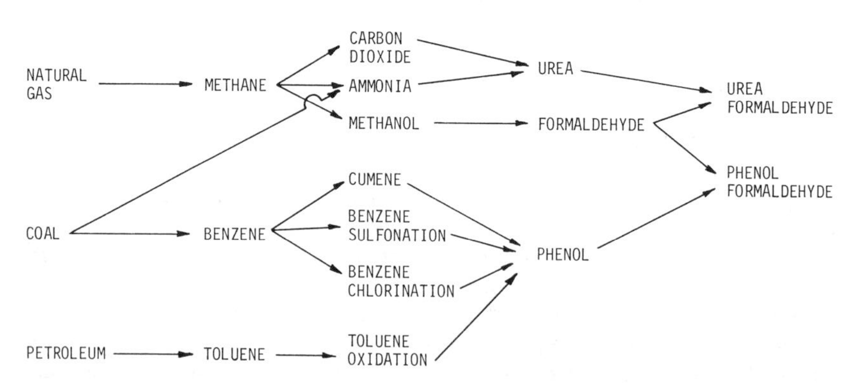

Figure 12.1. *Schematic flow chart of phenolic and urea resin feed stocks (Binder 1974).*

Other demands for the raw materials for producing urea and phenol resins come from the energy and agricultural industries. Natural gas supplies the methane for the methanol from which formaldehyde is produced. The formaldehyde is reacted with urea, phenol, and melamine to make the resins needed. Ammonia and carbon dioxide obtained indirectly from the natural gas are reacted together to provide the urea. Thus, natural gas is the basic source of almost all the material needed for making urea-formaldehyde resins. Melamine previously was traditionally produced indirectly from coal. However, in recent years the new plants have drawn upon urea as the basic raw material, which means melamine is also dependent at the present time upon natural gas.

The phenol comes from crude oil through the intermediate benzene, or toluene, or cumene. The problem faced here is that toluene may be required as part of the new fuel necessary to make the 1975 pollution-free automobile engines function properly. These new fuels cannot use lead and toluene is part of the substitution for the lead. Also, benzene is a critical raw material for use in pesticides which are badly needed for food production. At the present time there is about a 25% deficiency in pesticides around the world, and consequently there is a critical demand from the agriculture industry for this raw material.

With the great amount of urea production now going to fertilizer, and the demand for benzene and toluene by other products, the outlook at the present time is not bright for an expansion of the supply of the raw materials needed for synthetic resins. While there is a greater demand for these basic raw materials at the present time, part of the shortage problem has occurred because new refineries and new production facilities for resins have not been constructed at the level needed over the past ten years. The return on investments for the giant chemical complexes needed to generate these raw materials has not been high enough to encourage companies to make such investments.

Oil shortages, crude oil price increases, increased demand for petrochemical feedstocks, and lack of oil refining capacity, have all contributed to the price and availability problems of synthetic binders. An oil company has the choice, based on economics, of utilizing its crude oil supplies to produce gasoline, fuel oil, or petrochemical feedstocks. The raw materials for the petrochemical industry are the lighter hydrocarbons which are at the top of the so-called barrel of oil; but gasoline manufacturers and utilities also bid for these raw materials, high in Btu or calorie content and low in sulphur, because they are environmentally desirable as fuel. Energy demands are therefore also complicating the picture when considering synthetic resins.

With the recent worldwide increases in oil and gas prices, however, some new plants and expansions have been announced. It is anticipated that the methanol and urea problem will come into balance within the next few years. It is unknown whether the phenol supply will be sufficient for the proposed expanded composition board industry before 1980. The world supply has been at a deficit already, and large new plants are not planned so far as known; if they are planned, they will not be functional for a number of years. Contributing to some of the problem with the phenol shortage has been plant closures because of fires and explosions. At times these have taken a significant part of the world production capacity out of operation.

It would appear that developing countries that are rich in oil and gas resources would be wise to install complexes for generating methanol and benzene because of their need for food. This may not always be the case, however, as one oil-rich country has recently announced postponing a $500 million plant that was to produce methanol. The natural gas to be used as the basic raw material will be pumped back into the ground to force more crude oil up through the wells. Evidently they feel they will get more return on their investment by selling the oil around the world.

Urea Resins

As is well known, about 90% of the world's dry-process composition board products are bonded with urea resins. Consequently, urea resins have been studied thoroughly in research laboratories and by resin manufacturers. In the early days, the urea-formaldehyde resins normally used for particleboard were very similar to the glues used at that time for plywood. However, much research has gone into developing urea resins that have extremely fast cure rates in comparison to 15 years ago. About 1960 a typical ¾-in.-thick (19 mm) particleboard bonded with urea-formaldehyde resin would take about ten minutes for curing in the press. At the present time, depending on the system used, this board can be pressed in at least half the time and, in many cases, much less. At the same time, even with relatively sharp increases in resin prices over the last two years, the resin itself is lower in price than in 1960. Thus, a lower-cost resin is available now as compared to 1960, which can cure out at least twice as fast. Thus, without any consideration of inflation, the value of the resin has more than doubled, because of the increased production capacity of a plant based solely on improvements in the resin. This has been an extremely important development, pointing out that research indeed does have its place in this industry. This is one of the reasons for the rapid growth of the dry-process composition board industry.

To chemists the manufacture of synthetic resins is a relatively simple chemical reaction. However, it is not this simple reaction that is used in the industry. If it were simple, most large board plants would probably make their own resins. On the other hand, the ways to produce the sophisticated modern resins are trade secrets. Once in a while a resin manufacturer will take out patents on some of his recent developments, but much of the know-how or art of making the resin will not be revealed. Much of the tailoring of resins to fit a certain process or plant is indeed art based on science. Thus it is impossible to discuss the technical details of making resins because of the trade secrets involved. Consequently, resin technology per se will be dealt with in this discussion in general terms.

Urea-formaldehyde resins are relatively high-molecular-weight polymers of urea and formaldehyde dispersed in a water medium. The reactive end groups of these polymers are predominantly methylol groups. It is also a common practice of resin suppliers to add urea to the resin after the synthesis reaction with the excess formaldehyde. This reduces the level of unreacted formaldehyde in the liquid resin, but also greatly increases the amount of methylol end groups present in the resin. During the curing process of the hot press, a primary source of formaldehyde release from urea-formaldehyde binders is believed to be the deg-

NH_2-CO-NH-CH_2OH

monomethylol-urea

$HOCH_2$-NH-CO-NH-CH_2OH

or

NH_2-CO-N-$(CH_2OH)_2$

dimethylol-urea

$HOCH_2$-NH-CO-N-$(CH_2OH)_2$

trimethylol-urea

Figure 12.2. *Monomethylol-urea, Dimethylol-urea, and Trimethylol-urea: products from reaction between urea and formaldehyde (Kelly 1970).*

radation of methylol end groups which are not completely reacted. These low-molecular-weight methylol-ureas (Figure 12.2), are believed to contribute very little to the efficiency of the resin, but they do significantly reduce the formaldehyde odor of the liquid resin.

The predominant configuration of the urea-formaldehyde polymer molecules in a liquid resin has not been elucidated, although researchers have been studying these resins for a number of years. The proposed configurations range from a linear structure, to a branched structure and even to a possible cyclic configuration. Examples of three proposed configurations are presented in Figure 12.3. It is possible that these three configurations, as well as others yet to be proposed, are present in varying degrees in urea-formaldehyde resins.

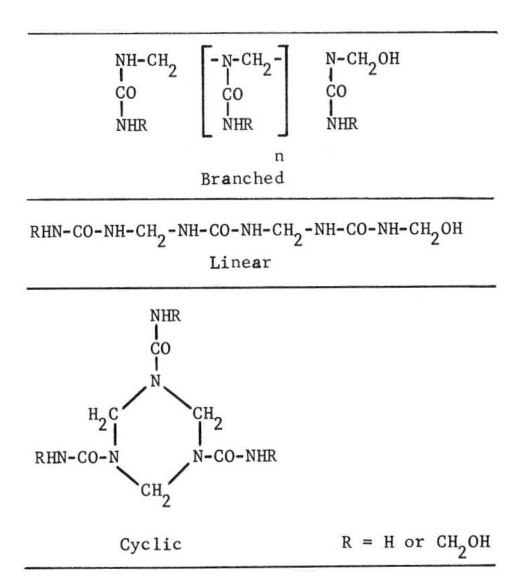

Figure 12.3. *Proposed polymer structures of urea-formaldehyde resins (Kelly 1970).*

Urea-formaldehyde resins are cured under acidic conditions; this acidity can be attained by using a catalyst, by the acidic nature of the furnish, or a combination of these two. Formaldehyde in the resin serves as a source of methylene bridges between the amide nitrogen of adjacent polymers. Without this formaldehyde, cure rate would be reduced and would be dependent upon the reaction between methylol end groups and amide nitrogens. For a methylene bridge to form between two polymers, these two reactive sites, the methylol end group of one chain and the amide nitrogen of an adjacent chain, would have to be in a favorable positon to each other for the reaction to occur. If these two sites were in a position such that mutual contact was difficult, the reaction would either be impossible or would not occur until the sites had moved to a favorable position; i.e., reaction rate would be dependent upon the rate of movement of the respective reactive sites of the polymers.

Cure of a urea-formaldehyde resin can also be effected by reaction between the methylol end groups of adjacent polymers. In this case, water can be split out, resulting in an ether linkage between polymers. This ether linkage can be converted to the more stable methylene bridge by splitting out formaldehyde. This conversion from an ether to a methylene bridge can occur immediately after formation or at some later time. Figure 12.4 presents the above two types of interpolymer linkages.

The presence of formaldehyde in the resin at the time of cure enhances the curing process. Cure without formaldehyde is possible, but not in the time most board producers have come to expect as normal press times.

Formaldehyde in the resin results in high formaldehyde odor and the resultant irritation to personnel in the vicinity of the press. This problem can be alleviated to some extent by adequate process controls. Among these are moisture content of the mat going into the press, length of the pressing cycle, and an adequate ventilation system in the pressing area. Atmospheric conditions such as temperature and relative humidity also have an effect upon the formaldehyde odor in the vicinity of the hot press. It is unlikely that the level of formaldehyde can be reduced to the extent that it is no longer bothersome to press personnel; however,

```
                                                     CH2OH
                                                       |
                                    -NH-CO-N-CH2-NH-CO-N-
                                           |
-NH-CO-N-CH2-NH-CO-NH-                     CH2
       |                                   |
       CH2                                 O
       |        CH2OH                      |
       |         |                         CH2   CH2OH
-NH-CO-N-CH2-N-CO-NH-                      |      |
                                -NH-CO-N-CH2-N-CO-N-CH2-NH-
   Methylene Bridge                    |
                                       CH2OH

                                           Ether Bridge
```

Figure 12.4. *Two possible interpolymer linkages in urea-formaldehyde resins (Kelly 1970).*

with suitable precautions it may be reduced to the extent that it no longer irritates other plant personnel.

There are several important interrelated production parameters governing the reaction of urea and formaldehyde to form a UF resin, such as the temperature and time of the reaction, the pH value, the concentration of the various reactants, and the mol ratio of the formaldehyde to the urea. Mol ratios of 1.4 to about 2.0 of formaldehyde to urea are reported. With lower ratios there is less free formaldehyde in the final board, but the curing rate of the resin will be slowed.

The formaldehyde odor is a significant problem in UF bonded boards. Formaldehyde in the air is toxic but it is so irritating to the eyes and nose that a person is driven from the area long before he is felt to be physically harmed. The least detectable odor of formaldehyde is reported as 0.8 ppm and the lowest concentration causing throat irritation is 5 ppm. A manyfold reduction in the formaldehyde around the press would be needed to completely eliminate this irritant. Recent developments have improved resins so that much of this problem has been relieved to a great extent in many plants. Formaldehyde is released slowly over long periods of time and this release is increased under hot and humid conditions. This is particularly noticeable in enclosed spaces such as closets which are lined with unfinished board. Good coating with clear finishes or paints, however, alleviates this problem. In overlayed boards, of course, the free formaldehyde is sealed off. Eventually the formaldehyde odor drops to an imperceptible level as the free formaldehyde is dissipated.

Returning to the manufacture of the resin, the reaction can take place about 5.0 to 5.5 in pH. The reaction is arrested by raising the pH to 7.0 to 8.0 and cooling the newly made resin. The concentration of the resin is about 40 to 60% nonvolatile solids. Some water is then distilled to increase the solids content to 60 to 65%. Some plants use the resin neat at this solids level while others dilute to a lower solids content (usually in the range of 50 to 60%) to improve spraying characteristics and coverage of the furnish with the resin.

Viscosities of the resin range as low as 30 centipoise for tack-free resins suitable for fiberboards to 200 to 300 centipoise for most particleboard applications. Higher viscosities are made to fit plants that need a more viscous resin.

A latent catalyst can be added, which acts once heat is applied to the mat in the hot press. Other catalysts such as ammonium sulfate or ammonium chloride (ammonium salts of strong acids) can be added at the board plant. These catalysts are not latent and, once added, the resin has a relatively short pot life. Another term used for catalyst is *hardener*.

The catalysts serve to provide acid which is needed for a fast cure. Some species such as Douglas fir are relatively acid (a pH of about 4.2), and in these cases sufficient acid is available in the wood to effect the cure. However, catalyst in the resin speeds the curing. Catalyst can be sprayed onto the furnish after applying the resin, and this is done in at least one plant.

Some woods or furnishes are so highly buffered (discussed later) that it is quite difficult to develop an acid environment in the furnish enabling faster resin curing to take place. An example is plywood trim bonded with PF glue. The alkaline PF glue results in a furnish that is very difficult to glue, and it took the resin companies a number of years to develop suitable resins for use with this furnish. A side note:

pressure-refined plywood trim appears to be a better furnish, possibly because some of alkalinity is removed during the refining.

Buffering of the resin is important when long storage life is important. Small amounts of ammonia or hexamine can be used. Urea and melamine can also serve as buffer. Resins used for board faces are usually more buffered to allow for consolidation of the mat before curing takes place. Curing, of course, takes place first in the faces in conventional operations because they are in contact with the hot press platens almost immediately. In most cases, it would be advantageous to use a highly catalyzed resin in the board to speed its cure. Ideally, both the face and core resins would cure almost simultaneously.

Mat moisture content, of course, is important as too high a level slows curing. In some plants, higher moisture levels are used in the surface layers to speed the cure of the core resin. Such extra moisture can also be sprayed onto the mat surfaces. This face moisture turns to steam and rapidly moves to the cool core of the board and, in doing so, carries heat to the core. This has been called the *steam shock* treatment.

The curing takes place by an increase in the resin viscosity followed by gelation and finally by forming of the rigid solid common to thermosetting resins. With UF resins, a core temperature of about 212°F (100°C) is needed for final curing. Face temperatures are much higher in the range of 300° to 370°F (149°–188°C) depending upon the press platen temperature. The key for fast press times is to elevate the board core temperature to a level sufficient for fast curing.

Melamine Resins

Melamines are high-cost resins that are better than ureas for exterior exposure but not as good as phenolics. Combinations of urea and melamine resins have been used for products that will be in damp but not severely exposed situations. Fortifying of urea resin with melamine is of value up to an addition of about 20%.

The production of melamine-formaldehyde resin is chemically similar to that of urea-formaldehyde resin. For many years the resin was supplied as a powder because of stability problems. However, recent developments have made it possible to deliver aqueous solutions of the resin. Most applications at present call for a combination of urea and melamine in the resin. This is a particularly important combination for use in the medium-density fiberboard (MDF) industry. These boards range up to about 1 ½ in. (38.1 mm) in thickness at board specific gravities in the medium-density range. Pressure-refined fiber is the furnish used.

Urea-formaldehyde concentrates or a melamine-urea-formaldehyde *in situ* formulation have been developed as binder systems. These binders have to have the proper tack at the right time to prevent the fibers from rolling up into balls and causing problems in handling and forming. Moreover, such balls usually have an excess of resin on the outside, and this will show up as resin spots on the board surfaces. Very low viscosity is also necessary to allow for uniform distribution of the resin over the fibers, which have very high surface area for a given weight.

One particular binder system developed for fiberboard is essentially an unpolymerized water solution of methylolureas, methylol-melamines, formaldehyde, and an acid-forming salt. It is called an *in situ* binder because the complete reactions occur in the hot press, whereas conventional resins are partially polymerized when manufactured. Proportions of the urea and melamine in the mix are varied, depending on whether the board produced is for interior applications such as furniture or where more water-resistant bonds are required.

Pope & Talbot, Inc., in Oakridge, Oregon, USA, has developed its own system for adding urea resin *in situ* to fiber. Part of the urea is added to the wood before fiberization with the remainder of the resin chemicals added afterwards. The reaction between the chemicals takes place in the hot press as with the melamine-urea-formaldehyde system.

Phenolic Resins

Phenolic resins have always cost more than the urea resins. In addition, it takes somewhat longer for the resin to cure out in a given board as compared with urea resin. Catalytic systems have been developed recently that have significantly decreased the press time when using phenolic resins. Of more importance in comparison with urea resins at the present time is the previously mentioned shortage of the raw materials needed for producing phenolics. Many plants in the United States have had their phenolic resins allotted at about 65% of their normal use during recent periods of high demand.

Phenol is characterized by the presence of a hydroxyl group attached to an aromatic nucleus (benzene). The major source of phenol is through synthesis from benzene by one of the three processes: sulphonation, chlorinaton, and cumene. The cumene process is the most important process at this time. Two types of phenolic resin are made: the resol or one-stage and novolac or two-stage. Phenolic resins have three further stages. The A-Stage is a polymer liquid or a liquid dehydrated to a solid. The solid is soluble in simple organic solvents and is fusible (thermoplastic). The B-Stage is the resin solid which is insoluble but can be swelled by solvents. It is infusible but can be softened by heat. The C-Stage occurs when the polymer is infusible, insoluble, and not softened by heat or swollen with solvents. It is a cured thermoset resin.

To make a liquid resol-type phenolic resin, the necessary reactants are first charged into a resin kettle and reacted using an alkaline catalyst. The mol ratio of phenol to formaldehyde ranges from about 1:1.8 to 1:2.2. The reaction is carried to a predetermined point and then slowed by reducing its temperature. These resins are not particularly stable since the reaction has only been retarded. Storage life may range from hours to months. They will continue to cross-link until the state of a completely reacted insoluble and infusible resin is reached. Storing at low temperatures is preferred. The resin as discharged from the kettle is thermosetting or heat-reactive and only requires further heating, a change in pH, or a combination of both, to complete the reaction to an infusible, insoluble state. The change in pH is not normally used in the wood industry. The curing is under the action of the alkali present in the solution. The high alkalinity of such resins can be detrimental in that more water than desired can be absorbed by such phenolic-bonded boards when placed in service.

The novolac resins have been around for over 50 years. They have been commonly used in molding processes and are the type used for powdered resins in some particleboard plants. They are made by reacting phenol and formaldehyde with an acid catalyst, although alkaline catalyst can be used. The mol ratio of phenol to formaldehyde ranges from about 1:0.8 to 1:1. To produce a hard resin all of the moisture is removed. These resins are permanently fusible or thermoplastic when removed from the kettle. For liquid resin production, an organic solvent can be added while the molten resin is still in the kettle. The hard resin is pulverized for use. Further formaldehyde is added usually in the form of hexa (hexamethylenetetramine) in the amount of 10 to 15%. The hexa breaks down under heat and forms formaldehyde and ammonia. The formaldehyde combines with the resin and converts it into a thermosetting product while the ammonia serves as a catalyst.

Novolacs are quite stable under normal conditions. They are hygroscopic or water seeking and should be stored in a dry place.

The phenolics are slower curing than the ureas. They are more heat stable and require higher press temperatures. Temperatures in the range of 250° to 300°F (121°–149°C) in the core are needed. Platens range up to over 400°F (204°C) in temperature. Catalysts to speed pressing time, such as resorcinol, can be used but this particular catalyst is quite expensive. The final cured resin is heat and chemical resistant.

Phenolic resins for bonding are fairly high in molecular weight. They stay on the particle surfaces and develop durable, rigid, strong waterproof bonds between the particles. Another type of phenolic resin is of interest to board manufacturers: aqueous solutions of low molecular weight. These resins have the ability to penetrate and swell the cell walls of the wood. When cured under heat, these resins strengthen the wood and impart excellent dimensional stability. A particleboard developed at Washington State University called Flapreg uses up to 30% of this resin and has excellent physical properties. It has a high density but requires only sanding and buffing to achieve an excellent finish. This durable product, similar to Compreg, which is made of resin-impregnated veneers, has been demonstrated to be an excellent furniture material. It can be made on standard production lines with the exception that the highest quality panel has to be compressed under 1200-psi (8.27 MPa) pressure in the press. Lower pressures will yield lower-density panels with concomittant lower properties which are still better than those of conventional boards. A technique that will improve board quality without complete impregnation is to use the impregnating resin in the faces only.

Similar approaches have been used to produce high-density fiberboards, particularly of the type used in high-density plastic laminates.

Isocyanates

Brief mention should be made of a more expensive but extremely interesting binder, namely isocyanates. In simple terms, these result in urethane chains. The hydroxyl parts of wood are chemically connected with this binder system resulting in extremely good bonds. Such bonds have good resistance against water, diluted acids, and liquors. This binder is approved for building board in the Federal Republic of Germany.

The advantage of this binder is that no water is contained in the system. All of the binder applied is used as an adhesive. Higher-moisture-content furnish can be used, thus costs of drying are reduced, and the "blue haze" found when drying to low moisture content is eliminated. Less of this adhesive is needed in a given board as compared with phenolic resin. Reportedly, board density does not have to be as high as with phenolic resins because of better bonds and thus there is a saving on wood costs. Bonds are very resistant to water and thus have good exterior exposure characteristics. These binders work well with difficult-to-glue material such as straw. Straw boards so bonded have been made that match phenolic-bonded wood particleboards. Urethanes are hydrophobic, thus some or all of the wax may be eliminated from the board.

Disadvantages include the aforementioned higher cost, which may be balanced by savings elsewhere in the production process. Further, isocyanates adhere firmly to aluminum and some steel, causing problems with bonding to caul plates and press parts. Some release agents, such as glycerine, have been found effective. Another solution to this problem has been to use amino or phenolic resins in the faces, thereby eliminating contact with metal by the isocyanates. The isocyanates must be handled carefully in production to avoid worker health problems.

NATURAL BINDERS

Some natural adhesives, such as casein, soy bean flour, and spray-dried beef and pork blood, have been used for many years for bonding plywood. The U.S. softwood plywood industry used tremendous amounts of such natural adhesives in the past. However, 80% of this production is now made with phenolic resin. Tannins have been successfully used commercially for plywood and particleboard, especially in Australia where tannin has been used for a number of years. Quebracho, a tannin product, is being used extensively in Argentina and Finland.

Recent Indian research has developed a tannin adhesive using a very small amount of formalin. This adhesive has been improved by the addition of a protein. Work has also been conducted on the use of lignosulfonates from spent sulfite liquor in Europe and Canada. Because lignin is a natural substance, this type of binder is being designated as natural in this discussion.

For particleboard and dry-process fiberboard, the same tannins and lignosulfonates used, or being considered, for plywood, are, or may be, utilized. Success has been achieved by replacing all of the synthetic resin with tannins. A simple lignosulfonate system appears to be reaching fruition in Canada for bonding waferboard. Availability of the lignosulfonates may be a problem as there is demand for them by other industries. Some of the tannin formulations require some small amount of formaldehyde and possibly some fortifying resins; nevertheless, the amount of synthetic materials required is dramatically reduced.

Tannins

Natural polyphenols, known generally as *tannins* and suitable for exterior-type glues, are found in a number of wood species and barks. The best-known sources to date have been mimosa wattle (*Acacia mollissima*) from South Africa and quebracho (*Schinopsis* spp.) from Argentina. These are being used commercially, mostly for plywood, in many places, including South Africa, Finland, Argentina, Brazil, Australia, the USSR, the United Kingdom, and India. Other natural polyphenols are presently being research, and hopefully commercialization of them will occur in the near future.

Quebracho is extracted from wood or bark and, when reacted with formaldehyde, forms an insoluble water-resistant resin. Because of its high molecular weight, a smaller quantity is needed to produce a resin than with synthetic phenol. Ordinary and bisulphated quebracho are used. Quebracho used with synthetic phenolic resins can improve the bond quality and strength, reduce pressing temperature and time, and reduce the amount of formaldehyde required. Quebracho has great reactivity with formaldehyde; therefore, problems with pot life and viscosity can occur. Both acceleration of cure and improvement of bonding strength are reported in Finland where quebracho makes up 10 to 20% of the total glue solids for plywood manufacture.

A new particleboard plant in Argentina plans to use quebracho formaldehyde binder exclusively. Quebracho can also be added to urea-formaldehyde resins to capture the free formaldehyde which normally evolves from the board after it is in place. About 22% of spray-dried quebracho based on urea-formaldehyde solids is added, and practically no free formaldehyde remains in the board.

At the present time adhesives based on wattle extract are a relatively simple mixture of aqueous solutions of tannin with formaldehyde in the form of paraformaldehyde as the cross-linking agent. The reaction between the two main ingredients takes place in the hot press.

In Australia a wattle tannin-formaldehyde binder with synthetic resin fortifier is used for producing a high-quality flooring-grade particleboard. A typical formulation calls for about 35% tannin, 2% fortifying resin, and 3% paraformaldehyde. Under plant conditions longer than normal press times of 9 to 10 minutes are usual for ¾-in.-thick (19 mm) boards. A balancing economic factor, however, is that higher mat moisture contents are necessary with levels of about 25% for surface layers and 17% for the core. Thus, far less energy is needed for drying of the particles. Furthermore, the risk of fire in the dryer would be minimized and potential dryer pollution problems may be reduced.

Other research and development efforts are going on in Australia and New Zealand, with particular emphasis on tannin from the bark of *Pinus radiata*, which is a major exotic species in those countries.

In India, tannin adhesive formulations of *Mimosa, Acacia mollissima,* and *Acacia mearnsii* have been used in bulk production of different grades of plywood and proved to be suitable. The new tannin-formaldehyde resin adhesives are easily mixed from ingredients without the need for any special manufacturing operation. It is recommended that formalin be used instead because of availability and lower cost.

It has been found that the addition of a proteinaceous material to tannin substantially improves the bond strength, presumably due to a reaction between the tannin and protein. A new catalyst has permitted a reduction in press temperatures to 212° to 248°F (100°–120°C). This adhesive has been adapted for particleboard, and successful tests have been conducted.

Research is in progress at the California Forest Products Laboratory on the possibilities of using bark extractives from four species available in quantity in the western United States and Canada. These include white fir (*Abies concolor*), ponderosa pine (*Pinus ponderosa*), Douglas fir (*Pseudotsuga menziesii*), and western hemlock (*Tsuga heterophylla*). The yields of bark extracts ranged from 320 to 370 lbs of extract solids per ton of oven-dry bark (161 to 187 kg/mt). In addition, much work was done in the western United States in years past on bark adhesives, and it is expected that these efforts will be revived because of the probable problems with supply and cost of synthetic phenolics.

In the California work, conventional three-layer flakeboards were produced with each bark extract using 8% of concentrated extract. After the blending of the extract with the furnish was completed, 1% of powdered paraformaldehyde was added to the blend. During hot pressing, the formaldehyde released reacts *in situ* with the polyphenolic bark components to form the resin system.

Press times at 356°F (180°C) platen temperature were three and eight minutes respectively for the ⅜- and ¾-in.-thick (9.5 and 19 mm) boards. Strength and moisture tests indicated the board met U.S. standards for phenolic-bonded boards; however, durability was not established.

Sulfite Liquor

Reacted waste sulfite liquor from the pulping process has been suggested for more than 20 years as is attested to by the patent literature. Laboratory and pilot-plant-made particleboard and building fiberboards were evaluated and found to have satisfactory or, in some instances, superior properties.

The system developed by Dansk Spaanplade Kompagni A/S in Denmark was brought to fruition during the late 1960s. Plants were built in Denmark and Finland. About 13% of waste liquor at 50% solids is first added to the furnish. The formed mat is pressed for 1.5 minutes per mm of board thickness at 356°F (180°C). The board is then placed in an autoclave at a temperature of 392°F (200°C). The pressure is slowly raised until it reaches 10 atmospheres. The entire process takes about 100 minutes. Boards made by this process are water resistant and meet minimum physical property requirements.

The plants reportedly are no longer operating because of economic reasons. Press times were too long as compared to fast-curing phenolic resins and, coupled with the autoclave post conditioning, the resulting production costs were too high.

The Eastern Canadian Forest Products Laboratory has pursued research on finding a practical, technical solution for the possibilities of using spent sulfite liquor as a base for adhesives for exterior plywood and as a binder for particleboard. The alternative approach they have used, compared to others tried previously, has been with acidified spent sulfite liquor (SSL). It is used either in liquid or powdered form in a conventional hot-press operation to form a waterproof bond. This process requires only a simple acidification of the spent sulfite liquor to convert the binder under heat and pressure into a water-insoluble bond in the particleboard. The laboratory work has shown that the acidification of the SSL, whether of calcium, magnesium, sodium, or ammonium base, can be carried out in several ways. Among these, the addition of a strong mineral acid, preferably sulfuric, appears to be the most practical and economical. Press time is similar to that required for a similar board made with phenolic resin. The finished board meets the Canadian Standards Association requirements (CSA–0188–68) for exterior particleboard. The long-term durability of the adhesive bond has not yet been completely established. This resin reportedly can provide tack which powdered phenolic resin does not. Thus, caulless lines can possibly use this resin successfully. Sticking to metal parts reportedly may be a problem that will have to be overcome with suitable release agents.

The final board may have to be reduced in acidity by a post-treatment. Otherwise, there may be a long-time destructive effect on the wood and corrosive effects on metal fasteners. As a result of the promise shown by current research, a plant trial using SSL powdered binder at a waferboard plant was completed recently with promising preliminary results.

RESIN DESCRIPTIVE TERMINOLOGY

A number of descriptive terms are used when discussing resins particularly the major ones for board—urea and phenol formaldehyde. Some understanding of these terms is necessary as they play important roles in the blending, forming, prepressing, and pressing stages of board manufacture.

Storage Life

The self-explanatory term, *storage life*, covers a very important resin parameter. Most liquid resins will have a much longer life if stored under refrigeration and, indeed, some phenolics must be frozen in order to be stored for a few weeks. Adhesive refrigeration, of course, is extremely difficult when manufacturing, transporting, and storing thousands of gallons of resin. Therefore, it is necessary to manufacture resin so that under "normal" conditions it will remain the same for a few weeks. Differences in ambient temperatures, e.g., between summer and winter can make such a task difficult.

In recent years, storage has become even more of a problem because resins have been improved to cure out at rates less than half of those in 1960. Such "hot" resins may be difficult to store because of the latent catalyst they contain, which may tend to start the curing reaction. The viscosity gradually increases as the resin goes "bad," and it finally becomes a solid. Storage tanks full of solid resins cause considerable problems.

For long storage periods such as when shipping overseas, the resins are spray-dried and only the solids are shipped. The resin must then be reconstituted for use.

Some dry phenolic resin is used in board manufacture. It also has storage problems in that ambient conditions can cause it to become lumpy. In such cases the resin must be reground to powdered form if it is to be used properly.

Tack

The term *tack* as applied to board resins carries such a variety of possible meanings it is useful to divide this important property into stages to discuss its function accurately. A subjective test is used for evaluating these stages. It is known as the cotton pick test and it is a widely used and fairly simple test for characterizing resin properties. It consists of drawing down a 1.5-mil (0.038 mm) film of any liquid resin on a ground glass plate, then pressing the film lightly with a 1-in. (25.4 mm) wad of surgical cotton or fiber tissue paper at frequent intervals. A fresh cotton surface is used for each contact.The time is noted when initial adherence occurs, when adherence under light pressure fails, the amount of pressure required for tack at that point, and the point at which no reasonable amount of pressure yields tack. Due to the effects of temperature, humidity, and air movement on test results, a control resin with known tack properties should be included in each test run.

Fluid stage: The liquid resin film prior to tack development. The resin wets but does not pull fibers in the cotton pick test.

Tack Point: The point in time when a film of resin ceases to be fluid and becomes intensely tacky, as indicated by intial fiber adherence in the cotton pick test.

Tack Loss Point: The point when a film of resin has dried sufficiently to cease being intensely tacky, as indicated by lack of fiber adherence under light pressure in the cotton pick test.

Residual Tack Level: The degree of tack retained by a dried film of resin for one to several hours after the tack loss point, as indicated by the amount of pressure required to cause fiber adherence to the dried resin film.

Dryout Point: The point where residual tack is lost. No amount of pressure will then cause adherence to the resin film.

Another version of this test is to draw a 3-mil (0.076 mm) film on a glass plate, using a Boston Bradley Bar. A cleansing tissue is used for the test rather than the cotton wad. The resin film is gently rubbed every two minutes with the tissue. The test notes four stages, A through D. Stage A is just after the film is formed and is wet and non-tacky. In Stage B, a little fiber is pulled from the tissue as some moisture is being evaporated from the resin. At Stage C, a firm touch is necessary for fiber to be pulled. The test must be performed quickly as heat from the hand will soften the resin film. When Stage D is reached, no fiber will be pulled from the tissue. Because of the subjectivity of this test, differences will occur between the individuals conducting it.

Relating these terms to a typical board process, for example, resin sprayed onto a fiber or particle should remain fluid and not develop tack in the blender. Instead, tack may develop at the former, at the prepress, or even at the loader depending on the requirements of the process. It has been found that tack loss can occur ahead of the former, in the hot press, or at any point between these operations.

Since residual tack governs the integrity of prepressed mats, it must also be controlled. Too little and the mat sloughs or sifts or fractures. Too much and it sticks to prepress rolls, platens, and hold-down bars.

Udvardy and Plester (1972) have worked out a system for numerically rating tack on a scale from 1 to 10 to assist resin manufacturers and plant operators in rating the tack and its profile throughout the production process.

To optimize the performance of any particleboard line, the tack requirement of each processing step must be assessed. Then a resin having this particular *tack profile* is developed for its use. This is important as, in most cases, it is necessary to tailor a resin for a particular plant depending upon its process, requirements, species, ambient conditions, etc. For example, different resins may have to be used for winter and summer operations.

Resin Flow

Flow is another property which is often interpreted differently. Flow means that ability of a resin to remain fluid under heat and pressure, wetting new surfaces and accommodating particle movement before being immobilized by loss of solvent and polymer growth.

Optimum flow is a valuable resin property, because it permits maximum board consolidation and internal bond strength with a minimum loss of surface hardness or strength due to precure. It is especially valuable in the manufacture of thick boards.

Until recently, resin flow was usually gained at the expense of curing speed. However, recent developments in both phenolic and urea resin design have made it possible to gain significant curing speed with little or no loss of flow under most normal processing conditions.

It is easily possible to have too much flow. This causes loss of board strength through excessive resin penetration into the furnish and retreat of adhesive solids from contact points of maximum pressure. This can be a significant problem in relatively porous species such as aspen. Particle and mat moisture content plays an important part in determining resin flow in a given system. The particle moisture must be balanced with the design of the resin for optimum results in the process.

Buffering

Buffering is an easily misunderstood term as it can be confused with pH. The fact that a resin has a high pH does not necessarily mean it is highly buffered, or vice versa. Actually, there is little connection between the two resin properties of buffering capacity and measured pH.

Buffering is defined as the degree to which a resin will tolerate contact with more acidic or basic materials without changing pH. Thus, it is helpful to buffer a urea resin for storage stability when it is desirable to have the pH remain near neutral (7.0 pH) where these resins store well. However, this becomes a distinct hindrance after the resin is applied and the acidity of the wood chip or a catalyst is necessary for lowering the resin pH and speeding the cure.

Buffering can be made to occur at virtually any pH level to serve a particular purpose. Figure 12.5 shows three examples of buffering curves. The average or center of the buffering levels are at pH 7.5, 5.5, and 3.5.

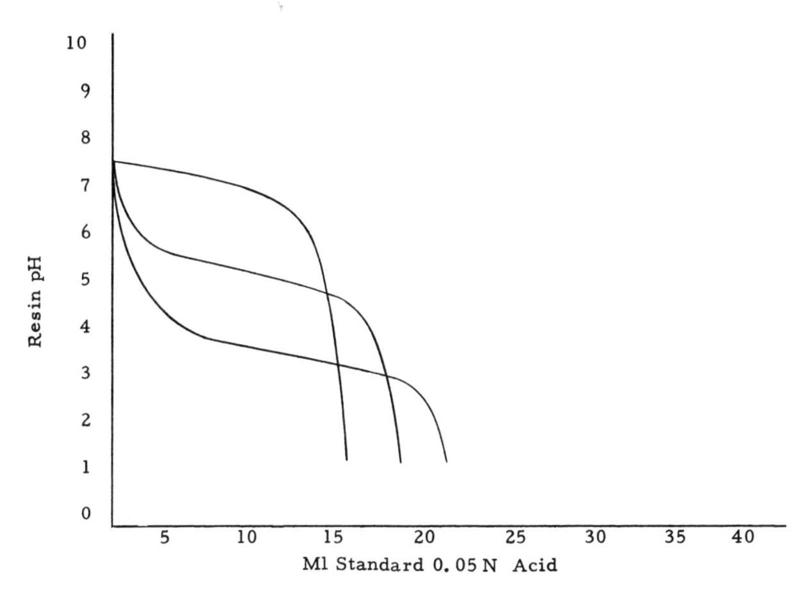

Figure 12.5. *Electrometric titration curves of UF resin buffered with different systems (Lambuth 1967).*

If three different portions of the same urea resin are buffered, as shown in the graph, to average pH levels of 7.5, 5.5, and 3.5, all three can be described as heavily buffered because of the quantity of standard acid or base required to change their pH significantly. However, the performance characteristics of the three buffered resins are vastly different.

The resin buffered at pH 7.5 will have an excellent storage life, but probably cannot be brought to curing pH by the action of wood acids alone. It will not cure rapidly, or possibly at all, unless an acid catalyst is added.

The resin buffered to pH 3.5 will have a very poor storage life, because of possible pH slide below the neutral range, even if initially adjusted to 7.5. However, it will cure rapidly at moderate press temperatures on even the least acid of wood particles.

The resin buffered to pH 5.5 will neither store well nor cure rapidly, regardless of its initial pH. It is buffered at a pH which is too low for good storage and too high for fast cure. For the resin at pH 7.5, the wood acids probably cannot overcome its buffering capacity near pH 5.5, and an acid catalyst will be required for normal cure times. Nothing can be done for its poor storage life.

These oversimplified examples show that buffering by itself has little meaning or purpose. It must be related to other resin performance properties. An experienced resin technologist can, by studying the wood pH level and the buffering data, determine through his experience which resins will work best. Unfortunately, no tried-and-true simple formula exists for exactly determining the "proper" resin. Only an experimental trial with the selected resin will determine its suitability.

Buffering is a two-way property, operating to hold pH constant in a given range whether the resin is being lowered in pH with acid or raised with base.

Resin pH

Once it was generally accepted that the higher the initial pH of urea resins, the better they would store. However, resins are available today which store well at pH 6.8 or 7.0, while others store poorly at 7.5 or 8.0. The mechanism of storage failure may not even be resin advancement in the normal sense, but an alkaline rearrangement called *pasting*. The storage capabilities of any specific binder resin should be judged on its merits and not on the amount by which its pH is above neutral. Under heat in the press, the resin pH will drop so that an acid condition results for curing the resin quickly.

Phenolic binder resins, unless separately acid catalyzed, will continue to be quite high in pH when stored and used. These resins derive their curing speed in large part from alkali content, but a pH check of finished phenolic particleboard will usually show wood acids have neutralized alkalinity during cure.

Viscosity

For board use, two aspects of resin viscosity are important, each for different reasons, and should be considered separately. One is the measured viscosity of the product as received and stored. The other is the fluid viscosity at the moment of spraying. Since resin viscosity is indirectly related to other performance properties, it is desirable to be able to produce board resins within the range of about 100 to 500 centipoise at 70°F (21°C). This permits the tailoring of other resin properties to the board process. Within this viscosity range, the resins should pump, store, and mix with reasonable efficiency, even when cold.

By contrast, resin viscosity should be low for application. Within reason and in general terms, the lower the viscosity the better. The thinnest resins in the range given above can easily be applied neat, while thicker resins must be warmed or diluted to reduce their viscosity for spray application. In many cases, the addition of wax emulsion alone will thin the resin sufficiently. Wax emulsions at lower solids content are routinely mixed with the resin for application to provide water resistance. With some of the newer short-retention-time blenders, however, high viscosity resins seem to function better.

Extenders

Extenders such as wheat flour are common in plywood glues. Their use is not widespread in composition board resins, but starch extenders are reported as additives in face resins in Europe.

Curing Speed

Curing Speed is another self-explanatory term. A conventional *gel* test to determine curing speed is conducted with 10 cc of the resin in a test tube set in a boiling water bath. The resin is agitated rapidly for one minute, using a small rod to bring the resin up to temperature, and then the time it takes the resin to gel is read. The resin in agitated occasionally during the test. A "hot" core resin may gel in five to ten minutes, while a highly buffered face resin will take much longer.

Urea-formaldehyde resins cure out at about the temperature of boiling water. Some parts of a mat being pressed into a board never get much hotter, with conventional press temperatures used with UF resins. This is particularly true when considering the board core around the periphery of the mat.

The curing of resin in a board is accomplished under different conditions than the gel test. Experienced personnel can correlate the gel test with actual pressing conditions. It may be necessary to place a small amount of catalyst in the resin under test to match board pressing conditions.

PF resins, as noted previously, require higher curing temperatures. Thus longer press times are required to allow time for the removal of moisture, which holds the core temperature of the board at about 212°F (100°C) until the moisture is dissipated.

Some like to classify curing speed as the number of seconds it takes to cure a millimeter of board expressed as thickness. An example would be 20 seconds per mm of board thickness. This can be misleading, however, as the catalyst and press temperature must be known for complete understanding of this description.

COMPARISON OF PAST AND PRESENT RESINS

Past Resins

It is of interest to review the resins used in the particleboard industry of the mid-fifties and to realize the improvements made. Particleboards were mainly homogeneous, with some layering through mechanical particle separation. In each case, resin requirements were basically uncritical with respect to tack control and other present-day refinements. In many cases the board line was changed to accommodate the resins available.

For instance, when resins were found to lack the residual tack to prevent the edges of mats from sloughing, at least one manufacturer solved the problem by simply bending up the edges of his cauls to support the mat edges (such cauls are still in use). The fines sifted through the mat, resulting in a two-layer board. But the board sold, probably bottomside up. In other cases, water and even urea solution was sprayed on the edges and on top to increase mat strength and for better consolidation.

The production rate was leisurely by today's standards. Typical press times for ¾-in. (19 mm) rough boards were 12 to 15 minutes at 300°F (149°C) for urea resins and 14 minutes and over at 350°F (177°C) for phenolics.

Degradation of hot board strength properties was a serious problem with urea resins, partly because of these long press times and partly due to hot stacking the boards themselves. Except for occasional *stickering*, cooling devices were seldom used. In fact, hot stacking had its purpose as an extension of phenolic resin cure cycles. Quickly cooled phenolic boards were much lower in strength properties than those which had been hot stacked.

The particleboard resins of the mid-fifties were basically conventional ureas and phenolics, reflecting the resin technology and board process requirements of the day.

Present Resins

Today's urea resins are considerably more sophisticated in performance than those used initially. They reflect the tremendous growth of the industry in terms of volume and diversity of individual plants. They are also lower in cost, even with recent price increases due to the raw material shortage problems, since they have over twice the curing speed. Resin tailoring for individual plant requirements is almost universal now. Among the resin properties most frequently optimized for individual process requirements are: viscosity, tack, flow, reactivity, pH, buffering, storage stability, foam control, odor, and wax compatibility. These properties are sometimes even adjusted to help overcome existing but basically undersirable operating conditions, such as foam caused by faulty transfer pumps, or poor distribution caused by improper spray heads.

Thus, today's ideal UF particleboard resin is in reality many different resins, each optimized for a particular board-forming process and furnish type. For example, today's resins, within certain limitations, can be designed to accomplish the following typical options:

1. Resin tack can appear or disappear at almost any point in the forming line for the purpose of making satisfactory mats from most any particle configuration and particle size distribution.
2. Resin reactivity can be adjusted over a wide range to accommodate virtually any reproducible cure condition, wood species, or board construction.
3. Allowances can be made for slow or fast press closing times without undue sacrifice of curing speed.
4. Viscosity can be adjusted for optimum dispersion, whether sprayed neat or dilute, hot or cold, alone or with wax.
5. A variety of storage requirements can be successfully met with respect to both resin thickening and settling.
6. Low formaldehyde odor resins can now be provided for most, if not all, board processes. Very significant advances have been made in this respect.
7. Significantly improved hot board degradation resistance or "mid-exterior" durability can be provided in urea binders at a moderate premium cost.

Some properties of urea and phenolic resins tend to be mutually exclusive, meaning they cannot be optimized in the same resin at the same time. Instead, they are generally reduced to a workable compromise. Examples of these opposite pairs of properties are: reactivity and storage life, late tack development and early tack loss, early tack loss and low odor, curing speed and durability.

A technical solution to at least one familiar pair of opposed properties, speed and flow, has been worked out for both types of resins, permitting faster press times with good board consolidation. Solutions to other mutually exclusive pairs of properties may well result from continuing research on binder resins.

The present phenolic particleboard resins also represent significant technical advances. The changes in resin performance may not seem as dramatic as those of urea resins, due to inherent differences in their performance properties. However, they still reflect real progress in developing improved performance characteristics for board use. As with the urea resins, technology has been developed for phenolic resins which permits selective modifications for particular fiberboard as well as particleboard applications.

To name a few improvements, the press times of today's phenolic board resins are roughly half of those of the early days, or even less, with catalytic systems now developed for even faster press times. Applied resin costs have also been reduced even with recent price increases. The problem of wax compatibility has been significantly reduced, if not overcome. At one time wax emulsions could not be blended with phenolic resins because the alkaline-based phenolic would break down the wax emulsion, causing fluid lines and spray nozzles to plug. The forming tack of some phenolics approaches that of urea resins. Adhesion to over-dried particles is significantly better.

CATALYSTS

Catalysts can be used to hasten curing speed. In some cases they may be blended with the resin immediately before spraying, or special mixers can blend the resin and catalyst right at the nozzles. The use of such a technique is important as plant breakdowns do not result in the loss of extensive quantities of catalyzed resin. Such added catalysts are acid salts for ureas. Catalysts can be a part of the prepared resin and not a simple addition of acid salts as noted previously. Such catalysts are termed *latent*, because they do not become active until heat is applied in the press. In the highly competitive resin business, the latent catalytic systems, as well as resin designs and tailoring schemes, are closely guarded secrets.

Catalysts are definitely not a performance cure-all. They cannot improve production rates and board strength properties simply by an arbitrary addition. The proper level of catalyst may vary tenfold or more from one process to another. For instance, from 0.05 to 0.5% ammonium sulfate (a widely used catalyst) or its equivalent in other materials may be required, depending on exact plant conditions. As a case in point, among plants in present operation, several are successfully using catalyst levels that would be harmful to a number of others.

The latent catalysts are much more complex because of the shipping and storage problems. Under these conditions, the catalyst must be able to remain in contact with the resin for extended storage periods without effect, and then be quickly activated when placed in the press.

One simple method for catalyzing liquid phenolic resin is to add resorcinol, but this is expensive. Paraformaldehyde blended with wax emulsion has been successfully used as a catalyst for powdered resin. This is of importance in speeding the resin cure in phenolic-bonded waferboard. Other catalysts are being used but their makeup is not known. Not as much research has been devoted to phenolic resin as urea because of the heavier use of urea. However, research on phenolic is expected to increase as composition board moves more into the structural field.

POWDERED RESIN

One other special category of binder, mentioned previously, is the powdered or spray-dried products. Both urea and phenolic resins are presently available in dry powdered form, generally containing an appropriate catalyst. Technically, these are known as B-Stage resins; they melt and flow in the press for good distribution before curing.

Spray-dried urea resins have been used primarily in the granuplast or wood flour molding industry rather than in the particleboard industry. One plant is reportedly using powdered urea resin. Powdered phenolic resins, generally classified as industrial resins because of their many non-wood applications, are used as binders for waferboard and certain specialty particleboards.

Powdered resins are also used in conjunction with liquid resins to provide increased binder solids in particular layers or areas of both particleboard and hardboard. For this purpose, each powdered resin type, UF and PF, is compatible with almost any liquid resin of its corresponding type. Spray-dried UF resins generally equal the curing speed of UF liquid resins, while powdered PF resins are considerably slower curing than today's liquid phenolic resins. Cross-mixtures of phenolic and urea resins in liquid or powdered form have not been practical because each interferes with the curing mechanism of the other. However, as mentioned previously, a new European development combines these plus melamine into an exterior resin.

The principal advantage offered by powdered resins of either type is the freedom to use wetter furnish, to a maximum of about 20% moisture content in conventional equipment or up to 50% in special presses. Conversely, powdered binder resins also permit very low moisture input levels, relative to binder solids, particularly in high-density board products and wood molding compositions where blistering is an extreme problem.

Among the disadvantages of powdered resins is a tendency to sift through mats while forming and handling, and much slower cure times for the phenolics compared with present liquid phenol-formaldehyde resins. Also, their price is considerably higher due to the added cost of spray-drying or casting and grinding.

Today's fast-curing liquid phenolic resins do not lend themselves readily to spray-drying or powdering. Even on moist furnish, the removal of solvent upsets their flow properties beyond any practical point. However, other techniques are now available which may permit the conversion of these reactive liquid resins to dry powders with little change in their performance properties.

SPECIALTY PRODUCTS

There are a number of "specialty" resin binders available or emerging commercially. Several of these special binders designed for "nonstandard" particleboard performance requirements are discussed below.

Phenolic resins which impart termite, mold, and rot resistance to particleboards for semitropical applications are available. Termite-resistant urea resins, although technically feasible, are basically inappropriate for this type of end use because rot and termites thrive in a warm, moist environment, which itself tends to degrade urea binders and thus limit the success of the application for purely mechanical reasons. Phenolic resins are far more resistant to this gradual, long-term degradation. Indeed the wood degrades before the resin.

High-temperature and moisture-resistant urea binder resins are presently available which make possible "mid-exterior" durable particleboard products. These products are considerably faster-curing than phenolic resins and are generally similar to ureas in handling properties. They give ASTM D 1037 accelerated aging test results of about 25 to 35% strength retention, compared with the 50% retention required for exterior particleboard binders. By contrast, straight urea resins fail completely in the first part of the first test cycle. The relative price of these resins, like their durability, is about midway between urea and phenolic resin prices.

A special application of these mid-exterior resins is their use in the core of layered phenolic particleboards, replacing the normal phenolic core binder. They permit much faster cure rates than those of present all-phenolic boards with near exterior durability. High-density boards made in this way are excellent core stock for high-temperature repressing with overlays. By actual exposure test, they may also prove to be sufficiently durable for certain exterior and wet end uses.

OTHER ADDITIVES

A number of other additives are used in composition board of which the most notable is wax or, in other words, a sizing agent. Wax is used in all boards in the United States and in boards of many other countries. However, some countries do not use it for certain products; for example, Canada does not require wax for furniture board. Other additives used in minor amounts are preservatives, fire retardants, and impregnating resins. Various dimensional stabilizers have been tried over the years, but no effective ones outside of the impregnating resins have been found to be economical, and these are economical only in special cases.

Wax

The use of sizing agents, such as wax emulsion in the production of board, appears to be a direct outgrowth of procedures followed in the paper, insulating board, and gypsum board industries. The primary need for a sizing agent is to provide the finished products with resistance to aqueous penetration. In composition board, wax emulsion provides excellent water holdout and dimensional stability in the board upon wetting. These properties are important both to subsequent secondary gluing operations, and to provide protection upon accidental wetting of the board during and after construction. In cases such as with siding, some wetting is to be expected, probably by a wicking action from the bottom side of the siding during rainstorms, and protection is provided by a sizing agent.

The water repellency provided has practically no effect upon dimensional changes or water absorption of boards exposed to equilibrium conditions. For short terms, however, as epitomized by the standard 24-hour water soak test, the presence of wax is very significant. Such repellency is only in the case of water as the absorption of water vapor is not hindered. Many investigators have found

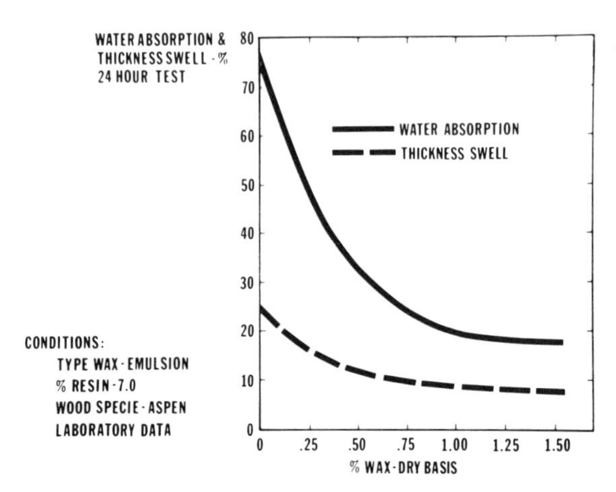

Figure 12.6. *Effect of wax on particleboard sizing–water absorption and thickness swell data for aspen (Albrecht 1968).*

that the addition of wax retarded water absorption temporarily, but had little effect when the board was exposed to equilibrium condition by water soaking or high humidity. Therefore the presence of wax primarily influences the rate at which the board reaches equilibrium. This property offers the potential advantage of extending the period of time for stresses to relieve themselves in structures.

Another function of wax emulsion in the production of board is to provide slip characteristics to the wood surfaces, thus reducing the tendency of blow pipes, forming stations, blenders, etc., to plug. Controversy surrounds this benefit but it reportedly has enabled plants to increase production without a major equipment expenditure.

Theory of Sizing: Figures 12.6–12.8 illustrate the effect that wax plays in providing resistance to water penetration. In the 24-hour soak test and 7-day humidity test, unsized particleboard is definitely inferior to the sized product. The species of wood involved has a strong influence on the performance of sizing agents. Boards produced from Douglas fir commonly use 0.25 to 0.5% wax solids. Aspen, on the other hand, usually requires 0.75 to 1.25% wax. As noted, wax emulsion does impart sizing characteristics to board. However, a great deal is yet to be learned concerning vapor resistance, although much has been learned about water absorption patterns and characteristics.

The initial penetration of liquids in board is through the voids between the wood particles. These, in effect, are small, irregularly shaped capillaries. The creation of new liquid surfaces, which must occur in water penetration, requires work because cohesive forces between the molecules in the liquid tend to limit the surface of the liquid to a minimum. This work is accomplished by capillary pressure plus whatever external pressure may be involved. Capillary action is illustrated in Figure 12.9. When the liquid wets the surface (contact angle $<90°$) the liquid rises in the capillary tube. When the liquid does not wet the surface (contact angle $>90°$), a depressed meniscus results.

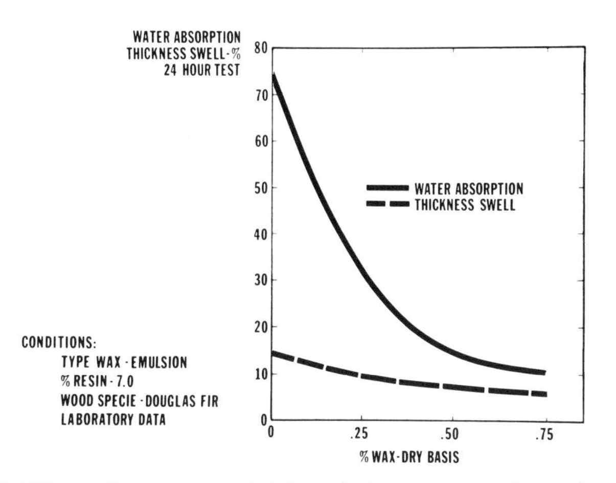

Figure 12.7. *Effect of wax on particleboard sizing–water absorption and thickness swell data for Douglas fir (Albrecht 1968).*

The amount of free water that enters a board is normally inversely related to its density. A low-density board, for example, usually has larger interstices between fibers. This assumes all other variables are constant. For example, a 55-lb/ft^3 (0.88 sp gr) dry-process hardboard made from aspen may absorb less water than a 42-lb/ft^3 (0.67 sp gr) particleboard made from Douglas fir.

Sizing, in providing resistance to water penetration, is a rate phenomenon. The rate of liquid flow through a capillary is a function of radius, pressure, and depth of penetration. Although a decrease in the radius of a capillary increases the pressure (height of capillary rise), frictional forces are even greater and rate of flow decreases. A demonstration of this phenomenon is shown in Figure 12.10. Under uniform conditions of board production, an increase in density results in improved

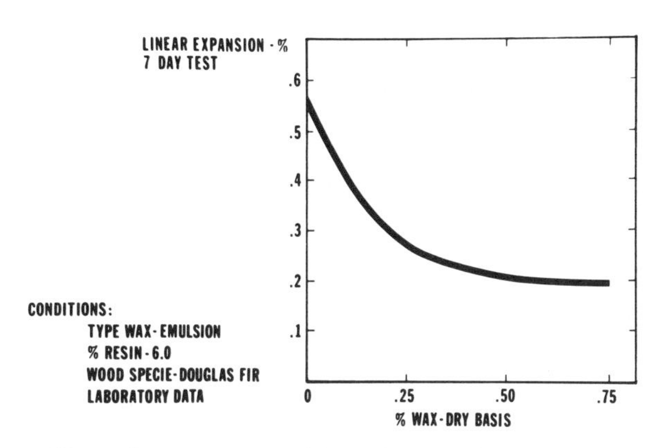

Figure 12.8. *Effect of wax on particleboard sizing–linear expansion data (Albrecht 1968).*

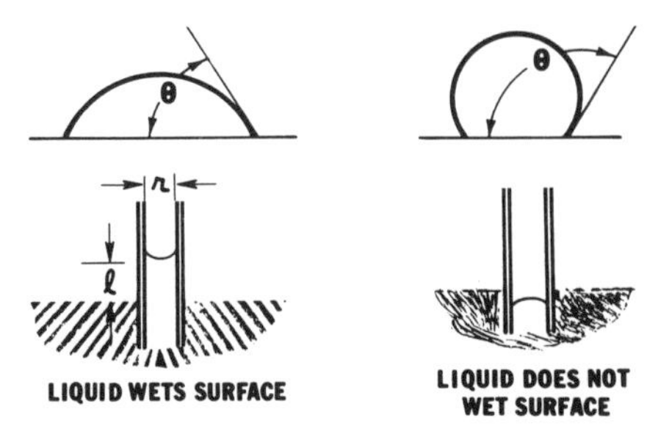

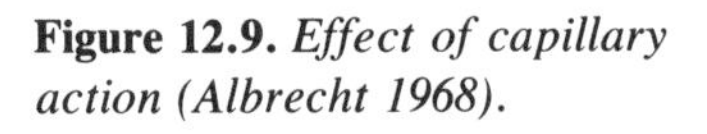

Figure 12.9. *Effect of capillary action (Albrecht 1968).*

water-absorption properties. This is seen constantly in board as it is very difficult for water to penetrate into a more densified product in 24 hours. Taken to equilibrium, as mentioned before, the curve will reverse itself as the higher-density board has more wood and therefore more potential for swelling. This is a good example of the short-term value of sizing.

The pressure developed in the capillary increases with decreasing radius and decreases with increasing contact angle. The primary effect of sizing is to make the fiber surfaces less wettable by the penetrant. In observing penetration under the microscope, the pattern does not appear to be the simple flow of liquid through a tube. It seems to jump from one void to another. It is possible the wood fiber surface to wood fiber surface bonds block the capillaries, making it necessary to cross these in order to continue penetration.

It is unlikely that water repellency can be attributed to the coating of fibers with wax. For example, 0.25% wax cannot cover the large surface area involved. The exact mechanism of sizing is often a controversial subject. There is general agreement, however, that the presence of heat is necessary to obtain good sizing. It is assumed that during the press cycle two things occur: (1) steam distillation of the wax occurs to some degree within the mat, resulting in vapor distribution of the wax, and (2) liquefaction of the wax permits it to wick over a large area, similar to a penetrating oil.

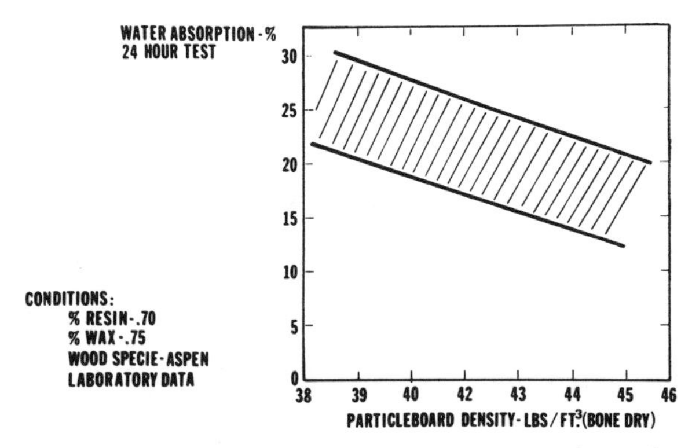

Figure 12.10. *Effect of particleboard density on water absorption test results (Albrecht 1968).*

In order to obtain efficient sizing of board, it is necessary to coat as much surface area as possible with a thin film of wax. Factors influencing this efficiency are spray application, blending, hot pressing, and hot stacking. Inefficient spraying and blending techniques can many times be overcome through long press times and board stacking periods. An ineffective spray application may result in large volumes of wax being applied to a small surface area. However, if press and hot stacking times are of sufficient length at temperatures above the melting point of the wax, flow-out occurs resulting in more efficient sizing.

An example of this is the relative efficiencies of molten waxes versus wax emulsions. Laboratory work has demonstrated that equal dry pound applications of both products result in equivalent sizing, if sufficient temperature and time are allotted for equilibrium conditions to be obtained. The primary reason wax emulsions have replaced molten wax is the higher efficiencies obtained through improved volumetric distribution of wax emulsion particles over larger surface areas. Under most plant operating procedures, insufficient time–temperature conditions prevail for effective utilization of molten waxes. Mathematically, increasing emulsion volume at constant dry-basis furnish levels should lead to more efficient sizing. This is one important reason for the incorporation of the wax emulsion in the resin furnish.

It should be noted that many plants used molten wax in the pioneer days of the particleboard industry but that few do now. The most noticeable users appear to be waferboard plants where the sticky wax helps the dry powdered resin adhere to the wafers. The press times are slightly longer and hot stacking is practiced, which helps use the wax efficiently.

Wax emulsions consists of minute wax particles of about 1 micron in diameter. Each wax particle is a homogeneous blend of all the waxes used in the original formulation. Each particle is enveloped by a surfactant containing an ionic charge that repels other similarly coated wax particles.

If the emulsion works properly, the water in this droplet is absorbed by the particles as soon as they come in contact. The wax, surrounded by the surfactant, remains on the surface. For the wax to flux or spread out and adhere to the wood fibers, the surfactant must release the wax. Surfactants cover a wide range of materials. Ideally, the surfactant selected should become ineffective when subjected to heat and therefore not function as a residual rewet agent. The heat sensitivity of the emulsion is critical. Utilization of the properties of the wax depends upon the more or less complete destruction of the surfactant in the hot press cycle.

Once freed from the surfactant, the wax particle fluxes and attaches to the wood fibers. How well this is accomplished depends upon the characteristics of the waxes as well as the surfactant. Figure 12.11 shows how a droplet of one emulsion will remain on a chip pretty much as it was deposited after being heated, while a droplet of another spreads out over an incredibly large area of fibers.

The fantastic array of waxes available to the petroleum refiner provides an almost infinite variety of choices in formulating an emulsion. Figure 12.12 represents the wax portion of a refinery's distillation column, arbitrarily separating paraffinic waxes, intermediate grades, and microcrystalline waxes. This column represents only the waxes in one crude oil; and the proportions of various waxes

Figure 12.11. *The fluxing property of droplet varies from one emulsion to another (Berrong 1968).*

and, in fact, the characteristics of each wax vary depending on the crude oil source. There are literally hundreds of waxes available to the laboratory developing a sizing emulsion.

The melting point of a wax is related to its boiling range as indicated on the distillation column. This relationship is particularly true for the family of paraffinic waxes, since their molecular composition consists of normal paraffin-type hydrocarbons varying only in chain lengths. In addition, paraffin waxes are all tested by the same method—melting point based on the release of latent heat as crystal formation occurs during cooling.

Intermediate waxes and microcrystalline waxes contain a complex mixture of non-normal hydrocarbons. They do not have a sharp melting point and therefore are tested under a different procedure than that used for paraffinic waxes. It is this difference in test procedures that prevents a direct correlation between boiling point and melting point.

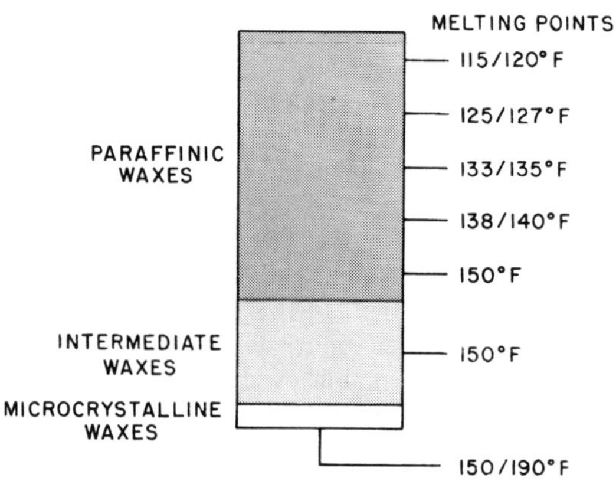

Figure 12.12. *Distillation separates waxes into three basic families (Berrong 1968).*

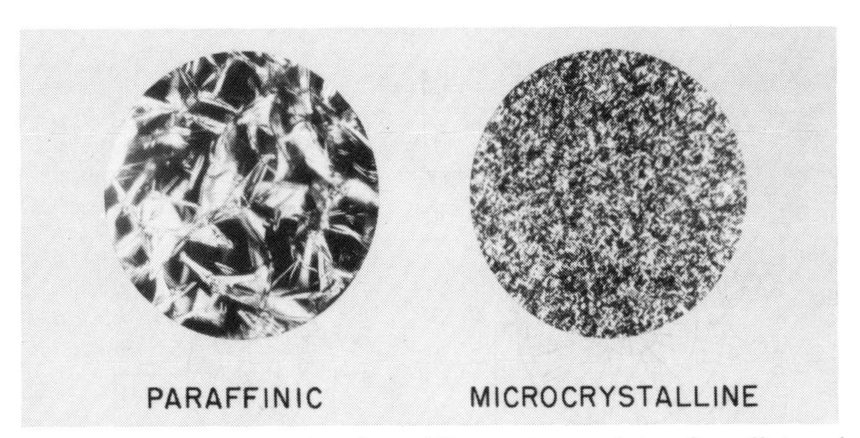

Figure 12.13. *Microscopic view shows difference in crystal size of paraffinic and microcrystalline waxes (Berrong 1968).*

A wax at the bottom may have as low a melting point as one in the intermediate range. And waxes with the same melting point can have entirely different sizing capabilities. This picture is further confused because melting points of different kinds of waxes are determined by different test methods.

Sizing properties are directly related to the type of crystals present in the wax. Paraffin waxes have long needlelike crystals that provide the best water resistance. The short, side-branched crystals of microwax are relatively ineffective for use as a sizing agent.

A microscopic view of a paraffinic wax and a microcrystalline wax (Figure 12.13) shows how different waxes are in their physical structure. Not surprisingly, this difference is reflected in sizing capabilities. Yet certain properties of both are useful in a board-sizing blend.

The larger crystals provide better moisture resistance. But some of the smaller crystal waxes are desirable to enhance handling characteristics and minimize paintability problems. They help prevent oil bleeding, which would interfere with enamel painting, and they minimize the tendency of large crystals to slow up lacquer drying.

There are exceptions to any generalization relating the physical characteristics of wax to sizing capabilities. These characteristics—melting point, boiling point, oil content, molecular weight, chemical structure and so forth—give the laboratory formulator useful clues to how a particular wax will contribute to effective board sizing, but, in every case, the final formulation is made by trial and error combinations.

Slack Wax: A term used loosely in the industry is *slack wax*. Formerly, *crude scale wax* was a term used to identify unrefined waxes having an oil content up to 6% and the term slack wax covered waxes of higher oil content (Figure 12.14). This nomenclature had significance to refinery people since it identified more than the oil content—it also indicated whether the wax was a by-product of light neutral or heavy neutral lubricating oils. The properties of the

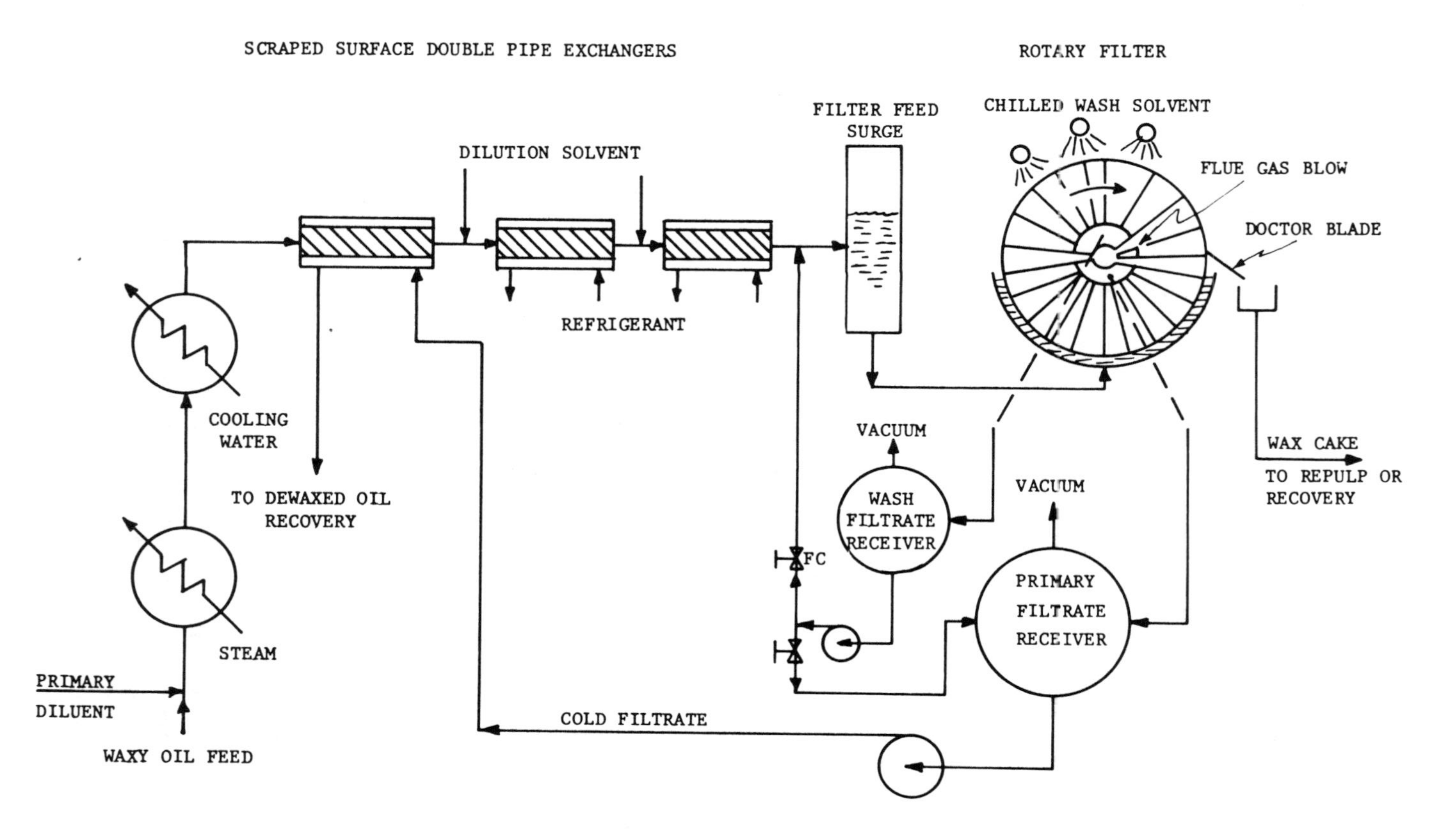

Figure 12.14. *A typical schematic of wax production. (General information, courtesy Mobil Oil Corp.)*

wax from each stream are considerably different, particularly the crystalline structure. Over a period of years, however, the term slack wax has become generally used by various industries when referring to the total family of unrefined wax.

In respect to sizing of composition board, slack wax can be used in either the neat form or applied in the form of a water-based emulsion. The price per pound strongly favors the use of neat wax, and its usage was fairly common in particleboard prior to 1960. As technology on emulsions improved, it was usually found that equal sizing could be obtained with one-half to one-fourth of the wax formerly applied in the neat form. In terms of cost per thousand square feet, the economics of emulsions versus neat wax currently vary from a break-even situation to a definite cost reduction for emulsions.

Other factors also enter the picture. Neat wax requires a high initial capital investment for storage tanks, pipes, and pumps that must be heated and insulated. This same system requires more maintenance and is more prone to failures than an emulsion system. The design of the blender is also important since neat wax can seldom be sprayed more than a distance of 12 to 18 in. (305–457 mm) without the molten droplets solidifying while airborne. If solidification occurs, positive attachment to individual particles is not obtained, resulting in loss of controlled distribution onto the wood.

Since consideration should also be given to other properties such as paintability, it should be noted that wax can adversely influence both lacquer drying time and the gloss properties of enamel. The type of wax becomes involved, but the most important point is the quantity of wax that exists on the surface of the board. The use of neat wax creates two potential problems. First, a higher percentage is needed to size the board. Second, there is a tendency for wax deposits to build up in the blender. These deposits periodically break loose and eventually become hot pressed into the mat. The economics and avoidance of operating problems have discouraged most particleboard and fiberboard plants from using neat slack wax. One major producer continues to use slack wax in all his plants. It is estimated that it is not used in more than 10% of the dry-process board produced in this country.

Wax Emulsions: Generally speaking, the more fully refined and purer wax grades impart more effective sizing characteristics to board. However, the process steps required in removing impurities, such as oils, increase their unit cost. Wax emulsions are petroleum products. Crude oil is distilled, giving gases, gasolines, fuel oil, paraffin distillates, and residue. Paraffin distillates provide the major source of wax used in board.

Predominantly crude scale wax or slack wax is used in the production of wax emulsion for board use. Manufacturers of wax emulsion use a variety of emulsification systems. These systems may be classified as: anionic, cationic and nonionic.

The type of emulsifier utilized is dependent upon the end-use requirements. Conditions of resin compatibility, mechanical stability, cost, retention, etc., determine the most optimum system. The emulsificaton system makes the hydrophobic wax suspended in water as a colloidal dispersion and such a material is easier to handle in the board plant's process. In some cases, sizing properties may be enhanced by the emulsifying system. However, certain emulsifying agents can

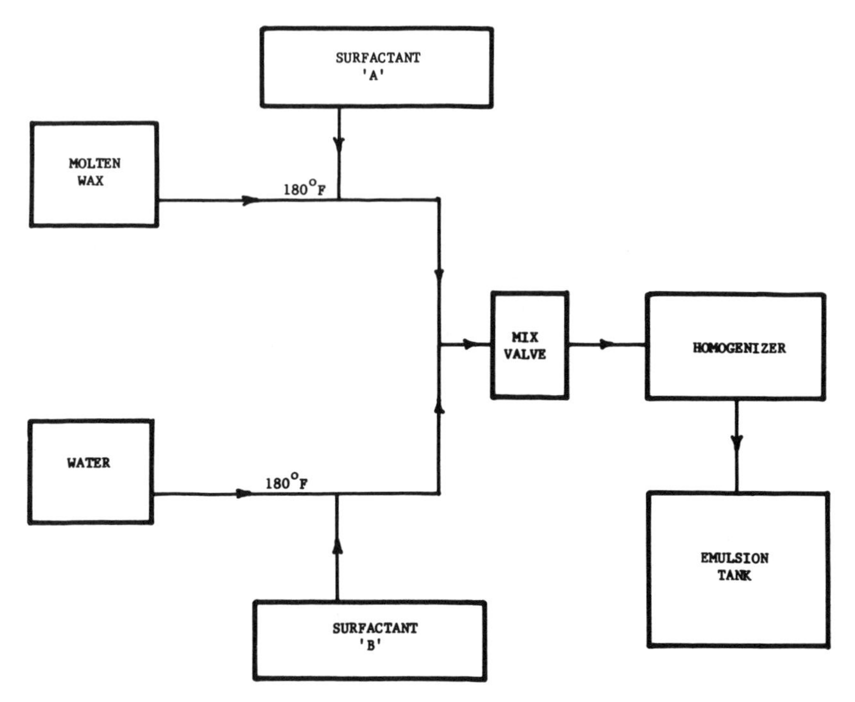

Figure 12.15. *Diagram of in-plant production of wax (Mohr 1968).*

detract from sizing, making it necessary to minimize their use in the wax emulsions used in board processes.

Most of the emulsions are supplied ready to use by large chemical manufacturing firms. However, wax emulsion can be manufactured at the board plant. This is rarely done in the United States because of economics. In the European industry, this has been an accepted part of the design concept of many board plant systems.

Growth of in-plant systems has been hindered by many factors. Some of these are: reservations concerning performance of a "home-made" emulsion, marked deficiencies in early equipment proposed as suitable for production, and lower commercial prices for wax as competition increased for this lucrative market. Keeping abreast of changes in emulsion technology is also a problem.

Concern has been expressed about three areas of performance: handling stability, sizing efficiency, and the paintability factor. Stability was of concern because even some of the commercial emulsions have had difficulty in this respect. Sizing efficiency was also a doubtful area, although the variation between different commercial emulsions is narrow. The paintability area was an unknown factor since there is no conclusive evidence that such a problem exists at modern and relatively low wax addition levels.

One simple process in use is outlined in Figure 12.15. Molten wax is metered from a hot tank and surfactant *A* is metered into this stream. Water is metered from a small surge tank and surfactant *B* is metered into this stream. Both streams then merge, pass through a mixing valve, and then through the homogenizer. Both wax and water are brought up to 180°F (82°C) by system heat exchangers before the surfactant is added. This discharge of completed emulsion is piped into an emulsion tank. Each material is pumped by a separate pump and the flow rate is monitored electrically. Variation from the set flow level of any material will stop the entire unit.

Figure 12.16 is a diagram of the homogenizer head. It functions somewhat like a three-piston metering pump and is capable of discharge pressures up to 5000 psi (34.5 MPa). The raw material comes into the intake manifold, is drawn into the pump chamber, and is discharged into the exhaust manifold. The emulsion is formed when this discharge is forced through the spring-loaded disc orifice at high pressure.

Figure 12.17 is a diagram of the storing and handling system. No attempt is made to cool the emulsion as it comes from the machine. It is stored and handled hot, and no change in sizing efficiency as a result of this has been noted.

Other Factors Affecting Sizing: Many waxes and wax emulsions are available, and a wide variety of sizing action is possible. Two other factors are involved, assuming the wax is uniformly distributed over the particles, and these are species and hot stacking.

The wood species used in board production determines the amount of wax required to obtain a given degree of sizing. Factors involved are wood specific

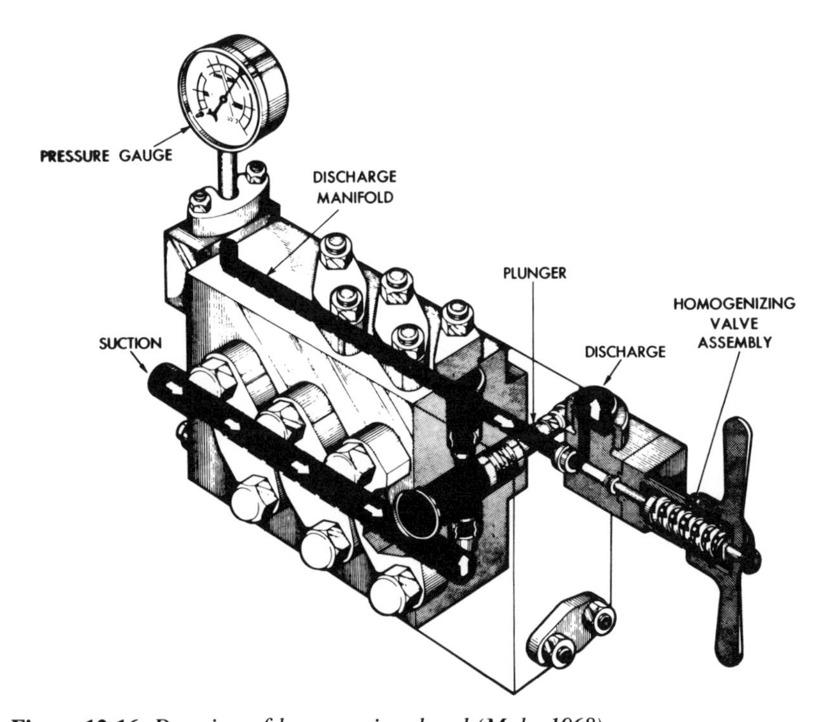

Figure 12.16. *Drawing of homogenizer head (Mohr 1968).*

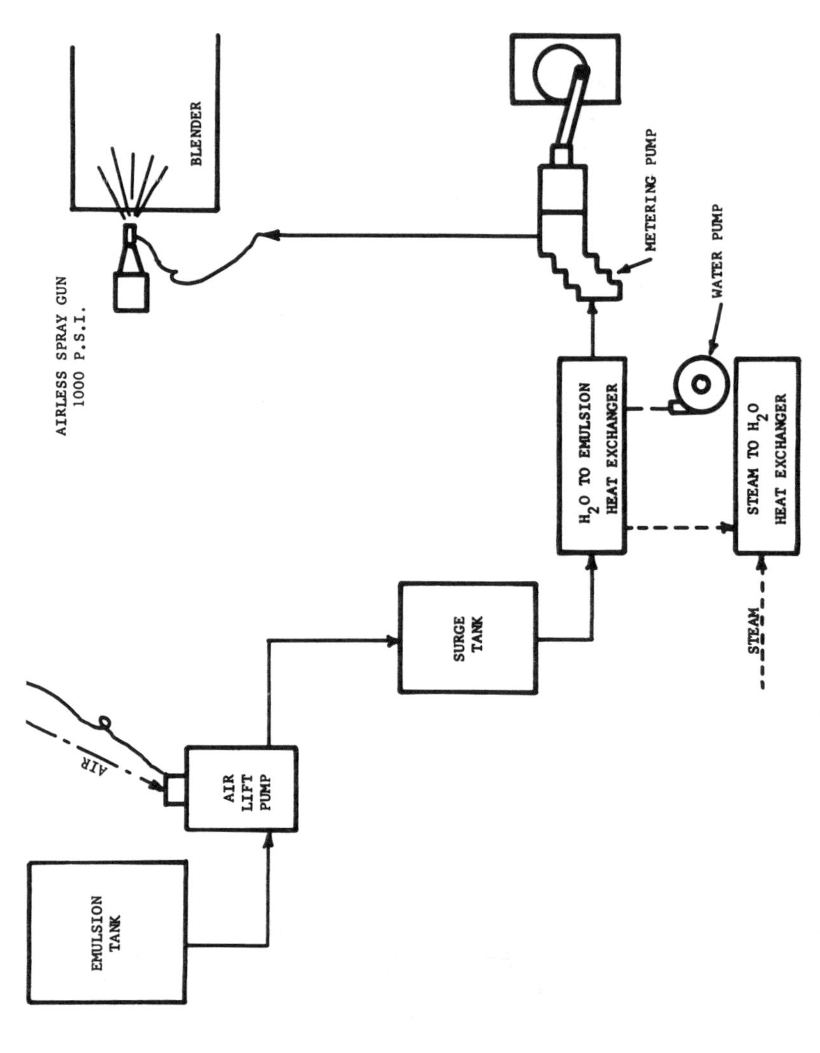

Figure 12.17. *Diagram of a storing and handling system (Mohr 1968).*

gravity, pH, resin content, internal structure of the species, and particle geometry resulting from plant processing. In addition to structural differences among species, some have natural resinous or waxlike extractives which impart a measure of water repellency to the boards made from those species.

Differences in thickness swell and water absorption are shown for Douglas fir, southern pine, and aspen in Figures 12.18 and 12.19. Twenty-four-hour water soak data are always difficult, if not impossible, to interpret. Note that southern pine has greater thickness swell than Douglas fir but that their water-absorption values are virtually identical. Also fir is better than the other two species in thickness swell without any wax in the boards. This is a good illustration of the species effect. Douglas fir is one of the best, if not the best, commonly used species as far as water-resistant qualities are concerned. It can easily be seen that mixtures of species, especially if the mix is unknown, can cause problems with the amount of size needed in the processing.

Linear expansion is primarily related to orientation of the fiber within the mat. Fibers oriented horizontally result in minimum linear expansion. An opposite extreme is represented by the extrusion method of producing board, where vertical orientation of fibers results in severe linear expansion.

One concern is that the inclusion of wax may lower some of the strength properties below the minimum values of the appropriate specification. Many researchers have worked in this area, and the consensus seems to be that wax quantities of 1% or less have little or no effect on the strength properties of the boards. As wax levels become higher, strengths are sometimes lowered, and this must be compensated for by additional resin, an increase in density, or a change in particle geometry.

Strength degradation from use of wax is normally quickly detected in routine production-control tests, and necessary corrections are made. Other possible side effects from the use of wax may not appear until the board is fabricated into its end use. These are more troublesome as the properties may not be evaluated by the manufacturer of the board. Two problems of this type that have been mentioned are interference in secondary gluing and difficulties in priming or painting. Terminology such as *wax migration* to the surface and *wax on the surface* have been applied to these problems. Perhaps these were problems at one time, but modern technology seems to have solved these problems. It is claimed, however, that the wax in the furnish helps to keep the board from sticking to the caul plates or hot press. In some cases, improved machinability of the board has also been reported.

The effectiveness of a wax emulsion in sizing board is primarily one of time and temperature. With steadily increasing production rates, hot stacking became more of a factor with respect to sizing efficiency. Hot stacking is the predominant requirement for effective sizing under these conditions. Board maintained at temperatures in excess of 120°F (49°C) for periods of one to four days allows for migration of wax over larger surface areas. However, the process of hot stacking has two prime deterrents: (1) possible loss of strength properties resulting from overcure of the resin, and (2) need for increased warehouse space. Most plants, therefore, now have board coolers. This has resulted in an apparent loss in sizing due to elimination of the benefits of hot stacking. Corrective measures have therefore had to be taken in the manufacture of the wax emulsion.

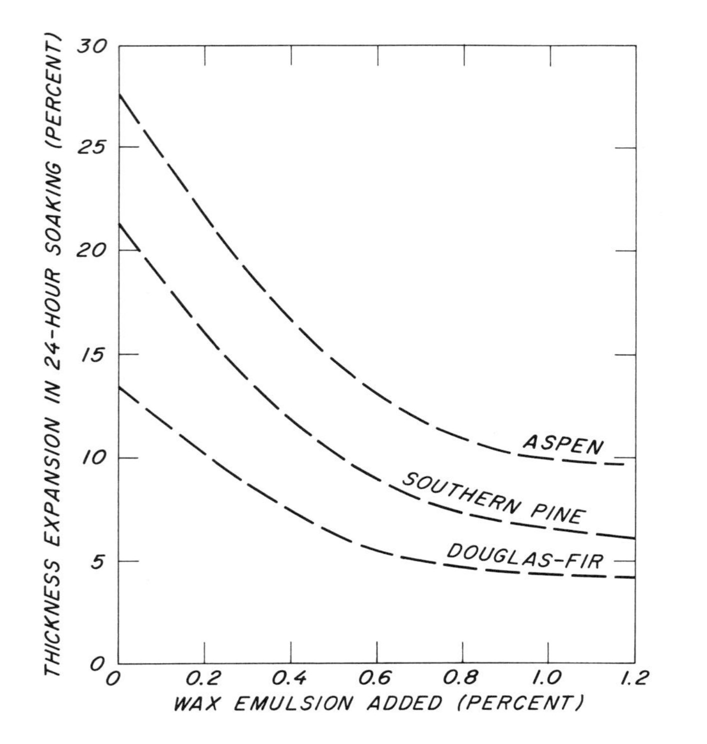

Figure 12.18. *General relation between amount of wax in particleboard and thickness swelling in 24-hour water immersion (Heebink 1967).*

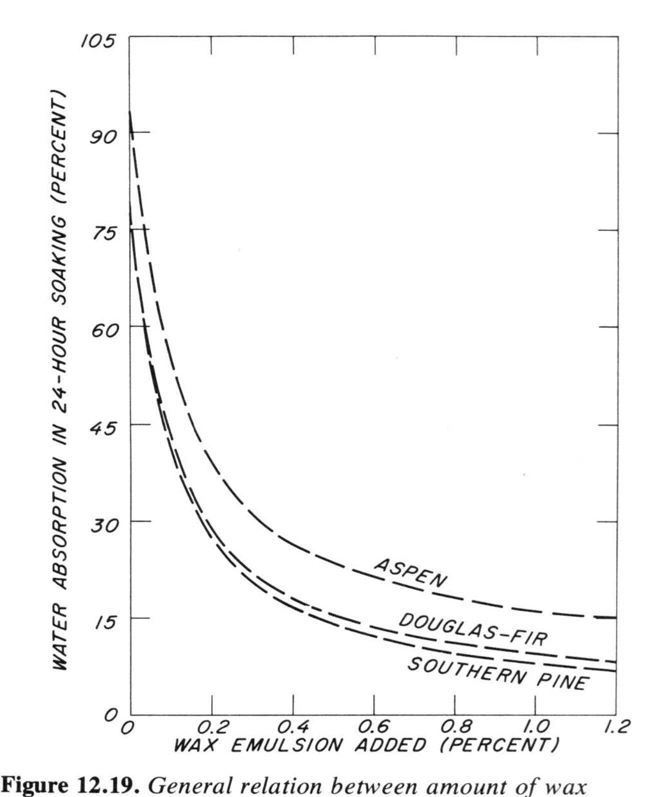

Figure 12.19. *General relation between amount of wax in particleboard and water absorption in 24-hour water immersion (Heebink 1967).*

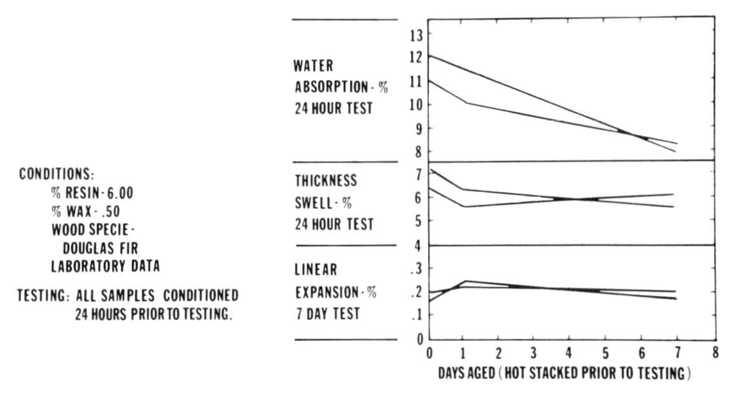

Figure 12.20. *Effect of hot stacking in water and/or moisture-absorption properties on ½-in. (12.7 mm) homogeneous particleboard (Albrecht 1968).*

Figures 12.20–12.22 demonstrate the effect of hot stacking board at temperatures of 125° to 140°F (52°–60°C). Figures 12.20 and 12.21 represent data collected on board produced in the laboratory. Boards produced in this work were cooled and then hot stacked at 125°F (52°C). Prior to testing all boards were conditioned for 24 hours. As can be seen, water-absorption results continued to improve over a seven-day period when hot stacked. Thickness swell and linear expansion results leveled off after 24 hours of hot stacking. Figure 12.22 demonstrates the effect of hot stacking on plant-produced ⅝-in. (15.9 mm) homogeneous particleboard. Four samples of board were hot stacked (after cooling) at 125° to 140°F (52°–60°C) for periods ranging from zero to eight days in duration. Water-absorption data continued to improve during the entire test period. These results were obtained with one type of emulsion. No attempt was made to determine the influence of different emulsions under hot stacking conditions.

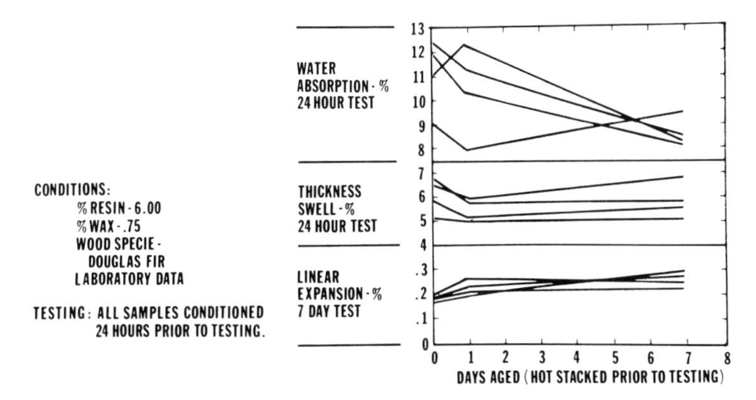

Figure 12.21. *Effect of hot stacking on water and/or moisture-absorption properties on ¾-in. (19 mm) homogeneous particleboard (Albrecht 1968).*

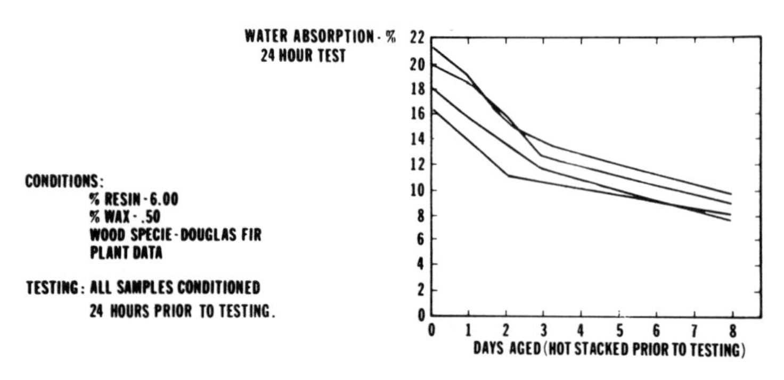

Figure 12.22. *Effect of hot stacking on water absorption of ⅝-in. (15.9 mm) homogeneous particleboard (Albrecht 1968).*

As mentioned in the discussion of testing, water soak evaluations are conducted under an equilibrium condition of 68°F (20°C) water temperature. Figure 12.23 illustrates the effect of water temperature on absorption test results. As can be seen, a few degrees change in temperature influences this test drastically. Fortunately when board is accidentally wetted, the water temperature is usually lower than 65°F (18°C).

The sensitivity of a board's water-absorption properties versus water temperature is also strongly related to the type of wax involved. If the wax contains components that melt at 90°F (32°C) it will obviously absorb water rapidly at this temperature.

Preservatives

Preservatives are not used widely in the United States. German standards call for such treatment in exterior boards. The popular preservative, pentachlorophenol, used in amounts of about 1.5%, has been found effective in

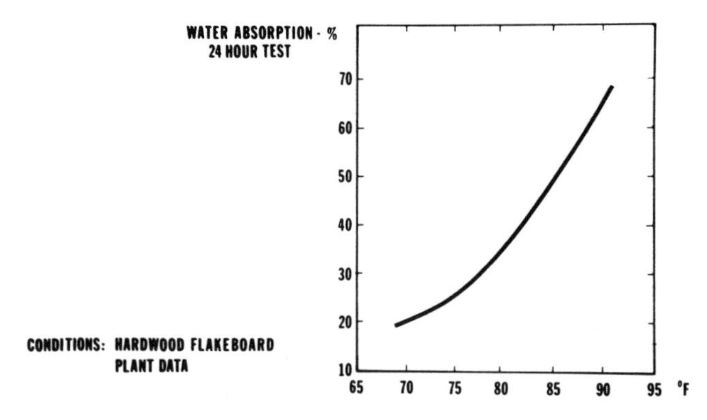

Figure 12.23. *Effect of water tenperature on absorption test results (Albrecht 1968).*

preventing decay and termite attack. It can be made into an aqueous solution of the sodium salt and mixed with the resin, or free pentachlorophenol can be mixed with the resin liquid. The addition of the resin to the furnish provides an even distribution of the preservative over the particles. In production, care must be taken to see that pentachlorophenol fumes are removed during hot pressing.

Other preservatives, such as chromated copper arsenate-type C and ammoniacal copper arsenite, can also be used.

The beauty of treating composition board during the manufacturing process is that little is added to the production cost outside of the cost of the preservative. Treating the panels after production calls for expensive pressure treating.

Fire Retardants

Fire retardants are used in some boards now, but their use is not widespread. Increased interest in safety, particularly in nursing homes, hospitals and multifamily dwellings, is increasing the interest in fire-retardant-treated boards. Fire retardants usually greatly reduce the production of wood smoke; however, some fire retardants contribute their own fumes which often result in an overall higher smoke contribution. It has also been observed that resin can contribute to smoke.

Other properties can be imparted by retardants that are not desirable. One of the most important is associated with smoke reduction, because smoke also contains hazardous fumes. Fume emission usually results from the breakdown of chemicals when heated or in contact with glowing charcoal. Acids of phosphorus and sulfur often form when fire retardants containing these compounds are used. Halides often give their corresponding acids under similar conditions. The use of carbon tetrachloride has been known to form the gas, phosgene ($COCl_2$). Thus consideration of possible fume emission is important when considering various fire-retardant systems. Additionally, inorganic salts, which under the above conditions break down into strongly acidic compounds, can have a very deleterious effect on wood strength.

Hygroscopicity is another property of many fire retardants that should be considered. Some very excellent fire retardants will tend to deliquesce in the wood structure, and this can be so severe under the proper conditions that the specimen will actually drip and small stalactites will form on the lower edge.

Preservative value of the fire retardant is one of the additional benefits possible. In the old proprietary mixtures used by lumber treaters, there were usually several different salts used; and one of these was usually a good preservative against biological attack.

However, in board production, where the system is sensitive to complexities, one attempts to keep the fire-retardant system simple and limited to only one salt. Among the salts which have been used in the past, and which are still used for this purpose, are zinc, arsenic, and copper. Others include borax, boric acid, and borate-containing materials as well as organic fire retardants.

Adding Fire Retardants: There are four principle areas where one can add a fire retardant during the board-making process, and one after the process. To begin the discussion in the reverse direction, let us start with area (*E*) on the diagram shown in Figure 12.24.

Figure 12.24. *Schematic drawing of a typical board process (Draganov 1968).*

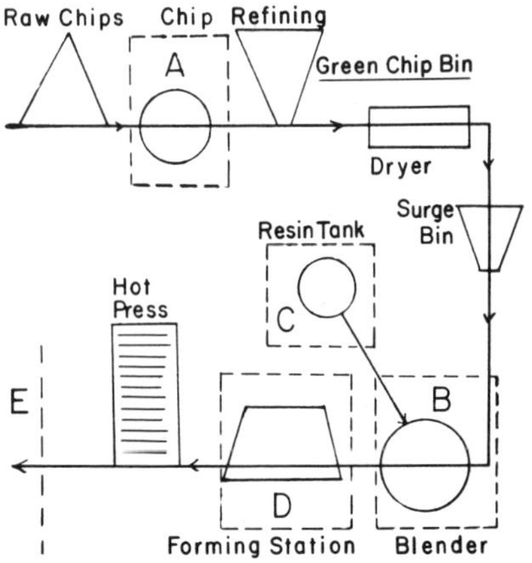

Until a few years ago, all fire-retardant board was prepared by pressure impregnation with a proprietary mixture similar to those used in lumber and plywood. The material to be treated was sent to a treating facility where it was placed in an autoclave and pressure impregnated with an aqueous solution of fire-retardant salts. The product was then removed, kiln dried, returned to the board plant for refinishing, and then sent to the customer. The advantage of this method was that it involved no effort by the producer of the board other than the use of a phenolic resin. It produced good fire-retardant board; however, its disadvantages outweighed its advantages. The principal disadvantage was cost. This method of treatment, in many cases, nearly doubled the cost of the product. The requirement to convert to a phenolic resin of course was quite a problem to many firms. The fire retardant used at that time had many of the poor characteristics mentioned earlier. The board was badly discolored, it was corrosive to hardware and suffered from a hygroscopic nature and, in addition, many of these boards carried high smoke contributions. Generally speaking, it was not a popular product.

Fire-retardant paints containing phosphates can be applied in thick coats over board substrates to provide some degree of fire retardancy.

A system has been recently developed in Canada in which fire-retardant chemicals are first applied to the finished particleboard surfaces and then driven into the board by hot pressing. As far as is known, this process is yet to be commercialized.

One of the first attempts to prepare a fire-retardant board during the manufacturing process was carried out in the forming station (area *E*, Figure 12.24). This has been used intermittently during the years. The method is simply to feed the fire retardant into the furnish distribution system in the forming station. The result is a heterogeneous mixture with particles of fire retardant interspersed with the furnish. One process operated with either granular boric acid or monoammonium phosphate. The pH of either material is such that the resin system is not seriously affected. The product produced had some reduced physical properties. The forming line was one that used a caul plate.

This method had the advantages of simplicity and minimal equipment investment. It also did not degrade the furnish as do some processes. It was, however, limited to a certain type of forming station. Problems would occur in adapting it to the modern stations which incorporate air-classification. The method of addition almost precludes satisfactory mixing of the retardant and furnish. If such a fire retardant as boric acid is used, 44% water contained in the boric acid is added to the furnish. This extra moisture should not be added just before pressing because resin bonding could not be achieved.

Fire retardant can be added to the furnish in the blender (area *B*, Figure 12.24) in powdered or liquid form. Such addition is limited to acidic constituents which will not seriously affect the cure of the urea-formaldehyde resin. Phenolic resins could be expected to have fewer curing problems than the ureas. The method has limitations as to the quantity of material that can be added in normal processing. During the normal treatment of furnish with resin-wax in the blender, only 5 to 7% of powdered fire retardant can be added, as this is all that can be reasonably expected to adhere to the furnish. If higher levels are used, fallout of the extra retardant will usually occur.

This problem can be largely circumvented by incorporating 10 to 15% of a powdered urea-formaldehyde resin along with the powdered fire retardant. The powdered resin in contact with the moisture in the furnish becomes tacky and allows more fire retardant to adhere to the wood furnish. Loadings, sufficient to earn a Class A or I flame spread classification, may be achieved by this method. However, the more resin added as a powder, the higher the smoke contribution is likely to be. This method could also be used with the more modern forming stations.

Addition of fire retardant as a powder has not been as effective as liquids. It has been found that fire retardant added as a solution and allowed to penetrate the particles is approximately 30% more effective than if the fire retardant is applied as discrete particles mixed into the furnish.

Fire retardants can be mixed with liquid resin in the resin tank (area *C*, Figure 12.24). In this method, a finely powdered fire retardant is mixed in with the resin until a suspension is formed. With sufficient stretching of operating factors, as much as 12 to 14% boric acid has been placed in a board in this manner. To do this requires more resin, some added water to reduce the viscosity of the mixture, and larger orifices in the application nozzles for handling the more viscous resin-retardant mix.

Polymeric fire retardants can be added as liquids in standard blenders. Normally, this is a separate processing step which is not incorporated with resin blending. High moisture contents after retardant blending may require further drying. However, if the retardant is added prior to drying, the redrying step can be eliminated.

Treatment at the raw material end is another possibility (area *A*, Figure 12.24). In this process, treatment is by solution or as a solid salt blended with extremely wet particles. This process makes use of the rapid diffusion characteristic of some fire retardants, especially the borates. The salt is diffused by the concentration gradient into the interior of the particles within a very short time. Rapid drying removes the moisture, leaving much of the fire-retardant salt in the interior of the wood particles. The furnish is then transferred to the blender and treated in a normal manner.

This method has the disadvantage of requiring more capital investment than those previously discussed, if more dryer capacity is needed. The moisture added to the green particles will increase the demand placed on the dryer and, in addition, it requires a plant layout which has a green holding bin to allow for a brief diffusion time.

The advantages of this method are: it allows the incorporation of very large quantities of fire retardant; it permits the addition of neutral or alkaline retardants; homogeneous products are reported tha display good physical properties.

Fire-Retardant Systems: A number of fire retardants have been developed; however, none have found widespread use as yet. Quite a number can be used but they will not all be reviewed here.

One system loads the resin up with about 12% of boric acid which acts as thc fire retardant. Usually more resin than normal is needed in such a system. A borate in aqueous solution mixed with green particles and acidified with a mineral acid makes up another system. In another approach, disodium octaborate in a strong solution is added to green particles, which are then dried and treated in the blender with a resin containing boric acid. Still another system uses a bromine treatment which supposedly combines with the lignin liberating hydrogen bromide. The furnish must be neutralized with borax or sodium carbonate to remove the hydrobromic acid formed as a by-product.

A chemical composition of water, sodium fluoride, urea, and mono-ammonium phosphate is used in another approach. However, this is a cumbersome system as the liquor is applied to the furnish in a slurry which means a great amount of liquid must be removed before board manufacture can take place. A mono-ammonium phosphate system in powdered form has been approved for use in another system.

A process used in U.S. manufacture calls for a mixed borax/boric acid composition (disodium octaborate tetrahydrate). An aqueous solution is added to the green particles or, if the furnish is extremely wet, powdered retardant can be added directly, and it diffuses into the wood quickly as noted previously. In board production with urea resin, it is sometimes necessary to add catalyst to cure the resin as the retardant tends to neutralize the chemical field of the furnish.

The system developed in Canada for hot pressing the fire-retardant chemicals into the particleboard uses ammonium dihydrogen orthosphosphate and liquid ammonium polyphosphate as the fire retardants.

With any fire retardant, it is important that it is nonleachable. Otherwise in exposed conditions, or in areas of accidental wetting, the retardant is slowly leached out. A relatively new retardant being developed is designed to be converted to a nonleachable state in the wood after impregnation. It was developed by Koppers as an all-weather fire-retardant formulation marketed as NCX. The fire retardant is composed of a group of water-soluble monomers, which polymerize to an insoluble, leach-resistant state when subjected to elevated temperatures as would occur in the drying and pressing operations. The polymerization action is similar to that which occurs in most resins used in the plywood and particleboard industry. Application of the retardant takes place in a conventional blender followed by drying, after which normal board processing takes place.

Manufacturing Fire-Retardant Board: In considering fire-retardant board, the type of board required must first be determined (Class A, B, C, D or I, II, III, IV). Experience has shown many prospective manufacturers are not even aware that there are several classifications.

There is a need to become familiar with fire codes and the extent and variation in the sales area. It probably will be necessary to market a product bearing an Underwriters Laboratories label. All pertinent information must be gathered and evaluated first.

The fire-retardant system should be the one most suitable for the particular plant. Of importance is to plan the installation carefully, particularly if one wishes to obtain label service from firms such as the Underwriters Laboratories Inc. In that case, the manufacturer will be required to standardize his procedures, and changes later can be awkward and costly.

Once experimental boards have been produced, they will have to be fire tested. Usually, a simple test such as the Fire Tube or Schlyter will suffice for developmental purposes. As the experimental board begins to approach the design properties desired, specimens should undergo experimental tunnel tests. Promising products must then undergo official testing, which is the only way to prove the boards meet the desired classification level. Similar efforts will have to be accomplished in getting ratings for fire-door material.

SELECTED REFERENCES

Albrecht, J. W. 1968. The Use of Wax Emulsion in Particleboard Production. *Proceedings of the Washington State University Particleboard Symposium, No. 2.* Pullman, Washington: Washington State University (WSU).

Anderson, A. B. 1975. Bark Extracts as Bonding Agents for Particleboard. Background Paper No. 67, third World Consultation on Wood-Based Panels, FAO, New Delhi, India, February 1975.

Anderson, A. B.; A. Wong; and K-T. Wu. 1974. Utilization of White Fir Bark and Its Extract in Particleboard. *Forest Products Journal,* Vol. 24, No. 7.

Anderson, A. B.; K-T Wu; and A. Wong. 1974. Utilization of Ponderosa Pine Bark and Its Extract in Particleboard. *Forest Products Journal,* Vol. 24, No. 8.

Becker, G., and M. Gersonde. 1975. Preservation of Wood-Based Panels Against Fungi and Insects and Testing Its Efficiency. Background Paper No. 114, third World Consultation on Wood-Based Panels, FAO, New Delhi, India, February 1975.

Berrong, H. B. 1968. Efficiency in Sizing of Particleboard. *Proceedings of the Washington State University Particleboard Symposium, No. 2.* Pullman, Washington: WSU.

———. 1973. Wax Sizing Agents for Particleboard. *Proceedings of the Washington State University Particleboard Symposium, No. 7.* Pullman, Washington: WSU.

———. 1975. Private correspondence on slack wax. Mobil Oil Corp.

Binder, K. A. 1967. Resin Application and Quality in Particleboard Manufacture. *Proceedings of the Washington State University Particleboard Symposium, No. 1.* Pullman, Washington: WSU.

———. 1974. The Outlook for the Supply of Resins. *Proceedings of the Washington State University Particleboard Symposium, No. 8.* Pullman, Washington: WSU.

Blais, J. F. 1959. *Amino Resins.* New York: Reinhold Publishing Corp.

———. 1962. Amino Resin Adhesives. *Handbook of Adhesives.* New York: Reinhold Publishing Corp.

Cagle, C. V., ed. 1973. *Handbook of Adhesive Bonding.* New York: McGraw-Hill Book Co.

Cazes, J. 1976. New Insights into Particleboard Resin Composition by Gel Permeation Chromatography. *Proceedings of the Washington State University Particleboard Symposium, No. 10.* Pullman, Washington: WSU.

Christensen, R. L. 1972. Test for Measuring Formaldehyde Emission from Formaldehyde Resin Bonded Particleboards and Plywood. *Forest Products Journal,* Vol. 22, No. 4.

Deppe, H-J., and K. Ernst. 1971. Isocyanate als Spanplattenbindemittel (Isocyanates as Adhesives for Particleboard). *Holz als Roh-und Werkstoff,* Vol. 29, No. 2.

Draganov, S. M. 1968. Fire Retardants in Particleboard. *Proceedings of the Washington State University Particleboard Symposium, No. 2.* Pullman, Washington: WSU.

Duncan, T. F. 1974. Normal Resin Distribution in Particleboard Manufacture. *Forest Products Journal,* Vol. 24, No. 6.

Duwin, W. 1975. El uso del Quebracho en los adhesivos para madera terciada y placas de madera aglomerada. Background Paper No. 107, third World Consultation on Wood-Based Panels, FAO, New Delhi, India, February 1975.

FAO. 1976*a*. Basic Paper II. Raw Materials for Wood-Based Panels. *Proceedings of the World Consultation on Wood-Based Panels,* held in New Delhi, India, February 1975. Brussels: Published in agreement with the Food and Agriculture Organization of the United Nations by Miller Freeman Publications.

———. 1976*b*. Basic Paper IV. Technology and Techniques in the Manufacture of Wood-Based Panels. *Proceedings of the World Consultation on Wood-Based Panels,* held in New Delhi, India, February 1975. Brussels: Published in agreement with the Food and Agriculture Organization of the United Nations by Miller Freeman Publications.

Froede, O. 1975. Synthetic Resin Adhesives for Wood-Based Panels. Background Paper No. 50, third World Consultation on Wood-Based Panels, FAO, New Delhi, India, February 1975.

George, J. 1975. Tannin Adhesives for Wood-Based Panels. Background Paper No. 14, third World Consultation on Wood-Based Panels, FAO, New Delhi, India, February *1975.*

Halligan, A. F. 1969. Recent Glues and Gluing Research Applied to Particleboard. *Forest Products Journal,* Vol. 19, No. 1.

Harkin, J. M.; J. R. Obst; and W. F. Lehmann. 1974. Visual Method for Measuring Formaldehyde Release from Resin-Bonded Boards. *Forest Products Journal,* Vol. 24, No. 1.

Heebink, B. G. 1967. Wax in Particleboards. *Proceedings of the Washington State University Particleboard Symposium, No. 1.* Pullman, Washington: WSU.

Hemming, C. B. 1962. Phenolic Resin Adhesives. *Handbook of Adhesives.* New York: Reinhold Publishing Corp.

Hine, J. M. 1975. Natural Adhesives for Panel Products. Background Paper No. 37, third World Consultation on Wood-Based Panels, FAO, New Delhi, India, February 1975.

Hse, C-Y. 1975*a*. Fast-Cure Phenolic Resin Binder for Exterior Flakeboard of Mixed Southern Hardwoods. *Proceedings of the Washington State University Particleboard Symposium, No. 9.* Pullman, Washington: WSU.

———. 1975*b*. New Developments in Phenolic Adhesives for Particleboard. Background Paper No. 73, third World Consultation on Wood-Based Panels, FAO, New Delhi, India, February 1975.

Humphreys, K. 1975. Trends in the Availability, Production and Prices of Synthetic Resin Adhesives for Panels Based on Wood and Other Fibrous Materials. Background Paper No. 8, third World Consultation on Wood-Based Panels, FAO, New Delhi, India, February 1975.

Jenks, T. E. 1975. Urea Resins in Hardboard Manufacture. Background Paper No. 68, third World Consultation on Wood-Based Panels, FAO, New Delhi, India, February 1975.

Johansen, T. A. 1971. Particleboard Bonded with Sulphite Liquor. *Proceedings of the Washington State University Particleboard Symposium, No. 5.* Pullman, Washington: WSU.

Johnson, R. E. 1968. Plant Experiences and Problems with Fire Retardants. *Proceedings of the Washington State University Particleboard Symposium, No. 2.* Pullman, Washington: WSU.

Kelly M. W. 1970. Formaldehyde Odor and Release in Particleboard. *Proceedings of the Washington State University Particleboard Symposium, No. 4.* Pullman, Washington: WSU.

Keutgen, W. A. 1969. Phenolic Resins. *Encyclopedia of Polymer Science and Technology,* Vol. 10. New York: Interscience Publishers.

Knapp, H. J. 1971. Continuous Blending 1948-1971. *Proceedings of the Washington State University Particleboard Symposium, No. 5.* Pullman, Washington: WSU.

Kulvik, E. 1975. Chestnut Wood Tannin Extract in Plywood Adhesives (PCF Adhesive). Background Paper No. 142, third World Consultation on Wood-Based Panels, FAO, New Delhi, India, February 1975.

Lambuth, A. L. 1967. Performance Characteristics of Particleboard Resin Blenders. *Proceedings of the Washington State University Particleboard Symposium, No. 1.* Pullman, Washington: WSU.

Lehmann, W. F. 1964. Retarding Dimensional Changes of Particle Boards. *Information Circular 20* (August). Forest Research Laboratory, Oregon State University, Corvallis, Oregon.

Maloney, T. M. 1970*a*. Fire-Retardant Particleboard. *Quest* (October). Pullman, Washington.

———. 1970*b*. Resin Distribution in Layered Particleboard. *Forest Products Journal,* Vol. 20, No. 1.

Mohr, G. D. 1968. In-Plant Production of Wax Sizing Agents. *Proceedings of the Washington State University Particleboard Symposium, No. 2.* Pullman, Washington: WSU.

Myers, G. C., and C. A. Holmes. 1975. Fire-Retardant Treatments for Dry-Formed Hardboard. *Forest Products Journal,* Vol. 25, No. 1.

Neusser, H., and M. Zentner. 1968. Causes and Elimination of Formaldehyde Odor from Building Materials Containing Wood, Particularly from Particle Boards. *Holzforschung und Holzverwertung,* Vol. 20, No. 5. (Translated by Joint Publications Research Service, U.S. Department of Commerce, December 1969.)

Nicholls, W. H. 1968. Phenolic Resin Binders. *Particleboard Manufacture and Application,* Chapter 2:3. Great Britain: Novello & Co. Ltd.

Parker, R. S. R., and P. Taylor. 1966. Phenolic Resin Based Adhesives. *Adhesion and Adhesives.* Elmsford, New York: Pergamon Press.

Peters, T. E. 1972. Manufacture of Products from Comminuted Wood. Patent No. 3,649,397. United States Patent Office (March).

Plomley, K. F. 1966. Tannin-Formaldehyde Adhesives for Wood. Division of Forest Products Technological Paper No. 39. Commonwealth Scientific and Industrial Research Organization, Melbourne, Australia.

———. 1975. The Development of Wattle Tannin-Formaldehyde Adhesives for Wood by CSIRO. Background Paper No. 13, third World Consultation on Wood-Based Panels, FAO, New Delhi, India, February 1975.

Plomley, K. F.; J. W. Gottstein; and W. E. Hillis. 1964. Tannin-Formaldehyde Adhesives for Wood. *Australian Journal of Applied Science,* Vol. 15.

Rayner, C. A. A. 1966. Resin Adhesives in Wood Chipboard—Some Comparative Features. *Wood* (July).

———. 1968. Amine-Formaldehyde Resins. *Particleboard Manufacture and Application,* Chapter 2:2. Great Britain: Novello & Co. Ltd.

Schwarz, F. E. 1971. Advances in Detecting Formaldehyde Release. *Proceedings of the Washington State University Particleboard Symposium, No. 5.* Pullman, Washington: WSU.

Senn, S. 1975. Equipment for and Experiences in Mixing Resin with Catalysts, Extenders, and Other Ingredients. *Proceedings of the Washington State University Particleboard Symposium, No. 9.* Pullman, Washington: WSU.

Shen, K. C. 1974*a*. Acidified Spray-Dried Spent Sulfite Liquor as Binder for Exterior Waferboard. *Proceedings of the Washington State University Particleboard Symposium, No. 8.* Pullman, Washington: WSU.

———. 1974*b*. Modified Powdered Spent Sulfite Liquor as Binder for Exterior Waferboard. *Forest Products Journal,* Vol. 24, No. 2.

———. 1975. Spent Sulfite Liquor as Binder for Particleboard. Background Paper No. 63, third World Consultation on Wood-Based Panels, FAO, New Delhi, India, February 1975.

Shen, K. C., and D. P. C. Fung. 1972. A New Method for Making Particleboard Fire-Retardant. *Forest Products Journal,* Vol. 22, No. 8.

Shepherd. J. W. 1974. Isocyanate Binders for Particleboard. Presented at the 8th Washington State University Particleboard Symposium.

Stashevski, A-M., and H-J. Deppe. 1973*a*. The Application of Tannin Resins as Adhesives for Wood Particleboard. *Holz als Roh-und Werkstoff,* Vol. 31, No. 11.

———. 1973*b*. Tannin Resins as Bonding Agents in the Manufacture of Chipboard. *Holz als Roh-und Werkstoff,* Vol. 31, No. 11.

Syska, A. D. 1969. Exploratory Investigation of Fire-Retardant Treatments for Particleboard. *USDA Forest Service Research Note FPL-0201* (August).

Talbott, J. W. 1957. *Flapreg. Technical Report 5.* Washington State University. Pullman, Washington: WSU.

Talbott, J. W. 1971. High-Resin Particleboards for Structural Uses. *Reprint No. 294.* Washington State University. Pullman, Washington: WSU.

Udvardy, O., and G. Plester. 1972. Tack and Its Interpretation. *Proceedings of the Washington State University Particleboard Symposium, No. 6.* Pullman, Washington: WSU.

White, J. T. 1974. New Developments Relative to Wood Binder Systems Created by the Energy Crisis. *Proceedings of the Washington State University Particleboard Symposium, No. 8.* Pullman, Washington: WSU.

Widmer, G. 1965. Amino Resins. *Encyclopedia of Polymer Science and Technology,* Vol. 2. New York: Interscience Publishers.

Winding, C. C., and G. D. Hiatt. 1961. Polymers Formed by Condensation Reactions. *Polymeric Materials.* New York: McGraw-Hill Book Co.

13. RESIN/WAX APPLICATIONS AND BLENDERS

Blending of resin, wax, and other additives with the furnish was relatively simple in the pioneer period of the industry. Pouring the resin onto agitated furnish was practiced much the same as mixing cake batter. Little was known about the effects on board properties caused by the many resin parameters already discussed, by resin distribution, resin droplet size, species, and particle size and geometry. Much research since then, however, has revealed the fundamental information needed for at least a partial understanding of optimum blending. Sophisticated control systems and blenders have been developed for implementing this information. The fundamental information will be discussed first in this chapter, followed by descriptions of some of the control equipment and the major blenders. Emphasis will be on the resin and wax application. Preservatives and fire retardants can be mixed with the resin or applied separately. In general, resin blenders can also be used with preservatives and fire retardants.

RESIN APPLICATION

Many factors are involved in applying resin for optimum use. In addition to the ones discussed in the resins chapter are the actual practices of storing and application. Massive amounts of resin are required to run large board plants; thus storage becomes a significant factor. Such a facility must be well designed to handle the resins, run trials, and determine the amount of resin going into the process at any given time. Next a precise method for controlling the ratio of wood to resin entering the blender is of paramount importance; otherwise, the board properties will vary widely.

It should also be pointed out that one of the more necessary features of any good resin-handling system is a quick and efficient means of cleaning all of the equipment, from the storage tank through the spray nozzles or centrifugal atomizers. Storage tanks should have easy access for workmen to get inside and clean the tank. Tanks should have a bottom drain or resin outlet, a valve on the resin line just outside the tank and/or a valve on the line just ahead of the next storage tank to insure against the possibility of noncompatible resins and additives intermingling in the line. Easy-flush facilities and direct-line access for cleaning all lines should be provided if possible. Readily accessible and easily cleaned strainers or filters should be installed in at least two places before the resin applicators to prevent this equipment from being clogged by dirt, gelled resin particles, and other foreign material.

To store the resin within the proper temperature range, resin storage tanks should be installed to allow proper storage temperatures in both winter and summer. A storage temperature of approximately 60°F (16°C) is ideal in that the resin advancement or aging is retarded and the resin viscosities are not too high for handling and pumping. Lower temperatures are rarely needed since resin usage is high and the resins used are usually no older than two weeks.

Provision for easy handling of tote bins or drums for trial runs of new resins and/or special resins and additives for new products should be included and carefully designed for ease of handling and minimum interference with normal production.

Provision to heat the resin in the range of 80° to 105°F (27° – 41°C) before spraying is generally found to be desirable in producing better atomization of the resin. Resin viscosity is very temperature sensitive. Thus, relatively uniform temperatures are needed when spraying resin to keep the resin droplet sizes within the specified range. This heating system should preferably include jacketed mix or use tanks with adequate constant agitation in order to add chemicals or additives to the resin. These tanks can also be used to double-check the hourly or daily resin usage rates, in addition to those meters which are used to record flow rate data.

Spraying or other application systems should incorporate quick-change connections to reduce the time necessary to change spray heads for regular maintenance or in case of a foul-up in the system. All of these features are useful in reducing downtime and in cutting costs on normal maintenance cleanup periods.

Pumping systems should be sized to provide rapid handling with complete flexibility to move resin in or out of any tank, to recirculate and provide blending when desired, and to operate with the least possible maintenance.

In the board industry, the quantity of resin used and the method of application of the resin are paramount determinants both of cost and quality of the finished product. Proper application of the binder is extremely critical in developing quality adhesive bonds.

Most plants apply their resin in a liquid state onto the furnish as it passes through the important piece of equipment known as the *blender*. Wax emulsion is normally mixed with the resin, or it can be dripped or sprayed upon the furnish as it is screw-conveyed to the blender, or it can be sprayed onto fiber as it is air-conveyed through blowpipes. The average blender is a piece of equipment designed to agitate the wood particles while the resin and/or other additives are being either sprayed or applied by centrifugal atomization onto the furnish. The object is to disperse the resin droplets uniformly over the wood particles. Each wood particle should have adequate exposure to the resin and/or additives, and should be coated with sufficient resin to attain the desired properties in the final board.

Most of the particleboard plants in this country have applied the resin by a spray system; however, a newer type of blender is gaining favor where resin is distributed by centrifugal force from a rapidly turning hollow shaft. It should be recognized that a resin and its additives may be satisfactorily introduced into the blender by air spray, airless spray, or the disc or centrifugal distribution. All three methods have their advocates, and it is possible that only one type may be preferable in certain types of blenders.

Of interest is the fact that the airless sprays, which operate under high pressures of 600- or 800-psi (4.14 or 5.52 MPa) to 1400- or 1500-psi (9.65 or 10.34 MPa) pressure, introduce less air into the blender. On the other hand, airless systems will usually have smaller orifices in the nozzles, and may pose a greater problem with flooding and in keeping the nozzles free from dirt and other contaminants which interfere with the proper spray pattern. Also, as no extraneous air is admitted into the blender during spraying with the airless system, mechanical agitation in some blenders may have to be increased to gain equally uniform distribution of the resin onto the particles.

Whether the spray system is one requiring a low pressure of 20 to 60 psi (138–414 kPa) on the resin as well as 40 to 100 psi (276– 689 kPa) of airstream to atomize the resin which is ejected from the nozzle, or whether an airless spray system is used, there are some important features to be kept in mind.

There are two primary factors which apply when the spraying method is considered. One factor is the fineness of spray and the other is selection of a suitable time period of application. Both factors are of cardinal importance. When the spraying method conforms to good practice relating to these factors, then uniform distribution of resin on the particles will be maintained.

It is of prime importance in the application of any resin through the spray system to be sure that there are sufficient nozzles available to do the job, and that the placement and direction of the nozzles are such that the best possible distribution of resin on the wood particles is obtained. Another factor of great importance is that there must be easy access to the spray heads. Any good system will incorporate a quick-change method to replace nozzles which become clogged, or malfunction in some other manner, since this reduces the efficiency of the resin application.

Also critical in the design of the resin-handling system is the consideration of whether the resin will be sprayed on the particles as received, usually a 65% nonvolatile (solids) solution in water for urea or 40 to 53% solids for phenolic, or whether the resin is to be diluted and/or mixed with water, wax emulsion, catalyst, or other additives.

The addition of any of the aforementioned additives to the resin is in essence diluting the resin solution. Theoretically this could give rise to better resin efficiency; however, additional moisture produced by diluting the resin solution might yield furnish with an excessive moisture content. Dilution could also lead to flooding (overloading) of the spray nozzles due to the increase in volume of the resin solution. Thus the apparent value of resin dilution could be negated by poor distribution of the resin over the furnish. Such dilution could also cause the resin to penetrate the wood and thus be lost for bonding the particles together.

All of these factors are important, not only in designing the resin-handling system, but in influencing the ease of application and controlling the method of application of the resin and additives. If the system is not properly constructed, operational difficulties are often encountered. This could result in inconsistent or low-quality board being produced when any one part is not under the proper control and functioning in the manner intended.

Discussion of resin application by the disc or centrifugal application methods will be dealt with in detail later. However, application is still designed to have small discrete droplets of resin uniformly applied to the furnish. These types of applicators are airless in that compressed air is not needed for atomizing the resin. Compressed air can have the disadvantage of causing overspray of the resin and the air supplied must be exhausted from the blender, although the latter is not always a problem.

In only a few plants in this country are resins applied in other than a liquid form, and, in these cases, the resin is generally applied as a spray-dried or finely divided powder.

When powdered resin is used, it is normally metered on a continuous basis along with the particles into an inclined rotating drum. Here the resin and wood particles are mixed as they travel from the inlet to the outlet end of the blender. In these cases, the wax in the form of an emulsion or in the molten state is applied ahead of time. One plant with finely divided furnish blends the wax with the furnish in the grinding operation performed in an attrition mill. Others spray the wax on before the resin is added.

Detrimental effects that could happen with the application of powdered resin are: (1) a poor distribution uniformity which leads to indeterminate dusting loss if levels equivalent to liquid phenolic are used (5 to 7%); (2) segregation of the resin from the particles which could lead to caul sticking; (3) furnish surface moisture content has to be kept at a sufficiently high level to obtain adhesion of the powdered resin to the wood particles, which reduces the cost of drying but may cause blowing problems; (4) proper precautions have to be taken against dust explosions and fires; and (5) the cost factor is higher since most powdered resins are expensive phenolics.

In contrast, manufacturers of fiberboard use at least five systems to apply the resin to the fiber. First, in the wet-process hardboard mills, the fiber is generally carried in a low-consistency water slurry. The resin is introduced to the furnish as a liquid and is precipitated upon the fibers by the use of an acid or acid salt which reduces the pH of the slurry to a point where the resin solids will no longer remain in solution. Second, the resin may be added through the center shaft of the attrition mill or refiner in one of two ways: In method *(a),* proper resin coverage of the wood fiber particles is accomplished in the attrition mill simultaneously with the generation of the furnish. This furnish is then dried; therefore, phenolic resins are used exclusively as ureas would cure out in the dryer. For method *(b),* resins are blended with fiber after the fiber has been dried in attrition mills equipped with the *devils tooth* blending plates (Figure 13.1) or regular grinding plates. Third, in some cases the resin is simply added to the wet or semi-wet fiber in the blow line between the refiner and dryer or in the blow line out of the dryer, using air turbulence for resin dispersion. Phenolic resins are used in the blow line to the dryer; however, UF or MUF resins can be used in the blow line out of the dryer. Fourth, resin is sprayed onto the fiber in much the same manner as is commonly done in particleboard plants, usually in short-retention-time blenders. Fifth, a combination system is used where urea is added to the fiber as it is being ground in the attrition mill; and, after drying is finished, the formaldehyde and other resin constituents are added in a short-retention-time blender.

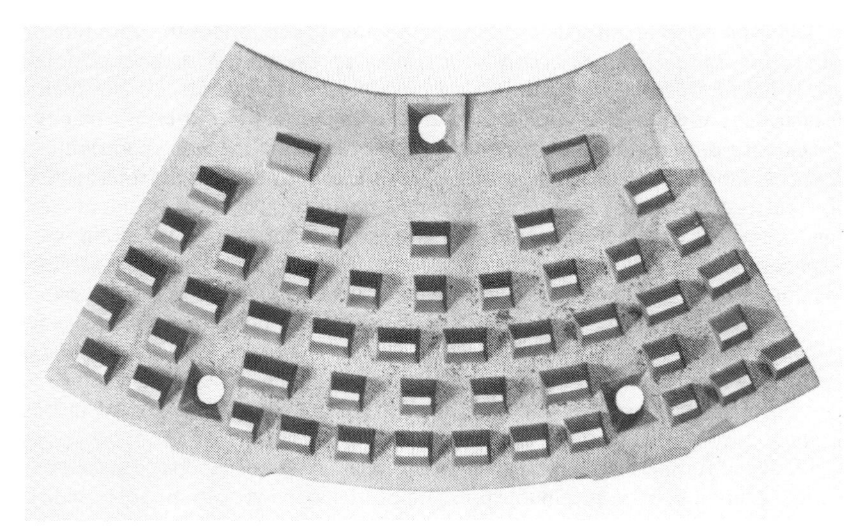

Figure 13.1. *A segment of a pin plate disc (McNeil 1967).*

Mixing Additives Versus Separate Additions

It can be said that the system which is geared to introduce the additives separately adds a degree of flexibility to the system. However, some plants believe there are advantages if certain or all of the additives are mixed before being introduced directly into the spray system. These advantages include: (1) lower initial cost of equipment, (2) possibility of less human error, and (3) less maintenance. It should be recognized that careful measuring of the ingredients into the blending or mixing tank is just as important as accurate metering into the spray nozzles. Therefore the possibility of error in such a system is just as great as that in which the additives are added through separate systems. It must also be realized the mixing or blending of the various ingredients prior to introduction into the spray system may give less flexibility in raising or lowering resin or wax content, in the event it becomes necessary to change the properties of the board. By proper engineering, either method can be made to work accurately and efficiently with no noticeable difference in results between the two methods.

Use of External Catalysts

Perhaps the addition of external catalyst to promote cure of urea-formaldehyde resins should be considered as a separate problem. This is because the addition of any acid or acid salt type of catalyst, exclusive of the latent-type catalysts which are sometimes integrated within the resin formulation, will lower the pH and accelerate the polymerization of the resin. This will result in a shorter storage or "pot life," and has been known to cause problems in plants which either are not properly equipped to handle the addition of external catalysts or whose personnel are not fully aware of all the problems involved.

The total wood acidity, as mentioned previously, determines the cure rate of urea resins, and aids immeasurably in determining press time. With the effect total acidity has on press times, it appears very probable that it is not the pH value of the furnish alone affecting press times, but rather a complex of pH, buffering capacity, and existing or potential total free volatile acid contents of the wood contributing to the overall effect on board press times. As little is known about the total acidity of wood as defined by pH, buffering capacity, and free volatile acid content, and how these factors function in the pressing process, further investigation with respect to their effects on press time, and to the factors which affect them prior to and during actual hot-pressing, seems to be merited. Knowledge of the value and variability of total wood acidity as defined above is of further help in solving some of the problems correlated to press time. This knowledge can also serve as a very useful guideline to the resin manufacturer in attempting to tailor a resin to a specific board process. Thus, it can be seen that much art is involved in the production of suitable resins for board use.

Woods or furnish low in total acidity require an external catalyst to accelerate the resin cure to realize acceptable press times. If a wood very low in total acidity is used, and the degree of total acidity difference between batches of wood particles is of sufficient magnitude, adverse effects may result. If the acidity were to vary over a wide range, it would be necessary to vary the amount of catalyst or contemplate a change in the resin being used. Excessive use of a strong catalyst could lead to precure of the resin on the particles and, if present in sufficient amount, could weaken the wood fiber on the bonding surface, or cause the glue bonds to deteriorate, thus influencing bond failures.

The addition of external catalyst is fairly common in particleboard plants where wood species have relatively high pH and thus low total wood acidity. Some plants, on the other hand, use wood species which contain enough wood acids and have low enough pH to lower the pH of the resin to a level where the resin cures rapidly in the hot press. As plants have grown larger, it has become necessary to use a wider variety of species, each usually with different characteristics as far as resin addition is concerned. Thus the resin manufacturer's job has become much more complicated, since it is difficult to determine the species mix as it comes in randomly as comminuted wood.

In normal commercial practice, the addition of external catalyst is accomplished by the addition of a dilute acid or an acid salt such as ammonium sulfate or ammonium chloride in a 20 to 30% solution. The catalyst is used in amounts ranging from about ¼% to about 1% of solid catalyst based on the liquid resin. The catalyst may be buffered with ammonium hydroxide to promote longer pot life and offer some precure resistance. The catalyst may be added to the resin with thorough agitation in a mix tank, through an in-line blending system just prior to the spray nozzles, or after blending. Adding the catalyst in solution before or after blending is unusual, but this is successfully being done. Three advantages accrue to this method: (1) there is no possibility of the resin's setting up in the pumps, pipes, nozzles, or resin application tubes; (2) the system can be easily adjusted to fit the species mix in the furnish; and (3) mixing and mixing errors are reduced because the catalyst is not mixed with the resin. The philosophy is to adjust the wood pH of species other than Douglas fir to the same level as Douglas fir, which contains sufficient acid to speed resin curing. Thus the control of catalyst level is independent of resin. Troubles may occur when resin levels are lowered if the resin and catalyst are mixed, because the mix is proportional. Such lowering may result in addition of too little catalyst, thus affecting bonding.

With any of these systems, proper precautions have to be taken to insure adequate mixing of the resin and catalyst. Facilities should be included to rapidly flush out any remaining resin in the lines to the blender in the event of a shutdown. It must be remembered that the use of an external catalyst is generally justified only if it will produce shorter press times.

Resin Usage in the Various Grades of Board

Resin levels and the levels of most other additives are calculated on what is known as the oven-dry basis of the wood, and only the solids level of the additive is considered. As an example, the formula for a certain manufacturer's product may be as follows:

	All parts by weight	Parts by weight including moisture
	100 oven-dry wood	102.0 (2% moisture content)
	7 (solids of resin as percentage of OD wood)	10.8 (65% solids resin)
	1 (solids of wax as percentage of OD wood)	2.0 (50% solids resin)
Totals	108	114.8

The calculations can be interpreted in several ways. For one, it can be said 7% of resin was used as shown in Column 1 but this is really 6.5% on a true percentage basis. The moisture can be considered as shown in Column 2, and thus the calculation would show the resin level to be 6.1%. The oven-dry basis method is the simplest and quickest to use but the various interpretations should be remembered when discussing *resin content*.

In the case of the few plants which manufacture homogeneous-type board, the resin content is the same throughout the board. On the other hand, plants which are equipped to produce three-layer or multi-layer board have to be able to control and vary the percentage of resin in the surface and core layers. Such board has a higher percentage of resin in the surface, because the fines have a greater surface area for a given weight. In a blender adding resin to all particles simultaneously, fines always pick up more resin than the coarse material. Thus, care must be taken to insure that the fines do not pick up too much resin. Fines must have higher resin levels because of their greater specific surface area, particularly when used in three- or multi-layer boards. The higher resin level reduces strip-off of laminates applied in secondary finishing and reduces the amount of telegraphing. Lower-grade board may contain as little as 4% resin in the core, while specialty boards may contain up to 8% resin in the core, with resin content in the surface layers ranging from 6 to 12%. Some very fine particles may have more than 18% resin. Dry-process hardboards, as noted earlier, will range from about 2 to 6% in resin level depending usually on board density (less resin in high-density boards and more resin in medium-density boards). In essence, the amount of resin needed in an explicit board type may be taken as that resin level needed to attain a specific level of board properties at a specified board density.

When less than 4% resin is used in the core layer or layers, it is generally found that the board is deficient in properties such as internal bond and screw holding. Most of this may be attributed to the fact that the distribution of the resin on the wood particles is not optimum. Some plants find that the inclusion of a certain amount of fines containing more resin is necessary to obtain the proper strength.

Resin contents in exterior-grade board will usually range from 5 to 10%, depending upon the end use intended. Waferboards made with powdered phenolic resin may have resin levels as low as 1½%. Boards containing phenolic resins require longer press times and produce higher strengths than those containing urea resins, when using equal percentages of resin. However, the cost of phenolic resins has been traditionally higher than that of ureas; and most manufacturers have made an effort to produce a board with acceptable strength, while attempting to keep the phenolic resin usage cost in the same range as urea resins. This results in phenolic resin usage in the board of one or more percent less than is used in urea resin boards. It has been proven of course, that phenolic-bonded boards have less deterioration than urea-bonded boards under exterior-use conditions.

Specialty particleboards are just exactly that and are intended for a specific market where particleboard competes with other materials. These are boards of higher densities, up to and over 60 lbs/ft^3 (0.96 sp gr), containing enough resin to produce the desired physical and surface properties.

Factors Affecting Resin Distribution and Quality

Resin distribution must be controlled in all types of boards to produce a uniformity of resin coverage on the wood particles, if a manufacturer is to produce a board with maximum strength while utilizing a minimum of resin.

There are two separate schools of thought regarding resin coverage of the wood particles. One advocates distributing the resin in a uniformly thin continuous film, while the other suggests depositing the resin in uniformly discrete droplets on the surfaces of the wood particles. The commercial blenders are based on the latter principle, as usually there is not sufficient resin available throughout the board to form a continuous film. Obviously the blenders are doing an adequate job, but none seem to provide an optimum distribution of the resin on the particles. Optimum distribution is obtained only when the desired, or necessary, amount of resin coverage is present on all fractions of the particles from the finest to the coarsest sizes. Optimum distribution, however, must be balanced with economics.

Laboratory blenders can be set up with very fine sprays resulting in boards with excellent properties. But the time required for such application is very long. Short-retention-time blenders, as will be discussed later, pass the furnish through in seconds rather than minutes. Nevertheless, significant strides forward have been made in blender design, and further ones are anticipated with the motivational force behind such developments being the rapidly increasing cost of resins.

In all blending systems the fine particles, on a weight basis, almost always receive a disproportionately large amount of the resin, and the coarse particles end up with a lesser amount of resin. This is because the fines have the largest exposed surface area for a given weight and therefore are subjected to more contact with the atomized resin. Resin pickup in many blenders has been determined to bear an almost straight-line relationship to the exposed surface area of the particles involved.

No single blender apparently will produce the desired distribution of resin on fines and coarse particles for a layered board when both size particles are run simultaneously through the same blender. In a homogeneous board, poor distribution is largely negated by the intermingling of various-size particles in the finished board.

Many plants producing layered boards have overcome this problem by installing separate blenders for face and core portions of the board. This produces resin distribution which more closely approaches the optimum, since resin coverage on core and surface material can be better controlled. Also resins specifically tailored for faces and cores can be used. For example, as mentioned previously, slow-curing resins can be used in the faces to overcome problems with precure, while highly catalyzed resins can be used in the core to effect faster curing in this area which is heated last.

Depending on the specific type of board-making process, certain blending systems produce better results on given types of particles. This effect is lessened in a homogeneous board-making operation, and when separate blenders are used in a layered board-making process.

In the blender itself, the degree of atomization of the resin is important. For those with spray guns, atomization is affected by the number and size of spray heads used, their placement, the pressure used, and the type of fan or spray that is obtained. Distribution in the blender is also affected by the amount of air introduced by the spray system. Because these vary widely from plant to plant, each plant must determine the combination that produces the best results. Those with disc or centrifugal atomizers have the droplet size adjustable by the factors controlling the atomizer, such as rotor speed and feed rate.

The tremendous difference in surface area between typical coarse particles and typical fine particles is shown in Figure 13.2. A 1-g sample of the $-6+9$ screen fraction had about 12 in.^{2} (0.008 m^2) of surface area, while the -42 screen fraction had about 100 in.^{2} (0.065 m^2) of surface area. Thus the small fraction had almost eight times as much surface area requiring resin.

In addition to the blender, which plays a primary role, there are other factors which influence resin distribution and efficiency. The ability to produce a large number of adequate bonds is dependent upon such factors as particle geometry, the ratio of various mesh sizes of particles present, moisture content, fiber structural integrity, density of particles, resin efficiency, and density of the finished board, as well as variables in the mat-forming process and pressing conditions.

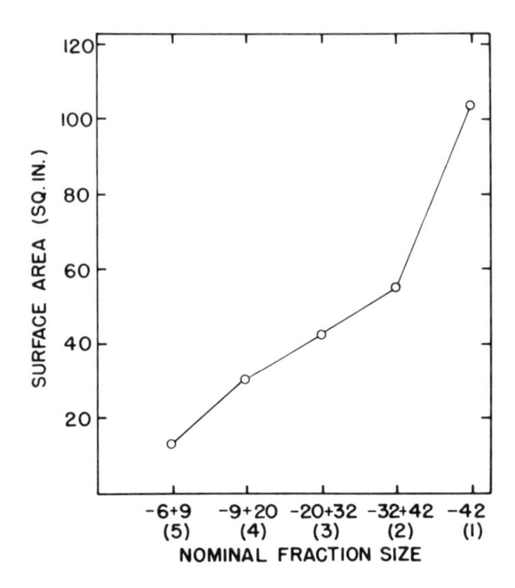

Figure 13.2. *Surface area of five screen fraction sizes based on one-gram samples (Maloney 1970).*

Particle size and form are definite factors in controlling board properties. The resin probably approaches most closely the desired maximum efficiency when applied to surfaces that manifest the lowest possible surface fiber damage. These intact surfaces allow better bonding of particles at lower pressures than do the rougher, surface-damaged particles. This fiber structural integrity allows the resin to function properly.

The glue-bond efficiency of the resin used in boards produced with a fine degree of atomization will be more uniform than discrete droplets from a coarse spray. In long-retention-time blenders, generally there has been little transfer of resin from one particle to another, even though prolonged mixing may take place. Some distribution by rubbing apparently takes place in short-retention-time blenders. Therefore, each board manufacturer must optimize the degree of atomization. Better spraying may sometimes be obtained by dilution of resin, but care must be taken to insure that such factors as lower tack, excessive penetration, higher moisture content or other side effects do not interfere with the best resin efficiency. Many times, conditions in the forming station will dictate what may be done. Pressing conditions also play an important role here, as do processing variables. Coupled with the importance of the individual components, one readily sees that all of these contributory factors are closely interrelated.

Recent experimentation has shown a possible change in resin transfer characteristics, as transfer of resin from particle to particle has been found. Perhaps the changes in resin formulation through the years are responsible for this transfer.

As mentioned, uniform distribution of the resin is of paramount importance. Regardless of the type of board produced, the most decisive factor is the degree of utilization of the resin. Properties of the board are reflected by the quality and quantity of glue bonds formed. Properties of the board improve with an increase in resin content, up to the point where the resin approaches filling all voids. But increasing resin content is not always the salvation it seems to be; because maximum resin efficiency must be balanced with economics, and this dictates minimizing the amount of resin used, particularly nowadays with the cost of resin increasing significantly. If resin usage is to be kept at the minimum, then other factors will have to be optimized.

One of the factors to be optimized, if the dictates of defined resin efficiency are to be followed, is the production of high-quality furnish. In making homogeneous board, the particles should only be refined to the size necessary to achieve ultimate board properties. This means that excessive fines should not be generated as they would, in turn, create excessive surface area to be bonded, resulting in poorer efficiency and utilization of the resin. If the particle geometry in a homogeneous board is similar between large and small particles, then an equal distribution of resin, on a surface area basis, between the particle sizes could conceivably take place.

Board properties are influenced not only by particle geometry and inherent properties of the wood species used, but also by resin content, the compression ratio of the species as determined by specific gravity, and moisture content. Increasing resin content of the board is the easiest way to improve its properties, but only to a maximum level. It must be borne in mind that the better properties obtained by increasing the resin level involve considerable expense. There is more resin at higher levels of addition that is not doing any significant bonding. This once again brings about the need for fiber structural integrity within the board so the resin can function effectively.

As noted previously, some hardboard operations apply the phenolic binder resin in attrition mills before the furnish is dried. However, U.S. particleboard plants dry the particles to a low moisture content, which allows for the moisture gained in the application of resins and other additives to give the desired final mat moisture content into the hot press. Some are finding that any moisture needed to adjust the moisture content should be added before the resin application as resin "strike in" or absorption by the furnish is reduced. There is no post-drying of the resin-blended furnish except for that resulting from the heat in the furnish out of the dryer and from exposure to air between the blender and the form station.

The temperature of the furnish directly from the dryer can, and does, affect the efficiency of the resin. Care must be taken to insure that the temperature of the furnish is not excessively high. If high furnish temperature is present, dryout of the resin can occur, resulting in loss of tack, too much resin penetration, and, in extreme cases, some precure. This excessive dryout can be amplified by using hot cauls. All of these should be avoided. Furnish coming to the blender should probably be kept below 100°F (38°C). If it is known that high temperatures exist, however, it is often possible to design a resin specifically to overcome or alleviate these problems.

One of the more irritating problems in blenders is the buildup of resin and particles within the blender, which markedly diminishes the efficiency of the blender and causes periodical maintenance shutdowns. This is because the build-up of resin and wood particles which adhere to the sides of the blender and the agitators dries out, and eventually becomes a hard mass which interferes with normal operation. This may occur at regular intervals ranging from as little as three or four hours up to several days. Factors which influence this are choice of resins, size of particles, type of agitation, and many of the same factors already mentioned. Some blenders may be more prone to this type of problem than other blenders under the same conditions. Newer blender designs are now helping to solve this problem, and these will be mentioned later.

Resin application, its efficiency, and its effect on the quality of the finished board, at its simplest, involves a myriad of complex interrelationships between all the integral constituents of the board, plus the influences of the many process variables.

WAX APPLICATION

The application of wax emulsion can be done before, with, or after applying the resin. No particular problems occur with urea resins; however, some decrease in sizing quality may occur when blending emulsions with phenolic resins. Emulsions could rarely be mixed with phenolics at first, but most problems have now been overcome. Thus, wax can be added where desired in the blending process with little or no deleterious effects.

Making emulsions work as they should is primarily a matter of maintaining their original quality. They require a well-designed handling system that avoids all the conditions that can make the wax separate before it should.

The best system is the one that requires minimum operating attention and maintenance. Plant personnel should not have to become experts in handling emulsions, as the system itself should take into account all possible variables that could cause trouble. It should be possible to apply emulsion without constant attention to the system, and this goal has been realized by the research and development activities of wax manufacturers and board plants.

Temperature control is important in shipping as well as in storage. Below 60°F (16°C), emulsions have less resistance to shear and are susceptible to damage and deterioration when flowing through pumps and valves. Temperatures below 40°F (4°C) result in separation that cannot be corrected by reheating or agitation. Allowing temperatures to rise above 110°F (43°C) can result in a coarseness of the wax-water mixture. Furthermore, rapid evaporation, which leaves a crust on the surface of the emulsion, can take place. Ideally, emulsion temperature should be maintained between 70° and 80°F (21°-27°C).

Insulated tank cars and trucks usually restrict temperature changes to less than 1°F (0.56°C) per hour while an emulsion is in transit, which is especially needed for protecting the emulsion for one to two days in subfreezing weather. Freezing, of course, breaks down the emulsion.

Outside storage tanks must be insulated and heated for protection against freezing temperatures. They may be heated with either a hot water piping loop at the tank bottom or by electric heating cables wrapped around the outside of the tank. Electric heating cables must be arranged and insulated to avoid hot spots that might cause localized boiling of the emulsion. Steam cannot be used because hot spots cannot be avoided in a steam system.

Spraying or atomizing by centrifugal force will be covered under "blenders," since the resin and wax are usually mixed together. However, as with resin, it is important to distribute the wax uniformly over the particles.

Solid wax must be heated for liquefaction before it can be sprayed. This calls for a completely hot system, including the spray nozzles. A problem can occur if the spray pattern is too long for applying the wax in the blender, as the wax will solidify before striking the furnish and thus it may not adhere properly.

INSTRUMENTATION AND MEASURING

For any blender to function efficiently, good control must be exercised over its operation. Accurate control of the ratio of resin, wax, or other additives to the furnish is necessary. Relatively simple equipment is used for preparing the additives prior to blending in U.S. plants, mainly because many of the U.S. resins are internally catalyzed. In Europe, however, it is common to mix resin, catalyst, ammonia, and even starch extender for face resins in the board plant. The resin preparation areas, which are called *glue kitchens* in Europe, are usually rather messy places. It is, however, a very important part of the plant, as errors here only cause problems with the board properties if too little resin or wax is used. On the other hand, excessive errors on the side of adding too much additive will increase the cost of producing the board significantly.

The simplest method of preparing the additive mixtures is to measure out each material in buckets and then mix them together in a tank. This system allows for easy changes in additive formulas, but is subject to frequent operator errors.

A common metering system is shown in Figure 13.3. It consists of a number of transparent graduated plastic cylinders. The level of the liquid in each cylinder can be varied by the positioning of magnetic switches, which means a plant can easily change its additive formulas. There can be problems with accurately measuring wax emulsion, as it may tend to foam. If a starch mixture is used in the face resin, it has been found that it frequently segregates in the cylinder and sometimes plugs the drain valve. The plastic cylinders used with emulsion also lose their transparency, and visual control of the flow is not possible.

Another system uses scales for measuring the amounts of the various additives. Limit switches are used to interrupt the flow of the fluids at preset scale readings. The use of punched cards makes changes in glue formulation relatively easy. The tolerance of the scale, however, does not allow metering of large and very small amounts of material on the same scale. Consequently, a two-range scale is used, or the main portion of the resin is metered into the mixing tank by a flowmeter or level indicator. The scales work with electronic controls, and at times the maintenance on them is not done efficiently. Stirring equipment is needed to obtain a homogeneous mix of all the materials going into the solution to be applied to the furnish in the blender. Investment costs for such an installation are high.

Proportioning pumps for metering the various materials for blending have been used in the industry for many years. All of the components are stored in containers above or near the pump unit. A pump unit consists of a drive motor, the gears for the pumps, and the respective pump heads. The measuring pumps are oscillating

Figure 13.3. *Graduated cylinder for measuring additives (Senn 1975).*

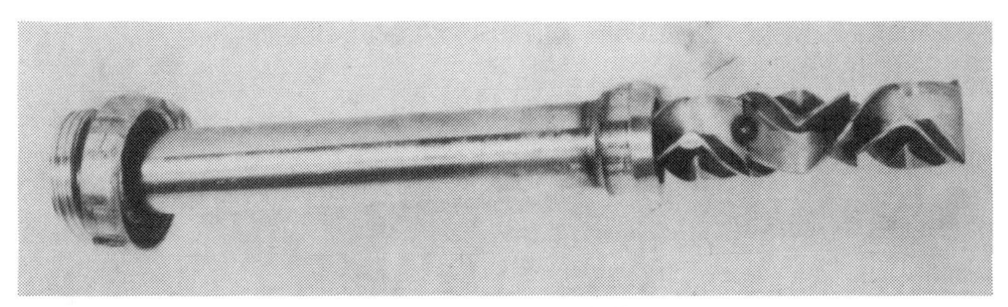

Figure 13.4. *One type of static mixer (Senn 1975).*

positive compressors in plunger-piston head construction. The quantity of fluid conveyed can be varied by changing the length of the stroke and/or the frequency. Such a pump is a conveyor and measuring device all in one. Units with speed transmitter and indicator are suitable for supplying ready-to-use resin mixes directly to the blender. Changes in the resin formula are achieved within seconds by changing the stroke of the respective pump. The proportioning pump system allows very accurate adjustment to binder requirements. The measured materials can be conveyed into a conventional mixing tank, or through a static mixer for homogenization.

The pump valves, the packing, and sometimes even the heads have to have a certain degree of maintenance. The overall costs over a year are higher than for the measuring or dosing cylinder equipment. The maintenance, however, is purely mechanical and may be performed more efficiently than with hydraulic or electric equipment. The investment costs, however, are high, particularly if a high degree of automation with respect to safety in cleaning is incorporated.

In those cases, particularly out of the United States, where the final mix of the resin and other additives is performed in the board plant, the aforementioned mixing tanks and static mixers are used for thoroughly mixing or homogenizing the additives. Stirring devices in the tanks include rotating discs on the bottom, vertical paddles, or blade stirrers. This mixing of the additives can cause problems when air is dispersed into the mix during the stirring operation. If this is the case, it must be recognized and handled in the plant as the volume of the additive mix will not accurately reflect the weight of material being delivered to the blender, since it can contain as much as 20% by volume of air.

The static mixers have to be used in connection with metering or dosing pumps. These mixers continuously blend the different fluids in parallel flows. Only the fluid is moved — not the mixer. The mixing energy is supplied by the metering pumps, and the result is an even distribution of the different materials throughout the fluid. These are closed systems; and since no high-speed agitation is needed, no air is emulsified into the additive mix. Thus flow accuracy is improved.

One type of static mixer is composed of several mixing elements, which are usually staggered in a pipe (Figure 13.4). Each element consists of perforated plates or spirals which are arranged in a way to form channels crossing each other. As the fluid flows through the mixer, it is divided into a great number of single streams which are then reunited at various points throughout the mixer. Since the fluid is forced to flow in a certain direction, the mixing process is independent of

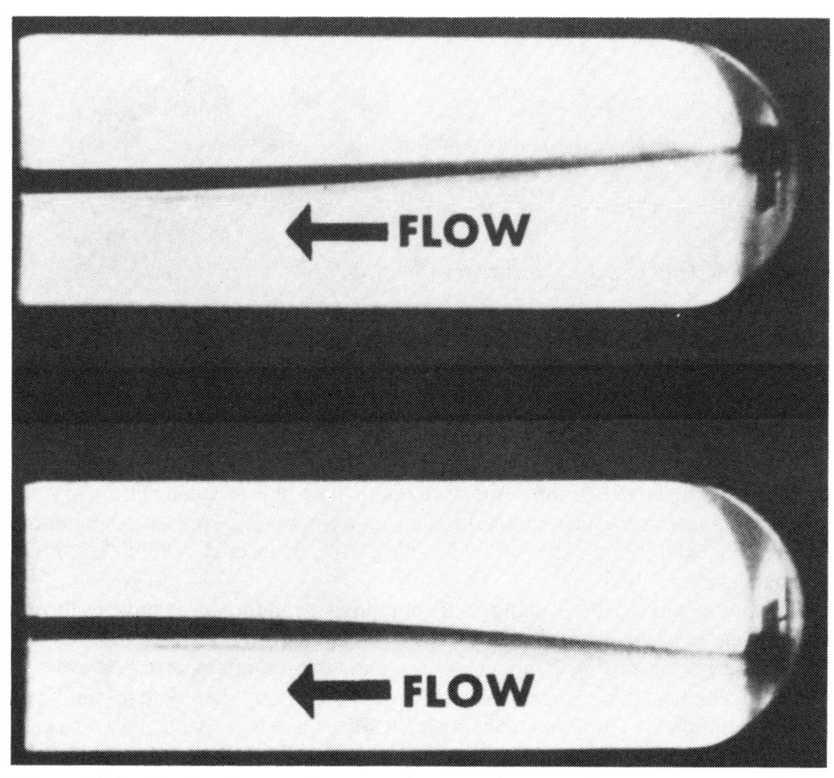

Figure 13.5. *Strobe photos showing vibrating element of in-line mixer in action. The liquid is flowing from right to left (Jacobs 1969).*

the kind of current, which can be either laminar or turbulent. Another type of mixer is shown in Figure 13.5. This mixer operates simply by the force of the liquids passing through it. The rectangularly shaped blade in the mixer will vibrate as many as 500 times a second with a pressure drop of 20 to 30 psi (138–207 kPa).

Plants have had difficulties with the first type of static mixer when operating without a retention valve at the inlet for each component. The unit is very compact if no intermediate tanks are installed after the mixing takes place. Some plants, however, use intermediate tanks in order to avoid downtime on the forming line if a valve in the pump fails. Others do not use the tanks in order to avoid any precure of the resin that may occur, solidifying the resin in the tanks. This is particularly important if a major failure occurs in the plant, shutting down the process line.

Maintenance of this type of mixer is negligible, and it only has to be dismantled perhaps once a month and cleaned mechanically. Some plants have eliminated the use of a flowmeter leading to the blender, and count the strokes of the glue metering pump to determine the amount of glue being delivered.

The magnetic flowmeter is one instrument apparently proven for metering and controlling additives, although there is not complete agreement on this point.

Coupled with a continuous measurement of the weight of the furnish entering the blender (mechanically or by nuclear gauge), the magnetic flowmeter enables the design of systems that will control the ratio of additives to wood both accurately and economically.

The magnetic flowmeter operates in accordance with Faraday's law of electromagnetic induction, which states that the voltage induced in a conductor of length moving through a magnetic field is proportional to the velocity of the conductor. By design, all variables are constant except velocity of the chemical through the flowmeter. The velocity is directly related to flow volume through the meter and is measured by a small electrode which generates a microvolt output. The microvolt output is linear with respect to volume through the meter. This microvolt signal is an accurate indication of the chemical flow volume through the meter.

In the pneumatic ratio controller used, the reference value for the ratioing of chemicals is the wood weight signal. Chemicals are ratioed according to this signal by use of levers and gears. Proper ratioing is accurately maintained by the feedback signal from the magnetic flowmeter. A scale running from 0 to 100% can be used to record chemical and wood usage. As an example, wood usage runs from no wood (0%) to 13 tons (11.8 mt) of wood per hour (100%) and chemical usage runs from no chemical (0%) to 10 gallons (37.9 liters) per minute (100%). Visually, it is possible to see resin and wood usage on a recording chart. Also, totalizers for recording chemical and wood usage are normally included for calculating resin and wax percentages.

Some of the characteristics of the continuous chemical/wood ratio controller include:

1. Accuracy is good when properly maintained. (Clean air is a must for proper operation.)
2. The maintenance schedule is easy to follow. Plant personnel check the system weekly for flowmeter accuracy and linearity. The ratio controller representative calibrates the system periodically.

If many ingredients must be measured, as in Europe, investment costs are higher than with metering or dosing cylinders.

A typical wood-to-resin ratio control system (Figure 13.6) is based on a single ratio controller. The wood flow is measured on a belt feeder. The wood flow rate becomes the primary measurement to the ratio controller, and the resin flow is adjusted to meet a predetermined resin-to-wood ratio. The actual modulation of resin flow is accomplished with a variable-speed-drive pump, or a constant-speed pump with a control valve downstream of the magnetic flowmeter. A "test-run" switch is provided to allow the operator to simulate wood flow during startup or to check the ratio controller and resin loop.

The system in Figure 13.6 may be duplicated with small-case pneumatic recorders and controllers (Figure 13.7). This small, light equipment allows for more compact panel boards when many loops are to be controlled from a central room. Furthermore, recording and controlling functions are separate, so that a failure in one will not influence the operation of the other. The record itself, while

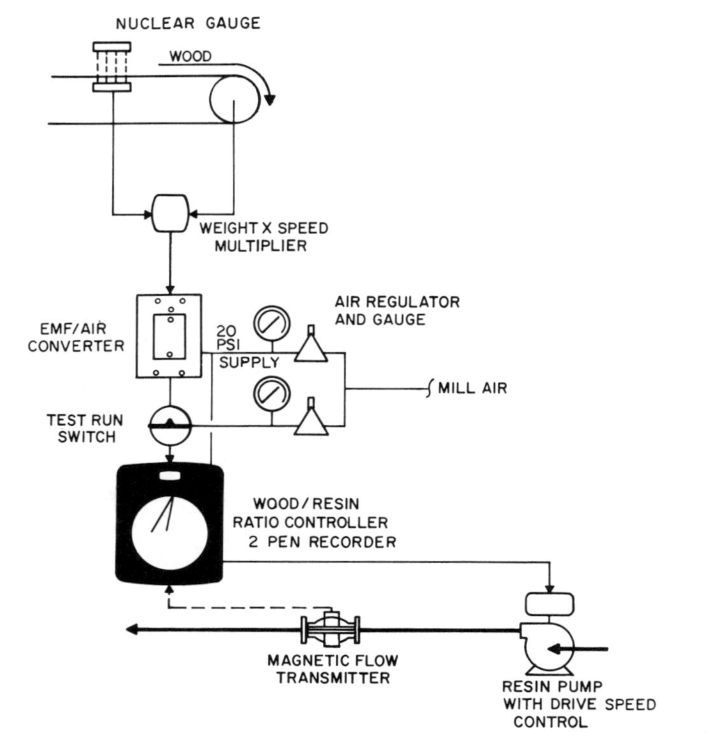

Figure 13.6. *A pneumatic wood/resin ratio control system (Ballard and Schlavin 1969).*

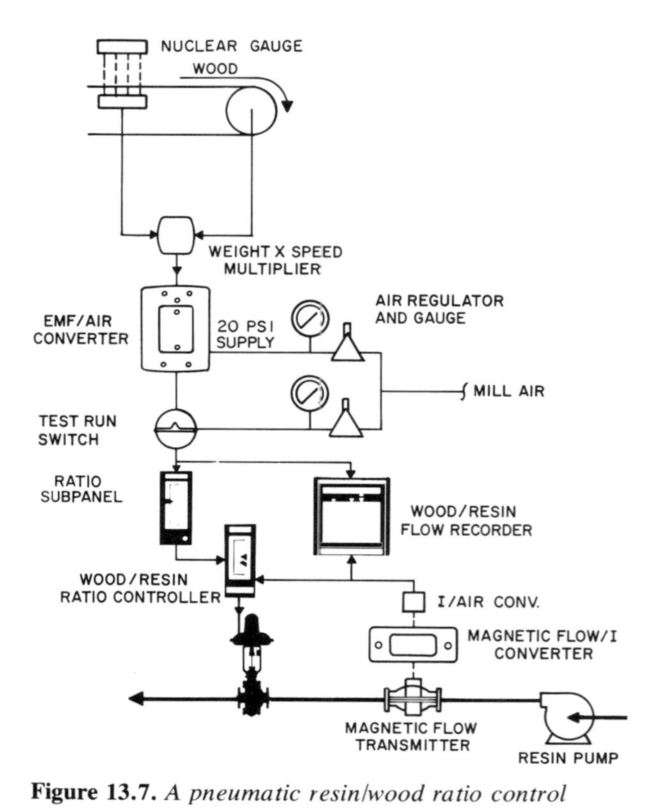

Figure 13.7. *A pneumatic resin/wood ratio control system with small-case pneumatic instrumentation (Ballard and Schlavin 1969).*

apparently smaller than the circular records, offers a scale of equal size and readability. This system is pneumatically actuated, and electronic-to-pneumatic converters are required.

A third system with identical control objectives is made up entirely of solid-state electronic instrumentation (Figure 13.8), with the exception of the final operator. Like the miniature pneumatic system, this equipment consumes little space. Recording and control functions are separate. The convertors are entirely electronic and have no moving parts. No instrument air is required.

Where the additive or additives are to be ratioed to the total product flow instead of wood flow alone, a system such as that shown in Figure 13.9 has been applied successfully. The resin here is ratioed to the total materials flow (wood plus resin) so that the ratio setting can be in percent of total.

The density of the blended product is measured with a gamma gauge. This signal multiplied by belt speed is the mass flow of the total product. The desired ratio of resin to total is manually set by the operator on the ratio controller dial. Any deviation between the ratio set on the controller and the ratio determined by a comparison of resin flow to a total product flow results in a change in controller output to the pump speed controller.

The instrumentation shown in this system is entirely solid-state electronic. A magnetic clutch drive is employed on the resin pump. The varying 10- to 50-milliamp DC signal produced by the controller increases or decreases pump speed to provide resin in the ratio as set on the ratio controller dial. Possible additional features of this sytem include digital flow integration on both the resin and total product flows. Since both measurements may be converted to proportional 10- to 50-milliamp DC signals, and those signals converted to digital pulse signals, the pulses may be totalized continuously in a resettable or nonresettable electromechanical totalizer. The nonresettable totalizer is frequently preferred where the information is to be used for accounting or plant balance purposes.

Additional ratio control loops are all that is required to expand the system in Figure 13.9 for multiple additive control such as shown in Figure 13.10.

The wood measurement can be done mechanically or with a nuclear gauge. These devices can be used elsewhere in the plant, and this description should cover other applications as well. The mechanical weight-o-meter is a system counterbalanced with weights as shown in Figure 13.11. A lever system is used to transfer the wood weight to the control box. The belt carrying the wood turns the integrating drive wheel. A chain attached to the integrating drive wheel provides the power for the mechanism in the control box. The integrating drive wheel also allows for varying belt speeds as the wood load changes.

To fully understand how the weight-o-meter operates, an examination of Figure 13.12 is in order. This shows the mechanism inside the control box. The weight is transferred from the bottom of the control box with levers and fulcrums to the dash pot. The counterweights counterbalance the wood weight and the dash pot dampens the system. The pendulum arm is connected securely to the horizontal member in the top of the box, so that it swings to the left as wood weight is applied. The pendulum arm is vertical when no weight is on the scale. As weight is applied, the levers connected to the base of the pendulum move the measuring wheel off the exact center of the integrating wheel. This wheel, as explained

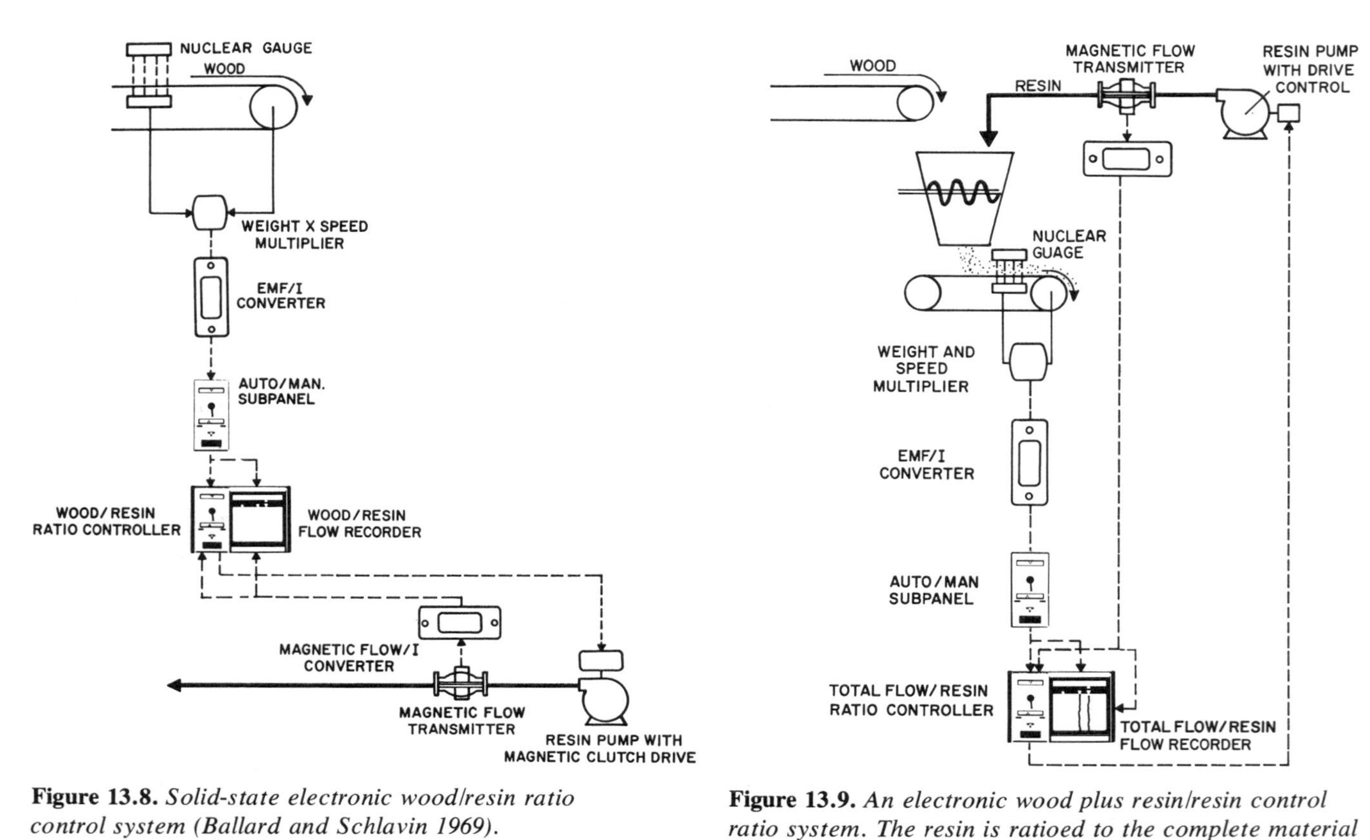

Figure 13.8. *Solid-state electronic wood/resin ratio control system (Ballard and Schlavin 1969).*

Figure 13.9. *An electronic wood plus resin/resin control ratio system. The resin is ratioed to the complete material flow of wood plus resin (Ballard and Schlavin 1969).*

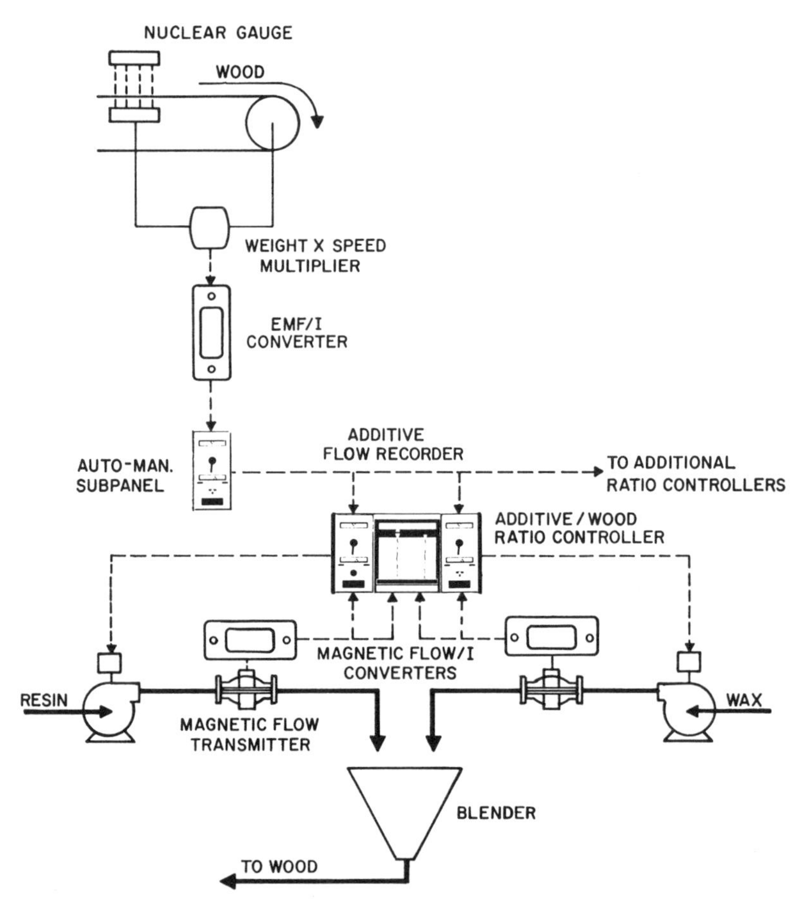

Figure 13.10. *An electronic/wood/resin/wax ratio control system. This is the same as the system shown in Figure 13.9 except for the addition of a wax ratio control based on the complete material flow of wood, resin, and wax (Ballard and Schlavin 1969).*

earlier, is driven by the belt carrying the wood. The measuring wheel starts to turn as it is forced off the exact center of the integrating wheel when wood weight is applied. As more weight is applied, the pendulum arm moves further to the left, forcing the measuring wheel to turn faster. The measuring wheel, in turn, is connected to a totalizing unit which indicates the amount of wood as read in pounds.

To obtain an electrical signal for the wood/chemical ratio control system, a cam and limit switch are used to pick up a pulse signal from the measuring wheel. The pulse signal is then converted to a 10- to 50-milliampere signal for the wood/chemical ratio controller. This signal is used as the wood usage signal for the ratioing of chemicals.

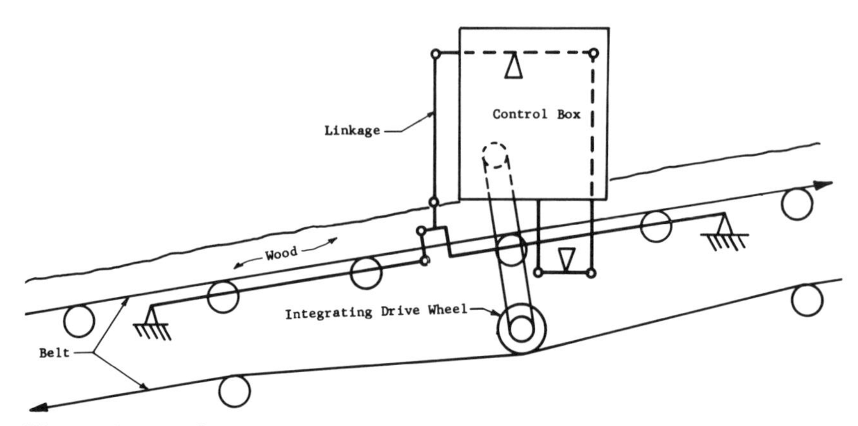

Figure 13.11. *Schematic drawing showing linkage for continuous wood weight-o-meter (Carter 1969).*

Some of the characteristics of the continuous wood weight-o-meter include:

1. Excellent accuracy when properly maintained.
2. An accurate, reliable signal is obtained even under adverse dusty conditions.
3. The maintenance schedule is easy to follow. Plant personnel check zero and accuracy with three different chains every week, and a scale representative checks the system periodically.

Belt conveyor scales are also available using a linear variable differential transformer to provide the weight signal to an integrator. Solid-state electronic

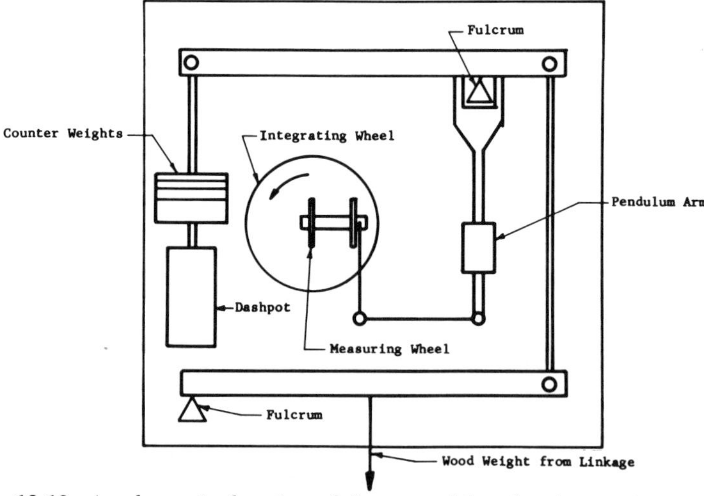

Figure 13.12. *A schematic drawing of the control box for the continuous wood weight-o-meter (Carter 1969).*

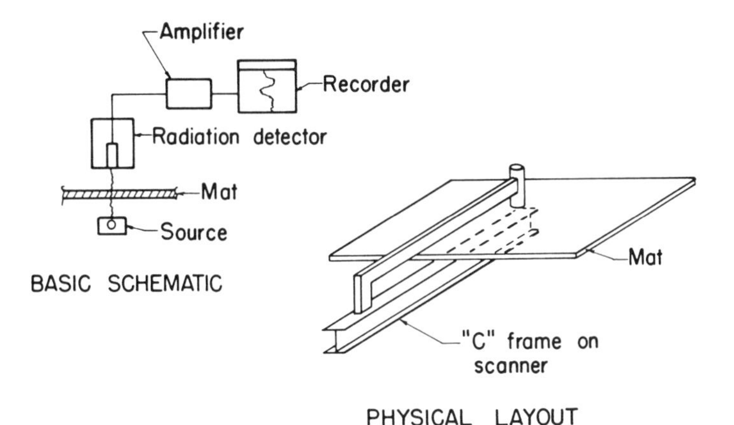

Figure 13.13. *Point measurement geometry (Rowe 1969).*

circuits which are less susceptible to plant environmental problems are used in such systems.

Nuclear gauges are becoming routine control instruments in many industries, including those producing board. For process control these are used to measure total mass. If a "point" measurement geometry is used (Figure 13.13), the gauge calibration becomes "weight-per-unit-area," which is the multiple of product density times the product thickness. Thus if the product thickness is maintained constant, the gauge calibration can be interpreted directly as density, and conversely, if the density is constant, as it is in metals and some plastics, the output is termed *thickness*.

Absorbing of radiation using beta rays and higher energy gamma photons (over 40 KeV energy, approximately), is based on the following equation:

$$R_{\text{output}} = R_{\text{air}} \times e^{-upt}$$

where (p) is density, (t) is thickness and (u) is the absorption coefficient, which, for higher energy gamma and beta radiation is proportional to Z/A, the atomic number divided by the atomic weight. For variations in wood fiber chemistry and water, the *Z/A* ratio remains constant enough so that these gauges are relatively insensitive to changes in fiber species and moisture content. Thus water weight is measured as just so much additional mass.

When it is desired to measure mass flow of wood fiber, use of strip geometry is required to insure measurement of the total-weight-per-unit-length along a conveyor. Thus, as shown in Figure 13.14, the variations in width of product as well as its weight-per-unit-area along a unit of length can vary the amount of radiation reaching the gauge in a linear fashion, thus measuring total-weight-per-unit-length. If speed is measured also, the strip gauge output is multiplied by this speed signal, the result is mass flow rate. And when mass flow rate is integrated with respect to time or distance, the total weight is determined (Figure 13.15).

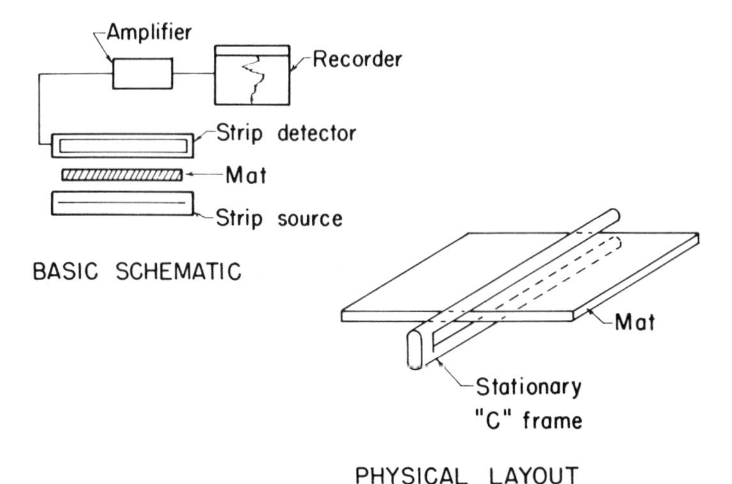

Figure 13.14. *Strip measurement geometry (Rowe 1969).*

The term *beta gauge* is a misnomer in that beta radiation as such cannot penetrate the mass-per-unit-area involved in the process. Beta sources can be used to create a secondary low-energy gamma radiation, called *Bremsstrahlung,* and this radiation has been used in the past to measure wood fiber mass. Its energy spectrum is quite broad, which is not helpful as too much energy of an excessively high level causes poor signal-to-noise ratio. These gauges do not need energies over about 70 keV for most applications on fiber mats, and all beta/Bremsstrahlung sources have spectra extending well over 150 keV. With the advent of Americium 241, and its accessibility in the market, a relatively monoenergetic source, the detector can be tuned for maximum output at that energy, leading to a high current output capable of higher speed measurement. As detector output increases, smaller input resistors can be used in the electrometer amplifier, permitting shorter time response in the amplifier.

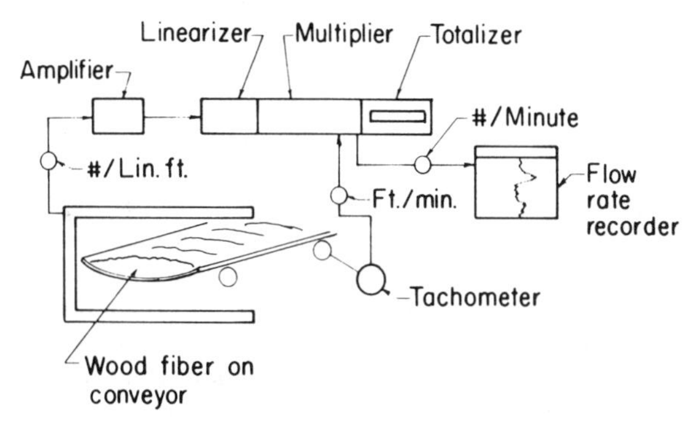

Figure 13.15. *Mass flow and total weight schematic (Rowe 1969).*

Sources are made in double-sealed stainless steel housings designed against leakage of the active material caused by corrosion, heat, or physical shock. Source holders are to be made to have enough shielding to limit the radiation fields to under 5 milliroentgens (MR) at 12 in. (305 mm) from the gauge head in most applications. Under such conditions no extra shielding or posting of the area as a *radiation area* is required. This level of 5 MR/hour is such that a man can stand a foot from the gauge head for 20 hours a week without exceeding the Atomic Energy Commission (AEC) accumulated dosage level.

Wipe testing of the source holder surface is required every six months (three years for certain types of isotopes), to insure that no leakage has occurred. This is a very simple procedure which is usually performed by the user. An inexpensive wipe-test contract is available to supply the wipe-test kit and to analyze the results. Figure 13.16 shows a typical installation on a conveyor belt.

Figure 13.16. *Ray-Weigh scale on wood chip conveyor belt (Rowe 1969).*

Nuclear wood measuring devices can also be used elsewhere in the production process, particularly for controlling the delivery of the wood in the mat former, as will be mentioned later, and as level gauges on fiber bins.

BLENDERS

Over the years a number of different types of blenders have been developed for use in the board industry. Some of them, no longer used extensively, will be mentioned briefly in order to gain a historical perspective as well as to indicate approaches that were rejected after experience in the field. Blenders are essentially of two types: long-retention and short-retention time. The first type may take minutes to blend the additives with the resin while the latter type performs the job in seconds. The first type is large in size and the second one small (considering the same capacity). In general, the objective of blending is to achieve an even and uniform distribution of resin, wax, and other additives over the particles. With resin, the goal is to glue the particles together with as little resin as possible. Thus, very accurate metering of the wood, resin, wax, or other additives is essential. The measuring system, therefore, is a very important part of the blending unit used in the board industry.

It is also difficult to work with a nonuniform size wood material which can run from very fine particles to large flakes. Theoretically, the more uniform the particles according to size, the better resin distribution should be. Since there usually is a very wide distribution of particle sizes within the furnish, it is necessary to compromise the blending operation to arrive at the most economical method of blending. Fine and coarse particles may be blended separately, for example. Speed of blending is important in an economic sense. At one time poor distribution of resin was overcome by adding a little more resin. However, the recent resin shortages and attendant price increases have made this an undesirable solution. Consequently, blending is moving to the forefront as a segment of the production process that should be optimized. The importance of such work can also be gathered from the fact that a conventional shavings-type ¾-in. thick (19 mm) particleboard has four to five times as much resin as an equivalent piece of plywood. Composition boards probably will replace plywood in many applications because of plywood shortages. However, having to use excessive amounts of resin will hinder this important development.

Laboratory experimentation has shown that industrial blenders do not perform near optimum conditions. Thus, important developments can yet be made in this critical production step.

In early efforts with liquid resin, the resin was simply dripped upon the furnish for blending. This method was quickly supplanted by the spray nozzle. Many of the early blenders were of the batch type and they worked very well and were reasonably controlled. But these systems could not be used efficiently in large production units. Batch blenders have been applied to a large production unit in Germany where they had four batch blenders on a turntable. One was being filled while two were blending, and one was dumping blended furnish. This turntable kept going around and around, supplying material to the felter. The filling and the discharge of the batch blender and the timing obviously has certain limitations.

About 1953–1954 in the United States, continuous blending became the way to operate as the plants grew in size. Two of the first blenders of the continuous type were the Grenco and the Miller Hofft, both developed in the United States. These two demonstrate the primary difference between the two types of blenders being used today.

One has a short-retention time as exemplified by the Grenco. In this blender, the resin is sprayed or applied by centrifugally operating tubes into a turbulent airstream containing a thin layer of the wood particles. Excellent control in metering the resin and furnish is imperative because of the short blending time. In the long-retention-time blender, such as the Miller Hofft, furnish is slowly blended with the resin, which statistically provides for a greater opportunity for good resin distribution, particularly if the resin/wood metering system is not functioning properly.

In the late 1950s, the wind-sifting blender (a long-retention type) was developed in the Federal Republic of Germany. Both fine furnish and coarse furnish were blended simultaneously. However, the fine furnish was quickly blown to the outfeed end of the blender and was exposed only briefly to the resin spray. Coarser particles remained in the blender for a longer period of time. Even with this system, the fines received more resin than the coarse particles, and the terminology used to describe such uneven distribution was that the fines *hogged* resin. This particular blender, however, was very popular in the United States through most of the 1960s.

Some plants always used the short-retention-time blenders and a move back to them started in the late 1960s. At this time, the short-retention-time blender is the most popular in the United States, having replaced most of the long-retention-time blenders. Its move to the forefront has been so strong that refinements have been made to improve this type in the USA, and new designs have been developed in the Federal Republic of Germany. Indeed it is safe to say that this type of blender is the premier one available in the world today.

Long-Retention-Time Blenders

Long-retention-time blenders, as noted, are slow-speed machines that treat the furnish gently. Resin is usually applied by spray nozzle in fine discrete droplets. The theory is to expose all particle surfaces to the atomized resin so that an even resin distribution is achieved. The longer the blending time and the finer the resin droplet sizes, the more efficient the blending. Such blenders can be quite large; and, because of their size, extensive operating and maintenance problems are possible. Blender lengths of about 16 ft (4.88 m) are known.

Long-retention-time blenders can be generalized under the terms roller, paddle or shovel, rotary, and wind sifting. A number of variations of these types of blenders are possible, and it is not intended to cover all of them in this discussion.

Roller Blenders: One of the first blenders was designed for use in the Fahrni or Novopan system. This blender was intended for use with extended resin mixes that were relatively high in viscosity. One version of this machine is shown in Figure 13.17. The furnish is first divided into fine and coarse, with the fines being exposed to resin only on the lower glue-applying roll to prevent hogging of the

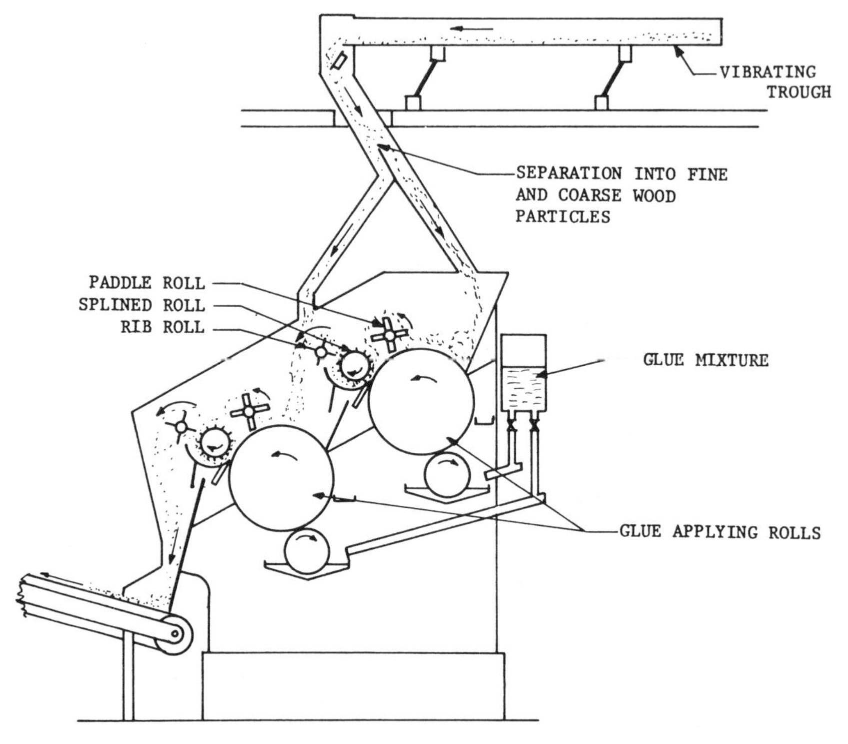

Figure 13.17. *Roller gluing machine, Fahrni system (Kollmann 1957).*

resin by the fines. The particles are rubbed onto the glue-applying roll by means of the paddle roll. The particles back up in front of this roll and are thoroughly blended with the additives before being carried over either to the next gluing roll or to the outfeed belt.

Paddle-Type Blenders: The early blender known as the paddle type is typified in the Miller-Hofft design developed in the 1950s. This blender consists of a long trough with a paddle arrangement inside that propels the furnish through. The agitated furnish is exposed to a number of spray nozzles set along the length of the blender. Figure 13.18 is a photograph of a small blender of this type showing the paddles.

A sophisticated version of this general type of blender has been developed that gently handles the furnish due to its low speed of operation. As shown in Figure 13.19, one central nozzle operating under higher pressures of up to 70 atmospheres is used for applying the resin or additives. Small droplets of resin are developed by this spraying system, resulting in satisfactory coating of even small particles. No undesirable wind currents are introduced into the blender because of the airless spraying; thus, the flow of the furnish is not disrupted. A centrifugal accelerator is used to form a conical curtain of particles in the blender. This style of blender is used throughout the world, particularly in the Novopan system.

A German design of another such blender took a different approach. Inside a cylindrical drum is mounted a revolving shaft which has heavy steel arms called shovels mounted axially (Figure 13.20). The furnish is picked up by these shovels, which are traveling fast enough to form a revolving envelope or ring of particles within the blender. Blending is accomplished in one and one-half to two minutes. The shovels are shaped to interchange the particles continuously from the outside of the ring through the center of the blender and back again as shown in Figure 13.21.

A different approach for applying the additives is used, namely high-speed centrifugal atomizers (Figure 13.21). No air, therefore, is necessary to achieve atomization of the additives. Each atomizer has about 125 holes for dispersing the additives and each one takes the place of several spray nozzles.

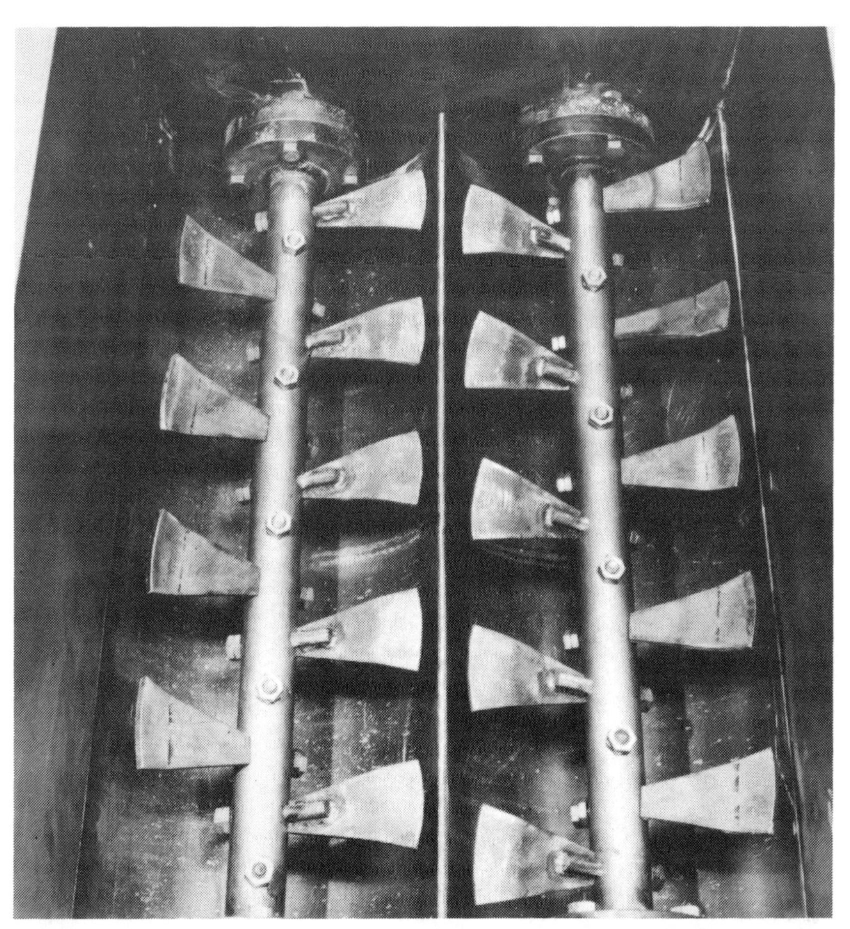

Figure 13.18. *Pilot mixer interior, looking down into blender. (Courtesy Miller-Hofft, Inc.)*

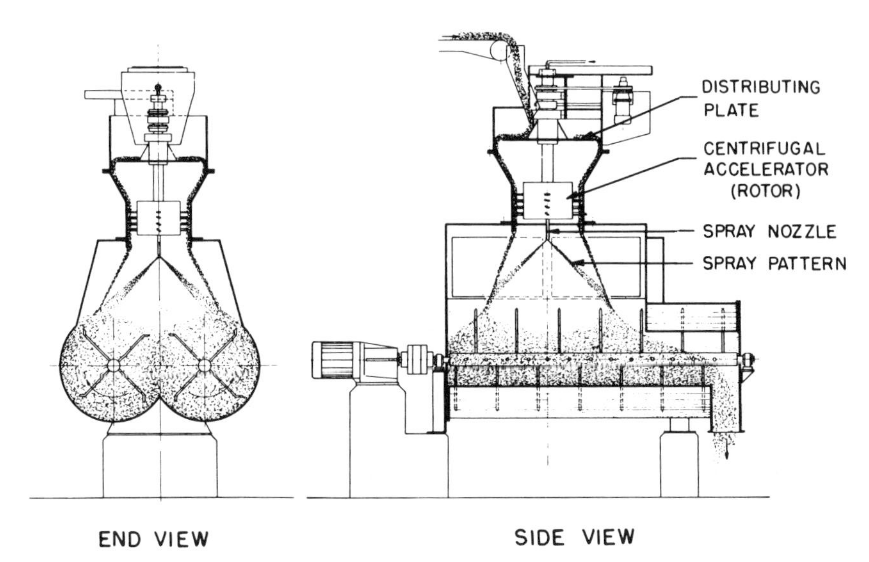

Figure 13.19. *Principle of the Fahrni gluing machine with one centrally placed special nozzle. Scale about 1:50. (Courtesy Fahrni Institute Ltd.)*

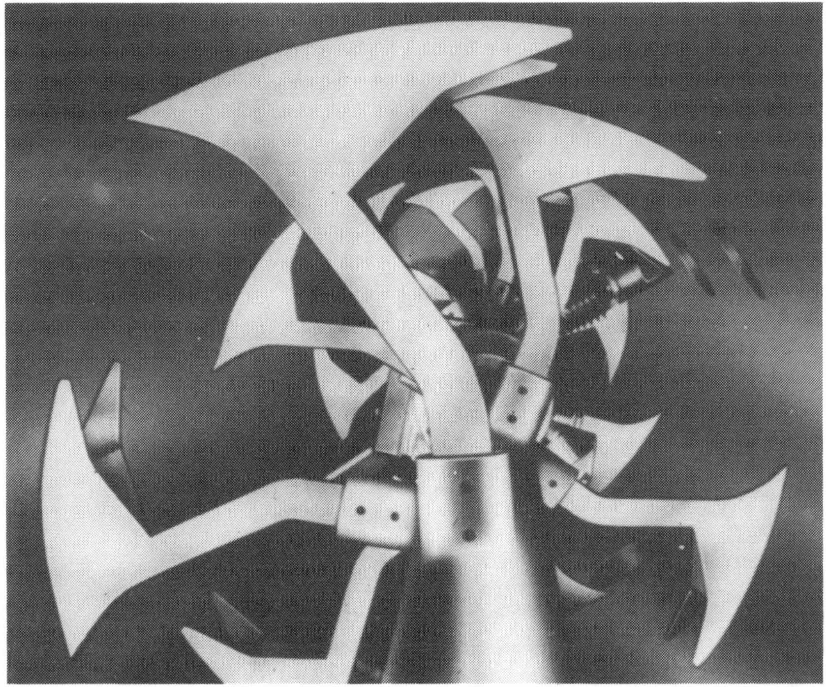

Figure 13.20. *View of the shaft and shovels in a Lödige Blender (Young 1967).*

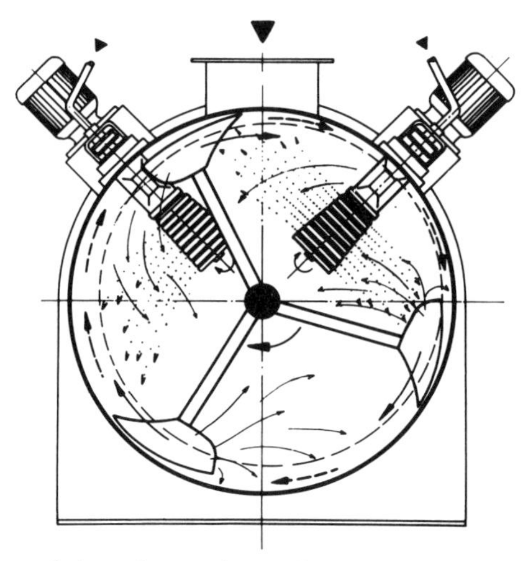

Figure 13.21. *View of ring of particles and disc atomizer (Young 1967).*

A blender for blending all of the furnish simultaneously is shown in Figure 13.22. Figure 13.23 is a view of a more sophisticated blender for separating the furnish by particle size before blending. Figure 13.24 is a drawing of the inside of the upper chamber shown in Figure 13.23. The furnish is screened with the coarsest furnish entering the blender so that it is exposed to the spray coming from all of the atomizer heads. The finer material is delivered to the blender at points

Figure 13.22. *Overall view of standard blender (Young 1967).*

Figure 13.23. *Overall view of sophisticated blender for preferential blending of furnish by particle size (Young 1967).*

beyond the first atomizer. This approach prevented at least some of the hogging of the resin by the fines.

Rotary Blenders: Probably the most obvious technique for blending is to use a rotating drum for tumbling the furnish and to spray the resin upon this tumbling mass. A number of designs have been developed and used, and this system is a popular one in research and development.

In one design, inside the revolving drum conveying the furnish through, is mounted a boom parallel to the axis of the drum. On this boom are mounted spray nozzles, the resin and wax manifolds, and the compressed air lines. The additives are sprayed at a right angle to the drum axis. Figure 13.25 illustrates this blender. The particles are exposed to resin spray the full length of the blender. The tumbling particles form several "curtains" of material so that overspray losses are minimized. It is important that spray nozzles on blenders can be observed and easily

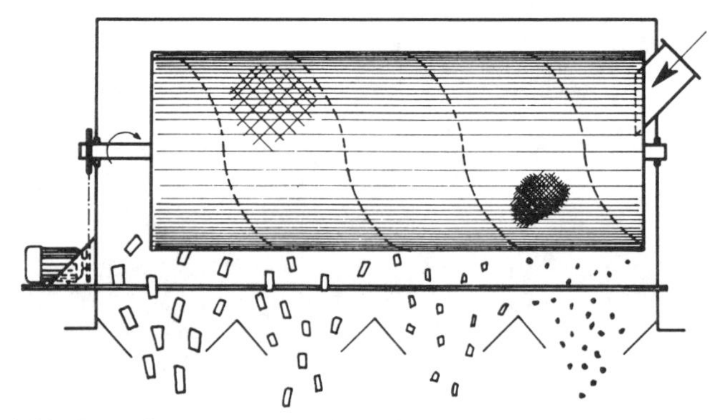

Figure 13.24. *Particles separating above blender (Young 1967).*

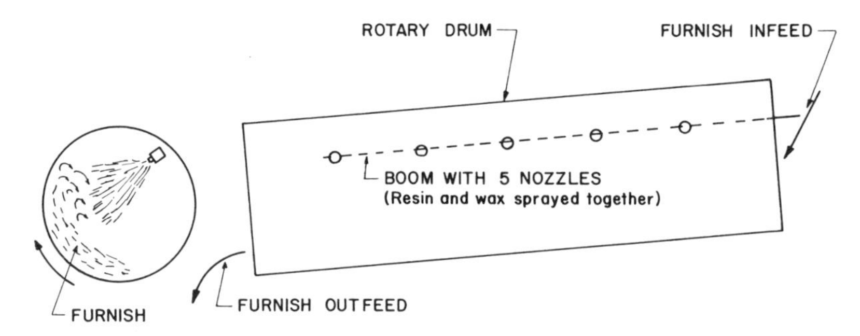

Figure 13.25. *Schematic drawing of a rotary blender with the spray nozzles set radially to the long axis of the blender drum. (Courtesy Washington State Univ.)*

maintained. However, in this blender they cannot be seen and spray nozzle problems can only be determined after the fact. Figure 13.26 shows the inside of one of these blenders during a maintenance shutdown.

Another design shown in Figure 13.27 has the additives sprayed parallel to the tumbling furnish. In this model, the spray nozzles are easily accessible so that malfunctions can be quickly rectified.

Wind-Sifting or Air-Classifying Blenders: This blender, which gained great popularity in the United States, was developed by Bähre Metallwerke (now Bison-Werke) in the 1950s. This manufacturer's first efforts with particleboard equipment were for its own use in its furniture manufacturing complex. With success in its endeavors, Bison expanded in the field not only with its own board plants but in manufacturing equipment and complete systems worldwide. One of

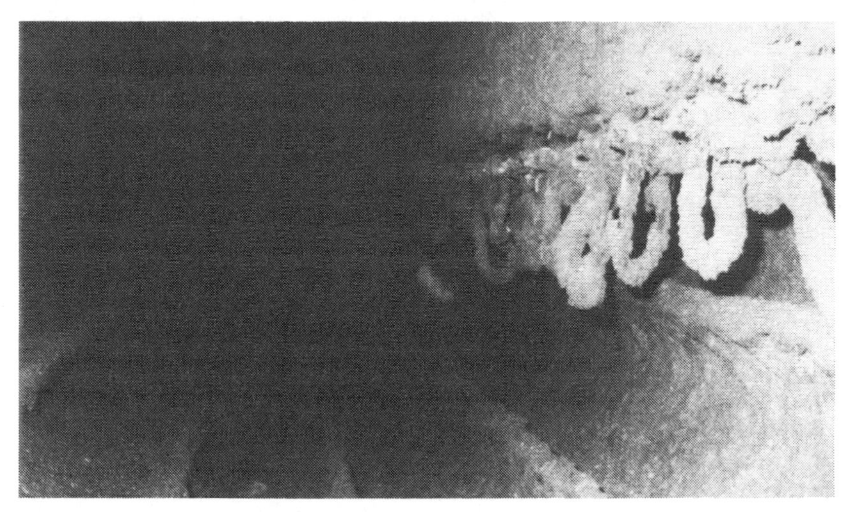

Figure 13.26. *The inside of a rotary blender, showing spray nozzles and flights (Thompson 1967).*

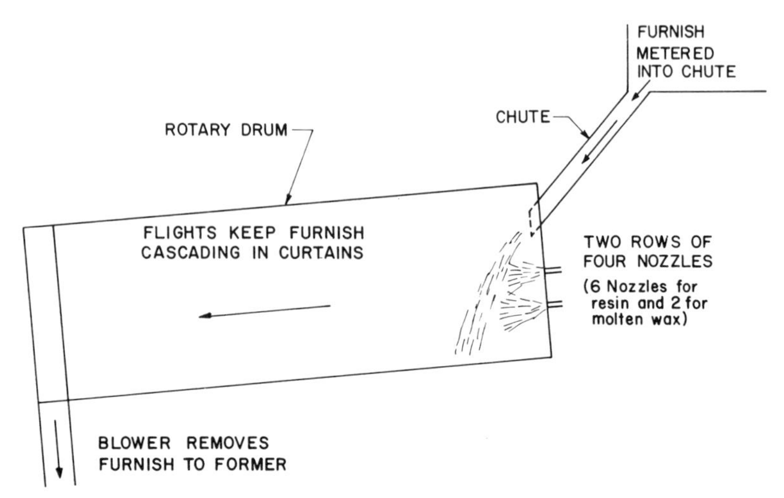

Figure 13.27. *Schematic drawing of a rotary blender with the spray nozzles set parallel to the long axis of the blender drum. (Courtesy Washington State Univ.)*

the pillars on which it has built its reputation is air-classifying of the furnish, particularly in blenders and forming stations.

Reportedly, the concept was first recognized while one of the individuals charged with coming up with a new approach was standing in a plant yard sifting dirt through his fingers on a windy day. He noticed the fine particles were blowing farther than the coarser particles. At the time a trough-type (paddle) mixer was being used in the operation. The air-classification idea was adapted to the paddle concept; the Bison air-classifying or wind-sifting blender and former were born.

A number of variations were developed for different applications. Figure 13.28 illustrates the basic principles. The furnish is dropped into an airstream at the inlet end of the blender. The coarse and therefore heavier particles drop straight down into the blender and are exposed to all of the spray nozzles. The rest of the particles are classified by size in a gradation from coarse to fine. Medium-size particles fall into the blender through the center of the blender and the very lightweight fines are propelled to the far end before being exposed to the resin spray. Rows of picker rolls or paddles propel the furnish through the blender thereafter. Resin is applied by two rows of spray nozzles mounted on top of the blender (shell) as shown in Figure 13.28. While the fines are not exposed to many of the spray nozzles, they still pick up a higher percentage of resin than the coarse particles. However, they would have an even higher resin content if exposed to all the spray nozzles. Thus this blender is capable of effectively blending various sizes of particles with additives simultaneously. These are large blenders as shown in Figure 13.29.

The furnish is fed into the blender off a weight scale on top of the blender. The weighed material is dumped into the surge bin on top of the blender and is then fed out of this bin at a controlled rate of feed. A picker roll or leveling device smooths the furnish out before it is fed into the blender proper.

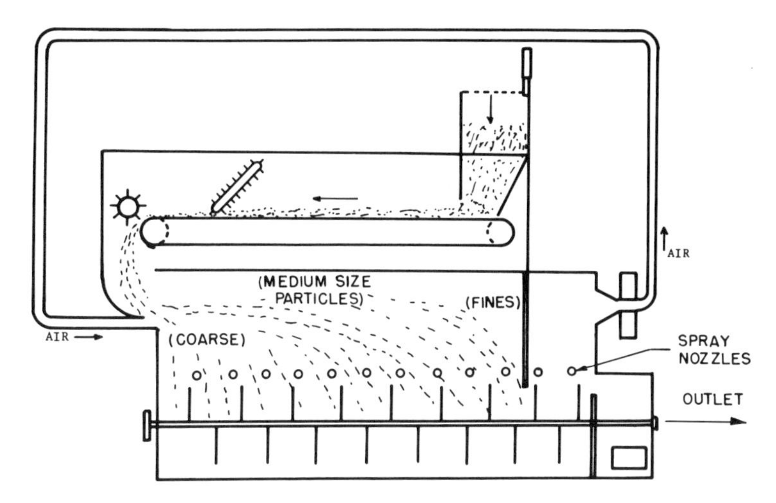

Figure 13.28. *Schematic drawing illustrating basic operating principles of a wind-sifting blender (Haigh 1967).*

Figure 13.29. *Long-range view of a wind-sifting blender (Haigh 1967).*

Short-Retention-Time Blenders

Short-retention-time blenders generally operate at high speeds as compared to the long-retention-time type. Blending, as mentioned previously, is accomplished in seconds. Four general types are available: blowline, attrition mill, vertical, and paddle. These blenders are relatively small and compact and their smaller size should reduce maintenance problems.

Blow-Line Blending: The simplest blender is the application of the additives into the blow line leading from the attrition mill to the dryer or from the blow line out of the dryer which is transporting the furnish to the former. The droplets strike the rapidly moving furnish, and further blending is done by the violent tumbling of the fiber in the blow line. This system is used in dry-process fiberboard plants.

Attrition-Mill Blenders: Attrition mills (normally double-revolving disc types) can be used for blending in two ways. In one, the additives can be blended with the furnish simultaneously with the grinding of fiber. Additives can be metered in with the basic wood raw material or they can be metered through a hollow shaft holding one of the grinding discs into the "eye" of the mill. In the other, special devil's-tooth or peg plates (Figure 13.1), or for that matter any other plates with the desired configuration, can be used to blend the additives with the fibrous furnish. The gap setting between the plates is much greater than when grinding fiber. This is usually a fiberboard blending technique; however, at least one particleboard plant blends wax and powdered resin with planer shavings in an attrition mill. The attrition mill also reduces the planer shavings in size simultaneously.

Vertical Blending: In a vertical blender, as shown in Figure 13.30, the furnish is dropped from a metering bin into a tower. Spray nozzles mounted at the top of the tower spray into the falling furnish. The resin and wood then spiral together onto a conveyor belt leading to the former. Vertical blenders have also been developed which spray the resin upwards through the falling furnish, but their use has been very limited.

Paddle-Type Blenders: Paddle-type blenders are the usual ones thought of when discussing short-retention-time blenders. The paddles are also called other names, such as *shovels* and *plows*. Long-retention-time blenders try to expose all particles to a fine mist of resin and other additives. This, as noted, takes a considerable length of time. A different approach is taken with the short-retention-time blenders.

Figure 13.31 is a schematic drawing of one type of these blenders. The furnish streaming through the inlet has a high resin level on those particles directly exposed to the spray while the others have little or no resin. The uniform distribution takes place during the violent subsequent tumbling action in the blender. A concentric ring of material forms in the blender as the furnish transits the blender, although there is some evidence that the furnish passes through in a helix or spiraling pattern. In general, depending on blender diameter and length, the rotor may turn between 700 and 1400 rpm. It is important that these blenders run loaded with furnish to insure that resin transfer takes place by rubbing particles together.

These are much smaller blenders than the large retention types illustrated in Figure 13.32. Because they are small, these blenders are much easier to operate and maintain. Malfunctions can be quickly corrected. Adjustments in resin and

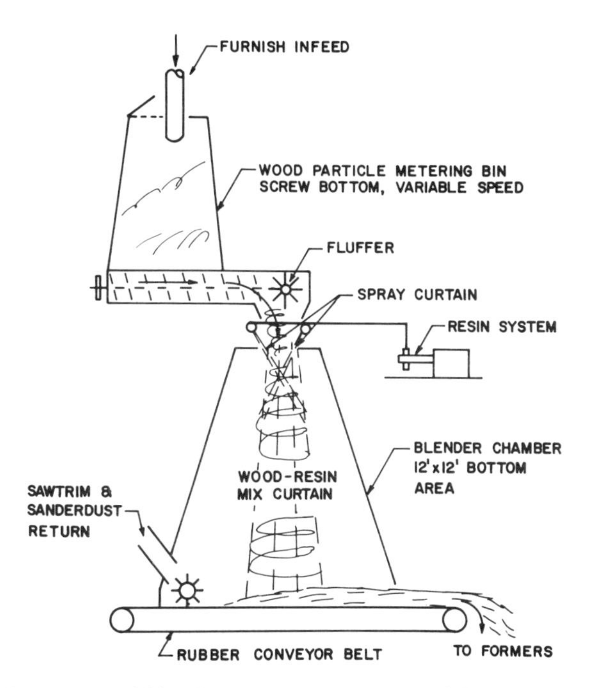

Figure 13.30. *A vertical blender. (Courtesy P. R. Walsh)*

other additive levels are fast to accomplish, which is important when changing production to a different product.

The first blenders of this type reportedly were used to blend molasses with alfalfa for cattle feed. They were first used for board in about 1948 in the development of the dry-process hardboard system. Later on they were used in early board plants of this type but were then supplanted by the introduction of simultaneous grinding of fiber and blending of additives in the attrition mill.

The first particleboard plants to use this blender were built in 1954. Others followed suit but the blender did not receive widespread use until about 1968. At

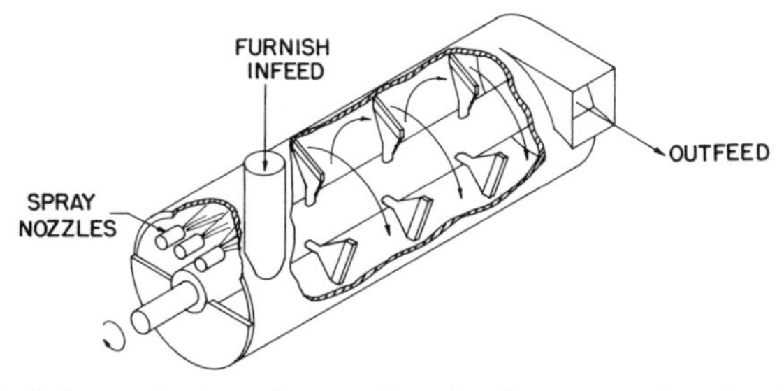

Figure 13.31. *Schematic view of operation of a short-retention-time blender using spray nozzles (Maloney 1975).*

Figure 13.32. *Continuous Drais Turboplan blender superimposed over a long-retention-time blender. (Photo composite courtesy Draiswerke GmbH.)*

that time, many U.S. plants using long-retention-time blenders started converting to the paddle-type short-retention-time blender. Today it is the dominant type used. Grencos and Keystones are made in the United States. In recent years, Bison, Drias, and Lödige (Littleford builds the Lödige in the United States) have developed competitive models.

These blenders have been used to blend all types of furnish. Resin can be applied by air and airless spray guns (Figure 13.33) dripped upon the furnish entering the blender or by centrifugal force within the blender. In this application resin is delivered into the blender through a rapidly rotating hollow rotor shaft.

The retention time in the blender depends upon several factors: blender size, shaft rpm, angle of the paddles, and furnish feed rate. The furnish passes through the blender along its periphery and is expelled tangentially, usually onto a belt.

Because of the relatively high operating speed and multiple-paddle rotor, the blender in a non-loaded condition also acts as an axial-type fan and can discharge high air volumes, thus creating a dusting problem. Blender loading and air intake have a significant effect on the amount of air discharge. If intake air can be restricted on the inlet side of the machine, then the air discharge is similarly restricted. The high-speed shaft action also ejects the furnish with sufficient force to make it self-conveying for a distance of 6 to 8 ft (1.83–2.44 m). Air-dusting can also be reduced by reducing the rotor speed and the paddle pitch.

The most popular type of discharge appears to be directly into a plenum-chamber-type housing over a belt conveyor feeding the blended furnish directly to the forming machine. In some installations, a return loop duct vents this plenum chamber back into the inlet side of the blender, thereby recirculating fines through the blender. In other installations, the plenum chamber over the conveyor may be directly vented into the suction side of the forming machine air system. Each installation seems to have its own peculiarities. There are also installations with a suction fan pickup for air-conveying of the blended furnish to the forming machine. Such installations should not be directly coupled to the blender since this additional suction tends to pull a portion of the resin spray through the eye of the blender right into the suction blower system which creates buildup in the fan piping and blower system. Air pollution problems are possible if the air is vented to atmosphere.

The capacity of these blenders can vary over a considerable range depending on style and manufacturer. Initially, when these machines were first considered for blending the new pressurized-refined fiber, the low bulk density of the fiber created concern that the short-retention-time blender would plug. This proved not to be the case. The concern was in error because the bulk density of the fiber was rated on a static basis instead of a dynamic basis. At rest, pressure-refined fiber may have a 2 lbs/ft^3 bulk density (0.032 sp gr). However, when this fiber is injected into the short-retention-type blender, the centrifugal action of the rotor separates the fiber from the static air in a similar manner to that in which a centrifuge separates solids from liquids. Thus, the dynamic bulk density within the blender may be 10 lbs/ft^3 (0.16 sp gr) or higher. That this occurs was readily evident in the blender installation at one medium-density fiberboard plant where, in operation, all the fiber transited along the periphery of the shell and the center of the machine was void of any material.

Experiments with the Grenco blender have been conducted comparing shaft injection of resin versus spray application. It was found that no significant difference in blending occurred. Thus, spray application has usually been chosen in the past as the optimum method because plugged nozzles can be observed immediately, and cleaning and maintenance can be performed easily. Some plants

Figure 13.33. *Spray nozzles applying additives in Grenco blender (Knapp 1971).*

Figure 13.34. *View of the rotor in a Keystone blender (Frashour 1967).*

also apply wax emulsion by shaft injection, as well as extra water needed for tack control in the formed mat.

Grenco and Keystone blenders use a paddle arrangement in the blender which has adjustments for paddle length and pitch (Figure 13.34). Removable inserts on the paddle ends can be replaced when excessive wear has taken place.

Figure 13.35. *View of rotor in a Lödige-Littleford blender (Campbell 1974).*

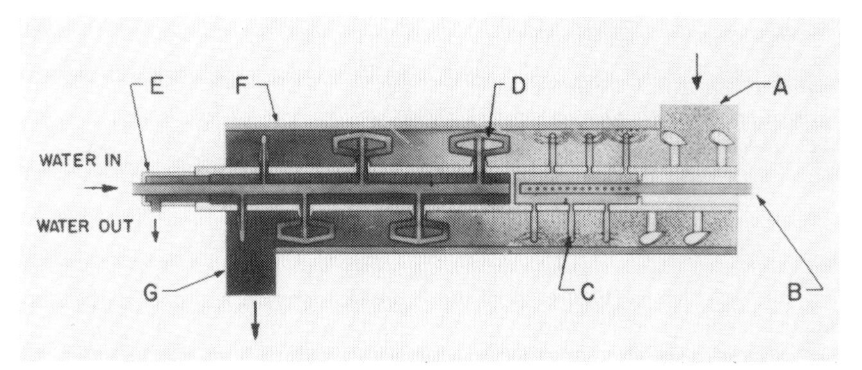

Figure 13.36. *Centrifugal application of resin in a Lödige blender (Campbell 1974).*

The Lödige-Littleford equipment shown in Figure 13.35, illustrates its shaft design. Figure 13.36 shows a slightly different approach to the blending action, compared to the models previously described. Particles are fed into the blender at *(A)* and formed by shovels into what the manufacturer calls the *chip-ring*. The resin is introduced through a distribution pipe *(B)*. Note that this is not a pressurized system. The resin runs out of the distribution pipe into the glue arms which extend from the hollow shaft *(C)*. About 96 holes for applying the resin are in the trailing edges of the resin-application tubes. These holes are about 0.100 in. (2.54 mm) in diameter; thus, small particles in the resin are not apt to plug the holes. Such resin application also eliminates the air introduced with air-atomizing spray nozzles. The resin is thrown from the glue arms into the chip-ring being propelled through the blender. Final blending is accomplished by action of the mixing plows *(D)* mounted on the shaft *(E)*. The hollow shaft and the mixing plows are water cooled. These blenders are jacketed and water cooled *(F)*. This water cooling results in a water-condensate layer, which inhibits the buildup of resin and furnish. The shaft, the tools, and the inner shell wall usually remain 100% clean. The cooling water, maintained at 53°F (12°C) or cooler, results in a heavy film of condensate on the surfaces of the shaft, tools, and inner shell wall. A counterbalance discharge flap *(G)* serves as a retention control system and prevents surging of discharge material.

The special design of the Lödige-Littleford blender has proven effective in providing good resin distribution and the elimination of resin spots. Competitive models from other manufacturers can also be found with water cooling. Since no air is added, problems with air pollution are eliminated.

An earlier version of a Lödige short-retention-time blender which is not emphasized at present is shown in Figure 13.37. In this design there are four cylindrical chambers with the resin added through centrifugal dispersion tubes in the second chamber. The furnish is driven tranversely to the machine axis through the four chambers. The furnish is dispersed into a ring in the first chamber, blended with resin in the second, and post blended in the third and fourth chambers (a fifth one is also available).

Figure 13.37. *A four-chamber short-retention-time blender (Schnitzler 1971).*

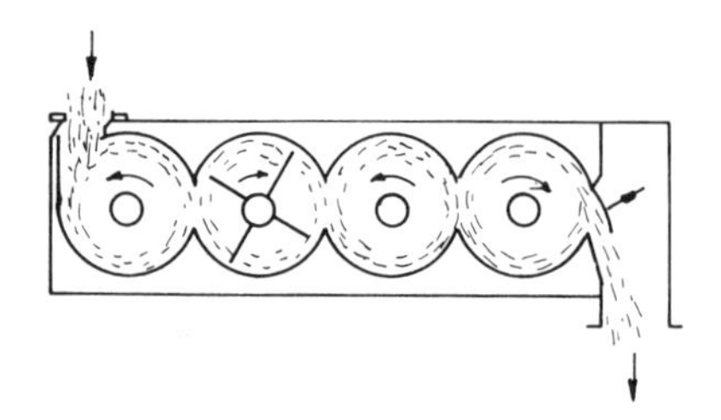

The Drais blender (Figure 13.38) shows the same general approach as the other single-chamber short-retention-time blenders. Of particular interest is their precision chain-belt-weigher for accurately metering the furnish into the blender. The resin is applied through a hollow shaft and "slinger" tubes on the shaft in the blending zone. Cooling of the shaft and housing is stressed.

In this blender, the resin is fed out of the ends of the resin-applicating tubes rather than out of holes in the trailing edge as in the Lödige-Littleford. The first set of resin-applying tubes injects the resin into the inner rim of the concentric ring of furnish, the second set of tubes applies resin into the center of the ring of furnish, and the third set of tubes applies resin into the outer rim of the furnish.

Problems can occur if the resin sets up in the hollow rotor shaft of this type of blender. If all piping and pumps are kept below the level of the resin-delivery pipe inserted into the hollow shaft and if the holes in the delivery pipe are always in the upright position, insignificant amounts of resin will be deposited in the hollow shaft if the blender is shut down. Problems with the setting up of resin in the hollow shaft can thus be negated.

With some relatively fragile furnishes, the violent action of short-retention-time blenders may cause some particle damage. However, research with nominally 0.030-in. thick (0.76 mm) flakes (wafers) of Douglas fir has shown relatively

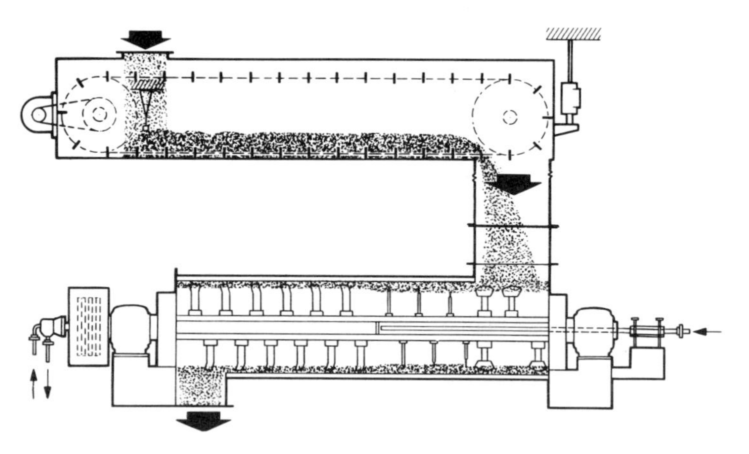

Figure 13.38. *Schematic drawing of Drais Turboplan blender. (Courtesy Draiswerke GmbH.)*

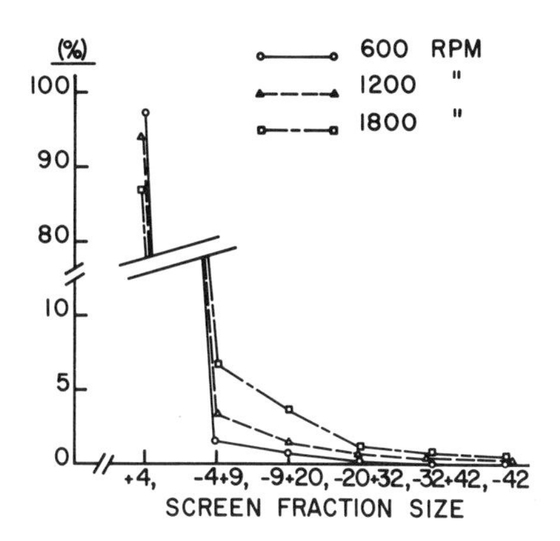

Figure 13.39. *Screen analyses of large-flake furnish processed through a short-retention-time blender at three different speeds (Maloney 1975).*

little breakage of these large flakes under conditions appropriate for good blending. Figure 13.39 shows a screen analysis on such flakes blended under these different conditions of blender rotor speed.

SUMMARY

As can be seen after the previous discussion, much effort has been devoted to the development of sophisticated, efficient blenders. Such development is very important nowadays with the increased cost and possible scarcity of resins. Work is underway on further improvements. It must be noted that even the most sophisticated blenders fall far short of the efficiency found with laboratory blenders. Thus, the optimum blender is yet to be built.

SELECTED REFERENCES

Albrecht, J. W. 1968. The Use of Wax Emulsion in Particleboard Production. *Proceedings of the Washington State University Particleboard Symposium, No. 2.* Pullman, Washington: Washington State University (WSU).

Ballard, G. A., and L. E. Schlavin. 1969. Instrumentation for the Particleboard Industry. *Proceedings of the Washington State University Particleboard Symposium, No. 3.* Pullman, Washington: WSU.

Berrong, H. B. 1968. Efficiency in Sizing of Particleboard. *Proceedings of the Washington State University Particleboard Symposium, No. 2.* Pullman, Washington: WSU.

Binder, K. A. 1967. Resin Application and Quality in Particleboard Manufacture. *Proceedings of the Washington State University Particleboard Symposium, No. 1.* Pullman, Washington: WSU.

Borgevink, G. 1975. Private communication on fiber blending. Medford Corp.

Buikat, E. R. 1969. Clean Dry Air for Instruments and Plant Use. *Proceedings of the Washington State University Particleboard Symposium, No. 3.* Pullman, Washington: WSU.

Burrows, C. H. 1961. Some Factors Affecting Resin Efficiency in Flake Board. *Forest Products Journal,* Vol. 11, No. 1.

Campbell, C. C. 1974. An In-Plant Study of Two Types of Particleboard Resin Blenders: Type I — Sprayed Resin Application; Type II — Through-the-Shaft Resin Application. *Proceedings of the Washington State University Particleboard Symposium, No. 8.* Pullman, Washington: WSU.

Carroll, M., and D. McVey. 1962. An Analysis of Resin Efficiency in Particleboard. *Forest Products Journal,* Vol. 12, No. 7.

Carter, R. K. 1969. Weighing Devices and Instruments Used in Particleboard Manufacture. *Proceedings of the Washington State University Particleboard Symposium, No. 3.* Pullman, Washington: WSU.

Christensen, R. L. 1974. Resin Efficiency and The Effect of Post-Blending in Particleboard Manufacture. *Proceedings of the Washington State University Particleboard Symposium, No. 8.* Pullman, Washington: WSU.

Crawford, R. J. 1968. Production Experiences in a Large Plant Using Several Species for Furnish. *Proceedings of the Washington State University Particleboard Symposium, No. 2.* Pullman, Washington: WSU.

Draiswerke GmbH. 1970. Drais Chip and Glue Mixing Machines for Batch and Continuous Operation. Bulletin FSP/K-FSP/SL. Mannheim, Federal Republic of Germany.

———. 1975. Drais Turboplan. Mannheim, Federal Republic of Germany.

Engels, K. 1971. High-Speed Chip and Resin Mixers with Axial Throughput. *Holz-Zentralblatt,* No. 124.

Fahrni Institute Ltd. Novopan. Zürich, Switzerland.

FAO. 1976. Basic Paper IV. Technology and Techniques in the Manufacture of Wood-Based Panels. *Proceedings of the World Consultation on Wood-Based Panels,* held in New Delhi, India, February 1975. Brussels: Published in agreement with the Food and Agriculture Organization of the United Nations by Miller Freeman Publications.

Frashour, R. G. 1967. The Forrest Industries Blender. *Proceedings of the Washington State University Particleboard Symposium, No. 1.* Pullman, Washington: WSU.

Haigh, A. H., Jr. 1967. The Bison Blenders. *Proceedings of the Washington State University Particleboard Symposium, No. 1.* Pullman, Washington: WSU.

Halligan, A. F. 1969. Recent Glues and Gluing Research Applied to Particleboard. *Forest Products Journal,* Vol. 19, No. 1.

Jacobs, J., III. 1969. Automatic Handling of Chemicals — Inline Glue Mixing. *Proceedings of the Washington State University Particleboard Symposium, No. 3.* Pullman, Washington: WSU.

Johnson, R. E. 1974. A Development in Production, Sales, and Use of Structural Particleboard. *Proceedings of the Washington State University Particleboard Symposium, No. 8.* Pullman, Washington: WSU.

Kollmann, F. 1957. Resin Application and Equipment for Blending Resin with Raw Material Particles. Technical Paper No. 5.30. *Fibreboard and Particle Board Technical Papers.* International Consultation on Insulation Board, Hardboard and Particle Board, FAO, Geneva, Switzerland.

Knapp, H. J. 1971. Continuous Blending, 1948-1971. *Proceedings of the Washington State University Particleboard Symposium, No. 5.* Pullman, Washington: WSU.

Lambuth, A. L. 1967. Performance Characteristics of Particleboard Resin Binders. *Proceedings of the Washington State University Particleboard Symposium, No. 1.* Pullman, Washington: WSU.

Lehmann, W. F. 1965. Improved Particleboard through Better Resin Efficiency. *Forest*

Products Journal, Vol. 15, No. 4.

———. 1970. Resin Efficiency in Particleboard as Influenced by Density, Atomization, and Resin Content. *Forest Products Journal,* Vol. 20, No. 11.

Lehmann, W. F., and F. V. Hefty. 1973. Resin Efficiency and Dimensional Stability of Flakeboards. *USDA Forest Service Research Paper FPL 207.* Forest Products Laboratory, Madison, Wisconsin.

Maloney, T. M. 1961. Composition Board Mills. Division of Industrial Research Report. Washington State University, Pullman, Washington.

———. 1970. Resin Distribution in Layered Particleboard. *Forest Products Journal,* Vol. 20, No. 1.

———. 1975. Use of Short-Retention-Time Blenders with Large-Flake Furnishes. *Forest Products Journal,* Vol. 25, No. 5.

Maxwell, J. W. 1969. Automatic Handling of Chemicals in Particleboard Plants. *Proceedings of the Washington State University Particleboard Symposium, No. 3.* Pullman, Washington: WSU.

McNeil, W. M. 1967. Defiberizors for Particleboard. *Proceedings of the Washington State University Particleboard Symposium, No. 1.* Pullman, Washington: WSU.

Mohr, G. D. 1968. In-Plant Production of Wax Sizing Agents. *Proceedings of the Washington State University Particleboard Symposium, No. 2.* Pullman, Washington: WSU.

Rowe, S. 1969. Nuclear Gages for the Forest Products Industry. *Proceedings of the Washington State University Particleboard Symposium, No. 3.* Pullman, Washington: WSU.

Schnitzler, E. 1971. New Methods of Chip Gluing. *Holz als Roh-und Werkstoff,* Vol. 29, No. 10.

Schoick, E. V. 1975. Private communication on adding resin catalyst to furnish. Evans Products Co.

Schwarz, F. E.; R. L. Anderson; and A. G. Kageler. 1968. Resin Distribution and How Variations Affect Board Quality. *Proceedings of the Washington State University Particleboard Symposium, No. 2.* Pullman, Washington: WSU.

Senn, S. 1975. Equipment for and Experiences in Mixing Resin with Catalysts, Extenders, and other Ingredients. *Proceedings of the Washington State University Particleboard Symposium, No. 9.* Pullman, Washington: WSU.

Surdyk, L. V. 1975. Private communication on short-retention-time blenders. Collins Pine Co.

Thompson, T. A. 1967. The Westvaco Blender. *Proceedings of the Washington State University Particleboard Symposium, No. 1.* Pullman, Washington: WSU.

Vajda, P. 1967. Blenders, An Introduction. *Proceedings of the Washington State University Particleboard Symposium, No. 1.* Pullman, Washington: WSU.

Walsh, P. R. 1975. Private communication on blender system design.

Wentworth, I. 1969. New Designs for Single-Opening Presses. *Proceedings of the Washington State University Particleboard Symposium, No. 3.* Pullman, Washington: WSU.

Wilson, J. B., and M. D. Hill. 1976. Resin Efficiency of Commercial Blenders for Particleboard Manufacture. *Proceedings of the Washington State University Particleboard Symposium, No. 10.* Pullman, Washington: WSU.

Young, D. R. 1967. The Lödige Blender. *Proceedings of the Washington State University Particleboard Symposium, No. 1.* Pullman, Washington: WSU.

14.
CAUL AND CAULLESS SYSTEMS

Before discussing mat formation and formers, prepressing, and hot pressing, it is necessary to first discuss the caul and caulless systems used in board production. It is difficult to separate mat formation, mat transportation, prepressing, and hot pressing, without first gaining an understanding of the two types of systems used in board plants. There are many possible ways that a forming and press line can be put together. There are a number of different formers available for use with different geometric shapes and sizes of particles. There are different types of prepresses and single-opening, multi-opening, and continuous hot presses. In this chapter, the forming press line will be presented in general, and in following chapters mat formers, prepressing, and pressing will be handled in detail.

CAUL SYSTEMS

The caul and caulless systems refer to the type of transport system used for moving the mat through the former and into and out of the hot press. The oldest system used in the board industry is the one using solid metal plates, usually called cauls or caul plates (see Figure 14.1). These plates about 3/16 in. (4.8 mm) in thickness are made of aluminum, steel, or alloys. A new system uses very thin flexible metal cauls of a special alloy, which can be heated quickly because of their thinness. It is reported that caul expansion takes place before full pressure is applied in the press, thus reducing problems of caul deformation caused by heating while being restrained.

The main advantage of the caul system is that it is relatively simple, as once the mat is formed it remains on this plate until after hot pressing is completed. Technological restrictions as to the particle geometry, the amount and quality of the resin, the tack of the resin, and particle moisture content are not as great as when using the caulless system.

With the caul system the mats are formed on the caul plates; the mats are cut to length at the juncture between caul plates, prepressed if desired, and then loaded into the hot press. After pressing, the finished boards are removed from the caul plates and the caul plates are returned around the forming-pressing line back to the infeed end of the former. Sometimes it is necessary to cool the caul plates on the return line so that the residual heat will not precure the resin on the first particles laid upon them on their next pass through the former.

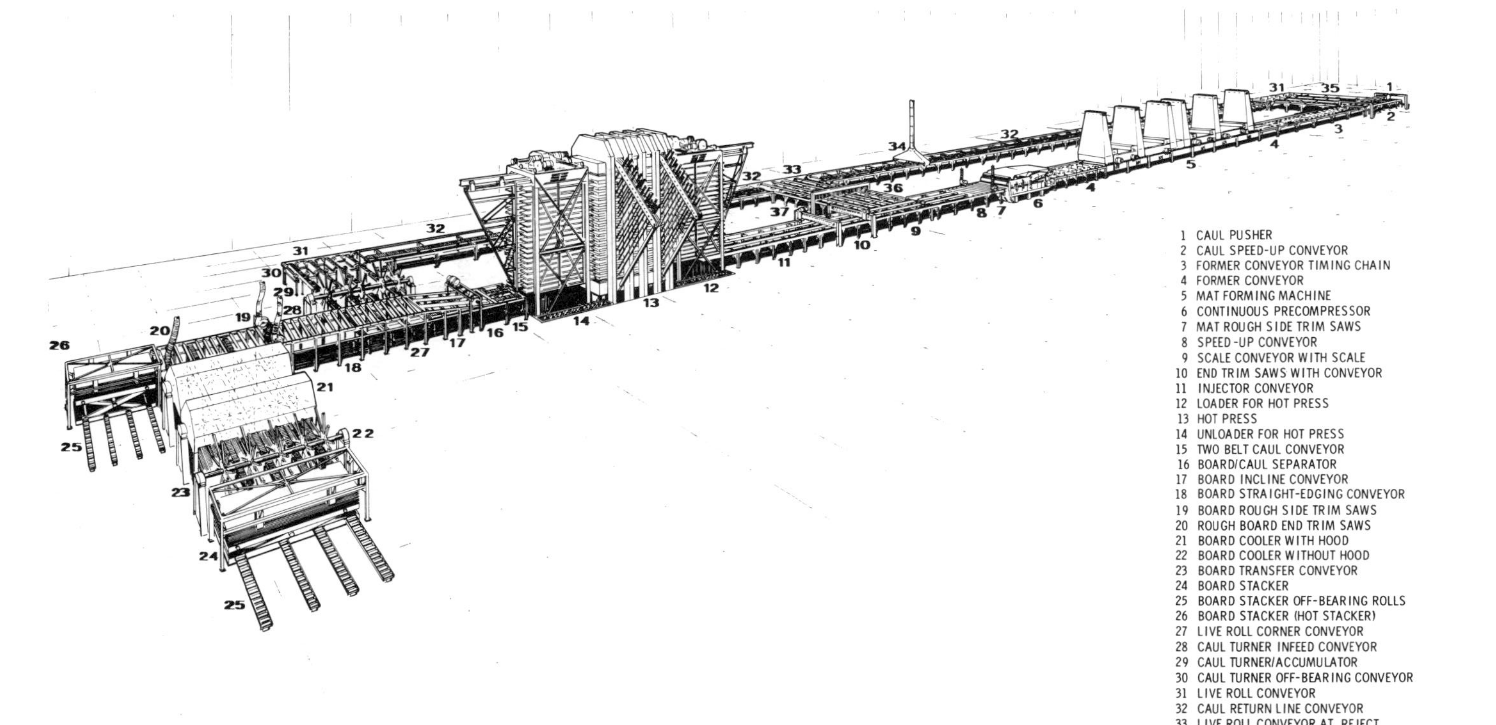

Figure 14.1. *Particleboard caul-type forming and pressing line. (Courtesy Tallman Machinery Co.)*

Normally a caul plate is not used on top of the mat. However, one may be used in certain cases where sticking is a problem (the board or particles stick to the caul plate and not to the press platens) or where a caul plate on top helps keep the mat from blowing apart as the press is rapidly closed.

A relatively recent development, similar to the caul-plate system, is one where a tightly woven metal mesh cloth is used in place of the caul plate. This flexible carrier is made of special woven steel wire 0.055 in. (1.4 mm) thick and is shown in Figure 14.2. The development of this system had the objective of eliminating the disadvantages of the caul system and the disadvantages of the caulless system, while at the same time combining their advantages. This system, called the Flexoplan process, has the following features:

1. After the mats are once formed they are transported into the hot press without any disturbance of the mat, in contrast to the caulless or semi-caulless processes. Prepressing can be used but it is not necessary.
2. The Flexoplan mesh metal cloth is dimensionally stable under pressure. This cloth can be heated up faster than a solid metal caul plate, and thus the heating time in the press can be reduced.
3. The thickness tolerance of the mesh metal cloth is determined by the tolerances of the steel wire. The wire varies less than ± 0.002 in. (0.05 mm) in diameter and is very resistant to changes in dimension under the influence of pressure and temperature in the press, which is contrary to caul plates.
4. In contrast to the caul system, this system requires much less floor space as the mesh metal cloths are returned under the press and the forming line.
5. After pressing, because of the structure of the weave, the mesh cloth quickly gives off heat to the ambient air so that no special cooling equipment is necessary in the return system.
6. During the time the hot press is being loaded, the mesh cloth insulates the bottom of the mat temporarily and prevents precuring of the bottom surface. This reportedly permits the use of higher press temperatures in the press as contrasted to the caulless system. In the caul system, of course, the bottom of the mat is not heated until the caul plate itself is heated, while at the same time, once the press is closed, the top of the mat is being heated immediately. In the caulless system, where the mat is deposited directly upon the hot platens, the bottom of the mat surface is immediately heated before the press is closed and there is the possibility of precure. The objective of the Flexoplan process in using the mesh cloth carrier is to control the heat applied to the mat in the press to such an extent that the heat is transferred to both faces of the mat at the same time and at the same temperature.
7. The very fine weave of the cloth makes it possible to lay down a very fine particle on the cloth without any sifting through the screen openings.
8. A very fine imprint is made on the bottom face of the board. This imprint is sanded off in the final dimensioning step.

The systems noted so far are normally used for multi-opening presses. For single-opening presses, the mats can be formed on a thin-gauge steel belt (a type of caul), prepressed if desired, followed by the belt moving the mat into the hot press.

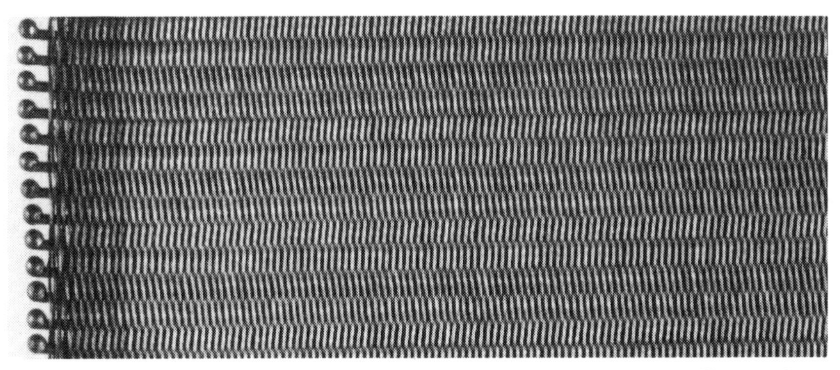

Figure 14.2. *Photograph of mesh metal cloth (Strisch and van Hüllen 1969).*

The steel belt, of course, stays in place underneath the mat during hot pressing. It is also possible to form the mat on steel wire mesh or woven fabric belts, prepress the mat, and then transfer it to a steel belt or wire mesh belt for transport into the hot press.

Continuous presses have been developed where the mat is formed on thin-gauge steel belts which then carry the mat through the hot press.

CAULLESS SYSTEMS

The caulless system uses either flexible or rigid plastic sheets as the transport medium through the former. It is then necessary to prepress the mat, with the exception of one system as described later. Afterwards the mat is deposited into the hot press for pressing into a board. The plastic sheet is returned to the head of the former to receive the next mat. A tightly woven wire screen (fourdrinier) or cyclone herringbone wire screen belt can also be used in the caulless system, particularly with fiber where vacuum formers are used.

The caulless system was developed for a number of reasons. There are problems with the caul system, particularly with the cauls themselves. They have a relatively violent treatment in being heated and cooled over and over in fairly rapid succession and they are also subjected to rather rough handling on a process line. Consequently they have a tendency to warp, twist, change their dimensions, and become defaced. The development of the caulless system meant eliminating the problems with the cauls themselves, plus eliminating the caul return system and caul cooler. Another improvement was the ability to employ slightly shorter press cycles, since the cauls did not have to be heated in the hot press prior to transmitting heat into the bottom face of the mat. The elimination of the cauls was also designed to contribute to improved pressing accuracy as far as thickness is concerned, because variations in caul thickness and imperfections in their faces were eliminated. Thus, the stock removal required during sanding can be minimized, which is important in that usually the face material of a three-layer or multi-layer board has the highest resin level and resin is quite expensive. The press must be well maintained to accomplish this objective.

Because the mat is usually prepressed in the caulless system, part of the compression as far as mat thickness is concerned has already taken place prior to hot pressing. Thus, the hot press daylight can be reduced slightly. Because the mat has already been formed into a relatively solid cake, the press can be closed quite rapidly. If the mat was not prepressed, the rapid closing of a press, particularly of the multi-opening type, could cause a great deal of what is called *blowout*. The rapid expulsion of air from the press openings blows the surface particles off of the mats and out of the press, thus ruining the final board product. Simultaneous closing devices on the press, as will be discussed later, lessen this problem. However, the blowout problem is minimized when mats are prepressed.

Another saving in raw material is possible with the caulless system because less trimming is needed with the mats and when cutting the pressed board to size. With a well-prepressed mat, it is possible to trim the mat edges and ends closer to the final board dimension. This trim material, because the resin has not been cured, can be reintroduced into the board former, thus conserving wood, particle quality, resin, wax, and other additives. In the case of the caul system, relatively wide strips, perhaps 2 to 3 in. (50.8–76.2 mm) or more in width, must be trimmed from the pressed board in order to obtain sound edges. This material can be milled back into particles and reintroduced into the furnish going to the blender, but this increases production costs since the resin has been cured and thus lost.

The caulless systems do have some technological restrictions that call for greater care on the production line. The mat must be one that can be held together after prepressing for deposition into the hot press. As such, particle geometry configuration has to be appropriate for holding the prepressed mat together. Greater amounts of resin might be required, or more precise control of the mat moisture level might be necessary. If a slightly higher mat moisture level is needed, problems with certain species and board blowing may occur. Tack in the resin is definitely necessary. Prepressing (with the one exception described later) is mandatory, and this adds a further cost to the production line.

It should be noted in passing that there are also three general types of prepresses. The first is a platen type where the formed mat is stopped momentarily for prepressing. The second is a belt type where the formed mat passes continuously underneath a belt compressor at the same linear speed as the former line. In this case the mat is gradually compressed as it passes through the prepress. The third type is similar to a giant caterpillar tractor tread; it operates much like the belt type and continuously compresses the formed mat.

Other problems have occurred with ambient conditions where the resin tack has suddenly disappeared at the wrong time, destroying the mat integrity. Also, certain species are apparently not suitable for use with the caulless system unless precautions are taken. One of these species, ponderosa pine, is relatively low in density and tends to develop a board with rather impervious faces. Plants running exclusively with ponderosa pine, or with a high proportion of this species in the furnish, have been ones having problems with blisters or blows. One plant using ponderosa pine almost exclusively for the raw material was originally built as a caulless plant. It was forced to convert to a caul line so that the moisture content could be lowered in order that good boards could be manufactured with a minimum number of blows and mat fractures.

Particle geometry, of course, is significant in assisting with the development of easy-to-handle prepressed mats. The high degree of fibrous material, particularly on the faces, will result in a mat much easier to handle than ones made up of small granular-type material. The fiber helps provide a good mat integrity. In fact, an all-fibrous furnish, as will be mentioned later, does not need any tack to assist in holding the mat together after prepressing.

Let us now consider some of the individual types of caulless lines that have been developed. They include the tray-belt, tablet, flexible plastic caul, continuous belt, and wire screen systems.

Tray-Belt System

The essential element of the tray-belt molding system is a retaining box, or mold, similar in principle to a deckle box with rigid side retaining walls. This mold, shown in Figure 14.3, has three features. First is an integral conveyor belt assembly which serves as the bottom surface covering the full area inside the frame. Second is a front gate which pivots from a point near the upper portion of the frame. Third is a movable rear wall located inside of the rigid stationary wall of the frame. This movable rear wall is fastened to the integral conveyor belt in the bottom. This is the system that does not require a prepress.

The tray-belt mold arrives at the press full of the board furnish laid down in the forming line. It enters the hot press with its front gate closed. During the press-charging portion of the loading cycle, the front gate pushes the board that has just been consolidated in the press into the unloader. Then, at a preset point in the forward overtravel of the mold, a mechanical latch opens the front gate permitting the deposition of the mats onto the press platen.

Figure 14.4 illustrates the principle with a single-opening press. Up to the point of maximum forward travel of the mold, the bottom belt assembly has been at rest relative to the mold frame. As the mold is withdrawn from the press, the bottom belt conveyor is activated to deposit the mat on the press platen. Withdrawal of the frame from the press is back toward the press loader. The conveyor belt is not independently powered, but derives its relative movement from the movement of the press loader. The function of the traveling rear wall inside the mold frame is to contain the rear edge of the mat during the period of its discharge from the frame. Part of the reset cycle to prepare each frame for its next orbit through forming and pressing is to return this movable rear wall to its original position inside of the

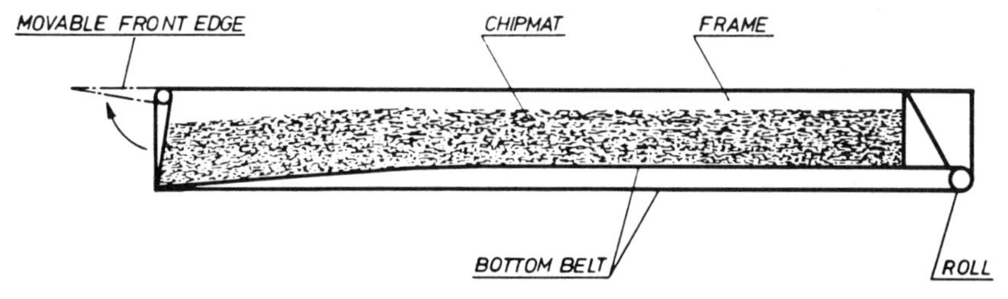

Figure 14.3. *The principle of the tray-belt molding frame (Steck 1970).*

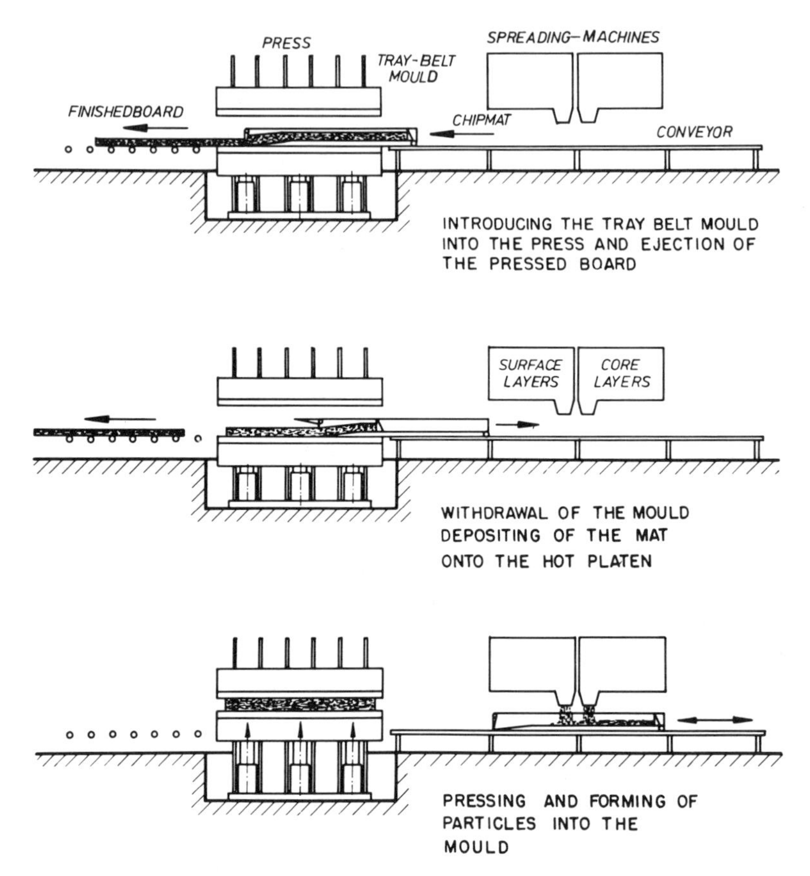

Figure 14.4. *The sequence of operations in the tray-belt molding system (Steck 1970).*

fixed wall of the mold. The front gate must also be closed again before the mold receives the next charge of raw material.

The mat makes its vertical drop from the belt surface to the platen surface by means of a prestretched, thin, and very flexible belt material. It is possible to use as little as a ⅛-in. (3.2 mm) radius at the nose end of the conveyor, while still retaining mat support beneath the entire belt surface. The approach to the nose is a long and very gradual mechanical decline to the point of mat transition from conveyor to platen.

Figure 14.5 illustrates various ways in which tray-belt molding plants may be arranged. In case of the single-opening press line, the forming station is normally directly in front of the press. Thus, the line is very compact and mechanically simple. During the pressing cycle, the tray-belt mold passes underneath the forming machines in a forward and reverse direction. In making a three-layer board, the bottom of the conveyor belt is first spread with face layer material followed by the first layer of core material. The reverse occurs when the travel of the molding frame is reversed under the forming line.

The center diagram illustrates a multiple-opening press line. Here a press loader is located in front of the press, and the frames are fed into the press by a loading boom. The shortest version of this system is one in which each tray-belt mold leaves the loading station, has a mat formed into it, and returns to the loading cage in sequence. This eliminates a merry-go-round system where the molds circle around and through the former and back into the loader.

Another operating variation is shown in the lower diagram in which a two-level conveyor is used to transport the tray-belt molds through a forming line and back to the press loader. While a filled mold is entering the press loader at the forming level, an empty mold is simultaneously removed from the press loader at a lower level. A lifting platform at the end of the forming conveyor accomplishes the transition from one elevation to another.

The system therefore can accommodate a relatively wide range of plant capacities, although the majority of tray-belt molding press lines built have employed presses with four or fewer openings and have been equipped with the single-level, self-reversing forming conveyor. Board sizes as large as 7 x 42 ft (2.13 x 12.8 m) are being produced with a version of this system.

By changing molds, board size can be varied in both length and width within the limits of the hot-press design. Since no prepress is used and a minimum of mechanical support equipment is required to complete the line, a captive plant with as little as 30 tons (27.2 mt) of daily capacity has been found to be economically practical.

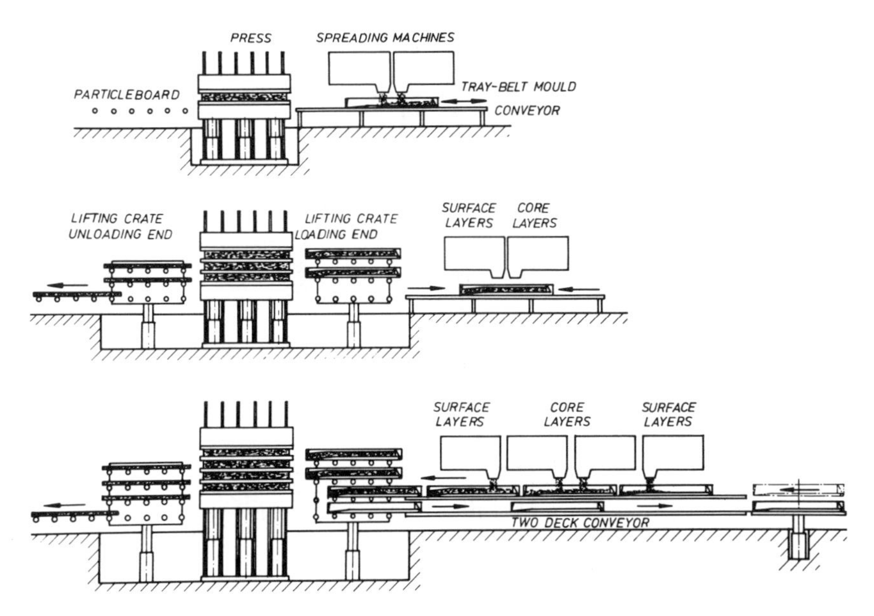

Figure 14.5. *Particleboard manufacturing with tray-belt molding (Steck 1970).*

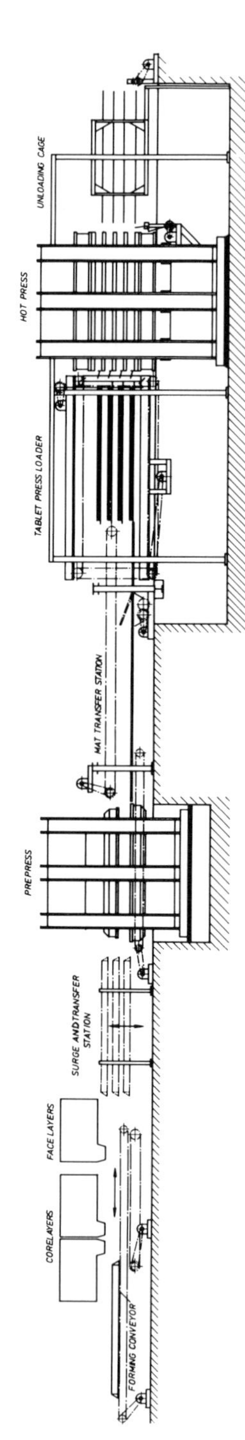

Figure 14.6. *Tablet loading particleboard system with self-reversing line (Steck 1970).*

As with all systems, the tray-belt mold has its limitations. The transporting frame assembly is a relatively complex mechanism. Also, the timing necessary for the press loading, loader indexing, and forming sequence functions must be handled by accelerating and decelerating speed controls in the frame transport system. There is no opportunity to reject the mat before pressing. The normal maximum capacity of this system is about 160 to 170 tons (145–154 mt) per day. Attractive aspects of tray-belt molding are the range of product thickness capabilities, including thin boards, efficient utilization of floor space, and large board size capabilities.

Tablet System

The tablet system also has a rigid frame in which the mat is formed. There is also some similarity in the flow of the frames through the mat forming station. The primary difference occurs in the floor for the frame, which is a rigid plate or tablet made of reinforced plastic instead of the integral conveyor belt used in the tray-belt system. The frame is not attached to the floor plate, but instead rests upon its edges with mechanical provisions to keep the frame properly centered on the plate. There the similarity between systems ends, because the tablet system centers around the use of a stationary prepress. In its prepressed state, the mat can be handled without damage by sliding it over smooth surfaces onto the hot platen.

Figure 14.6 illustrates a self-reversing tablet line. The surge and transfer station moves one frame on its supporting plate, or tablet, to the forming line at the left while a second tablet and frame assembly is simultaneously moving into the transfer station from the single-opening downward-acting platen prepress. The third level of the elevator is the surge station to compensate for the time the frame requires to make its forward and reverse path under the formers without a time penalty to the prepress.

Experience indicates that the prepress pressure should approximate the pressure of the hot press, or be at least in the 350- to 400-psi (2.41–2.76 MPa) range. Also, a distinct holding time in the neighborhood of 15 to 18 seconds for the total prepress cycle time is required. The prepressing technique consists of a rapid approach to the mat with a speed reduction before actual contact. Mat contact is followed by relatively slow pressing. The prepress platen fits inside of the frame within close tolerance. Before full pressure is exerted on the mat, the frame is stripped off the tablet surface in an upward direction so that lateral expansion of the mat is not restrained under the final thrust of the prepress platen. This prevents the mat from binding inside of the frame. Normally, the prepress platen is the same size as the pressed board after trimming.

A conveying mechanism inside the prepress slides the cake off the original conveying tablet to the mat transfer station to the right of the prepress shown in Figure 14.6. The mat is either passed through or dumped into a reject mat station, whichever is appropriate. Finally it is slid into the press loader. The loader is equipped with a series of thin, flat metal plates, usually aluminum, which carry the mats into the press openings. In the top diagram of Figure 14.7, the last prepressed mat is being pushed by the preloader onto the transport plate of the press loader. In the center diagram, the loader plates are conveying the mats into the press and pushing out the finished boards. In the bottom drawing, the loader plates are being withdrawn from beneath the mats, after restraining crossbars (one for each press opening) have been moved into the press opening to a position on top of the loading plates and behind the mats to hold them in place. At that point in the loading cycle, the plates are pulled back into the press loader. The restraining crossbars then move out of the press openings and return to their "at rest" position.

The frictional forces between the mat and the carrier plate determine the pressure exerted by the trailing edge of the mat on the restraining crossbar, and it is predominantly the length of the mat which determines the specific pressure between crossbar and mat. If the force is too great, the mat structure will crumble. In practice, mat length is normally restricted to between 12 and 14 ft (3.66 and 4.27 m) in the tablet system. In these lengths, the resin content is usually no higher than would be required by the raw materials in any other system, caul type or caulless.

A different approach on one arrangement of this type of line is to form part of the mat on the return conveyor, in order to limit the internal shear stresses in the board which some feel occur when the mat is laid in only one direction (Figure 14.8). In this merry-go-round system, the first forming machine for a three-layer

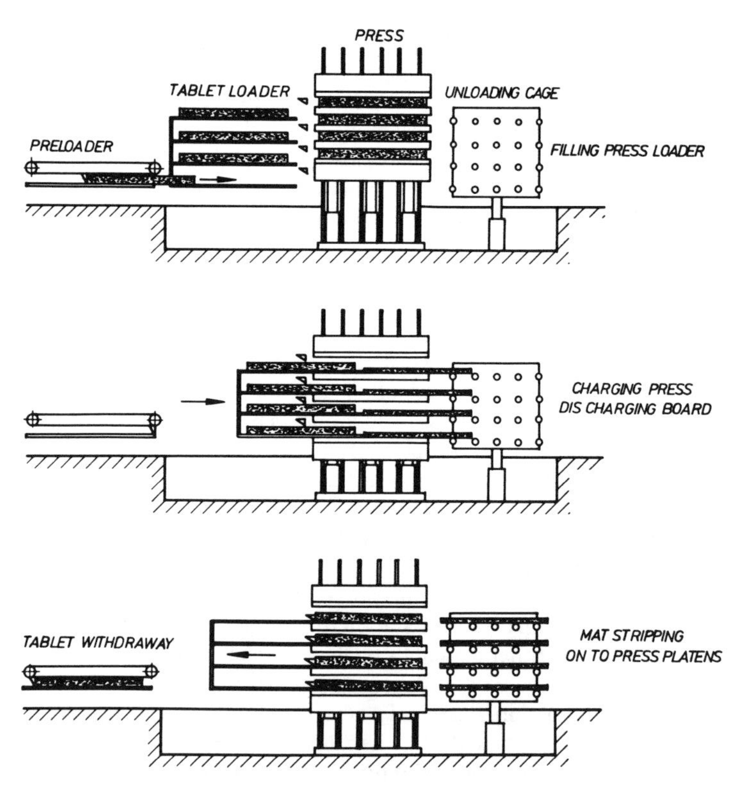

Figure 14.7. *Tablet loading particleboard system (Steck 1970).*

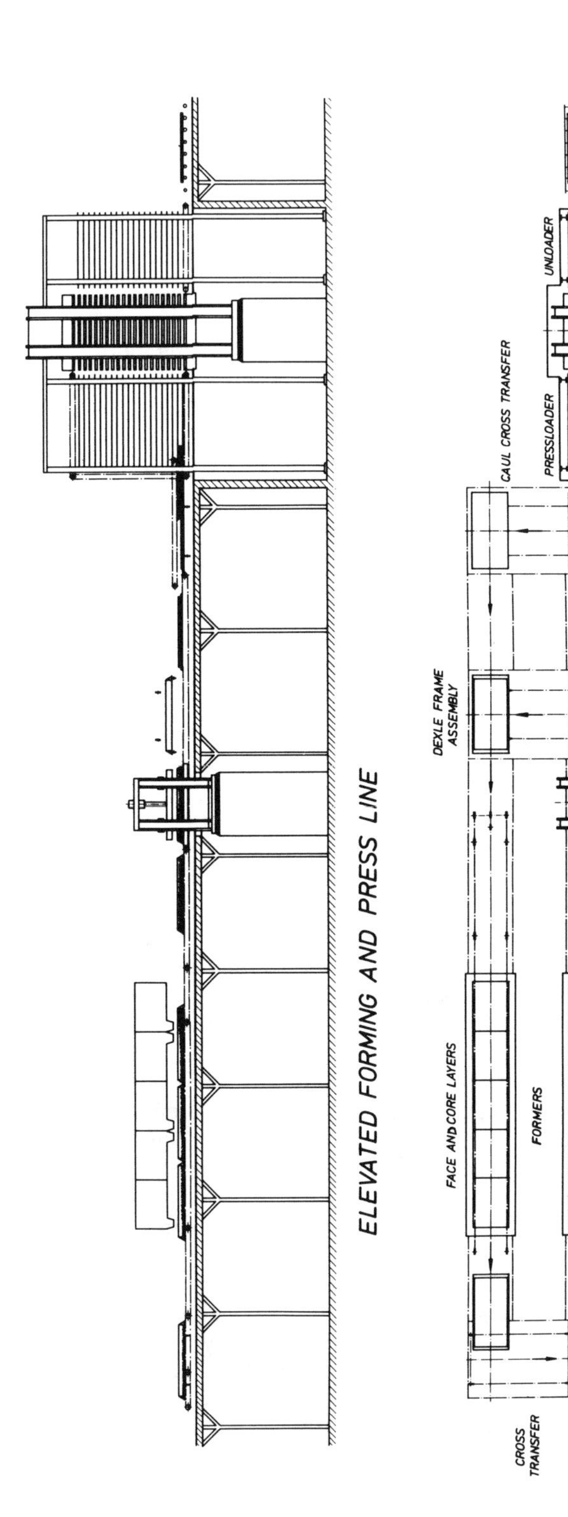

Figure 14.8. *Tablet loading merry-go-round system (Steck 1970).*

board on the return line is a face layer former. Next come one or more core layer forming units, depending on the capacity of the plant. The partially filled frame assembly is then cross-transferred to a position in line with the remaining forming heads leading to the prepress. In this fashion, the lower half of the mat is laid with particles oriented in one direction and the upper half of the mat with particles oriented in the opposite direction. In the self-reversing line, the same structure of mat buildup occurs. Even in homogeneous board, the reverse orientation of the particles in each mat is a desirable structure.

As in any system, the tablet-type plant has certain limitations. One already mentioned is that of board length. Another is minimum board thickness. The normal minimum considered to be safe is ⅜ in. (9.5 mm) with any dry refined wood as the basic raw material. Under certain selective conditions, that minimum might be as low as ¼ in. (6.4 mm), but a careful analysis of both raw material and the properties required in the finished product would be essential before positive commitment to such a product. The prepress is a machine of major magnitude and represents a significant portion of the total cost of the line. Economically, its cost can be justified, but mechanically, it represents the critical path in the board-manufacturing process. Since a specific holding time in the prepress is required to solidify each mat, the prepress is the limiting factor in the productive capacity of the system, especially when producing thin board.

Flexible Plastic Caul

Compared with previously mentioned caulless systems, the flexible plastic caul system is the most popular in the United States. It is a combination of using a wind-sifting former, a platen prepress, and multi-opening hot press. A flexible plastic forming sheet is used as the mat transport on which the mat is formed. A popular material for the sheet has been polyvinylchloride-covered belting. On the head end of the sheet is a metal puller bar that extends 6 in. (152 mm) beyond each edge of the sheet. The forward movement of the forming sheet is initiated when lugs on forming and speed-up chains alongside the forming line engage the puller bar extensions. As the formed mats leave the mat-former, a flying cutoff saw cuts the continuously formed ribbon of furnish into individual mats by making a cut directly over the forming sheet puller bar. This is the same type of cut used on a caul line to divide the mats as to length. After the flying cutoff saw has performed its function, the puller bar is engaged by a speed-up chain lug, and the individual mat is then accelerated to a speed approximately four times that of the forming-line speed.

The mat is rapidly moved ahead into the platen prepress where it is then stopped and prepressed for a few seconds. By the time the prepressing is completed, the next newly formed mat has arrived for prepressing. The top platen in the prepress is usually heated to about 130°F (54°C) in order to prevent buildup of the tacky furnish on the platens, but prepresses of this type are operating without heat.

A typical cycle that has been used for prepressing takes about 17 seconds. A fast approach speed and two pressing speeds are used. During the approach phase, if too high a speed is used the air currents developed disturb the top surface of the mat. In the second phase, most of the air is evacuated from the mat. In this case, if an excessive speed is used there is some rupturing of the mat, particularly along the edges. The final mat consolidation occurs during the third phase where the final pressure of about 350 to 400 psi (2.41–2.76 MPa) is held for three seconds. A tightly stretched curtain is used over the bottom of the top platen (the moving platen). Polyethylene sheet can also be stretched over this curtain to serve as a release agent.

After the prepress is opened, the consolidated mat is advanced onto a fabric belt. As this action takes place, the flexible forming sheet is separated from the prepressed mat and travels downward underneath the forming line and back to the start of the forming station to receive another mat. At this time the prepressed mat has a high degree of integrity and is formed into a fairly hard biscuit. To get this type of integrity, it is necessary to have a resin with tack that will assist in holding this partially consolidated mat together as it proceeds down the forming-press line. The mat is carried forward on a series of individually driven wide belt conveyors where final end trimming and side trimming are performed.

The mat ready for pressing is then injected into a press loader tray which is equipped with belts for loading and unloading as shown in Figure 14.9. When the press is ready for a new load, the press loader moves onto the trays of the unloader

Figure 14.9. *Caulless press loader. Each loader tray is in effect a conveyor equipped with four belts. The belts move forward at the same speed that the loader is withdrawn from the press and deposits the mats directly upon the press platen (Peinecke 1967).*

shown in Figure 14.10. Compressed air can be released from the nosepiece of the loader as it passes into the press. This air blows all loose particles off the press platens. As the loader is withdrawn from the press, the belts on the loader trays move forward at the same speed as the trays are withdrawn, depositing the mat onto the press platens (Figure 14.11).

Changes in ambient temperature have caused problems in this type of line as well as others. At lower temperatures a phenomenon called *peel-off* has been seen. This is a separation of the bottom face of the mat from the core. This separation has occurred as the mat leaves the prepress at the time the forming sheet is being separated from the mat. *Lift off* is the term used by some when the top layer sticks to the top platen on the prepress and is then lifted off when the prepress opens.

As noted previously, problems have also been exhibited with resin tack by changes in mat temperature and humidity. The resin companies have been tailoring their resins for different seasons of the year in order to assist with the proper operation of caulless lines, although newer resins may have overcome this problem. In the prepress it is necessary to keep the mats from falling apart as they move down the line into the press loader and finally into the press. Problems are found when rapidly accelerating the mats out of the prepress, decelerating and finally stopping in the press loader. At least one company makes a conscious effort

Figure 14.10. *Caulless press unloader. No caul-board separator or caul-return system is required (Peinecke 1967).*

Figure 14.11. *Photograph of mat being deposited on platens. (Courtesy Robert K. Carter, Boise Cascade Corp.)*

to keep the plastic sheets on which the mats are formed at a constant temperature. This is done by spraying steam and water upon the sheet just before it passes through the former. They try to maintain an approximate temperature of 75°F (24°C).

CONTINUOUS BELT SYSTEM

In recent years, there has been a trend toward larger board sizes and plants of larger capacities throughout the world, as larger boards can better serve certain specialized markets. Edge trim and cut-to-size losses decrease in percentage of raw board produced as the panel size becomes larger. In general, plant cost per ton of productive capacity in a single-line plant becomes lower as the output of that line increases.

Two obstacles had to be overcome to make plants of very large capacity feasible. The first was industry's acceptance of the practicality of making board wider than 4 or 5 ft (1.22 or 1.52 m). The press manufacturer was not the only one involved. The forming machine producers, the resin scientists, and the process engineers all had to work together. The second major problem was one arising from the discontinuous pressing process. It was obvious that, when the press was consolidating and curing a load of boards, a number of mechanical devices were needed to transport and store the mats that were being formed in the meantime.

A system for continuous forming with minimum mat loss is schematically depicted in Figure 14.12. In this system the mat remains at the same elevation from the beginning of the forming line until it is in the press loader.

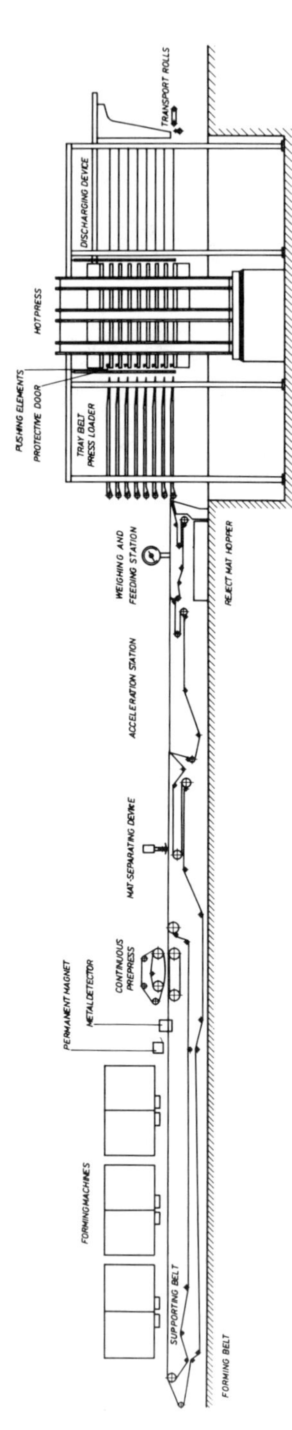

Figure 14.12. *Continuous forming belt system (Steck 1970).*

The method of mat separation is shown in Figure 14.13. This figure shows three conveyors in tandem. The forming conveyor is at left. For a specified board thickness and density, its speed is constant. The conveyor section in the middle is the acceleration station. It must run at two speeds: (1) the speed of the forming conveyor during the period of time a single mat is in process or transfer from the forming conveyor until that mat is totally on the accelerating conveyor and (2) at a higher speed to create the time separation between mats required to accommodate the capacity of the total press line. At the right of the diagram is the preloading conveyor which receives the mats from the acceleration conveyor and delivers those mats to the press loader. The preloading conveyor must similarly run at two speeds: (1) at the high speed of the acceleration conveyor during the period of mat transition, and (2) at the speed required to load the press loader in balance with the capacity of the total press line.

In the top diagram (Figure 14.13), one mat is on the forming conveyor, one on the accelerating conveyor, and one on the preload conveyor. In the second diagram, the higher velocity of the separating conveyor has taken effect and has created a larger separation between the mats. Meanwhile, however, the leading edge of the mat on the forming conveyor has advanced, still at its constant forming speed. To prevent a gap between the forming and mat-separating conveyors, the front nose of the forming conveyor has advanced at the same velocity as the forming belt conveyor. The forming belt take-up system has actually permitted the forming belt conveyor surface to become longer.

In the third diagram, the nose of the forming belt conveyor has advanced even further, but at this point, the previous mat has traveled beyond the mat-separating belt section. The second and third diagrams show that, as the nose of the forming conveyor was advanced, the trailing edge of the mat-separating conveyor was similarly advanced. With the mat off of the separating belt, the leading and trailing edges of these two conveyors can be brought back toward the original starting

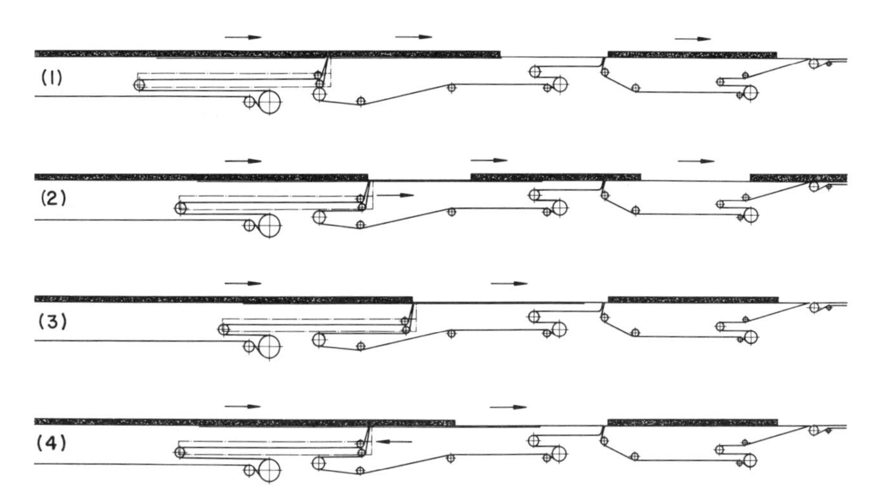

Figure 14.13. *Mat movements in the continuous forming belt system (Steck 1970).*

position. That movement is occurring as shown in the fourth diagram and establishes the starting position once again. Transition from the separating station to the press loader prefeeding station is always at the speed of the prefeeder. Transition from the forming belt to the separating station is always at the speed of the forming line. It is only the speed of the separating station which changes to either the speed of the preceding or the following conveyor. Since all are mechanically locked to a common drive, and the relationship between sections always remains constant, perfect synchronization is automatic.

The functioning of the press-loading mechanism is important since it could easily destroy the well-prepared mat. The press-loading action is shown in Figure 14.14. A minimal vertical drop through a long gradual slope to a relatively sharp nose and full support for the belt is the key. Normally, a separate pushing mechanism is used to eject the pressed boards out of the hot press far enough to be picked up by nip rolls in the press unloader.

Wire Screen System

When working with dry-process lines that use fiber for the furnish, the caulless system is relatively simple. Fibrous furnish can be laid down by a number of different formers, but one of the more popular ones is the vacuum former which will be described later. In this system, the fiber is laid down onto a wire mesh screen and then passes through one or two belt prepresses. The mat is then cut to length by flying cutoff saws and placed in a press loader of the same type discussed under the flexible plastic caulless system. Pressing is also accomplished with much the same type of press. With this line, tack is not as important as it is with the granular types of furnish. Indeed too much tack would create problems with the fiber, causing it to roll up into balls which would cause problems in the conveying system and in the former itself. Figure 14.15 shows this type of forming line ahead of the press loader.

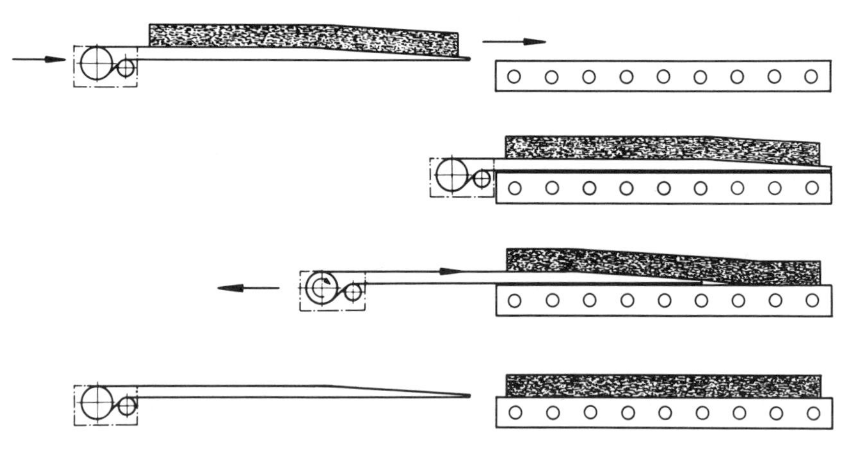

Figure 14.14. *Press loading in the continuous belt system (Steck 1970).*

Figure 14.15. *General arrangement of the forming line in a typical dry-process fiberboard plant (Peters 1968).*

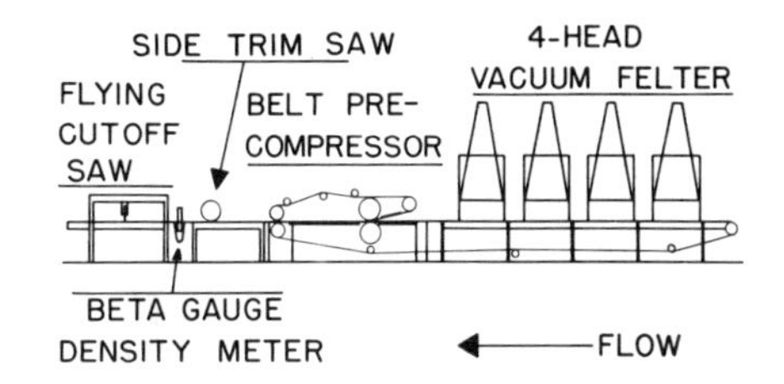

CONCLUSION

As discussed, there are many ways to put a forming-pressing line together. For any new plant using a multi-opening press, the differences between the old reliable and relatively unsophisticated caul system, the metal cloth system, and the various types of caulless systems have to be carefully evaluated on the basis of the raw material available, production rate desired, the climatic conditions, and the technical expertise that can be secured for proper operation of the plant. The cost of investment and operating will be different, of course, for each system. It would be of great benefit to have a sophisticated formula developed to determine the best system for a particular location. So many factors must be evaluated, however, that it would be very difficult to come up with such a universal formula.

SELECTED REFERENCES

Achenbach, A. 1975. Problems and Aspects of Economics in the Particleboard Industry. Background Paper No. 44/8, third World Consultation on Wood-Based Panels, FAO, New Delhi, India, February 1975.

Carlsson, B. J. 1975. Presses and Pressing Techniques in Wood-Based Panel Industries. Background Paper No. 105, third World Consultation on Wood-Based Panels, FAO, New Delhi, India, February 1975.

FAO. 1976. Basic Paper IV. Technology and Techniques in the Manufacture of Wood-Based Panels. *Proceedings of the World Consultation on Wood-Based Panels,* held in New Delhi, India, February 1975. Brussels: Published in agreement with the Food and Agriculture Organization of the United Nations by Miller Freeman Publications.

Maloney, T. M. 1961. Composition Board Mills. Division of Industrial Research Report. Washington State University, Pullman, Washington.

Peinecke, R. 1967. Prepressing and the Caul-less System. *Proceedings of the Washington State University Particleboard Symposium, No. 1.* Pullman, Washington: Washington State University (WSU).

Peters, T. E. 1968. The Vacuum Former. *Proceedings of the Washington State University Particleboard Symposium, No. 2.* Pullman, Washington: WSU.

Schwiertz, H. F. 1974. Same Line Presses Fiber- or Particle-board. *World Wood* (December).

Steck, E. F. 1970. Caulless Pressing Systems for the Manufacture of Particleboard. *Proceedings of the Washington State University Particleboard Symposium, No. 4.* Pullman, Washington: WSU.

Strisch, H., and S. van Hüllen. 1969. New Systems and Recent European Developments. *Proceedings of the Washington State University Particleboard Symposium, No. 3.* Pullman, Washington: WSU.

15. MAT FORMING AND FORMERS

Felting, forming, or spreading of furnish all refer to the formation of a uniform mat for consolidation in the press into the final board. Some feel this is the most important piece of equipment in the process line. There are many types of forming machines in use today around the world, and not all can be considered here. Most are manufactured by equipment companies; however, in addition, there are many custom-designed forming machines in a number of plants. No other area in the manufacture of board has such a large variety of machines available. Also, many problems have been found in this area. As in the rest of the process line, many factors have to be considered in deciding which former to use. The type of furnish used and the type and quality level of the board all have to be considered. Some machines require more raw material preparation than others. Some machines require constant adjustment and higher-caliber operators. Some are rather simple while others are quite sophisticated.

Some of the factors involved in mat formation will be explored first, followed by a review of some of the popular types in use employing different operating principles. Company names are used for easy identification. Several companies may be taking similar approaches to mat forming.

IMPORTANCE OF MAT FORMING

The preparation of a consistently uniform mat of furnish is one of the most important parts of the board manufacturing process. A lack of uniformity in this distribution will result in physical property variations, wide density variations, and excessive thickness springback in the high-density areas. Even if mechanical properties were not of prime importance, the uneven mat formation would result in a board more susceptible to warp. Poor mat formation may affect appearance and, therefore, the saleability of the board due to poor surface and edge characteristics. Unevenly formed mats can also cause damage to the press.

PROBLEMS ENCOUNTERED IN MAT FORMATION

The greatest problem encountered in the process of mat formation occurs because a volume of wood of varying bulk density is metered into the former with the expectation that an even weight distribution will result in the mat. The bulk densities of the prepared furnish will constantly change with the variety of conditions under which the furnish is prepared. Its properties will also be changing depending on the source from which it comes and, perhaps, the season of the year. Further compromises must be made depending on the desired product to be manufactured, such as homogeneous or layered board for different end uses calling for high and low physical properties. Most plants will be wanting to make several grades or types of board, thus compromises must be made all through the production process. Perhaps the raw material will impose certain limitations, while advantage can be taken of other characteristics. For example, fibrous furnish cannot be used successfully in some formers but can be in others.

Bulk density and its variation is consequently of prime importance. If a highly refined fiber is allowed to fall freely, in all probability the mat will have a relatively uniform bulk density. However, this bulk density will change dramatically when conveyed in a screw conveyor or if stored in a large high-surge bin.

Carefully prepared flakes will also have a uniform bulk density, since all of the flakes have a similar shape and size. This will change as a result of manufacturing circumstances, such as breaking of the flakes if conveyed through fans.

Most of the raw material used in particleboard is not, for one reason or another, as well defined as those just described. In fact it covers everything from wood flour to large chips and slivers. Fiber is more uniform but wide variations are also found in its sizes. Periodic screen analyses of the furnish will show changes from hour to hour, day to day, and season to season. The particle geometry and size range will also change as knives, plates, or hammers become dull in the particle-generation process. The moisture content of the log, slab, or shavings will vary causing a further change in the particle. The resultant particle will also vary, depending on whether it is from a kiln-dried or air-dried source, as well as on the characteristics of the planer which produced it. Planer shavings may be thicker when planed from lumber produced in a circular sawmill than in a band mill.

With the wide bulk-density variations found, the approach is to try to establish average conditions and then to minimize wide fluctuations. This is another example of why the manufacture of board is an art as well as a science, since the human element must correct for and control fluctuations as well as other constantly changing factors.

FORMING MACHINE DESIGN

All forming machines meter the fiber or particles based on an average bulk density and spread it uniformly across the width of the machine. In some cases the furnish is weighed just before it is metered out to fall on the caul or other transport medium. Sometimes there is a leveling device to assist in forming a uniform mat as it leaves the forming machine. All forming systems have certain things in common, although the mechanics of each may vary. They all have a surge area, which may be integral or slightly remote, and usually a weighing section incorporated with a final surge or metering section. All have some method of dropping the furnish as individual particles onto the caul or transport medium. The fall is by gravity in almost all cases. The surge bin may be slightly remote as in the vacuum felters, on top, or behind the former. The primary purpose of this surge area is to insure a continuity of material flowing into the machine. It is also important for spreading the wood uniformly across the width of the forming area. A uniform level in this bin must be maintained for bulk-density control. If the bin is too full, the density will rise; if it is too low it will decrease from material weight alone. With tacky resins, bin level change also causes tack change, which affects some former designs more than others.

Since this surge area is usually large, many, but not all, forming machines meter the fiber into a batch-weigh section to help overcome varying bulk densities. The batch-weighing machine can be preset for a given weight and number of weighings per minute. The scale is interlocked with the feed conveyor. The batch-weigher discharges into a final small surge area for final metering. This final surge area is small to minimize density-variation problems. From this final surge area, the wood is metered out on a volumetric basis by means of rakes, picker rolls, or similar device. Some, such as the Microfelter, use a bank of metering rolls. Other machines use only a single surge area but keep it low and long instead of high, so as to control bulk density in an attempt to be very accurate in metering.

Very few of the machines let the particles fall freely onto the caul or other transportation medium. Originally, forming machines were raised high above the cauls, and the material fell freely in the belief that the long fall would result in an even mat. Unfortunately, the particles, instead of falling down evenly, were affected by natural air currents and drifted to one side or the other being heavily influenced by the final flow direction of the air current created by the falling material itself. The forming machines were then lowered as close as possible to the transport medium. In some forming machines, screens were placed in the path of the fall furnish to break the material fall and to help the particles, particularly flakes, fall flat. In others, small screens were placed in the former to allow fines to fall onto the transport medium first, with coarser material on top. Other machines used picker rolls to spread the material out in the direction of caul transport movement. Still another machine dropped the material into a high-volume, low-pressure air duct to give a graduated layer effect. Another forming machine fed the material in the direction of the rotation of rollers and thus gave a grading effect from each forming head.

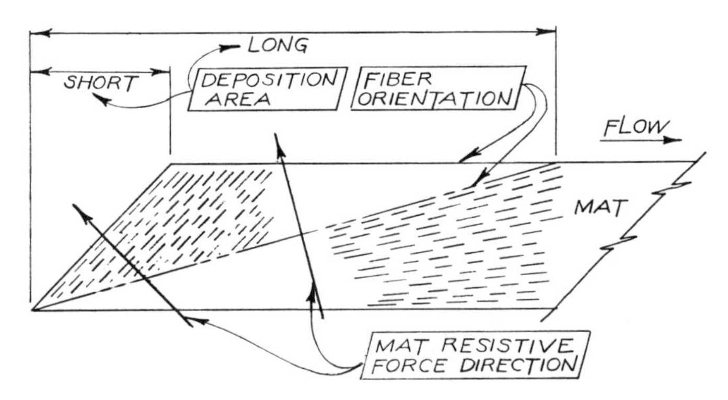

Figure 15.1 *Mat-deposition angle (Newman 1967).*

Originally, many mats were formed in a small area with the flakes or particles at a steep inclination (Figure 15.1). This was satisfactory with smaller presses. However, as presses became larger, this steep angle created a horizontal component to the pressing force which actually moved the press crown off its longitudinal centerline. It then became necessary to flatten or minimize this particle-deposition angle. The devices for dropping the particles were sophisticated in design to lengthen the mat-formation area, lessening the particle-deposition angle.

In spite of all of the elaborate mechanisms, some units still require leveling screws, picker rolls or shave-offs to level the final mat.

TYPES OF FORMING MACHINES

There are many different types of forming machines, each designed for specific types of raw material and end products capable of doing a variety of jobs within reason. With multi-opening presses, the former is stationary. There are some single-opening press line plants, however, where the forming machines must move. In these a single steel belt transits the former and press. Consequently, while the press is closed, the belt cannot move. In this case, the forming machine moves over the board area laying a mat down. Thus, the forming is intermittent and not continuous. Other small single-opening plants use individual cauls and a stationary forming machine.

In larger plants, an individual forming head has definite capacity limitations. In these cases, it is desirable to have multiple forming heads for continuous forming. Individual forming head capacities should not be pushed to their limit since, almost without exception, they will form a more uniform mat at lower rates. As a general rule, it can be said that for a given volume of wood, the more forming heads or drop points used, the more uniform the resultant mat. Further, with a given quantity of material, the more forming heads employed, the lower the particle-deposition angle.

With multiple heads, advantage should be taken of the opportunity to separate the finer material and place it on the faces. It is now also possible to work effectively with different particle-moisture contents and resin levels and resin formulations–for example, more moisture and resin in the faces and slower curing resin in the faces. This also means specialized equipment in the milling area of the plant for efficient separation of various-size particles. Some plants make three separations and use five or six forming heads. The two end heads are used for fines, the center one or two for coarse particles, and the two intermediate heads for a semi-coarse particle.

The capacity of a forming head should not be overextended. An additional head is inexpensive compared to the cost of rejects and sales complaints. Multiple-forming heads with a given amount of raw material will ease quality-control problems and make the plant generally more flexible to serve a variety of markets.

INSTRUMENTATION

The importance of accurate metering of furnish to the former has already been noted. Different approaches will be noted in the discussion of the different types of

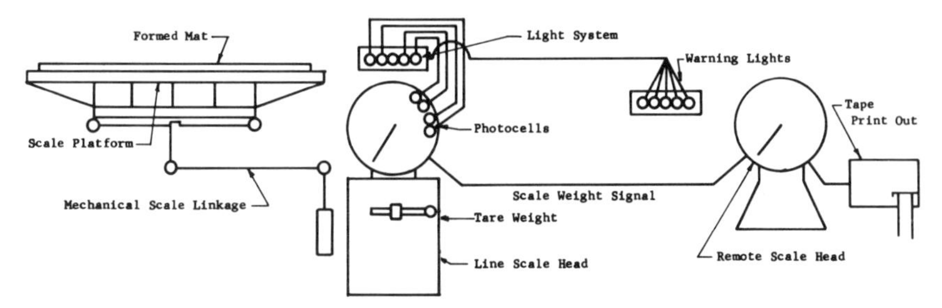

Figure 15.2. *A schematic drawing of a mat-weighing system (Carter 1969).*

formers available. Recently, nuclear gauges have been incorporated into the metering system on some formers. It is important also to weigh the finished mat after forming to insure uniformity. As noted, this is extremely important if boards of consistent quality are to be produced. Two systems are used: (1) weighing the entire mat with a conventional scale and (2) continuously determining the weight by means of a nuclear gauge.

Figure 15.2 is a diagram of a conventional mat-weighing system. The lever system transferring the mat weight to the scale-control box is similar to the continuous wood weight-o-meter used in blending. Mat weight is read directly from the scale face in one-pound increments.

For a superior mat-weighing system, some attachments have been added to the scale. Since the operator is located at a remote distance from the scales, a remote scale head is installed for close observation of mat weights by the operator. Attached to this scale head is a printed tape readout system to allow mat-weight trends to be monitored for correction before reject limits are reached.

An auxiliary light system is also included. The purpose of this attachment is to actuate automatically the mat-reject system if mat weights are too high or too low. Four photocells are used on the scale face in order to detect the position of the weight-indicator arm. Two of these photocells actuate the reject system, and the other two photocells indicate "accept over" and "accept under" mat weights. No photocell is needed to show accept values on mat weights.

Figure 15.2 shows the part of the auxiliary light system above the scale face. This part of the system, connected to the warning lights, may be used by the operator to control mat weights in case the remote scale head ceases to function. As can be seen, five lights are above the scale head. If either of the two outside lights (red) flash, the mat is rejected. If the second or fourth lights (yellow) flash, mat weights are running slightly high or low. If the center light (green) flashes, mat weights are right on target. The accuracy of this system has been reported to be excellent. The mat scale itself is accurate to better than ± 5 lbs (2.27 kg). The control of the mat weights is ± 15 lbs (6.82 kg). Much of the control of mat weight depends on the furnish bulk density and the skill of the operator.

By using a strip nuclear gauge, measurement of mat-weight-per-unit-length across the machine can be done on the fly, without altering the basic conveyor speed. This eliminates the static weigh station and speed-up device. Short-term variation in mat weight from foot to foot in the machine direction is evident by observation of the strip-chart readout. Alarm contacts can be installed to monitor excess variations. Long-term trends are also evident on the strip chart.

For nuclear gauge measuring of mat weights, the basic weight or weight-per-unit-area, usually expressed in pounds per square foot, is used. Measurement is by point source and point detector geometry (see Figure 13.13). These elements are mounted on a C-frame, which in turn traverses a base beam, or on an O-frame wherein they ride parallel ways, which are held in line precisely by lead screws or cables. The C-frame is less expensive, requires less maintenance, but it does require aisle space equal to the width of the line, when it is backed off the mat. Newer models are temperature-stabilized, which prevents drift. The C-frame has the further advantage of complete removal from the line; while the O-frame, while it saves aisle space, remains at all times on the conveyor, where it conceivably could be in the way of the maintenance people. The C-frame has been popular.

It is difficult to use such gauges on caul lines as waviness of the cauls and variation in caul thickness cause problems. Consequently, the popular use of nuclear gauges is on caulless lines. In the caulless system, it is popular to mount the basis-weight-gauge C-frame before the prepress (Figure 15.3). In this line the gauge "sees" only the mat as no caul plate is involved. When the air space appears between the individual mats, the gauge changes its reading markedly. To prevent this wild reading and subsequent one- or two-second delay in reading on the next mat, the gauge can be arranged for zero center calibration, and the amplifier input can be zeroed by a photocell, sonic switch, or other external sensor when the gap appears between the mats. The zero center calibration creates easy operator interpretation; since he does not have to remember the absolute value of target-mat-basis-weight, and can be told to take action only when the gauge deviates a given percentage from the zero center.

A typical calibration for such a gauge would be ± 0.5 lbs/ft^2 (± 2.44 kg/m^2). The precision of measurement and the resolution of 1% of full scale is ± 0.01 lbs/ft^2 (54 g/m^2) for this application. Figure 15.4 shows a gauge in operation.

Metal detection and removal are of extreme importance at all points of the production line because of damage to the equipment that can occur. If sparks result from metal striking parts of the equipment, fires and explosions can occur. Rocks also cause these problems, but they have to be handled with rock traps.

The performance of any metal detector is quite variable as it does not detect all metal. Because of this, it is necessary to develop a metal detection and removal

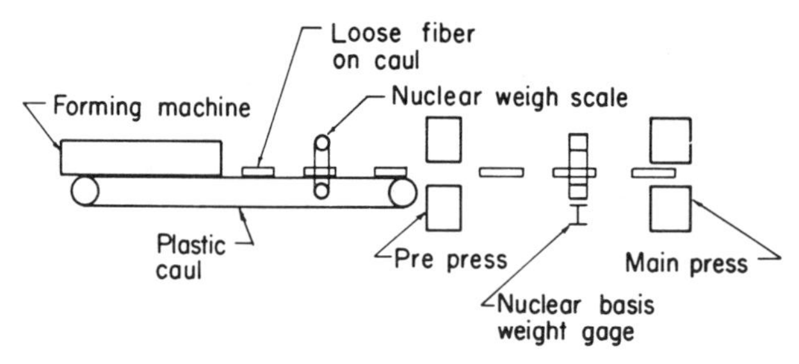

Figure 15.3. *Nuclear gauge application on a caulless line (Rowe 1969).*

system consisting of magnets, permanent magnets, electromagnets, magnetic-head pulleys, air-system traps, vertical air pickups, chipwashers, as well as metal detectors. Some have discontinued the use of metal detectors, because of variable performance, and depend instead upon the other methods of removal for handling metal and rocks.

After the mat is formed, a metal detector is needed to keep metal out of the press where expensive-to-repair platen damage can occur. Additionally, the metal would end up in the final board, where further damage to equipment could happen when the board is sawn either in the plant or in the field. When metal is detected in a mat, it is normal to reject the mat and not to try to reclaim the furnish.

The principle of operation of a metal detector is to use a radio frequency (RF) generator and RF receiver (Figure 15.5). The radio generator emits an RF magnetic field and the RF receiver is adjusted so it matches the RF generator field. Thus, the generator and receiver are in balance. This balanced condition is disturbed any time a ferrous metal passes between the generator and the receiver. Any change is monitored by the receiver which trips a relay, thereby sounding an alarm or operating a mat-dump gate.

Figure 15.4. *Typical installation for basis-weight-gauge (side view) (Rowe 1969).*

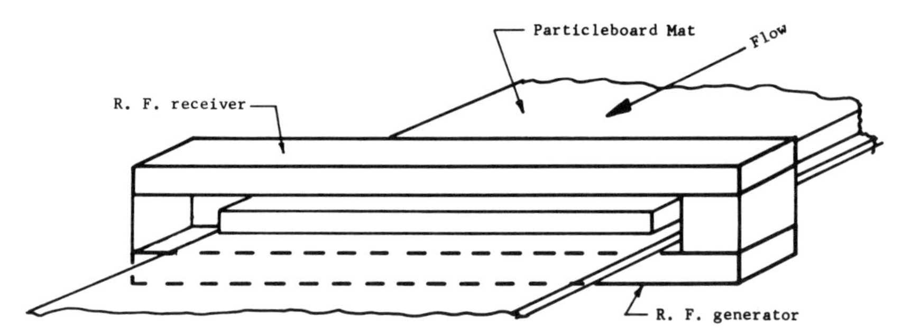

Figure 15.5. *A schematic drawing of a metal detector (Carter 1969).*

FORMERS

Wind-Sifting or Air-Classifying Forming

One type of forming machine uses the principle of air separation or wind sifting to lay up a graduated or multi-layer board. It was originally developed for single-opening presses and is used extensively around the world. It has been particularly popular in the United States in particleboard plants. It is now undergoing revision for use with fiber.

This former produces a mat with fines on the faces and without distinct layers, unless distinct heads are used for faces and core. The particle size is graduated from coarse material in the center of the mat to the finest material on the faces. This is done automatically, regardless of the furnish being fed to the forming machine, thus no large or thick particles end up on the surfaces of the mat, nor is there a distinct fines face layer subject to sand-through, thereby showing coarse particles. With this system, therefore, the classification of particles is not as critical, and rejects through poor surface characteristics are reduced. Experience has shown that the Bison former can perform satisfactorily with a wide range of furnish (e.g., large and small flakes, coarse, fine and very fine particles).

In this former, the furnish is metered from a surge bin at the top and drops in front of a high-volume, low-pressure air system designed to give a uniform airflow. The effect of the horizontal airflow is to blow the finest particles to the extreme ends of the former, while the heavier particles drop almost vertically. The exact distance the particles travel before deposition is a function of the relation between their mass and falling velocity and the air velocity, except with thin, large flakes with a high "sail" surface. These tend to be carried along with the fine material. Figure 15.6 illustrates how this former functions.

The first part of the former to consider is the surge bin on top of the unit (Figure 15.7). It is long instead of high, to minimize wide bulk-density variations, and meters the furnish by reels which gradually reduce the height of the metered area. In the type shown, some material is always recirculating, which is only a problem when using tacky or highly catalyzed resin. In this case, the furnish must be felted without delay or problems will occur further down the processing line. Other types run without recirculation to eliminate this problem.

On the formers for single-opening press lines, there is one surge bin dropping between two banks of air ducts (Figure 15.8). Each duct is offset from the opposing one to allow the particles to pass through. Each individual duct has its own air-control valve. There are also vertical screens below and behind the air system, to assist in mat leveling. This former is also used in a few smaller multi-opening plants.

The original Bison forming machine used in the United States separated this air duct and reversed the fan system as shown in Figure 15.9. A common fan was placed in the center with a side air intake. This fan discharged in both directions, separating the material as it was metered from two separate surge bins. This unit has a greater capacity than the previously mentioned forming machine and produces a more uniform mat.

It should be mentioned that the original forming machine used in the United States was in a plant with a multi-opening press. Newer plants have not used this former design as it has a limited capacity. As plant capacity increased, more places for dropping the furnish were required. It was also learned that the air direction of the single-fan discharge resulted in different board surface characteristics. On one side, the air discharged in a downward general movement and, on the other side, in an upward movement. The air direction characteristics were different in each side of the machine.

The next innovation was to separate the surface sections and use a separate fan for each. A single surge bin similar to those on the top of the former was placed in

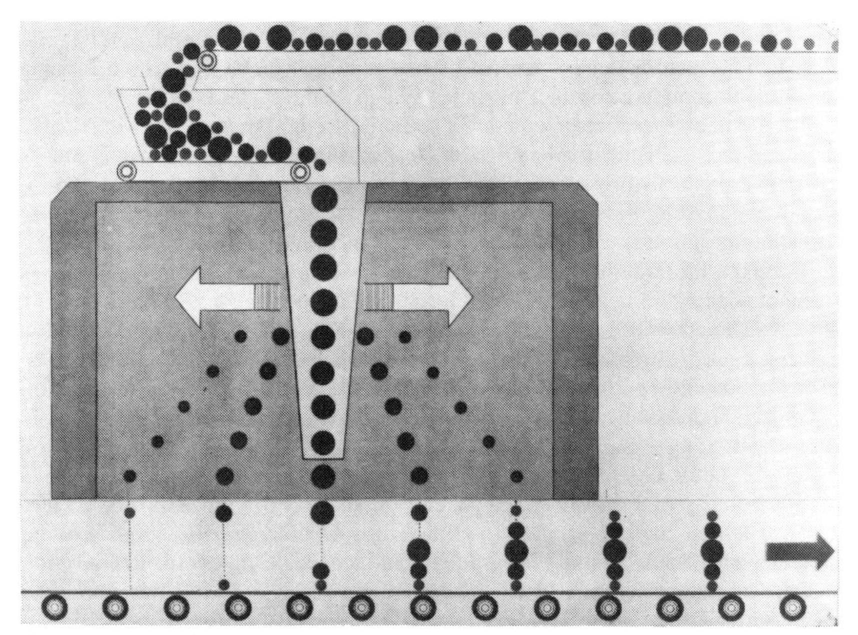

Figure 15.6. *A detailed illustration of the Bison-Werke method of forming a particleboard mat (Wentworth 1969).*

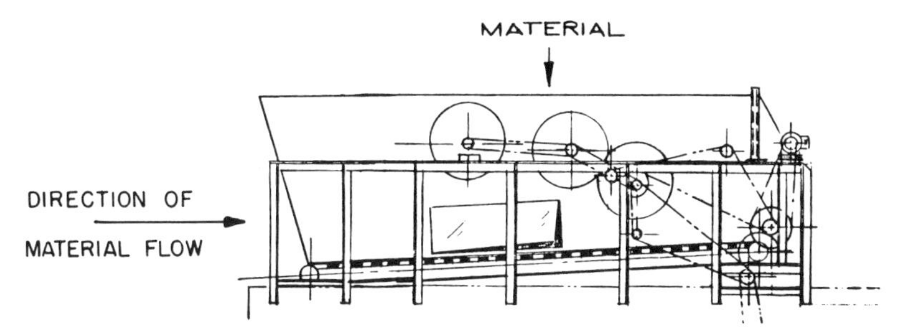

Figure 15.7. *Surge bin on top of a Bison forming machine (Newman 1967).*

between the fans for accurate metering of a homogeneous core section. This in effect added versatility to the machine. A separate resin level as well as different moisture contents could be used in the core. Further, a different type of flake or particle could be separated out for use in the core section only. At the same time, a series of baffles were added to the surface sections for better air-movement control. A version of the more sophisticated former is shown in Figure 15.10.

As with all formers, the forming action is taking place within a large box. Since the operation is a dusty one, particularly with fine furnish, it is difficult to observe what is going on inside. This is a disadvantage when the former is not functioning properly and the operator is trying to adjust the machine. The larger machines, laying faces and cores separately, have had a recent revision in design making it easier to determine where the malfunctioning of the equipment is taking place. In

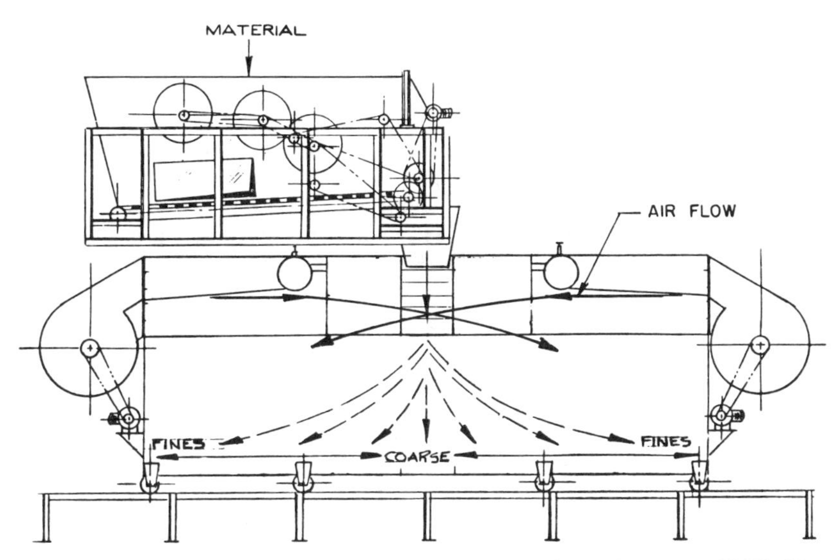

Figure 15.8. *Bison former for single-opening press lines (Newman 1967). Note rails and wheels for moving this former back and forth.*

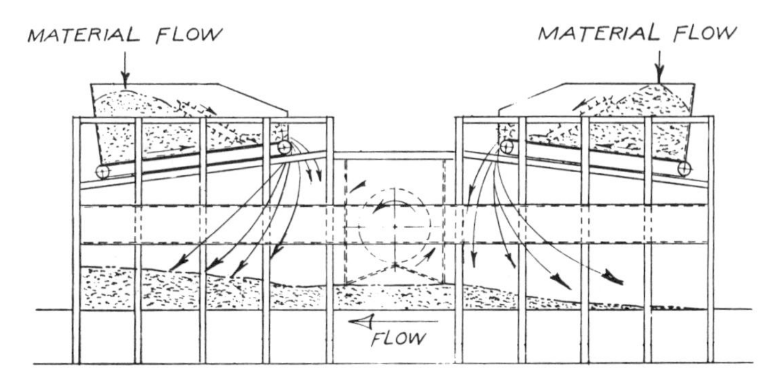

Figure 15.9. *Original Bison forming machine used in the United States (Newman 1967).*

this design, the face formers are separated from the core former. Thus, the operator can see which part of the former is giving the trouble. Additionally a scalper (a picker roll device that shaves the mat off to a uniform height) has been placed at the outfeed of the core section to level the mat before the top face is laid.

This leveling of the mat before forming the top face was handled differently in the Bartev design, where the mat was prepressed before the top face was laid.

Fahrni Spreaders

Novopan plants throughout the world have been successfully using the older type Fahrni former. The former works on the principle of superimposing multiple layers one on top of the other. Several heads are used to accomplish this spreading, one of which is shown in Figure 15.11. The multiple spreading is designed to insure that the particles are oriented mostly in the horizontal direction, which results in a compact mat. This in turn assists in reducing the need for prepressing and narrows the size of the press opening "daylight" required.

Face and core furnish is fed to separate silos above the forming line. The furnish is then metered onto a vibrating horizontal tray which feeds the particles under a comb roll for accurate metering to a pair of flinger rolls. The gap between

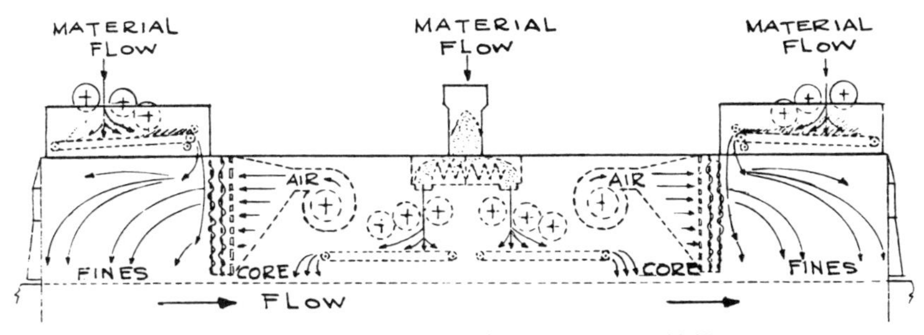

Figure 15.10. *A large Bison forming machine (Newman 1967).*

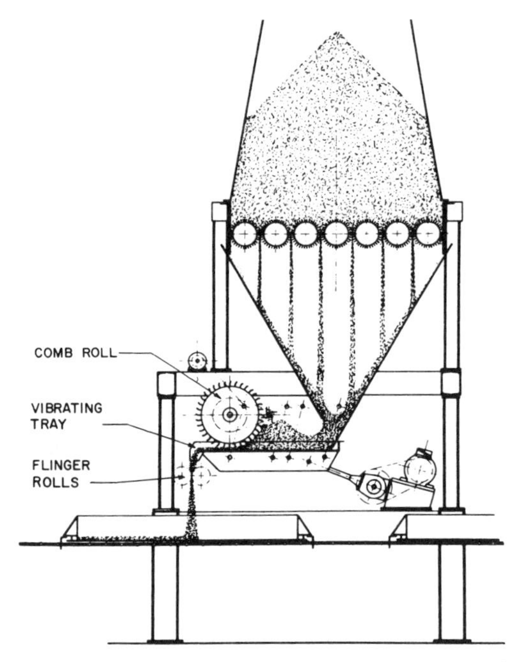

Figure 15.11. *Principle of the Fahrni spreader with horizontal tray and dosing system. (Courtesy Fahrni Institute Ltd.)*

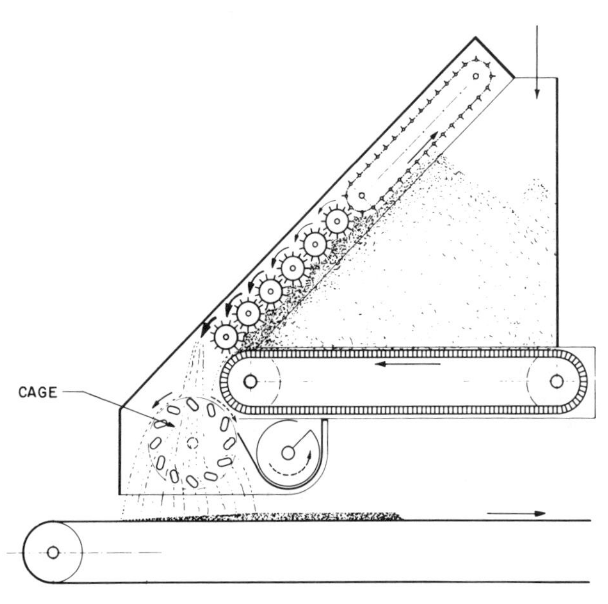

Figure 15.12. *A new type of Fahrni former. (Courtesy Fahrni Institute Ltd.)*

the vibrating tray and metering roll can be adjusted to suit the desired mat height. The metered furnish then falls between the flinger rolls which deposit the particles onto the mat carrier.

A newer former has been developed specifically for lines requiring a greater throughput. This former is shown in Figure 15.12. The vertical silo or hopper is replaced by a horizontal one. This design prevents fines from running through out of control. Other refinements not shown on this drawing include a pocket feeder roll between the bin and cage for accurately feeding the furnish to the cage.

The cage consists of steel tubes held between two rotating steel discs. The furnish is dropped through the cage for distribution upon the mat carrier.

Schenck Formers

A number of different approaches to forming have been taken by Carl Schenck Maschinenfabrik GmbH. The one shown in Figure 15.13 has a single spreader head. Several are normally used in series, depending on the capacity required. Several dosing bunkers (surge bins) are arranged ahead of the forming station for metering the material to the spreading head. The furnish is discharged volumetrically with the weight of the material flow controlled by a belt-weighing system installed in series.

A gravity spreader head (Figure 15.14) has a roll fitted with steel or plastic spikes for spreading the furnish. The speed of the gravity spreading roll is adjusted according to the desired classification effect (heavier particles are thrown farther), the required quantity of particles to be formed, and the type of particles. A prerequisite for this operation is the precise adjustment of the point where the particles come into contact with the spreader roll in relation to the mat speed. Figure 15.15 shows a typical line equipped with four of these spreader heads.

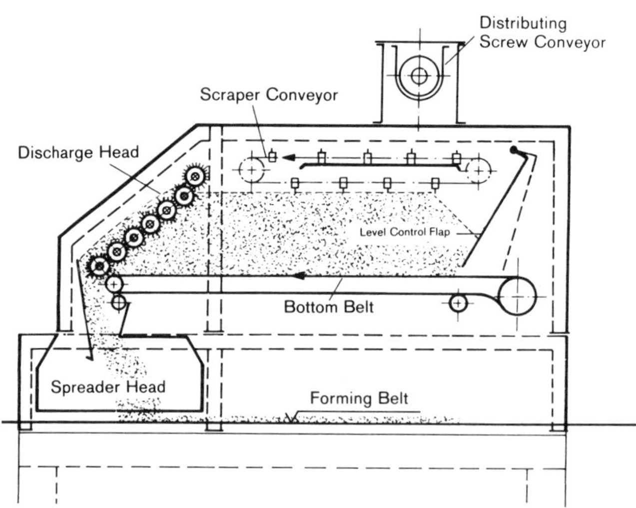

Figure 15.13. *A single-head Schenck spreader (Strisch & van Hüllen 1969).*

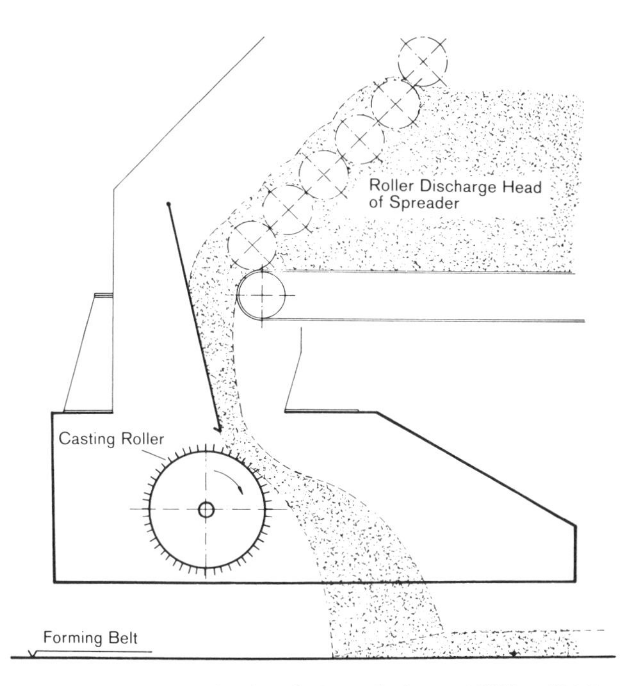

Figure 15.14. *A gravity spreader head (Strisch & van Hüllen 1969).*

Figure 15.15. *A typical Flexoplan forming line equipped with four Schenck spreader heads (Strisch & van Hüllen 1969).*

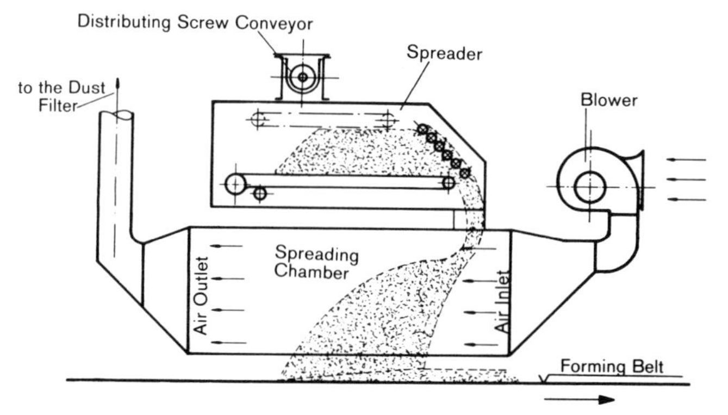

Figure 15.16. *An air grading spreader (Strisch & van Hüllen 1969).*

The Schenck version of an air-classifying mat former is shown in Figure 15.16. Here the particles coming down from the dosing feeder fall into a controlled airstream. The fines are given the longest trajectory, while the coarse particles have the shortest trajectory. This results in a mat formed with graduated layers. The exhaust air is filtered to the atmosphere.

A mat former with a two-roller disintegrating head is illustrated in Figure 15.17. These two rollers rotate in opposite directions and are fitted with long plastic spikes. The direction of rotation of the rollers is opposed to that of the curtain of

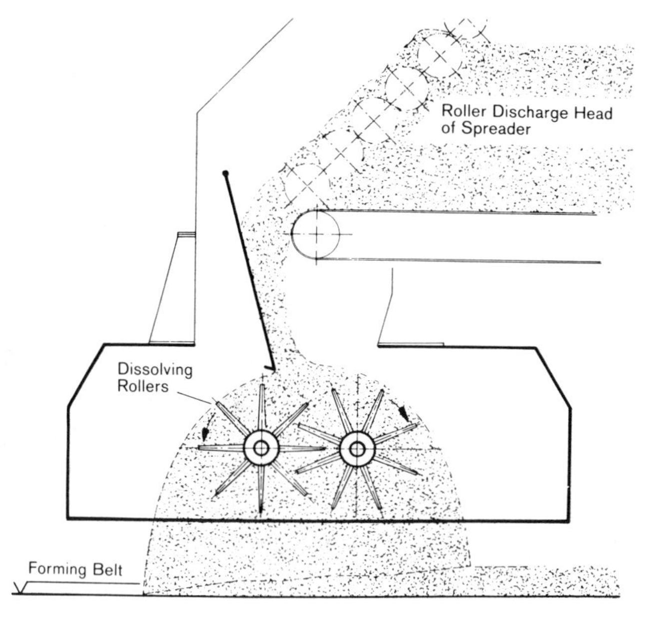

Figure 15.17. *A twin-roller spreader head (Strisch & van Hüllen 1969).*

material being fed to the rollers. This impedes the motion of the particles as they fall. The counter-pushing action of the rotating spikes causes the particles to fall down in small, different curves of descent. The wide disintegrating area of the particles being formed into a mat develops a low particle-deposition angle, the wide veil concept. This low angle is important in insuring that particles or flakes are in a flat position in the mat, as mentioned previously.

Figure 15.18 shows a triple-roller disintegrating mat former which is designed similarly to the head described in Figure 15.17. The speeds of the three rollers are infinitely variable and, as can be seen, the roller frame is slightly inclined. The first spiked roller transfers part of the particles to the next two spiked rollers, whereas the remainder of the particles fall between the first two rollers. These three rollers spread particles over a great length, resulting in a very flat mat on the mat transport medium. The former is now mostly used for core spreading, although it may also be used for surfaces, unless the desired grading effect is too great.

In Figure 15.19 is shown a specially developed spreader head which is used for spreading fibrous and superfine particles for surface layers. It uses a large spreading roll and two rake rollers covered with plastic spike rings. The spreading roll and the rake rollers rotate in opposite directions. The rake rollers are separately adjustable as to rpm with fixed-speed ratios. Because of the faster circumferential speed of the spreading roll as compared to the first rake roller, the fibrous particles are dammed up to a certain extent. The spreading roll itself combs out a uniform portion of the particles to the second rake roller. The circumferential speed of the second rake roller is slower yet, which again causes the spreading roll to comb out a uniform portion of particles. Thus, the fibrous particles are thoroughly unraveled and the proportioning action of the rollers insures that the required rate of fall of the particles is achieved.

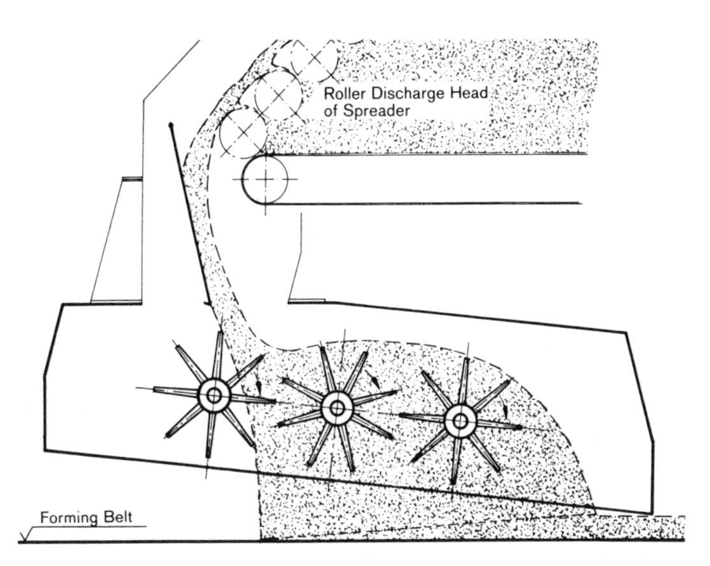

Figure 15.18. *A triple-roller spreader head (Strisch & van Hüllen 1969).*

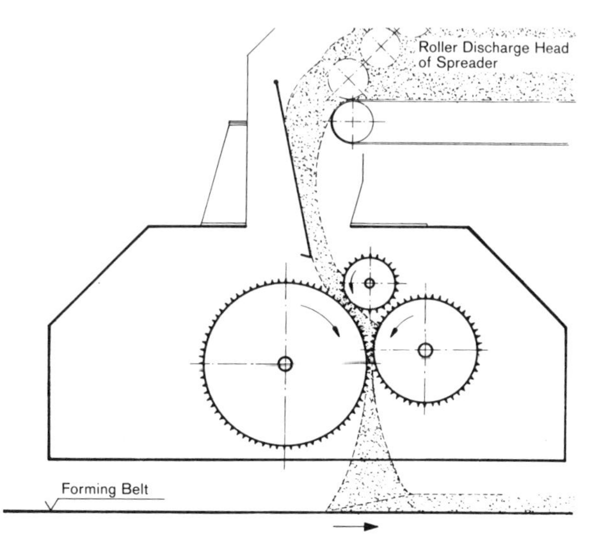

Figure 15.19. *A special spreader head for fine-surface material (Strisch & van Hüllen 1969).*

Durand Microfelter

The Durand Microfelter (originally known as the Columbia Microfelter) was designed to be used with free-flowing particulate types of furnish. Fibrous particles will "ball up" and jam between the metering rolls as will be seen shortly.

This forming machine (shown schematically in Figure 15.20) operates upon the principle that, in volumetric metering, a constant volume is discharged only when the head and all other conditions at the metering point are constant. In design, the machine is divided into two sections: the upper for primary metering, and the lower for precision metering and mat formation.

The upper metering rolls are so mounted and the design so made that special profiled teeth will form a live bottom to the upper hopper, which is sloped outward toward the bottom to eliminate hang-ups. When running, these upper feed rolls meter an average quantity of material from the hopper to the lower forming bin. The distribution across the machine width is reasonably uniform, and the tooth action tends to break up any particle bundles.

The seven lower metering rolls form an inclined bottom to a constant-volume recirculating material hopper from which they draw a uniform volume across the width of each roll. To insure that the volume of material is constant in this section, the upper feed rolls are adjusted to overfeed the mat requirements. The flight conveyor system sweeps the material along to fill any voids beneath the sweep bars, and carries the excess over into the return screw conveyor across the end of this lower hopper. This excess material is returned via a screw lift or conveyor to the upper hopper.

The seven metering rolls drop seven curtains of material from the former (Figure 15.21). As mentioned, fibrous furnish does not work well in this former.

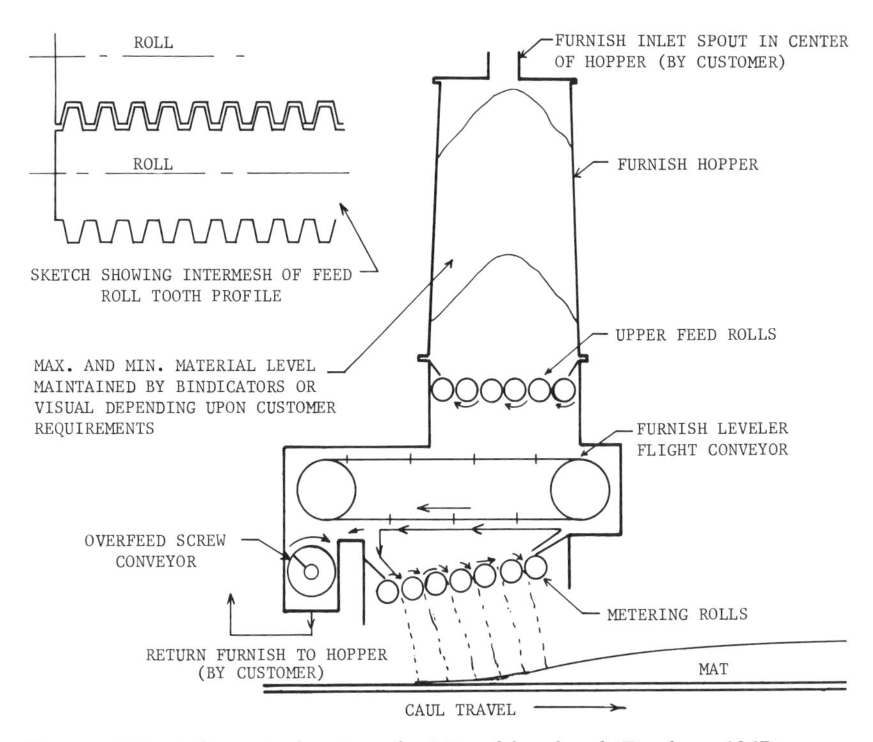

Figure 15.20. *Schematic drawing of a Microfelter head (Frashour 1967).*

Figure 15.21. *Curtains of particles falling from the former (Frashour 1967).*

Also tacky or high-moisture-content furnish will tend to cause problems between the rolls. Polished metering rolls or special coated ones will alleviate some of this problem. The forming is affected by air currents and these must be controlled to prevent malformed mats. For producing layered boards, several of these former heads are placed in series; for example, one for each face and two for the core. Width of mat can be easily adjusted by moving side deckles. The unwanted edge portions of the mat are deflected into return conveyors leading back to the infeed of the former.

Würtex Former

The Würtex former controls the forming process by weight and volume and is normally used in multiples of two. As shown in Figure 15.22, the material is fed into the loading hopper (1) of the spreader, preferably by a scraper belt to insure uniform distribution across the full useful width of the machine. The material is then fed from the loading hopper (1) to the weighing or dosing hopper (3) by means of a steep-angle conveyor (2). The kick-back roller (4) insures that a uniform layer is fed into the weighing hopper. As soon as the preset batch of furnish has been delivered to the dosing scale, the feed hopper (5) is closed pneumatically by two magnetic valves while at the same time the steep-angle conveyor (2) is stopped by the action of an electromagnetic clutch until the weighing hopper (3) has delivered the batch and has returned to the closed position.

The batches of furnish delivered at preset intervals by the weighing hopper (3) are fed to the discharge conveyor (7) by a distribution cell wheel. Thus, the latter takes care of the preliminary equalization of the batches discharged after each

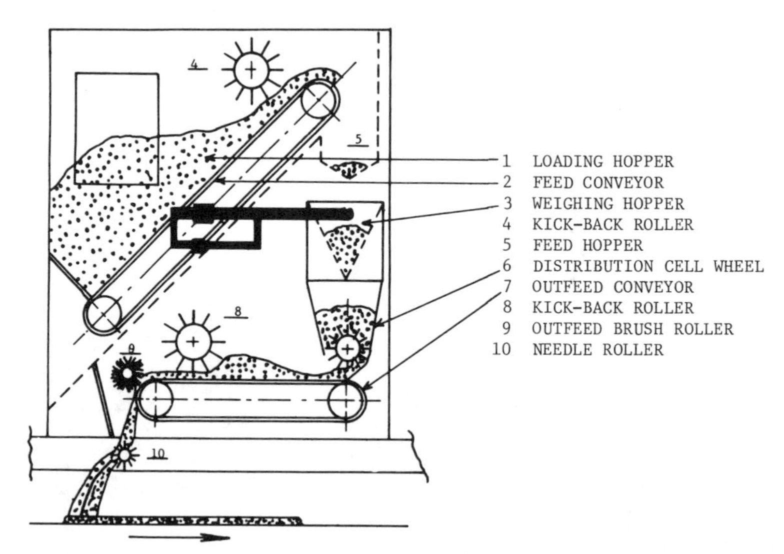

Figure 15.22. *A Würtex forming station for conventional particleboard (Young 1967).*

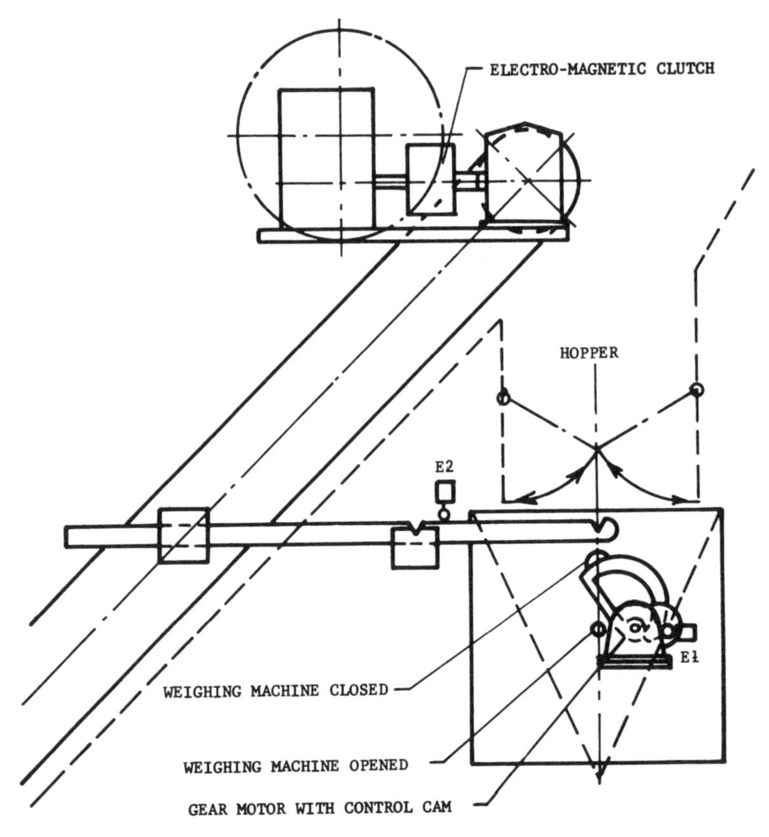

Figure 15.23. *Control of the weighing device (Young 1967).*

weighing cycle. By vertically adjusting the equalizing or scalping roll (8), the volume of the material is kept at a constant level. The speed of the discharge conveyor (7) is infinitely variable so that the material weighed during the preset time can be handled independently of its bulk weight; thus combined weight and volume batching is insured.

The brush roll (9) with vertical and angular adjustment feeds the furnish to the needle roller (10) which then spreads the furnish on the cauls or belt passing underneath the forming station.

The weighing hopper is a container with two flap-type gates which are opened and closed by a cam as shown in Figure 15.23. When the weighing machine is closed, the motor is stopped by limit switch (E1) in preparation for the weighing of the next batch. The gates of the feed hopper are then opened via two magnetic valves and the electromagnetic clutch is activated to feed furnish to the weighing machine by the means of the steep-angle conveyor. Once the proper weight of material is fed into the weighing hopper, the hopper is pneumatically closed by weighing beam switch (E2) via two magnetic valves and the conveyor is stopped by disengagement of the electromagnetic clutch. The weighing machine gates are then opened, dropping the batch of furnish into the distribution cell wheel.

The way the furnish falls upon the transport system passing under the forming station can be varied by the speed and position of the needle roller. By proper adjustment, the larger particles can be thrown farther than the fines so that, when two forming stations are used, a graduated three-layer board can be formed. When four forming stations are used, the first and last formers form the bottom and top surface layers and the center two formers lay down the core layer.

An air-classification system has been built into this type of former by using negative air pressure rather than positive pressure as in the former described previously. In other words, the classifying is done by sucking rather than blowing. Very light, thin flakes have a tendency to fly to the outer extremities of the former when air-classified and thus end up on the board faces. A clever scheme has been devised in this former for preventing such an occurrence when fines only are desired on the faces. A vibrating screen with large holes placed in the airstream screens out the large, thin flakes, causing them to drop into the core of the mat being formed. The fines pass through the screen and are deposited on the face in the normal manner.

In operating the Würtex former, one user has found that the level of furnish in the loading hopper has to be controlled by an upper and lower bin control. If the level is allowed to rise too high or drop too low the upper kick-back roller cannot do its job properly. A close relationship between belt speeds and the amount of material being passed through the former must be maintained, which requires a skilled operator.

One advantage of the Würtex is that the flakes are thrown with some force onto the mat being formed resulting in a compact and relatively thin mat.

Wafer Former

Wafers as noted before are large flakes. If they are produced according to the original concept, the ends are tapered; however, this is not the case normally as the wafers are usually of relatively uniform thickness. These large, flat particles are difficult to transport and handle with equipment designed for "conventional" board processes. Each wafer is a potential small wedge and a number of these can quickly work together and jam mechanical parts of processing equipment. Use of air conveyors usually will result in breaking the wafers.

A former developed for use with wafers is shown in Figure 15.24. Wafers are fed into a revolving distributor head (17) at the top of the former from a belt conveyor. The distributor head drops the wafers into the open ends of a series of tubes arranged in a circle (27). The configuration of the tube arrangement at the bottom of the former is in the form of rows placed across the width of the mat to be formed. The single stream of wafers entering the top of the former is therefore separated into a number of small streams which deposit wafers evenly across the caul plate traveling beneath the former. With this former no jamming of the wafers occurs, as a belt is used for feeding and the passage through the former is by gravity. A number of these formers or heads can be aligned in series to form the amount of furnish desired.

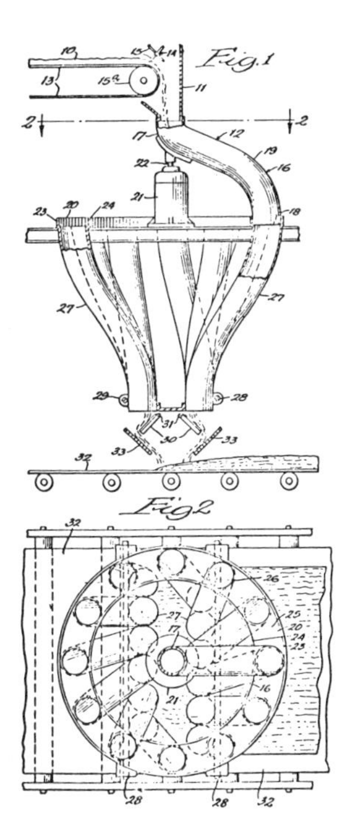

Figure 15.24. *A former used for laying waferboard mats (Gunn 1972).*

The Würtex former and a modification of the previously described Durand Microfelter are also being used successfully for forming mats of wafers. In the Microfelter, the wafers are metered upon a spreader roller in a manner somewhat similar to that shown in Figure 15.14.

Vacuum Formers

Vacuum formers are used when forming dry or semi-dry (about 25-30% moisture content) fibrous furnish. To date their use has been in the hardboard and medium-density fiberboard segments of the industry; however, some may find this approach of value in some of the process lines using more particulate-type furnish.

Vacuum forming is a means of producing a continuous mat by the passage of furnish-laden air through a moving fourdrinier wire screen or a cyclone herringbone wire screen. The furnish enters the former from the top and may pass through a device such as a squirrel cage, shaker screen, metering rolls, or swing-vent to disperse the fiber across the former. Hammermills can also be used. A negative pressure is applied in a chamber below the forming wire. The passage of particle- or fiber-laden air to the chamber traps the furnish on the wire.

One of the more popular formers drops the furnish into the former pneumatically through a swing-vent sometimes called an *elephant trunk*. This swing-vent oscillates perpendicularly to the mat travel. The vent is rectangular with the long

Figure 15.25. *A vacuum former head (Peters 1968).*

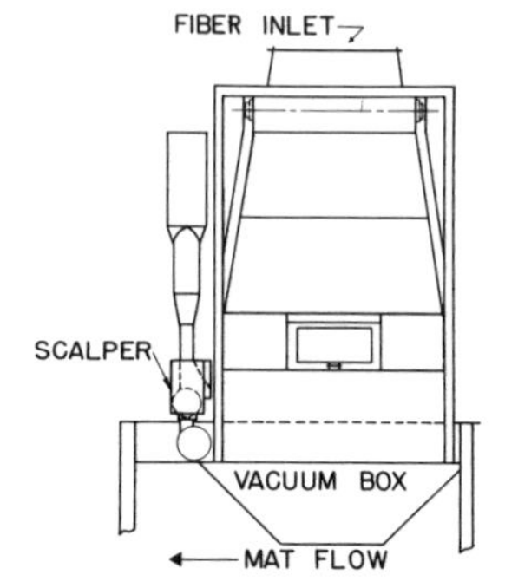

dimension parallel to the mat flow. The action of the swing-vent is analogous to the motion of a reciprocating spray gun carriage in a spray booth for continuously applying a coating to the work passing on a conveyor. Figure 15.25 shows a drawing of one head of such a former.

The air velocity at the swing-vent discharge in some dry-process hardboard felters varies from 4000 to 6000 fpm (1219–1829 mpm). Some users claim that these velocities are necessary for adequate dispersion since fibrous furnishes require high velocities to minimize agglomeration. These high velocities, most likely, would not be necessary for the forming of the nonfibrous furnish used in most particleboard plants. Lower velocities are desirable to prevent disruption of the mat surface in each succeeding felter head.

The distance from the swing-vent to the mat surface tends to decrease the velocity because of the increased volume of furnish in the felter head. Air introduced in vents at the top of the felter also decreases the furnish velocity since the air-to-furnish ratio increases.

As the accumulation of the furnish takes place, the passage of the air through the mat is restricted. Areas of low weight permit greater airflow; thus, there is a greater chance that low-weight areas will subsequently receive additional material. This aids in obtaining uniform mat density. The passage of air through the mat also draws the individual particles into closer relationship with each other. The impact of fibers hitting the mat tends to integrate the fibers with each other. Also, alternate release and application of negative pressures such as occurs between felter heads tends to work the furnish into new interparticle matrixes.

The frequency of swing-vent oscillation depends on the speed at which the forming wire is moving and the length of the pattern discharged from the nozzle. Each oscillation across the mat felts a layer of furnish that overlaps the preceding oscillation. In practice the speed of oscillation may vary from 40 to 120 oscillations per minute.

At the discharge of the felter head, a shave-off roll or scalper rotates in the opposite direction to the mat movement. The shave-off roll removes the excess furnish through a vacuum hood which closely surrounds the scalper rotor. The excess furnish is returned to the fiber bins or is reintroduced into the top of the felter. The shave-off roll is vertically adjustable, which permits accurate reduction of the mat to the desired mat height. The vacuum box under the former is extended

Figure 15.26. *Fiber mat emerging from a vacuum former (Peters 1968).*

under the shave-off roll to prevent mat springback while shaving the mat to a precise height. Automatic control of the weight of each layer may be had by controlling the shave-off height with a radiation gauge and closed-loop instrumentation. Figure 15.26 shows a mat emerging from a vacuum former after it has been scalped to the desired height.

Additional felter heads may be added, depending on the requirements of the board manufacturer such as production rate, furnish, species, and product design. Figure 14.15 showed a schematic of a typical forming line and Figure 15.27 shows a typical forming line in operation.

Deckles (thin side plates) of several types are used on vacuum formers. The fixed deckle is common but a moving deckle may be used. The deckles are side

Figure 15.27. *A Washington Iron Works dry-process forming line (Peters 1968).*

forms controlling the width of the mat formed. Some machines deliberately overshoot the deckles with furnish and the excess furnish is removed from the side-deckle area by a vacuum fan. An adjustable deckle which permits different mat widths is a desirable feature for some processes, because the excess furnish from each head can be reintroduced into the same felter head. Such an arrangement reduces the mixing of dissimilar furnish in the system.

The maintenance of a slight negative atmospheric pressure within each felter head is essential to prevent blowout of the furnish. In order to maintain a negative pressure in a felter head as the weight and density of the mat increase, the horsepower requirement must be increased to remove the same air volume because the mat acts as a filter or damper. For example, one particular felter requires 40 HP (30 kW) in the first head to maintain 1 in. (25.4 mm) of water differential pressure in the vacuum chamber; however, the number four head uses approximately 150 HP (557 kW) to remove the same amount of air at a negative pressure of 25 in. (635 mm) of water.

This system of forming allows for close control over mat weight. The mat is also compressed by the action of the negative air pressure pulling the fibers against the fourdrinier screen. The bulk density of the mat is much higher than the bulk density of the fiber. This provides a mat with better integrity which is easier to prepress and one that is easier to handle on caulless lines. A roll prepress is normally used in this system and will be described later.

Vacuum felting permits flexibility in mat formation. A layered or homogeneous mat can be formed, depending on the product requirement. Also different wood sources or species can be selectively placed in the face or core layers. This selectivity is also applicable to maintenance of different layer moisture percentages or resin contents. Furthermore, face and core furnish may be felted according to particle geometry. Uniformity of the mat is enhanced by controlled buildup and precise shave-off.

The felter is not a self-contained unit. A separate metering system is required; thus, the accuracy is also affected by the metering system feeding the felter. The former has a relatively high horsepower requirement which again adds to the investment and the operating cost. The system may reduce urea resin tack to an unacceptable level because of frequent pneumatic transfer and recirculation.

The vacuum felter is based on volumetric metering. This principle assumes that the furnish bulk density changes slowly or not at all. Consequently, plants with variable raw materials, i.e., species, wood geometry, or wood source, may experience erratic weight control of the mats.

Rando-Wood Former

Recently a new former has been developed to handle fiber called the Rando-Wood MDF former. This machine is manufactured by the Rando Machine Corporation, which has used techniques involved in its equipment made for the non-woven and textile industries.

Figure 15.28 shows this new former, which is now being used in the board industry. The former is divided into two parts: the feeder and the former. Figure 15.29 is a photograph of the complete former undergoing tests.

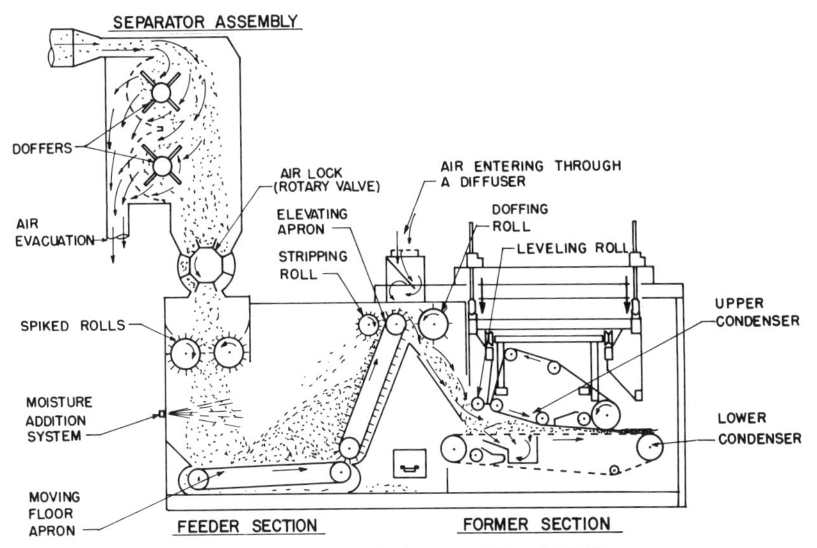

Figure 15.28. *The Rando-Wood MDF former (Wood 1976).*

The resinated fibrous furnish is supplied to the feeder via a mechanical or a pneumatic delivery system which insures a constant and uniform supply of material to each forming station. The furnish flows through the machine shown in Figure 15.28 as follows. The fiber enters the forming unit via the pneumatic separator where the furnish is separated from the air-conveying system and retained in the chute attached to the first stage of the forming equipment. A fiber beater system consisting of a pair of spiked rolls fiberizes any clumps of fiber and provides a uniform supply across the width of the hopper. Moisture can be added to the fiber as it passes from the beater rolls to adjust the moisture content. The fiber falls upon a moving delivery floor apron, which introduces the fibers to the inclined apron equipped with specially designed slats and pins. This apron picks up the fibers from the creeping delivery floor apron and raises the fibrous mass to a stripping conveyor. At this point, the excess material is removed by the stripping conveyor and returned to the rear of the feeder. This action also reduces any tendency for the fibers to form clumps or balls of material. The material remaining on the incline apron pins is thus distributed across the width of the former at a constant volumetric flow. This remaining material held by the pins passes over the apex of the apron into the mat-formation section.

At the apex of the incline, an airstream is created which strips the fiber from the pins of the elevating apron. The free fall of the fiber is also influenced by the negative air pressure developed in the mat-formation section. In the mat-formation section is an "air bridge" which has a wedge-like cross section. Situated on the lower side of this wedge is a perforated condenser assembly with an inner suction chamber. The wedge developed by the condenser provides an ever-decreasing path for the fibers so as to provide a slight pressing action to the mat structure formed within the air bridge.

The fibers are compacted by the suction pressure drawn through the condenser assembly. As the condensers are rotated forward, the surface movement, combined with the suction pressure, causes the fiber to move forward in the air bridge. Wherever there is a weak packing of fibers, a strong airflow is established; and fibers are immediately pulled into it, thus building up a uniformly dense mat.

The output of the former is measured volumetrically. The packing pressure is dependent upon the intensity of the suction chamber pressure in the condenser and is directly related to the static pressure of the airflow. Output can be controlled by the surface speed of the condenser assembly (all other variables held constant) by increasing or decreasing the cross-sectional distance of the throat, and by increase or decrease in speed of the hopper section's supply aprons.

Each former develops a single ply of mat with the previously mentioned intertwined arrangement of the fibers. The formed mat is laid upon a conveyor for transport down the press line. For thick mats, a number of formers can be used in tandem, laying up a multiple lamination of thinner mats. It is reported that a single former can lay up to 150 tons (136 mt) of mat per day at a maximum mat width of 100 in. (2.54 m).

Figure 15.29. *A Rando-Wood former undergoing tests. (Courtesy Rando Machine Corp.)*

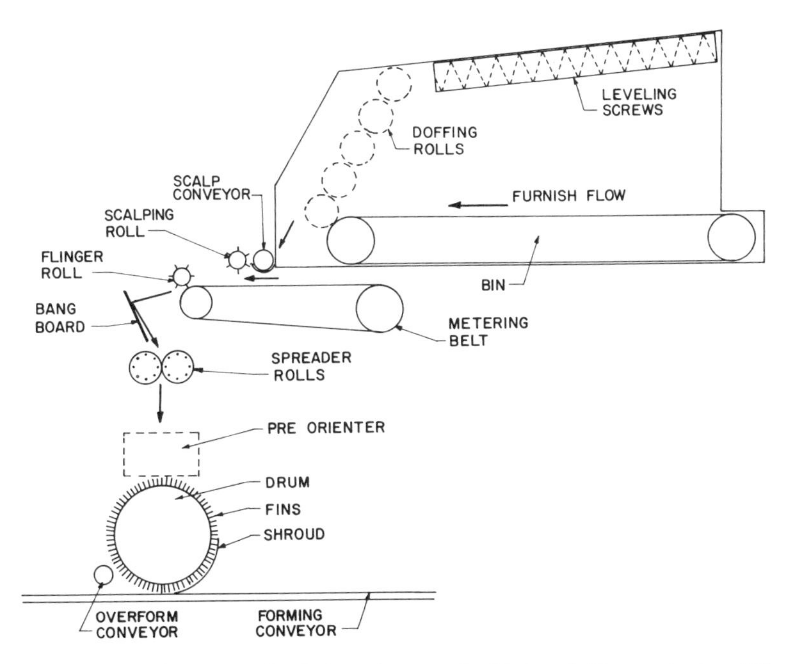

Figure 15.30. *Schematic of a flake orienter, the Potlatch Crossorienter. (Courtesy Potlatch Corp.)*

Orienting Formers

Orienting or aligning particles in boards is gaining in importance, as stated previously. Two ways of orienting are now available: electrical and mechanical.

With the electrical means, particles are dropped between charged plates (one positive, the other negative). The electrical field aligns the particles as they fall between the plates onto the mat carrier. A wide range of particles from large flakes to pressurized-refined fiber can be aligned by this method. A wide-ranging variety of particle sizes in the furnish can be aligned simultaneously.

A mechanical orienter now in use is shown in Figure 15.30. This machine was developed for aligning flakes. The flakes are fed from a bin onto the metering belt. The flakes are fed through the system into a preorienter. Reciprocating fins in the preorienter align the flakes and drop them into pockets formed by steel fins on a large-diameter drum. The fins hold the flakes in the aligned position. The drum is rotating and deposits the aligned flakes onto the mat carrier. A shroud keeps the flakes from falling out of the pockets until they are dropped onto the mat carrier.

SELECTED REFERENCES

Buikat, E. R. 1969. Clean Dry Air for Instruments and Plant Use. *Proceedings of the Washington State University Particleboard Symposium, No. 3.* Pullman, Washington: Washington State University (WSU).

Carter, R. K. 1969. Weighing Devices and Instruments Used in Particleboard Manufacture. *Proceedings of the Washington State University Particleboard Symposium, No. 3*. Pullman, Washington: WSU.

FAO. 1976. Basic Paper IV. Technology and Techniques in the Manufacture of Wood-Based Panels. *Proceedings of the World Consultation on Wood-Based Panels,* held in New Delhi, India, February 1975. Brussels: Published in agreement with the Food and Agriculture Organization of the United Nations by Miller Freeman Publications.

Fischbein, W. J. 1957. Continuous Press Manufacture of Particleboard — Economics, Quality, Production. Technical Paper No. 5.35. *Fibreboard and Particle Board Technical Papers*. International Consultation on Insulation Board, Hardboard and Particle Board, FAO, Geneva, Switzerland, January-February 1957.

Frashour, R. G. 1967. The Columbia Microfelter. *Proceedings of the Washington State University Particleboard Symposium, No. 1*. Pullman, Washington: WSU.

Gunn, J. M. 1972. New Developments in Waferboard. *Proceedings of the Washington State University Particleboard Symposium, No. 6*. Pullman, Washington: WSU.

Hansen, S. K. 1974. Brochure on Durand Microfelter. Durand Machine Company. New Westminster, B.C.

Maloney, T. M. 1961. Composition Board Mills. Division of Industrial Research Report. Washington State University, Pullman, Washington.

Newman, W. M. 1967*a*. The Bähre Bison Forming Machine. *Proceedings of the Washington State University Particleboard Symposium, No. 1*. Pullman, Washington: WSU.

———. 1967*b*. Principles of Mat Formation. *Proceedings of the Washington State University Particleboard Symposium, No. 1*. Pullman, Washington: WSU.

Peters, T. E. 1968. The Vacuum Former. *Proceedings of the Washington State University Particleboard Symposium, No. 2*. Pullman, Washington: WSU.

Raddin, H. A., and D. E. Wood. 1973. The Rando-Wood Process for the Manufacture of Continuous Thin Fiberboard. *Proceedings of the Washington State University Particleboard Symposium, No. 7*. Pullman, Washington: WSU.

Rando Machine Corp. The Model 110 WF Rando-Wood Former. Macedon, New York.

Rowe, S. 1969. Nuclear Gages for the Forest Products Industry. *Proceedings of the Washington State University Particleboard Symposium, No. 3*. Pullman, Washington: WSU.

Strisch, H., and S. van Hüllen. 1969. New Systems and Recent European Developments. *Proceedings of the Washington State University Particleboard Symposium, No. 3*. Pullman, Washington: WSU.

Talbott, J. W. 1974. Electrically Aligned Particleboard and Fiberboard. *Proceedings of the Washington State University Particleboard Symposium, No. 8*. Pullman, Washington: WSU.

Wentworth, I. 1969. New Designs for Single-Opening Presses. *Proceedings of the Washington State University Particleboard Symposium, No. 3*. Pullman, Washington: WSU.

Wood, D. E. 1976. Pneumatic Web Formation and the Rando-Wood Medium-Density Fiberboard Former. *Proceedings of the Washington State University Particleboard Symposium, No. 10*. Pullman, Washington: WSU.

Young, D. R. 1967. The Würtex Forming Station. *Proceedings of the Washington State University Particleboard Symposium, No. 1*. Pullman, Washington: WSU.

16. PREPRESSING

In order for the caulless system to operate without using the tray-belt approach, a prepress is needed to consolidate the mat for subsequent handling. Prepressing is not required when using a caul or metal mesh cloth or steel belt for transporting the mat into the press. However, prepressing reduces the "daylight" or press opening needed, particularly with thicker boards, and allows for faster press closing without blowing the top surface of the mat apart. It is claimed that fast press closing reduces surface precuring and increases the core density with prepressed mats, which in turn provides a higher internal bond. The opposite is true when pressing unprepressed mats, as will be noted later. Because of these advantages, some multi-opening caul-type and single-opening press lines also use a prepress. Pressures of up to 560 psi (3.86 MPa) have been reported to improve the pressing conditions and board properties. Higher pressures have reportedly been used in production plants, but this is unusual. Higher pressures have been shown in laboratory work to decrease board strength properties using conventional pressing. Either damage to the particles or driving the resin from the particle surfaces into the particles, which removes the resin from the bonding areas, appear to be two reasons for this drop in board strength properties.

Two types of prepress are used: a platen prepress which has to have the mat stopped for operation and a continuous prepress which consolidates the mat as it moves from the former.

PLATEN PREPRESSES

The platen type previously mentioned in the discussion of caulless lines operates with either a heated or unheated top platen. The top platen moves in such a press. Consequently, care must be taken that hydraulic oil leakage does not spill onto the mats. Figure 16.1 shows one of these prepresses.

For the platen prepress to function, it is necessary to move the formed mat quickly into the prepress and consolidate it before the next mat arrives. This calls for rather sophisticated belt-transfer systems and acceleration and deceleration controls. It is the most commonly used type in the United States for particleboard production.

Figure 16.1. *A mat entering a caulless prepress (Peinecke 1967).*

CONTINUOUS PREPRESSES

In a continuous prepress, the mat is compressed while moving through it on the same belt on which it was formed; consequently no change in transport-belt speed is required. The prepressed mat is then cut to length and quickly moved ahead, into the press. The continuous presses use belts or metal tractor-type treads to convey the mat through the pressure section.

The belt prepress also goes under the name of a *roll prepress*. A typical one used in dry-process fiberboard plants is shown in Figure 16.2. The mat entering the prepress is first quickly consolidated by a pair of nip rolls and then held at that thickness through a holding section. The prepressed mat springs back to some degree upon leaving the prepress. However, a considerable reduction in mat thickness takes place. One of these prepresses is shown in Figure 16.3.

In some cases, when very thick fiberboards are being made, several prepressed mats are stacked on top of each other. Otherwise, it would be impossible to fit the mats into the press. Unprepressed mats could easily be over a foot in thickness.

Another version of a belt prepress is shown in Figure 16.4. This prepress is operating on a caulless particleboard line where the mat has been formed with an air-classifying former. A nip roll is not used in this press to gain immediate compression of the mat to final prepressed thickness. Particulate-type mats would probably have too much disruption with nip rolls. Fibrous furnish, on the other hand, can be successfully prepressed initially with nip rolls.

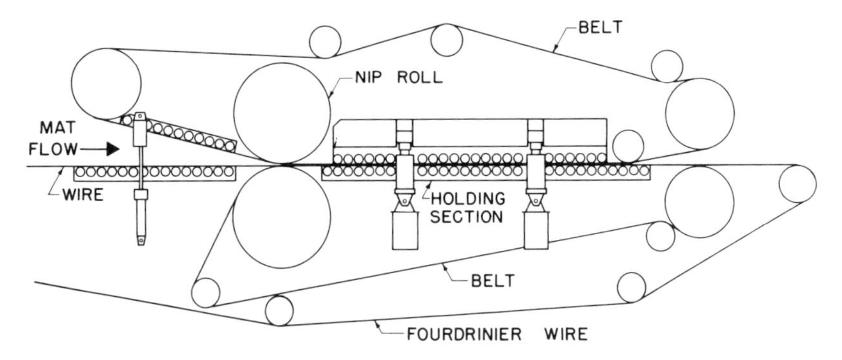

Figure 16.2. *Schematic drawing of a roll prepress (Peters 1968).*

The type of prepress using a heavy metal tread is shown in Figure 16.5. An endless plastic-faced belt runs through the forming machine and on into the opening to the prepress. This belt carries the mats through the prepress and returns to the entrance of the forming machine, leaving the prepressed mat to be carried forward to the traveling cutoff saw. The pressure plates, located at both top and bottom of the press, are segmented and are rotating around the end drums. The pressure plates are covered, both top and bottom, by endless plastic-coated belts, so there is not an imprint of the pressure segments on the mat.

As the mat enters the pressing area, the first pressing plates are under a roll pressure of about 100 psi (0.69 MPa). As the mat progresses further, the pressure is increased gradually until 500 psi (3.45 MPa) is exerted on the mat. The drives of this prepress are variable to fit the line speed of the former.

Figure 16.3. *Precompressor (Johnson 1973).*

Figure 16.4. *Belt precompressor. (Courtesy Siempelkamp Corp.)*

As the mat leaves the prepress, it is monitored by a gamma density gauge which indicates whether it is acceptable or not. If it falls outside of the limitations set, then the mat is rejected after it leaves the traveling cutoff saw. After the mat has been crosscut, a speedup conveyor section moves the mat over a tipple arrangement. If the mat is not acceptable, the tipple pops up, and the mat is rejected through a spiked breakup roll into a return system which relays it back to the forming machine for reuse. Figure 16.6 is a view of a typical trimmed prepressed mat showing the tight edges.

Different users will have different needs with a prepress. Whether the platen or continuous type is the better will depend upon the process line and desires of the manufacturer.

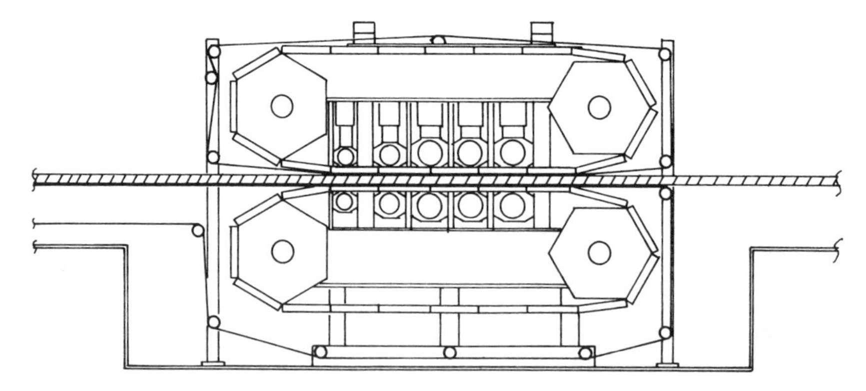

Figure 16.5. *Schematic drawing of high-pressure prepress (Wentworth 1972).*

Figure 16.6. *The trimming section cutting the prepressed mat to length (Wentworth 1972).*

Preheating the Mat

Of interest is the ability to preheat the mat before pressing. Such preheating is not sufficient to cure the resin but does impart heat to the mat before pressing in order to shorten the pressing cycle. Just a few seconds less time in the press means a considerable increase in press production over a year's time.

An early approach to this preheating of the mat was the use of high-frequency heating just before the mat entered the Bartrev continuous press. Recent press lines have also been built using an RF tunnel preheater to elevate the mat temperature before hot pressing.

A heated-platen prepress has been developed for use with single-opening presses. The mats are consolidated with this low-pressure, heated prepress using the same cycle as in the main press. The top and bottom platens of the prepress have individual temperature control; thus, the prepressed mat can be brought up in temperature as desired, reducing the final curing time required in the main press. In many single-opening press lines, without a prepress, a certain dead time is required for heating up the bottom steel belt when it carries a fresh mat into the press. The top platen cannot be brought in contact with the mat surface until this is accomplished. Now, by controlling the bottom prepress platen temperature at a different level than the top platen in the prepress, this dead time is eliminated. A decrease in hot-pressing times of 33% has been reported by the industry, using the heated prepress.

SELECTED REFERENCES

Bucking, H. G. 1974. Exterior Structural Board Products Manufactured in Single-Opening Presses with Heated Prepresses. *Proceedings of the Washington State University Particleboard Symposium, No. 8*. Pullman, Washington: Washington State University (WSU).

Carlsson, B. J. 1975. Presses and Pressing Techniques in Wood-Based Panel Industries. Background Paper No. 105, third World Consultation on Wood-Based Panels, FAO, New Delhi, India, February 1975.

FAO. 1976. Basic Paper IV. Technology and Techniques in the Manufacture of Wood-Based Panels. *Proceedings of the World Consultation on Wood-Based Panels,* held in New Delhi, India, February 1975. Brussels: Published in agreement with the Food and Agriculture Organization of the United Nations by Miller Freeman Publications.

Fischbein, W. J. 1957. Continuous Press Manufacture of Particleboard—Economics, Quality, Production. Technical Paper No. 5.35. *Fibreboard and Particle Board Technical Papers*. International Consultation on Insulation Board, Hardboard and Particle Board, FAO, Geneva, Switzerland, January-February 1957.

Gefahrt, V. J. 1975. Zur Physik and Technik der Spänevorwärmung mit Hochfrequenzenergie (About Physics and Technique of Preheating Woodchips with High Frequency Energy). *Holz als Roh-und Werkstoff,* Vol. 33, No. 5.

Johnson, D. R. 1973. Panel Discussion on Experiences and Problems with Pressure-Refined Fiber. *Proceedings of the Washington State University Particleboard Symposium, No. 7*. Pullman, Washington: WSU.

Peinecke, R. 1967. Prepressing and the Caul-less System. *Proceedings of the Washington State University Particleboard Symposium, No. 1*. Pullman, Washington: WSU.

Peters, T. E. 1968. The Vacuum Former. *Proceedings of the Washington State University Particleboard Symposium, No. 2*. Pullman, Washington: WSU.

Steck, E. F. 1970. Caulless Pressing Systems for the Manufacture of Particleboard. *Proceedings of the Washington State University Particleboard Symposium, No. 4*. Pullman, Washington: WSU.

Wentworth, I. 1972. Recent Advances in Prepresses for Single-Opening and Multiple-Opening Particleboard Plants with Automatic Stacking for Cut-up Operations. *Proceedings of the Washington State University Particleboard Symposium, No. 6*. Pullman, Washington: WSU.

17. HOT PRESSING AND PRESSES

HOT PRESSING

The most expensive single piece of equipment is the press, and speeding the press time makes better use of it. Press times have been shortened dramatically since the 1950s through resin and press improvements. For example with ¾-in.-thick (19 mm) board, press time has dropped from 15 minutes to about 3 to 5 minutes at present when using urea resin. Thus, production through the press has been more than doubled.

In pressing, a number of factors are involved, including the resin type, resin catalyst, press temperature, wood species and particle geometry, mat moisture level and distribution and heat transfer, press closing time, pressure, and its relation to the board-density profile, vapor pressure within the board during pressing, and precure and postcure of the resins. As is usual in the board process, all of these factors interact, making it difficult to provide easy-to-use formulas when considering hot pressing. These factors will be considered separately and references made to major interactions; however, all interactions will not be discussed in detail since this part of the production process is still part of the web of interrelated variables that are difficult to sort out and understand completely. This will be followed by a discussion of the major types of presses.

Resin Type and Catalyst

As noted before, urea and melamine-urea resins are much faster curing and less expensive than phenolics. Thus, phenolics are used only when absolutely necessary because of the slower press times. The amino resins are catalyzed for the fastest cure. Recent developments in phenolics have also made it possible to speed the cure; however, the aminos are still faster.

Press Temperatures

High press temperatures, from about 360° to over 400°F (182°–204°C) are needed with the phenolics. The aminos are also speeded in cure rate by high press temperatures, but lower press temperatures can be used successfully. In earlier times, many plants operated with press temperatures for ureas at about 290°F (143°C), and this level of press temperature is not unknown now. However, newer presses are usually designed to operate at much higher temperatures, in particular the single-opening types where press time is of extreme importance. Heat from the press is also supplemented from the heat developed by the condensation reaction of the resin curing. One study showed that as much as 20% of the total heat needed for polymerization came from this reaction, although 10 to 15% may be a more reasonable level.

Wood Species and Particle Geometry

Wood species and the configuration of the particles are important factors affecting board consolidation. Easy-to-compress species such as redwood require lower pressures for consolidation than harder-to-compress species such as oak (considering that both species are being increased in density in the final board proportionately). Easy-to-compress species have a tendency to "seal" the surfaces of the board, resulting in blows because vapor cannot escape through the face during the latter part of the pressing time.

Particle geometry affects pressure because larger particles that must be "crushed" one into another require greater consolidation pressures than smaller ones that mesh together, such as −20 screen fraction sizes and some small fibers. However, small fractions of high-density species are hard to compress into the smooth surface similar to those with small comminuted particles used in typical softwood particleboards. Voids are readily apparent between high-density small particles used in the faces as they do not compress as well and flow together. In some developmental work, it has only been possible to develop suitable surfaces for overlaying by using a high percentage of small particles (−42 screen fraction size) or by using high face moisture contents (16–18% for plasticizing the wood).

Mat Moisture Level and Distribution

Probably the two most significant factors governing the properties of a given board (considering reasonable resin levels, press times, and press temperatures, etc.) are moisture content and its distribution and press closing time.

The moisture content of a mat entering the hot press is of great importance in pressing composition board. The moisture in the furnish comes from three sources. First, there is the moisture contained in the furnish after it has been dried in preparation for blending with the resin. Depending on the process line, this will range from about 2 to 5%, based on the oven-dry weight of the wood. This level is somewhat on the high side when powdered resin is to be used. Keeping the furnish moisture content as high as possible when blending with resin may help prevent resin "strike in" and its subsequent loss as a binder.

Second, there is the water that enters the furnish as part of the liquid resin used. Urea-formaldehyde resins are normally supplied at a solids content of about 65%. Resin may be applied at this solids level or for production reasons, where better atomization can be achieved, diluted to a lower level of 55 to 60%. A number of different solids levels are found with phenol-formaldehyde resins. A low resin solids level of about 20 to 25% is normally used in the dry-process hardboard-type plants to obtain better distribution. Since the phenolic resin is normally applied in the attrition mill, the excess moisture can be removed in the dryer. For particleboard, the phenol-formaldehyde resins were manufactured at a 40% solids level for many years, thus 60% of the resin was moisture. In many cases this excessive amount of moisture raised the final mat moisture content above the 10 to 11% maximum that many plants can tolerate when higher resin levels were needed. Since post-drying is not done in particleboard plants, this severely limited the amount of phenolic resin that could be added to the furnish. However, in recent years, newer formulations of phenol-formaldehyde resin have been developed with a solids content around 53%. This solids content is not all phenolic; another additive has been used to lower the moisture in the resin and to aid in its handling and curing characteristics. The phenolic solids content has remained at about 40%.

The third source of moisture in the mat comes from the condensation reaction when the resin cures. In general, two resin molecules combine and thereby split off one water molecule resulting in a small amount of extra moisture. One report notes that on a 6% resin solids addition of urea-formaldehyde, a maximum of 0.9% water can be added to the furnish in this way.

A fourth type of "moisture" can be a problem in pressing board. This is not truly moisture but vapor that is generated during hot pressing from the extractives in the wood itself. A problem species of this type is western red cedar. More on vapor pressure will be mentioned later.

As with any gluing operation, excessive moisture can hinder or prevent the adhesion of two pieces of wood. Most particleboard or fiberboard can be bonded at moisture contents ranging from 2 to 18%. Semi-dry hardboard may go as high as 30% in mat moisture content. At the high-moisture-content level, it may require an excessively long press time. At the low-moisture-content level, problems can be encountered in closing the press so that the desired board thickness cannot be achieved. A highly significant aid to press closing is the moisture in the mat which converts to steam during hot pressing. This steam plasticizes the wood particles, enabling the press to consolidate the mat with relatively low pressures as compared to cold pressing. Indeed, many presses in use would not be able to consolidate the mats to the desired thickness without the use of heat because of pressure limitations.

Boards are pressed at moisture contents ranging between 7 and 16%, depending on the particular plant's operating parameters. Troublesome species like the aforementioned western red cedar and ponderosa pine are difficult to press at moisture-content levels around 11 to 12%. Such a moisture level is relatively high in the United States, but higher levels are common elsewhere in the world. Single-opening presses are dependent upon very short pressing times for achieving economical production. Thus, moisture contents of the mats are held somewhere around 7%. Higher pressing pressures are needed to accomplish board consolidation, but at the same time excessive amounts of moisture do not interfere with fast bonding of the resin.

As the mat is heated in the press, the moisture changes to vapor and migrates to cooler areas, most notably at first to the core of the board. Within a short time after the press is closed, a temperature gradient exists between the faces of the board and the core; and also between the center of the panel and the edges. Most people recognize that such a gradient exists between the faces and the core but some forget about the gradient between the center and the edges. These gradients, of course, are constantly changing as the temperature of the various zones of the mat being consolidated into a board changes. Because vapor pressure is related to temperature, the pressure gradients will follow the temperature gradient during the initial part of the cycle.

If the board is left in the press for a long time, the pressure gradients and temperature gradients will eventually come to equilibrium, or perhaps even reverse themselves. However, this does not come to pass because boards are taken out of the press long before such an equilibrium occurs. Once the entire core of the board reaches about 212°F (100°C), the final cure is effected within about 30 seconds with urea resins. A somewhat longer press time is needed with phenolic resins. Core temperatures at the panel corners will rarely get above this level unless the board is left in the press for a long period of time.

Many studies have been conducted showing the temperature of the core of the boards at various time periods during hot pressing; however, it can be noticed when running a study to determine the shortest press time that the last places in a board to cure are at the corners. Moisture is always working its way out of the board to the edges during pressing. The corners are the coolest parts of the board because of their exposure to atmosphere on two edges. Consequently, these are the last areas to get above the boiling point of water, and thus are the last areas in which the resin is cured. In press time studies on a ⅝-in.-thick (15.9 mm) board, for example, it may be that at 3 minutes the board is not cured in the core at all. At 3 minutes and 30 seconds there is some curing in the core. At 4 minutes the board is completely cured except at the corners. This is noticeable in that the corners are much thicker, and the core at the corners is not bonded together at all while the core at the midpoints of the panel edges is cured. Within about 30 seconds afterwards, however, the core at the corners will be completely cured.

In the early days of this industry, it was felt that most of the moisture was vented from the board during hot pressing. Since with ¾-in.-thick (19 mm) board press times ranged between 10 to 12 minutes there was a very good chance that such was the case, particularly with boards 4x8 ft (1.22x2.44 m) in size. However, with the increase in board sizes up to 8 ft (2.44 m) wide to more than 60 ft (79.2 m) in length and the very significant decrease in press time due to improvements in the resins, not much of the moisture can be evacuated from the board during hot pressing. Board moisture content after pressing may be 5 to 7% and not oven-dry as some might believe. Thus, only a small amount of the moisture escapes through the edge of the board. The other moisture that has been turned to vapor during pressing is released through the faces of the board upon opening of the hot press. If the face of the board is relatively impervious because of large flakes or highly densified, resin-rich, small particles, blows occur. Lowering of the moisture content is one possible solution to alleviating this blowing problem.

The moisture level and its distribution is closely related to the heat transfer in a standard contact heating press. The high-frequency heating system is different in that the entire mat can be heated simultaneously, whereas with contact heating, heat is transferred from the mat faces to the core. One study has shown that the heat-transfer process in the hot-pressing cycle involves two interrelated components. The first component is the transfer of heat by conduction of water vapor flow from the interface of the hot platen and mat to the mat core. The second component is the transfer of heat by vapor flow from the interior of the board to the exposed edges. Permeability and relative vapor pressure are much more dependent upon the board variables than are the heat conduction parameters. Consequently, the rates at which the interior temperature changes are regulated by the flow of water vapor through the porous structure of the board. Vapor flow from the faces to the board core accelerates the temperature rise, while vapor flow towards the edges of the board retards temperature rise by consuming thermal energy as heat of vaporization.

Centerpoint temperatures in low-density board experience a rapid rise and level off quickly near the boiling point of water, while those in high-density boards have slower initial rises but continue to rise steadily throughout the pressing time. Increasing the moisture content hastens the initial rise and causes the low-density board core temperatures to level off faster and the high-density board core temperature to increase more rapidly. Low-density panels benefit from vapor flow from the faces to the core early in the pressing and are penalized by lateral flow toward the board edges later. High-density boards, on the other hand, get little assistance in increasing the core temperature at the start of the press cycle but tend to increase in temperature later in the cycle without the restricting influence of the lateral flow of the vapor. Rates of temperature increase ranged as high as seven times or greater for ½-in.-thick (12.7 mm) board as compared to 1½-in.-thick (38.1 mm) boards. A thin board, of course, has a rapid temperature increase in the core during pressing because of the core's close proximity to the hot platens.

Moisture content of the mats, as has been discussed so far, can be used advantageously and can also cause problems. It is easier to close the press quickly at the higher mat moisture contents, and the thickness of the board can be more easily controlled. Presses can be built lighter because less pressure is needed for consolidating the mats into boards. High moisture, particularly on the faces, speeds the transfer of heat from the hot platens to the core of the mat, as is shown in Figure 17.1. This use of moisture for rapidly transmitting heat to the core is also called *steam-shock*. However, the cure time of the resin, particularly in the core of the board, can be impeded by an excess of moisture. The higher mat moisture contents of faces aid in plasticizing the wood particles, providing tight, hard faces. Levels of about 16% in moisture content are excellent for this purpose. This high face moisture has the further advantage of retarding the cure of the face resin, thus reducing precure problems. However, higher mat moisture contents will result in lower-density board cores, lower internal bond strengths, the possibility of blows or delaminations and caul-to-board sticking problems. As will be mentioned later, the press closing time and the various moisture contents and distributions of moisture throughout the mat have a pronounced effect on the final board-density profile and physical and mechanical properties. It may be necessary to lower the mat moisture content if the species or particles compress too easily.

Advantages of lower mat moisture content are: higher general strength properties, in particular internal bond; shorter press times; freedom from caul sticking; and a more uniform density profile. Problems that can result from low mat moisture content are: higher water absorption in the boards, rougher faces due to a

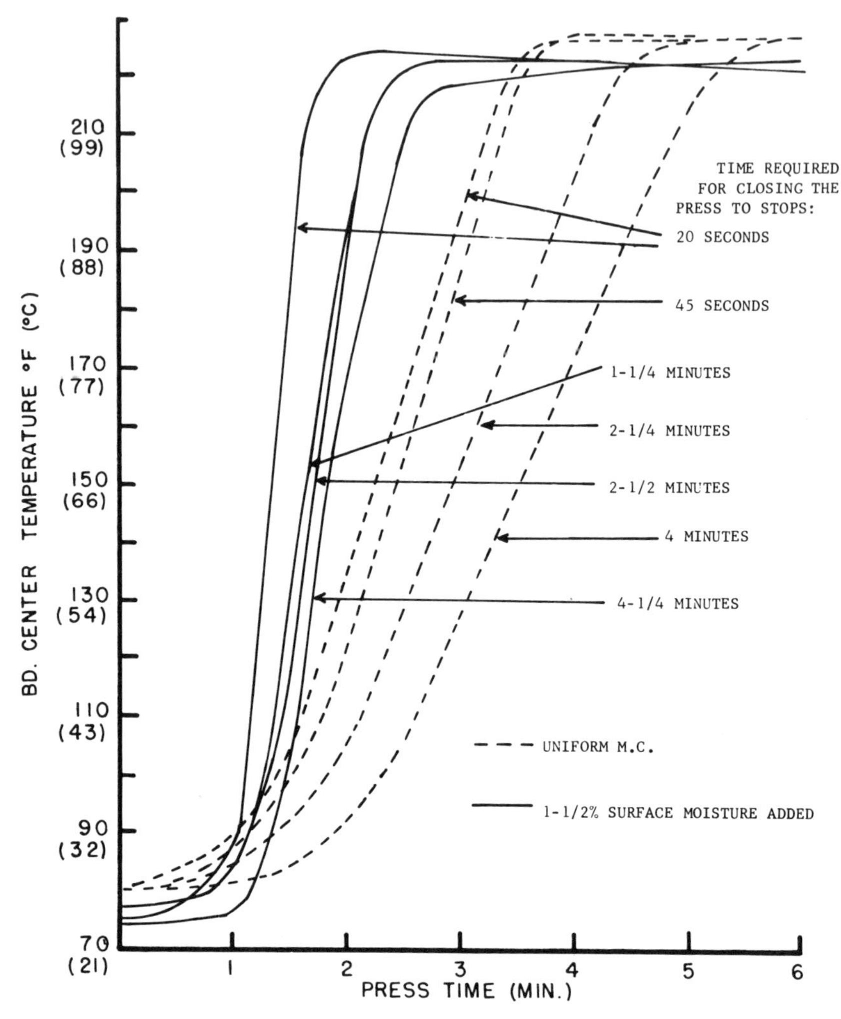

Figure 17.1. *Differences in time required for the development of similar core temperatures between boards pressed with a uniform moisture content and boards pressed with 1% additional moisture added to the surface (Crawford 1967).*

decrease in the wood plasticization; slower heat transfer to the mat core; slower close time to stops at a given pressure, and possible problems with tack. However, in recent years, newer developments with resins have made possible the development of tack at relatively low furnish moisture content.

It is difficult to define optimum conditions when considering mat moisture content and distributions thereof. It appears 10 to 12% is the optimum moisture content if this alone is considered in pressing board. However, single-opening press lines are operating at levels of 6.5 to 7.0%. Problems of insufficient moisture to accomplish good resin bonding when using "standard" resins must be overcome by resin tailoring. Higher mat moisture contents may require longer press times, although resin advancements appear to have overcome some of the problems. Variations in total moisture content between boards in a plant will result in differences in final board thicknesses resulting in rejects because of "thick" or "thin" panels.

Vapor Pressure Within the Board

The internal bond of the board in the hot condition as it leaves the press must be sufficiently high to withstand any vapor pressure that exists within the board when the press is opened. Because of the relatively porous structure of composition board, vapor pressures are usually relieved quickly, with the aforementioned exceptions that can cause blows. Since vapor pressure is associated with temperature, it is important to recognize that the relatively low temperatures in the cores of boards being compressed in normal cases does not generate a very high vapor pressure. At 212°F (100°C), a vapor pressure equivalent to atmospheric pressure is found. As the core temperature rises the vapor pressure increases rapidly. If the core temperature of the board exceeded 300°F (149°C), it would probably be necessary to keep the board in the press long enough to evacuate all of the moisture through the edges to prevent blowing. For example, at 300°F (149°C) the vapor pressure would be 67 psi (462 kPa). At 350°F (177°C) the vapor pressure would be 136 psi (938 kPa). Fortunately, the usual case is one where the core temperature is on the low side and thus the vapor pressure does not become extremely high.

The increase in vapor pressure can cause an interesting problem. In one particleboard study where boards of 0.80 sp gr (50 lbs/ft^3) were being made at 375°F (191°C), with catalyzed urea resin, it was shown that boards could be pressed in 1 minute. However, if the board was left in the press for another 30 seconds, the board blew because the additional heat added to the board increased the vapor pressure above that of the internal bond of the hot board. Thus, additional press time was needed to relieve the vapor pressure before the board could be removed from the press without blowing. This same experience has occurred in the field with boards as low in density as 0.70 sp gr (44 lbs/ft^3). Boards can also blow shortly after leaving the press because of excessive vapor pressure.

There is some confusion existing on the difference between delaminations and blows. A delamination can be defined as a definite separation between the particles in the core of the board. This can be the result of low platen temperatures, insufficient cure time, insufficient pressure, low resin levels, or a combination of these factors. Some of these same conditions can cause springback. In such cases, the thickness of the board leaving the press is much greater than that to which it was compressed in the pressing operation. It is still bonded together but the bonds are not of good quality.

The blows in the board are caused when vapor pressure of the volatiles (moisture and vapor from extractives) exceeds the internal bond strength of the board. The blow normally occurs near the center of the board because this is where the board is the weakest. Many parts of the board can be well bonded with the blows distributed throughout. Some people would call the blow a blister, since there will be definite bulges in the board faces where the blows occur.

Previously the sealing of the board surfaces was mentioned as a contributor to blow problems. It should also be noted that the wood density and board specific gravity or density are also causes of the problem. As the board density increases, the sealing effect becomes more pronounced. Thus, it may be impossible to produce higher-density boards from some species without special precautions. Problems have been noted with ponderosa pine and with some of the cedars.

Press Closing Time, Pressure, and Density Profile

Press closing time is one of the most significant factors in determining board properties; and closing is closely associated with moisture content, as well as a number of other factors. *Press closing time* is defined as the time it takes to compress the mat to final board thickness once the press platens make contact with the mat surfaces. A number of density profiles can be developed in a board by manipulating these factors. Figure 17.2 shows the typical density profiles that have been observed. Of course, once an imbalance of heat, pressure, or moisture content occurs, the profile is no longer symmetrical but skewed. Such a board is to be avoided since its properties will be affected and it will probably be warped as well. The drawings show the density of the board from one face to the other. Another term used when discussing density profiles is *layer density* or, in other words, the density of the various layers in a board.

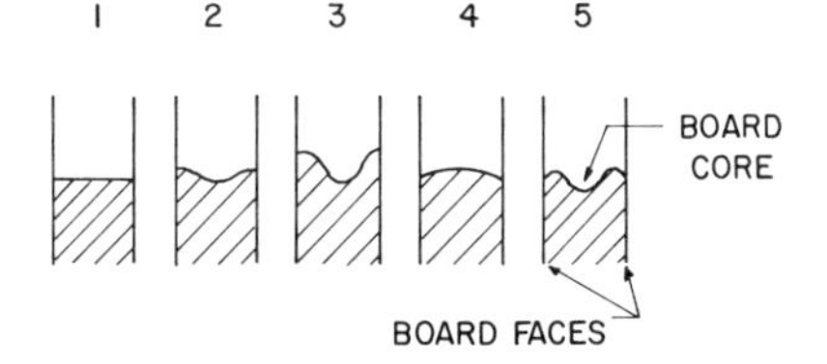

Figure 17.2. *Typical density profiles. No. 5 is typical of many particleboards (Polovtseff 1961).*

In Strickler's classic research on press cycles and mat moisture content, ¾-in.-thick (19 mm) boards were made with Douglas fir flakes using various mat moisture contents and variations thereof and six press closing times of 0, 1, 2, 4, 6, and 12 minutes. It should be noted that his research was conducted at a time when it took 10 to 12 minutes to cure urea resin in boards of this thickness. Figure 17.3 shows the press cycles that he studied. Cycle here does not include the time for loading and unloading the press, which is normally included in the board industry's measurements.

It is immediately obvious that much higher pressures are needed with fast press closing times. This occurs because these mats are essentially cold during consolidation.Once the heat from the platens starts to penetrate the mass of wood, two things happen which are critical factors in determining final density profile (along with pressure). First, there is a transmission of heat from the mat faces to the core.

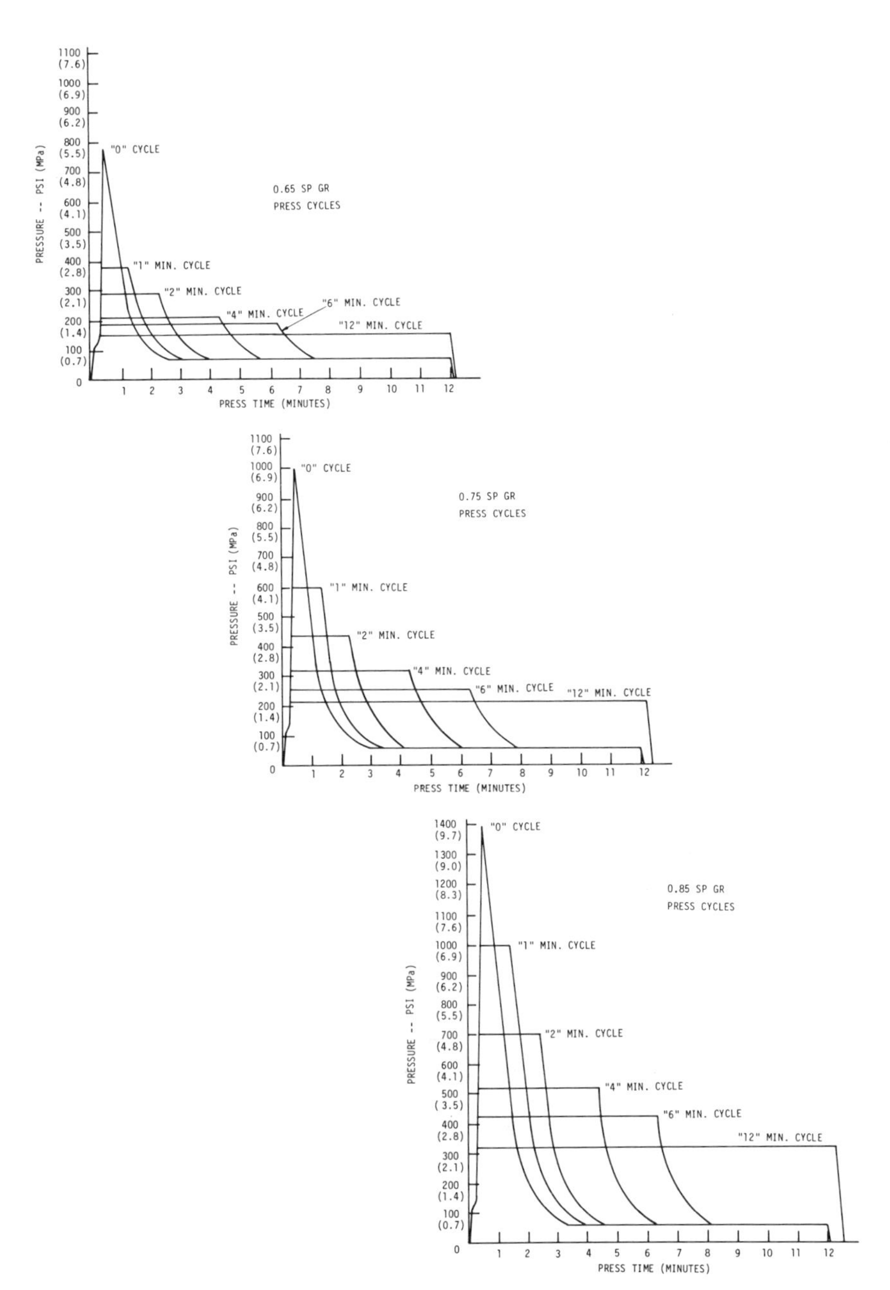

Figure 17.3. *Shown here are press cycles that produce ¾-in.-thick (19 mm) Douglas fir flakeboards of three density levels–0.65, 0.75, and 0.85 sp gr–at 12% moisture content. Press cycle designations (''1'' minute cycle, etc.) indicate minutes at maximum pressure (Strickler 1959).*

Second, the moisture turns to steam and the steam plasticizes the wood, making it easier to compress. This plasticizing takes place first in the faces and with the faster closing times, where the mat is almost at final thickness within a minute or less, the plasticized faces "collapse" to a higher density under the pressure. The core is still cold and is resisting the pressure. When sufficient heat reaches the core for plasticization, the faces are already set at higher density and the resin has cured to hold these layers at the higher density. As a result, there is insufficient wood left in the core to permit compression into a higher density at a given thickness.

At slower press closing times (i.e., using less closing pressure) there is less pressure and thus less compression in the face layers. Also the resin in the face layers may be cured before mat consolidation takes place. Thus the face layers are relatively rigid and at a lower density during the latter part of the closing time, and the plasticization and compression of the furnish under pressure while the resin is uncured takes place at intermediate layers or in the core.

The density profile, however, can be altered by prepressing. Prepressing the mat to final thickness would tend to eliminate density variations throughout the board thickness. This would require high pressures, and some research has shown possible deleterious effects on board properties with such high-pressure cold prepressing. However, by consolidating the mat in prepresses to about 0.30 in sp gr (given a 0.70 sp gr board), it is possible to close the hot press faster and provide a board with a more uniform density profile. A single-opening hot press is able to consolidate a mat into a ⅝-in.-thick (15.9 mm) board in about 13 seconds. In this example the pressing pressure is built up to 500 psi (3.45 MPa) within a period of 8 seconds.

It should be noted that once the mat has been compressed to final thickness, pressure must be reduced or otherwise the mat will be consolidated further. In Strickler's study, boards were pressed to caliper or thickness without stops or gauge bars in the press. He therefore shows the pressure required to hold the board at the intended thickness after consolidation was accomplished (Figure 17.3). It is readily apparent that only about 50-psi (344.7 kPa) pressure was needed to hold caliper during the latter part of all but the 12-minute press-closing time (in this case it took the entire time to close the press). Less pressure is required at the end of the press time because much of the resin is cured and holding the particles in their "set" position. The plastic deformation and set of the wood particles themselves is also a factor in reducing the pressing pressure. This may be demonstrated by hot pressing a mat without resin.

Most presses use stops to control final board thickness. If press pressure is maintained at the level needed for closing, it can be seen that a tremendous pressure is built upon the stops which can cause damage to the stop system and perhaps to the press itself. Therefore, even with a stop system, pressure is reduced during the latter part of the pressing time.

Figure 17.4 shows the density profile found when using 0-, 2-, 6- and 12-minute press-closing times, three moisture-content levels and three board-density levels. The important observation is that fast press closing results in high-density board faces and low-density cores. Thus, board properties such as bending and stiffness are enhanced while internal bond is diminished because of the density profile. The reverse is true when a long press closing is used.

The density profile can also be adjusted by an uneven moisture content between faces and core. Higher face moisture contents developed by adjustments in blending or by spraying water on the mat faces can accentuate the density profile obtained by fast press closing. Reversing the face and core moisture gradients produces a more uniform density profile as shown in Figure 17.5.

Press temperature is also involved with the press closing and resulting density profile. Higher press temperature increases the core density while decreasing the face density because of faster heat transfer to the core. Similar pressing approaches to low-, medium-, and high-density boards result in different density profiles although the profiles in general are similar (Figure 17.4). This is a complicated subject, and other factors can come into play providing a different density profile than anticipated. Species has been one of these. The important point is that by manipulating all of the factors involved with density profile, it is possible to control board properties to the manufacturer's advantage.

Face quality can be improved by post-pressing board at temperatures around 700°F (371°C) for a few seconds as shown in recent experiments. This high temperature setup would call for an additional press, but it would not need to apply high pressures.

Precure and Postcure of Resin

Figure 17.5 also shows the characteristic found in many boards of a slightly lower face density on the outer part of the board faces, probably because of complete resin cure before final board thickness is reached in pressing. Much of this "softer" (low-density) surface is removed in the sanding step. These soft surfaces are related to the *precure* or dryout of the resin caused when the hot platens first come in contact with the mat. At the normal temperatures of 325° to 375°F (163–191°C), resin curing is almost instantaneous at the faces and happens before the mat is consolidated to final thickness. Although the precure can be alleviated by the use of higher face moisture levels, this may extend the press time.

It is interesting to recall the relatively slow closing of some of the early multi-opening board presses. Very soft surfaces occurred and sanding off as much as 0.125 in. (3.2 mm) of material from faces was not uncommon. Many early multi-opening presses were plywood presses modified for use with composition board. These closed from the bottom so that the bottom opening closed first and the top opening last. In a 16-opening press, this resulted in each board having a different density profile and also different properties. This problem was rectified by the development of simultaneous closing systems so that all boards are pressed identically.

Postcuring can be useful or deleterious. First, it is of advantage to hot stack phenolic-bonded boards to complete the cure. Second, hot stacking of urea-bonded board can break down the resin bond by hydrolysis, which is a function of temperature, moisture content, and time. Hydrolysis is where the resin molecule is cleaved but the effects on the board are being argued. It is claimed by some that the effects are exclusively in the wood because it is more sensitive to strength loss. Others feel the breakdown is in the wood-resin interface.

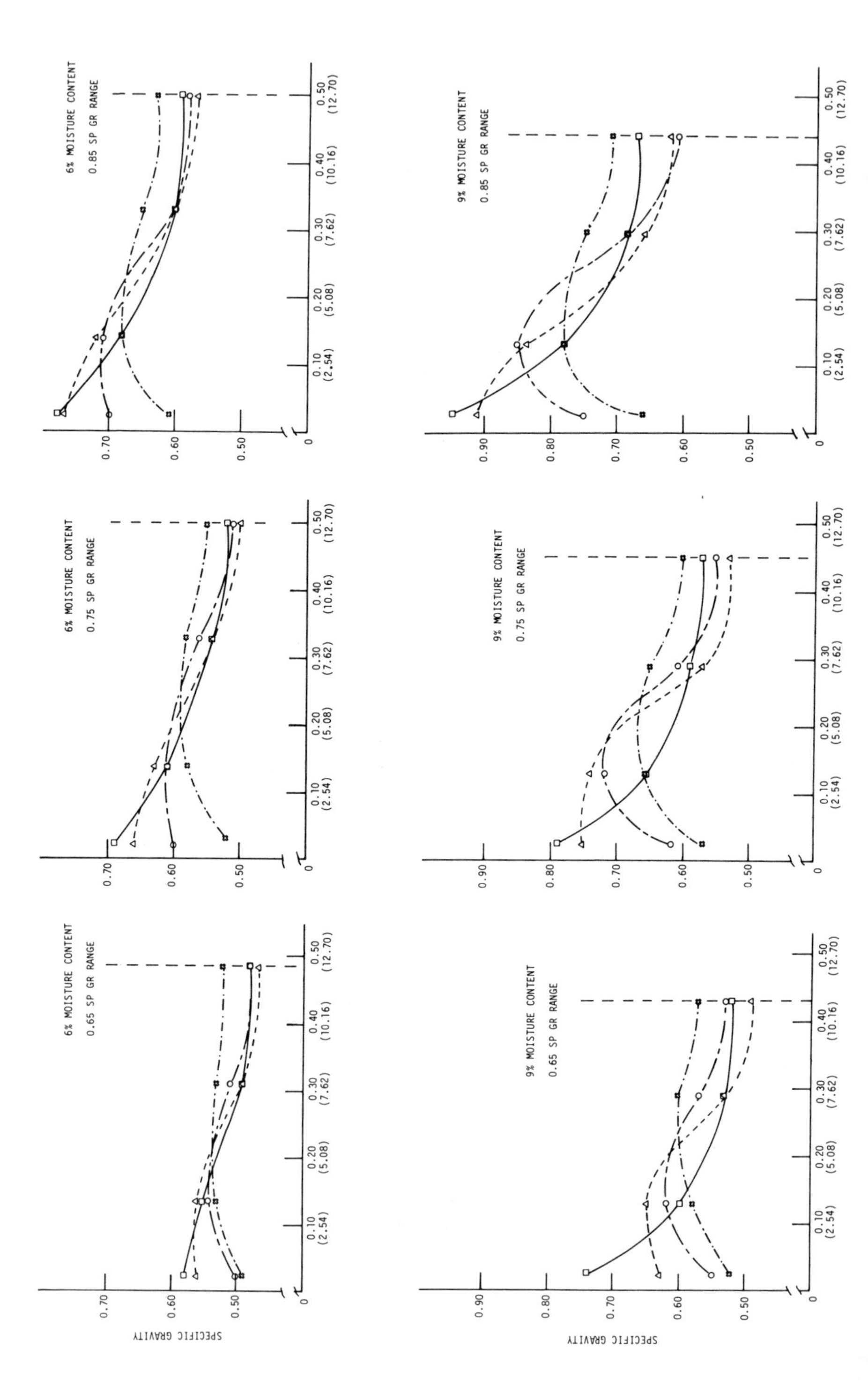

6% MOISTURE CONTENT
0.65 SP GR RANGE
6% MOISTURE CONTENT
0.75 SP GR RANGE
6% MOISTURE CONTENT
0.85 SP GR RANGE
9% MOISTURE CONTENT
0.65 SP GR RANGE
9% MOISTURE CONTENT
0.75 SP GR RANGE
9% MOISTURE CONTENT
0.85 SP GR RANGE
SPECIFIC GRAVITY
0.10 (2.54)
0.20 (5.08)
0.30 (7.62)
0.40 (10.16)
0.50 (12.70)
0.50
0.60
0.70
0.80
0.90
0

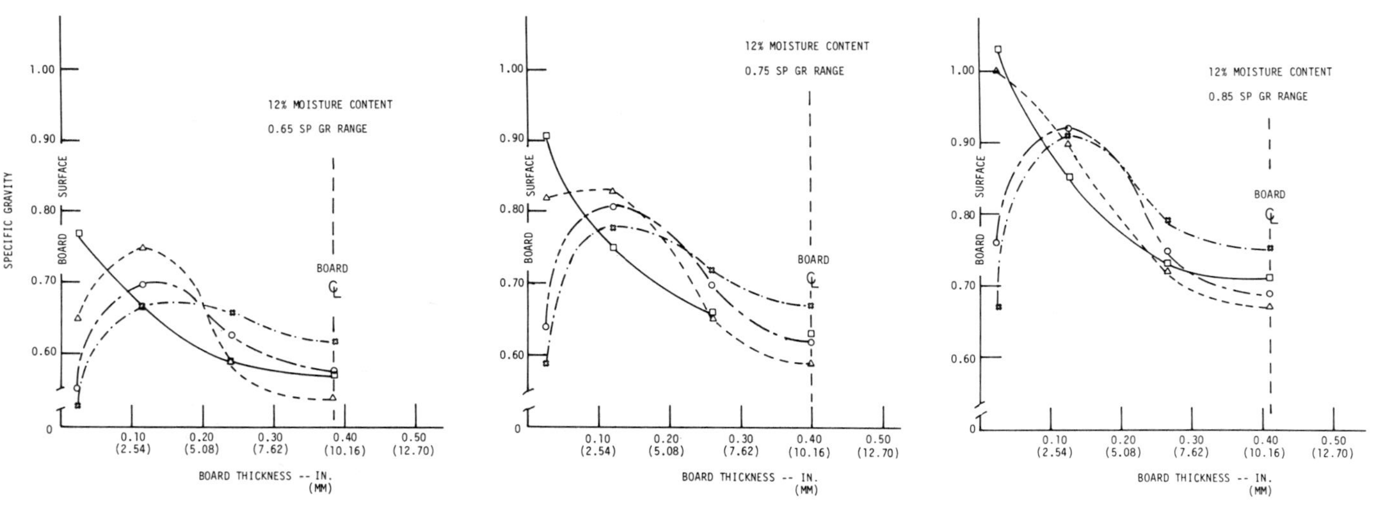

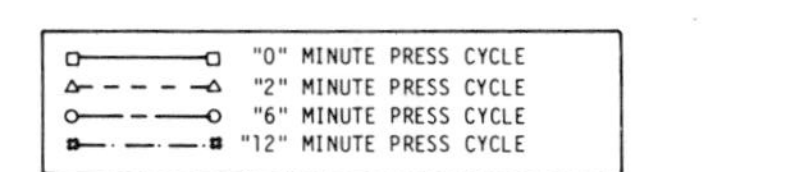

Figure 17.4. *Adjusted layer density curves from board surface to center line for four press cycles, three moisture content levels, and three density levels. Curve points indicate only the average density for each layer (Strickler 1959).*

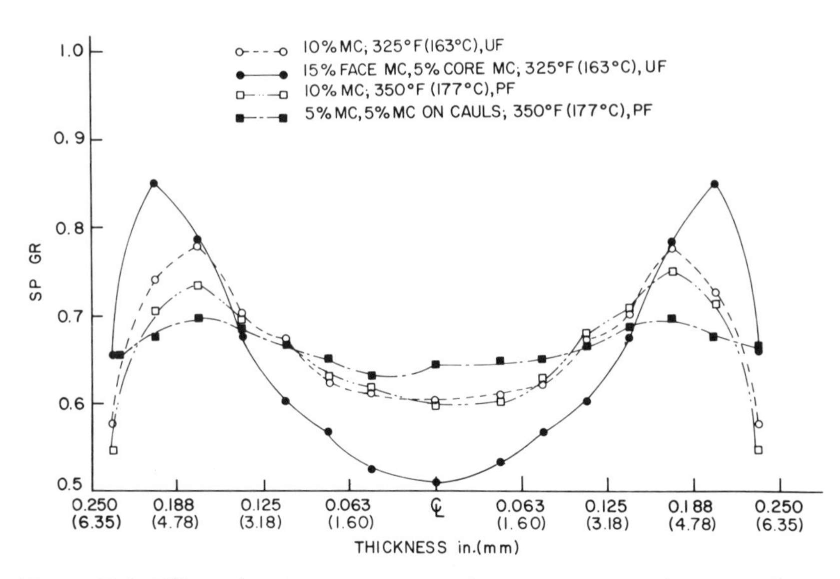

Figure 17.5. *Effect of various moisture and temperature conditions on density gradients in ½-in. (12.7 mm) particleboard. Each line is average measurement of two panels (Heebink, Lehmann, and Hefty 1972).*

PRESSES

Presses can be divided into platen (of which there are multi- and single-opening types) and continuous types. The type of press to use must be carefully evaluated for each individual case. Advantages and disadvantages may not be the same for each. Multi-opening presses are dominant in the industry at present; however, many single-opening types are in operation. Continuous presses are being installed worldwide with the development of a system for making thin board. Few such presses for thicker boards are now in operation, although they were developed in the 1950s.

Important factors to consider when evaluating the pros and cons of any press line are the quality of the board to be made; the input of the raw material (wood, resin, energy) in relation to the produced quantity of boards; the length of the scheduled down times for maintenance, changing of stops or thickness control devices, cleaning, and the length of the unscheduled down times that can be expected.

Early multi-opening presses were adapted from the plywood industry. These were not suitable for pressing dry-process fiberboards and particleboards; consequently, the multi-opening types were developed and now are the most widely used.

Further development in the industry has made possible the construction of presses for producing larger and larger board sizes in both single- and multi-opening presses. Both multi- and single-opening presses are as wide as 8 ft (2.5 m). Multi-opening presses are as long as 33 ft (10 m) with single-opening presses being offered up to 100 ft (30.5 m) long. The longest known in operation is 68.8 ft (20.96 m). Flexibility in cutting-to-size resulting in lower trim losses is a particular advantage with large boards. There are difficult production conditions in making and handling such large boards, and these must be considered in the plant design. Investment costs are higher per unit of produced board.

Multi-opening presses operate with the caul, metal cloth, or caulless systems. Most of these have many openings running as high as 40, although 14 to 18 are more usual. For high production capacities, multi-opening presses are used. Capacities of 572 to 715 tons (519–649 mt) of board per day are not unique. The greatest capacity found is about 1300 tons (1179 mt) per day.

The single-opening presses operate with either a steel belt or the special metal mesh cloth used to transport the mat into the press, which is simple and reliable. Single-opening lines have lower production capacities than the large multi-opening lines; however, single-opening lines have found great favor for smaller plants. Some plants run with more than one such line, with perhaps the extreme example being a plant reportedly with nine single-opening lines. Technological progress in single-opening press lines in recent years has made these systems more attractive as alternatives to multi-opening systems, even for large production capacities where two or more presses are used.

Single-opening press lines are now available in sizes up to approximately 592 ft^2 (55 m^2) of pressing area, corresponding to an output of 286 to 358 tons (400–500 m^3) of board per day, based on a reasonable thickness; but considerably larger presses are being discussed. There are a number of reasons why single-opening presses are favored for use in board plants. It is possible to insert the new mat into the press in 10 to 15 seconds, depending upon the press length. It takes only a short time for building up the full pressure upon the mat, particularly if the mat is prepressed. One firm claims approximately 10 to 15 seconds for such an accomplishment. High press platen temperatures ranging around 390°F (200°C) can be utilized with a minimum of surface precuring because of the fast charging and closing of the press, which also reduces the press time. It is also possible to preheat the steel belt carrying the mat into the press and to preheat the formed mat. Both of these procedures contribute to shortening the pressing time. For three-layer boards with a density of 41 lbs/ft^3 (0.66 sp gr) and with a thickness of ⅝ in. (15.9 mm) after sanding, a total press time of 150 seconds is reported. This is compared to a total pressing time in multi-opening presses of about 250 to 350 seconds. A further reduction of 50 seconds in press time with single-opening presses is anticipated with the use of high-frequency preheating of the mat.

It is claimed that there is a better thickness tolerance in pressing a single mat rather than many, as in the multi-opening presses, and thus less sanding is required, saving wood, resin, and other additives. If sanding can be minimized—and this applies to all types of presses—there is less risk of exposing coarse particles in layered-type boards. This is important in cases where the boards will be used for secondary processing such as printing or laminating. The single-opening-panel sizes are quite large, and thus there is less loss in edge trimming, compared with the smaller boards produced in multi-opening presses. Single-opening presses are less complicated, and the mat loading system is also relatively simple, compared with multi-opening presses.

Because of the relatively long length of single-opening presses in comparison to multi-opening press lines, however, the equipment—including the former, pre-press and press—requires a much larger building area than the multi-opening presses. To date, the total single-opening press line is generally more expensive in capital investment on a production unit base. In large plant operations where more than one single-opening line is involved, more operators are usually required. Single-opening presses, because of the large size of the board, usually require mat moisture contents of about 7%. This is several percentage points lower than normal for multi-opening press lines. Such dry mats can require a greater pressure for closing the press to the final board thickness. Since the closing takes place in a relatively few seconds, there is little, if any, plasticization of the mat during the press closing time. The requirement for extra pressure capability in these types of presses also can add to the capital cost.

Another approach has been taken with a system that does not require as much space as a single-opening press line but is designed for production capacities similar to the single-opening lines. It generally uses a 6-opening press of a size which has a total pressing area about equal to that of the larger single-opening presses. This reduces the space requirements considerably, and this multi-opening press is not as complicated as those with many more openings. Indeed, simultaneous closing may be eliminated; but problems with blowing mats out of the press during closing may occur, depending upon the furnish type or whether prepressing is performed.

Single- and multi-opening presses are unproductive during the loading-unloading, closing, and opening part of the press cycle. This dead time becomes extremely important when pressing thin board. In broad terms, it is claimed that economic feasibility of a continuous press is limited to a process where the cycle time is one minute or less. It may be possible to go as high as three minutes in press time depending on the product. The reason three minutes is the cutoff point is that no big multi-opening press with 20 or 30 openings can charge the loader, insert the mats into the press, and close the press in less than three minutes on a reliable basis. The dead time in a single-opening press may be 15 to 20 seconds. Thin boards can be cured in less than 30 seconds, thus the dead time can be about 70% of the pressing time. The elimination of this dead time has led to the development of continuous presses for thin board.

A new system called the Mende process for producing a continuous thin board is being used widely over the world. In this system a continuous steel belt runs through the forming area and partially around a 10-ft-diameter (3.05 m) heated pressing drum. This belt carries the formed mat through the continuous press. The machines are supplied in five widths, ranging from 4 to 8 ft (1300 mm–2500 mm).

Another design of continous press has been developed by Washington Iron Works. It consists of two heated steel belts with a pressure-holding section about 36 ft (11 m) long. A set of two large nip rolls, located at the beginning of the pressure section, first presses the mat to nearly its final thickness. This quick compression of the mat is different from the Bartrev press developed in the 1950s which consolidated mats continuously into board, using a stainless steel band running over a caterpillar backing track. The nip roll pressure is approximately 2000 to 3000 lbs/linear in. of width of the rolls (909–1364 kg/25.4 mm) Between 30 and 100 lbs of pressure/in.^{2} (207 to 689 kPa) is then applied in the holding section which is backed up by lubricated slides. The mat is pressed and cured in less than 30 seconds depending upon the thickness of the board and the resin catalyst used. This is a flat straight-through press, and the board is cured in the flat position, whereas the panel in the Mende system is mostly cured as it passes around the 10-ft-diameter (3.05 m) heated pressing drum. The development stage for this newer press was completed, and it was ready for production application when this book was prepared for publication.

The advantage of crosscutting the endless board to any desired length is not agreed upon by all authorities. The possibilities of making boards more economically in a 16.4- to 24.6-ft (5–7.5 m) long multi-opening press or a single-opening press 45.9 to 49.2 ft (14–15 m) long, where cutting losses can be less than 10% on the full length, must be evaluated to determine whether the continuous press is indeed better for any installation.

Another pressing system which is producing thicker board makes a continuous panel employing a single-opening press to press board on a step basis. In this process, a mat of unlimited length is pressed one step at a time. A special heating arrangement in the press platens prevents the resin in the mat at the entrance of the hot press from curing. A number of these systems are in operation.

Presses are now run with automatic controls, although manual operation is possible. In this age of automation, programmed controls are common, and it is easy to change pressing cycles simply by changing the punched control card.

Press Heating Systems

The most common way of heating a press is to introduce some type of heating medium into the press platens. The most commonly used medium in the United States is steam because boilers can operate off plant residuals. Other systems widely used in the world are hot oil and hot water. There is some use of high-frequency pressing, and its popularity has gained in the last several years. With the exception of the high-frequency, the heat is applied to the mat in the hot press through conduction. With the high-frequency or, as this is sometimes called, *radio-frequency* type of heating, the mat is placed in an alternating electric field which is set up between press platens. In this field molecular vibration occurs, which leads to heat generation for curing the resin. A new system using electric heat has been recently announced, and a unique system using direct heating by gas flames has been used for many years.

Steam: In steam-heated presses, the steam as generated by boilers is run through a passageway in the press platens. The steam temperature is directly related to the steam pressure. Thus, pressure control valves are used for adjusting the press temperatures. These controllers are operated by air controllers, making it imperative to have clean, dry compressed air which is not contaminated by water or compressor oil. Otherwise the controllers will not function properly and the press temperature will be out of adjustment. In this system there must also be proper steam trap operation, and the strainers must be checked frequently to insure that they are not plugged with pipe scale or boiler water additives. There can be a difference in temperature between the top and the bottom of a press platen, due to a film of condensate forming on the bottom of the passageway in the press platen. Condensate lines must be laid out properly so that they will drain properly.

Hot Water: Hot water systems have been popular because of the relative ease in controlling the heating of the water and the final water temperature; however, they are not normally used in the United States. A boiler is used to generate steam for heating the water in a hot water generator. The maximum press temperature that can be practically achieved with hot water is about 420°F (216°C). Hot water systems eliminate steam traps and their associated leakage problems. The heat transfer is uniform throughout the platen, varying only with flow velocity and temperature. More interest is being expressed in this heating medium nowadays when the fuel used is expensive fossil types because savings in fuel costs of more than 19% have been demonstrated in comparison with steam. Problems with hot water heating include the need for treating the water for pH and oxidizing problems and leakage. The main disadvantage is that the hot water has to be pumped through the system under very high pressure.

Hot Oil: When heating with hot oil, it is quite easy to go to relatively high temperatures such as 450 to 500°F (230–260°C) with the present heating systems. Higher temperatures are possible for special cases. Electrical means are used for heating and controlling the hot oil temperatures; however, recent developments have resulted in a way to use heat from wood waste incineration for heating the oil. The hot oil is pumped through the platens continuously. There must be sufficient flow of the hot oil to insure that enough heat is delivered to the platens, eliminating wide fluctuations in press temperatures. A big advantage with hot oil is that only the pressure necessary for pumping the oil is required. In the case of steam-heated platens it is necessary to treat them as pressure vessels. There is a fire danger with hot oil, however, in case of leaks; and it is important to keep the pressing area clean of wood particles, dust, and fiber which can easily catch on fire in case of an accident.

High Frequency: A number of advantages are possible with the use of high-frequency heating in pressing board, all of which have to be evaluated in the total concept of the plant before deciding whether or not to use it. Some of these advantages are: a short press time; press time is proportional to mass in the press, not the heat and vapor transfer properties of the mass as with conventional pressing; and low platen temperatures. Typically, during pressing ¾-in. (19 mm) boards, about half of the heat utilized is from the hot platens and about half from the high-frequency source. The heat from the high frequency appears immediately throughout the full thickness of the mats and results in faster cure cycles simply because the rate of energy input is greatly increased compared to hot plates. Other advantages when producing MDF include: the use of low-cost *in-situ* resins; the use of the same adhesive formulation, no matter what the thickness and density of the final board product; the elimination of board-cooling equipment after pressing, since the boards can be stacked immediately because of lower board temperatures; and a more uniform density profile in the board. Better overall physical properties are developed, particularly in the thicker fiberboards. The storage time between board pressing and finishing can be reduced, since the moisture content of the panels is more or less equilibrated as the board comes out of the press.

A number of improvements of the fiberboard properties and characteristics are claimed with the combination of contact and high-frequency heating. They include better machinability, better coatability, superior dimensional stability and higher internal bond, improved edge screwholding and edge hardness. Because the faces of the finished board will probably be lower in density than conventionally pressed boards, a slightly lower bending strength and stiffness can be expected. This is not true if the boards pressed with contact heating have soft surfaces.

A number of plants have been built in the medium-density fiberboard field using the high-frequency type of pressing. There also have been a number of plants built using only the conventional type of heating in the press platens. Both types of plants appear to be operating successfully in the marketplace. However, it does appear that the high-frequency type of resin curing is better for boards that are ¾ in. (19 mm) or over in thickness. It should be mentioned that a few small single-opening press lines producing "conventional" particleboard also use RF heating.

High-frequency heating has been finding more favor in the medium-density fiberboard lines because of the speed of curing the resin during pressing, particularly with thicker boards over ¾ in. (19 mm). High frequency assists in producing thicker boards with good hard surfaces. It has been difficult to do this with conventional platen heating systems because of precure or overcure of the surfaces. Figure 17.6 shows production versus generator power output at four levels.

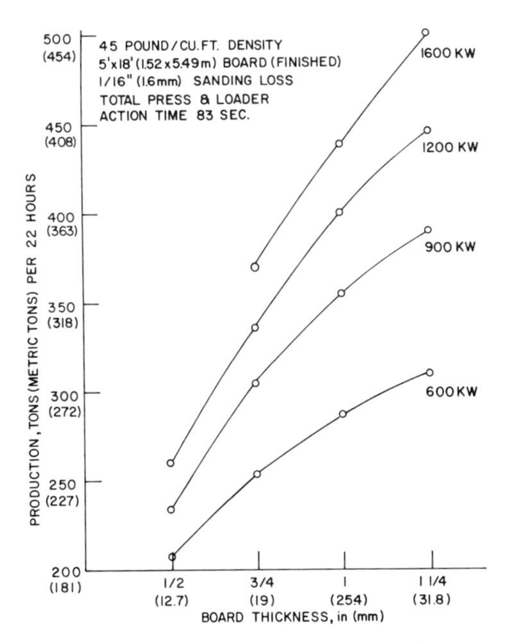

Figure 17.6. *Production vs generator power output levels in a six-opening press. (Courtesy Votator Division, Chemetron Corp.)*

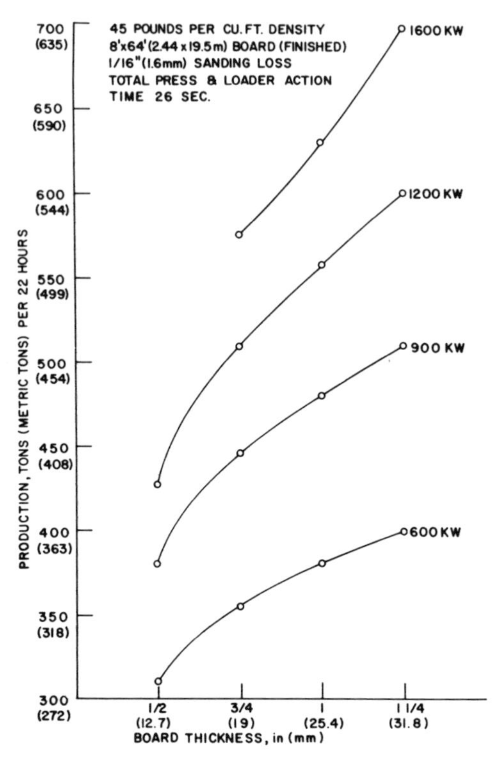

Figure 17.7. *Production vs board thickness at four high-frequency power levels in a single-opening press. (Courtesy Votator Division, Chemetron Corp.)*

Figure 17.7 indicates the increased production that can be expected when making boards using either 600, 900, 1200, or 1600 kW high-frequency power levels. This information is based on *in situ* UF resins. The heat cycle includes time necessary to cure the resin which is not a function of RF power input. Thus, if other types of UF or PF resins are used—types which cure slower or require higher temperatures—the heat cycle will be longer and the production less than that shown in the figures. A large single-opening press is in operation producing boards 8 ft wide by 65 ft long (2.44x19.8 m) using high frequency (1600 kW) as part of the curing system. Multi-opening presses, however, to date have been limited to eight openings at 5x24 ft (1.52x7.32 m).

High-frequency units are used in conjunction with platen heating systems because, if the platens are not heated, heat generated in the mat during pressing will rapidly move to the cold platens, thus insufficient heat will be in the mat surface for cure of the resin. A hot platen also prevents the moisture from moving to the cold surfaces where it would condense and cause wet, soft surfaces as well as possible rusting of the platen and mottling of the board surface. When pressing, the high-frequency power is not turned on until the mat has been compressed to approximately the final board thickness.

With high-frequency heating, the particles are uniformly heated. Unless the mat surfaces are in contact with a hot platen, they lose heat to the platen. The moisture in the wood aids heating and, when vaporized, is dispersed throughout the mat. This can wet the particle surfaces and further improve the distribution of the resin as well as assist with the plasticization of the wood furnish.

As is well known, wood is a poor conductor of electricity. However, the moisture in the mat, which ranges from about 8 to 13%, assists with the absorption of the high-frequency energy during pressing. The system is based on the absorption of energy by the mat when it is placed in an alternating electrical field. At frequencies of about 6^{10} hertz, molecular vibration occurs, which leads to the heat generation within the mat. This power comes from oscillator tubes which are self-excited. This power is transmitted via conventional hot press platens, which form electrodes, to the mat or mats in the press.

The heating of wood with high frequency takes from about 1.2 to 2.5 kW minutes of RF power per 2.2 lbs (1 kg) of wood depending on the temperature rise.

A single-opening press requires insulation of one of the platens. The multi-opening presses with intermediate pressing platens require insulation of the guides and hangers on these platens. If the number of openings is even, it is not necessary to insulate either the top or the bottom platen. Figure 17.8 illustrates these layouts. Figure 17.9 shows the connection of a 900-kW generator to a multi-opening press.

Since it is necessary to insulate presses that are using high frequency, it is difficult to use high frequency on continuous presses except as a source of heat for preheating the mat. Such preheating can speed up the pressing time in a conventional system and has been used for many years with the Bartrev system. Newer systems are also using high frequency for preheating the mat before it enters the hot press.

Complete closure, or shielding of the high-frequency system inside the press, is necessary because the platens can serve as a radio frequency transmitter. Thus, without shielding, radio and television transmission could be interrupted. Figure 17.10 shows a typical arrangement for shielding the press. It is nothing more than a thin sheet of sheet metal which serves to confine the electric energy. This means doors for loading and unloading the press must be used to seal off the press after loading the mats.

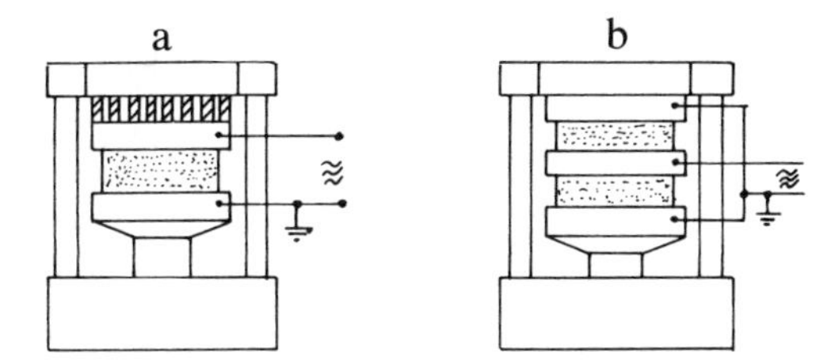

Figure 17.8. *Schematic diagram of the connections for a single- or two-daylight press from the HF generator. One side of the single-daylight press (a) must be insulated from the HF energy, whereas the two-daylight unit (b) only needs insulation for the guides of the middle plate (Hafner 1964. Courtesy Brown Boveri Corp.)*

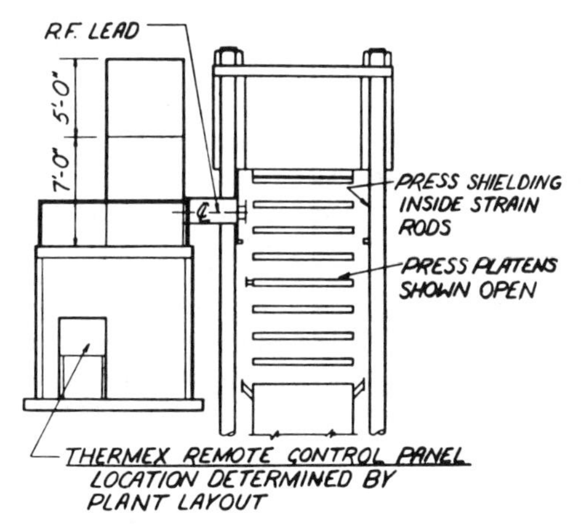

Figure 17.9. *End view of a 900-kW generator and press (Hopkins 1970).*

It is also necessary to insulate the pipe connections for the platen heating medium on all ungrounded platens with a high-frequency generator. This can be done with insulated hoses or metal pipe which is wound into a spiral. This spiral acts as a choke coil which prevents the escape of high-frequency signals. Platens connected to ground do not need such insulation. If stops are required between the press platens to control board thickness, they must be made of material other than metal. Suitable plastic materials can be recommended by manufacturers of the high-frequency units. Also, all metal parts involved in the actual electrical field should have the sharp edges rounded off to prevent corona and random arcing at these points. A serious problem occurs if any metal gets into the mats since an arc can be created that pits the faces of the platens. This can eventually require the removal and resurfacing of the platens. At last one supplier has a device, called an

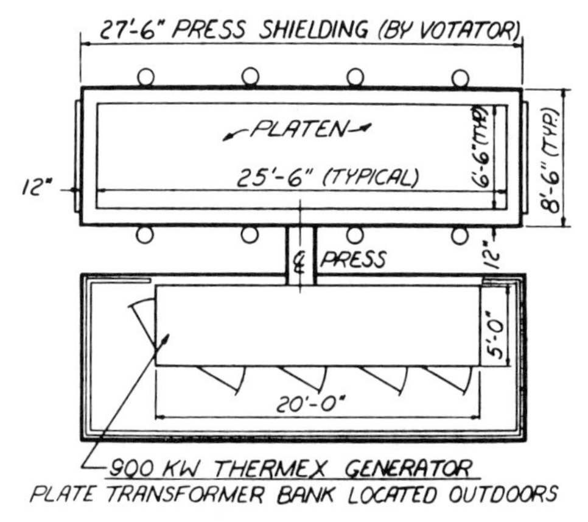

Figure 17.10. *Plan view of a 900-kW generator and press (Hopkins 1970).*

arcex, which senses the arc and stops the generator before such damage can occur.

According to the rules and regulations of the Federal Communications Commission, each high-frequency installation has to be measured for spurious radiation. The field strength limit has been set at 10 microvolts per meter measured at one mile from the equipment, which has not been difficult to achieve with proper press shield design.

With high-frequency generators, more than the rated power can be delivered because the generator only delivers power during the curing time of the press cycle. During press loading and unloading the generator does not deliver power and is thus able to cool down. However, there are limitations on how long the generator can deliver excess power, and the details must be worked out with the manufacturer to prevent damage to the equipment.

The oscillator tubes are cooled by a closed-loop system, using distilled water. The heat exchanger, which is cooled by plant water, is included (as is the storage tank) to maintain the cooling water at the proper temperature. Figure 17.11 is a photograph of such an installation.

There are many safety provisions for such equipment both for the personnel and for component protection. Safety grounding devices are used where lethal voltages exist. Interlock switches are used both for personnel protection and to provide proper sequencing of operations. For example, the use of interlocks in the generator circuit insures press closing and high-frequency application in the proper sequence. Pilot lights are also usually used to indicate whether control, filament, and plate circuit functions are in operation. They also indicate operation of crowbar and other protective devices. A bank of small indicator lights is used by one manufacturer to locate trouble spots such as open doors (there are 15 in this system), unactuated flow switches, actuated temperature switches, and so forth. For component protection, rapid-acting vacuum switches de-energize the plate transformer primaries within 10 ms after operation of the electric crowbar.

The so-called crowbar has as its principal purpose the prevention of an oscillator tube from being destroyed when internal arcs occur during operation. In addition to this, the crowbar also provides protection for other components in the high-frequency section of the equipment, especially the vacuum condensers.

The crowbar is a brute-force device. When the oscillator is operating correctly, an ignitron, which is connected across the DC supply terminals, acts like an open circuit (Figure 17.12). When an arc occurs in the high-frequency section, the ignitron is fired by the crowbar control circuits and shorts the DC supply within microseconds of the occurrence. This diverts the electrical energy from the fault location to the ignitron and prevents damage to the component in which the arc occurred. The crowbar firing circuit is a single-pulse device. Its fault-sending circuit utilizes several relays to make sure the crowbar is ready to operate before the high frequency is turned on. Failure of any crowbar relay to close its contacts will prevent operation of a high-voltage power supply. In addition to the fail-safe relay, a test button for checking the crowbar firing circuit is provided.

Each manufacturer is cognizant of the potential problems with safety and has taken precautionary steps, such as described here for one manufacturer, to eliminate the hazards of using high-frequency generators.

Figure 17.11. *Thermex generator (250 kW) with a solid-state rectifier. It has a distilled water to plant water heat exchanger (Hopkins 1970).*

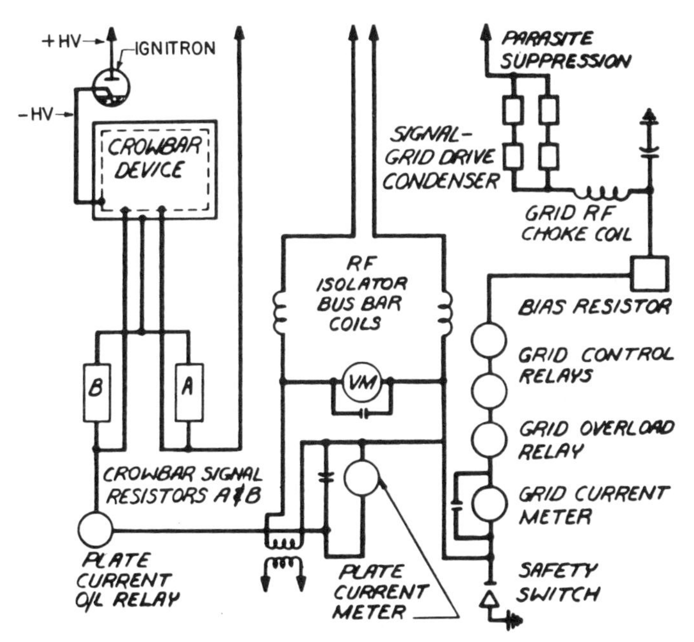

Figure 17.12. *Diagram of the crowbar installation (Hopkins 1970).*

At least one manufacturer provides door interlocks and high-voltage grounding switches at all access panels and doors. These open the high-voltage circuit when the door is open and also ground the high-voltage circuit to further insure its safety. Such devices are normally sufficient for personal protection but can be disabled if sufficient efforts are made. Such practice should be highly discouraged as the voltages within a generator can be lethal.

Electric: Electricity is commonly used for heating small laboratory presses. Such energy is uncommon in production processes; however, a recently unveiled system uses electricity to heat special caul plates. The first plants are being proposed for producing thin board. Mats on special, thin caul plates are stacked one upon another and then inserted into a single-opening press. Electrodes are attached to the ends of the special cauls, and as soon as the press is closed, the electrical system is energized, heating the cauls and the mats. It is claimed that no low-density faces develop because heat is not applied until the mat has been subjected to maximum pressures. Thus, the soft surfaces found because of immediate contact between the mat surface and hot-press platens in conventional heating are eliminated. Near perfectly formed mats are needed, otherwise the boards do not have uniform thicknesses. At first it might be anticipated that higher pressing pressure would be needed since plasticizing of the wood by the steam formed would not occur until after the press was closed. This is not too different from some fast-closing standard presses, which close in about 20 seconds. However, consolidation of the mat to final caliper does not occur until heat is applied. Figure 17.13 illustrates this system.

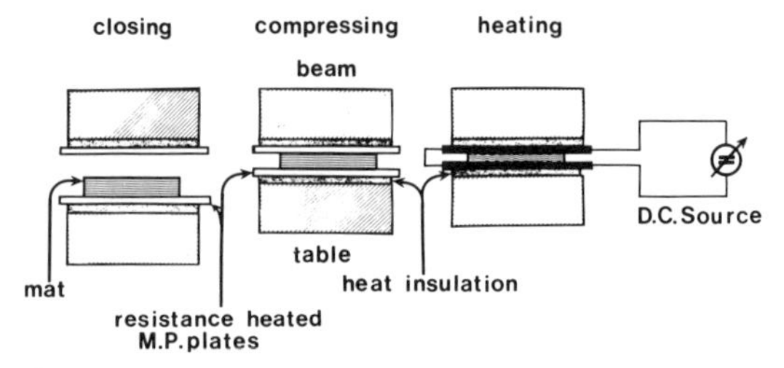

Figure 17.13. *Electrical heating system for board (Axer 1975).*

Gas Flame: As will be described later under "stack presses," heat is applied to platens in a press by direct gas flame. Continuous presses also use this heating system to heat the heavy links in the "caterpillar" tread which transmits the mat through the press. The heat stored in the links is used to heat the mat. This can be a hazardous operation unless care is taken to keep combustible material such as fine particles away from the open gas flame.

Thickness Measuring

An important measuring gauge of value to a press operator is a thickness gauge measuring board caliper out of the press. Excessive thickness is an indicator of low resin cure, shortage of resin, or moisture content that is out of control, etc. A continuous data reporting system allows the operator to adjust the blender, former, or press to possibly overcome the problem. Automatic marking of the thick panels allows for special handling of them later.

One version of a continuous thickness gauge is shown in Figure 17.14. The electronics are solid-state and use linear variable displacement transducers (LVDTs) or other electronic devices to determine thickness. The sensing wheels need to be held together with sufficient force to overcome any warp. All components must be dirtproof and waterproof.

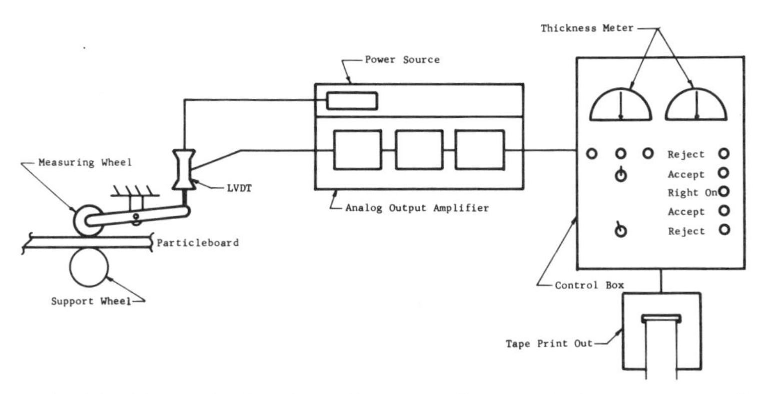

Figure 17.14. *Schematic drawing of an out-of-press thickness gauge (Carter 1969).*

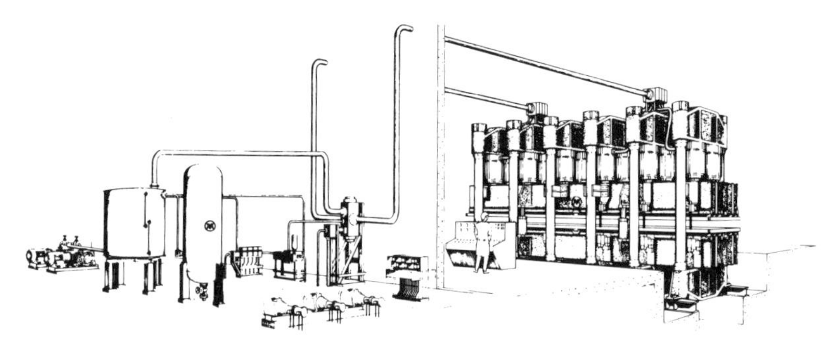

Figure 17.15. *A single-opening press. (Courtesy AB Motalla Verkstad.)*

Platen Presses

By far and away the most popular type of hot press used is of the platen type. A further breakdown is into multi-opening and single-opening (Figures 17.15 and 17.16 illustrate these two types of presses). Another type of platen press is what can be termed the *stack*, which will be dealt with separately at the end of this section. These types of presses are discontinuous in that the mats are introduced into the press and are held there during the pressing operation. The speedup systems in the mat formation line are needed in order to gain time for the actual

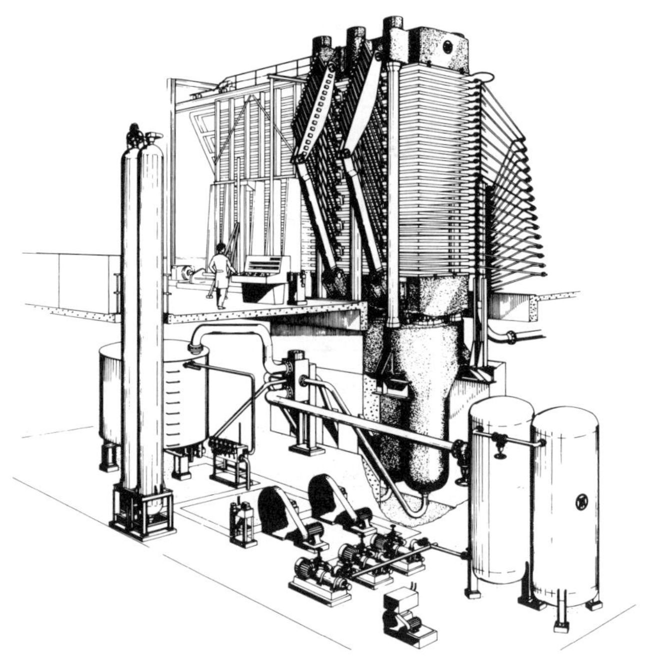

Figure 17.16. *A multi-opening press. (Courtesy AB Motalla Verkstad.)*

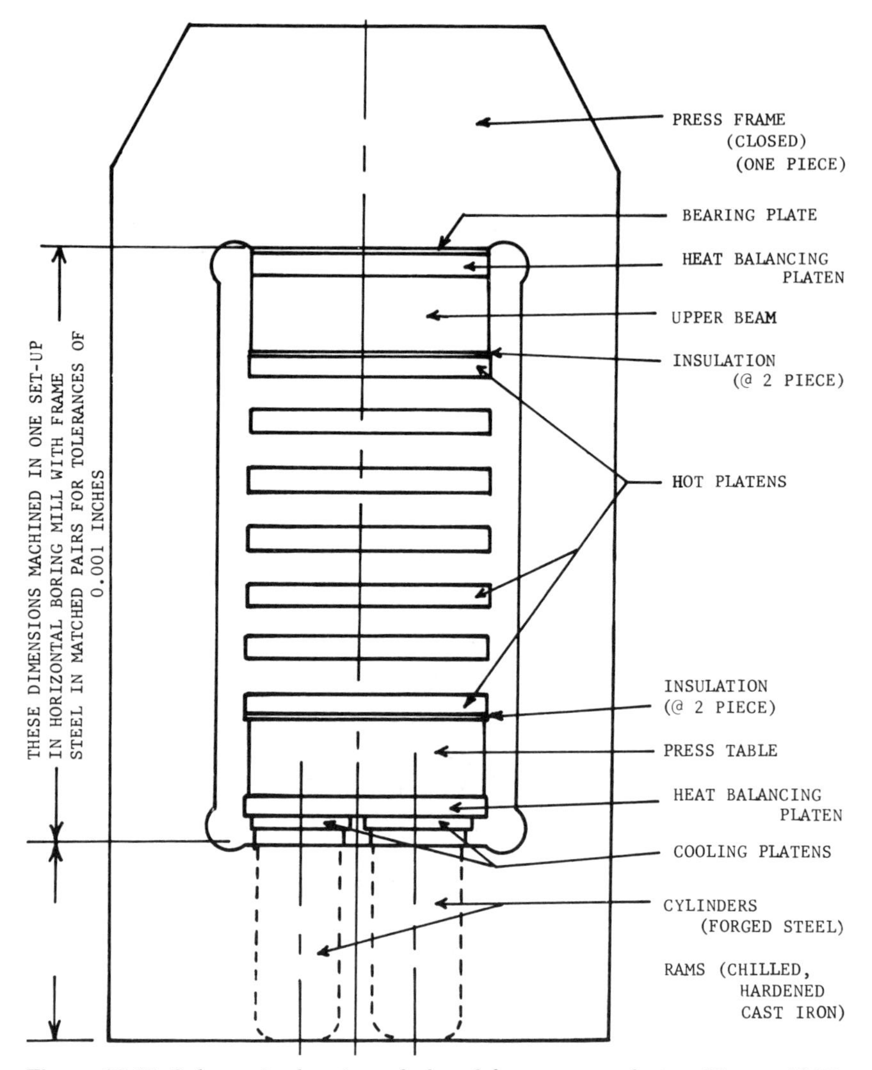

Figure 17.17. *Schematic drawing of closed-frame press design (Ogran 1967).*

pressing of the boards. For the most part, the multi-opening presses are upward-acting presses while the single-opening presses are downward-acting presses. In other words, the hydraulic system operates from either the bottom or the top of the press. Presses must be designed to handle the maximum bending moments, tensile forces, and shear forces which can occur in pressing.

Sophisticated press loaders have been developed for placing the mats into the multi-opening presses. A typical system was shown in Figure 17.15. With a downward-acting, single-opening press, it is not necessary to have any type of press loader outside of the steel belt or wire mesh belt that brings the mat into place between the hot platens.

The optimum action of any platen press, of course, is one in which the horizontal members, the upper beam, bottom press table or platen and the intermediate hot platens all remain flat, level, and parallel throughout the complete press cycle. Smooth press travel, accelerating, and decelerating means minimum wear and maintenance and longer press life.

Press Type: Two different types of platen presses are produced: frame and column. Figure 17.17 shows a typical frame press and lists its parts. Other drawings of frame presses are shown in Figures 17.18 and 17.19, which also illustrate ways to install the cylinders. The frame for the presses is made of thick sheets of steel, out of which an opening is cut for the platen or platens. The best frames should be made from single plates and not welded from four strips. It is a simple operation to manufacture individual plate units, intersperse them with spacer-rings and clamp them together to form a press frame. Each frame is composed of two vertical pieces of steel plate matched in pairs with web plates between to form a rigid and stable component. When cutting the frame from the sheets of metal, it is important that rounded corners be cut in the plates to reduce the stress concentrations that would occur in a right-angle corner. It is desirable to round or chamfer the outside corners of the frames. Sometimes these frames are set close together as shown in Figure 17.18 for assembling a press. Because of the number of frames a heavy upper beam is not needed. However, because the frames are relatively close together, there is a somewhat crowded condition for attaching parts to the press platens and for general maintenance.

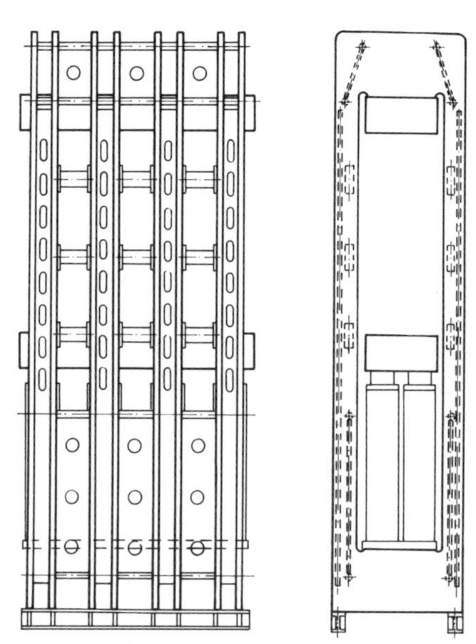

Figure 17.18. *Frame press with eight independent cylinders on base plate. (Courtesy AB Motalla Verkstad.)*

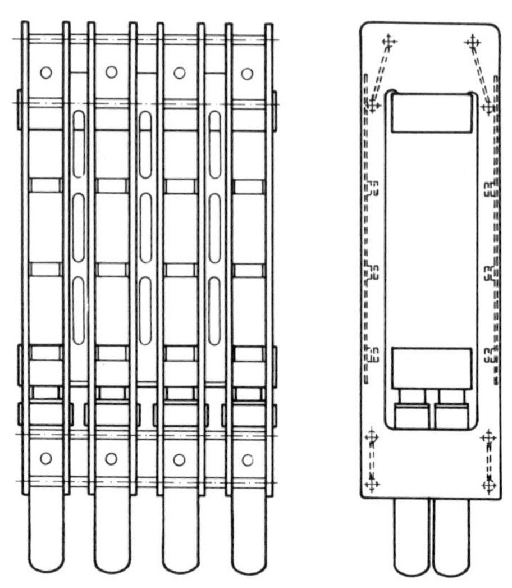

Figure 17.19. *Frame press with eight suspended cylinder units. (Courtesy AB Motalla Verkstad.)*

The number, size, and arrangement of the frame sections are dependent upon the platen length and width, the compression characteristics of the materials being pressed, the specific maximum pressure on the platen area, the hydraulic design requirements, the ram force distribution, the bending and stress analysis of the horizontal components, the press accessibility for operators and maintenance

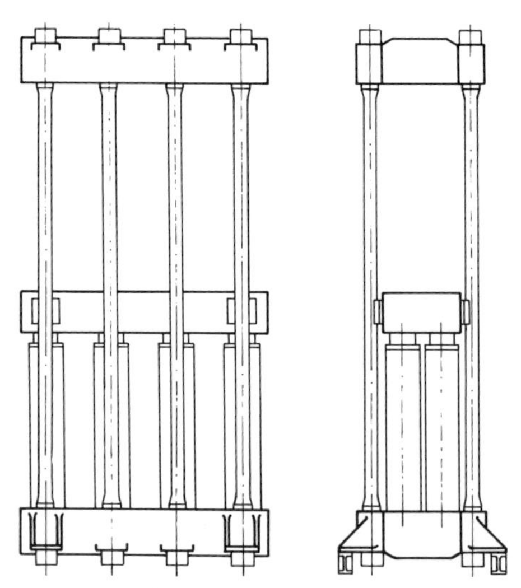

Figure 17.20. *Column press with eight independent cylinders on base plate. (Courtesy AB Motalla Verkstad.)*

personnel, the press auxiliary equipment which includes simultaneous closing units, and the material and fabrication utilization and economics. In the frame press it is normal to use a number of smaller rams, usually in pairs, rather than large rams. Two or more of the main rams are used as jackrams first and then as main rams. With these smaller rams, fast press closing can be accomplished without the need for jackrams or accumulators.

With column presses, as shown in Figures 17.20–17.22, the wider spacing between the columns and the heavier top platen or crown is shown. For example, one column press for producing 8x24-ft (2.44x7.32 m) boards has a crown that weighs 120 tons (109 mt). As can be seen, there is more space between the columns in this type of design. Columns can be cylindrically ground columns or welded frames. In column presses larger cylinders are used than in the frame presses and they are mounted as shown in Figures 17.20 and 17.21. The cylindrical columns are built with threaded ends. Large takeup nuts are screwed onto these threads, and it is important that the threaded area at the bottom of the screw area is not less than the diameter of the column proper; otherwise the column is weakened at this point and stress failures will occur. These columns are prestressed when put in place. After the installation is running, the columns are checked periodically and the nuts adjusted so that the proper stress is kept on the columns. Such presses are very rigid, whereas frame presses have more "give" in them when subjected to stresses that could occur from eccentric loads, such as unevenly formed mats.

With such large rams, fast closing of the press becomes a problem that must be solved. As shown in Figure 17.23, jackrams (fast-acting rams) can be used to close such a press at high speed while the main cylinders are opened to the tank for prefilling by gravity. The main rams are then used to apply the high pressure on the mats being consolidated into boards.

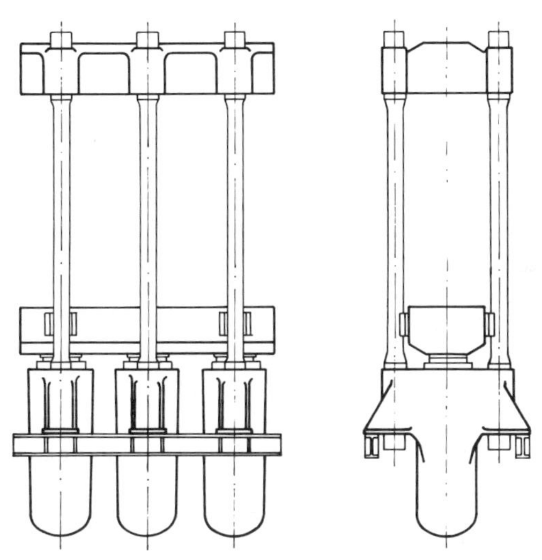

Figure 17.21. *Column press with three cylinders directly connected to the columns. (Courtesy AB Motalla Verkstad.)*

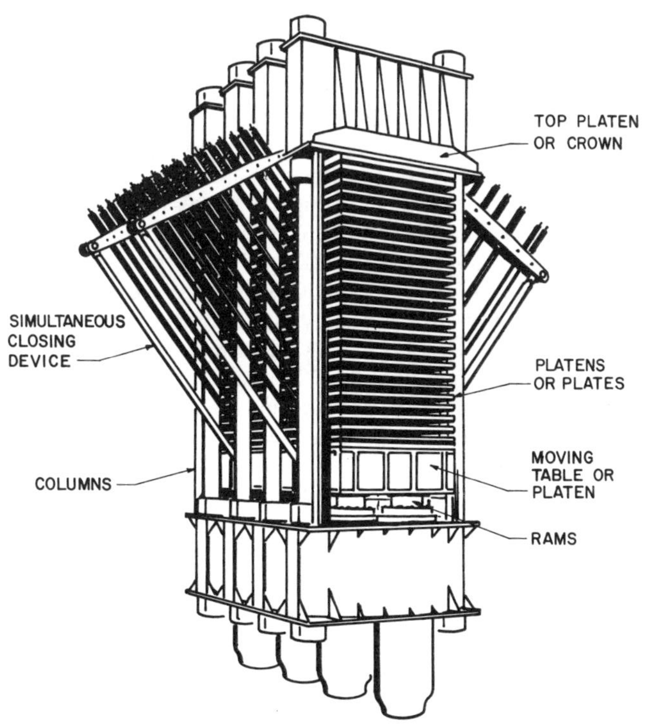

Figure 17.22. *A 5x8-ft (1.52x5.48 m) hot press with simultaneous closing (Morse 1967).*

Another method of achieving fast closing is found with low-pressure accumulators which are capable of rapidly filling the main oil or water lines and raising the press at the required closing speed. The type of accumulator used on board presses is shown in Figure 17.24. It is not an accumulator in the true sense as a bladder or piston does not separate the gas and hydraulic fluid. The bottle is pressurized with compressed air for water systems. Nitrogen is used for oil systems to avoid explosion. Air or gas will be absorbed in the fluid and so refilling of the compressed air or nitrogen is required. As a simplified illustration, the pump first pumps the hydraulic fluid into the accumulator until the desired pressure is attained. For fast press closing, valves 1 and 2 (Figure 17.24) are opened, which relieves the pressure in the accumulator. The pressure built up in the bottle then rapidly forces the hydraulic fluid out of the accumulator and into the cylinder, thereby closing the press quickly. The pump then takes over, applying high pressure for final consolidation of the mat into a board. During the time pressing is taking place, the pump also recharges the accumulator with hydraulic fluid in preparation for the next press closing. Several accumulators can be used with a press, and a somewhat complicated valving system is needed.

High-pressure accumulators reduce the pump capacity required, but the direct high-pressure pump is less complicated and more reliable.

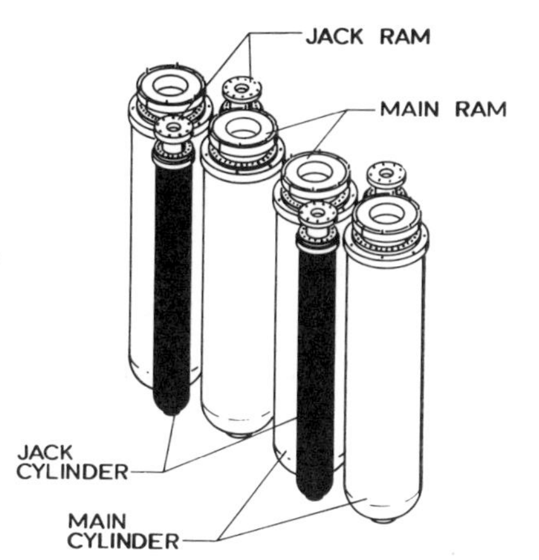

Figure 17.23. *Illustration of main rams and jackrams (Morse 1967).*

Hydraulic Fluids: Water can be, and is, used as the hydraulic fluid. It has long been used because large quantities can be made to flow quickly through small-diameter piping. The susceptibility of the system to corrosion has been overcome by using water softeners and additives such as soluble oil. It is low in cost and, if leaks occur, the water evaporates without leaving dirty residues. Additionally, there is not the fire hazard with water that there is with some oils.

Despite these advantages and the fact that nonflammable hydraulic oil is expensive, oil drives are now the most popular. Oil systems are capable of giving

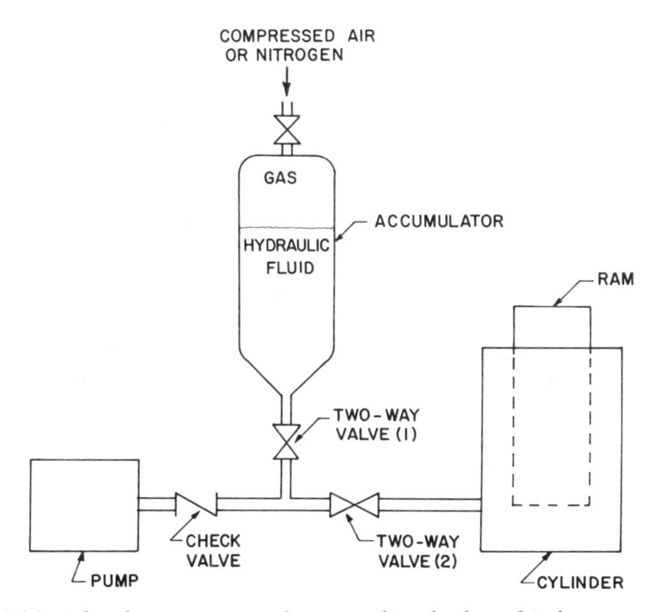

Figure 17.24. *A bottle-type accumulator used in the board industry.*

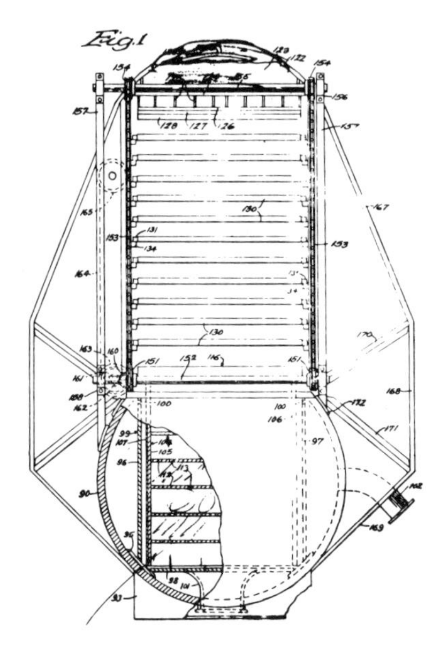

Figure 17.25. *Design of a water hydraulic press for waferboard (Gunn 1972).*

high outputs with relatively small tank and pump units compared to water hydraulic systems. Clean, acid-free machine oil used in these systems is the best preservative for metals, and its good lubricating properties minimize wear at seals and valves.

Normally, the diameter of the rams totals much less than the total surface area of the press platen. Advantage is taken of Pascal's principle that an increase in pressure at any point in a liquid results in a like increase at every other point in the liquid. Once the press has closed, it is possible for a large force exerted over a small distance (the rams) to be obtained by exerting a small force through a large distance (the pump). Thus relatively small rams in comparison to the total platen area can be used, and these are standard items. However, another approach has been taken on at least two presses. This design (Figure 17.25) has a large ram which is about the same size as the platen. At the top is a semicircular tank enclosed by the top member of the press which is pipe-connected to the lower tank holding the ram. The concept is that the pressure acting on the mats during the pressing cycle is equal and opposite at all times as the upper tank adjusts position to accommodate pressure differentials. Water is used for the hydraulic medium, and, since the ram is so large, only pressure slightly more than needed for consolidating the mat into a board on a one-to-one basis need be applied by the pump. Cost of the press is therefore claimed to be lower because of reduced costs for pumps, piping, and control valves. Furthermore, such a press is much lower in weight. Only two of this type are known to be in use in the industry.

Deflections: The press platens must be made to resist deflections under a number of cases. The deflection of the structural members of a press can never be totally eliminated, only minimized to acceptable limits. Figure 17.26 illustrates the type of deflections that can be found on hot presses. The main consideration of any press design regarding deflection is to build into the members sufficient strength to limit the amount of deflection to certain specifications. The most common figure used is to limit the combined transverse and longitudinal deflection within the area of the hot platen size to 0.010 in. (0.25 mm) when the maximum rated pressure is applied to the largest mat size. If equal hydraulic pressure is applied to mats of smaller size, the deflection will increase as the load is distributed over a smaller area that is further away from the points of support for the platens. When mats smaller than maximum are pressed, it is possible to build up a higher than designed pressure upon the mat. Thus it is important to check and insure that such higher pressures are not applied to the smaller mat, as the press can be severely damaged.

Another cause of deflection in the press table and upper beam is the unequaled heating of these units causing more expansion in them. Cooling plates and a layer of thermal insulation material are used to minimize this deflection as shown in Figure 17.17.

Platen damage can be divided into two main types. The first is permanent bending of the platens that results from pressing to full pressure when a mat is missing from one or more openings. This applies only to presses that use thickness stops or gauge bars located at the outer edges of the hot plates as shown in Figure 17.27. Also, the narrower the mat, the more damage to the platens because the load becomes concentrated more in the center. The second type of damage is localized deformation of the surfaces caused by tramp metal such as bolts, nuts, wrenches, and so forth.

The repair of platens damaged by pressing without mats is done by using the press hydraulic system to bend the platen past its yield point in the opposite direction in order to flatten it. Perfect flatness is hard to achieve without removing the plate and having the surfaces remachined. In repairing small localized damage, the usual way is to build up the spot with welding rod and grind the excess weld off to give a smooth surface.

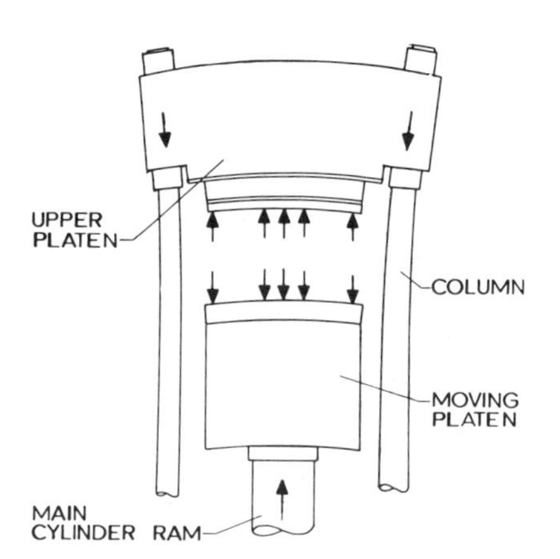

Figure 17.26. *Some possible press deflections to be found on hot presses (Morse 1967).*

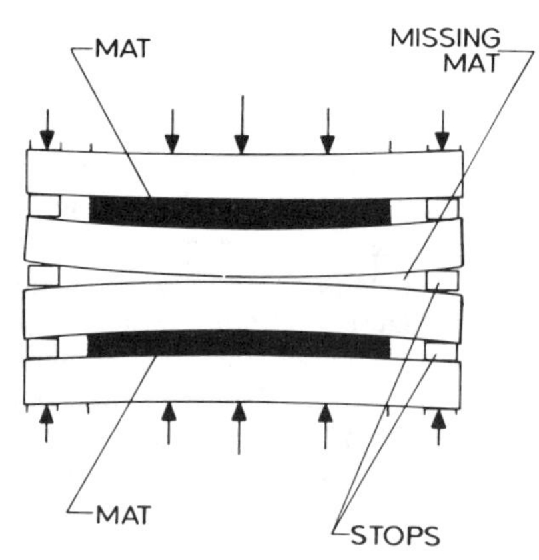

Figure 17.27. *Platen deflection damage possible when a mat is missing (Morse 1967).*

Proper maintenance of a press is very important, not only to prolong its useful life but to prevent major breakdowns. Periodic checks should be made by personnel who are well qualified and properly trained. These checks should locate loose bolts, improper lubrication, excess wear on parts, damaged platens, misadjustment of simultaneous closing units, hydraulic leaks, improper hydraulic cycles, and so forth.

Stops: Some method of holding the press platens at the right spacing during the pressing period is necessary to insure that the finished boards are of the correct thickness. In a small laboratory-type press, this can be done in a relatively simple manner by using dial gauge indicators. However, no matter what the size of the press, uneven mat formation will make it difficult to hold the press opening at the correct or desired dimension over the entire area of the press platen.

The simplest means for controlling thickness is the placing of stops or gauge bars between the hot platens. Thus, once the platens are on position against the stops, the correct thickness of the final board is assured unless very poorly formed mats are used. In that case, there will be extreme springback in the areas of high density, and the final board will have a marked variation in thickness. Indeed, in some older presses in operation, some of the boards do come out wedge-shaped, either because of wear on the press or because the mat-formation systems are not as sophisticated as the newer ones. In operations using such stops, it is necessary to take into consideration the thickness of the caul plates being used, if cauls are used, and the amount of springback that can be expected in the board due to species, moisture, and particle geometry effects. Such stops are relatively simple to use but are subject to peening caused by pressure from the platen or from having foreign matter such as particles crushed between the stops and the platen. In a large, multi-opening press, the usual system in changing such stops is to have a number of the plant staff work together to quickly change the stops by hand; however, systems are available for automatically changing stops.

Some of the multi-opening presses use linear variable displacement transducers (LVDTs) to assist in controlling board thickness. The LVDT is used to generate a voltage signal in a secondary winding in proportion to the displacement of its core from center. This signal is used as a command to an electrohydraulic servo control. This control system adjusts the variable-volume hydraulic pumps so that just enough pressure is maintained in the hot press to keep the press on the stops; or if position control is being used (as will be discussed later), the platens maintain the proper opening for producing the boards desired. In the case of multi-opening presses using stops, this control is set for the entire package of boards. Figure 17.28 illustrates this type of control. Thus, when all the platens are set upon their stops, the LVDT will be in contact with a preset gauge bar. If the LVDT moves away from its contact point, a command will be sent to the pumps to supply more oil to the hydraulic cylinders until the LVDT is back in its proper position. Less pressure is needed at the ending of a press cycle because the resin is curing out in the board and holding it in its consolidated position. The LVDT system, therefore, automatically compensates for the need for less pressure at the end of the press time.

With the stop system used in multi-opening presses, extra space is needed to accommodate the stops. As a consequence, many tons of extra weight can be

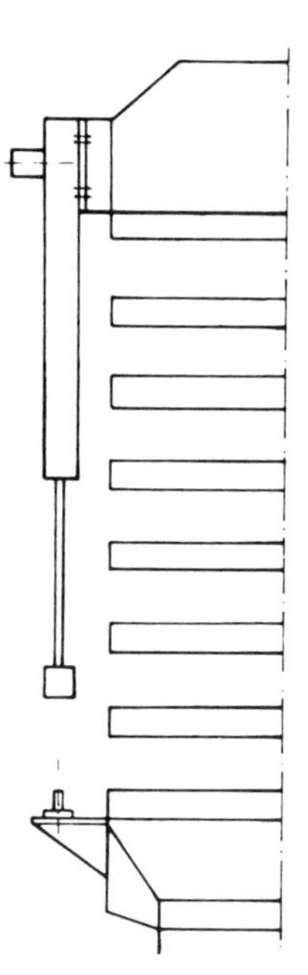

Figure 17.28. *Schematic drawing of an electronic position sensor (Strisch and van Hüllen 1969).*

added to the press for the stop systems. Stops can also prevent proper venting of the steam or vapor coming from the edges of the boards being pressed. If the hydraulic system goes out of control, the stops between platens can cause severe deflection and possible damage to the platens.

Thus, it is of advantage in some cases to be able to press without stops, particularly with multi-opening presses. One such system utilizes what is called *four-point control*. It is necessary that the press table at the bottom of the hot press be held in a perfectly horizontal position during the closing process to avoid horizontal movement of the heating platens. A table rod guide is used by one company to accomplish this, and these guides insure parallel position of the pressing table and the head or crown of the press. The rods are fixed at the table and guided by the head and are made rigid enough to keep the table horizontal should it try to move to an inclined position. The rod is also resilient enough to yield to abnormal pressure, as illustrated in Figure 17.29. The table is measured at all four corners by electronic position sensors, such as shown in Figure 17.28. The values determined electrically are simultaneously transmitted to the hydraulic control of the press. The four corners of the press are guided by the four-point control acting on four independent hydraulic cylinder systems. These four hydraulic systems have their own decompression and compression valves and prefilling. All systems work jointly during the closing process until they reach their preselected nominal positions. At that point, a changeover takes place, and the individual systems operate independently and hold their positions in accordance with the individual electronic signals. Such pressing is not possible without precision in forming the mats.

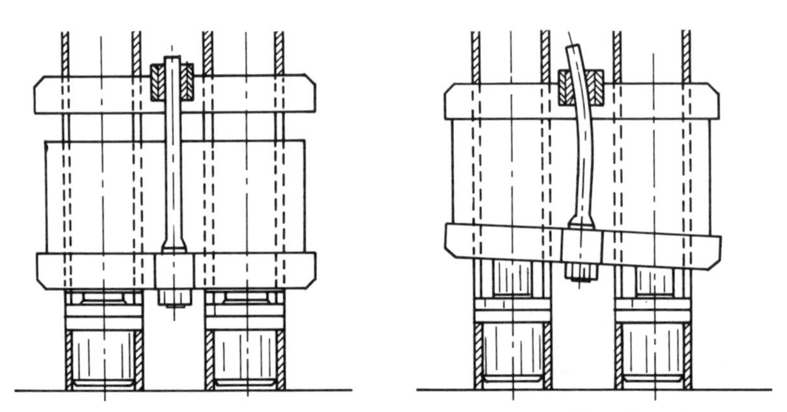

Figure 17.29. *A schematic drawing of the action of the table rod guides (Strisch and van Hüllen 1969).*

With single-opening presses, the use of stops has been popular, although position control is also in use. One system places the stops about 5 ft (1.52 m) on center on both sides of the press, but about 10 in. (254 mm) above the surfaces of the hot plates when the press is closed. Thus, they are not contaminated by blown-out particles from the mat and finished board. It is also possible to change the stops quickly, as they are readily accessible. Some feel that the thermal deflection of the top and bottom of the single-opening press is of such magnitude that thermal deflection takes place in spite of the best precautionary measures. While these variations are not of too great significance with multi-opening presses, they are magnified on single-opening press lines. Thus there is the possibility of major variations in the final board thickness.

One company uses spacer plates mounted on swiveling devices so they can be swung in between the table and head of the single-opening press outside of the actual area of the heating platen. These spacer plates, as with the previously mentioned system using gauge bars, do not lie on the pressing area. The swivel device makes it possible to change stops quickly.

However, a newer design has replaced the stop systems with a hydraulic thickness-control system. In this single-opening press, the press cylinders are controlled individually by direct-acting valves giving the desired thickness over the entire board area, irrespective of normal variations in the mat itself. The position valves control the hydraulic pressure in the cylinders so that the back pressure of the mat at the respective cylinder location is counterbalanced to give the correct thickness of the board. In areas of the mat where it has been evenly formed and has the desired moisture content, the thickness is achieved at the prescribed point in a press cycle; and pressure reduction commences. In areas of the mat which are unevenly formed or the moisture content is out of control, the hydraulic pressure required for proper consolidation is maintained for either a shorter or longer period until the desired board thickness has been achieved.

Steel stops cannot be used between platens when high-frequency heat is used. Special plastic stops or position control must be utilized.

Multi-Opening Presses

Sophisticated loaders for multi-opening presses have been discussed previously under prepressing. Some simply push the caul plate holding the mat into the hot press. Other systems used for caulless production lines deposit the prepressed mat directly on the hot platens by means of belt systems. For the Flexoplan system, a press loader can either carry the wire mesh screens holding the formed mats into the press, where the bars at the leading edge of the wire are engaged by a holding device after which the press loader is withdrawn from the press; or preferably a chain system engages the bars at the leading edge of the wire-mesh screen and pulls the entire assembly into the hot press.

It is imperative to use simultaneous closing with a multi-opening hot press with many openings. The early board presses were adapted from the plywood industry, where simultaneous closing is not needed. With this system, the bottom opening was closed before the top opening and the problems with variation in density profile occurred. Simultaneous closing devices have made it possible to close each opening or "daylight" at the same speed as any other. Thus the density profiles of each board are identical. In addition, simultaneous closing makes it possible to close the press far faster than previously. In one example, a 24-opening press with simultaneous closing will have the rams traveling at about 400 in. (10.2 m) per minute. However, this translates into about 17 in. (432 mm) per mintue per opening. This type of operation enables modern multi-opening presses to close to final position within about 30 seconds for boards ⅝ to ¾ in. (15.9–19 mm) thick. At first glance, it would seem that the simultaneous closing would not be necessary if the press would indeed close at 400 in. per minute. However, this rapid closing sets up tremendous wind forces around the mat because of the air being exhausted from the press openings. These forces would be so severe that the entire mat would be disrupted, making it impossible to press boards of any quality.

Figures 17.30 and 17.31 illustrate one approach for providing simultaneous closing. Figure 17.32 shows that not every opening has to be equipped with simultaneous closing but that they can be closed in pairs. Figure 17.33 illustrates

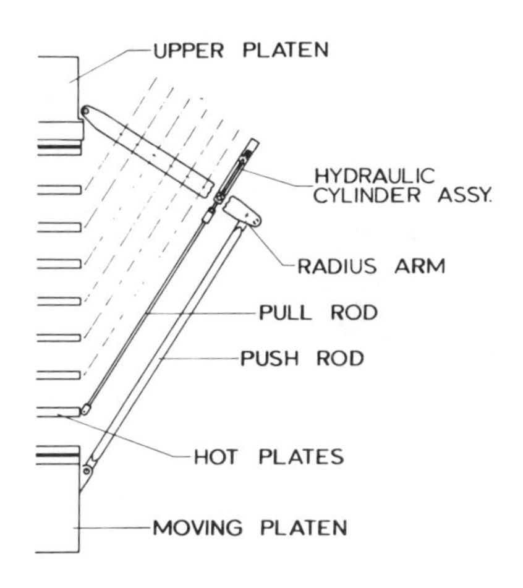

Figure 17.30. *A simultaneous closing system (Morse 1967).*

the comparison between an open and closed press equipped with simultaneous closing. Simultaneous closing is not used with some presses with a few openings because of well-prepressed mats or the type of furnish used.

The press unloaders are simply receiving elevators that take the finished boards and bring them to the level of the board finishing line (Figure 17.34). Boards can be pushed into the unloader by the nosepiece of the press loader. One system uses drive rolls located at the lead end of the unloader to remove the boards from the press once the unloader pushes them far enough out of the press for the rolls to engage. This enables the use of compressed air for blowing loose particles out of the press before the next press load is inserted.

Another system used with cauls has a draw-out beam which grips special notches in the lead end of the caul plates and pulls the cauls and boards out of the press into the unloader. The unloader itself can be equipped with belts or idler wheels to facilitate unloading.

Figure 17.31. *Simultaneous closing arrangement on a multi-opening press. (Courtesy Washington Iron Works.)*

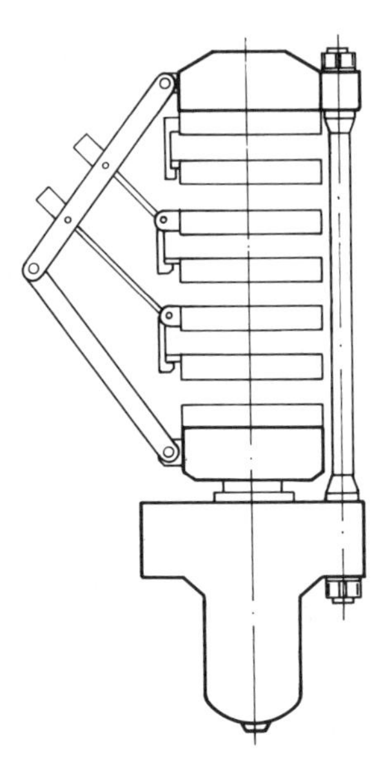

Figure 17.32. *A simultaneous closing system for every two openings. (Courtesy AB Motalla Verkstad.)*

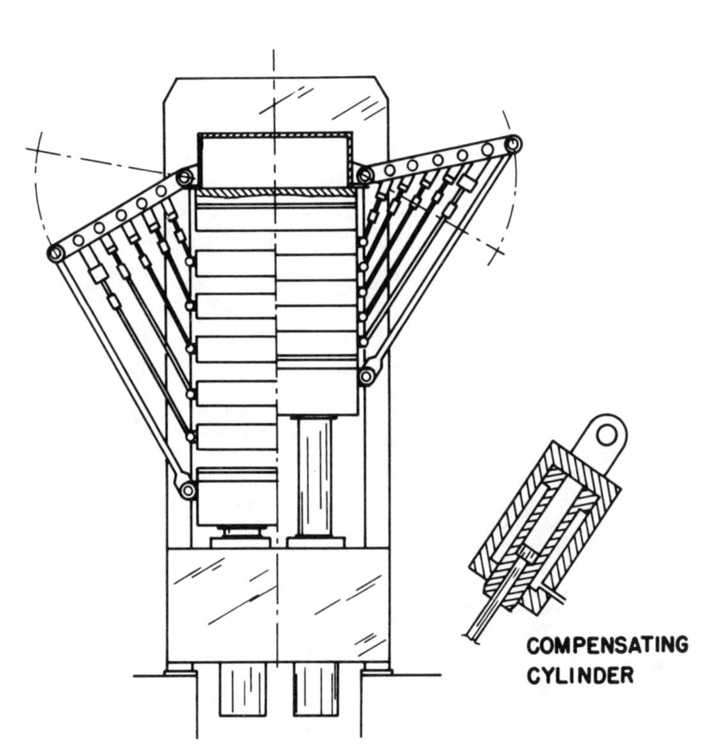

Figure 17.33. *A simultaneous closing device for hydraulic presses (Strisch and van Hüllen 1969).*

Because of the tremendous heat generated in a multi-opening press, heat shields are used to insulate the press loader and the mats it contains and the finished boards in the unloader from the press. These are simply large sheets of metal that are rolled into the space between the loader and unloader and the press during the pressing time. Such shields can also be used on the outfeed side of the press. Their main purpose is to keep the boards from "floating" out of the press when the press is opened. The vapor venting through the bottom face of the boards provides an air cushion on which the boards can easily move out of the press into the unloading elevator due to the forces of gravity. Unfortunately, the boards may enter the elevator improperly and jam the mechanism.

In the caul and caulless systems used with multi-opening presses, loose particles fall from the mats being inserted into the press or drop off the edges of finished boards being removed from the press. These particles can be between the stops (if they are used), and the press between the caul plate and the press platen, or between the surface of a freshly deposited mat and the press platen in the caulless system. All of these occurrences are troublesome. In presses using cauls, compressed air released through a manifold system can be used to blow these particles out of the press. In caulless systems, compressed air for blowing these particles out can be released through the nosebar of the press loader trays as they pass through the press openings carrying the new mats into position for deposition upon the press platens.

Figure 17.34. *A Flexoplan press and unloader (Strisch and van Hüllen 1969).*

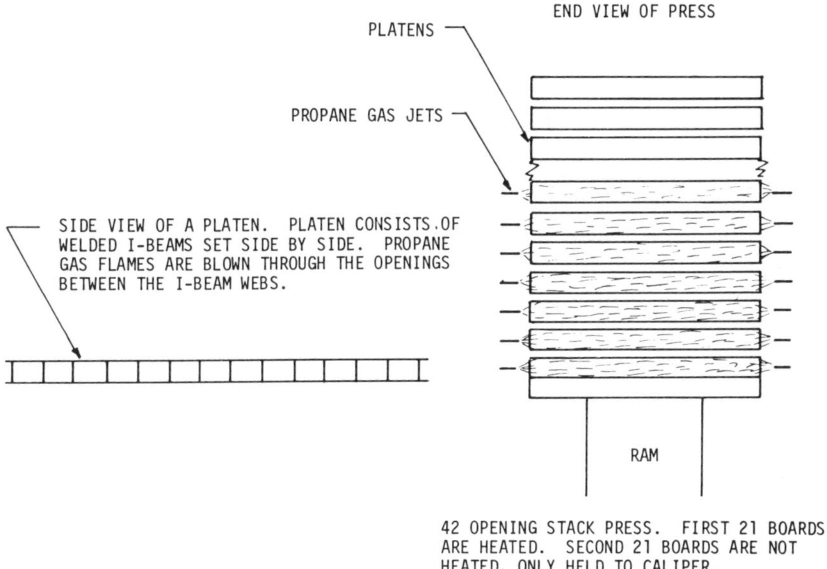

Figure 17.35. *Schematic drawing of the Chapwood stack press installed at the Philomath, Oregon, plant (now out of business).*

Stack Presses

A particularly interesting press that has been used for many years in the industry is a stack press as shown in Figure 17.35. The concept is illustrated in a 42-opening press as described in the following paragraphs.

The particular board is made with wet-lap fiber faces and a shavings core. Drying and pressing of the board takes place simultaneously. The formed mat on a caul plate is first placed on a *platen* as illustrated in Figure 17.35 and the entire package is inserted into the bottom of the press. Then the ram drives the package upwards until it is held in place by catches or dogs. At that time all of the 41 packages of platens, caul plates, and boards above are resting on the newly inserted mat. This dead weight supplies the pressing pressure, although the total weight can be adjusted by using dogs for each platen. As one package is inserted into the bottom of the press, another is taken off of the top and brought down to floor level by means of an elevator.

A temperature of more than 400°F (204°C) is held throughout the first half of the cycle. The heat is introduced in a unique way by means of gas jets, set in manifolds along the sides of the press, that blow flames through the entire width of the platens. The platens themselves are made of I beams welded side by side. The gas flames go through the openings between the I-beam web. One-half the jets operate from one side and one-half from the opposite side of the press. No heat is applied in the upper half of the press. The final board is quite dry at the time it leaves the press.

Single-Opening Presses

Multi-opening presses have been the most popular for board pressing because of their high productivity. Increased production can be had by increasing the number of openings in the press. This also allows for some flexibility in production, such as increasing press time without a great loss in production. The larger presses, however, require sophisticated loading systems such as a preloader for the press loader in order to have sufficient mats ready for injection into the press. The forming line also has to run at a high speed in order to lay down a sufficient number of mats. Such speed lends itself to poorly formed mats. In such cases, it is advisable to determine whether two press lines are advisable. The addition of a single-opening line can provide the extra capacity that is needed for quality board manufacture.

Because of their massive size (one company offers presses 8 ft 9 in. wide up to 100 ft long (2.67x30.48 m), metal caul plates are not used in single-opening presses. The tremendous expansion and contraction of such large pieces of metal plus the difficulty in handling them, along with the massive caul return lines that would be required, make them uneconomical. The single-opening presses use either thin, steel belts or woven wire mesh for the carrier system.

For those presses using the thin, steel belts, two variations are available. In the first one, the belt extends under a form station through the prepress (if one is used), through the hot press, and back around beneath these stations on the production line to the head of the form station. The mat is formed by a traveling former, which is unlike the conventional multi-opening system. While one mat is being consolidated into a board in the hot press, the traveling form station lays down a new mat upon the section of the steel belt beneath it. In this type of system, the mat forming is therefore discontinuous. When the press opens after consolidating the board, the steel belt moves forward, ejecting the finished board and moving the freshly formed mat into the hot press.

A platen-type heated prepress is now used with some of the single-opening operations. The advantage of such prepressing is that the press time is significantly reduced. In the normal single-opening press line, without the prepress, a certain dead time is required for heating up the bottom steel belt when it carries a fresh mat into the press. The top platen should not be brought in contact with the mat surface until this heating is accomplished. Now, by controlling the bottom platen temperature in the prepress at a different level than the top platen, this dead time can be eliminated. To further balance the application of heat in this system, a top rotating steel belt is used in the press itself. This belt is three times the length of the press and moves one length at a time in sequence with the movement of the bottom steel belt as it removes the finished board from the press and moves a new mat into the press. The bottom of a mat inserted into a press has heat applied to it immediately. In this press it is also possible to have different heat levels in the top and bottom press platens for applying the heat in a precisely balanced manner to the mat undergoing consolidation into a board, resulting in a perfectly symmetrical density profile in the board. One of these press lines is shown in Figure 17.36.

The second type of single-opening press line, as shown in Figure 17.37, utilizes three different belts in the forming and press line. The mat is first formed upon a wide belt conveyor using a stationary former. The mat then passes through a metal detector and a 600-lb/linear-in. (10.7 kg/linear mm) continuous prepress to a cutoff saw at the end of the forming conveyor. There the mat is transferred to a second wide belt conveyor. When the length of mat transferred to the second conveyor corresponds to the press length, a cutoff saw is activated and trims off a section of the continuously formed mat. The cutoff mat is then accelerated ahead and transferred to the steel belt conveyor that runs through the hot press. The steel belt carries the mat into the press and positions is for final pressing, as with the system utilizing a continuous belt throughout the entire system. After pressing, the board is moved out onto a roller conveyor by the steel belts, while another mat is simultaneously being carried into the press.

Prepresses were initially installed to allow for a faster closing of the hot press without blowing off the top surface layer of the mat; however, they were found to have other advantages. It has been found that the prepressed mat has less resistance to compression in the hot press. Thus, as soon as the press platens touch the mats, it is possible to build up the pressure much faster. It is also claimed that the prepressed board, as such, ends up with a better internal bond with this system of pressing, because the mat is compressed to thickness quickly and the high-temperature platens provide heat rapidly to the board core, resulting in a density profile that is relatively even throughout the board cross section as compared to boards pressed at lower temperatures.

The advantages of keeping the steel belt that passes through the hot press warm is twofold: less time is needed in the hot press to cure the board, since the steel belt is already partially hot, and the thermal stresses in the steel belt that occur during the hot pressing are reduced. In extreme cases, the belt itself can be folded if some part of the mat was formed too thin by accident.

The use of a heated prepress has already been mentioned. Another technique for speeding press curing is to heat the mat by means of a high-frequency-heated

Figure 17.36. *A single-opening press line with a heated prepress (Bucking 1974).*

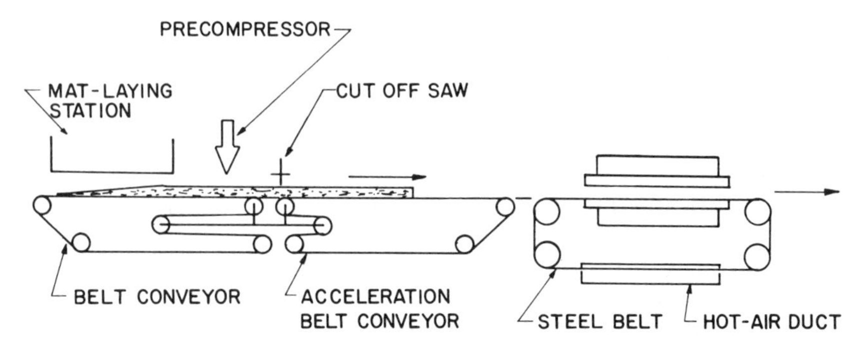

Figure 17.37. *Three-belt feeding system for the single-opening press line (Carlsson 1973).*

tunnel (Figure 17.38). This type of heating has been used for many years in the continuous Bartrev pressing system. It is now being introduced with single-opening press lines for heating the prepressed mat from approximately 80° to 160°F (27°–71°C). Thus, the required temperature increase in the board core to 212°F (100°C) is reduced to about 50°F (10°C) from 160°F (71°C).

The total length of this type of single-opening line is greater, as the preheater takes up about the same amount of space as the press. Conversely, the installed high-frequency power is less, as compared to using high frequency in the press, as the preheater is working continuously at a lower level and not intermittently at a higher level as it does in an RF-heated hot press. In addition, the shielding to prevent the electromagnetic radiation is less complicated and less expensive in the preheater. It is reported that a mat for conventional board about ⅝-in. (15.9 mm) thick, preheated to 160°F (71°C) requires about 50 seconds less time in the hot press than a mat that enters the press at room temperature. Since, in the single-opening press lines, preheated mats go into the press soon after preheating, such preheating should not cause problems with advancing the resin cure, as might be expected with multi-opening press lines where a number of mats are accumulated before they are inserted into the press.

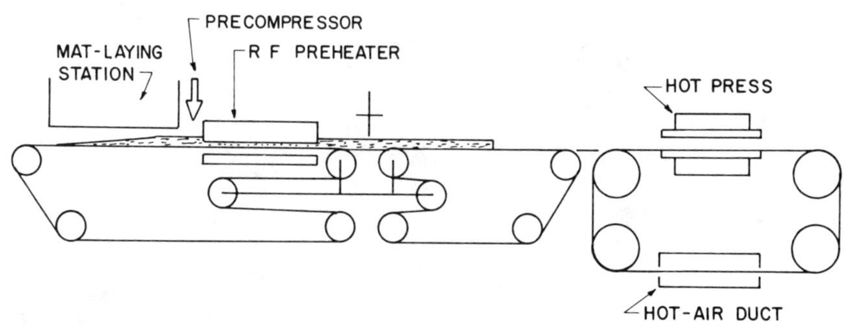

Figure 17.38. *RF preheater installed on the press line (Carlsson 1973).*

Single-opening press lines at present can produce up to 250 tons (227 mt) of ⅝-in.-thick (15.9 mm) board per day. Further improvements are expected to increase the production to about 350 tons (318 mt) a day. By using the preheating system, it is anticipated that capacities of upwards of 500 tons (454 mt) a day may be reached within a few years.

A third type of single-opening system uses the aforementioned woven wire mesh. This, in actuality, is a discontinuous pressing system that produces an endless board. It uses the same type of wire-mesh cloth as the Flexoplan system. This particular wire cloth has a good dimensional stability, with a maximum thickness variation of 0.004 in. (0.10 mm) over the full dimensions of the belt in use. Damage to this type of belt can usually be rectified easily within the production plant, whereas damage to the thin, steel belts used in other systems may be more difficult to repair. The wire cloth, however, has to be in perfect condition over its full length as it does not index to a given point each cycle. The steel belt is in sections and does index to a given point; so if damage occurs in an area, only the one-length section of the press size where the damage occurred need be replaced, instead of the whole belt. This type of metal cloth adapts itself to the temperatures and high pressures encountered in the hot press without developing particularly high stresses within itself.

In this system, the forming station is mobile and moves back and forth, forming the mat on the mesh cloth. The cloth carries the mat into the press and returns to the head of the forming station beneath the press and forming line after the board is pressed. Since the mat is not precut before it enters the hot press, the hot press itself has a special design to prevent precuring of the mat at the entry of the press. It is also necessary to have this hot press designed in such a way that the endless mat is not fractured at the juncture between the loosely spread mat and the finished board. A smooth transition from the mat to the finished board is produced by a combination of two techniques, as shown in Figure 17.39. In this figure, it is

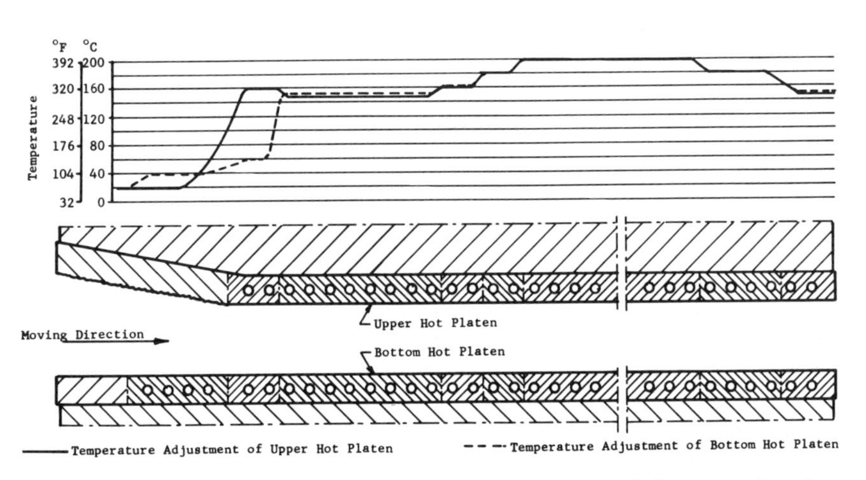

Figure 17.39. *Diagram showing temperature adjustment of platens and wedge-shaped opening at the feed end of the press (Strisch and van Hüllen 1969).*

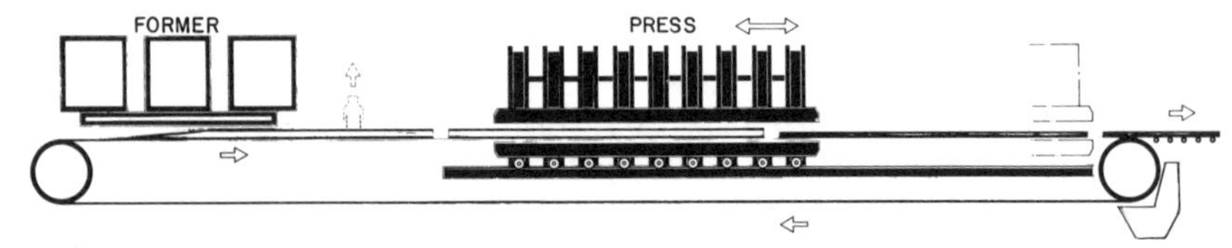

Figure 17.40. *Moving single-opening press. (Courtesy Raute, Inc.)*

shown that the temperatures in the hot press are not applied uniformly over the whole length of the heating platen, but that the temperatures are stepped down at the inlet and at the outlet of the hot press. The temperatures in the top and bottom platen are held at about the same level from one end to the other. The mat is stepped through the press in such a way that the part pressed first in the inlet zone of transition has final pressing accomplished in the outlet zone of transition. Thus, the mat is pressed twice in the zones of transition and with reduced temperature, which reportedly gives, in the end, the same quality of board as that pressed in the main temperature zone. At the inlet of the press, the temperature is slowly reduced in the critical stage below a temperature of 212°F (100°C); since at that temperature the resin is curing out, particularly in the case of urea, but final curing is not effected.

Another special feature of this press is that the top platen is not pressed perpendicularly on the mat at the infeed end of the press, but presses down in the shape of a wedge, as shown in Figure 17.39. To avoid shifting of the different layers of the mat with accompanying tearing, this wedge has a wavy surface on its bottom, where it contacts the top of the mat.

A fourth type of single-opening press developed in Finland is quite interesting in that the press moves back and forth, as illustrated in Figure 17.40. The mat is formed upon a continuous closed-mesh wire belt which moves at a constant speed. A section of the formed mat corresponding to the press length is next separated by a traveling cutoff saw. As soon as the mat is in the press, the press closes and consolidates the mat while traveling forward at the same speed as the wire belt. As soon as pressing is completed, the press opens and travels rapidly back toward the former as the next mat enters the press.

Continuous Pressing

Continuous presses have always captured the imagination of those in the industry. With continuous pressing, there is no need for handling much of the production line with batch processes. Quick acceleration of mats down forming lines, followed by deceleration as they are loaded either into press loaders or presses, is not needed. The discontinuous method has called for extensive and sophisticated engineering to overcome all of its problems. However, continuous pressing has its own problems, which have prevented its widespread use, except in manufacturing thin board.

Continuous presses for laminated lumber products, plywood, and composition board have been proposed and patented for at least 50 years. Conceptual feasibility, technical feasibility, and commercial feasibility are all interrelated. The economics of making a profit with the system is the governing factor, determining whether a continuous pressing system can indeed be successful.

Almost every conceivable type of pressure backing—including direct steam pressure, direct fluid pressure, direct air pressure, rollers, circulating bearing balls, double bands, caterpillar sections, friction sections, nip rolls and combinations thereof—has been patented for continuous-band-type presses.

The first continuous press actually used in production was the Bartrev, which was developed in the United Kingdom over 25 years ago. This particular system produces 50-in.-wide (1.27 m) particleboard at standard thicknesses. A typical production press is shown schematically in Figure 17.41. The press itself, shown in Figure 17.42, is a massive piece of equipment. As the mat enters the press, it is first heated by high frequency which shortens the time necessary for complete curing of the resin in the press. Stainless steel bands run over the caterpillar backing track. The caterpillar tracks are heated by gas panels and the track links store the heat for use in curing the board.

In each one of the track links are special high-pressure bearings, which call for frequent greasing. There are also special high-pressure, heat-resistant roller bearings running on the guide rail inside the press, as shown in Figure 17.43. It is immediately apparent that the large number of bearings require an enormous amount of maintenance, and this was felt to be one of the difficulties with the original system. A few of these systems were built and some are still in operation. One has been particularly successful in Sweden and the maintenance cost has been much lower than anticipated. However, its use has not become widespread.

Recently a modernized version of this type of continuous pressing has been developed, using the caterpillar-type tread or lug prepress. The first system of this type is now in operation in the Federal Republic of Germany. Figure 17.44 is a drawing of this recent development.

In the 45-ft-long (13.7 m) press, the thin, steel belts go through the press at the same speed as the segmented platens. Indeed, the steel belt runs free except

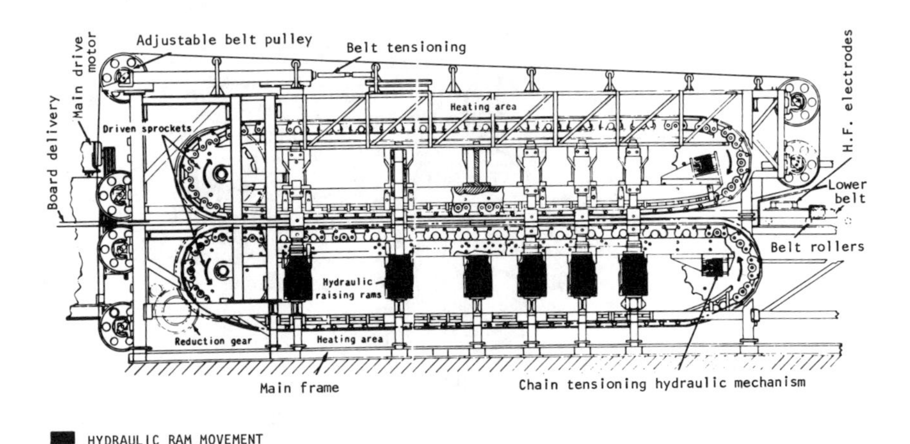

Figure 17.41. *A Bartrev press 50 in. (1.27 m) wide (Knapp 1973).*

where it is necessary to run the lower belt around a pulley to provide for transferring the mat as received from the former. Thus the platens and belts are not dragging against each other. The drive rolls are the small ones inside the press, which also control the board caliper. This contrasts with the large drive sprockets at the outfeed end of the Bartrev.

Both single- and multi-opening platen presses are unproductive during the loading-unloading-closing-opening part of the press cycle. This unproductive time becomes critical when pressing thin boards. In a large multi-platen press with 20 to 30 openings, experience has shown that a minimum of three minutes is required for charging, closing, opening and discharging the press. Since a thin board can be cured in less than a minute, a fraction of the total press cycle that a multi-platen press is unproductive becomes prohibitively large when producing thin boards.

Figure 17.42. *Side view of Bartrev press under construction (Knapp 1973).*

Figure 17.43. *A view of the track or chain-tightening mechanism (Knapp 1973).*

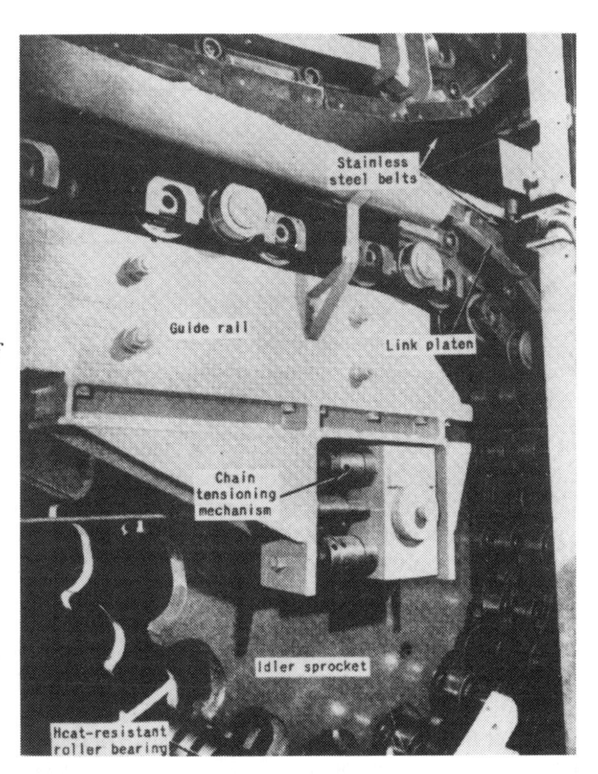

This has been the impetus for developing the continuous thin board presses now in use in industry plants.

About 50 of the Mende type are now in operation. It is an outgrowth of a press developed initially for the continuous curing of rubber sheeting or laminating thermoplastic films. Other versions are also undergoing development with one ready now for the market.

Their capacity ranges up to about ¼ in. (6.4 mm) in thickness with widths between 4 and 8 ft (1.22 and 2.44 m). Line speeds can vary from 15 to 90 ft (4.6–27 m) per minute depending on board thickness. Resin levels can range from 5 to 12% depending upon the particle geometry and the board properties desired. When using particulate-type furnish, board densities of about 46 lbs/ft^3 (0.74 sp gr) have been the maximum. However, experience now with pressure-refined fiber is showing that the easier-to-compress fiber allows compression of boards into final densities of about 55 lbs/ft^3 (0.88 sp gr). It appears quite likely that boards made of fiber at these higher densities will meet minimum specifications for hardboard. Particle geometry is still an overriding factor in developing good bending and stiffness properties in composition board. Thus, the fibrous material enables the production of higher-strength boards; whereas the particulate-type furnish, such as planer shavings, results in boards with much lower bending and stiffness properties. As mentioned, it has not been possible to make higher-density boards out of

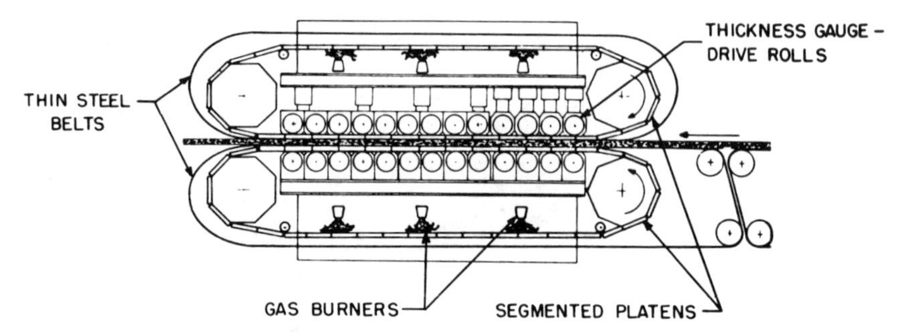

Figure 17.44. *Continuous press employing gas flame to heat pressing treads. Note high-pressure section at entrance of press (Wentworth 1975).*

the shavings-type furnish in this system. Thus, density cannot be used to compensate for the lower physical properties of these boards. In some countries using flakes as part of the furnish, board properties intermediate between the shavings-type boards made in parts of the United States and the fiber-type boards are found.

The Mende press is shown in Figures 17.45 and 17.46. First there is a 10-ft-diameter (3.05 m) heated press drum and a heated pressure roll at the infeed, which holds the steel belt at a given distance from the press drum and establishes the caliper or thickness of the board. In essence, this roll operates as a nip roll. Calibrating rolls further along in the pressing section hold the steel belt at the proper distance from the heated press drum. The finished board passes out either back over the forming station (which is usual) or over the top of the press station itself. Infrared heaters are placed within the press assembly to assist in maintaining the press temperature on the outer side of the continuous belt.

A high-grade Swedish steel belt is normally used in this press; however, a stainless steel belt has been used recently with success. Belts can be changed in 24 hours. A length of new belt is first fed through the machine and then the two ends are welded in place. Wide belts such as needed for 8-ft (2.44 m) board must be welded together out of two narrower bands of steel at present because of the unavailability of wider material.

The 10-ft-diameter (3.05 m) press drum and the calibrating rolls are crowned. In initial designs, some problems were observed with boards thicker in the center

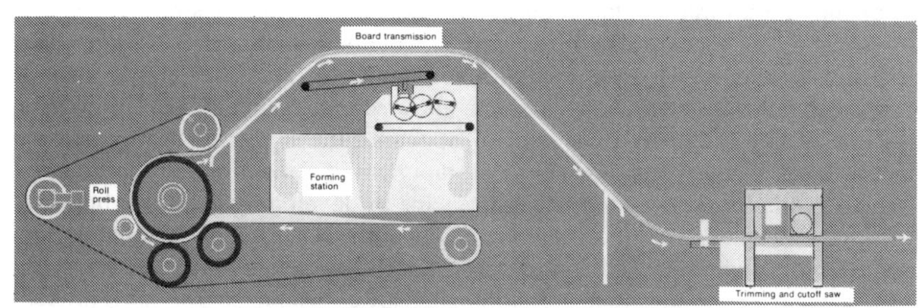

Figure 17.45. *Schematic drawing of a Mende press. (Courtesy Bison-Werke.)*

than at the edges. By crowning the rolls (the infeed pressure roll is not crowned), this problem has been overcome.

As can be readily seen, the hot board leaving the press experiences a severe curvature. However, when all things are equalized and carefully controlled in the production process, it is possible to end up with flat board. The board is cut to length and trimmed to width within a short time after leaving the press. Thus, the final set of the hot board takes place in the flat position.

The heating medium for the drum in the system can be either hot water, hot oil, or steam. The manufacturer is suggesting that hot oil be used to provide the greatest flexibility in temperature adjustment for manufacturing either interior- or exterior-type products and to eliminate the boiler.

At the present time, the system is being refined to overlay the board simultaneously with thin plastic sheets. This will eliminate a separate step for doing such overlaying.

The other continuous-band-type press now ready for production application consists of two horizontal steel belts, which are arranged in close proximity for pressing the mat into the final thin board in the flat position. A drawing of this press is shown in Figure 17.47. A particle or fiber mat is fed continuously between these two steel belts and the initial rolls serve to nip or press the mat to its final caliper. A pressure of 2000 to 3000 lbs/linear in. of roll width (35.8–53.7 kg/linear mm) is maintained. The nip rolls compress the mat to a thickness which is within about 25% of the final thickness of the board. Initially, experimentation showed that the mat could be overcompressed so that it was thinner than desired.

The final compression is achieved by having the two steel belts and the compressed mat passed between two heated horizontal steel platens, as shown in Figure 17.48. This figure shows the basic concept of the press, in which the horizontal flow of the mat passes through the nip roll section for calipering and then into the holding section, which is 36 ft (11 m) long. The holding section consists of four subsections, the first two of which are 6 ft (1.83 m) long and the second two of which are 12 ft (3.66 m) long. Each one of these sections is heated.

Figure 17.46. *A continuous Mende press in operation. (Courtesy Bison-Werke.)*

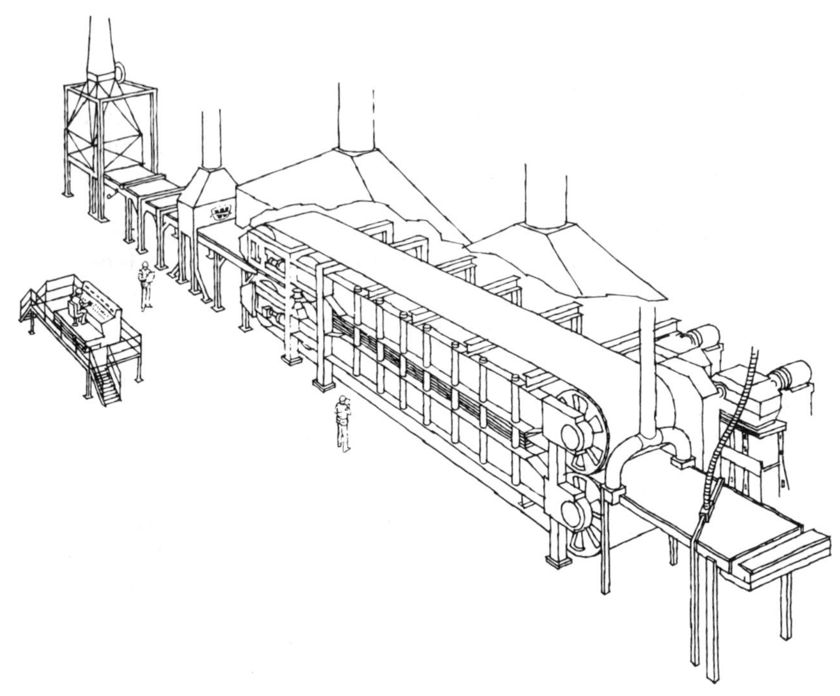

Figure 17.47. *Continuous band hot press (Johnson and Martin 1975).*

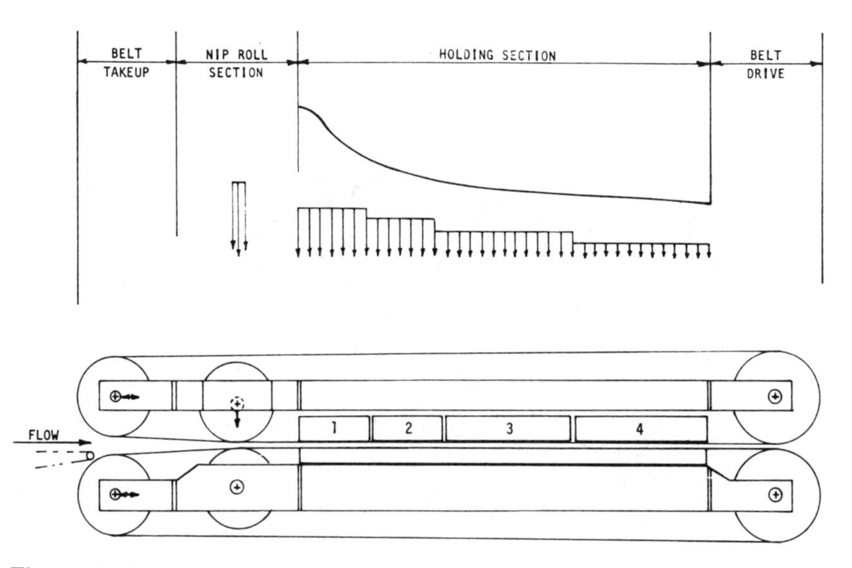

Figure 17.48. *Continuous pressing concept (Johnson and Martin 1975).*

The pressure to maintain board thickness does not exceed 100 lbs/in.^{2} (689 kPa) on the mat. Depending on board thickness and resin, as well as press temperature, the resin sets in between 15 and 50 seconds.

It has been found that the preferred temperature ranges between 300° and 390°F (150–200°C), as the effective conditions of dynamic rolling are best within this press temperature range. The temperature and moisture conditions used in the high-pressure dynamic rolling operation are for the purpose of causing the plastic flow and compression of the fibers to take place without setting up or curing the resin to any appreciable extent. The fibers are compressed quickly and continuously to the required density and thickness. It has been found that the holding section of a press provides the cure at a relatively low pressure of about 50 psi (345 kPa) and at a press temperature of 350°F (176°C).

Figure 17.49 shows a pilot plant model and Figure 17.50 shows a cross section of the holding section of this press. The section works against a stop beam to guarantee the thickness of the final board, although it is felt that this may not be necessary. On the top of the stop itself is a cooling plate and then a course of insulation between the cooling plate and the hot plate. The antifriction shoes are located between the hot plate and the steel belt which runs on top of the shoes. The upper belt is separated from the upper hot plate by another layer of antifriction shoes. Another course of insulation separates this hot plate from a cooling plate and the pressure section. The antifriction shoes consist of a metal backing plate with an antifriction material applied to the top of it. This is a dry lubricant which has the proper coefficient of friction for correct operation of the continuous press. However, in experimentation to date there has been shown a tendency for the shoe surface to glaze over, because of contamination. This problem has been eliminated by the introduction of a small amount of liquid lubricant for the purpose of purging the shoes and keeping them clean.

Figure 17.49. *Pilot plant model of continuous band press (Raddin and Wood 1973).*

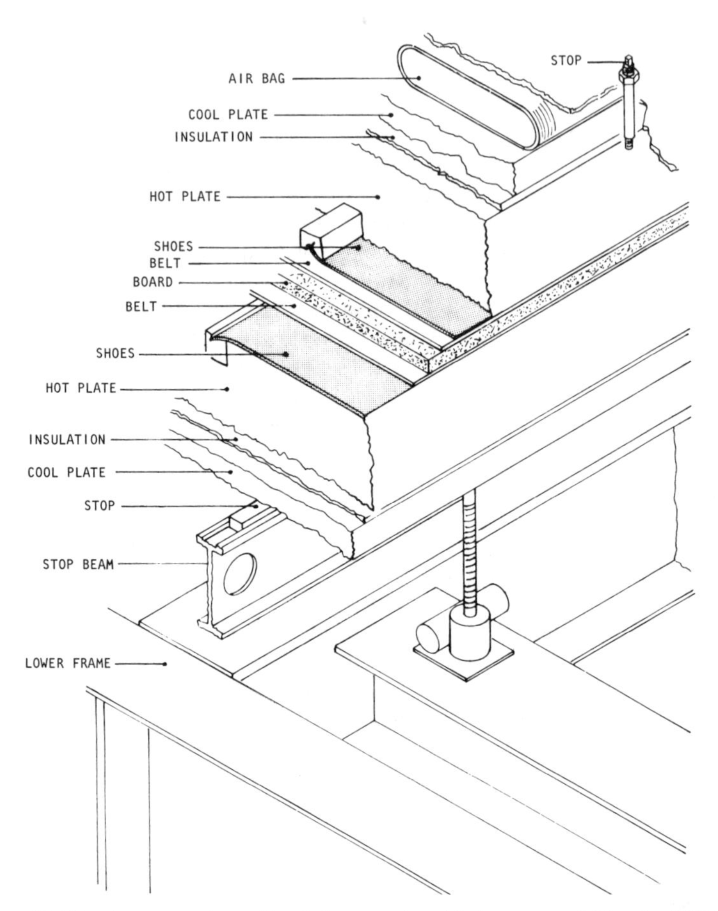

Figure 17.50. *Cross section of holding section (Johnson and Martin 1975).*

With any type of belt press, there are a number of problems of maintenance. A system has been devised for changing both belts in an eight-hour period on this particular press. Experimentation has also determined a belt that has the necessary properties to withstand the stress required of it as it is wrapped around the driving drum and pulled through the pressure section. All calculations show that the belt has sufficient yield strength and fatigue resistance to operate for a long time without a need for changing.

The initial press is to produce a 4-ft-wide (1.22 m) panel. This press is designed for producing 100 tons (91 mt) of ⅛-in.-thick (3.2 mm) board per day, at a line speed of 66 ft (20.1 m) per minute and a maximum horsepower of 220 (164 kW) per belt for a grand total of 440 HP (329 kW).

On experimentation to date, the density profile of the thin boards produced is relatively uniform throughout the cross section of the board. Figure 17.51 shows the density profile of boards made with hardwood fiber and softwood fiber.

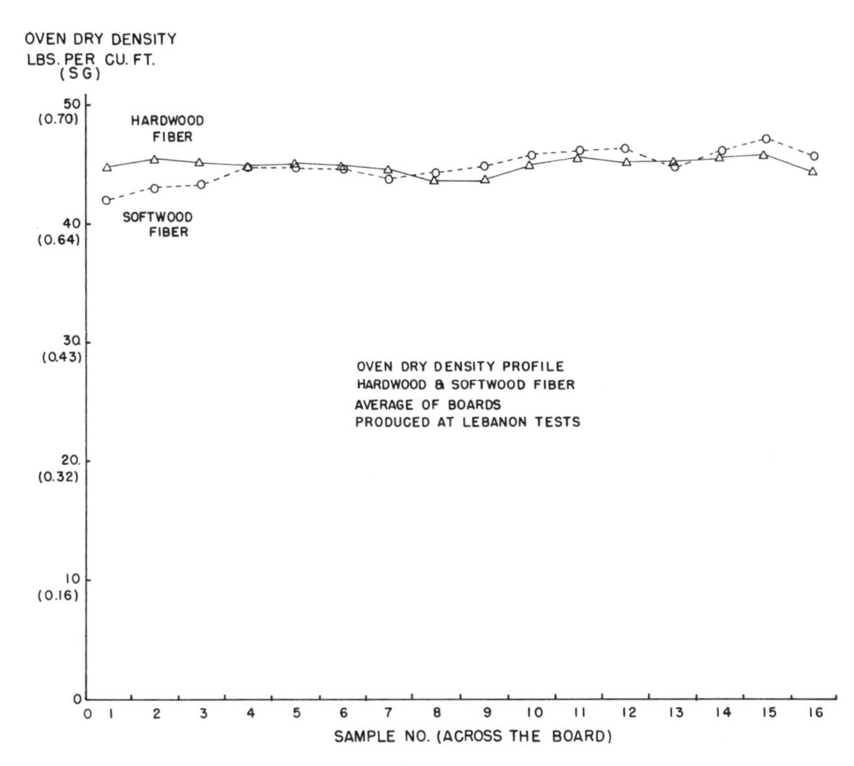

Figure 17.51. *Density profile of fiberboards made on a continuous band hot press (Johnson and Martin 1975).*

SELECTED REFERENCES

Achenbach, A. 1975. Problems and Aspects of Economics in the Particleboard Industry. Background Paper No. 44/8, third World Consultation on Wood-Based Panels, FAO, New Delhi, India, February 1975.

Advantages of High Frequency Heating in Fiberboard Pressing. 1971. Richmond, Virginia: Bulletin printed by Miller Hofft, Inc. (November).

Allan, D., and B. Polovtseff. 1961*a*. The Effect of After-pressing Conditioning on Wood Chipboard Stability. *Wood Chipboard* (July). Weybridge, Surry, England: The Airscrew Co. and Jicwood Ltd.

———. 1961*b*. The Effect of Hydrolysis on Wood Chipboard Quality. *Wood Chipboard* (July). Weybridge, Surry, England: The Airscrew Co. and Jicwood Ltd.

Axer, H. 1975. New Concepts for the Production of Particleboard. *Proceedings of the Washington State University Particleboard Symposium No. 9*. Pullman, Washington: Washington State University (WSU).

Board Mill Designed for 250,000 m³/Year. *World Wood* (May).

Bosch, J. M. 1969. Experiences with Closed Circuit Television in a Particleboard Plant. *Proceedings of the Washington State University Particleboard Symposium, No. 3*. Pullman, Washington: WSU.

Bowen, M. E. 1969. Heat Transfer in Particleboard During Hot Pressing. Ph. D. Thesis, Colorado State University, Fort Collins, Colorado.

Bucking, H. G. 1974. Exterior Structural Board Products Manufactured in Single-Opening Presses with Heated Prepresses. *Proceedings of the Washington State University Particleboard Symposium, No. 8*. Pullman, Washington: WSU.

Buikat, E. R. 1969. Clean Dry Air for Instruments and Plant Use. *Proceedings of the Washington State University Particleboard Symposium, No. 3*. Pullman, Washington: WSU.

Carlsson, B. J. 1973. Large Size Single-Opening Press Lines. *Proceedings of the Washington State University Particleboard Symposium, No. 7*. Pullman, Washington: WSU.

———. 1975*a*. Presses and Pressing Techniques in Wood-Based Panel Industries. Background Paper No. 105, third World Consultation on Wood-Based Panels, FAO, New Delhi, India, February 1975.

———. 1975*b*. Private communication on single-opening pressing and accumulators. AB Motala Verkstad.

Carter, R. K. 1969. Weighing Devices and Instruments Used in Particleboard Manufacture. *Proceedings of the Washington State University Particleboard Symposium, No. 3*. Pullman, Washington: WSU.

Crawford, R. J. 1967. Pressing Techniques, Problems, and Variables. *Proceedings of the Washington State University Particleboard Symposium, No. 1*. Pullman, Washington: WSU.

Dupu, M; T. Oradeanu; G. Badanoiu; and I. Alexandru. 1960. *Industria Lemnului,* No. 3. (Translated from the Rumanian by Research Information Service, 1962. Translation No. 473.)

FAO. 1976. Basic Paper IV. Technology and Techniques in the Manufacture of Wood-Based Panels. *Proceedings of the World Consultation on Wood-Based Panels,* held in New Delhi, India, February 1975. Brussels: Published in agreement with the Food and Agriculture Organization of the United Nations by Miller Freeman Publications.

Fischbein, W. J. 1957. Continuous Press Manufacture of Particleboard—Economics, Quality, Production. Technical Paper No. 5.35. *Fibreboard and Particle Board Technical Papers*. International Consultation on Insulation Board, Hardboard and Particle Board, FAO, Geneva, Switzerland, January-February 1957.

Freidman, J., and W. M. Teller. 1975. Why Convert from Steam to Hot Water Heating? American Hydrotherm Corp., New York, New York.

Gunn, J. M. 1972. New Developments in Waferboard. *Proceedings of the Washington State University Particleboard Symposium, No. 6*. Pullman, Washington: WSU.

Hadekel, R. 1954. *Hydraulic Systems & Equipment*. Cambridge: At the University Press.

Hafner, T. 1956. Radio-Frequency Heating in the Woodworking Industry. *The Brown Bovari Review,* Vol. 43, No. 5.

———. 1964. High-Frequency Heating in the Manufacture of Chipboard. *The Brown Bovari Review,* Vol. 51, No. 10/11.

Hedin, B. V. 1970. Motala Hardboard Presses—Development of Presses and Appertaining Equipment. AB Motala Verkstad, Motala, Sweden.

———. 1975. Single Opening-Press-Lines for the Manufacture of Particleboard and Medium-Density Fibreboard. Background Paper No. 103, third World Consultation on Wood-Based Panels, FAO, New Delhi, India, February 1975.

Heebink, B. G.; W. F. Lehmann; and F. V. Hefty. 1972. Reducing Particleboard Pressing Time: Exploratory Study. *USDA Forest Service Research Paper FPL 180*. U.S. Dept. of Agriculture, Forest Products Laboratory, Madison, Wisconsin.

Hesch, Rolf. 1969. Single & Multi-Daylight Presses. *Board Manufacture* (June).

Hopkins, B. R. 1970. High-Frequency Curing of Fiberboard. *Proceedings of the Washington State University Particleboard Symposium, No. 4*. Pullman, Washington: WSU.

Hurtienne, R. C. Electrohydraulic Servo Control for Board Presses. *Proceedings of the Washington State University Particleboard Symposium, No. 3*. Pullman, Washington: WSU.

Jenks, T. E. 1974. Process Engineering Research on Medium-Density Fiberboard. *Proceedings of the Washington State University Particleboard Symposium, No. 8*. Pullman, Washington: WSU.

Johnson, A. L. and R. T. Martin. 1975. The Washington Iron Works Continuous Thinboard System. *Proceedings of the Washington State University Particleboard Symposium, No. 9*. Pullman, Washington: WSU.

Knapp, H. J. 1973. Continuous Pressing—Past, Present, and Future. *Proceedings of the Washington State University Particleboard Symposium, No. 7*. Pullman, Washington: WSU.

Knox, L. C.; A. Schmitz-LeHanne; and R. R. Maddocks. 1973. FHL-Dryliner—A Thin, Continuously Produced Dry Fiberboard. *Proceedings of the Washington State University Particleboard Symposium, No. 7*. Pullman, Washington: WSU.

Kollmann, F. 1957. Effect of Moisture Differences in Chips before Pressing on the Properties of Chipboard. *Holz als Roh-und Werkstoff*, Vol. 15, No. 1.

Lehmann, W. F. 1960. The Effects of Moisture Content, Board Density, and Press Temperature on the Dimensional and Strength Properties of Flatpressed Flakeboard. Master's Thesis, North Carolina State College, Raleigh, North Carolina.

Lehmann, W. F.; R. L. Geimer; and F. V. Hefty. 1973. Factors Affecting Particleboard Pressing Time: Interaction with Catalyst Systems. *USDA Forest Service Research Paper FPL 208*. U.S. Dept. of Agriculture/Forest Service, Forest Products Laboratory, Madison, Wisconsin.

Maloney, T. M. 1961. Composition Board Mills. Division of Industrial Research Report. Washington State University, Pullman, Washington.

Morse, E. H. 1967. Washington Iron Works Presses. *Proceedings of the Washington State University Particleboard Symposium, No. 1*. Pullman, Washington: WSU.

———. 1972. Particleboard Press Maintenance. *Proceedings of the Washington State University Particleboard Symposium, No. 7*. Pullman, Washington: WSU.

Neusser, H., and W. Schall. 1970. Experiments for the Determination of Hydrolysis Phenomena in Particle Boards. *Holzf. und Holzv.*, Vol. 22.

———. 1971. Hydrolysis Phenomena in Particleboard. *Chemical Abstracts*, Vol. 74, No. 24.

Oehler, G. 1962. *Hydraulic Presses*. London: Erward Arnold, Ltd.

Ogran, E. 1967. Siempelkamp Closed-Frame Multiple-Opening Press for Particleboard. *Proceedings of the Washington State University Particleboard Symposium, No. 1*. Pullman, Washington: WSU.

Perry, W. D. 1969. Thickness Gauges for Particleboard. *Proceedings of the Washington State University Particleboard Symposium, No. 3*. Pullman, Washington: WSU.

Polovtseff, B. 1961. Practical Aspects of Wood Chipboard Densification Patterns. *Wood Chipboard*, (July) Weybridge. Surry, England: The Airscrew Co. and Jicwood Ltd.

Raddin, H. A. 1967. Hi-Frequency Pressing and Medium Density Board. *Proceedings of the Washington State University Particleboard Symposium, No. 1*. Pullman, Washington: WSU.

Raddin, H. A., and D. E. Wood. 1973. The Rando-Wood Process for the Manufacture of Continuous Thin Fiberboard. *Proceedings of the Washington State University Particleboard Symposium, No. 7*. Pullman, Washington: WSU.

Rowe, S. 1969. Nuclear Gages for the Forest Products Industry. *Proceedings of the Washington State University Particleboard Symposium, No. 3*. Pullman, Washington: WSU.

Schubert, D. F. 1975. Continuous Presses for Board Products. Background Paper No. 79, third World Consultation on Wood-Based Panels, FAO, New Delhi, India, February 1975.

Shen, K. C. 1974. Improving the Surface Quality of Particleboard by High-Temperature Pressing. *Forest Products Journal*, Vol. 24, No. 10.

Smith, P. H. 1975. High-Frequency Curing of Medium-Density Hardboards and Particleboards. Background Paper No. 77, third World Consultation on Wood-Based Panels, FAO, New Delhi, India, February 1975.

Stainless Steel Belt Cuts Downtime in Georgia-Pacific's Thin Panelboard Process. 1975. *Forest Industries* (April).

Strickler, M. D. 1959. Effect of Press Cycles and Moisture Content on Properties of Douglas Fir Flakeboard. *Forest Products Journal*, Vol. 9, No. 7.

Strisch, H., and S. van Hüllen. 1969. New Systems and Recent European Developments. *Proceedings of the Washington State University Particleboard Symposium, No. 3*. Pullman, Washington: WSU.

Suchsland, O. 1967. Behavior of a Particleboard Mat During the Press Cycle. *Forest Products Journal*, Vol. 17, No. 2.

Suchsland, O., and G. E. Woodson. 1974. Effect of Press Cycle Variables on Density Gradient of Medium Density Fiberboard. *Proceedings of the Washington State University Particleboard Symposium, No. 8*. Pullman, Washington: WSU.

van Hüllen, S. 1975. Private communication on hot-water heating in hot presses. Becker & van Hüllen. Federal Republic of Germany.

Vermaas, C. H. 1975. New Development Related to the Continuous Manufacture of Dry-Process Hardboard. Background Paper No. 113, third World Consultation on Wood-Based Panels, FAO, New Delhi, India, February 1975.

Waferboard Uses Cross Grain Chips. 1956. *Lumberman* (June).

Wentworth, I. 1969. New Designs for Single-Opening Presses. *Proceedings of the Washington State University Particleboard Symposium, No. 3*. Pullman, Washington: WSU.

———. 1971. Production of Thin Particleboard in a Continuous Ribbon. *Proceedings of the Washington State University Particleboard Symposium, No. 5*. Pullman, Washington: WSU.

———. 1972. Recent Advances in Prepresses for Single-Opening and Multiple-Opening Particleboard Plants with Automatic Stacking for Cut-Up Operations. *Proceedings of the Washington State University Particleboard Symposium, No. 6*. Pullman, Washington: WSU.

———. 1975*a*. Manufacture of Particle- and Fiber-boards with Flat and Rotary Continuous Presses and Continuous Lamination of Thin Boards. *Proceedings of the Washington State University Particleboard Symposium, No. 9*. Pullman, Washington: WSU.

———. 1975*b*. Private communication on hot water heating of press. Bison-Werke.

18. FINISHING BOARD

For this discussion finishing is being separated into primary and secondary. Primary will cover the cooling, sanding, grading, cutting, and shipping of the raw board. Secondary will briefly discuss filling, painting, laminating, and edge finishing of board. However, this is a lengthy subject and space will not allow a full discussion. Much finishing is done elsewhere, such as in special finishing plants or in furniture factories. However, more and more finishing is taking place in board plants to add value to the product, which hopefully increases the profit.

PRIMARY FINISHING

Cooling

Immediately upon leaving the press, the board is hot and will not reach its maximum properties until it has cooled. High-density hardboards usually are quite dry and have to be humidified to adjust their moisture contents before finishing. Particleboard is not normally humidified. Medium-density fiberboard may or may not be humidified.

Most board, because it is urea-bonded, is cooled upon leaving the press. The familiar *star* or wicket cooler shown in Figure 18.1 has been used for many years. These coolers hold about two press loads of board, and heat is dissipated from the boards while they are in the cooler. With urea-bonded board such cooling reduces the board temperatures to about 130°F (54°C). Such temperature reduction is necessary to keep the urea resin from breaking down. A hot-stacked pile of boards can maintain a much higher temperature for days because of wood's good insulating property.

Newer types of coolers are enclosed chambers such as shown in Figure 18.2, where the boards are transported on edge through the cooler. Air of controlled temperatures can be passed through these coolers to assist with the lowering of the board temperatures. Coolers can be massive pieces of equipment holding several press loads of board, some of which are 8 ft (2.44 m) wide and over 25 ft (7.62 m) long. With adequate cooling, sanding can be done immediately without tearing out the surfaces, which can occur when sanding "hot" boards.

Most all board is cooled; however, it can be advantageous to hot stack phenolic-bonded board to complete the resin cure. Some urea-bonded board is also hot stacked when the urea resin system is so formulated as to benefit from hot stacking. Some plants also find improved resistance to water absorption and thickness swell in boards where the hot stacking assists with the distribution and flow of the wax in the boards.

Hot stacking assists with final curing of phenolic resins and this has been a common practice with such boards. Retail yards have reported that packages of phenolic-bonded boards are still warm upon receipt, illustrating the insulation found with wood.

Cooling is needed to prevent hydrolysis from occurring in the board. Hydrolysis has already been mentioned as a problem in hot pressing. The temperature of the board, the moisture distribution with the board, and the vapor pressure are constantly changing during the pressing operation. This dynamic situation continues until the board cools after leaving the press. Asymmetrical stress changes consequently can occur, resulting in warping of the panel. Cooling is designed to provide symmetrical cooling of the panel and therefore balanced stresses from face to face, as well as to prevent hydrolysis.

Panels can be stacked by automatic stackers before further processing or they can go directly to sanding, trimming-to-size, and on to cut-to-size sawing.

Sanding

Panels should be relatively cool for sanding and final trimming to size. Hot panels have not achieved their full strength, and sanding such panels may not result in the best finish. The same is true with trimming-to-size, and some plants trim the panels to a rough size followed by a final trimming later on.

Figure 18.1. *Board turner and cooler. (Courtesy Washington Iron Works.)*

Figure 18.2. *Boards leaving an enclosed board cooler. (Courtesy Washington State Univ.)*

In the early days of the industry, drum sanders were used since they were about all that was available. These sanders consist of a number of sandpaper-wrapped drums which, because of their limited contact with the panel, tend to leave *chatter* marks (slightly corrugated surface) which is poor for overlaying and surface finishing. Some of these sanders are still in use. A narrow-belt sander traversing across the panel is used to try to minimize the chatter marks.

The development of the wide-belt sander, however, has made it possible to quickly dimension particleboard and develop a fine finish in the same machine. Modern sanders fall into two basic types: high-speed, multiple-head machines and slow-speed, single-head machines. Sanders will now sand panels up to 8 ft 5 in. (2.57 m) wide at line speeds to about 170 ft (52 m) per minute. Modern boards do not require as much sanding as some of the pioneer boards. Cases are known of having to sand off about 0.125 in. (3.2 mm) of surface because of poor soft surfaces and poor dimension control at the press.

The high-speed multiple-head machines can range from a two-head dimensioning machine to a six-head machine that will abrasively plane the panel to dimension with the first two heads and put on a fine finish with the last four heads. Fine control can be exercised over each head. Such a machine is shown in Figure 18.3. The slow-speed single-head sander is used on finished lines for buffing or finishing cut-up stock and banded stock. Some sanders can be purchased in modules, for example, putting in four heads now and two more at some time in the future.

Figure 18.4 contains schematic drawings of some of the sanding-head configurations showing the basic parts. The opposed-head configuration is designed for maximum stock removal and dimensioning or sizing of the panel. The platen head or smoothing bar is used for finishing. As shown in the drawing, the sander belt in this case contacts the panel over a wide area. This removes the chatter marks that occur from the dimensioning step. Also shown in Figure 18.4 are the combination head and combination four roll, both of which are designed for smoothing the final surface of the panel being sanded. Acceptable final board thickness after sanding is about ± 0.003 in. (0.08 mm).

Most sanders today have all of the sanding heads contained in one machine or in modules designed for close coupling. However, there are versions where the sanding heads are separated by several feet. With these sanders there is usually a brush before each sanding head to clean all debris from the surface being sanded. Brushes are also used in the close-coupled machines.

In a sander, the first heads are designated as primary. As mentioned, they are for sizing the panel to the desired thickness. The opposed-head configuration, with the top and bottom steel contact rolls directly opposing each other, provides the most accurate dimensioning. Secondary heads are usually of the combination type, using both a contact roll with a smooth surface or spiral serrations and a finishing platen or smoothing bar. The secondary head is used to dimension the board to final caliper and to remove any grit and rotation marks caused by the primary contact rolls. These secondary heads can be opposite each other, or opposite a platen or a back-up roll, which is called a *billy roll.* The finishing heads can be either a combination head or a platen only. The platen-type head consists of two belt carrier rolls and a pressure platen. This type of head removes very little stock but removes all of the imperfections developed by the primary and secondary heads, resulting in a very fine finish.

Single-head machines can be used for sanding on a production line. A few of these have been used, but they must run at slow speeds and the panels must pass through the sander several times.

Good feed systems have been developed for transporting the panel through these sanders. In earlier sanders this was quite a problem, resulting in poorly

Figure 18.3. *A six-head sander. (Courtesy The Coe Manufacturing Co.)*

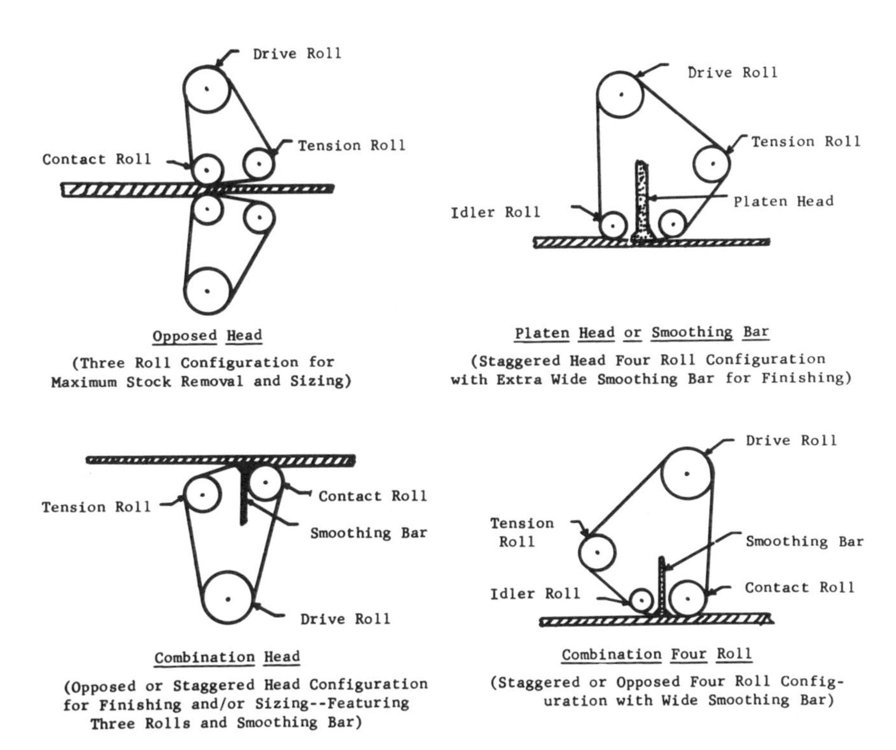

Figure 18.4. *Some sanding-head configurations (Murdock 1968).*

finished panels which would have end or edge snipe. In present-day sanders, the bottom feed rolls are the only ones driven and these rolls are covered with polyurethane to provide a good grip upon the panel. The upper or pinch rolls are usually equipped with soft rubber or polyurethane covers. Some operators like to use knurled steel feed rolls wherever possible. There apparently are more maintenance problems with the polyurethane-covered rolls, and these are used only where the conveying is critical. The billy rolls opposite the contact drums are either rubber-covered or smooth steel. Since these are powered feed rolls, the rubber-covered ones are superior in controlling the panel feed. The overhead or hold-down rolls are normally spring-loaded rather than driven as mentioned. Spring tension on these rolls must be sufficient to prevent slippage of the panels and to provide enough friction to the lower drive rolls to insure constant end-to-end feed. The contact drums can be either knurled steel or covered with rubber. Knurled-steel drums are relatively maintenance free. Rubber-covered contact drums have a maintenance problem from wear in the areas where the edges of the sander belts oscillate. This problem is enlarged in sanders which use both wide and narrow belts. Steel drums are utilized for an aggressive sanding operation.

The sander belts oscillate back and forth across the panel as sanding is accomplished. Consequently sophisticated sander belt tracking systems have been developed. Some of the techniques employed are: air jets against limit-switch arms, electronic eye systems, and air jets from idler rolls. The air jets, when covered or uncovered, cause differential pressure in the instrumentation which controls the automatic belt oscillation. This method of signaling the instrumentation has provided excellent control of the oscillation.

It is important to have proper belt tensioning in a sander as the belts tend to stretch during their operating life. Pressure gauges are used to monitor the tension, and adjustments can be made by the operator.

The polishing platens require guides and adjusting mechanisms with close-fitting tolerances. Their end-to-end adjustment must be extremely accurate to prevent sanding a taper across the panel. Polishing platens have covering material, normally of polyurethane or hard-pressed felt covered with graphite cloth. Minor defects in this covering must be monitored and corrected, if they occur, since a poorly applied covering can cause streaks on the sanded panels. It may be necessary to have frequent maintenance and changes of the polishing platen material.

Large-capacity air systems are needed to remove the dust generated by the sanding heads. If this is not done properly, the dust can be left on the surface of the sanded panels and this dust can then contaminate the bearings, guides, motors, and controls. Thus, close-fitting hoods, large pipe sizes, and high-capacity fans are necessary for efficient cleaning of both the machine and product. A problem does occur with the dust-collecting system in that every once in a while a sander belt "explodes" or breaks apart in the machine. When this happens, it is very likely that a great number of sparks are generated; and these sparks can pass through the dust-collecting ducts into the dust-collecting system and baghouse, causing an explosion and fire. Explosion-suppression systems are therefore important in the sander dust collecting system.

Sander belts are usually the most expensive item in the finishing department. Many variables are involved in determining the life span of the sander belts. These conditions vary between plants in many ways, all of which affect the production and belt expense. Some of these conditions are: difference in species; panel temperature and panel size; variation in panel thickness, requiring different amounts of stock removal; density of panel surfaces; storage and handling of belts; maintenance of the equipment; and the experience of the sander operator (which is a very important factor).

The sequence of sandpaper grits and the type of sandpaper used varies between operations. Sander design dictates, in many cases, the use of high-cost belts rather than medium-cost belts and low-cost paper-backed belts. With few exceptions, silicone carbide grain is the abrasive material used to sand board. The toughest resin bonds and the strongest backings available are combined with the silicone carbide grain to produce these belts. Similar belts are used for cutting glass, plastics, leather, ceramics, and nonferrous metals. It has been noted that no material requires more horsepower per inch of belt width, more aggressive contact wheels, or more massive machine construction than does this particleboard sanding operation.

The type of grit used on sanders varies between the sander and the product being sanded. Underlayment, for example, might not require any finer grit than 80, since an extremely smooth finish is not needed. Furniture-type panels however, requiring finer finishes, can be sanded in sequences of 50 to 100, 60 to 100, and 20 or 80 to 150 grit. Prefinishing plants, however, may go as high as 400 grit with the sandpaper. With the coarser-grit paper, the belt speed can range from about 550 to 6500 ft (168–1981 m) per minute. When using finer-grit paper, slower speeds are found to be better for providing a better surface with speeds running in a range between 400 and 2000 ft (122–610 m) per minute.

Open-coat belts (determined by the amount of cutting grains glued to the belt) are used for heavy sanding, as the grain spacing minimizes the buildup of material on the belt. Open-coat belts will have about 50 to 60% of the belt covered with grains while close-coat belts will be completely covered with grains. Close-coat belts are appropriate for finish-type sanding.

Belt life can be shortened by high humidity; inadequate dust exhaust; a heavy rate of cutting; a high sander feed speed; a high board resin level (especially in the faces); high board specific gravity (as we have learned, faces are normally higher than the average board density); metal or impurities in the board; the wood species itself; and extractives such as resin in the wood.

Belts should be stored on special racks before use in a controlled atmosphere of 50% relative humidity and 70°F (21°C). With proper storage and prevention of mishandling, such as getting the belts wet, maximum belt life can be expected.

Grading

After the panels leave the sander, it is normal for a grader to check the board surfaces visually for imperfections. An oblique light is used to show up these imperfections. A board turner can also be placed in the line so that the operator can grade both sides. Figure 18.5 shows one of these turners in operation. An

Figure 18.5. *A board turner in operation. (Courtesy Washington State Univ.)*

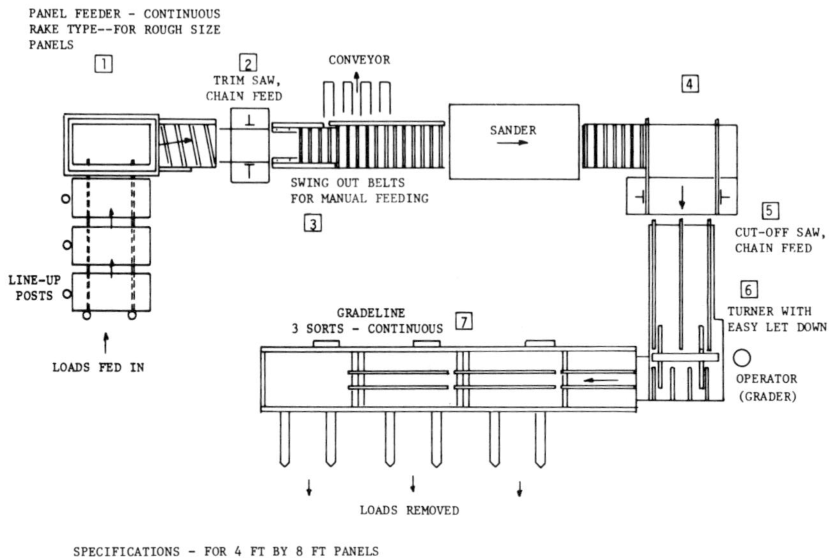

1. FEEDER, HOOK TYPE, CONTINUOUS CYCLE NOT REQUIRING RELOAD DOWN-TIME. INCLUDES AN INFEED BUNDLE CHAIN FOR LOAD STORAGE. INCLUDES AN OUTFEED ROLLCASE WITH A LINE-UP STRAIGHTEDGE.
2. SKINNER SAW.
3. ROLLCASE, POWERED ROLLS TO SEND PANELS FROM SAWS INTO SANDER.
4. PANEL TRANSFER, DESIGNED TO PUT PANELS ONTO CUT-OFF SAW CHAIN BEAMS.
5. CUT-OFF SAW.
6. POWER TURNER, FOR RAISING BOARDS TO VERTICAL POSITION FOR GRADING PURPOSES.
7. PANEL STACKERS WITH ACCUMULATORS FOR CONTINUOUS STACKING AND COUNTERS.

Figure 18.6. *A typical grading, sorting, and stacking line. (Courtesy Keystone Machine Works Inc.)*

automatic stacking system then follows so that several sorts of the material can be made. This is automatically controlled by the grader. Figure 18.6 illustrates a typical grading, sorting, and stacking line.

Cutting

Saws: A basic two-pass trimming system has been standard in the plywood and other panel industries right from the start. This consists of a first-pass or skinner saw unit with two side saws for edge trimming, a 90-degree corner transfer to a second-pass or trim saw with two saws as shown in Figure 18.7. The first evolutionary step for this system was to add one or more splitter saws for the first-pass saw and one or more saws for the second-pass unit. This was the beginning of the cut-to-size operation as illustrated in Figure 18.8. The primary and best use of this type of line is to break large panels down to sizes that are easier to handle and sand or cut up further. This is especially important with the monstrous-sized panels that are being produced today. From this line, which can be either in the board press line or in a separate line, the panels can be cut to order on a quick-set type island saw with either chain- or slat-bed feed. A slat-bed saw is shown in Figure 18.9.

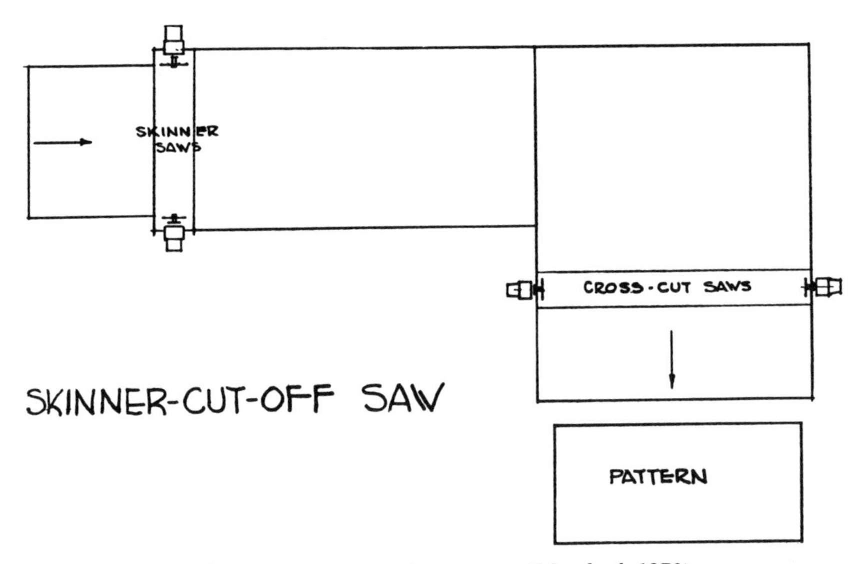

Figure 18.7. *The basic two-pass cutting system (Murdock 1970).*

The slat-bed saws come in two types with the most common one using a wooden bed into which the saw cuts. This bed is a replaceable item, and how often it is replaced depends on the number of size changes. The other type uses an aluminum slat with a replaceable neoprene-covered slat. On this type the bed dips under the saw line so that the bed is not cut.

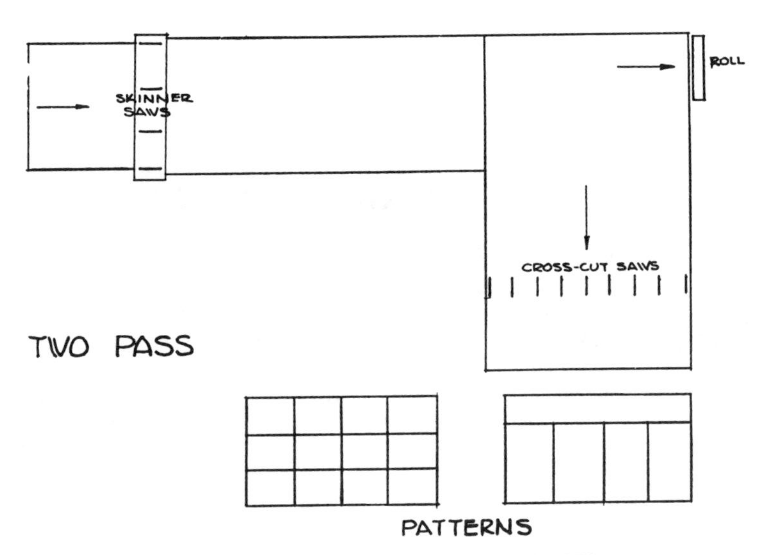

Figure 18.8. *The basic cut-to-size operation (Murdock 1970).*

Figure 18.9. *Dip chain slat bed with overhead aligning devices (Murdock 1970).*

An interesting new saw has been developed for cutting to length the continuously pressed board as it leaves the press or the very long boards while they are traveling down the production line. The saw traverses at an angle across the moving board at a synchronized speed so that the board is cut off square to the edges. One of these saws is shown in Figure 18.10. Other saws cutting the thin, continuously pressed board momentarily stop the board with clamps so the cutoff can be made. The thin board continuously coming from the press buckles up for an

Figure 18.10. *A trim/crosscut/splitting unit for continuously pressed or very large boards cut while moving. (Courtesy Mereen Johnson Machine Co.)*

instant against the clamp restraint while this cut is being made. A clipper- or guillotine-type saw has also been used with thin board.

If further breakdown is needed for drawer parts and so forth, ripsaws and double end machines are used for ripping, bullnosing, grooving, and sanding of these parts. These machines can be connected together for high production or fed individually.

The current trend is to more sophisticated systems using punch cards or tapes to control the saw. With these systems orders can be programmed at some central location and transmitted to plants where tapes are punched, inserted in the machine, and the order cut utilizing the best pattern for the primary board size. Large companies are doing their own programming, but there are independent firms also offering this service.

The new sawing systems are of three types. The first uses multiple-panel feeding with the panel stack initially moving into position and being clamped. The cuts are then made by traveling saws for both the rip and crosscuts. The second type uses a combination of the traveling saw and positioned saw arrangements. Here the multiple-panel stack is moved into position, clamped, and a rip cut made by a traveling saw as shown in Figure 18.11. After this rip cut is made, the stack is indexed for the next rip cut. The previous rip-cut stack is moved into a conventional-type crosscut saw with saws positioned for a desired length and cut. These units can be fully automatic with tape record control or they can be manually controlled.

The third type of sawing system is called the three-pass. This system for a high-speed, accurate cut-up operation uses a normal skinner-splitter saw and two crosscut saw assemblies arranged side by side with a residual off-bearing assembly at the third saw as shown in Figure 18.12. One system of this type can be tape controlled so that a complete change in setup can be made in 1¼ minutes while making 33 changes in the line. The large amount of electronics required for the

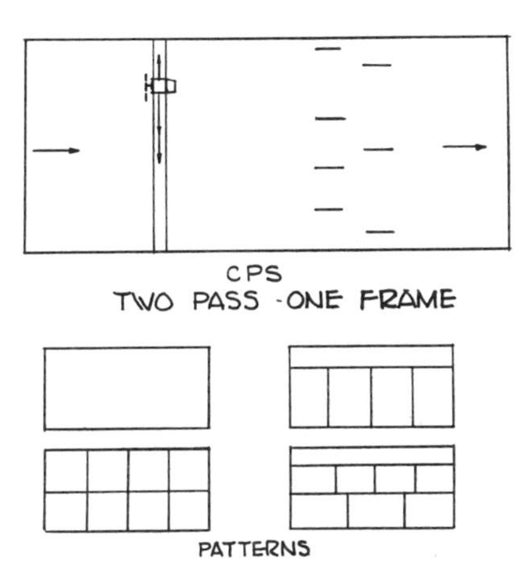

Figure 18.11. *Combination cutting with a traveling saw and positioned saws (Murdock 1970).*

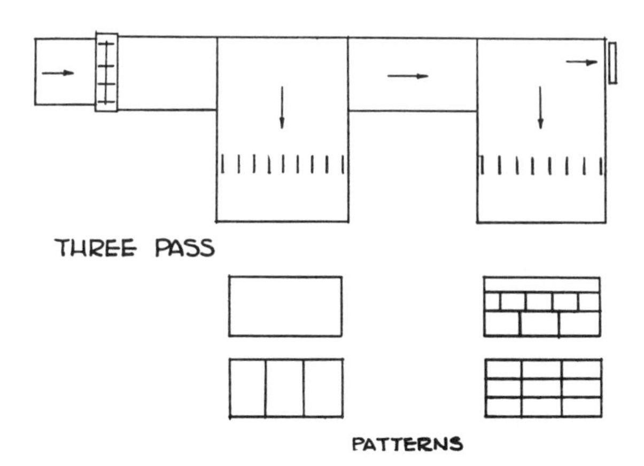

Figure 18.12. *The three-pass cut-to-size arrangement (Murdock 1970).*

three-pass system is of course expensive. A more practical approach is a combination of manual set and tape control. This consists of zero saw, stops and guides being manually controlled, and 17 saws, chain beams and hold-downs being tape controlled. This cuts the cost substantially but still allows the normal setups in a very short time. A completely manual setworks depending upon the skill of the operator, would take 10 to 15 minutes for a 17-saw setup to be adjusted. This time interval includes the time it takes the operator to measure and check the size of the parts being cut.

The higher-production panel saws cut the material to size for the customer from large panels, either directly out of the press or later from panel storage. The scheduling task in a cut-to-size operation is a very complex and difficult one because of the interactions between all of the factors involved. However, the computer is presently being used successfully in two areas: daily scheduling and planning. By using a systematic method of cutting, the plant can consistently meet its cutting schedules and shipping dates. This method assures that orders will be continuously completed and shipped throughout the week. Orders may be changed, added, or deleted up to one day before the cutoff date for each period. One of the most powerful features of the computer scheduling service is its consistency, since the computer can analyze and solve the problems quickly and is never forced to take short-cut approaches to meet the deadlines, which manual schedulers often must do. Historically, a computer service can improve board yield over those prepared manually by 3 to 10%. As a result, primary board utilization can be maintained at 91 to 93%, leaving 7 to 9% incurred off-fall in wastage. Since it is usually possible to recover about 1 to 2 percentage points of the off-fall, the system can come within 2 to 4 percentage points of the perfect 97% utilization (considering saw kerf).

The cost of a computer scheduling service is low in view of the benefits mentioned. Considering board yield alone, computer scheduling should save the user more than three times the cost of the service. For example, $1 spent for the service saves $3 on material costs alone. The highly sophisticated new saws can cost up to $1 million, which is far more than the basic skinner-type saws initially used in the industry.

Board breakers are used to break the trim into small pieces. These are refined further for reuse as furnish or as fuel.

Sawblades: It has not been possible to use conventional cutting tools for particleboard and fiberboard. Tools employing tungsten carbide or modified grades of tungsten carbide for the cutting edges have become the best way to cut composition board. High-speed steel has not been successful because the resin in the board, the natural resins in some of the wood species, and the high silica content in some of the wood species (because of their geographic location) have made the board extremely abrasive. The use of high- and low-pressure overlays, vinyl coverings, fillers, and base coats has further complicated the cutting practices required for composition board.

The selection of the proper grade of carbide is a matter of careful analysis. It is necessary to obtain the most wear-resistant grade of carbide that will do the cutting job without causing chipping or flaking away of the carbide. The correct grade selection is most often made from experience. A grade of carbide must be hard enough for the material being cut but must also have the characteristic or property in its makeup to withstand the shock of impact when cutting. Usually a non-alloyed grade of tungsten carbide will give the longest tool life. In some instances a micro-grain grade (a finer grain) of tungsten carbide is used, particularly for high-density particleboards. Using proper tool design will allow a reduction in cutting pressures, which aids in optimizing the economy of operation.

There have been many recent developments in the style and configuration clearance angles of sawblades and in the grades of carbide required for the various jobs. The combination of these advances has provided the best cutting tools yet for cutting or machining board. The selection of sawblades for cutting the board can usually be determined by the density of the board. The coarseness of grain, particle size, and geometry, must also be considered in a sawblade selection to insure the most effective results. Ideally the design of the sawblade must match its rpm with the feed rate of the material through the cut, so that the desired chipload per cutting tooth will be maintained. If this is not done chip-out, fiber tear-out, and a decrease in the length of tool life will occur.

The end use of the material will determine the order of importance for the quality of the cut or life of the cutting tool. Both factors are vital points to consider in the selection of sawblades. The use of carbide-tipped sawblades is a must in order to achieve even moderate tool life when cutting. Sawblades with maximum shear and sufficient teeth, exerting the least possible pressure on the board as it is being cut, will assure the most effective results.

No matter what type of cutting is being done, the material must be supported properly, right up to the edge of the cut. This will help to avoid edge chipping or chip-out. In some cases special devices must be designed to provide proper material support. However, there are some applications where full support is not possible. An example of this is found when cutting against the feed on slat-bed machines (Figure 18.13). Cutting forces tend to lift the material away from the machine bed and hold-downs cannot be set close enough to the sawblade. Reversing tool rotation to a climb-cutting configuration improves the edge finish (Figure

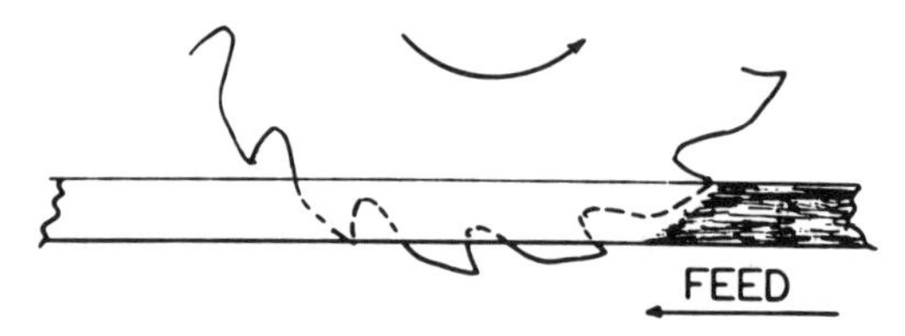

Figure 18.13. *Cutting a panel from above with a power cut (Jones 1968).*

18.14). If tool rotation cannot be reversed, the cutting should be performed as close as possible to the hold-downs and a zero or positive five-degree hook angle sawblade used to reduce the lifting force on the material. Cutting pressure can also be reduced by adding more teeth. Also, greater shear angles can be designed into the tooling to change the cutting force direction.

As the board becomes more dense, the leaving edge becomes more likely to chip or break out. The use of an alternate top bevel tooth form (Figure 18.15) can provide enough shear for a good top and bottom edge finish as well as good leaving end cuts that do not have breakout. The addition of alternate face bevel added to the alternate top bevel design will help avoid any leaving end breakout that still may occur (Figure 18.15). The alternate face bevel acts as a shearing tooth delivering a slicing movement to the cut, which reduces the cutting pressure.

As mentioned, a factor to consider when cutting board is tool life. The alternate top bevel design features a sharp point that dulls easily because it is subjected to high impacts as it enters the material first. In this case shorter tool life results. The use of a triple-chip grind design will give added tool life over the sharp alternate top bevel design. A tooth with 45-degree corner angles opens a path through the hard abrasive board as shown in Figure 18.16 (1). It is followed by a flat tooth which takes a light finishing cut on the sides of the material, shown in Figure 18.16 (2). Since the 45-degree tooth opens the path in the cut, the regular tooth only has to make the smooth side cut. The 45-degree "lead-tooth" is also extended so that it always cuts deeper than the raker tooth, thus protecting the sharp corners of the raker tooth and extending tool life. Such a cutting action is not only beneficial in producing longer tool life, but also produces a fine side finish.

Some very dense, hard panels are so brittle that the raker tooth chips the edges of the material. When this occurs an alternate face angle is sometimes added to the raker teeth to provide a slicing, shearing, side-cutting action which will not chip the side (Figure 18.17). This then changes the standard two-tooth design to a three-tooth design pattern.

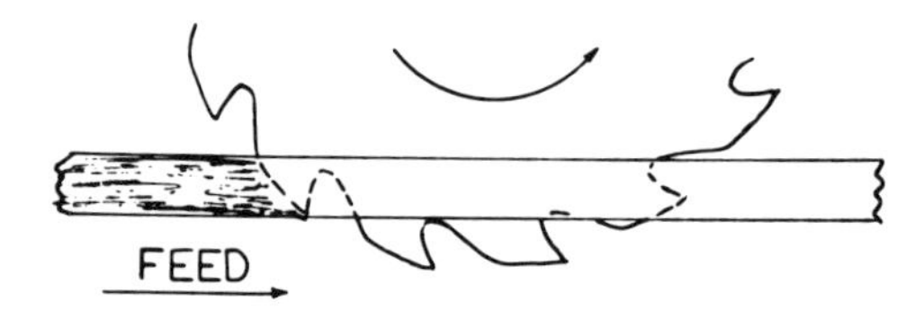

Figure 18.14. *Cutting a panel from above with a climb cut (Jones 1968).*

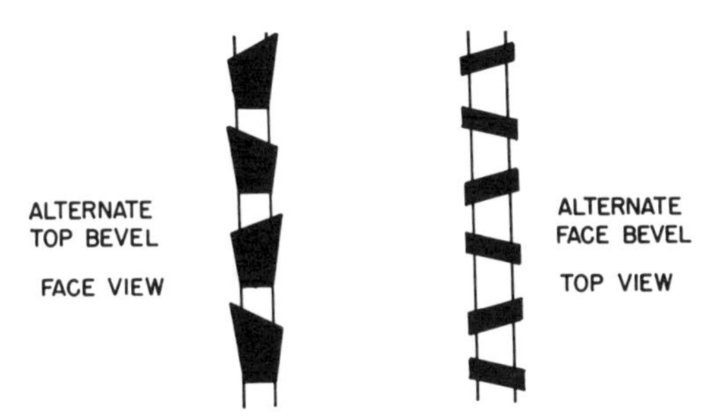

Figure 18.15. *The alternate top bevel and face bevel tooth form used to prevent edge breakout (Kovich 1972).*

In addition to the triple-chip design with alternating face bevel, there are other modifications to the standard alternate top bevel design which, in specific applications, help increase the tool life while still maintaining the sharp cutting point. One of these is the use of a raker tooth along with the alternate top bevel teeth, which helps increase the tool life (Figure 18.18). By adding alternate face bevels to

Figure 18.16. *The triple-chip grind with flat raker tooth design (Kovich 1972).*

alternate top bevels, the leaving edge breakout can also be avoided. This combination will provide a finer finish with a reasonably extended tool life. However, the triple-chip design with alternate face bevel on the raker teeth will generally provide the longest tool life.

There are five more important design features to consider, when cutting board, which directly affect any tooth style used. These factors are: clearance angles,

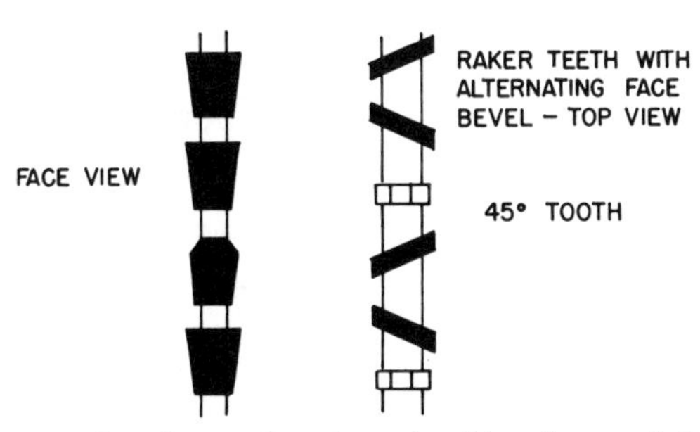

Figure 18.17. *Forty-five-degree bevel tooth with raker teeth having alternating face bevel design (Kovich 1972).*

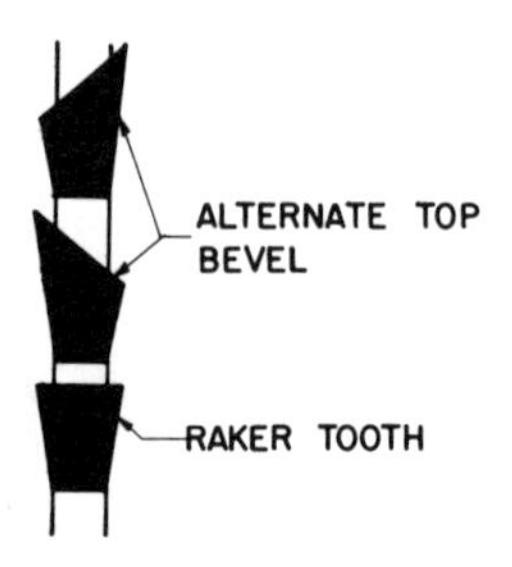

Figure 18.18. *A raker tooth added to the alternate top bevel tooth design (Kovich 1972).*

hook angles, tooth approach angle, body thickness, and tensioning. Clearance angles have an important effect on blade design. In cutting board, high clearance angles are needed due to the abrasiveness of the resin binders. Also, some wood species are more abrasive due to the higher silica content in the wood fiber. One of these, southern pine, is used extensively for producing particleboard.

The resin binder tends to build up rapidly on the cutting teeth, resulting in dulling of the cutting edge and shorter tool life. Increased tip-to-body-clearance, side or back clearance, a radial clearance, and an outside-diameter clearance (Figure 18.19) must be designed into the sawblade. These increased clearance angles must be incorporated into the design of the cutting tool to allow for the *flow-back* of the material being cut. The flow-back of the material is the degree to which a material will compress before penetration of the tooth occurs. Low-density fiberboards will compress and bend ahead of the tooth (Figure 18.20). When the bent fibers are cut, the sudden release of pressure causes the fibers to flow back in a wave on either side of the tooth. If this wave contacts the sides of the tooth because of insufficient angular clearance, frictional heat develops, not only producing drag on the blade, but also burn marks on the board.

The *hook* or *rake* angle of a cutting tooth affects the ease in which the tooth penetrates the work being cut. The greater the hook or rake angle, the less power required, and the greater the shearing action. However, tool life is shortened decidedly. The less the hook or rake angle, the more power required. Consequently, a loss of shearing action occurs but the tool life is substantially increased. There are three basic hook or rake angles to consider in sawblade selection as shown in Figure 18.21.

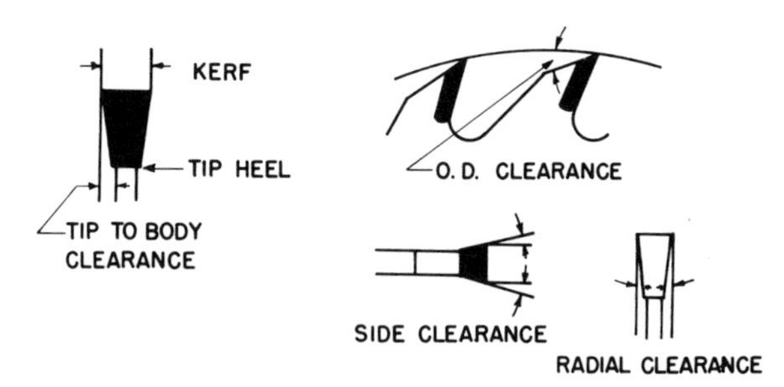

Figure 18.19. *Clearance angles considered for sawblade design (Kovich 1972).*

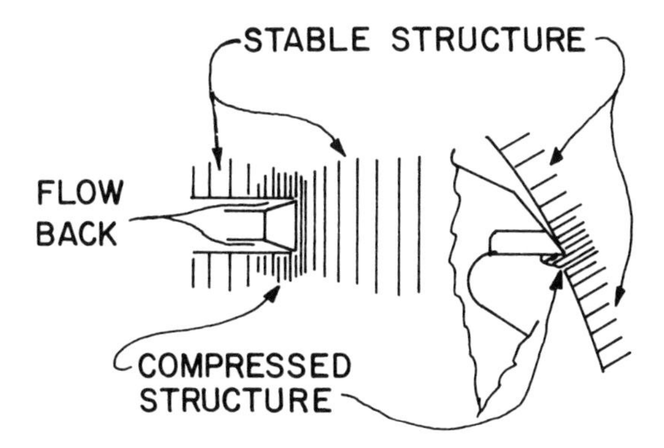

Figure 18.20. *Illustration of compression of fibers in a low-density fiberboard being cut. (Courtesy North American Products Corp.)*

To understand the importance of each hook angle, it is first important to consider the tooth approach angle that the sawblade is under in its cutting application. The relation of the blade position to the work affects the tooth approach angle. This in turn affects leaving-end breakout, edge chipping or chip-out, and thin panel cutting. There is a direct relationship between hook angles and tooth approach angles. The basic importance of tooth approach angles is in the control of cutting force direction. In thin panels, the sawblade should run low in the cut and barely project through the material being cut (Figure 18.22, Position 1). This directs the cutting force as nearly parallel to the plane of the material as possible; thus, the uncut portion acts as a support and edge chipping is controlled.

When work is fed into the saw or cutter near the centerline of the arbor (Figure 18.22, Position 2) there is more of a tendency to have a pounding effect resulting in more chipping.

On machines where climb cutting is necessary, the hook angles of the tool must be reduced so that the teeth do not grab the material and whip the sawblade across the table. The reduced hook angle will resist the tendency of the tool to cause chip-out and also resist the tendency of the tool to pull into the work. This can occur with such force that the tool will lunge forward, causing damage to both the tool and machine (Figure 18.22, Position 3).

The body thickness of a sawblade is another important factor for careful evaluation. The cutting width of a sawblade, as well as tooth form, hook angle, and size of chipload taken by each tooth, influences the cutting pressure. In general,

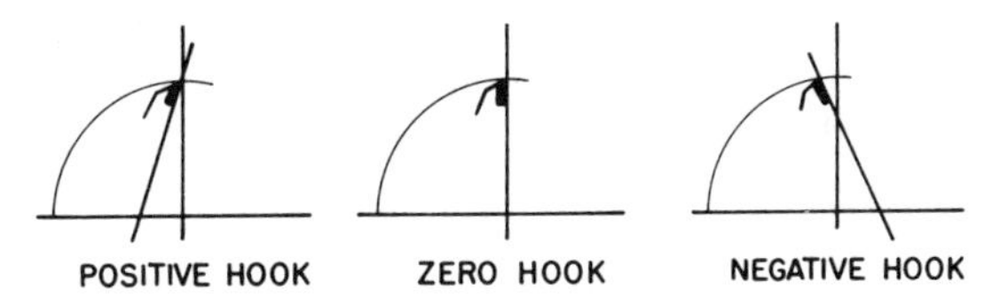

Figure 18.21. *The three basic hook or rake angles (Kovich 1972).*

Figure 18.22. *Three blade positions for cutting board (Kovich 1972).*

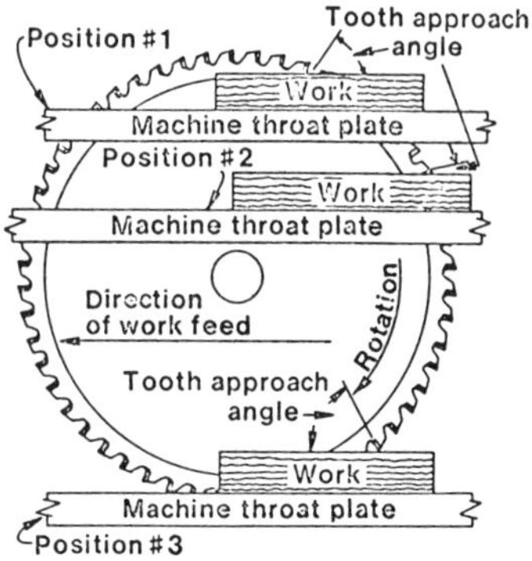

the thinner the kerf width of the saw, the lower the cutting pressure. However, as the kerf is decreased, the saw body must be made thinner and therefore becomes weaker. The sawblade body thickness is related to the diameter and, therefore, it is advisable to select the smallest-diameter blade commensurate with the thickness of the material being cut and the number of teeth and the tooth spacing required for the feed rate. For fixed-speed direct-drive saws, the rim speed of the blade and, therefore, both the aerodynamic noise and blade body vibration increase as the blade diameter is increased. On fixed-arbor machines where the blade cannot be raised or lowered, large-diameter blades should be equipped with supporting collars as large in diameter as possible. Saw collars should be of the hollow ground design and equal in diameter to provide equal support on both sides of the blade at the outer edge of the collars.

The use of a thin-rim-body sawblade is not always the answer to solving a quality or cutting pressure problem. At today's higher rpm, various sawblade applications, and increased feed rates, the stronger the saw body, the longer the saw life will be, and the more stable it will be (i.e., true running or body run-out) under full cutting load. In more operations, thin-rim sawblades will not be stable and run true when cutting particleboard. Therefore, the use of a stronger, heavier saw body is recommended. However, the use of thin-rim sawblades definitely has its application in cutting particleboard.

Proper tensioning is extremely important not only for the quality of the cut but also for blade stability and life as well. A properly tensioned sawblade for the correct rpm is very critical in controlling noise as well. The rim speed of a sawblade at 3600 rpm is approximately 180 to 200 (290–322 km) miles per hour. The centrifugal force acting on the thin-blade body is very large at such speeds and the internal stresses, due to heat treating, are very sensitive in the thin-sawblade body to such a force. During the processing of the sawblade, the body is straightened and tensioned by means of pressure rolls or saw hammers so that the steel in the blade body is uniformly stretched in a manner that will allow the blade body to retain its true circular shape at running speed.

Three or four equally spaced expansion slots, usually about one-sixth of the blade radius, are cut into the rim of the blade body at the bottom of the saw gullet as an aid to obtaining the proper straightening and tensioning of the blade body.

These slots permit the metal of the steel body in the rim area where the stresses are greatest, to be stretched in a more uniform manner.

Since the centrifugal force acting on the blade body is related to the rim speed and blade body thickness, the amount of tension must be adjusted to these two factors. Large-diameter thin-body saws require more tensioning than small-diameter thick-body saws.

Improper tensioning can cause severe vibration resonance in the blade body. This, in turn, will not only cause the blade to cut much wider than it should, but will also cause the blade to develop a great deal more noise. Sawblades that have too much tension will eventually produce cracks in the gullet and expansion slots of the blade; in severe cases, they may cause cracks in the collar area.

It is important to note that sawblade body cracks, excessive saw noise, and poor surface finish can also result from running a blade that is misaligned with the feed direction of the material. This latter condition is all too prevalent in machines where preventive maintenance is not a part of the company procedure.

Too many expansion slots can prevent the blade from running true under cutting conditions. As a result the saw will generate noise as well as yield poor-quality cuts. Therefore, it is very important for the proper number of expansion slots to be designed and engineered into a sawblade. Too many expansion slots only try to cover up problems due to improper tensioning. There are other factors that can cause a sawblade to scream. Sometimes, regardless of how much care is taken in tensioning properly, the problem can be built in at the time of manufacturing the steel blank itself (i.e., uneven cooling of the steel). To insure continued proper tensioning, the sawblade must be checked each time it comes in for servicing and maintenance.

In general, any sawblade, regardless of tooth form, can be made to cut. However, the blade may only cut for a short time before a complete breakdown, or it may leave a very poor finished cut. The reason one should understand the limitations and advantages of tooth form, hook angles, chipload, and the other factors covered here is that these basic factors can be put together into various combinations to achieve the desired results. Since the success of these combinations is dependent upon the type of machine used, work support provided, size of machine collars supporting the blade, saw tensioning, and the type, thickness, and density of the board as well as the kind of cut being made, it is obvious that it takes a combination of these factors to insure the best possible cutting results.

Recently, because of the limitations placed on saw noise by the federal government and administered under the Occupational Safety and Health Act (OSHA), the factor of noise has had to be considered with saws. It is not that noise problems did not exist before but that they have increased directly with the demands of the industry for faster production. Saw and cutting tool manufacturers have been able to do a commendable job with noise reduction on the rigid-body type of cutting tools. The development of spiral-tooth and staggered-tooth cutters in several design styles, which replace the simple, solid straight-tooth cutterheads, has provided good results. These tools are more expensive, however. More work has to be done for some of the very high-speed operations in meeting acceptable sound levels along with maintaining long tool life and appropriate surface finish requirements.

There have been many research studies conducted on the various types of sawblades used throughout the woodworking industry. This has been an extremely difficult problem to solve, particularly in meeting the proposed OSHA maximum allowable noise levels. It has been very difficult, because of the complex data developed, to come up with clear-cut solutions to circular saw noise. Great strides have taken place but much more work has to be done. There has also been a great deal of trial and error work performed by the industry in attempting to solve this relatively new problem.

It has been pointed out, however, that saws must be made to very close specifications and that manufacturing quality control must be high. In addition, the saws must be used as designed and the servicing of the saws must be done promptly and correctly.

Different types of saws have been evaluated, particularly those of the laminated type. With this sandwich-plate construction, two plates of equal thickness are laminated together with a viscoelastic material in between to make up the sawblade body. When either system undergoes flexure in the dynamic condition, the interface is subject to differential stress that converts some of the mechanical energy into strain energy which is finally dissipated as heat. This serves as a damping mechanism which suppresses some of the saw noise.

It is expected that the saw itself is operating properly in order to suppress whatever noise is being generated. Other techniques to assist in decreasing the noise level are coating the inside of the saw hood with sound-dampening material, enclosing all openings around the hold-down devices between the blade guard and the panel being cut, and using as large a saw collar as possible. A plastic polymer has also been used as a damping medium between the saw collars and the blade.

Much additional work will have to be done to meet the proposed OSHA standards as shown in Figure 18.23. A firm has been engaged to perform research on this subject. Its area of expertise is in aerodynamics and the researchers are to develop new blade designs (if possible) and to develop a system to absorb sound generated. They are to perform their research in six steps.

1. Identify the most used blade configuration, so when the noise is quieted, the most significant impact in the workplace will be possible.
2. Seek out the predominant noise sources, especially the aerodynamic ones, and attack them where they originate.
3. Use analytical noise-prediction models so they can look at their models and ask what effect they will get if they make a certain change.
4. Identify various possible noise-control methods both by making modification to blades, and by designing retrofit packages to absorb sound already generated. This phase will include cost/performance studies.
5. Prepare comprehensive design manuals to help reduce noise, predict how much reduction, and forecast the cost/performance impact.
6. Write a final report with all the technical information of the study and make it available.

Since this is a very difficult problem, it probably will take several years to solve to everyone's satisfaction.

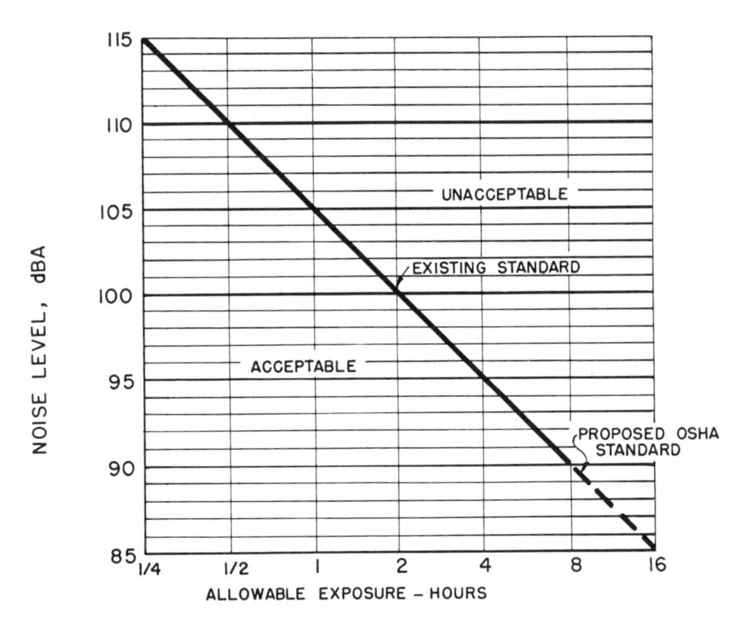

Figure 18.23. *Noise exposure limit (Crawford 1975).*

Shipping

Shipping composition board is an important facet of the industry. It is the last time the product is inspected prior to being used by the consumer. The shipping department of any plant assumes the position of being the "unseen salesman" and is a behind-the-scenes partner of the sales department. It is responsible for providing shipments of board to the customer, which are perfect in appearance, free of damage and which are easily unloaded. This calls for careful preplanning, packaging, tallying, and proper preparation of the boxcars or flatcars for loading and shipping. Truck loading is not as complicated but care must also be exercised here; otherwise the product will be damaged.

The preplanning for loading a boxcar becomes an intricate process when the multitude of small, cut-to-size pieces that can be shipped are taken into consideration. Standard sizes such as 4x8-ft (1.22x2.44 m) panels do not require extensive preplanning. However, with complex cut-to-size pieces and larger panels, detailed diagramming of the boxcar in preparation for loading must be performed. Indeed, it has been found that a single diagrammer may take up to one full working day in preparing his diagrams. Computer programs have also been developed for determining how a car should be loaded for a certain order. It is important that the final load is relatively level within the car and the units are tight together, which insures that the panels will not work loose and become damaged in transit.

The diagramming, whether done manually or by a computer, is the initial step when an order is received at a plant. This is done prior to cutting to size and unitizing, as the loading diagram for level, tight loading indicates the unit sizes and heights. In addition, consideration is given at this time to the use of cover sheets and the direction of applied bolsters to the units at the time of loading. The diagram provides direction to the car loader regarding the exact placement of the units within the car, enabling him to load quickly and with a high degree of efficiency.

The package method of shipping board provides a system of mechanically handling products throughout warehousing, loading and unloading, usually by means of a forklift. Although a cost of packaging is incurred during processing of the product, either after sanding or after cutting, this cost is substantially less than that for the hand loading of loose products into a car piece by piece. The potential damage to corners, edges, and surfaces is also much less with this system. Unit height varies due to the level loading requirements and the individual plant preferences. A common number of panels per unit as established by several particleboard producers, for example, results in a standard-size unit height ranging between 22½ and 26¼ in. (572-667 mm).

Proper strapping is used to provide the unit with sufficient strength to maintain its integrity while being loaded, throughout the transit period, and during unloading and handling in the customers' warehouses. Strapping manufacturers are able to specify proper thickness and width as well as strength of the types of strapping needed for packaging the various composition boards.

Corner protectors are used under strapping for those units that do not use wasteboard for cover sheets. Care must be exercised to insure that a corner protector is used which will prevent the strapping when under tension from damaging the edge of the board. Corner protectors are also necessary when using 1¼-in. (31.2 mm) banding to secure tiers of board during the car-loading process. Unit surface protection is provided on many board products when they are to have further coatings to protect coatings previously applied or to protect against panel surface damage caused by forks on lift trucks or other mechanical devices.

Waste board material is used when a rigid pallet is necessary to retain very small cut-to-size panels in a tight unit. Corrugated paper is used to give dust and mechanical protection to larger panels which provide their own rigidity. At times, coated slip sheets are used between individual panels of filled composition board to prevent sticking between panels during transit.

Bolsters for separation of units must be of dry material and of sufficient thickness to allow the entry and removal of forklift forks without damaging the panels of the unit. Dry Nominal 2x4-in. (51x102 mm) lumber and waste board strips are used for this purpose. Bolsters are strapped to the units by some particleboard manufacturers, providing a very tight unit with good spacing between the bolsters, which helps to prevent warpage during storage. Strapping bolsters to the units dictates that the bolster length be exactly the same as the unit width to prevent strap damage at the corners of the units. In some areas, a supply of bolster material is not sufficient to warrant cutting to exact lengths due to the waste of shorter than usable ends. Figure 18.24 shows a typical unit.

A tally or accounting of the product shipped is necessary to enable correct invoicing for an order. Several methods of tally are in use in the industry, such as the sheet tally, where a separate sheet is made out as the various products are loaded; the order tally, where the amount being loaded is indicated below that particular item on the order; and the use of tally tags. Tally tags are applied to a

Figure 18.24. *A package of particleboard showing the proper application of bolsters on the bottom of the package for efficient forklift operation (Horner 1969).*

product during manufacture and are used throughout processing. At the time of loading, a carbon copy of the tag is removed. The item is tallied on a separate sheet and the carbon copy is filed with the original order for any future reference. This is important in cases of complaints so that the processing parameters followed for producing the particular board in question can be accurately determined.

Extreme care must be exercised in preparation of boxcars for loading. All foreign material must be removed and sharp objects removed from the walls to insure that the product is fully protected from mechanical or weather damage. The ends of boxcars must be prepared for loading; since they are normally bowed and damaged, requiring custom-cut blocking for this area. Figure 18.25 shows one technique for such preparation of unlined end walls in box cars. Lined boxcars must also be prepared for loading in similar ways to those shown in Figure 18.25. The end blocking must be solid, strong lumber such as 2x4s or larger to sustain the shock loads during transit. All voids must be filled; the blocking must be vertical and of sufficient thickness to prevent the board from touching the end wall of the car. Vertical 2x4s nailed to the sidewalls are used to prevent the end unit from shifting to the wall and thus causing damage to the board at the corner of the bowed car end, as shown in Figure 18.26. Protection along the sidewalls is normally provided by nailing vertical waste board strips along the wall at the end of each tier. In some cases waste board sheets are used to completely line car walls when necessary to protect small units of cut-to-size panels or any products that are loaded tight to the wall.

The loads are placed in boxcars in several arrangements, such as: straight down the center without side blocking; staggered tiers with the ends blocked; and cars with one-half of the load at one end and on one side of the car and the other half of the load on the opposite side at the other end of the car.

Loading standard sizes straight down the center of the car without end blocking each tier has been successful and has provided a fast, economical method of

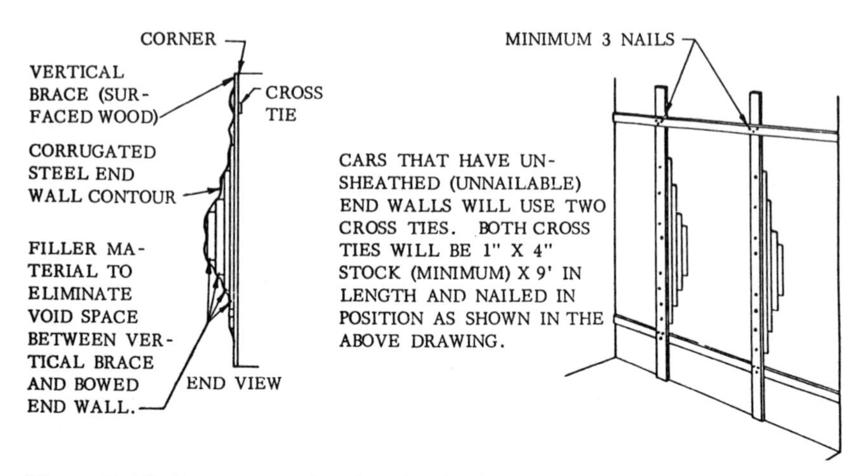

Figure 18.25. *Preparing unlined ends of a boxcar for loading of particleboard (Horner 1969).*

loading. The most important factor in this concept of loading is to have an extremely tight load which will transmit shock loads during transit from one end of the car to the other without disrupting any unit or tier in the load.

Strips of veneer or waste board should be used between tiers to prevent scuffing, or abrasion if the units should move or sway in transit. It is important to use a material for this purpose which will retain its integrity and will not crush or crumble when exposed to extreme pressures. In cars loaded with cut-to-size

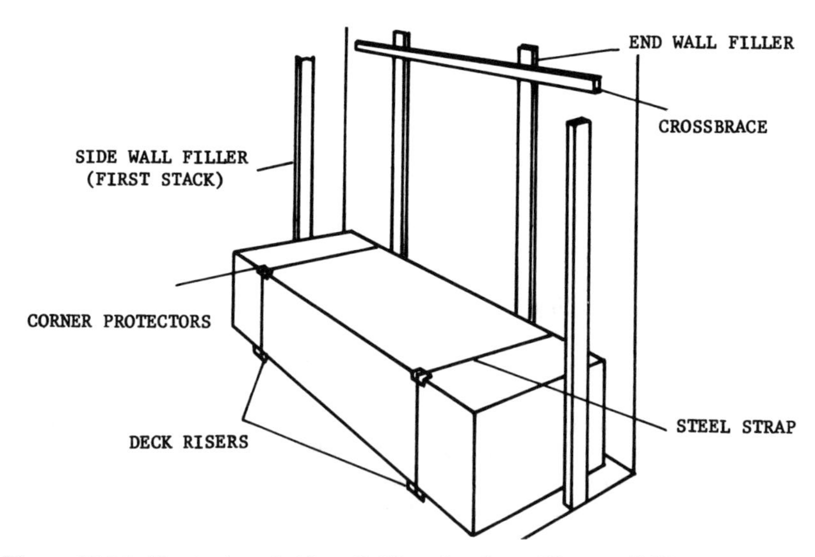

Figure 18.26. *Vertical and sidewall fillers in place (Horner 1969).*

material, high-tensile steel strapping is used to tie the tiers together at the ends of the cars and to the subsequent tiers throughout. This prevents the tiers from slipping sideways into the walls of the car.

Center blocking consists of lumber or rubber dunnage. Rubber dunnage has been used successfully. However, a large number of dunnage bags are necessary and at times they remain in the customer's warehouse and are not returned promptly. A combination of using rubber dunnage bags to align all tiers tightly and then installing lumber blocking in the remaining void has been a common practice of this industry and is illustrated in Figure 18.27. After the lumber blocking is firmly wedged into place, the rubber dunnage is removed and used again to assist in loading the next car. Thus only a few such bags are necessary and they never leave the manufacturing plant. It is important to install vertical uprights from the wedges to the ceiling of the car and to brace them across the car to prevent them from working loose, as shown in Figure 18.28. Vertical 2x4s are used between doorway tiers to give the necessary rigidity to the units and to prevent shifting (Figure 18.29). High-tensile steel strapping is used to tie all the doorway units together into one integral mass (Figures 18.30 and 18.31).

Recently, however, a new technique has been replacing both the rubber dunnage bag and lumber blocking systems. Paper air bags, which have polyvinyl liners, are used instead and are discarded by the customer when he unloads the boxcar.

When a change in the elevation of a carload is necessary, vertical pieces of lumber having minimum dimensions of 4x5 in. (102x127 mm) can be used to secure the highest units in place. Damage-free boxcars provide ideal conditions for changes in elevation, since the crossbars can be positioned to hold the top units. Steel strapping attached to the sidewalls of the car is also used to hold the

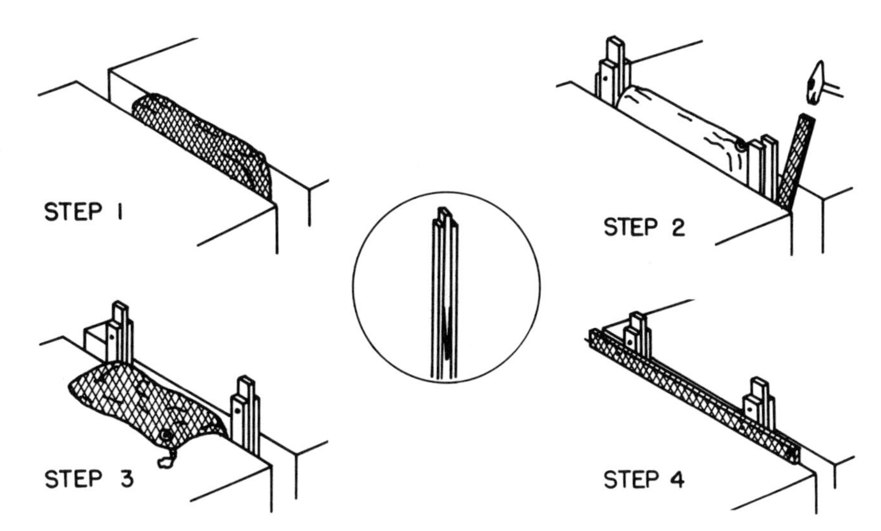

Figure 18.27. *The installation of center blocking by using rubber dunnage bags and lumber blocking (Horner 1969).*

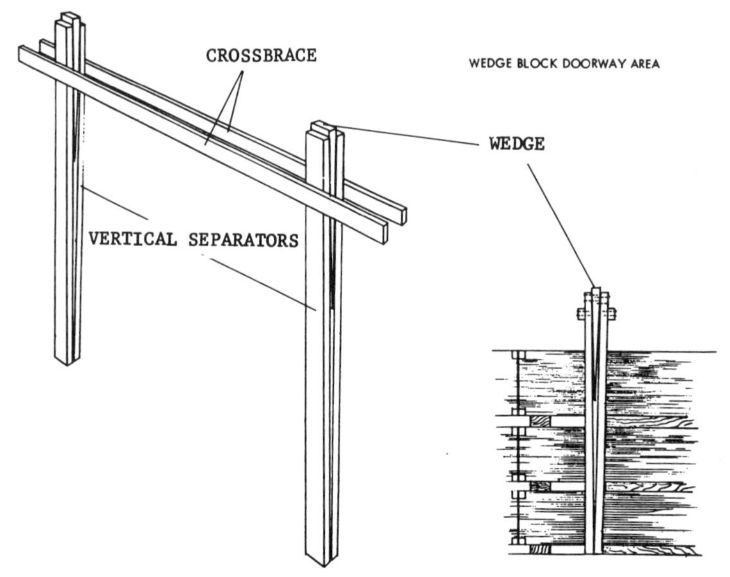

Figure 18.28. *A diagram showing the proper placement of vertical uprights (Horner 1969).*

top units in place. However, the more positive methods are vertical uprights and damage-free cars.

Many of the cars used for shipping are called *lot cars, stop-off cars,* or *cars that have two or more destinations.* The product loaded in the multiple-destination cars must be sectionalized to allow each customer to unload his portion and still have the load reach all other destinations without damage to the products. Steel strapping is used to tie together one-third or one-half portions of a car, for

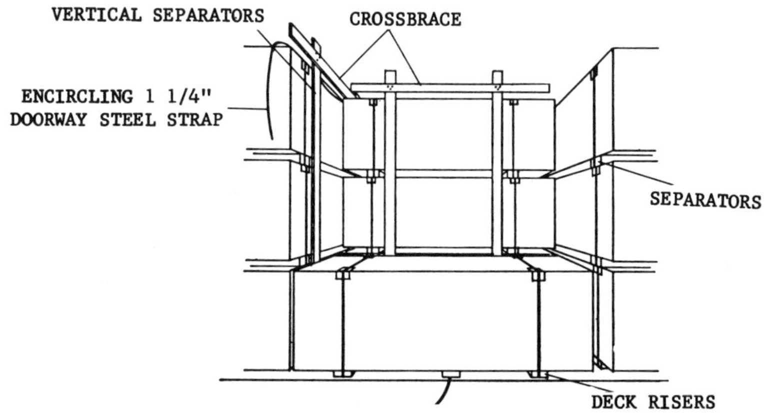

Figure 18.29. *Vertical separators between doorway tiers (Horner 1969).*

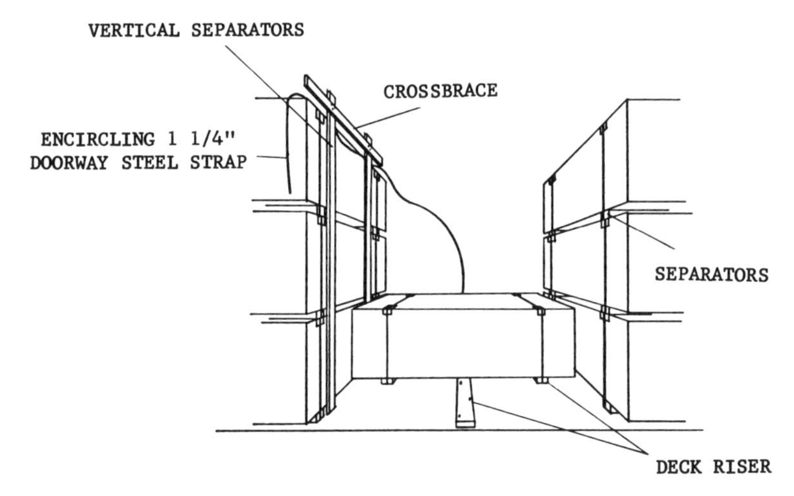

Figure 18.30. *Preparation for loading the doorway of a boxcar (Horner 1969).*

example; and, upon unloading a portion of the car, it is the customer's responsibility to reblock the car and ship it to the next destination. Damage-free cars are also used for this purpose, allowing reshipment to the next destination without a very large amount of reblocking.

Another method of separating individual lots in a car is to load the first lot the full length of the car as the bottom unit in each tier, the second lot the full length of the car as the middle unit of each tier, and the third lot as a top unit of each tier. This method forces each customer to completely unload and reload each car at his destination, which can cause handling damage to the product. However, it also provides full-length loading and reduces the hazard of having poorly blocked cars at subsequent destinations.

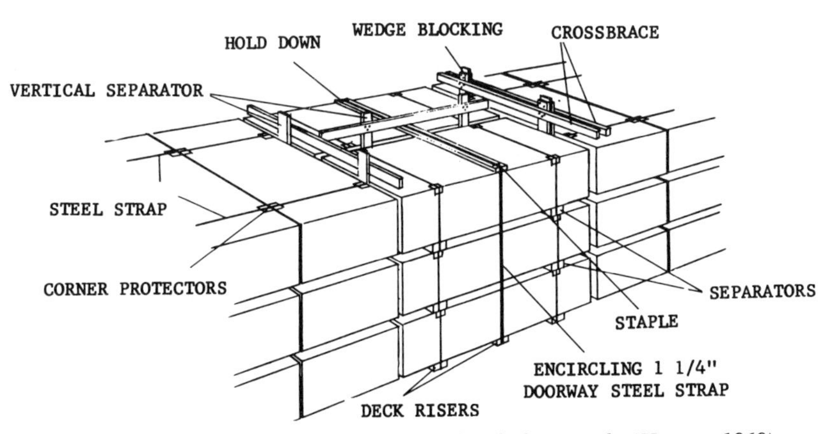

Figure 18.31. *Diagram of boxcar doorway loaded properly (Horner 1969).*

Flatcar loading of composition board is especially attractive when long lengths of 12 ft (3.66 m) or longer are shipped. The flatcars used must be sufficiently long to contain a load as in boxcar loading, and the bulkheads on the ends of the cars must be of sufficient height to support all units in the end tiers. The bulkheads are first covered with a 6-mil (0.15 mm) vinyl sheet before any units are placed on the car. Preparation of a flatcar for loading is similar to that of a boxcar in that nails, sharp objects, and other foreign objects must be removed.

Following the loading of the units on the flatcar, the bottom units across the car are strapped to the center units and, in turn, the center units are strapped to the top units of the load, producing one solid package of board across the length and height of the load. The center block is applied as in boxcar loading, and the entire load is covered with another 6-mil vinyl sheet. This vinyl sheet and the vinyl sheet on the bulkheads are then joined together at the end of the car using a wraparound method on two 2x4s. Then the 2x4s are nailed to the bulkhead, and a similar installation is made along the bottom edge of the length of the load. All joints, seams, or holes in the vinyl are taped to prevent air from entering and causing ballooning of the vinyl in transit, which can cause tears and removal of the covering.

Hazards of wind and mechanical damage, as well as vandalism, cause transient removal of vinyl in flatcar shipping. Of course at any time that the covering is damaged, the product is subjected to whatever the weather conditions are at the particular time of transit.

Truck shipments of board are popular, both for short local hauls and for long cross-country hauls when shorter transit times are important. Such hauling provides for more convenient unloadings, especially with small consumers of board who do not have railroad facilities of their own. The trucking firm normally supplies the cables, corner protectors, and tarpaulins as needed to hold the load in place and protect it from the elements.

SECONDARY FINISHING

Composition board products can be sold in the raw state, primed or completely finished with some type of overlay and edge finish. Surface finishing ranges in quality from a simple paint finish to high-density plastic laminates. Edge finishing is necessary for most particleboard furniture-type applications; while, as already mentioned, one of the virtues of the medium-density fiberboard is the minimizing or complete elimination of the edge treatment. Tempering is practiced with thin hardboard by dipping it in oil and then heating in ovens to polymerize the oil. Oil-tempered hardboard consequently has excellent dimensional stability and resistance to weathering and the pickup of water. It is also possible to apply the oil to the fibers during the board-making process and to polymerize the oil during hot pressing. Oil tempering, however, will not be dealt with in this discussion, as surface and edge treatment, not complete treating of the panel, is the subject.

Surface and edge finishing or overlaying panels improves their appearance and properties, resulting in high-value products. Overlays can be printed with various wood grains or other designs for top-quality applications. Some overlays, such as those of resin-treated fiber, serve as a substrate for providing an excellent paint finish. It is possible to eliminate painting by using pigmented acrylic films combined with resin-treated paper. Resin-treated paper overlays enable the use of some panel products for exterior applications, because the faces are protected. Particleboard, for example, tends to surface-roughen even though an excellent paint film is applied. Such roughening can provide a decorative effect; but, if a smooth surface is desired, overlays eliminate this particular problem.

In addition to providing satisfactory appearance qualities, protection for the substrate and, in some cases, upgrading low-quality surfaces, overlays offer a means for providing a number of specific characteristics beneficial for various end uses. Durable, hard, abrasion-resistant, and smooth surfaces can be developed. In the extreme cases, for very high-quality faces, fiber glass or metal can be used for the overlay.

Figure 18.32 from Guss (1975) shows the relative amount of the various types of finishing done in 1972 and the predicted amounts for 1977. These pie charts show the dominance of laminated-type finishes.

Surface finishes, classified as wet or laminated, consist of a variety of different types of paint, wood veneer, high-pressure decorative laminates, plasticized polyvinyl chloride film or foil, unplasticized polyvinyl film, melamine-resin-saturated paper, polyester-saturated paper (the last two are both low-pressure systems), resin-treated fiber overlays, polyvinyl fluoride film, vulcanized fiber, fiber glass reinforced plastic overlays, and metal. A variety of techniques is used for applying these finishes. Some of the pieces of equipment used are roller coaters, curtain coaters, roll printers, roll laminators, platen presses (hot and cold), pinch or nip rollers, postformers, and vacuum formers. Various means of curing liquid finishes, including ovens, ultraviolet light (UV), and irradiation are used. Microwaves, high frequency, and plasma arc are also suggested for curing purposes.

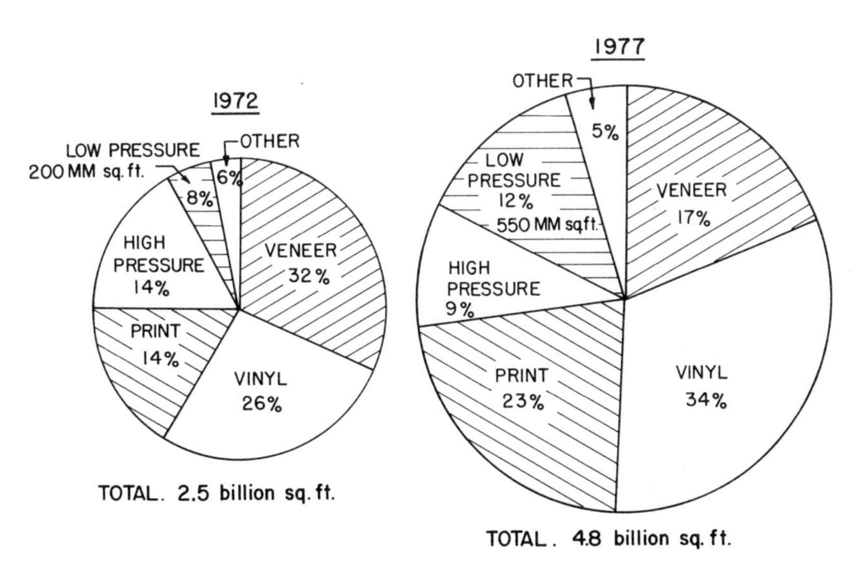

Figure 18.32. *Finishes used over particleboard (Guss 1975).*

An interesting surface coating originally designed for use with particleboard floor underlayment is a hot-melt wax-polymer. This coating provides a moisture barrier that improves the underlayment quality. However, a good market for this product is industrial shelving, where the wax finish makes it easy for workers to slide boxes.

Wet Finishing

The basic requirements for wet finishing composition board consist of three elements: coating equipment, drying equipment, and auxiliary machinery. The coating equipment can apply the finish by spray coating, curtain coating, direct-roll coating, knife coating, or reverse-roll coating. The drying ovens can operate as convection ovens using hot air, infrared ovens that are gas fired or electric, and the newly developed ultraviolet radiation-type drying equipment. The auxiliary equipment consists of panel feeders, brushes, sanders, burnishers, stackers, and conveying means for transporting the board panels through the finishing process. The importance of these auxiliary elements, particularly sanders and burnishers, for proper substrate preparation prior to coating cannot be ignored. Various types of finishing schemes for particleboard are shown in Figure 18.33. Many modern finishing lines now use ultraviolet for curing the filler and topcoat.

Face finish material now is far more sophisticated than that of just a few years ago. It ranges over solvent-free liquids, chemically cross-linked resins, water-base finishes, and irradiation-cured polymer systems. It is anticipated that greater advances are to come through research and development by the suppliers.

High-density hardboard can be finished directly with wet finishes, because of its dense, smooth surface. However, particleboard and medium-density fiberboard will normally need some type of preparation for final finishing, because of the voids between particles on the surfaces. Recently, more very fine material has been used on the faces of the boards that will receive high-quality finishes. This reduces the need for filling. However, with the wet finishes, some type of filling will usually have to be done. Boards over 45 lbs/ft^3 density (0.72 sp gr) are recommended.

A typical wet system for treating particleboard or fiberboard is: first fill the surface; then apply a base coat; print on the highlights of a pattern with a gravure press; print on the final design; and apply a top coat.

The filler is the foundation of the coating system. In addition to uniforming the surface, it must have excellent adhesion, be hard and tough, and hold out the subsequent coats so that a deep, full appearance can be obtained.

Solvent-based fillers used include lacquer, vinyl, water emulsion, alkyd, polyurethane, and urea-alkyd. Lacquer or vinyl-based fillers have good adhesion to the substrate, excellent sanding, very fast drying by conventional oven systems, excellent recoating characteristics, and they may be pigmented to give excellent hiding properties. They have a lower volume solids, however, resulting in lower filling action, higher cost, and only fair chemical resistance.

Water-emulsion fillers have high-volume solids, good adhesion to most substrates, excellent sanding, relatively fast drying with conventional oven heating, good chemical resistance, and they may be pigmented for hiding. They are usually

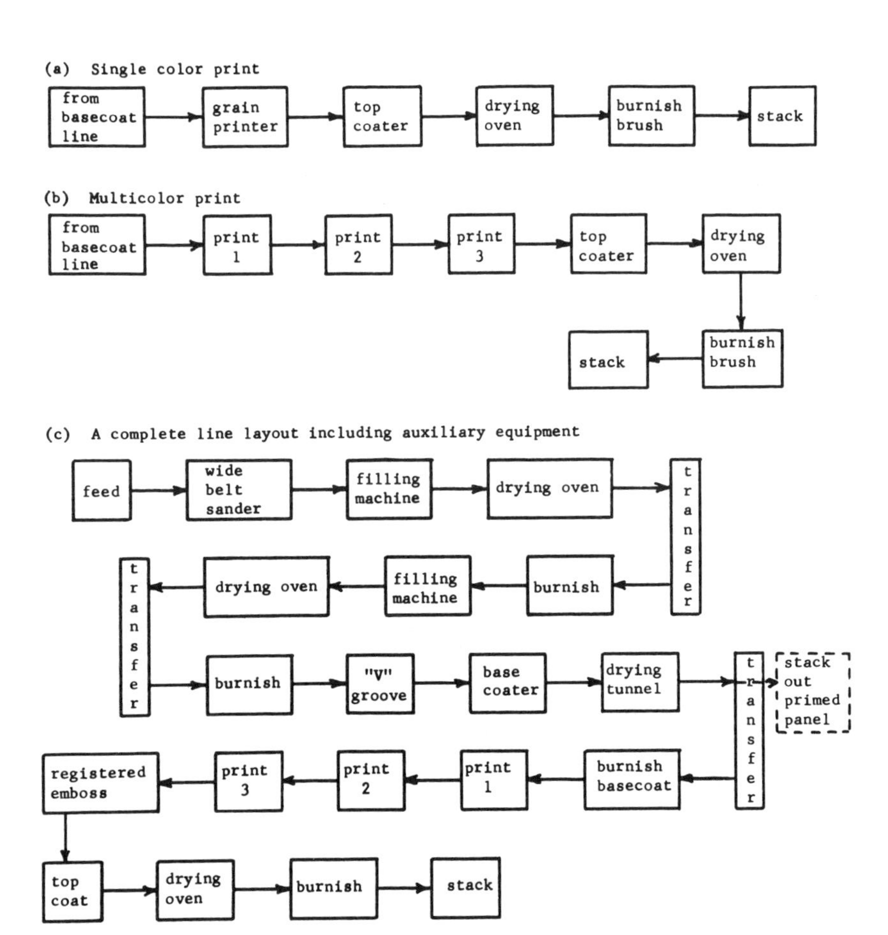

Figure 18.33. *Typical particleboard grain printing (Knapp 1968).*

sanded to smooth the surface before the next finishing coats are applied. They should receive at least one coat of a solvent-based material (e.g., basecoat, sealer, or topcoat) which will act as a moisture-vapor barrier. This will control flake swelling during the service life of the panel. Water-based fillers are a satisfactory means of reducing the air-polluting effluents on existing finishing lines.

Polyurethane and alkyd-based fillers have a high-volume solids level, good adhesion to most substrates, good sanding after curing, and excellent recoating characteristics. They have the disadvantage of relatively slow drying speeds.

Urea-alkyd fillers have high-volume solids, good adhesion, good sanding and recoating characteristics, and fast drying. However, they have limited pot life.

In the past few years, new fillers have been developed, based on photochemically reactive resins, primarily polyesters and acrylates. In place of a volatile solvent, they use reactive monomers (e.g., styrene, vinyl toluene, ethyl acrylate, methyl methacrylate) which become a part of the filler during curing.

A number of benefits are obtained with these ultraviolet light-curable systems:

1. Cure time is reduced by one-half or more compared with alkyd or polyurethane fillers.
2. Immediately out of the oven, the boards can be sanded to give a smooth, uniform surface suitable for printing.
3. Since there is no loss of volatile material, film shrinkage is much less than with conventional fillers. One coat of UV-curable filler will usually give a completely smooth surface.
4. Cost per square foot of polyester filler compares very favorably with alkyd and polyurethane fillers.
5. Since there is virtually no loss of volatile material to the atmosphere during the curing process, air-pollution problems are greatly reduced.

The UV energy is normally supplied by multiple groups of high-pressure mercury-vapor lamps. Each lamp consists of a quartz tube containing small amounts of liquid mercury. Upon striking an arc, the liquid mercury is vaporized by a rapid increase in temperature within the tube. The mercury vapor then emits specific wavelengths in the ultraviolet region, and this radiation provides sufficient energy to cause a rapid reaction of the photoinitiators in the coatings. The lamps also emit energy in the infrared region of the radiation spectrum, thus producing heat and speeding the reaction of the polyesters.

Ultraviolet radiation is hazardous, and direct exposure to the lamps can cause severe skin burns and eye damage. Thus, the units are shielded and equipped with electrical interlocks to prevent accidental opening while the lamps are burning. These units do have a good safety record.

There are two types of filling machines in use. They are the bottom-knife coater and the reverse-roller coater (Figures 18.34 and 18.35). The coating is applied to the bottom of the board in knife coatings. The filler is contained in a reservoir and is transferred to the board by a roller. The excess of coating on the surface of the board is removed by the knife, and this material flows back into the reservoir.

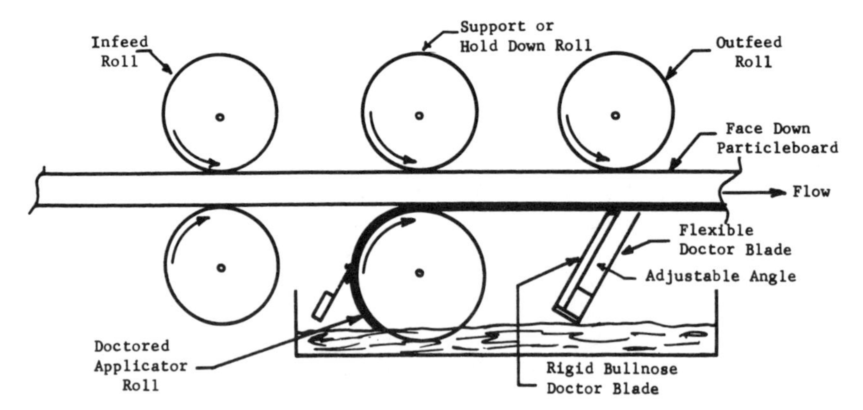

Figure 18.34. *Schematic drawing of a knife coater (Knapp 1968).*

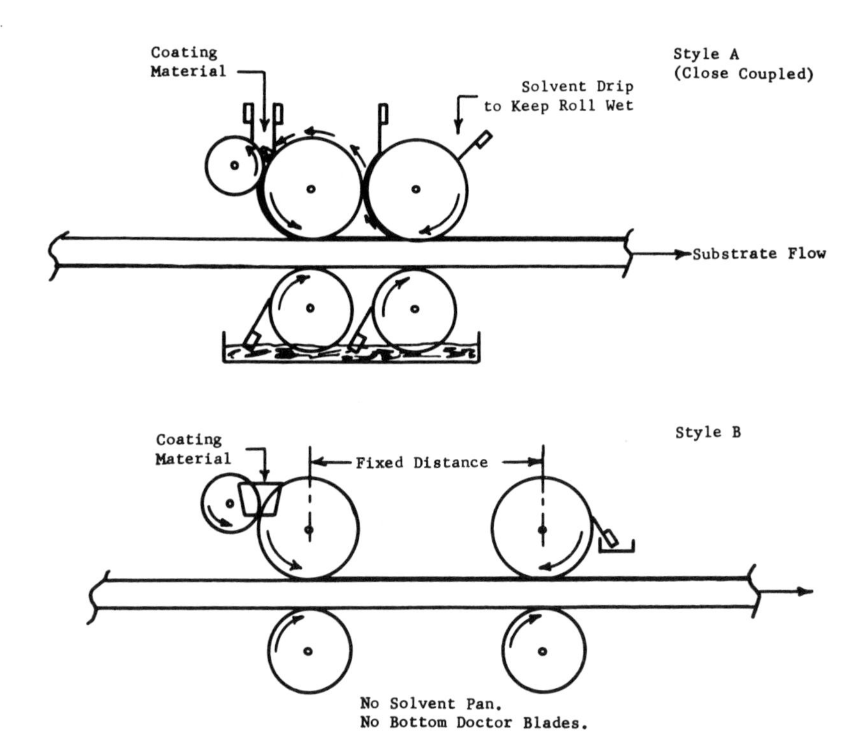

Figure 18.35. *Schematic drawings of two styles of reverse coaters (Knapp 1968).*

Normally, a knife coater is operated at comparatively low throughput speeds and with a high knife pressure. Because the knife is flexible enough to follow small variations in the board surface, there is very little surface buildup of the coating, while complete filling of the voids is achieved.

The upside-down operation makes it a little difficult for in-line inspection of the filling operation. Also, the possibility of damage to the filled surface is much greater than with the top side coating; so careful cleaning of conveyors and transfers is essential. Nevertheless, knife coaters are very effective where good filling with minimal surface build of the coating is required.

The reverse-roller coater is much more commonly used. The first section is essentially a direct-roller coater, followed by a doctor or metering roll. Support rolls are below the upper rolls (Figure 18.35). A controlled excess of coating is applied to the board by the application roll. The wiping roll is adjusted so that there is a snug fit between it and the pressure roll. As the board passes under the wiping roll, excess material is wiped from the surface and is removed from the roll by the scraper blade. The amount of material left on the surface can be controlled by varying the relative speeds of the application and reverse-wiping rolls. The faster the wiping roll speed, the less the amount of material left on the surface.

In newer machines the two rolls are closely coupled, so that the excess material is transferred directly from the wiping roll back to the application roll. There is essentially no change in the coating composition from the beginning of the run to the end of the run, and there is no concentration of contaminants.

After filling and curing, the fillers are often sanded with a fine-grit belt or a flap-wheel unit, which is a slotted aluminum hub into which coated abrasive segments are loaded. These flaps of sandpaper can be used as is, or they can be partially cut into narrow strips which provide extra flexibility. These flaps or strips then whip over the panel during sanding and are particularly effective in obtaining abrasive contact with out-of-level surfaces. The panels are then brushed or blown clean and proceed to the basecoating step. Some board plants ship filled panels for final finishing by the consumer.

The basecoat or ground coat is then applied over the filled surface with a direct-roller coater, precision coater, or curtain coater. A curtain coater is shown in Figure 18.36. Precision coaters are similar to direct-roller coaters, but they can also control the texture of the coating.

Basecoats are high-solids-pigmented materials that form an opaque film. These are usually lacquer- or vinyl-based, though water-reduced basecoats are finding increased use. It should be mentioned that such coating can also be done on paper-overlaid board, veneer-overlayed particleboard, low-pressure-laminated particleboard, or composition board that does not require filling. Several coats of the basecoat may be applied. The basecoat is cured in an oven before passing through a buffer which can be a sander or a nylon mesh roll impregnated with abrasive grains. This scuff sanding can improve the quality of the print by removing high spots. All dust must be removed from the panel after buffing.

The basecoat supplies the initial undertone color for the final finish. A grain print is then applied by engraved rollers. The printing can be done in several colors with each print keyed in register to introduce an interplay of color and create the illusion of depth, wood texture, and figure highlighting. The engraved rollers are usually steel with a copper engraving which is chrome-plated to resist scraping wear. Its design is based on the principle of using many small, etched wells in various depths to hold and deposit ink. Deep wells produce dark tones and shallow wells, light tones. Each additional color requires a separate cylinder. Triple printing is necessary for high-quality prints. Machines may run with a number of cylinders, or they may be ganged up to permit multicolor to be printed in a single pass. Combination top and bottom printers are available, which print grain on

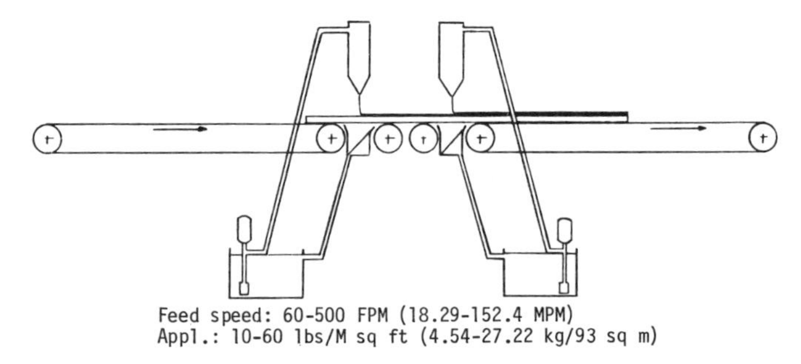

Figure 18.36. *Double-head curtain coater (NPA Bulletin 10).*

Figure 18.37. *An embossing roll. (Courtesy Washington State Univ.)*

both sides of the board with their cylinders synchronized to make the grain appear to show completely through. The surface can be embossed to accentuate the grain design. An embossing roll is shown in Figure 18.37. Such a roll also is used to apply ink at the bottom of the grain impressions simultaneously with the embossing. This is followed by the normal printing.

Figure 18.38 shows a one-roll printer. The ink container *(A)* keeps ink at a constant level. The rubber application roll *(B)* covers the engraved print roll *(D)* uniformly with ink. An oscillating doctor blade *(C)* removes the excess ink and so ink remains only in the depressions of the print roll. The rubber-covered transfer roll *(E)* picks up the ink in pattern form and transfers it to the panel.

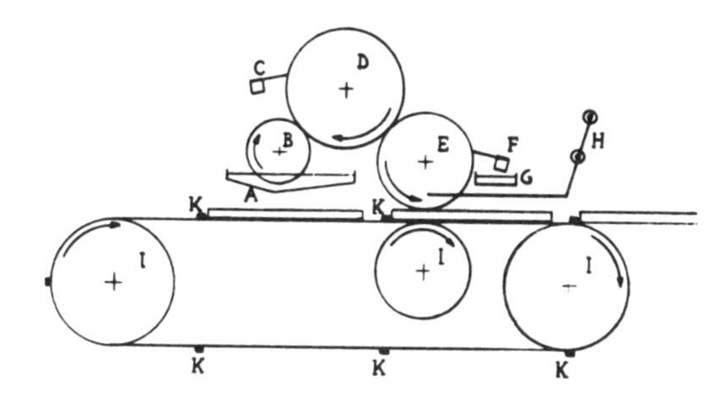

Feed speed: 30-150 FPM
(9.14-45.72 MPM)
Appl.:<1 lb/M sq ft (0.45 kg/93 sq m)

Figure 18.38. *One-roll printing machine (NPA Bulletin 10).*

An open conveyor can be used to insure that the drying-type inks are dry before sealing. Nondrying inks must pass through a dryer. The sealer or topcoat is then applied by roller coater, precision coater, or curtain coater. A sealer is used to protect the print of panels for furniture manufacture, where additional finishing steps are to be applied. Sealers are usually nitrocellulose- or vinyl-based. If the completed composition board panel is used as is, generally a heavier coating of a more abrasion- and stain-resistant topcoat is applied as the last step. Alkyd-ureas are most frequently used, combining high build and abrasion resistance, rapid cure response, good chemical resistance, and low cost. Nitrocellulose, catalyzed nitrocellulose, vinyl, and cellulose acetate butyrate (CAB) are also used.

Fairly recently, UV-curable polyesters and acrylics have been introduced, utilizing similar technology and curing systems to the UV-curable fillers. Another recent addition is the water-reducible topcoats based on acrylic or polyester resins, cross-linked with urea- or melamine-formaldehyde resins. Both UV-curable and water-reducible topcoats will see increased use in the future as the industry moves to improve its environmental impact.

Laminating

When it comes to laminates for overlaying board substrates, there are two types: self-bonding and adhesive-bonded. Adhesive-bonded laminates include wood veneer, the familiar high-pressure plastic laminate used for many years, paper with precondensed amino and phenolic resins, film underlays with phenolic resin or amino resin impregnation and coatings, vulcanized fiber with amino resin coatings, and thermal plastic films such as polyvinyl chloride. The self-bonding laminates are newer and include melamine and polyester resin-impregnated paper. Diathyl-phthalate-(DAP) impregnated paper is another but is not used in the United States.

For successful laminating with thin films, it is important to have very fine particles on the faces of the board panel. Very fine material, previously thought to be a resin hog in board manufacture, is being used more and more for achieving smooth surfaces. The smooth surfaces, as mentioned, are needed for preventing show-through of irregularities through the thin surface in papers. It appears that more attention must also be paid to the particle size and geometry in the board core to insure a high-quality end product. If a high gloss on a film is used, irregularities are highlighted. However, most of the thin films are matte, satin or "furniture" finished, which tends to diminish this problem. Particleboard with large particles on the faces needs some type of "barrier," "cushion," or underlayment sheet underneath the film surfaces, such as phenolic- or polyester-impregnated paper (kraft). The use of panels without faces of fine particles is highly unusual in the United States.

A very low moisture content of the board can cause difficulties when the board and surface film are combined. Too high a moisture content causes stains due to the increased vaporization and also runs the risk of a poorly closed surface. Moreover, such a condition can cause overlay "blisters" and an undue shrinkage in thickness. The optimum moisture content for good bonding ranges between 6 and 7.5%.

Dense panel faces can give problems with adhesion to the films. Poorly cured adhesive joints can let go because of high tensile and shearing forces developed in the laminate. Because most particleboard has a low surface density on the outermost faces, problems can occur because of the lower strength associated with lower density. It has been found that 0.8 to 0.9 sp gr on the panel surface appears satisfactory. Thus, low-density faces must be sanded off, particularly if this has not been accomplished on the regular production line. These faces were mentioned in the discussion of density profile. The density of board for conventional laminating normally exceeds 45 lbs/ft^3 (0.72 sp gr). Boards of lower densities have reportedly been used in short-cycle laminating but this is unusual in the United States.

Variations in specific gravity with thin boards are superimposed upon stresses generated by the laminating pressure. These stresses can become of such magnitude as the board shrinks and swells, due to changes in humidity, that stress cracks form in the laminates. Thin boards surfaced with melamine-resin overlays, ultraviolet-cured polyester underlays, and hard urea resin coatings are particularly susceptible.

It is also possible to laminate some overlays onto particleboard during the actual pressing of the particleboard. This is becoming of more importance with the thinner boards that are now being produced, particularly the continuous type where surface layers can be rolled upon the formed mat as it enters the hot press. In this case, almost perfect forming of the mat is necessary to keep a uniform density over the entire area of the final board. Let us now turn to the major types of overlays and their applications.

Wood Veneer: High-quality wood veneer overlays, of course, are a premium product as far as quality is considered in the marketplace. Particleboard and medium-density fiberboard have become the premier substrates for use with quality wood veneer. In fact, core-type hardwood plywood, which is a major product, consists of the hardwood veneer over the particleboard or fiberboard substrate. Most of the volume of the material is composition board, not hardwood plywood.

Wood veneer is also used for crossbands over the composition board substrate before high-quality veneer is applied to reduce telegraphing problems. Wood veneers such as tupelo, maple, and birch are laminated onto board to provide a base for subsequent printing or film overlaying. Veneers are applied to the substrate by conventional hot- or cold-pressing means developed over many years, and much experience has been gained in this type of laminating. Care must be taken to equilibrate the veneer and substrate to about 7% moisture content before gluing to minimize checking problems. Telegraphing is a problem with veneer laminating; consequently, exposing the substrate to high humidities before gluing and the use of glues with high water content should be avoided.

High-Pressure Decorative Laminates: High-pressure decorative laminates have been around for many years. They consist of printed papers which have been saturated with thermosetting resin, generally of the melamine-formaldehyde type and bonded to multiple layers of phenolic-impregnated kraft paper under high heat and pressure. This is a high-cost product with excellent impact and abrasion resistance, good chemical resistance, and durability. It should be used where the high-performance properties can be best utilized. These uses include tabletops and countertops and other horizontal surfaces on furniture, kitchen cabinets, and on bathroom walls.

The pattern sheet of the composite is a high-quality pigmented alpha-cellulose paper, which is printed generally in up to four additional colors for optimum appearance quality. The sheet is impregnated with melamine resin and a water-alcohol solution and then dried. A clean alpha-cellulose tissue highly impregnated with melamine is assembled over the dried patterned sheet, and these in turn are placed over several sheets of phenolic-impregnated kraft paper. The assembly is then bonded and cured in a press operating between about 250° and 280°F (120°–138°C). Pressing pressure ranges between 1100 and 1400 psi (7.58–9.65 MPa) for periods varying from 45 to 60 minutes and longer, depending on the type of material being made. These laminates must be cooled in the press and the cooling time may be as long as the heating period. The resultant high-pressure laminate is in turn bonded to the wood substrate using contact-type adhesives, polyvinyl acetate (white glues), hot melts, urea-formaldehyde adhesives, casein, etc., for interior or protected applications and resorcinol or similar waterproof adhesives for exterior use. Where high laminating pressures under heat are required, phenolic-bonded high-density board is normally used.

For interior-type applications, the contact-type adhesives offer an advantage which the others do not in that they are suitable for on-site installation and do not require clamping or platen pressing. They provide the most versatile means of installing decorative laminates and have been one of the major reasons for the use of laminates. These adhesives have also been the cause of many problems when they have been used improperly.

Polyvinyl Chloride: One of the best-known overlay materials is polyvinyl chloride (PVC) film. There are two types: plasticized film and rigid. When applied in a plant, the film is usually in a pure form and is bonded to the substrate with an adhesive in a roll-laminating process. This film can be printed and embossed. Prints on the outside of the film can wear away in areas receiving severe abrasion. This has been overcome by printing the pattern on the reverse side of the film, which is placed next to the substrate. The back print is then covered with a ground coat or color coat which masks the substrate and adhesive. With wood grain printed pattern, the back printing technique has a degree of realism very difficult to attain otherwise. It is also possible to put a clear protective film over any of the films where face printing is desired, and this is a common practice.

It is reported that polyvinyl fluoride overlaid PVC film is nearly equal to ceramic tile in stain-resistance and cleanability. This two-sheet laminated overlay is valuable for the manufacturer of vinyl-clad panels which are folded for cabinet or furniture parts. This system is used for making finished products of square, rectangular, or polyhedral shape. The first step consists of lamination of the board with the flexible plastic sheet. Then accurate miter-grooves are cut through the substrate in the desired locations. Adhesive is then metered into the grooves and the panel is folded into the desired shape, using the flexible plastic sheet as hinges. Figure 18.39 shows some typical configurations that are possible. When back-printed film is used, cutting into the film results in an undesirable and distinct white line. But, by using face-printed film (usually with a protective film on top), overcutting to a considerable depth is possible before it is visible from the face.

The back- or reverse-printed film is normally about 3 mils (0.08 mm) thick. The volume market for the face-printed film, which has a clear film laminated to the top, calls for 6-mil (0.15 mm) material (commonly 4-mil [0.10 mm] film for printing with a 2-mil [0.05 mm] overlay or a 3-mil [0.08 mm] film for printing with a 3-mil overlay).

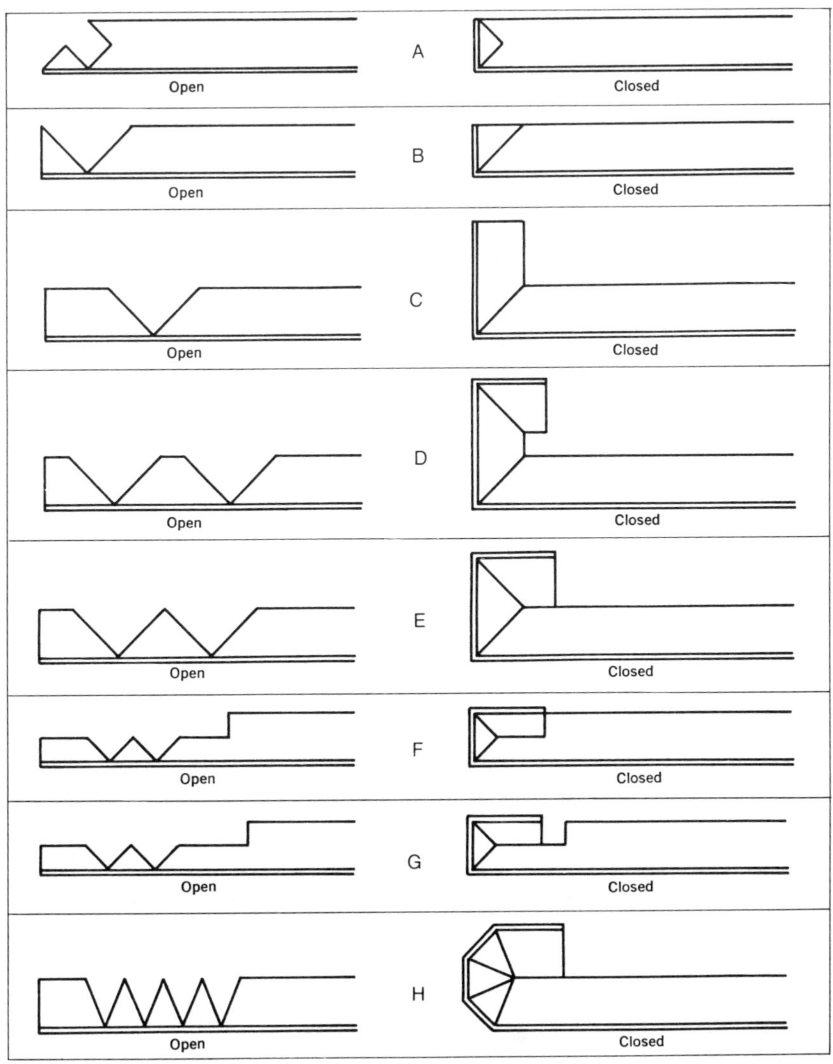

Figure 18.39. *V-grooving process permits many simple as well as complex forms to be accomplished in uncomplicated grooving, folding, and gluing procedures (NPA Bulletin 13).*

Adhesives are necessary for bonding vinyl film to wood-based panels and generally are of two classes: water-based (polyvinyl acetate) and solvent (polyester or polyurethane) which are applied to the substrate. Epoxy is being used more and more and has the advantage of being 100% solids, thus, there is not any solvent to evaporate. The solvent- based systems are two-part materials and are considered to produce the maximum overlay bond strength. The water systems are usually thermoplastic and are the types most commonly used. Whichever system is used, the adhesive must be resistant to the plasticizers that can migrate from the PVC film. The bond of the film to the panel should be sufficiently strong to resist film creep or shrinking away from the panel edges.

Application equipment for this system can be relatively simple (Figure 18.40). The essential parts consist of rolls to tension the vinyl sheet, a heating unit to soften the film and make it easier to handle, an adhesive applicator roll to apply the adhesive to the film, a heater to flash off the water and also soften the film, and nip or pressure rolls to provide intimate contact between the film and the panel substrate. In some cases, the adhesives may be applied to the panel substrate rather than to the film. Folding and tucking rolls may be used for the edge wrapping of the panel, as pointed out earlier. An embossing roll is normally used to emboss a realistic wood grain effect. Figure 18.41 shows a laminated panel leaving the press.

Vinyl films can also be applied to wood substrates, particularly those with three-dimensional characteristics, by vacuum lamination. The substrate material should be between 6 and 10% in moisture content. It is first shaped for profile to desired contours. The curves and corners are rounded, avoiding 90-degree angles.

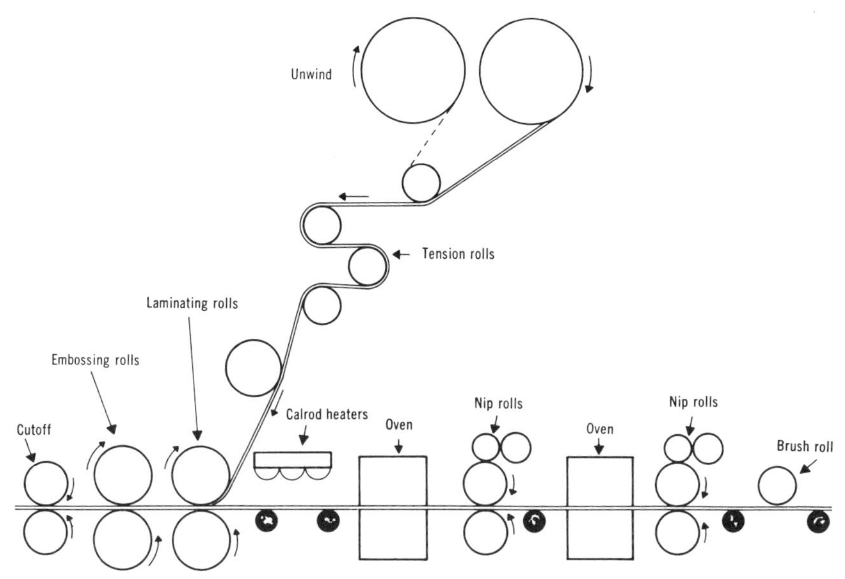

Figure 18.40. *A typical direct-roll coating system for application of vinyl film to particleboard (*Using Particleboard *1972).*

Figure 18.41. *A vinyl overlaid panel leaving the press (Wentworth 1972).*

Next, the part is spray-coated with a thermosetting resin. Only one coat of adhesive is necessary on dense substrate surfaces, whereas two coats may be necessary on low-density surfaces and edges. A part or a number of parts are then placed on the platen of a vacuum-forming machine. This platen is perforated with holes. Next, the vinyl film is placed in position above the workpiece and clamped into position. The film is then heated from above, while air is forced up from below, relaxing the film. The air trapped under the film is then evacuated to a reserve tank below the assembly, creating a vacuum below the film. Next, the film is drawn down to cover the part's surface and edges. More heat is then applied to the film until the glue line is activated. The heat is then removed and the part cooled while still under vacuum. Cycle times for this system are about four to five minutes in duration. The plasticized PVC film used on wood substrates is restricted to interior applications such as for wall paneling and furniture.

Fabric-backed or supported vinyl film, which is used in field applications, can be used for special cases in a plant because the greater thickness lends itself to deep embossing. Textured wood pattern is one example.

Unplasticized polyvinyl chloride film (rigid PVC) is formulated without a plasticizer and is normally thinner than the plasticized material. Thicknesses of 10 mils (0.25 mm) or more are known; however, 1.7- to 1.8-mil (0.04–0.05 mm) film is the volume market. The material is roll-laminated to board substrates. It is reported that a great market potential for this material is in residential siding products, where the chemical resistance, color durability, and low maintenance qualities of the exterior weatherable PVC can be utilized.

Low-Pressure Decorative Overlays: Of great importance in the industry today are the low-pressure decorative overlays. The first was diallyl phthalate (DAP) which was first developed (in the United States) about 1960. By 1964, an overlay material had been developed in straight polyester or melamine systems or combinations of the two.

Polyesters were most popular through the 1960s. Later on, European melamine-resin manufacturers came out with hot press and treating equipment packages for producing melamine-faced board panels in short cycles. A single-opening, fast-cycle system for 5x18-ft (1.52x5.49 m) panels was introduced in the United States around 1970, and several of these plants are operating or about to start operations. One will produce 8x24-ft (2.44x7.32 m) overlaid particleboards in one minute or less. These overlays can contain a decorative pattern or they can serve as a paint substrate.

Before pressing, polyester overlays are flexible and virtually tack free. Melamines are also quite tack free but are more brittle and may fracture if not handled carefully. Polyesters have a longer shelf life and can be successfully stored for six months or more at 65° to 70°F (18°–21°C). Melamines, however, slowly become more brittle and should be used within 30 to 60 days provided they are kept at 65° to 70°F (18°–21°C). Less advanced resins can be used in melamines to prolong their storage life in cases such as overseas shipments. Longer press times, however, are required.

The polyester-type overlay is alpha-cellulose paper treated with polyester resins ranging in resin content from about 52 to 62%. Most of the pressing at present is done in multi-opening presses at specific pressures of about 150 to 250 psi (1.03–1.72 MPa) and temperatures between 270° and 350°F (132°–177°C) at the board surface. In overlaying with both polyesters and melamines, press pads can be used between the press platens and pressing plates which are in contact with the overlays. West Coast U.S. plants, however, are mostly operating without press pads. Press or cushion pads tend to distribute the pressure uniformly over the panel being laminated and are particularly useful with panels of uneven dimension. Release papers are normally used to prevent sticking to the press platens and to provide the type of finish desired (smooth or textured). Pressing time in these presses is about one to two minutes for single-opening presses and five to eight minutes for multi-opening presses. This is time under pressure and does not include loading, unloading, press opening, and closing. The overlaid panels are hot when pulled from the press.

These polyester overlays are nonshrinking while curing and are consequently stress-free and nonwarping if cooled flat. Polyester is tougher and more resilient than melamine. In addition, polyester overlays exhibit less checking, greater impact resistance, and superior machinability.

The production of the overlays based on melamine-resin-saturated alpha-cellulose paper is rapidly growing in importance. The melamine-formaldehyde resin used has modifiers added to it to enhance the elasticity of the hardened melamine resins. Other materials are added to the resin solution, which later on improves the impregnation of the paper and helps prevent films from sticking to the platens.

The base papers for the decorative films are alpha-cellulose papers, which are used almost exclusively. For impregnation, these papers must have a high wet strength and the right porosity to accept the proper amount of resin. The board laminate consists of one to three papers: an overlay sheet, a decorative sheet and a barrier (or base) sheet. Barrier sheets are not known to be used in the United States. Most U.S. board has very smooth surfaces, eliminating the need for such sheets, which are used to cover porous surfaces. It should be mentioned that balancing papers are also normally used on the back of the board so that, after it is laminated, warping will not occur. In some cases, the same type of melamine overlay is used on both sides of the panel. In this way, if some minor defect happens to occur on one side, the other side is still perfectly acceptable for a top-grade face, and warping problems are minimized with identical papers on both sides.

The melamine-saturated papers usually contain 50 to 55% resin. The overlay or cap paper is an unfilled alpha-cellulose paper which, upon pressing, becomes transparent and protects the decorative paper lying underneath, particularly against wear and tear. The use of a cap sheet is popular in the United States on the wear side of panels. To improve abrasion resistance such cap sheets can also be used on polyesters. It has already been mentioned that the decorative papers are either pigmented solid colors or pigmented sheets with a wood grain or random design in one color or multi-color prints. It should be mentioned that these colors must be fade-resistant and compatible with polyester resins and solvents used in polyester and melamine resins.

Not much melamine overlay is applied in multi-opening presses in the United States. Special handling is necessary to prevent precure of the resin when loading this type of press. In the instances where such laminating is done, press times between two and eight minutes are reported at temperatures ranging from 280° to 330°F (138°–166°C). Pressures are higher than used for polyester, being about 350 psi (2.41 MPa). Release papers can be used to give smooth or textured surfaces.

Melamines shrink slightly during curing; consequently, they must be handled carefully if laminated to thin substrates below ½ in. (12.7 mm) in thickness. Warp problems, which will be covered later, must be handled. In comparison to polyesters, the melamine surfaces have better scratch resistance at the same resin level.

Formerly, relatively long times were always necessary for applying these types of decorative surface laminates. However, there has been a clear move to high temperatures at relatively short pressing times, made possible by the development of better resin systems.

A great step forward in this type of laminating is the previously mentioned short-cycle process with single-opening presses where lamination is accomplished in about one minute. High-gloss finishes so far have not been possible with this type of finishing. For most applications in the United States, high-gloss finishes are not desirable because any slight imperfection in the overlay or substrate is shown up. Resins with very short cure times (from 20 to 30 seconds or less) are now available; thus, very fast press times are possible. These fast-curing resins require, however, that the time between initial contact of the film with a hot heating platen or plate to the time it takes to reach a specific pressure of approximately 120 psi (0.83 MPa) must not exceed a value of 15 to 20% of the condensation time. This means that in the case of a condensation time of 30 seconds, loading of the press, closing of the press, and generating of sufficient pressure must not exceed a time of about 6 seconds. Therefore, once the board and overlays are inserted into the press, maximum pressure must be applied quickly or the fast-curing resin in the overlays will polymerize before good bonding to the substrate takes place (referred to as *precure*).

Polyester laminating is accomplished in about 1 minute (40 seconds for resin cure and 20 seconds for unloading and loading in the press). Pressures range from about 175 to 220 psi (1.21–1.52 MPa). Release papers are usually necessary.

Melamine laminating takes about 80 seconds (60 seconds for resin cure and 20 seconds for press unloading and loading). Pressing pressures range as high as 350 psi (2.41 MPa), and resin curing temperatures range from 310° to 350°F (154° to 177°C). Release papers may or may not be used. The use of polished chrome-plated pressing plates and melamine containing a release agent often eliminates the need for release papers.

It must be kept in mind that there are many variations of the low-pressure laminates available and that the figures given here are intended as guidelines.

Plants equipped to overlay with melamine can usually operate with polyesters, but the reverse may not be true because of the higher pressing pressures required.

One system in use loads the press by means of a tray-belt on which the composition board with top and bottom overlay is aligned, as shown in Figure 18.42. Unloading is done by a traveling vacuum lift, which travels into the open press and carries out the finished board; or the bottom caul travels out of the press and a vacuum lift behind the press carries the board via an intermediate storage station to the finished board stack. The press plates and press pads (if used) usually remain in the press. In case a vacuum lift is used for unloading a press, the bottom heating platen is usually controlled differently than the top heating platen with regard to temperature. This is necessary to avoid differential curing rates between the top and bottom overlays, as the bottom overlay is in contact with the heated platen longer than the top overlay.

Another type of short-cycle laminating which has been developed is a vertical laminater as illustrated in Figures 18.43 and 18.44. The substrate is cleaned and transferred to the vertical position by a special vacuum lift. The overlay papers are placed in positions and "tacked" there by means of a gluing process. The sandwich is then fed into the vertical press, and heat and pressure are applied simultaneously to both faces of the panel. In both systems the finished panel is then edge trimmed with conventional equipment.

Vulcanized Fiber: Vulcanized fiber has been successfully bonded to wood panel products for many years. It possesses some properties such as toughness not matched by the more common resin-treated fiber overlays, but it has not been able to compete economically with the medium-density overlay products. A paperlike sheet is made from chemically converted cotton cellulose generally treated with zinc chloride and is felted into a multi-layer product of whatever thickness is desired. It is very tough and dense and has excellent electrical properties. Many of its properties, such as dimensional stability, are quite similar to those of wood. Modified polyvinyl acetate and resorcinol adhesives provide durable exterior quality bond on wood-based panel products. An important use for vulcanized fiber has been as crossbanding in the furniture industry under fine hardwood face veneers applied to plywood, particleboard, or lumber core panels. The sheet provides a stabilizing influence on the thin hardwood overlay and resists telegraphing from the core panel.

Resin-Treated Fiber Overlays: Resin-treated fiber overlays are another important surfacing for wood-based panel products. They consist of cellulose sheets or papers which may be plain, bleached, dyed, or printed, saturated with a resin such as phenol formaldehyde or polyester, and then hot-press-bonded to the wood substrate by heat and pressure. Some systems contain sufficient resin to be self-bonding, while others require secondary adhesive.

High-density overlays are described as a surfacing that is hard, smooth, and of such a character that further finishing by paint or varnish is not necessary. These overlays consist of a cellulose-fiber sheet or sheets containing not less than 45% resin solids based on a volatile-free weight of fiber and resin, with the resin being a thermal-setting phenol or melamine type.

Medium-density overlay is a resin-treated facing that presents a smooth, uniform surface intended for high-quality paint finishes. It consists of a cellulose fiber sheet containing not less than 17% resin solids for a beater-loaded sheet, or 22% for an impregnated sheet, both values based on the volatile free weight of resin and fiber exclusive of glue line. The resin can be a thermosetting phenol type, or a combination of polymers that will pass performance standards of U.S. Product Standard PS 1-74 for Construction and Industrial Plywood. Commercial production is usually hot press laminating, but roll laminating has been done.

A light-weight, medium-density type overlay is manufactured particularly for use with particleboard and insulation board. This phenolic resin-treated paper weighs approximately 33 pounds per thousand square feet without the glue line. When bonded with phenolic adhesives and finished with conventional paints, it performs very well on products for which it is designed. Considerable production of a fiberboard siding product overlaid with this type of material and preprimed and/or prefinished at the factory has been performing well for several years.

Polyvinyl Fluoride: Polyvinyl fluoride overlay is supplied in a variety of thicknesses, up to about 2 mils (0.051 mm). It is commonly roll laminated onto products such as hardboard, utilizing a modified epoxy adhesive. The adhesive is applied to the board, the solvent flashed off, and the film nip rolled to the substrate. The overlay possesses excellent color retention, good chemical resistance, and has good abrasion resistance. However, it is expensive, which limits its use. It has been used on some residential house siding (particularly plywood), where its weathering properties are excellent.

Specialized Overlays: Specialized overlays, such as fiber glass, reinforced plastic, and metal, have not found widespread use in the composition board industry as yet. However, their success with certain plywood products foreshadows the possibility of more use of these types of products taking place in the future.

Figure 18.42. *A short-cycle press line for overlaying. The four photos show assembly of resin-treated papers with board substrates and press loading. (1) Laydown and positioning of bottom surface papers. (2) The top papers are*

positioned over the substrate. (3) The assembly is then clamped prior to being loaded into the press. (4) Finished board is ready for unloading. (Courtesy Siempelkamp Corp.)

Figure 18.43. *The panel in front is raised into vertical position by means of a movable vacuum table shown in rear. Resin-impregnated papers are then fixed to the panel. (Courtesy Becker & van Hüllen.)*

Warp

Composition boards must stay flat in service conditions, particularly when they are unsupported in modes such as cabinet doors. Wood products have tendencies to warp because of unbalanced construction either from natural causes, such as differential shrinking and swelling between summerwood and springwood, or from man-made causes in panels, such as particleboard where a skewed density profile is developed. Nothing can be done about the natural causes of warping found in lumber and, to some extent, in plywood. If the composition board is produced properly, however, such sources of warp should not exist. However, the man-made causes of warp must be considered and handled. A secondary manufacturer is at the mercy of the board manufacturer in that he has no control over the way the board was produced. However, a secondary manufacturer can induce problems such as a skewed density profile in the panels by sanding only one side of a board he receives, which can result in warp problems.

The type of unbalanced construction developed by sanding can be handled by proper care. Such a panel should not warp when subjected to the forces induced by moisture changes. However, upon applying decorative plastic laminates and backing sheets, differentials in response to the changes in moisture content can cause an imbalance within the panel. These stresses caused by the differential movement are a major cause of warpage in the field. Much of the warpage is created because of different stresses within the materials which are originally bonded together into the laminated composition board final product. There can also be an unequal stress as exerted by the laminate material on opposite sides of the board. This can be caused by having unequal thicknesses of the sheets of material or by having different types of material.

Temporary unbalance can also be caused by unequal rates at which face laminates and backing sheets change dimensions. When decorative laminates and backing sheets are exposed to high relative humidities, backing sheets have been found to attain their maximum expansion in about one day and decorative laminates in about one week. When panels of unbalanced construction are exposed to such moisture conditions, it is common for the warp to occur in the opposite direction to that logically expected, based on the unrestrained dimensional characteristics of the panel components. This reverse movement can be misleading if conclusions about the warp proneness of a particular construction are based on too short an exposure to different humidities.

It is important to condition all the materials to the same relative humidity conditions before lamination takes place. It is recommended that exposure to about 40 to 45% relative humidity for seven days will accomplish this objective, but local conditions must be considered. Ideally the backing sheets should exactly offset the longitudinal shrinking or swelling of the laminate on the opposite side in order to prevent warping. The ideal backing sheet not only responds to the moisture change just as the laminate does, but also transmits as much stress to the core as does the laminate. An ideally constructed panel would use the same laminate on both sides of the core. This may be feasible on thin boards that are exposed from both sides, such as used for cabinet doors. However, the laminate is more expensive than backing sheets and, whenever possible, backing sheets are used, running the risk of differential characteristics between the laminate and the backing sheet.

It should also be pointed out that laminates and backing sheets may have a machine direction and exhibit very different expansion and contraction rates in response to changes in moisture content. It may be found that linear movement across the machine direction of the sheets is twice as much as that in the machine direction. This must be taken into account when laying up the package of materials prior to final laminating. It has also been found that the elastic-type adhesives such as rubber-base-contact cements reduce somewhat the amount of stress transferred to the substrate by the face laminate, which reduces the tendency for warping.

Figure 18.44. *Finished panel leaving press. (Courtesy Becker & van Hüllen.)*

Some cabinet tops which are fastened down can be made with these adhesives and no backing sheets.

It is also important to consider stabilizing the back or bottom face of a panel that has been wet finished on the top. Thus some sort of back sealing can be used for providing the same type of restraint on both sides of the panel.

Edge Finishing

A number of different edging materials are used, particularly with the thicker products such as particleboard and medium-density fiberboard. Wood strips can be edge banded to the board and these strips can be profiled to the preferred shapes, using conventional routers and shapers. Plastics can be used to fill the edges, and these plastics can be later finished similarly to the top of the panel. A molded polyester system has been developed that uses a carved wooden design to make silicone rubber molds. Exotic patterns can be thus developed and the polyester edge banding on the wood paneling will reflect accurately this pattern right down to the wood grain and pores. Other materials used for edge banding are vinyl, resin-free wood fiber which needs final finishing or which can serve as the backing of vinyl films, resin-saturated wood fiber, high-pressure plastic laminates, and wet finishes. The most widely used continuous decorative edge banding is a polyester decorative paper combination, generally in the same pattern or solid color used on the low-pressure or high-pressure laminate surface.

The wet finishing is similar to that found with surface finishing where filling and the application of graining, ground coats, printed designs, and a topcoat take place. Miter folding or V-folding of vinyl simultaneously with its application to the surface of the panel has already been mentioned. Also previously mentioned was vacuum forming of vinyl over three-dimensional configurations on a panel. T-molded pieces of plastic can be used for edge banding where the vertical member of the T is fitted into a slot cut into the edge of the panel.

Another type of edge finishing takes place with postforming, where the edges and surface, as with V-folding, are covered with one sheet of laminating material. In postforming, a rigid plastic laminate is molded to the shape of the substrate. Heat is first applied until the plastic becomes pliable. This must be controlled precisely; otherwise, cracking will occur if the plastic laminate is too cool or blistering will happen if the laminate is too hot. The pliable part is then held by the postforming machine against the substrate until the adhesive has cured. This is the popular method for applying a laminate to a countertop where a curved or rounded front edge is desired.

SELECTED REFERENCES

Batey, T. E., Jr. 1970. Overlays for Wood-Based Panel Products. FAO Committee on Wood-Based Panel Products, Third Session, Rome, Italy, November 1970. FOI: WPP 70/6.1.

Best, M. K. 1975. Low Pressure Laminate Uses, Popularity Growing. *Wood & Wood Products* (February).

———. 1976. Overlays for Composition Board (Interior and Exterior Applications, Hot Pressed and Cold Pressed, Decorative and Industrial). *Proceedings of the Washington State University Particleboard Symposium, No. 10.* Pullman, Washington: Washington State University (WSU).

Brumbaugh, J. I. 1967. Overlaying of Particleboard. *Proceedings of the Washington State University Particleboard Symposium, No. 1.* Pullman, Washington: WSU.

Bryan, E. L. 1971. Modern Management Techniques for Particleboard Operations. *Proceedings of the Washington State University Particleboard Symposium, No. 5.* Pullman, Washington: WSU.

Crawford, D. R. 1975. Noise Problems and Their Potential Solution with Sawing Equipment in Board Plants. *Proceedings of the Washington State University Particleboard Symposium, No. 9.* Pullman, Washington: WSU.

Crawford, R. J. 1976. Introduction to the Surface Treatments of Composition Board. *Proceedings of the Washington State University Particleboard Symposium, No. 10.* Pullman, Washington: WSU.

Currier, G. M., and J. H. Clontz. 1975. Finishing and How It Relates to Today's New and Old Substrates for Furniture Manufacturing. Background Paper No. 88, third World Consultation on Wood-Based Panels, FAO, New Delhi, India, February 1975.

Edge Processing. 1974. *Wood & Wood Products,* Vol. 79, No. 5.

FAO. 1976. Basic Paper IV. Technology and Techniques in the Manufacture of Wood-Based Panels. *Proceedings of the World Consultation on Wood-Based Panels,* held in New Delhi, India, February 1975. Brussels: Published in agreement with the Food and Agriculture Organization of the United Nations by Miller Freeman Publications.

Graham, P. 1974. Working with Particleboard. *Furniture Production* (January-February).

Guss, L. M. 1975. Overall Implications of Panel Products in North America. *Proceedings of the Washington State University Particleboard Symposium, No. 9.* Pullman, Washington: WSU.

Harre, K., and U. A. Hartl. 1972. European Practices in Cutting Particleboard. *Proceedings of the Washington State University Particleboard Symposium, No. 6.* Pullman, Washington: WSU.

Heape, G. C. 1970. Computer Analysis for Cut-to-Size Production Control. *Proceedings of the Washington State University Particleboard Symposium, No. 4.* Pullman, Washington: WSU.

High Density Particleboard. 1974. Published by *Woodworking & Furniture Digest* in cooperation with the National Particleboard Association, Silver Spring, Maryland.

Horner, K. M. 1968. Plant Experience and Problems with Cutting and Sanding. *Proceedings of the Washington State University Particleboard Symposium, No. 2.* Pullman, Washington: WSU.

———. 1969. Shipping Particleboard. *Proceedings of the Washington State University Particleboard Symposium, No. 3.* Pullman, Washington: WSU.

Jenuleson, W. R. 1973. Flat Sheet Hardboard Finishing. *Forest Products Journal,* Vol. 23, No. 4

Jones, W. A. 1968. Principles, Problems and Techniques for Cutting Particleboard. *Proceedings of the Washington State University Particleboard Symposium, No. 2.* Pullman, Washington: WSU.

Kline, R. H. 1975. Mitre-Groove Components in the Laminated Vinyl Veneer Board Industry. Background Paper No. 66, third World Consultation on Wood-Based Panels, FAO, New Delhi, India, February 1975.

Knapp, H. 1968. Development, Principles, Problems and Techniques of Finishing Particleboard. *Proceedings of the Washington State University Particleboard Symposium, No. 2*. Pullman, Washington: WSU.

Koch, R. L., II. 1976. Finishing Board Via the UV Cure Process. *Proceedings of the Washington State University Particleboard Symposium, No. 10*. Pullman, Washington: WSU.

Koellisch, J. F., ed. 1973. Working with Hardboard. Chicago, Illinois: Wood & Wood Products and American Hardboard Association.

Kovich, J. T. 1972. New Developments in the USA for Cutting Particleboard. *Proceedings of the Washington State University Particleboard Symposium, No. 6*. Pullman, Washington: WSU.

———. 1974. Selecting Sawblades for Cutting Particleboard. North American Products Corporation, Atlanta, Georgia.

Kraemer, E., and A. Schmitz-LeHanne. 1971. European Laminating Practices. *Proceedings of the Washington State University Particleboard Symposium, No. 5*. Pullman, Washington: WSU.

Leary, P. E. 1974. Ultraviolet Curable Coatings for Industrial Finishings. *American Paint Journal,* Vol. 59, No. 15.

Maxwell, R. 1975. Noise Problems and Their Potential Solution with Sawing Equipment in Board Plants. *Proceedings of the Washington State University Particleboard Symposium, No. 9*. Pullman, Washington: WSU.

Murdock, R. G. 1968. Principles, Problems and Techniques for Sanding Particleboard. *Proceedings of the Washington State University Particleboard Symposium, No. 2*. Pullman, Washington: WSU.

———. 1970. Complete Panel System. *Proceedings of the Washington State University Particleboard Symposium, No. 4*. Pullman, Washington: WSU.

National Particleboard Association. 1970. Edge Filling of Particleboard. *Technical Bulletin,* No. 5 (April). Washington, D.C.

———. 1971*a*. Sawing Particleboard. *Technical Bulletin,* No. 6 (April). Silver Spring, Maryland.

———. 1971*b*.Edgebanding and Veneering of Particleboard. *Technical Bulletin,* No. 7 (November). Silver Spring, Maryland.

———. 1972. Filling Particleboard. *Technical Bulletin,* No. 9 (December). Silver Spring, Maryland.

———. 1973. Direct Printing and Finishing Particleboard. *Technical Bulletin,* No. 10 (December). Silver Spring, Maryland.

———. 1974*a*. The Handling and Storage of Particleboard. *Technical Bulletin,* No. 11 (March). Silver Spring, Maryland.

———. 1974*b*. Sanding Particleboard. *Technical Bulletin,* No. 12. (July). Silver Spring, Maryland.

———. 1974*c*. V-Grooving with Particleboard. *Technical Bulletin,* No. 13 (December). Silver Spring, Maryland.

New Molded Shaped Edge Process for Decorative Tops and Panels. 1969. *Furniture Methods and Materials* (April).

Plath, L. 1971. Requirements of Particle Boards for Resin Surfacing. *Holz als Roh-und Werkstoff,* Vol. 29, No. 10.

Powell, H. L. 1968. Sandpaper for Particleboard. *Proceedings of the Washington State University Particleboard Symposium, No. 2*. Pullman Washington: WSU.

Reichhold Chemicals, Inc. 1975. Application Bulletin. *Bulletin No. TFP-4f* (March). Tacoma, Washington.

Robertson, A. A. 1976. Printing and Finishing of Composition Board for High Grade Furniture. *Proceedings of the Washington State University Particleboard Symposium, No. 10*. Pullman, Washington: WSU.

Ryckman, J. D. 1973. Plastic Overlays for Woodworking. *Woodworking & Furniture Digest* (September).

Saul, J. D. 1974. Roseburg Enters the Prefinished Panel Market. *Wood & Wood Products* (February).

———. 1975. Your Good and Faithful Servant, the Machine. *Wood & Wood Products* (April).

Scanning Devices Should Help on the Grade Line. 1974. (Chapter 13 in the series "Plywood Plants of the Future.") *Plywood & Panel* (November). Special Reprint and Directory Edition.

Segal, A. R. 1960. What You Should Know About Carbide-Tipped Saws. *Carbide Engineering* (July, August, and September).

———. 1975. Noise Problems and Their Potential Solution with Sawing Equipment in Board Plants. *Proceedings of the Washington State University Particleboard Symposium, No. 9*. Pullman, Washington: WSU.

Siempelkamp Corp. 1975. Short Cycle Presslines for Decorative Laminating without Recooling. Atlanta, Georgia.

Steck, E. F. 1975. Private Communication on Short-Cycle Laminating. Siempelkamp Corporation.

Stensrud, R. K. 1968. Overlays for Exterior Particleboard. *Proceedings of the Washington State University Particleboard Symposium, No. 2*. Pullman, Washington: WSU.

Stephenson, J. A. 1968. Fillers for Particleboard. *Proceedings of the Washington State University Particleboard Symposium, No. 2*. Pullman, Washington: WSU.

———. 1975. Private communication on wet finishing. The Sherwin-Williams Co.

Storch, H. J. 1974. Quick Clamp Stack Veneering. *Wood & Wood Products*, Vol. 79, No. 12.

Surdyk, L. V. 1973. Some Causes of Warp in Particleboard Production. *Proceedings of the Washington State University Particleboard Symposium, No. 7*. Pullman, Washington: WSU.

Trotter, T. 1968. Sawblades for Particleboard. *Proceedings of the Washington State University Particleboard Symposium, No. 2*. Pullman Washington: WSU.

Using Particleboard. 1972. National Particleboard Association and *Wood & Wood products*. Chicago Illinois: Vance Publishing Co.

Wentworth, I. 1972. Recent Advances in Prepresses for Single-Opening and Multiple-Opening Particleboard Plants with Automatic Stacking for Cut-Up Operations. *Proceedings of the Washington State University Particleboard Symposium, No. 6*. Pullman, Washington: WSU.

———. 1975. Manufacture of Particle- and Fiber-boards with Flat and Rotary Continuous Presses and Continuous Lamination of Thin Boards. *Proceedings of the Washington State University Particleboard Symposium, No. 9*. Pullman, Washington: WSU.

Whaley, J. H., ed. 1974. Working with Particleboard. *Furniture Production*, (January and February), Vol. 37, Nos. 263, 264.

Wikstrom, G. G. 1976. A Secondary Manufacturer Views Problems with Overlaying and Composition Board Substrates. *Proceedings of the Washington State University Particleboard Symposium, No. 10*. Pullman, Washington: WSU.

19. NEW DEVELOPMENTS

The new technologies developed for the wood-based composites sector over the past 40 years are remarkable. Millions of tons of products are now produced annually from wood materials that had been mostly unused previously. Today, composites range from the familiar older products such as plywood and hardboard to large timber composites used in heavy construction. This chapter describes some of the newer technologies now being used in the wood-based composites sector of the forest products industry.

The field of wood-based panel materials has changed drastically since 1950. New particle types, new adhesives, adhesive improvements, equipment developments, process control developments, and inventions (including computers), as well as innovations in usage have created a whole new world of wood-based panels or materials in 1992. These materials are now better known as "composites." These composites include not only panels, but also molded products, lumber, large timbers, components, and materials made with combinations of wood and other materials.

The 1992 world of composites sees millions of tons produced worldwide annually. Plywood, inorganic-bonded materials, insulating board (noncompressed fiberboard), and hardboard (compressed fiberboard) were produced in significant amounts before 1950, but these products, plus all the newer composites, have become a major part of the forest products industry today. The 1989 FAO World Survey report (Secretariat 1990), for example, shows 125,883,000 m^3 of panel production throughout the world in 1988.

This chapter discusses the new technologies in the wood-based panels-to-composites sector. The subject of this chapter has been changed deliberately from wood-based panels to that of composites so that molded and lumber composite materials would be included—they are already important products and will become even more important in the future. Because of the size of the industry today, it is impossible to discuss the subject completely. Consequently, there will be omissions of what some will consider very important developments.

"Composites" is a word easily misunderstood. In this discussion, composites are defined as materials that have the commonalty of being glued or bonded together. Thus, glue, resin, or another binder is an important part of quality composites. Conventional glue and resin, based on petroleum and natural gas, are expensive and they will become more expensive as oil prices increase. A contributing factor to developing countries' problems with glues and resins is that the latter usually are imported and this can only add to these countries' costs.

It is a well-known fact that wood and wood-based composites are important materials that benefit humanity. The world's population is growing and wood and composite products are badly needed by all countries; however, the forest resource is shrinking even faster than it has in the past. On a per capita basis, the amount of wood available worldwide has decreased dramatically since 1950. Furthermore, the cost of raw materials to produce composites is increasing rapidly. Technology has advanced to the extent that more and more low-quality material is being used for composites, better equipment is being built, and operating crews are becoming better trained. The greatest challenges the industry faces are: learn to use more of the wood species found in the forest and also different raw material types, develop new economical adhesive systems, develop environmentally safe preservative and fire-retardant systems, and educate those who are not familiar with wood composites with proper techniques for using these composites in structures and furnishings.

This chapter discusses panels, molded products, lumber composites, and adhesives. Panels include softwood plywood, hardwood plywood, particleboard, blockboard, MDF (dry-process medium-density fiberboard), wet-process fiberboard (noncompressed and compressed), dry-process hardboard, OSB (oriented strand board) and waferboard, COM-PLY®, molded materials, and inorganic-bonded materials. Lumber or timber composites include laminated veneer lumber (LVL), COM-PLY, parallel strand lumber (PSL), and oriented strand lumber (OSL). The discussion of adhesives is followed by a section entitled "Specific Considerations" providing more details on the industry. Not discussed, except in a few cases, is the increasing importance of the marriage of wood and other materials such as plastics and metals. These resulting composites take advantage of the best characteristics of each material and the quality of appropriate or desired properties into these materials is vital to providing better composites as well as prolonging the life of the composites.

A last introductory, but important, comment relates to the economic impact of new technology. In North America, for example, the sale prices of the end products have remained relatively stable despite labor, wood, and adhesive cost increases. New technology has made the difference in lowering production costs—a dramatic economic point in basic wood products manufacture. Without new technology, product sale costs would be much higher today.

COMPOSITE PANELS

Softwood Plywood

Softwood plywood dominates the North American structural panel market. Significant improvements in production resulting in reduced production costs have enabled the industry to increase production and remain competitive with OSB. In 1992, however, a reduced timber supply in the western United States and increases in veneer log prices caused a number of plants to close. Plywood plants are now attempting to produce mostly higher-value products rather than sheathing (basic building panels), leaving much of the sheathing market to lower-cost OSB.

Softwood plywood manufacture has undergone a dramatic change in recent years. Some examples of the manufacturing changes include pretreatment of the veneer blocks, which has enhanced veneer quality and the amount of veneer produced; computerized X-Y peeler block chargers; ultrasonic veneer block scanning; powered backup rolls; powered nose bars; and linear-positioned lathe knife carriages, which have become part of the conventional lathe and further increase the amount of veneer produced. Also, the spindleless lathe, which peels blocks to cores of less than 2 in. (50 mm) in diameter, has been introduced. Changes in the gluing system, such as gluing veneer of higher moisture content and foaming the glue, have reduced the amount of adhesive required and, hence, have lowered production costs.

For softwood plywood, the development of phenolic glues for use with veneers at moisture contents of 10% or more instead of 3 to 4% is a major breakthrough. This has resulted in higher dryer productivity, improved press productivity, lowered glue spreads, better prepressing, better assembly time tolerance, improved veneer gluability, and fewer panel blows. Some plants have obtained savings of up to 22% in the glue cost while simultaneously saving on dryer and press costs.

Gluing high-moisture–content veneer (up to 15%) results in a product closer to the equilibrium value in actual use, thus reducing warp and dimensional changes. However, a plant must operate under strict process controls to be successful in gluing high-moisture–content veneer; otherwise, product quality will be reduced significantly.

Hardwood Plywood

Unlike softwood plywood, which is used mostly for structural applications, hardwood plywood is used both structurally and also for decorative panels. A number of developments have taken place over the years that would enable a plant to recover more veneer from a given log and to produce quality plywood more efficiently. Most of these developments are several years old, but they are just now coming into prominence. It must be remembered that in the United States veneer overlaid particleboard is also categorized as hardwood plywood.

The newer versions of lengthwise veneer slicers have been installed in many plants, particularly in east Asia and Australia. These slicers do not have the output of conventional slicers, but they work well in smaller plants. They allow for recovering difficult-to-process wood in conventional plants and for producing veneer from quality wood material in sawmills. Thus, sawmills gain the opportunity for a value-added product by producing veneer as well as lumber.

Some of the reasons for using lengthwise slicers include low investment, no foundation requirements, low energy consumption, complete or near-complete utilization of the flitches, automatic operation, and the ability to produce thin veneer with high-quality surfaces as well as thick veneer. The flitch can be any length depending upon the slicer capabilities, i.e., 3 to 12 ft (1.22 to 3.66 m). If more veneer is required, it is relatively inexpensive to add another slicer. The combination of a sawmill and veneer mill has resulted in significantly increased product recovery and profit. This slicer development also presents an opportunity for the softwood mills by providing thin, high-quality veneer for overlays, edgebanded composite materials, and laminated products.

A major competitor to hardwood plywood is thin MDF. Thin MDF with appropriate high-quality overlays or prints appears the same as hardwood plywood. This product has taken significant amounts of the hardwood plywood market.

Blockboard

Blockboard is often overlooked today, but its old technology, or variations on the old technology, could provide panel-building products in many places. Blockboard panels are made with strips of lumber in the core with faces of thin lumber, veneer, or hardboard. The core lumber strips and the lumber face material are glue-edge–jointed. This product line has been revitalized in Europe.

Particleboard

Particleboard remains the world's dominant furniture panel, although a considerable amount of its production also goes into structural applications such as manufactured home floors, roof sheathing, wall panels, and stair treads. The process lines are becoming quite sophisticated as computers, programmable logic controllers, and many types of modern measuring devices are now considered normal to the lines.

The basic raw material remains the same: plant residues or low-quality logs. Some recycled wood material, where it is economical to use such a substance, is now part of the raw material supply.

Preparation of particles is changing, however, not so much because of the development of new particle-generating machines, but because of the size of the particles. Competition from MDF, which has a relatively uniform edge amenable to clean profiling as well as having surfaces of fibers, is driving at least part of the particleboard industry in the United States toward the use of smaller and more slender particles. These finer particles result in a board more like MDF in edge and surface qualities. Such particleboard can be edge-profiled better and is not as susceptible to "fiber pop" on the board surfaces. When wetted with laminating glue for applying thin foils, even very small particles (less than 0.079 in. [2 mm] in length) will swell sufficiently (fiber pop) to show the outline of their shape through the foil. Also, even extremely low levels of glue (water-based) will result in fiber pop.

Continuous presses have attracted great interest recently for the manufacture of all types of composition board. Many different kinds of continuous presses have been developed over the years. There was limited success by the Bartrev, about 40 years ago, and a great deal of success by the Mende system in producing thin boards. Currently, newer, improved Mende presses are producing thin MDF very successfully.

The new generation of continuous presses has four major versions: the ContiPress manufactured by Küsters, the Hydro-Dyn produced by Bison Werke, the ContiRoll built by Siempelkamp, and the ContiPower System made by

Dieffenbacher. All four presses have been in operation for several years. About 50 of these presses are now running or soon will be. Also, steam-injection pressing has gone into production lines since 1989. It is being used also for MDF and oriented strand lumber (OSL).

Recycled wood is being used more widely as a raw material for composites. This type of wood has been used in South Africa and Germany for years. There is concern, however, about contaminants in this type of wood, particularly the wood that comes in as demolition waste or as preservative- or fire retardant–treated material. Recycled wood is being used more and more in North America.

Many different types of treatments have been used to improve the dimensional characteristics of particleboard. Most treatments are applied to products bonded with phenol-type resins. However, recent work has shown that dimensionally stable urea-formaldehyde–bonded products can be produced by a short steam pretreatment of the particles. This is done before board manufacture with steam treatments as short as one minute.

In the United States, there has been an extraordinary change in the use of particleboard since 1978. From 1973 to 1978, 40% of particleboard was used as underlayment in single-family homes or as decking in mobile homes and manufactured housing. However, waferboard and OSB have entered these markets recently and have pushed particleboard out. A panel called Sturd-I-Floor™ was developed as a combination subfloor and underlayment panel. Sturd-I-Floor can be made as plywood, waferboard, or OSB. Lower installation cost and improved wood yields in manufacturing made this product line very competitive and virtually eliminated the use of double-layer flooring systems in many parts of the United States.

By 1989, the flooring market for particleboard accounted for less than 12% of the total material shipped. However, it is interesting to note that the total shipments of particleboard in 1989 reached an all-time annual record. The reason for the increase in use and production of particleboard, despite a phenomenal loss in one marketing area, was the expanded use of industrial particleboard. In order to produce the higher-quality industrial particleboard, many of the manufacturing plants had to make major quality improvements in their production process. To handle the raw material species, size, moisture content, resin/wood blending, density within panels and between panels, and out-of-press thicknesses, a much tighter control of variations had to made. Capital expenditures were made on additional screening equipment, refining equipment, blenders, and forming equipment. Most plants now have automated continuous monitoring systems measuring the critical quality characteristics, including moisture, weight, and thickness. Also, computerized control of blending, forming, and pressing has been developed over the last number of years. The control systems now employed in all composite plants are a far cry from those used in the early 1980s.

There is still some consumer resistance to the use of particleboard and MDF. The low-quality consideration often goes back to the pre-1980 period when the lower-grade flooring and utility-type particleboards were considered typical. As already noted, most of this type of particleboard is no longer produced. Also, consumers, in the form of furniture manufacturers and cabinet manufacturers, often used the wrong quality particleboard and MDF (and often used poor fabrication and construction techniques). In addition, the emotional considerations involved with the formaldehyde issue provided the public with a bad perception of these products. It is now the purpose of the industry to educate consumers about the present-day quality of the product line, showing that, indeed, high-quality furniture and cabinets can be and are being manufactured with the right quality particleboard and MDF.

MDF

Many new MDF plants have been constructed since 1989 and more are in the planning stages. The plants produce thin and thick boards. Usually, thin MDF is produced by continuous presses—those presses mentioned under particleboard as well as larger and newer versions of Bison's Mende calendar press.

The technology for producing MDF has increased greatly in recent years. There are now over 100 plants worldwide. It appears that a mixture of species can be handled better in an MDF plant than in a particleboard plant. There are, however, still some problems, particularly in designing the resin system when there are significant chemical differences between the species. Urea-formaldehyde is still the dominant resin. The interaction between the chemistry of the wood species and the urea-formaldehyde is well known, and control of the species' chemistry is needed to ensure that quality MDF is produced consistently.

An MDF plant can handle a wide variety of raw material types from pulp chips to planer shavings to plywood trim to sawdust. However, all of these different raw material types have to be introduced into the process line at appropriate levels to ensure a consistent quality of fiber. The total wood raw material mix should include at least 25% pulp chips to produce the better-fiber furnish. Other nonwood materials such as bagasse can also make excellent MDF.

Refiners have been improved consistently over the years, leading to more efficient fiber production at a lower cost. New refiner design variations have become available recently. Due to greater throughput of material and ease of operation, virtually all MDF plants now use single revolving disc refiners, as contrasted in the past to double revolving disc refiners.

Drying is done through tube-type dryers that are less apt to "ball" fibers into clumps and are easier to control as far as fires are concerned. The blowlines from the refiners to these dryers are used for blending resin with the fiber, particularly for the face furnish.

Other developments with MDF include placing radio frequency or high-frequency preheating units at the entry of continuous presses making thicker panels. This reduces press times up to 40%. Also, formaldehyde-free MDF has been developed that is suitable for exterior uses, such as signs, billboards, and other sites subject to extremes in moisture level.

MDF lends itself well as a substitute for clear lumber. MDF does not have a grain structure, but finishes and overlays can be used effectively to provide an MDF product that, in appearance, looks like wood. The average United States specific gravity level of MDF is 0.72 (about 46 lb/ft^3). However, recent developments in New Zealand have led to the manufacture of MDF at specific gravity levels of 0.60 and 0.48 (37.5 and 30 lb/ft^3). The advantage of this particular product, called "Liteboard," as manufactured by Nelson Pine Industries, Ltd., is the reduction of density while at the same time maintaining good board surface quality, having acceptable board strength and stability, and providing a high-surface–density board product that can be applied with high-quality finishes. This kind of board is made from a combination of resin adjustments and pressing strategies. The pressing strategy calls for developing a high-density surface but a relatively uniform density throughout most of the interior or core of the panel. This particular product is being produced on continuous presses, which have adjustable features, allowing for this type of pressing.

The contribution of Owen Haylock at Canterbury Timber Products in Canterbury, New Zealand, should be noted in the development of the MDF industry. Haylock and his colleagues developed a pressing technique in their multiple-opening press resulting in MDF boards with relatively high surface densities and uniform core densities. The pressing strategy uses a step system, rapidly bringing the mat after insertion into the press to a thickness about 30% greater than the final targeted panel thickness. This produces a panel with a high-density surface. Subsequent steps are then taken to bring the panel to final caliper. This technique eliminates much of the soft-surface face found in the early production lines using platen heating only for pressing. Most of the earliest MDF plants used a combination of platen heating and high-frequency curing.

Fiberboard

Insulating Board (Noncompressed Fiberboard): Insulating board manufacture has been decreasing in size slowly. The need to clean the water effluent has been a major problem and competition from foamed plastic panels has had further impact on insulating board manufacture. There may be ways to eliminate the water effluent problem. Work is under way using enzymes or bacteria and nutrients that clean the water of the hemicelluloses (the major contaminants in the wastewater) and other solids in the wastewater.

One insulating board product still enjoying great success in the United States is Homosote. This product is made of waste newspaper and ground wood–paper publications waste. Manufacture started in 1916 (based on technology from Agasote Millboard in the United Kingdom). It is a structural product based on recycled wood fiber. Besides various panels, components for walls and roofs are produced combining the board with rigid urethane foams, foils, fiberglass, and other materials.

Hardboard (Compressed Fiberboard): There are three versions of hardboard. These are wet-process, dry-process, and semi-dry process. Wet-process hardboard continues to be an important panel product worldwide. As with insulating board, there are questions about cleaning the wastewater. Stricter environmental regulations can make ways to use wastewater that have been used previously (such as for irrigation) against the law in some countries. The work, mentioned previously, on developing enzymes and bacteria for cleaning water could be an important development for wet-process hardboard.

Wet-process hardboard has been affected in the United States by the development of OSB siding, as siding has been the major market for hardboard in the United States. Hardboard in the United States is also being used as a substitute for solid wood in trim applications.

Expansion of product lines has included roof-covering material in the form of tiles or simulated wood shingles or shakes. Such hardboard is untreated and is being manufactured and used in South Africa and the United States.

Dry-process hardboard is now becoming quite confused with thin MDF. The dry-process board is bound together with synthetic adhesives just as particleboard and MDF are. It usually has poorer properties under wet conditions than wet-process hardboard does because the hemicelluloses, with their affinity for water, are retained in the furnish rather than partially washed away as in wet-process hardboard. However, a significant development in dry-process hardboard technology is the production of door skins. This development will be discussed under molded products, but it is a major product line in the United States.

The semi-dry process prepares the fiber in the same way as in the wet- and dry-process systems. Some resin is added (usually phenolic) in the fiber-grinding process step. The fiber is then dried to about 30% moisture content at which time the forming and pressing is performed the same as in the dry-process system. The extra moisture in this system (at least at the higher board densities) permits some self-bonding to develop between fibers as in the wet-process system. Thus, in the semi-dry system it is possible to use a lower amount of resin than in the dry-process system.

COM-PLY

COM-PLY panels are made with veneer and particles/flakes. Three-layer panels have veneer faces and particle cores while five-layer panels have a veneer core (laid up with its grain direction at 90° to the face veneer grain direction). One plant is producing the five-layer board very successfully.

OSB/Waferboard

The development of OSB and waferboard in North America is well known. In the past few years, almost all of the production of these two panel products has shifted to OSB. Reportedly, only one out of some 40 plants now produces waferboard. Almost all of the panels are used in structural applications in the same way as plywood. Thus, making OSB with greater bending strength in one panel direction (usually the long direction) results in a product much like traditional plywood. OSB is usually produced with a thinner thickness, e.g., 7/16 vs. 1/2 in. (11.1 vs. 12.7 mm), than waferboard for comparable uses. Thus, there is a material savings in manufacturing OSB.

Most OSB is now made with 3-in.-long (75 mm) or longer strands in the surface layers. The core may be of smaller strands and may or may not be oriented. Today's OSB panels are made with oriented layers (mostly three-layer [some with a randomly laid core], but some five-layer constructions also) that are laid up similar to these in plywood.

The OSB segment of the industry has become an important part of the structural panel business in recent years. Its growth has been greatest in the United States and Canada, but OSB plants are operating now in Europe and are planned for South America as well as other parts of the world.

The OSB industry can use small irregular logs as the raw material, but it prefers ones that are relatively straight and about 14 in. (350 mm) in diameter. The reasoning behind this is that usually the logs are debarked with ring-type debarkers. Such debarking is very time consuming and inefficient with the smaller logs. However, a great amount of the logs used are smaller than 14 in. in diameter. Drum debarkers designed especially for smaller-diameter logs are being used for debarking the smaller logs although there is some damage to the wood in the debarking operation.

The industry has been developed on low-density hardwoods such as aspen; however, other species such as southern pine, lodgepole pine, jack pine, and Scotch pine are being used. High-density species are still difficult to use because boards produced from such species are also quite high in density. These boards are difficult to handle, cut, and nail, and they are more expensive to ship.

To cut quality flakes, the OSB process requires green logs. In cold climates, hot ponds (about 120°F [50°C]) are required to thaw frozen logs. After drying, the finer material is screened out and normally used for fuel. Most flakers are still of the disc type. New knife-ring flaker and disc machines have been developed that cut flakes from full-length logs without the need to slash the logs to lengths of about 36 in. (900 mm) long, as is required for feeding into the older disc flakers. These machines are moving quickly into dominant positions in the plants, and the continued improvement of flakers will help in the development of OSB.

Continuous pressing is also being introduced into OSB manufacture. In addition, steam-injection pressing, already used for particleboard, MDF, and PSL, is being considered. This would speed the pressing of phenolic-bonded panels, isocyanate-bonded panels, and thicker (over ¾ in. [19 mm]) panels.

Considerable concern is being expressed in the industry about excessive thickness swell, particularly with OSB. Processes for correcting this problem using steam treating are being proposed. Steam post-treating has been promoted before and has been used, reportedly successfully, in the industry.

Chemical modification of composites to provide dimensional stabilization has been proposed and performed, to a limited degree, for many years. In recent years, the United States Forest Products Laboratory has performed considerable research in this field, particularly with products designed for use under adverse environmental conditions. The lab's work involves chemically modifying the reactive components of the wood cell wall. Technology is emerging that can be of value to commercial board manufacturers.

In the general area of structural panels, there is the possibility of using either standard pulp chips or maxi-chips (over 2 in. [50 mm] long), ring-flaking them, and using electrical orientation to align the particles that are of varying lengths. A mixture of strand lengths can also be used to produce structural panels. This approach has great potential, particularly for locations where suitable logs are not available for cutting long strands.

Molded Materials

Molded materials of fiber or particles have a long history in the industry, especially in western Europe and the United States where molded door skins are a major product line. Molded parts for the automobile industry have taken an increasing share of the market formerly served by plastics. Combinations of wood and plastic fiber have made it possible to produce inside car door panels with deep draws or impressions—an impossibility with wood fiber alone. As many as twelve of these types of panels are now being put into some models of American cars. These panels are overlaid with various plastic foils.

The high strength-to-weight measure of wood fiber moldings has made their use quite attractive. Another successful molded product is pallets. Molded structural parts for building offer excellent opportunities in the future. Folded plate designs, for example, offer great strength with a minimum use of material. Perhaps molded wood products will find a new life if the cost of plastics continues to increase.

Inorganic-Bonded Panels

There has been considerable worldwide interest in the development of inorganic-bonded panels. New developments with fast-curing bonding materials and processes have sparked this interest. The greatest interest has been with the gypsum fiberboards. This product, developed in Germany, has captured a significant part of the German gypsum board market. Several equipment companies have developed process lines. The raw material is designed to be manufactured from all recycled material—waste paper for the fiber and gypsum reclaimed from coal-burning electrical generating plants.

Two plants were operating in North America—one in eastern Canada and one in the eastern United States—but the United States plant has been closed. This product is different from conventional gypsum board, which has gypsum between two thick sheets of paper. The gypsum fiberboard product has the fiber evenly blended throughout the board. Thus, the gypsum fiberboard has to be cut and applied differently from the conventional gypsum board. If the United States and Canadian construction trades accept this new product, there is a good potential for many new plants in North America.

Roof shingles and shakes are now being produced using wood fiber, sand, and cement. One such product, "Hardishake," is produced in New Zealand and is molded to look like wood shakes. It is produced in several colors. A new plant of this type has been built in California, and another plant is operating in Oregon producing roof shakes of wood particles bonded with cement.

LUMBER OR TIMBER COMPOSITES

The decreasing quality of the world's forest and the shrinking amount of timber available for producing either conventional lumber or high-quality lumber has led to a tremendous interest in producing composite lumber-like materials. Research has shown that such materials can be made and, indeed, plants are now producing these lumber-like products.

The products include laminated veneer lumber (LVL), COM-PLY, parallel strand lumber (PSL), and oriented strand lumber (OSL) (Figure 19.1). As of 1991, LVL had become the most important construction material. New plants are being considered, as are process improvements. Long press times of over 20 min-

Figure 19.1. *Examples of (left to right) LVL, PSL, Parallam™ and COM-PLY (in background). (Courtesy of Washington State University.)*

utes are needed to cure the adhesive in the 1½-in.-thick (40 mm) product. High-frequency (HF) curing of the glue lines can reduce press times dramatically, and two HF plants are under construction. Steam injection also works in reducing press time.

LVL products, so far, have been used primarily for structural applications. More recent growth has been in the development of processes for the production of LVL that is cut up into other products such as millwork. These products are made with colorless glues, so it is difficult to tell that the millwork is LVL and not solid wood. A new type of lathe is capable of peeling veneer at much greater thicknesses. In this type of LVL, fewer glue lines are needed.

The development of COM-PLY panels and lumber has been going on for many years. COM-PLY lumber is a material composed of random or oriented wood flakes or particles sandwiched between one or more layers of veneer. Panels were described earlier. For lumber, one or several layers of veneer are placed on the faces or edges of the lumber.

Panel-producing plants commenced production in the mid-1970s. A plant for producing COM-PLY lumber has been in operation in North Carolina, U.S.A. This plant produces an oriented flakeboard product that is cut into desired widths for making composite lumber. Several layers of veneer are then laminated onto each edge of the flakeboard to make the COM-PLY joist lumber. Veneer is laminated onto the flakeboard faces to make truss lumber. The veneer is oriented parallel to the lumber length. Different layers of veneer are used depending upon the width of the joist lumber. This product is being directed towards the truss market where lumber members of assured strength are desired and for the joist market where long straight members are desired.

Parallam, or PSL, is a development of MacMillan Bloedel, Ltd., of Canada. The operating company is now Trus Joist MacMillan. Many years of research and development led to the pilot plant production of PSL. In the pilot plant, this material is composed of oriented strands made from waste softwood veneer. Strand dimensions are about ½ in. and up to 39 in. long (12 mm–1 m). Exterior-type plywood glue is applied to the strands and the strands are oriented and laid up into a mat that is then processed through a continuous press. A microwave-type heating system was invented to permit the production of such a thick billet.

The billet is sawn into desired lengths, thicknesses, and widths. The resulting PSL is of high quality and has been designed to serve markets where high-strength lumber or timber materials are needed. Two commercial-size plants are now running in Georgia, U.S.A, and Vancouver, B.C., Canada. Veneer is being peeled in standard 100 in. (2.5 m) lengths and then cut into ½-in.-wide (12 mm) strips specifically for use in these plants. The commercial-size billets are 11 by 16 in. (279 by 406 mm) and 11 by 14 in. (279 by 356 mm).

Scrimber was a development in Australia for producing lumber. "Scrim" is a term from the textile industry that means loosely woven. The "ber" in the name Scrimber comes from timber. Small logs are crushed between rollers, producing the scrim. Then layers of scrim are dried, followed by the application of glue. The layers of scrim are laid up into a mat and consolidated in a hot press into lumber. The first commercial plant was put into operation in the late 1980s, but it failed. The Japanese have developed a similar product called Zephyrwood. Reportedly, it is ready for commercialization.

Trus Joist MacMillan has moved forward with the development of a lumber product called PSL 300. This is oriented strand lumber made with nominal 12-in.-long (300 mm) strands. In the process, round logs are barked and then cut into the 12-in.-long strands. The strands are nominally 0.030 in. thick by 1 in. wide (0.8 by 25 mm). The strands are dried, blended with isocyanate binder and then oriented in the long direction into a thick mat. The mats are pressed in an 8 by 35 ft (2.44 by 10.67 m) steam-injection press. The pressed billets can be produced from any thickness ranging from 1 to 5.5 in. (25 to 145 mm) and may weigh up to 6,000 lb (2,720 kg). The pressing in the steam-injection press results in a product with uniform density throughout and a relatively short press time. The first manufacturing plant is operating in Minnesota, U.S.A. The product is aimed primarily at the window and window frame market for now, but it is a structural material. Much of the product will be wrapped with vinyl, other plastics, or wood veneer.

OSL is being considered for manufacture by some OSB manufacturers; some is being produced already. These OSB plants have been built so all the strands can be aligned in one direction making it convenient to produce both OSB and OSL.

OSL faces a struggle in entering conventional lumber markets. Strength properties have to be measured and design values assigned to the lumber. This takes considerable time and effort; however, this effort is under way as OSL will be needed more and more in the future. Weyerhaeuser has produced OSL products for the furniture market.

A timber product now being produced under the name Cedrite is a railroad tie (sleeper) produced by recycling old railroad ties. This product, originally invented in the 1970s, is being manufactured in a plant designed to produce 1,600 ties per day. Old ties are chipped, the metal removed, the chips hammermilled and then blended with resin in preparation for production. The blended furnish is placed into metal molds. Hardwood stiffeners are placed in the center of the mats that measure about 30 in. (760 mm) high before being compressed to a final thickness of 7 in. (180 mm). Once a mat is compressed into a mold, a top plate is secured over the compressed mat. The mold then goes into an oven to cure the resin.

ADHESIVES

Urea-formaldehyde and phenol-formaldehyde resins are still the dominant adhesive systems used. Isocyanate binder has become important particularly in the OSB industry.

Formaldehyde emissions from urea-formaldehyde–bonded products have been reduced dramatically, meeting standards established in many countries in recent years. Additives can be used to reduce emissions further. Ammonia systems that can be used to remove free formaldehyde immediately after pressing are now being used in several countries. Reduction in formaldehyde emission still continues, although remarkable reductions have occurred since 1980.

In the United States, formaldehyde emissions from urea-formaldehyde–bonded wood panel products have decreased by 75–90% since 1980. In 1984, the United States Department of Housing and Urban Development established formaldehyde emission standards for particleboard and plywood used in the construction of manufactured homes. The industry has used this rule as the model for voluntary emission standards; it was incorporated into a national consumer standard in 1989. At present, the industry is working on developing a new voluntary standard for flooring that will have a much lower level than the previous standard. The previous standard called for a 0.30 ppm (parts per million) emission limit. A new standard will call for a 0.20 ppm emission limit. This new level will be compatible with the most stringent emission levels found anywhere in the world.

Isocyanate (4-4' diphenylmethane di-isocyanate known as MDI) has been used in the particleboard industry for many years. Usually, isocyanate is used as a substitute for the durable resins: phenolic and melamine. Isocyanate has also been used in combination with these resins. Some use has been made of isocyanate for furniture board when it was desirable to have board without any chance of formaldehyde emissions.

In Canada and the United States, there has been a breakthrough in the use of isocyanate resin. Isocyanate-bonded OSB is being produced in Canada and the United States, particularly for use in manufacturing siding. This product is overlaid with an embossed resin-impregnated paper and has its edges coated to prevent water absorption from rain. Some of this board is made with isocyanate in the board core and phenolic resin in the board faces. This practice eliminates problems with the isocyanate sticking to the press platens and cauls but still takes advantage of the faster curing speed of the isocyanate. Newer phenolic resins have reportedly been made to match the curing speed of isocyanate.

All of the resins and glues discussed are based on natural gas and/or petroleum. Tannin adhesives, particularly from wattle, have been developed at a high level of quality. This effort has been particularly successful in South Africa, Australia, and southeast Asia. Tannin-based adhesives are being used successfully in Brazil, as well as in other countries. The most success appears to have come from the wattle-based tannin adhesives.

Research has been conducted on modifying the chemistry of wood surfaces (flakes and fiber) to enhance adhesive bonding. Mechanical properties have shown a strong relationship to chemical treatments, but a complete understanding of the relationship is yet to be determined.

Work on producing composites without binders has proceeded for many years. Indeed, wet-process hardboard can be and is produced without binders. Shen has done a considerable amount of developmental work in this field. His "Auto-Bonding" technique produces an exterior-grade product by using a thermohydrolysis system to act chemically on the cellulose, lignin, and hemicellulose of the wood. Successful trial runs have been made in several countries. The disadvantages of the process include a darker-colored board, the need for high press temperatures, and the need for high-pressure steam.

Many research efforts are continuing with lignin and carbohydrate glue developments. These efforts are very important because new glue systems are needed, particularly if oil and gas prices increase again, as is fully expected. It should also be remembered that the animal and vegetable protein glue technology, developed in the 1930s and 1940s, is still available for use. Both interior and exterior glue systems for use in both cold and hot pressing were used for many years. Undoubtedly, further improvements are possible due to scientific knowledge gained since 1960 when most of the protein glues were replaced.

Problems with efficiently gluing different mixtures of species are still of concern. More research is needed to solve these problems, particularly for those countries who have species mixtures with no possibility of separating the species.

SPECIFIC CONSIDERATIONS

Standards and Grades

Standards and grades (found in Chapters 2 and 4) of different materials are always being reconsidered. Chapter 2 in this book presents tables of grades and properties of a number of different products as of 1977. Some of these grades and properties have been updated and they are being presented here. The first, Table 19.1, covers hardboards; Tables 19.2 and 19.3 cover hardboard siding (Table 19.2 covers physical properties and Table 19.3 lists thicknesses and tolerances). Table 19.4 presents the two tables of grade information from the American National Standard on Wood Particleboard (National Particleboard Association 1993); Table 19.5 provides information on the grades and general uses for each grade; and Table 19.6 covers medium-density fiberboard.

Table 19.1 *Classification of Hardboard by Surface Finish, Thickness, and Physical Properties*

Class	Nominal thickness (in.)	Water resistance (max avg per panel): Water absorption based on weight (%)	Water resistance (max avg per panel): Thickness swelling (%)	Modulus of rupture (min avg per panel (psi)	Tensile strength (min avg per panel): Parallel to surface (psi)	Tensile strength (min avg per panel): Perpendicular to surface (psi)
1 Tempered	1/12	30	25	6000	3000	130
	1/10	—	—			
	1/8	25	20			
	3/16	—	—			
	1/4	20	15			
	5/16	15	10			
	3/8	10	9			
2 Standard	1/12	40	30	4500	2200	90
	1/10	—	—			
	1/8	35	25			
	3/16	—	—			
	1/4	25	20			
	5/16	20	15			
	3/8	15	10			
3 Service tempered	1/8	35	30	4500	2000	75
	3/16	30	30			
	1/4	30	25			
	3/8	20	15			
4 Service	1/8	45	35	3000	1500	50
	3/16	40	35			
	1/4	40	30			
	3/8	35	25			
	7/16	35	25			
	1/2	30	20			
	5/8	25	20			
	11/16	25	20			
	3/4	—	—			
	13/16	—	—			
	7/8	20	15			
	1	—	—			
	1-1/8	—	—			
5 Industrialite	1/4	50	30	2000	1000	25
	3/8	40	25			
	7/16	40	25			
	1/2	35	25			
	5/8	30	20			
	11/16	30	20			
	3/4	—	—			
	13/16	—	—			
	7/8	25	20			
	1	—	—			
	1-1/8	—	—			

Source: ANSI/AHA A135.4-1982 "Basic Hardboard."

Table 19.2 *Physical Properties of Hardboard Siding*

Property	*Requirements*	
Water absorption, percent based on weight	12 (max avg per pannel)	
Thickness swelling, percent	8 (max avg per panel)	
Weatherability of substrate (max percent residual swell)	20	
Weatherability of primed substrate	No checking, erosion, flaking or objectionable fiber raising. Adhesion: less than 0.125 in. of coating "picked up"	
Linear expansion 30–90% RH (max percent)	Thickness range	Maximum linear expansion
	0.220–0.324	0.36
	0.325–0.375	0.38
	0.376–0.450	0.40
	over 0.451	0.40
Nail-head pull-through, lb	150 (min avg per panel)	
Lateral nail resistance, lb	150 (min avg per panel)	
Modulus of rupture, psi	1800 for ⅜- & 7/16- & ½-in. thick 3000 for ¼-in. thick (min avg per panel)	
Hardness, lb	450 (min avg per panel)	
Impact, in.	9 (min avg per panel)	
Moisture content,* percent	4–9 incl., and not more than 3% variance between any two boards in any one shipment or order	

Source: ANSI/AHA A135.6-1990 "Hardboard Siding."

*Because hardboard is a wood-based material, its moisture content will vary with environmental humidity conditions. When the environmental humidity conditions in the area of intended use are a critical factor, the purchaser should specify a moisture content range more restrictive than 4–9% so that fluctuation in the moisture content of the siding will be kept to a minimum.

Table 19.3 *Thicknesses and Tolerances for Hardboard Siding*

Nominal thickness (in.)	*min.-max. (in.)*
1/4 (0.250)	0.220–0.265
3/8 (0.375)	0.325–0.375
7/16 (0.438)	0.376–0.450
1/2 (0.500)	0.451–0.525

Source: ANSI/AHA A135.6-1990 "Hardboard Siding."

Until the 1980s, almost all structural panels used in the United States were plywood, particularly softwood plywood. They were manufactured under what is known as commodity or product standards. Basically, these are prescriptions on how the minimum product is to be manufactured that do not address product application. In contrast, a performance standard, or what some used to call an end-use standard, reverses the process in that the end-use of the product is considered first, and the standard does not prescribe by what means a product will be manufactured. Consequently, a performance standard must define performance criteria and test methods. The process has to ensure that the product line will perform successfully in the field.

Consideration of the performance standard took place first in the late 1970s as more of the particleboard, flakeboard, waferboard, and, later on, OSB products were manufactured for the marketplace. This wide variety of product lines brought forth the effort of the American Plywood Association (APA) to develop performance standards covering what came to be known as structural panels.

End-use categories, such as single-layer floors, roof sheathing, subflooring, structural exterior siding, and concrete forms, were established. The performance standard determines a baseline of performance. All products must meet the minimum baseline. However, the manufacturer is free to produce a product from his available raw material, such as veneer, flakes, strands, or particles, as long as the minimum baseline of performance is met. Thus, a considerable amount of developmental work went into creating what as now known as the APA Performance Standards and Policies for Structural-Use Panels, PRP-108 (1991).

The performance standard approach has been a resounding success. Most of the OSB-type production in the United States and Canada now goes to the marketplace under a performance standard. The use of a performance standard has made it possible for these relatively new products to move quickly to the market. This approach allows for innovation, and the quick use of new manufacturing technologies and modification of panel manufacture. Figure 19.2 shows the APA program for panel qualification and quality assurance for several products.

Currently, in Canada, a new product standard on OSB and waferboard is close to being approved. This standard, CSA 0437-92, proposes the grade property requirements shown in Table 19.7. Construction Sheathing, CSA 0325-91 (Canadian Standards Association 1991), is the Canadian performance standard harmonized to APA PRP-108 (noted previously) that is soon to become DOC PS-2-92. These two standards are being harmonized under the Canada–United States Free Trade Agreement.

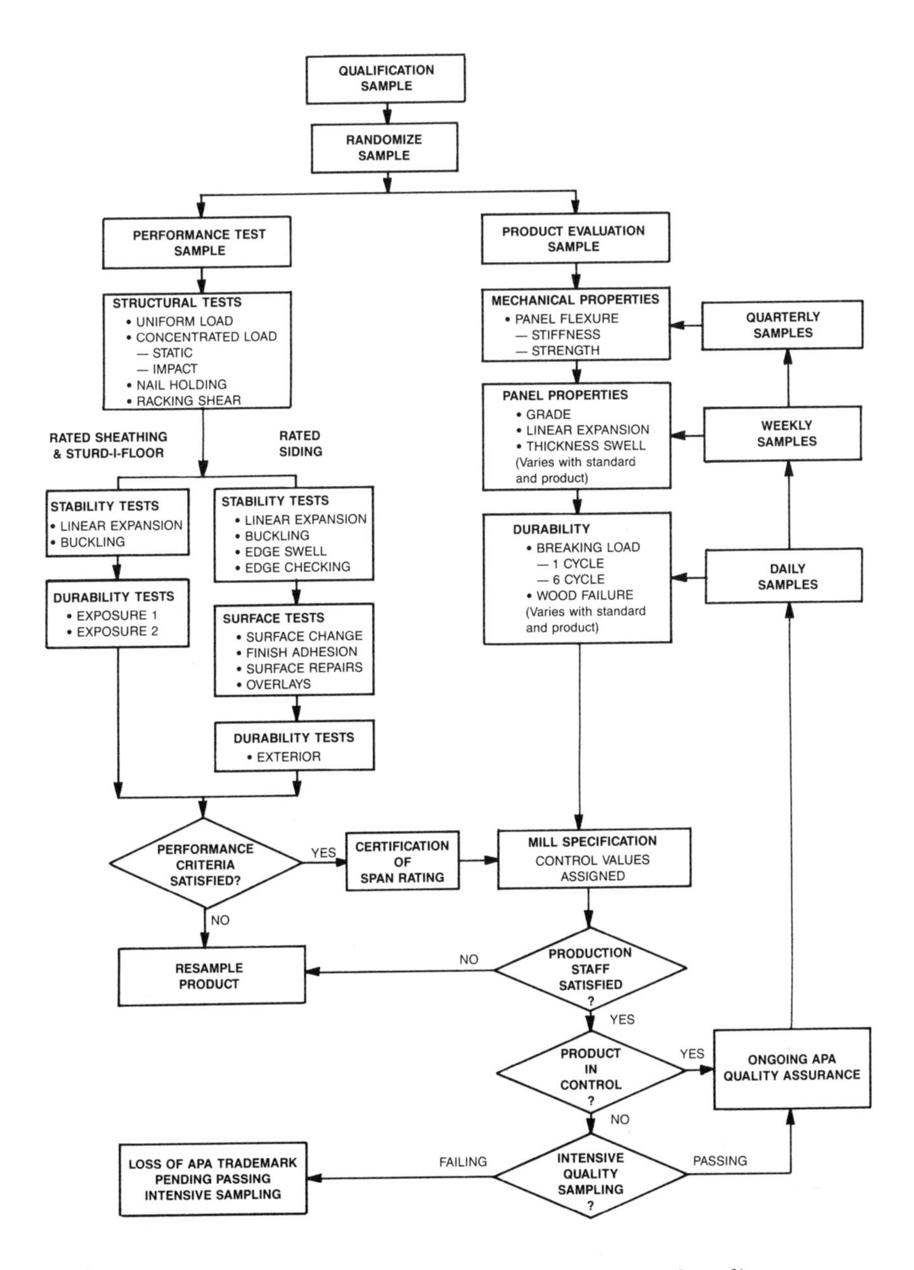

Figure 19.2. *APA performance-rated panel qualification and quality assurance program for APA rated sheathing, APA rated Sturd-I-Floor & APA rated siding. (Courtesy of APA PRP-108 Performance Standards and Policies for Structural-Use Panels.)*

Table 19.4 *Property Requirements of Wood Particleboard*

Requirements for Grades of Particleboard[a,b]

Grade[c]	Length & width tolerance mm (in.)	Thickness tolerance[d]: Panel avg from nominal mm (in.)	Thickness tolerance[d]: Variance from panel avg mm (in.)	Modulus of rupture N/mm[b] (psi)	Modulus of elasticity N/mm[b] (psi)	Internal bond N/mm[b] (psi)	Hardness N (lbs)	Linear expansion max. avg (%)	Screwholding: Face N (lbs)	Screwholding: Edge N (lbs)	Formaldehyde max. emissions (ppm)
H-1	±2.0 (0.080)	±0.20 (0.008)	±0.10 (0.004)	16.5 (2393)	2400 (348100)	0.90 (130)	2225 (500)	NS[e]	1800 (405)	1325 (298)	0.30
H-2	±2.0 (0.080)	±0.20 (0.008)	±0.10 (0.004)	20.5 (2973)	2400 (348100)	0.90 (130)	4450 (1000)	NS	1900 (427)	1550 (348)	0.30
H-3	±2.0 (0.080)	±0.20 (0.008)	±0.10 (0.004)	23.5 (3408)	2750 (398900)	1.00 (145)	6675 (1500)	NS	2000 (450)	1550 (348)	0.30
M-1	±2.0 (0.080)	±0.25 (0.010)	±0.15 (0.005)	11.0 (1595)	1725 (250200)	0.40 (58)	2225 (500)	0.35	NS	NS	0.30
M-S	±2.0 (0.080)	±0.25 (0.010)	±0.15 (0.005)	12.5 (1813)	1900 (275600)	0.40 (58)	2225 (500)	0.35	900 (202)	800 (180)	0.30
M-2	±2.0 (0.080)	±0.20 (0.008)	±0.10 (0.004)	14.5 (2103)	2250 (326300)	0.45 (65)	2225 (500)	0.35	1000 (225)	900 (202)	0.30
M-3	±2.0 (0.080)	±0.20 (0.008)	±0.10 (0.004)	16.5 (2393)	2750 (398900)	0.55 (80)	2225 (500)	0.35	1100 (247)	1000 (225)	0.30
LD-1	±2.0 (0.080)	±0.15 (0.005) −0.40 (0.015)	±0.15 (0.005)	3.0 (435)	550 (79800)	0.10 (15)	NS	0.35	400 (90)	NS	0.30
LD-2	±2.0 (0.080)	±0.15 (0.005) −0.40 (0.015)	±0.15 (0.005)	5.0 (725)	1025 (148900)	0.15 (22)	NS	0.35	550 (124)	NS	0.30

Requirements for Grades of Particleboard Flooring Products[a,b]

Grade[c]	Length & width tolerance mm (in.)	Thickness tolerance[d]: Panel avg from nominal mm (in.)	Thickness tolerance[d]: Variance from panel avg mm (in.)	Modulus of rupture N/mm[b] (psi)	Modulus of elasticity N/mm[b] (psi)	Internal bond N/mm[b] (psi)	Hardness N (lbs)	Linear expansion max. avg (%)	Formaldehyde max. emissions (ppm)
PBU	+0 (0) −4.0 (0.160)	±0.40 (0.015)	±0.25 (0.010)	11.0 (1595)	1725 (250200)	0.40 (58)	2225 (500)	0.35	0.20
D-2	±2.0 (0.080)	±0.40 (0.015)	±0.25 (0.010)	16.5 (2393)	2750 (398900)	0.55 (80)	2225 (500)	0.30	0.20
D-3	±2.0 (0.080)	±0.40 (0.015)	±0.25 (0.010)	19.5 (2828)	3100 (449600)	0.55 (80)	2225 (500)	0.30	0.20

Source: ANSI/A208.1-1993 "Wood Particleboard" (under revision).

[a] Particleboard made with phenol-formaldehyde–based resins does not emit significant quantities of formaldehyde. Therefore, such products and other particleboard products made with resin not containing formaldehyde are not subject to formaldehyde emission conformance testing.
[b] Grades listed in this table shall also comply with the appropriate requirements listed in Section 3. Panels designated as "exterior glue" must maintain 50% MOR after ASTM D 1037 accelerated aging (paragraph 3.4.3).
[c] Refer to Table 19.5 for general use and grade information.
[d] Thickness tolerance values are only for sanded panels as defined by the manufacturer. Unsanded panels shall be in accordance with the thickness tolerances specified by agreement between the manufacturer and the purchaser.
[e] NS = Not Specified

Table 19.5 *Explanation of Grades and General Uses of the Different Grades of Wood Particleboard*

Explanation of Grades

The particleboard grades in this standard are identified by a letter designation, followed by a hyphen and a digit or letter. Additionally, the two-part identification may be followed by a third letter, number, or term identifying a special performance characteristic.

The first letter designations have the following meanings:

H High density (generally above 800 kg/m³ (50 lb/ft³))

M Medium density (generally between 640–800 kg/m³ (40–50 lb/ft³))

LD Low density (generally less than 640 kg/m³ (40 lb/ft³))

D Manufactured home decking

PBU Underlayment

The second digit or letter designation indicates the grade identification within a particular density or product description. For instance, "M-2" indicates medium-density particleboard, Grade 2.

The optional third designation indicates that the particleboard has a special characteristic. For instance, "M-3-Exterior Glue" indicates medium-density particleboard, Grade 3, made with exterior glue to comply with the durability requirements in paragraph 3.4.3 of the standard.

Information to be provided

All particleboard which is represented as conforming to this American National Standard shall be identified with the following information:

a. Manufacturer's name or trademark and mill identification

b. "ANSI A208.1-1993"

c. Grade

d. Lot number or date of production

e. For D grade products only, the words "MANUFACTURED HOME DECKING"

f. For PBU grade products only, the word "UNDERLAYMENT"

g. For products complying with paragraph 3.4.3 only, the words "EXTERIOR GLUE"

General Use for Grades of Wood Particleboard

Use	*Grade*	*Table*
Commercial	M-I,S	A
Industrial	M-2,3	A
High-density industrial	H-1,2,3	A
Door core	LD-1,2	A
Exterior construction	M-1,2,3-Exterior glue	A
Exterior industrial	M-1,2,3-Exterior glue	A
High-density exterior industrial	H-1,2,3-Exterior glue	A
Underlayment	PBU	B
Manufactured home decking	D-2,3	B

Source: ANSI/A208.1-1993 "Wood Particleboard" (under revision).

Table 19.6 *Property Requirements for Medium-Density Fiberboard*

Nominal thickness (in.)	*Modulus of rupture (psi)*	*Modulus of elasticity (psi)*	*Internal bond (Tensile strength perpendicular to surface) (psi)*	*Linear expansion (%)*	*Screwholding Face (lb)*	*Screwholding Edge (lb)*
13/16 and below	3,000	300,000	90	0.30*	325	275
7/8 and above	2,800	250,000	80	0.30	300	225

Source: ANSI/A208.2-1986 "Medium-Density Fiberboard for Interior Use."

*For boards having nominal thicknesses of 3/8 in. or less, the linear expansion value shall be 0.35%.

Table 19.7 *Grade Property Requirements for OSB and Waferboard (Average Values for a Sample Consisting of Five Panels)*[a]

Group 1	*Limit*	*Units*	*Direction*[b]	*R-1*	*O-1*	*O-2*
Modulus of rupture	Min	MPa	//	17.2	23.4	29.0
			⊥	17.2	9.6	12.4
Modulus of elasticity	Min	MPa	//	3100	4500	5500
			⊥	3100	1300	1500
Internal bond	Min	MPa		0.345	0.345	0.345
Bond durability—MOR after 2 h boil	Min	MPa	//	8.6	11.7	14.5
			⊥	8.6	4.8	6.2
Thickness swell—24 h soak						
—12.7 mm or thinner	Max	%		15	15	15
—thicker than 12.7 mm				10	10	10
Group 2						
Linear expansion—oven dry to saturated	Max	%	//	0.40	0.40	0.35
			⊥	0.40	0.40	0.50
Lateral nail resistance[c]	Min	N	//	70t	70t	70t
			⊥	70t	70t	70t

Source: CAN/CSA-0437.0-92—June 1992 (under revision).

[a] In addition no individual panel in the five-panel sample shall have any property more than 20% below (or above in the case of thickness swell and linear expansion) the listed five-panel average value for that property.

[b] For OSB random waferboard, // means parallel to the longer dimension of the panel; ⊥ means perpendicular to the longer dimension. For OSB (aligned) board, // means parallel to the indicated direction of face alignment; ⊥ means perpendicular to the indicated direction of face alignment.

[c] t = nominal thickness in millimeters.

Processing Equipment

As would be expected, many different process lines and individual pieces of machinery have been developed over the last years. Only a few of them, highlighting some of the recent developments, will be discussed in this chapter. The concentration will be on particle and fiber generation, improved drying, improved blending, orienting formers and MDF formers, continuous pressing, and steam-injection pressing.

Particle and Fiber Generation: There have been many refinements in the machines for producing conventional particles and fibers. The big changes in machinery have been in the production of strands, fiber, and recycled wood.

With the rapid development of the waferboard/OSB industry since 1980 has come the introduction of debarkers into the composite industry in North America. Both single log and drum debarkers have been used. Prior to the development of the OSB industry, very few whole logs were used in the North American industry. At first, almost all plants that produced wafers, in particular, used disc flakers. These disc flakers were increased in size and, basically, dominated the industry for a number of years. Currently, the disc flakers require that logs be cut to a predetermined length, for example, 36 in. long, before being fed into the flaker. With the recent demand for longer strands now that the industry has shifted from wafers to strands, two new developments have occurred for cutting strands; both will take whole logs. The first one, the CAE development, is shown in Figure 19.3. The logs are placed in position alongside a disc flaker head (118 in. diameter [3 m]). The disc flaker head then moves across the log, or bundle of logs, cutting the strands.

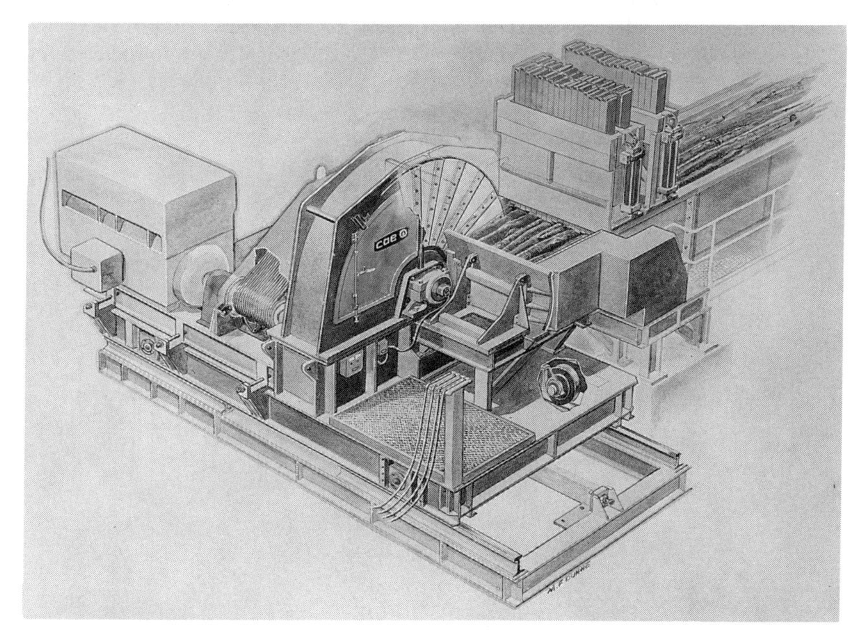

Figure 19.3. *The CAE long log flaker. (Courtesy of CAE Machinery Ltd.)*

The head then returns to its original position, the bundle of logs is indexed forward, and another cut of strands takes place. This type of flaker is also being used in the particleboard industry.

The second development by Pallmann is shown in Figure 19.4. With this type of flaker, logs are placed inside a knife ring. The knife ring then passes across the set of logs, or bundle of logs, cutting strands. The knife ring returns to its original position, the logs are indexed forward, and another cut is taken.

There have been many innovations in producing pressure-refined fiber for MDF plants. The early historical development was in the United States. Refiner capacity was typically 200 tons per day or less using a 40-in.-diameter (1 m), single–revolving disc machine. Thus, more than one machine was needed in the MDF plants being built at that time. These plants developed using separate refiners for producing core and face furnish and in doing so they could treat the face and core furnish differently with resin.

Now, there are refiners with 54-in.-diameter discs producing over 300 tons of fiber per day. Figure 19.5 shows a flowsheet for a modern European MDF plant. These plants, unlike almost all North American plants, run off of logs. North American plants use many sawmill residues (shavings, sawdust, etc.) with about 20% pulp chips. The pulp chips provide some long fiber. Refiners, for the most part, now use "side feeding" of the chips into the refiner (Figure 19.6). The digester is separated from the refiner infeed and a conical discharge metering screw at the digester feeds a high-speed ribbon screw going into the eye of the refiner. This works very well when only chips are being used. At this time it is not known whether a mixture of chips, shavings, and sawdust will work as well.

Now it is a well-known fact that the blending of resin and the drying of the fiber is all tied into the refiner. The fiber, as it leaves the refiner, goes through what is called a blowline on its way to the dryer. The resin and wax are introduced into

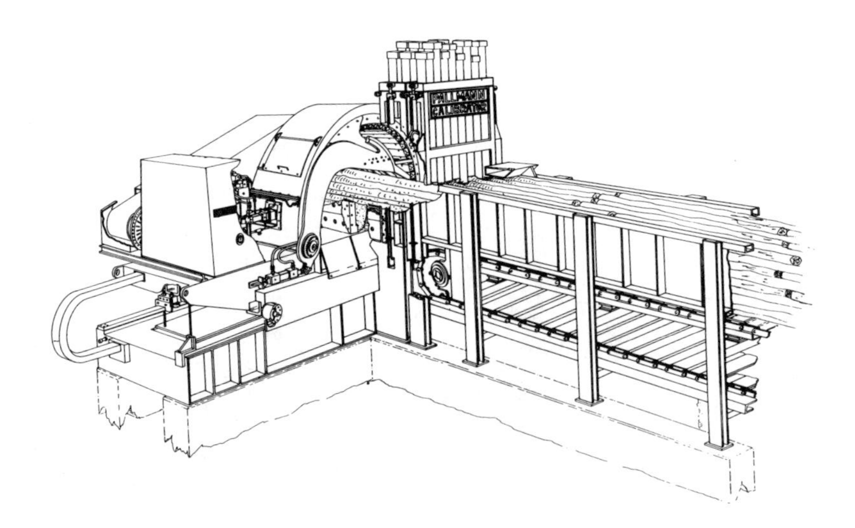

Figure 19.4. *The Pallmann long-log flaker system. (Courtesy of Pallmann.)*

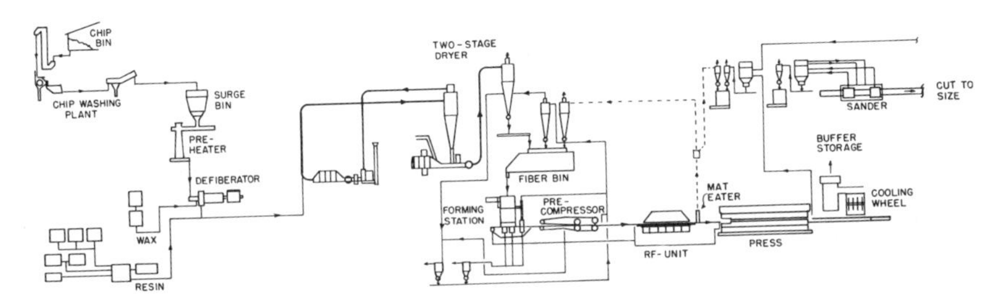

Figure 19.5. *Flowsheet of a modern European MDF plant (Weicke 1990).*

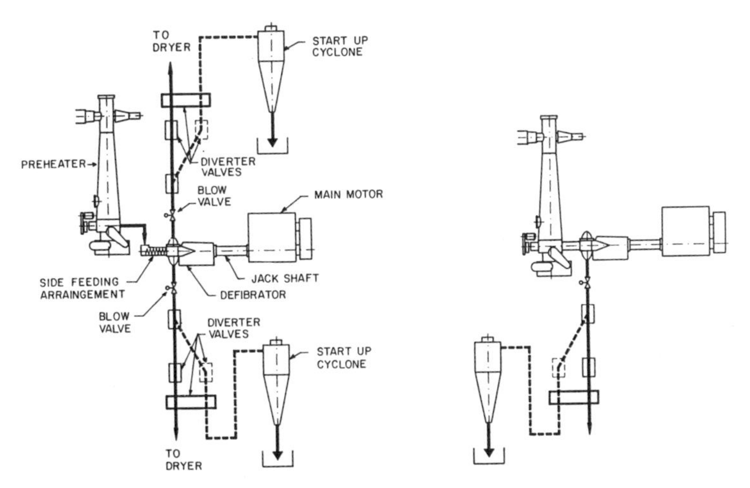

Figure 19.6. *Diagram of direct feeding (right) and side feeding (left) of the refiner (Ramos 1991).*

the blowline and they are blended intimately with the fiber as it passes through the blowline turbulently. Figure 19.7 illustrates mechanical versus blowline blending.

Drying: The tube or flash dryers used today must then dry the fibers at low temperatures so the urea-formaldehyde resin that is used normally is not cured.

Additionally, a number of two-stage dryers are used now. The first dryer is very short and built completely out of stainless steel because it operates with a relatively low air volume in one system. The moisture content of the fibers is taken down to an intermediate level between the final moisture content desired and the incoming moisture content. The second-stage dryer further reduces the fiber moisture content to about 8%. The return air off the first-stage dryer is taken to a heat exchanger, which reclaims most of the heat. The reclaimed heat is used to heat up the inlet air to the first-stage dryer. This system saves 12 to 14% of the dryer energy.

The first stage of the dryer is a relatively short pipe about 100 ft (30.5 m) long. The fiber stays in the pipe for approximately 2.5 to 3 seconds. The second stage can be a number of different lengths as it only does a very gentle drying, with an inlet temperature well below 200°F (93°C). A lump separator was devel-

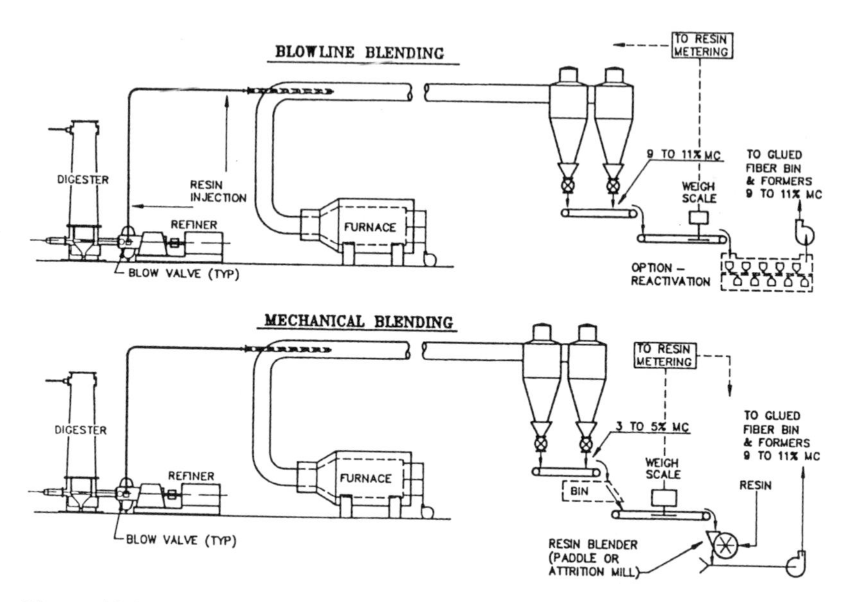

Figure 19.7. *Blowline versus mechanical bending (Frashour 1990).*

oped by one company to ensure that no large fiber bundles or resin clumps get through the process and into the mat former.

The major problem found in recent times with drying has been with toxic effluents being released into the atmosphere. Severe restrictions are being developed throughout the world for limiting the amount and type of effluents that can be released into the atmosphere from any manufacturing process. Some dryers, particularly those drying strands, may have inlet temperatures in the range of 1600°F (871°C). Usually, these types of dryers are much more difficult to control when they must meet environmental regulations. There is a trend in Europe to use more dryers and dry at lower temperatures to alleviate such problems. A number of developments for emission control in the United States include such devices as electrostatic precipitators and electrofilter beds.

Blenders: Blowline blending has been discussed in some detail along with the improvements in the refining of fiber. Some other information should be noted, however. Some plants are still using double-disc refiners for applying resin for MDF manufacture. They report very low resin consumption. There is also the technique call "belt blending." One plant has been using this for many years; it simply involves placing a number of resin spray guns over a wide belt conveyor and spraying the resin directly onto the fiber as it goes underneath the spray guns. The fiber then drops into the standard short-retention-time type of paddle blender, or any other type of mechanical blender, for the final distribution of the resin. Hammermills and similar devices can also be used to distribute resin.

Blowline blending was developed specifically to reduce resin spots on the surface of the panel. The system is not foolproof, as resin can build up within the first part of the dryer, and every once in a while the built-up resin can break off and cause some spots. Such build-up can also happen elsewhere in the system, such as in elbows. The resin formulation is critical to good performance of the blowline system. Naturally a resin has to be formulated so that it will not cure when passing through the dryer. Almost all of the resin used in such systems is diluted in resin solids to somewhere between 35 and 50%. This reduces resin viscosity so that it is easier to distribute the resin over the fiber.

A number of new blenders have been developed for use in blending resin, waxes, and other chemicals with traditional flakes and particles. These newer blenders are very sophisticated and have controls to limit variation in resin application and moisture distribution. In addition, these new blenders have been designed to limit the amount of particle destruction found in some of the older types of short-retention-time blenders. Work a number of years ago at Washington State University's Wood Materials and Engineering Laboratory showed that the short-retention-time blenders utilizing a centrifugal application of resin did not function as thought. As shown in Figure 19.8, these types of blenders apply the resin in a stream rather than an atomized spray. This was recognized by much of the industry and changes were made so that the resin, for the most part, was simply pumped into the blender through various tube systems. The newer blender systems have gone back to atomizing the resins.

Other research at Washington State University, as shown in Figure 19.9 shows how a semi-inverted cone or disc can atomize resins into very fine droplets as indicated by a number of resin and blender studies. A commercial development is shown in Figure 19.10 where a spinning disc is also used for finally atomizing resin. This development has enabled the isocyanate and liquid phenolic adhesives

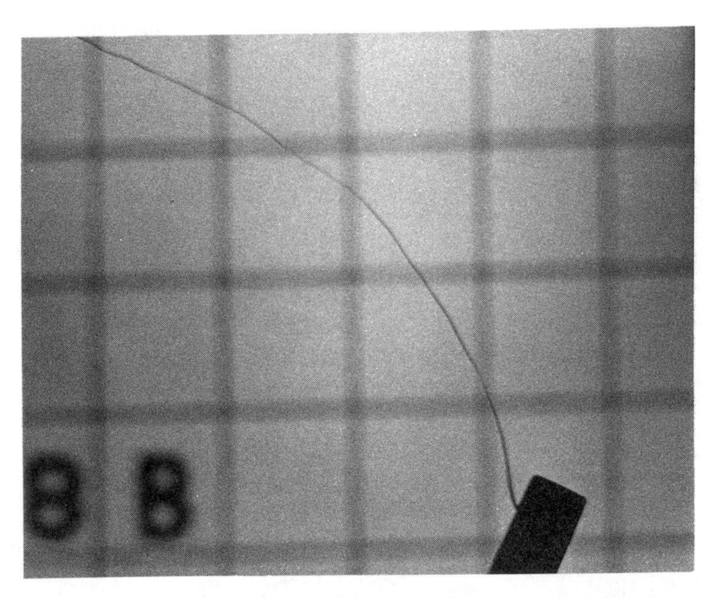

Figure 19.8. *Resin discharge from orifices in the trailing edge of the blending arm. Experimental parameters were: orifice: 0.120 in. diameter; feed rate: 1.7 lb/minute; rotation: 1100 rpm; resin viscosity: 415 cps. (Maloney and Huffaker 1984).*

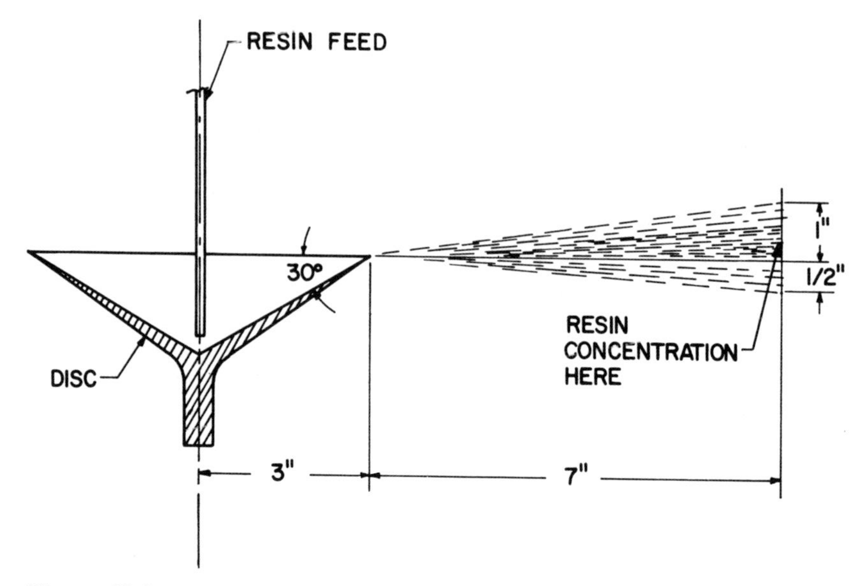

Figure 19.9. *Schematic drawing of inverted cone in operation. (Courtesy of Washington State University.)*

to be applied to strands and wafers at economical levels. Older spraying systems used in blenders did not apply resin well, therefore higher quantities of resin were needed to make a good product. The spinning cone development has reduced the amount of resin needed by providing better dispersion of the resin and better dis-

Figure 19.10. *Air driven spinning cone for atomizing resin (Coil and Kasper 1984).*

tribution of the resin over the strands. This development is now in trial in particleboard plants and it shows promise in reducing resin levels, as was also indicated in the Washington State University work.

Formers: With the advent of the oriented strand board business, there has been great progress in a number of different orienters for aligning the strands in making this product. The electrical orienting system (on page 172) has been developed to the commercial stage, but is not yet in production. It works extremely well with fibers through strands. This concept is illustrated in Figure 19.11. In pilot plant work making oriented strand lumber, bending properties that were as much as nine times greater in the oriented direction than in the cross-oriented direction were achieved.

One variation of a flake orienter is shown in Figure 15.30 (page 505). Figure 19.12 shows another version of a cross-orienter. The wood strands fall into the

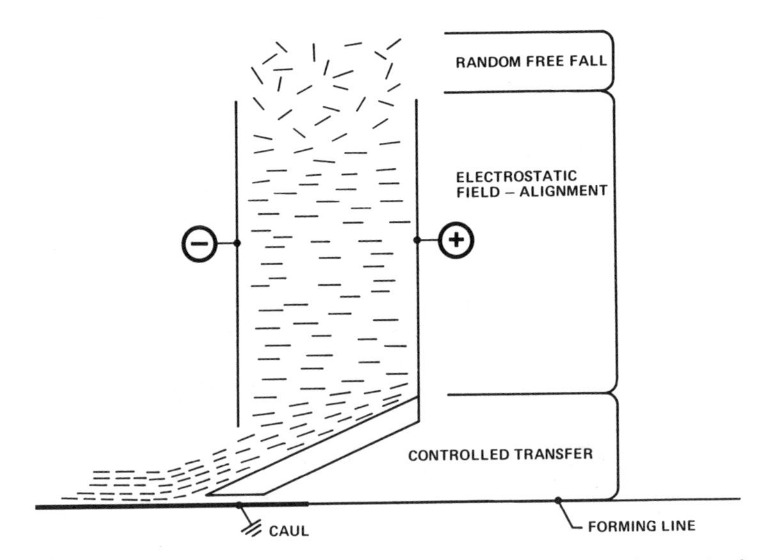

Figure 19.11. *Controlled transfer of oriented mat to forming line—single cell (Fyie et al. 1980).*

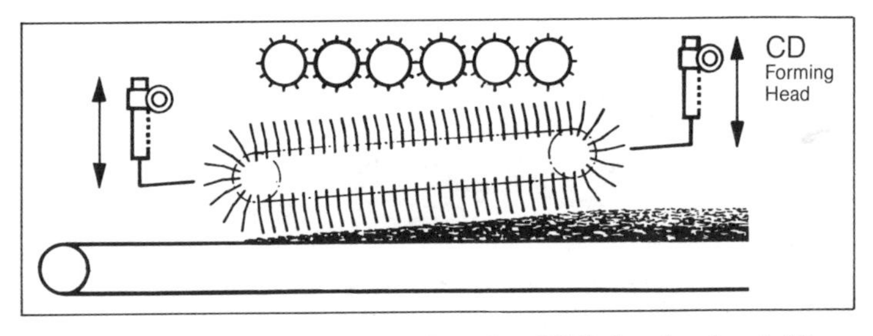

Figure 19.12. *The Siempelkamp cross-direction (CD) forming head. (Courtesy of Siempelkamp.)*

pockets of the orienter rolls as shown at the bottom of the diagram. The strands are then laid crosswise on the mat passing beneath much the same as in the machine in Figure 15.30.

There are different versions of machines for aligning the strands in the long direction. Figure 19.13 shows a disc orienter system for aligning particles. The strands are dropped on top of the disc shown at the bottom of the drawing (Figure 19.14). The discs force the strands to be aligned in the long direction and they are dropped onto the mat going beneath. These discs can also be organized in such a way that the bottom strands are deposited on the very bottom and the top of the mat passing beneath, with the shorter strands being aligned closer to the core of

Figure 19.13. *Shows the strands being dropped onto the disc head on the left side and traveling over the live discs toward the right side. (Courtesy of Siempelkamp.)*

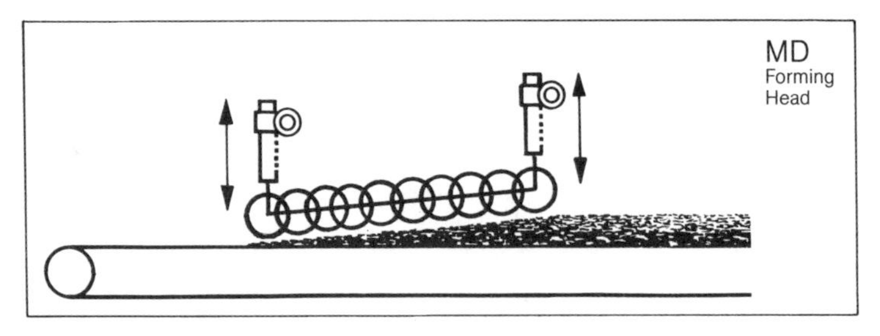

Figure 19.14. *The Siempelkamp machine direction (MD) forming head. (Courtesy of Siempelkamp.)*

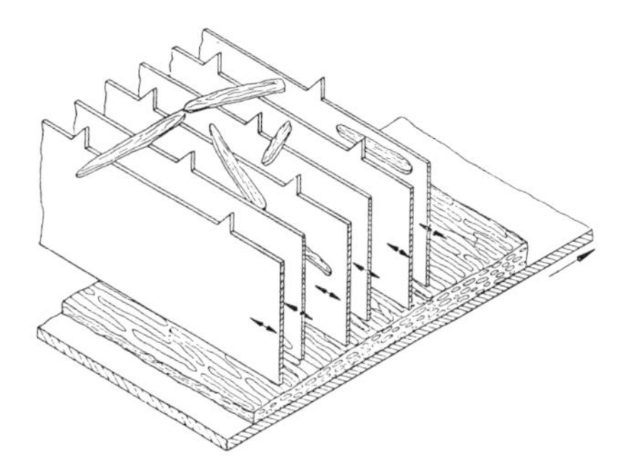

Figure 19.15. *The fin type of orienter. (Courtesy of Bison Werke.)*

the board. It is simply a matter of the spacing between the individual discs and shafts that determines the degree of alignment. A counter-moving fin system for lengthwise orienting is shown in Figure 19.15.

In the fiber arena, a number of new vacuum formers have been developed for producing MDF. Many sophisticated controls are part of the machines. The controls allow for a more uniform separation of the fiber going into the mat that results in a higher-quality pressed panel.

Presses: There have been a number of new press developments over the last several years. They include innovations in single-opening presses, continuous presses, and steam-injection presses. All of these have benefited from the rapid incorporation of computerized control systems, whereby the operation and the control of the press is far more sophisticated than it was only a short time ago.

In the single-opening area, there are now presses producing panels as large as 171 ft (52 mm) in length, 85 in. (2.2 m) in width, and ½ to 1.1 in. (12 to 28 mm) thick. Naturally, these panels are cut to size to fit the customer's needs after production. Producing such large panels dramatically reduces the amount of waste in the form of edge trim. For example, if a plant was producing a final board size of 4 by 8 ft (1.22 by 2.44 m) and trimming the edges of the panel by 2 in. (50.8 mm), the trim waste would be about 11.4%. However, a very large board has a significantly lower trim loss. The huge Dieffenbacher single-opening press at Egger in the United Kingdom has a platen format of 7.25 by 172 ft (2.2 by 52.4 m). A 2 in. (50.8 mm) trim on this board results in a trim loss of 5%—far less than the 11.4% found with a 4 by 8 ft (1.22 by 2.44 m) format. A 2-in.-wide trim, of course, is a rather wide trim, but this example illustrates the point.

Continuous presses have seen a spectacular improvement since 1980. The Mende line, as illustrated in Figures 17.45 and 17.46 (pages 564 and 565), is used widely for producing thin boards of the particleboard and MDF type. The heating drum has now been increased from the 10 ft (3.05 m) diameter to a diameter of 16.4 ft (5 m). Board widths range from 4 to 8 ft (1.22 to 2.44 m). In order to obtain increased production capacity and a better quality of product from these thin board presses, HF (high frequency) units have been installed between the forming machines and the infeed to the pressing drums. Also, the mat can be formed on fabric belts and then transferred to the steel belt running through the press itself. This eliminates the steel belt return from the press drums back underneath the forming drum and speeds the production process overall. Also, the steel belt maintains a constant heat by not being cooled down as it passes back under not only the whole system, but also under the former to pick up the next mat.

The great advance in continuous pressing in recent years has taken place in the presses that produce thicker boards. These presses are being used to produce particleboard, MDF, and OSB, LVL, and OSL (Parallam). Some of the advantages of these boards are: very close control of thickness tolerance, no precured surfaces, no cycle–dead time for loading/unloading as found with multi-opening and single-opening presses, quick change in producing different thicknesses of panels, better control of the thickness profile of the board, savings in wood and resin by greatly reducing the sanding required, and a reduction of the trim loss as illustrated earlier with the single-opening press. With the continuous press, however, there is no trim loss at the end of the sheet except for the width at the kerf of the sawblade. It is claimed that savings reach at least 20% of the material when board is produced using the continuous press.

The continuous presses have been developed by Küsters, Bison, Siempelkamp, and Dieffenbacher. All use steel belts to move the mat through the press as it is being consolidated into the final board. However, each press has its own design for supporting and moving the steel belts. This has been a difficult problem to solve because the presses are applying considerable pressure to the mat being consolidated into the final board.

For the Küsters design, there are a number of identical frames within this press, as shown in Figure 19.16, that are in the shape of I-beams. The steel belts are moved through the press on beds of roller-chains. These roller-chains, approximately 2.2 in. (55 mm) wide, are shown in Figure 19.17. A multitude of these chains set close together establish rolling beds in the press for moving the steel belt and the mat through the press. Figure 19.18 shows how these belts travel through the press and are returned through the press platen to the head of the press. Figure 19.19 shows a cross-section of the press, illustrating how the chains are heated by means of the internal heating platen and how the thickness is controlled. This press has, as do the others in the continuous field, an adjustable infeed arrangement, as shown in Figure 19.20, that assists with the control and produces the desired density profile through the board. This particular press system is also used for producing Parallam using Trus Joist Macmillan's proprietary Microwave Type Heating System.

The Bison system (the Hydro-Dyn continuous press [Figure 19.21]) has taken a different approach to moving the mat through the press. The upper heating platen is vertically adjustable and is suspended from a number of hydraulic pistons and

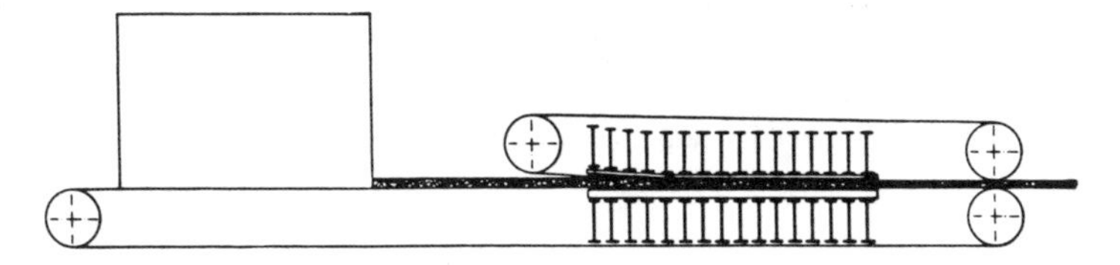

Figure 19.16. *General view of the Küsters Press® (Ahrweiler 1982).*

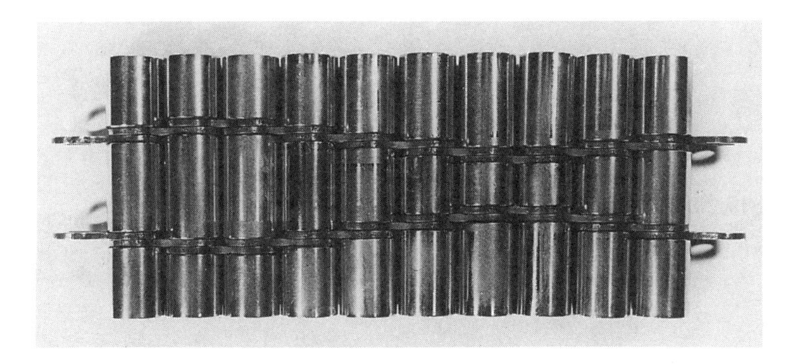

Figure 19.17. *Roller chain used in the press (Ahrweiler 1982).*

Figure 19.18. *The Küsters roller chains. (Courtesy of Edward Küsters.)*

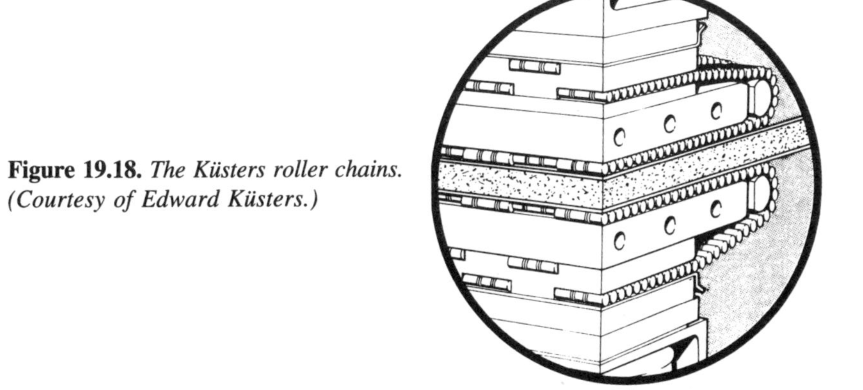

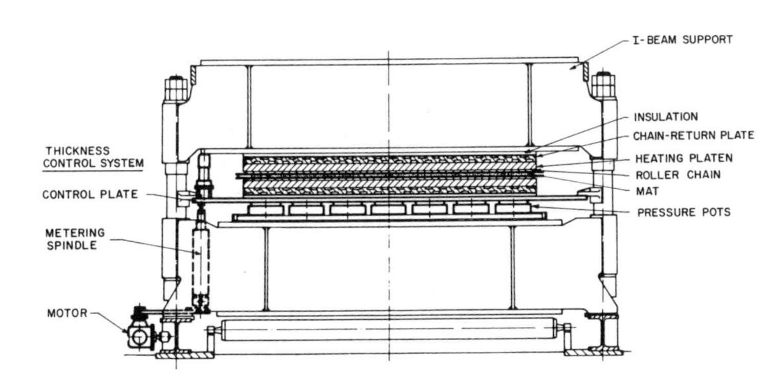

Figure 19.19. *System for maintaining uniform board thickness. Also shown is the heating system. (Ahrweiler 1982).*

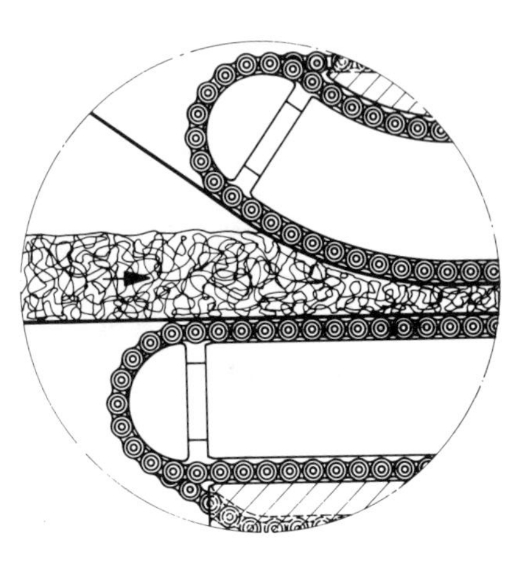

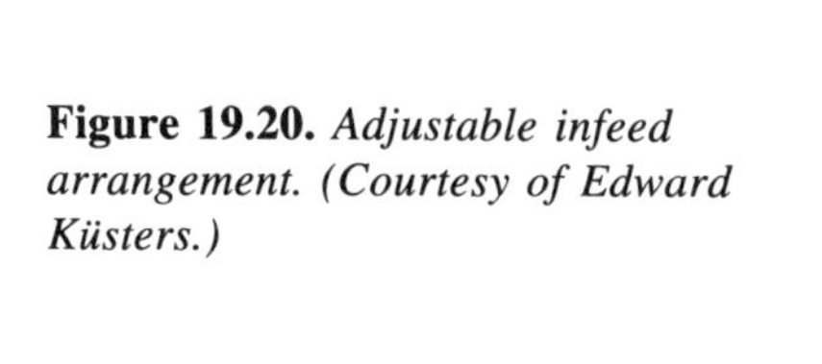

Figure 19.20. *Adjustable infeed arrangement. (Courtesy of Edward Küsters.)*

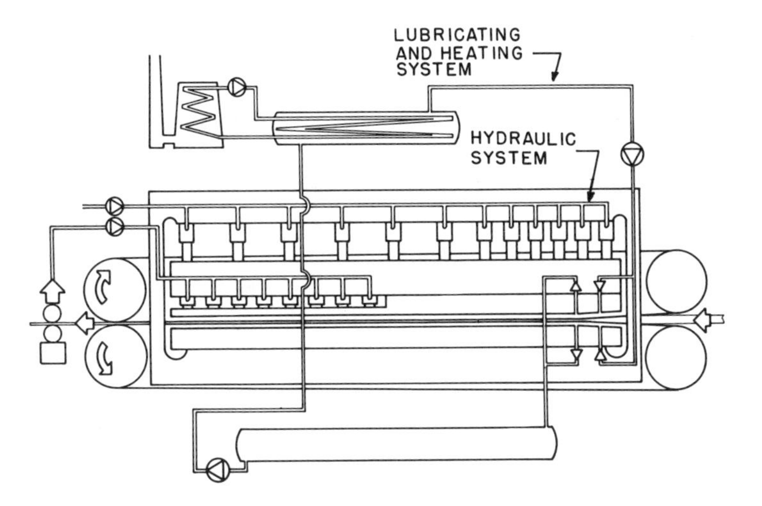

Figure 19.21. *Schematic of the ContiRoll Press operation (Sitzler and Vining 1986).*

cylinders that are adjusted to the board thickness required. The steel belts are preheated by the infeed drive drums. The top and bottom platens of the press have small holes drilled in the surface. It is here that hot oil flows, providing not only the heat for the steel belts, but also the lubrication for them to move freely over the surface of the platens. The holes are placed strategically for the infeed of the hot oil, so that at the out-feed, the cool oil is returned to the hot oil heat exchanger. As with all the continuous presses, there are different pressure zones in the press. The inlet is wedge shaped and adjustable, as mentioned previously, to allow for control of the preferred density profile of the board being produced. The inlet zone of the press is a high-pressure zone and this is followed by subsequent zones of lower pressure.

In essence, in actually consolidating the mat into a board, the continuous press works just the same as single-opening and multi-opening presses. The continuous press can apply high pressure on the mat immediately to densify surfaces, followed by a lower pressure when the board has reached its final thickness, or the pressing can be controlled to provide a panel with a relatively uniform density profile. The continuous presses can also be used to apply laminates on board surfaces.

The Siempelkamp system, known as the ContiRoll Press, is shown in Figure 19.22. Figure 19.23 shows a schematic of how the press functions. As is obvious with any of the continuing presses, the problem is to move the steel belt through the press easily, while simultaneously applying pressure and consolidating the wood mat into the higher-density final product. In the ContiRoll, both the upper and the

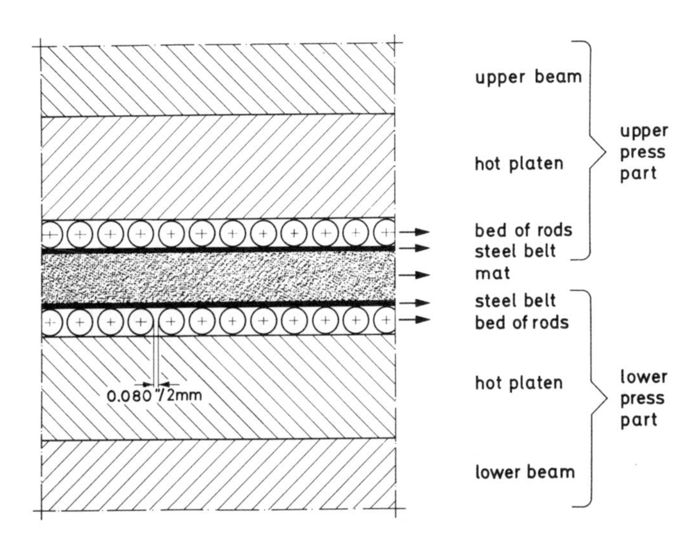

Figure 19.22. *Hydro-Dyn Press system (Krichel and Wentworth 1983).*

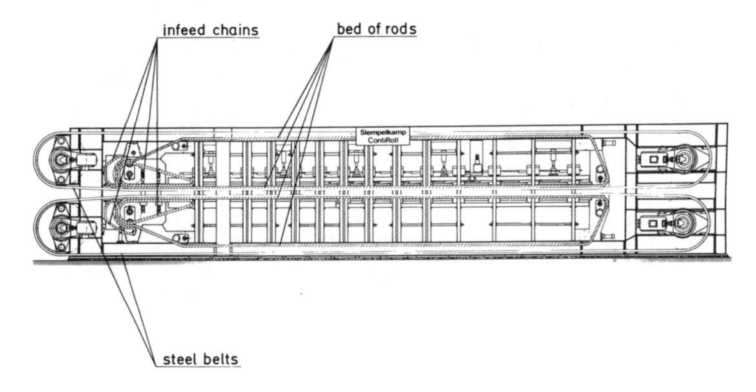

Figure 19.23. *The ContiRoll Press (Sitzler and Vining 1986).*

lower part of the press incorporate a bed of rods located between the steel belts and the hot platens to move the mat through the press. In the lower section, the rods return below the press to the head of the press; the upper bed of rods is returned to the infeed end in the upper section of the press. Figure 19.24 shows this bed of rods, also known as a "rolling grizzly," in the ContiRoll Press. Figure 19.25 shows a ContiRoll Press in operation.

The latest entry into the field of the continuous press was developed by Dieffenbacher. This system, known as the ContiPower System, is illustrated in Figure 19.26. It uses a wedge-shaped press inlet to control the density profile of the final product (Figure 19.27). This wedge positioner is heated, so both heat and pressure are being applied at the very beginning of the press cycle. This press system also uses rolling rods for moving the steel belts through the press easily. The platens are divided into five different heating circuits. Each of the heating circuits is individually controlled and each one is adjustable to different temperatures. This is illustrated in Figure 19.28.

Figure 19.24. *Bed of rods operating between the steel belt and the stationary hot platen (one each for the top and bottom) (Sitzler and Vining 1986).*

Figure 19.25. *The ContiRoll Press (Long 1989).*

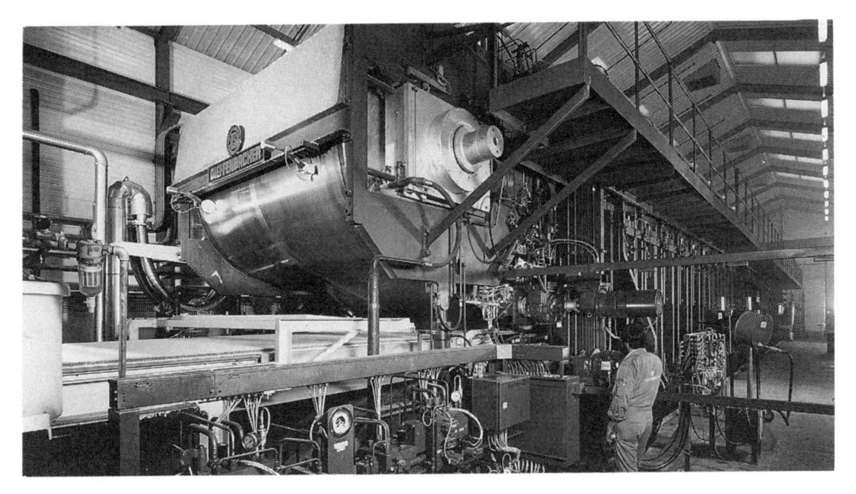

Figure 19.26. *The ContiPower continuous press. (Courtesy of Dieffenbacher.)*

The development of steam-injection pressing has resulted in the installation of these presses in a number of places throughout the world. The use of steam-injection pressing drastically reduces the amount of time in the press for curing the resin. Also, the injection of steam into the mat partially plasticizes the furnish, thereby reducing the closing pressure to as little as 40% of that needed in a conventional press. This process also creates a very flexible control over the vertical density profile or thickness profile of the board through proper choices of onset of the steam, duration of the steam, and the press closure rate. It has been shown that with

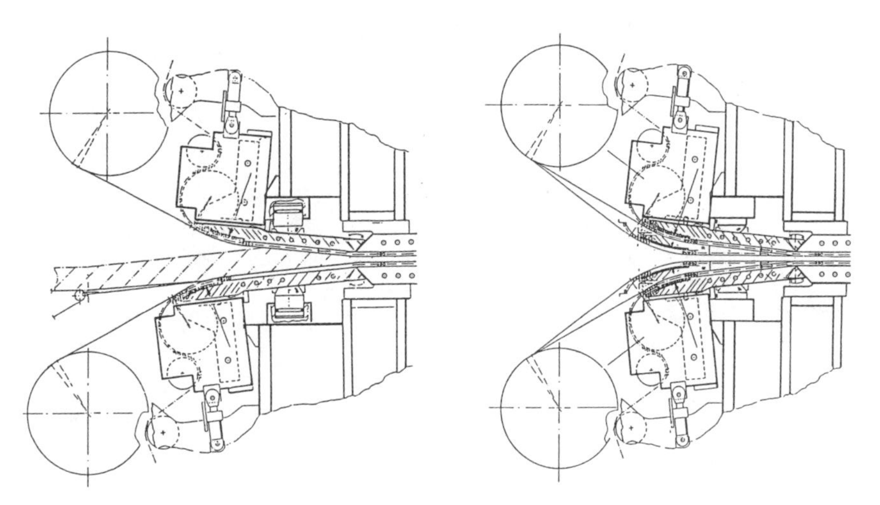

Figure 19.27. *Compression of the mat in the wedge positioner (Kohler et al. 1991).*

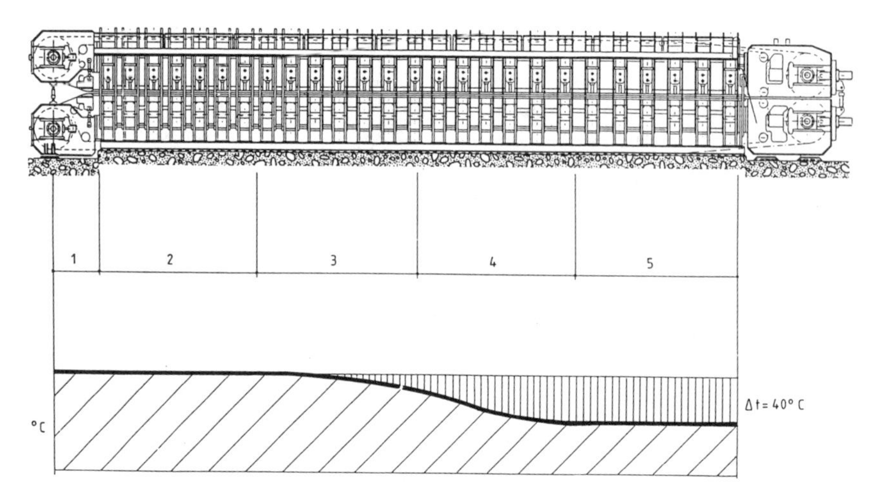

Figure 19.28. *The five heating zones of the CPS (No. 1 is the wedge positioner) (Kohler et al. 1991).*

phenolic resins the curing time of the resin can be reduced to about 90 seconds. With thicker boards, steam-injection pressing, therefore, becomes a very important possibility for anyone considering their production.

Figure 19.29 shows the concept of steam-injection pressing. A schematic of the steam-injection press system is shown in Figure 19.30. In simple terms, the press is closed, air is evacuated out of the mat, and steam is injected. It is important to remove the cool air out of the mat so that the steam can do its job properly and

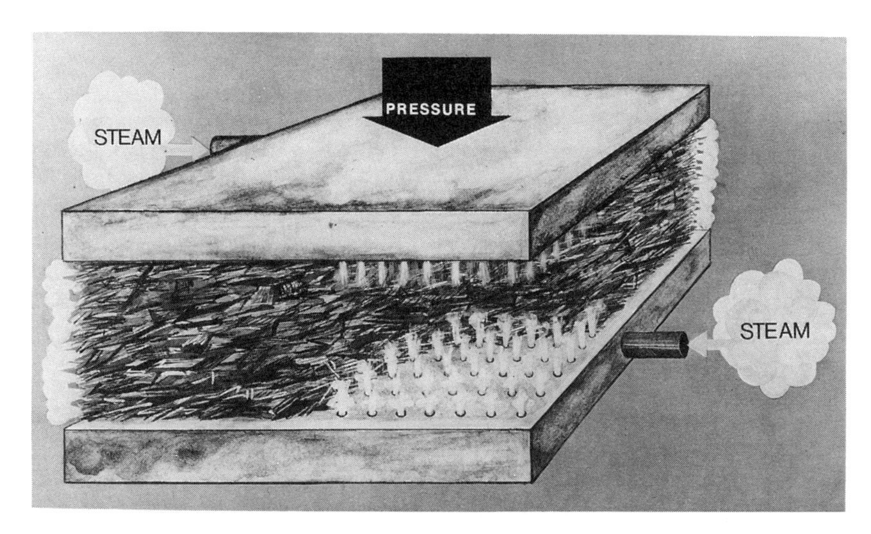

Figure 19.29. *Saturated steam is injected into the mat through top and bottom perforated platens during press closure (Geimer and Price 1986).*

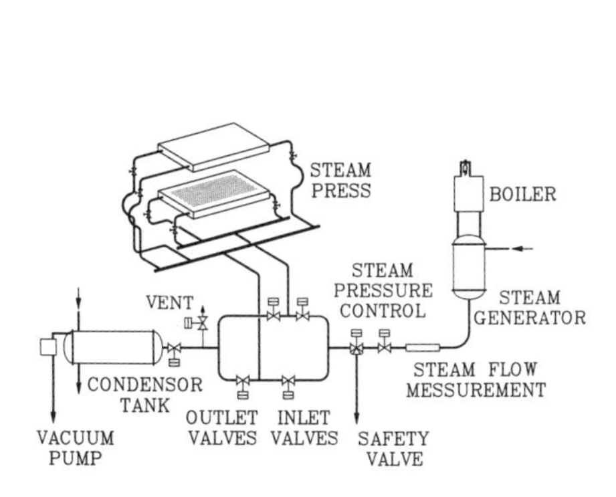

Figure 19.30. *The steam-injection press system (Ramos 1991).*

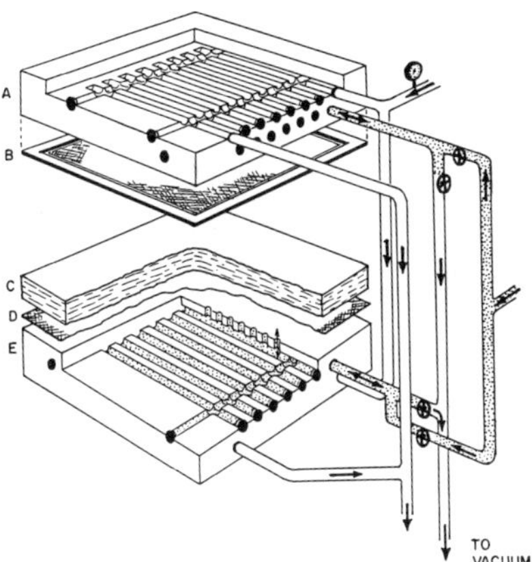

Figure 19.31. *Schematic of a self-sealing steam press (A = upper platen [heating channels shown], B = compression frame and upper screen, C = mat, D = lower screen, E = lower platen [injection channels shown]) (Hsu 1991).*

quickly. It is also necessary to then remove part of the steam before opening the press so the vapor pressure within the mat is reduced. This avoids blows.

A self-sealing steam process has been developed by W. Ernst Hsu at Forintek in Canada. In this process, a frame attached to the press platen is used around the mat being compressed into the final product as shown in Figure 19.31. This frame seals the edges of the mat, preventing the loss of steam. A reduction in press time, greater dimensional stabilization of the product, ability to cure higher-temperature-curing resins, good steam penetration even when large flakes or wafers are used, low loss of steam, low risk of explosion, and an easy control of density profiles are advantages claimed for this later generation of the steam-injection press.

Recycled Wood: A number of devices have beeh developed over the years to reclaim recycled wood. Recycling of waste wood is not new, going back many, many years in a large part of the world. However, in the last several years, a greater interest has been expressed in recycling of wood, because of the environmental pressures being placed on the disposal of wood. Also, in some areas, a shortage of wood material for board plants makes recycled wood a possible valuable raw material. Figure 19.32 shows a device for preparing particles from recycled wood.

As in most production processes, when relatively large pieces of material for recycling are brought in, it appears that at least a two-step process is needed. The first step is to reduce the recycled wood from its large size as either individual boards, pallets, packing cases, etc., to relatively large pieces; secondary reduction in industry terms reduces these pieces to a smaller size. Usually, after the first reducing step, the material is passed beneath magnets that remove much of the

Figure 19.32. *Machine for preparing particles from recycled wood (Fischer 1992).*

ferrous metal. Magnets also are used after the second reduction step for further removal of ferrous metals. The problem with any recycled material is that other materials have been thrown in with the wood because many people consider this to be garbage and anything can go into the waste stream. This provides great problems in removing other material, such as plastics. Dirt also has to be removed either by air classifying or screening.

CONCLUSION

Much has happened in the field of composites since 1950. Materials are now being manufactured in quantities that once were only dreams.

The composite products part of the forest products industry is positioned to economically provide vitally needed building materials and other materials to the world society. Composites have been responsible for the greater use of wood raw materials and this trend should increase. High technology makes it possible to supply reasonably priced composites.

The challenge to the composites industry is to not only develop wood composites to replace older, solid-wood products, but to develop wood composites that replace materials made of nonrenewable resources. Many of these composites will probably be marriages of several materials. Such developments are critical to enjoying high living standards in the 21st century.

SELECTED REFERENCES

Ahrweiler, K. 1982. The Continuous Küsters Press System. *Proceedings of the Washington State University Particleboard Symposium, No. 16*. Pullman, Washington: Washington State University (WSU).

Alvarez-Novoa, C. 1990. The Tafisa Plant in Spain. *Proceedings of the Washington State University Particleboard/Composite Materials Symposium, No. 24*. Pullman, Washington: WSU.

American Hardboard Association. 1988. *American National Standard—Basic Hardboard,* ANSI/AHA A135.4-1982. Palatine, Illinois: American Hardwood Association.

American Hardboard Association. 1990. *American National Standard—Hardboard Siding,* ANSI/AHA A135.6-1990. Palatine, Illinois: American Hardwood Association.

American Plywood Association. 1991. *Performance Standards and Policies for Structural-Use Panels,* APA PRP-108. Tacoma, Washington: American Plywood Association.

Andrews, R. 1988. Hardishake Roofing Development. *Proceedings, Composite Wood Products Symposium*. Rotorua, New Zealand: Forest Research Institute.

Anonymous. 1990. *PSL Newsletter*, Volume 1, Issue 1 (September) Plymouth, Minnesota.

Automated Builder. 1990. The Market, the 1989 U.S. Housing Pie. *Automated Builder,* Vol. 6.

Ball, G. W. 1981. New Opportunities in Manufacturing Conventional Particleboard Using Isocyanate Binders. *Proceedings of the Washington State University Particleboard Symposium, No. 15*. Pullman, Washington: WSU.

Borgrevink, G. 1980. Fabricating and Using MDF. *Proceedings of the Washington State University Particleboard Symposium, No. 14*. Pullman, Washington: WSU.

Canadian Standards Association. 1991. *Standards on Construction Sheathing,* CSA 0325-91. Rexdale, Ontario, Canada: Canadian Standards Association.

Canadian Standards Association. 1992 (under revision). *Standards on OSB and Waferboard,* CAN/CSA 0437-92. Rexdale, Ontario, Canada: Canadian Standards Association.

Coil, G. K., and J. B. Kasper. 1984. A Liquid Blending System Developed by Mainland Manufacturing. *Proceedings of the Washington State University Particleboard/Composite Materials Symposium, No. 18*. Pullman, Washington: WSU.

Detlefsen, W. P. 1987. Making Plywood from High-Moisture Veneer. Presented at the Forest Industries 1987 Clinic and Machinery Show. Portland, Oregon, March 4–6, 1987.

Donnell, R. 1990. Highland American Starts Gypsum Fiberboard Plant. *Panel World,* Vol. 11.

Ellingson, G. P. 1985. Production of Isocyanate-Bonded Particleboard. *Proceedings of the Washington State University Particleboard/Composite Materials Symposium, No. 19*. Pullman, Washington: WSU.

FAO. 1990. FAO World Survey of Production Capacities: Plywood, Particleboard, and Fibre Board 1989. Food and Agriculture Organization of the United Nations Committee on Wood-Based Panel Products, Rome, Italy, September 1990.

Fischer, K. 1992. Scrap Wood Recycling—Raw Material Source for the Board Industry. *Proceedings of the Washington State University Particleboard/Composite Materials Symposium, No. 26*. Pullman, Washington: WSU.

Frashour, R. 1990. Production Variables, Blowline, and Alternative Blending Systems for Medium Density Fiberboard. *Proceedings of the NPA Resin and Blending Seminar,* John Bradfield, Editor. Gaithersburg, Maryland: National Particleboard Association.

Fyie, J. A. 1980. Electrostatic Systems for Engineering Composition Panel Properties. *Proceedings of the Washington State University Particleboard Symposium, No. 14.* Pullman, Washington: WSU.

Fyie, J. A., T. E. Peters, and D. J. Henckel. 1980. Electrostatic Orientation Systems for Engineering Composition Panel Properties. *Proceedings of the Washington State University Particleboard Symposium, No. 14.* Pullman, Washington: WSU.

Geimer, R. L. 1982. Steam Injection Pressing. *Proceedings of the Washington State University Particleboard Symposium, No. 16.* Pullman, Washington: WSU.

Geimer, R. L., and E. W. Price. 1986. Steam Injection Pressing—Large Panel Fabrication with Southern Hardwoods. *Proceedings of the Washington State University Particleboard/ Composite Materials Symposium, No. 20.* Pullman, Washington: WSU.

Godfrey, J. 1991. Plant Experience in Gypsum Fiberboard Production in Rhode Island, USA. *Proceedings of the Second International Inorganic Bonded Wood and Fiber Composite Materials Conference,* University of Idaho, Forest Products Research Society, Madison, Wisconsin.

Groah, W. 1986. Hardwood Plywood Association. Personal communication. Reston, Virginia.

Heebink, B. G., and F. V. Hefty. 1968. Steam Post Treatments to Reduce Thickness Swelling of Particleboard. *Proceedings of the Washington State University Particleboard Symposium, No. 2.* Pullman, Washington: WSU.

Hsu, W. E. 1987. Post-Treatment Process for Stabilizing Waferboard. *Proceedings of the Washington State University Particleboard/Composite Materials Symposium, No. 21.* Pullman, Washington: WSU.

__________. 1989. Steam Pretreatment for Dimensionally Stabilizing UF-Bonded Particleboard. *Proceedings of the Washington State University Particleboard/Composite Materials Symposium, No. 23.* Pullman, Washington: WSU.

__________. 1991. A Practical Steam Pressing Technology for Wood Composites. *Proceedings of the Washington State University Particleboard/Composite Materials Symposium, No. 25.* Pullman, Washington: WSU.

Johns, W. E., G. C. Myers, and W. K. Motter. 1987. Bondability of Wood Surfaces. *Proceedings of the Washington State University Particleboard/Composite Materials Symposium, No. 21.* Pullman, Washington: WSU.

Kikata, Y. 1990. Zephyr Wood, a Network of Continuous Fibers: Defibration of Wood by Roller Crushing and Recomposition of It. Presented at XIX IUFRO World Congress, Division 5, Montreal, Canada, August 5–11.

Knokey, A. 1987. New Gluing Techniques Related to Veneer Drying. Presented at the Forest Industries 1987 Clinic and Machinery Show, Portland, Oregon, March 4–6, 1987.

Knudson, R. M. 1992. PSL 300™ LSL: The Challenge of a New Product. *Proceedings of the Washington State University Particleboard/Composite Materials Symposium, No. 26.* Pullman, Washington: WSU.

Koenigshof, G. A. 1983. Use of Electrically Oriented Core for COM-PLY Truss Lumber. *Proceedings of the Washington State University Particleboard/ Composite Materials Symposium, No. 17.* Pullman, Washington: WSU.

Kohler, E., H. Günzerodt, and R. A. Wood. 1991. New Control Flexibility for More Product Variability in Continuous Pressing of Wood Composites. *Proceedings of the Washington State University Particleboard/Composite Materials Symposium, No. 25.* Pullman, Washington: WSU.

Krichel, H., and I. Wentworth. 1983. Continuous Board Pressing Technology. *Proceedings of the Washington State University Particleboard/Composite Materials Symposium, No. 17.* Pullman, Washington: WSU.

Lambuth, A. L. 1985. Adhesives for Wood, Where From Here? *Proceedings, Wood Adhesives in 1985: Status and Needs.* Madison, Wisconsin: Forest Products Research Society.

Larsson, L. 1989. *MDF Technology Updated.* Sundsvall, Sweden: Sunds Defibrator AB.

Lehmann, W. F., and E. Roffael. 1992. International Guidelines and Regulations for Formaldehyde Emissions. *Proceedings of the Washington State University Particleboard/ Composite Materials Symposium, No. 26.* Pullman, Washington: WSU.

Lembke, C. A. 1986. Veneer Mill Extends Resource Value. *Australian Forest Industries Journal,* Vol. 11.

Long, P. J. 1989. Continuous Pressing of Medium-Density Fibreboard at Fibron Industries. *Proceedings of the Washington State University Particleboard/Composite Materials Symposium, No. 23.* Pullman, Washington: WSU.

Luckman, E. R., and H. F. Pagel. 1983. Manufacture of Formaldehyde-Free Plywood. Presented at the Third Annual International Symposium on Adhesion and Adhesives for Structural Materials, Pullman, Washington, September 27–29.

Maloney, T. M. 1982. Masonite Denies Hardboard's Fall with Introduction of Woodruf. *Plywood and Panel World,* Vol. 9.

__________. 1984. Electrically Oriented Composition Boards Made With Smaller Size Strands and Particles. *Proceedings, Comminution of Wood and Bark.* Madison, Wisconsin: Forest Products Research Society.

__________. 1987. Technology Extending the Resource. *Proceedings, Expert Consultation on Wood-Based Panels.* Rome, Italy: Food and Agriculture Organization of the United Nations.

Maloney, T. M., and E. M. Huffaker. 1984. Blending Fundamentals and an Analysis of a Short-Retention-Time Blender. *Proceedings of the Washington State University Particleboard/Composite Materials Symposium, No. 18.* Pullman, Washington: WSU.

Marenzi, L. 1987. Old Technology Still Valid: Blockboard. *Proceedings, Expert Consultation on Wood-Based Panels.* Rome, Italy: Food and Agriculture Organization of the United Nations.

Maylor R. 1988. Phenolic Resins for Gluing High Moisture Veneers. *Proceedings, Composite Wood Products Symposium.* Rotorua, New Zealand: Forest Research Institute.

Mayner, J. A. 1990. The Economics of Continuous Pressing. *Proceedings of the Washington State University Particleboard/Composite Materials Symposium, No. 24.* Pullman, Washington: WSU.

McCafferty, P. 1990. Reinventing Wood. *Popular Science,* Vol. 5.

McCredie, W. H. 1992. Formaldehyde Emissions from UF Particleboard, Voluntary Standards vs. EPA Regulation. *Proceedings of the Washington State University Particleboard/ Composite Materials Symposium, No. 26.* Pullman, Washington: WSU.

Moore, G. 1988. Triboard and Steam Pressing. *Proceedings, Composite Wood Products Symposium.* Rotorua, New Zealand: Forest Research Institute.

National Particleboard Association. 1992. *Standard for Particleboard for Mobile Home Decking,* NPA 1-82. Gaithersburg, Maryland: National Particleboard Association.

National Particleboard Association. 1992. *American National Standard Medium Density Fiberboard for Interior Use,* ANSI/A208.2-1986. Gaithersburg, Maryland: National Particleboard Association.

National Particleboard Association. 1993. *American National Standard Wood Particleboard* ANSI/A208.1-1993 (under revision). Gaithersburg, Maryland: National Particleboard Association.

Natus, G. 1991. Gypsum Fiberboard Production in Nova Scotia, Canada. Proceedings of the Second International Inorganic Bonded Wood and Fiber Composite Materials Conference, University of Idaho, Forest Products Research Society, Madison, Wisconsin.

O'Halloran, M. 1980. The Performance Approach to Acceptance of Building Products. *Proceedings of the Washington State University Particleboard Symposium, No. 14.* Pullman, Washington: WSU.

Pallman, H-I. 1987. A Revolutionary Concept for Flaking Random Length Logs and Slabs. *Proceedings, Structural Wood Composites: New Technologies for Expanding Markets.* Madison, Wisconsin: Forest Products Research Society.

__________. 1990. Improved Flake and Particle Generation for the Board Industry. *Proceedings of the Washington State University Particleboard/Composite Materials Symposium, No. 24.* Pullman, Washington: WSU.

Parker, D. J. 1987. Parallam Parallel Strand Lumber: Evolution. *Proceedings of the Washington State University Particleboard/Composite Materials Symposium, No. 21.* Pullman, Washington: WSU.

Pierce, J. 1984. A New Approach to Blending—Conical Atomizers. *Proceedings of the Washington State University Particleboard/Composite Materials Symposium, No. 18.* Pullman, Washington: WSU.

Pollman, D. 1990. Merritt Plywood Machinery Inc., Lockport, New York. Personal communication.

Post, H. 1990a. The Homosote Story. *Wood Based Panels International,* Vol. 11.

Post, H. 1990b. Lumbering On. *Wood Based Panels International,* Vol. 10.

Potter, E. E. 1975. Ced-Rite Particleboard. Background Paper No. 64, World Consultation on Wood-Based Panels, FAO New Delhi, India, February, 1975.

Ramos, J. M. 1991. The MDF Industry in Financiera Maderera S.A. *Proceedings of the Washington State University Particleboard/Composite Materials Symposium, No. 25.* Pullman, Washington: WSU.

Scholes, W. A. 1990. Scrimber: Another Way to Stretch Timber Supply. *World Wood,* Vol. 10.

Shen, K. C. 1988. Binderless Composite Panel Products. *Proceedings, Composite Wood Products Symposium.* Rotorua, New Zealand: Forest Research Institute.

Sitzler, H. D., and S. Vining. 1986. New Developments in Continuous Pressing of Panel Products. *Proceedings of the Washington State University Particleboard/ Composite Materials Symposium, No. 20.* Pullman, Washington: WSU.

Stevenson, J. 1990. Not a Bad Idea. *Timber Processing,* Vol. 10.

Sturgeon, M. G., and N. M. Lau. 1992. GoldenEdge Liteboard (600 kg/m^3) Manufactured at Nelson Pine Industries. *Proceedings of the Washington State University Particleboard/ Composite Materials Symposium, No. 26.* Pullman, Washington: WSU.

Sundin, B. 1990. Erfarenheter av Olika Formaldehydanalyser Utforda pa E1-Skivor. *Proceedings, Dynobels Skivdagar i Stenungsbaden.* Stockholm, Sweden.

Triolo, P. 1986. Molded Hardboard—Past and Present. Presented at the Fourtieth Annual Meeting of the Forest Products Research Society, Spokane, Washington, June 22–26, 1986.

Voss, N. A. 1991. Impact of Changes in Industrial Particleboard Manufacture. *Proceedings of the Washington State University Particleboard/Composite Materials Symposium, No. 25.* Pullman, Washington: WSU.

Wentworth, I. 1975. Manufacture of Particleboard and Fiberboard with Flat and Rotary Continuous Presses and Continuous Lamination of Thin Boards. *Proceedings of the Washington State University Particleboard Symposium, No. 9.* Pullman, Washington: WSU.

Wiecke, P. H. 1984. Will Continuous Presses Replace Existing Presslines for Wood Based Panels? *Proceedings, New Available Techniques.* The World Pulp and Paper Week, April 10–13. Stockholm, Sweden: The Swedish Association of Pulp and Paper Engineers.

__________. 1990. Worldwide Development with MDF. *Proceedings of the Washington State University Particleboard/Composite Materials Symposium, No. 24.* Pullman, Washington: WSU.

Woodworking International. 1985. Revolutionary Development in Veneer Slicing. *Woodworking International.* No. 1.

Youngquist, J. A., and R. M. Rowell. 1988. Can Chemical Modification Technology Add Value to Your Products? *Proceedings of the Washington State University Particleboard/ Composite Materials Symposium, No. 22.* Pullman, Washington: WSU.

APPENDIX

MEASURES AND BOARD THICKNESSES

Forest Products Measures

Product and Unit	*Cubic meters*	*Cubic feet*	*1000 board feet*
Panels			
1000 sq m (1 mm thick)	1	35.315	0.4238
1000 sq ft (1/8 in. thick)	0.295	10.417	0.125

Product	*Kilograms per cubic meter*	*Cubic meters per metric ton*
Particleboard	650	1.54
Fiberboard		
Compressed	950	1.053
Non-compressed	250	4

Source: Yearbook of Forest Products, Food and Agriculture Organization of the United Nations, Rome, Italy.

Board Thicknesses

in.	*mm*	*in.*	*mm*
1/8	3.2	9/16	14.3
3/16	4.8	5/8	15.9
1/4	6.4	11/16	17.5
3/8	9.5	3/4	19.1
7/16	11.1	7/8	22.2
1/2	12.7	1	25.4

UNITS AND CONVERSION FACTORS

*Basic SI Units**

Quantity	*Unit*	*Symbol*
Length	metre	m
Mass	kilogram	kg
Time	second	s
Electric current	ampere	A
Temperature	kelvin	K
Amount of substance	mole	mol
Luminous intensity	candela	cd

* The International System of Units (Système International d'Unités or SI)

Derived SI Units of Use in the Board Industry

Quantity	*Unit*	*SI Symbol*	*Formula*
Area	square meter	–	m^2
Density	kilogram per cubic meter	–	kg/m^3
Energy	joule	J	N·m
Force	newton	N	$kg{\cdot}m/s^2$
Frequency	hertz	Hz	(cycles)/s
Power	watt	W	J/s
Pressure	pascal	Pa	N/m^2
Quantity of heat	joule	J	N·m
Stress	pascal	Pa	N/m^2
Voltage	volt	V	W/A
Volume	cubic meter	–	m^3

Multiples and Submultiples of Base Units

Example– Base Unit = Meter

Multiplication factor	*Prefix*	*Symbol*	*Multiplication factor*	*Quantity*	*Symbol*
10^{12}	tera	T	10^{12} meters	terameter	Tm
10^{9}	giga	G	10^{9} meters	gigameter	Gm
10^{6}	mega	M	10^{6} meters	megameter	Mm
10^{3}	kilo	k	10^{3} meters	kilometer	km
10^{2}	hecto*	h	10^{2} meters	hectometer	hm
10^{1}	deka*	da	10^{1} meters	dekameter	dam
10^{0}			10^{0} meters	meter	m
10^{-1}	deci*	d	10^{-1} meters	decimeter	dm
10^{-2}	centi*	c	10^{-2} meters	centimeter	cm
10^{-3}	milli	m	10^{-3} meters	millimeter	mm
10^{-6}	micro	μ	10^{-6} meters	micrometer	μm
10^{-9}	nano	n	10^{-9} meters	nanometer	nm
10^{-12}	pico	p	10^{-12} meters	picometer	pm
10^{-15}	femto	f	10^{-15} meters	femtometer	fm
10^{-18}	atto	a	10^{-18} meters	attometer	am

*** Multiple and prefixes representing steps of 1000 are recommended. For example, show force in mN, N, kN, and length in mm, m, km, etc. The use of centimeters should be avoided unless a strong reason exists. An exception is made in the case of area and volume used alone. With SI units of higher order such as m^2 and m^3, the prefix is also raised to the same order; for example, mm^3 is $10^{-9}\,mm^3$ not $10^{-3}mm^3$. In such cases, the use of cm^2, cm^3, dm^2, dm^3, and similar nonpreferred prefixes is permissible.**

When expressing a quantity by a numerical value and a unit, prefixes should preferably be chosen so that the numerical value lies between 0.1 and 1000, except where certain multiples or submultiples have been agreed to for particular use. The same unit, multiple or submultiple is used for tabular values even though the series exceeds the preferred range 0.1 to 1000.

Conversion Factors Used Frequently

From	*To*	*Multiply by*
Length		
millimeter (mm)	inches (in.)	0.0394
meter (m)	in.	39.37
meter	feet (ft)	3.2808
mils or 0.001 in.	mm	0.0254
inch	mm	25.4
feet	m	0.3048
yard (yd)	m	0.9144
Area		
square mm	sq in.	0.0016
sq m	sq ft	10.7639
sq m	sq yd	1.196
sq in.	sq mm	645.16
sq ft	sq m	0.0929
sq yd	sq m	0.8361
Volume		
cubic mm	cu in.	0.00006
cu m	cu ft	35.3149
cu m	cu yd	1.3080
cu in.	cu mm	16387.1
cu ft	cu m	0.0283
cu yd	cu m	0.7646
Mass		
gram	ounce, avoirdupois	0.0353
kilogram	pounds (lb)	2.2046
metric ton	short ton (2,000 lbs)	1.1023
ounce, avoirdupois	gram	28.3495
lb	kilogram	0.4536
short ton	metric ton	0.9072
Density		
grams per cu cm	pounds per cu ft	62.43
kilograms per cu m	pounds per cu ft	0.0624
pounds per cu ft	grams per cu cm	0.0160
pounds per cu ft	kilograms per cu m	16.0256

Other Important Units

From	*To*	*Multiply by*
pound-force per sq in. (psi)	kilopascal (kPa)	6.895
pound-force per sq in. (psi)	Megapascal (MPa)	0.006895
kilowatt-hour	joule (J)	3 600 655
horsepower (electric)	watt (W)	746
horsepower hour	kilowatt hour (kWh)	0.746
horsepower day	kilowatt hour (kWh)	17.9
degree Fahrenheit	Celsius (C)	$t_c = (t_f - 32)/1.8$
degree Celsius*	Fahrenheit (F)	$t_f = 1.8\ t_c + 32$
degree kelvin	degree Celsius	$t_c = t_k - 273.15$
degree Celsius	degree kelvin (K)	$t_k = t_c + 273.15$
degree kelvin	degree Fahrenheit	$t_f = 1.8\ t_k - 459.67$
degree Fahrenheit	degree kelvin	$t_k = (t_f + 459.67)/1.8$
British thermal unit (Btu)(mean)	joule (J)	1 055.87
calorie (mean)	joule (J)	4.190 02

* The SI unit used is the kelvin. However, wide use is made of the degree Celsius (formally called the centigrade scale) in Engineering.

GLOSSARY

The following definitions are taken mainly from the references noted at the end of this glossary. Other terms included are common to this field and are defined from the author's experience.

ABPA. American Board Products Association.

ADDITIVE. Any special material incorporated in a panel in the course of manufacture to impart special properties. The term includes preservatives and water repellents and fire retardants, but not binders.

ADHESIVE. A substance capable of holding materials together by surface attachment. The term is used to cover the bonding of sheet material and is synonymous with glue. The term *binder* is used for materials concerned in the bonding of fibers, particles, etc., (*see* BINDER).

AIR-CLASSIFIER. A machine that uses airstream to separate particles on the bases of weight and surface area.

AIR FELTING. Forming of a fibrous-felted board from an air suspension of damp or dry fibers on a batch or continuous forming machine (sometimes referred to as the dry or semi-dry process).

ATTRITION MILL. A machine for reducing wood particles to fibers by grinding action.

BALANCED CONSTRUCTION. A construction such that the forces induced by uniformly distributed changes in moisture content will minimize warping. In practice, this means that layers of particles on either side of the center line are of the same density and thickness.

BELT CONVEYOR. A moving belt for conveying materials.

BINDER. A bonding agent used to bond together the particles in a particle board, or a material added during manufacture to improve the bonds in a fiberboard (*see* ADDITIVE and SIZE). Can be organic or inorganic.

BLEED-THROUGH. Glue or components of glue which have seeped through the outer layer of ply of a veneered product and which show as a blemish or discoloration on the surface.

BLENDING. (1) The application of binder and additives to particles. (2) Mixing materials properly.

BLISTER. A bulge on the surface of a panel due to separation of the layers and somewhat resembling in shape a blister on the human skin; its boundaries may be indefinitely outlined and it may have burst or become flattened.

BLOWS. Blisters or bloated spots on the surface caused by the accumulation of gases.

BOND, *n*. The union of materials by adhesives.

BOND, *v*. To unite materials by means of an adhesive.

BUFFER. To regulate the reactivity (pH) of a resin.

BULK DENSITY. The weight of uncompacted particle or fiber material. Usually expressed as weight per cubic foot.

CALIPER. An instrument for measuring diameters or thicknesses. Also used as the term describing board thickness.

CATALYST. *See* HARDENER.

CAUL. A flat metal plate on which particles are conveyed and pressed. Board or plastic sheets (either rigid or flexible) on which particle mats are formed. Formed mats are removed from this type of caul for hot pressing.

CAULLESS SYSTEM. A manufacturing process in which particle mats are formed and conveyed on moving belts or other mat carriers and then pressed directly between the press platens or plates without the use of caul plates.

CHIPS. Small pieces of wood chopped off a block by axlike cuts as in a chipper of the paper industry, or produced by mechanical hogs, hammermills, etc.

CLASSIFIER. Equipment for separating particles according to size or weight.

CLOSING TIME. Time interval to compress a mat from initial to final thickness in the hot press.

COMMINUTED WOOD. Wood reduced to small or fine particles.

COMPOSITION BOARD. Board made of communited wood or other lignocellulosic material, bound with adhesive and made under heat and pressure. Covers both particleboard and fiberboard.

CONDENSATION. A chemical reaction in which two or more molecules combine with the separation of water or some other simple substance. If a polymer is formed, the process is called polycondensation. *See also* POLYMERIZATION.

CORE (MIDDLE). The center layer in a composition board panel.

CURE, *n*. The physicochemical change brought about, either in thermosetting synthetic resins under the influence of heat and catalysts (polymerization), or in drying oils used for oil-treating board (oxidation). It sometimes refers to the setting of natural glues, generally by water or solvent evaporation. The term can be used either to denote the treatment itself, or its effect.

CURE, *v*. To change the physical properties of an adhesive or binder by chemical reaction, which may be condensation, polymerization, or vulcanization; usually accomplished by the action of heat and catalyst, alone or in combination, with or without pressure.

CURING TIME (CURING CYCLE). Specified minimum time for a resin to cure at a given temperature.

CURL. *See* PARTICLE.

DEFIBERIZE. Convert chips, sawdust, shavings or other relatively large particles into fibers by attrition or mechanical grinding.

DELAMINATION. Separation of a panel into layers.

DIELECTRIC SUBSTANCE. An electric insulator or nonconductor.

DILUENT. A substance that reduces the effect or concentration of another substance.

DIMENSIONAL STABILITY. The degree to which a material retains its dimensions when exposed to varying conditions of temperature and humidity.

DISC FLAKER. A flaking machine with knives on a rotating disc.

DOSING. Metering of material.

DRUM FLAKER. A flaking machine with knives mounted on a rotating drum.

DRY OUT. Loss of tack to a point where no amount of pressure will cause adhesion.

DRY SOLIDS. Weight of oven-dry solid material in a mixture or solution.

EFFLUENT (DRYER EFFLUENT). Volatile emissions from dryers in form of vapor, mist, or fumes.

EMBOSSING. A process by which the surface of a panel product is given a relief effect, usually by using a patterned caul in the hot press.

ENGINEERED BOARD. A (structural) board of dependable high strength and stiffness.

EXCELSIOR. *See* PARTICLE.

EXTENDER. A low-cost additive used with an adhesive or binder. The cost of a glue is thus reduced, since the minimum amount of adhesive may be used to achieve the desired qualities, while the necessary amount of glue for proper spreading is still retained. Cereal flours are often used for this purpose.

EXTRACTIVES. Soluble components of wood.

EXTRUDED PARTICLEBOARD. A particleboard made by extrusion through a die. The particles lie with their larger dimensions mainly perpendicular to the direction of extrusion.

FACE. That surface of a panel on which the grade or quality is chiefly judged. When both surfaces are of the same quality, both are described as faces.

FELTING. *See* AIR FELTING and WET FELTING.

FIBER. The slender threadlike elements of wood or similar cellulosic material, which when separated by chemical and/or mechanical means, as in pulping, can be formed into fiberboard. The term also includes fiber bundles.

FIBERBOARD. A sheet of material made from fibers of wood or other lignocellulosic material and manufactured by an interfelting of the fibers followed by compacting between rolls or in a platen press. Fiberboards are classified according to their density and their method of manufacture. Alternative spelling: fibreboard.

FIBER BUNDLES. Threadlike groups of wood or similar fibers, held together by their natural binders, suitable for coarse pulps such as are used for fiberboards. Also referred to as fibers.

FIBERIZE. *See* DEFIBERIZE.

FILLED PARTICLEBOARD. Particleboard having a factory-applied coating of filler on one or both faces to prepare the surface for further finishing by printing, lacquering, or painting.

FILLER. (1) An additive used with adhesive or binder to reduce too free penetration of the adhesive into the wood and to improve other characteristics such as gap-filling. Finely ground minerals, and wood and nut shell flour are commonly used (compare with extender). (2) A material used in finishing a wood-based panel to fill surface pores and irregularities. *See also* FILLED PARTICLEBOARD.

FIRE-RETARDANT TREATMENT. The application to a panel or its components of chemicals that reduce the rate of combustion and/or the spread of flame over the surface.

FLAKE. *See* PARTICLE.

FLAT-PLATEN PRESSED PARTICLEBOARD. A particleboard manufactured by pressing a mass of particles coated with an extraneous binding agent between parallel platens in a hot press with the applied pressure perpendicular to the faces. Compare with EXTRUDED PARTICLEBOARD.

FLOW. Movement of an adhesive in liquid phase during the bonding process, before the adhesive is set.

FLUFF. Convert clumps of fibers into a fluffy mass of free fibers.

FORMALDEHYDE. A reactive organic compound, HCHO.

FORMATION (FORMING)

Fiberboard definition: the making into a mat of felted fibers.

Particleboard definition: the layering of the blended mass of particles to form a mat.

FORMER (FORMING MACHINE). A machine that spreads particles to form a mat.

FREE FORMALDEHYDE. Release or emission of uncombined formaldehyde, usually from UF-bonded boards.

FURNISH. (1) The particles used for making board. (2) The blended particles, binders, and additives ready for the board-forming process.

GLUE. *See* ADHESIVE.

GRADED BOARD. Board produced from fines-surfaced mat without distinct layers, the particle size normally being graduated from coarse material in the center to the finest material on the surfaces.

GRADED DENSITY PARTICLEBOARD. Particleboard in which the density is made to vary through the thickness.

GRANULE. *See* PARTICLE.

HAMMERMILL. A machine for reducing chips or other particles to smaller particles, having pivoted hammers and a perforated screen.

HARDBOARD. Usually a fiberboard of 0.50 to 1.20 g/cm^3 (31.5–75 lbs/ft^3) density.

HARDENER. A substance or mixture of substances added to an adhesive to promote or control the curing reaction by taking part in it. The term is also used to designate a substance added to control the degree of hardness of a cured film.

HEAT TREATING. The process of subjecting a wood-base panel material (usually hardboard) to a special heat treatment after hot pressing to increase some strength properties and water resistance.

HIGH-DENSITY PARTICLEBOARD. Particleboard with a minimum density of 0.80 g/cm^3 (50 lbs/ft^3).

HIGH-FREQUENCY HEATING. Heating by high-frequency electrical energy.

HOT PRESSING. A process of pressing a mat between hot platens or hot rollers of a press to compact and set the structure by simultaneous application of heat and pressure.

HOT STACKING. Stacking boards when they are still hot. Used to improve properties, particularly for completing the cure of phenolic bonded boards.

HYDROCARBONS. Compounds containing only carbon and hydrogen.

HYDROXYL. –OH group.

IMPREG. *See* IMPROVED WOOD.

IMPREGNATE. Treat particles with low-molecular-weight resin (to about 30% resin content). *See also FLAPREG* under IMPROVED WOOD.

IMPROVED WOOD. A general term for wood that has been specially treated in various ways to reduce or retard warping and movement or to increase its strength or other properties.

Compreg. A form of improved wood in which the wood is impregnated with synthetic resins and compressed.

Compressed wood. A form of improved wood in which the wood is compressed.

Flapreg. A form of improved particleboard similar to compreg.

Impreg. A form of improved wood in which the wood is impregnated with synthetic resins.

INDUSTRIAL BOARD. Nonstructural board used for furniture.

INFRARED DETECTORS. Equipment for detecting hot material or glowing embers before flame occurs.

INTERNAL BOND (tensile strength perpendicular to surfaces). A measure of the bonding force between particles.

ISOCYANATE BINDERS. Urethane binders containing –N:C:O group.

LAMINATE, *n*. A product made by bonding together two or more layers of material or materials.

LAMINATE, *v*. To unite layers of material with adhesive.

LINEAR EXPANSION. Change in length with change in moisture content.

LIPPING. A strip of wood or other material applied to the edge of a panel.

LOW-DENSITY PARTICLEBOARD. Particleboard with a density up to 0.60 g/cm^3 (up to 37.5 lbs/ft^3).

MAT. Particles loosely laid together in preparation for being pressed into a board.

MAT-FORMED PARTICLEBOARD. A particleboard in which the coated particles are formed into a mat (having substantially the same length and width as the finished board) before being flat pressed.

MEDIUM-DENSITY FIBERBOARD. Normally a compressed fiberboard of about 0.40 to 0.80 g/cm^3 (25 to 50 lbs/ft^3) density, manufactured for a wide range of uses such as panel material for use in building, furniture, and similar constructions.

MEDIUM-DENSITY PARTICLEBOARD. Particleboard with a density of 0.60 to 0.80 g/cm^3 (37.5–50 lbs/ft^3).

MODIFIER. Any chemically inert ingredient added to an adhesive formulation that changes its properties. *See also* FILLER and EXTENDER.

MOISTURE CONTENT. Amount of moisture in a board normally expressed as percentage of dry weight.

MULTILAYER BOARD. A particleboard made of several layers of like material. Includes boards made from layers with particles of different shapes and sizes. It can also refer to some grades of fiberboard.

MULTI-OPENING PRESS. A press having openings for pressing a number of boards simultaneously.

NONCOMPRESSED FIBERBOARD (INSULATION BOARD). Fiberboards with a density of 0.40 g/cm^3 (25 lbs/ft^3) or less. This group may be divided into semi-rigid and rigid insulation board, primarily on the basis of density but also with reference to physical properties such as bending strength.

NONDESTRUCTIVE TESTING. Measurement of the strength of a board without breaking it.

NOVOLAK. A phenolic-aldehydic resin which, unless a source of methylene groups is added, remains permanently thermoplastic. *See also* THERMOPLASTIC.

NPA. National Particleboard Association.

OD. Oven dry.

ORIENTATION. Alignment of particles.

OVEN-DRY WOOD. Wood dried until all moisture is removed.

OVERLAID BOARDS. Boards having factory-applied overlays which may be resin-treated papers, high- or low-pressure decorative plastic laminates, plastic films, or wood veneers.

OVERLAY. A thin layer of paper, plastic, film, metal foil, or other material bonded to one or both faces of a panel.

PANEL. A sheet of plywood, fiberboard, particleboard, etc., of any type.

PARTICLE. Distinct fraction of wood or other lignocellulosic material produced mechanically. Used as the aggregate for a particleboard. Types of particles include:

Curl. Long, thin flakes manufactured by the cutting action of a knife in such a way that they tend to be in the form of a helix.

Fiber and Fiber Bundles. *See also* FIBER and FIBER BUNDLES.

Flake. Specially generated thin, flat particles, with the grain of the wood essentially parallel to the surface of the flake, prepared with the cutting action of the knife in a plane parallel to the grain but at an angle to the axis of the fiber.

Flax shives. Fine rectangular-shaped particles of lignocellulosic material obtained by longitudinal division of the stalk of the flax plant during scutching of the retted flax.

Granule. A particle in which length, width, and thickness are approximately equal, such as a sawdust particle.

Shaving. A small wood particle of indefinite dimensions developed incidental to certain woodworking operations involving rotary cutterheads usually turning in the direction of the grain; and because of this cutting action, producing a thin chip of varying thickness, usually feathered along at least one edge and thick at another and usually curled.

Sliver. A particle in which width and thickness are approximately equal and whose length is parallel to the grain of the wood and at least four times the thickness.

Splinter. An alternative form for *sliver*.

Strand. A relatively long (with respect to thickness and width) shaving consisting of flat, long bundles of fibers having parallel surfaces.

Wood wool (excelsior). Curly, slender strands of wood made usually by scoring and cutting knives moving along the grain of a block of wood, reducing it to narrow, thin ribbons.

PARTICLEBOARD CORESTOCK. Common name given to a particleboard manufactured for use as a core for overlaying.

PARTICLEBOARD FLOOR UNDERLAYMENT. A grade of particleboard made or sanded to close thickness tolerances for use as a leveling course and to provide a smooth surface under floor-covering materials.

PARTICLE GEOMETRY. Size and shape of particles.

PARTICLES. The aggregate components of a particleboard manufactured by mechanical means from wood, including all small subdivisions of wood. Particle size may be measured by screen mesh that permits passage of the particles and another screen upon which they are retained, or by the measured dimensions, as for flakes.

PENETRATION. The entering of an adhesive into an adherend.

NOTE. This property of a system is measured by the depth of penetration of the adhesive into the adherend.

PLATEN. A part of a press consisting of a rigid metal plate, usually heated, for exerting pressure on the mat. Intermediate platens in a multi-opening press are also called plates.

POLYMER. A compound formed by the reaction of simple molecules (monomers) having functional groups which permit their combination to proceed to high molecular weights under suitable conditions. Polymers may be formed by polymerization (addition polymer) or polycondensation (condensation polymer). When two or more kinds of monomers are involved, the product is called a copolymer.

POLYMERIZATION. A chemical reaction in which the molecules of a monomer are linked together to form large molecules whose molecular weight is a multiple of that of the original substance. When two or more monomers are involved, the process is called copolymerization or heteropolymerization.

POSTCURE. Additional curing after pressing.

POT LIFE. *See* WORKING LIFE.

PRECURE. Curing of a resin before pressing.

PREFINISHED PANELS. Panels having factory-applied decorative or protective coatings.

PRESERVATIVE TREATMENT. The application to a panel or its components of chemicals that inhibit attack by mold, bacteria, fungi and insects.

PRESSING TO CALIPER. Applying pressure upon the mat in the hot press until mat thickness is equal to a desired caliper or thickness reading.

PRESSING TO STOPS. Applying pressure upon a mat in the hot press until mat thickness is equal the thickness of "stops" (metal bars) interposed between the press platens at their edges.

PRESS TIME. Total time of pressing.

PRIME-COATED BOARD. A panel given an initial coating of a sealer or paint to satisfy a particular use requirement or to enhance its use in a certain way.

PROCESS VARIABLE. Factors of the manufacturing process which can be varied to change the properties of the product.

PROFILE (DENSITY PROFILE). Variation of density of a panel from face to core.

RADIO FREQUENCY (RF) HEATING. Heating with high-frequency radio waves. This term is used interchangeably with high frequency (HF).

RESIN. (1) A solid, semisolid or pseudosolid organic material which has an indefinite and often high molecular weight, exhibits a tendency to flow when subjected to stress, usually has a softening or melting range, and usually fractures conchoidally. (2) A bonding material for board in liquid or other form which can be heat-cured to a resin as defined in (1) above. (3) The most widely used is urea formaldehyde (UF), followed by phenol formaldehyde (PF); some use is made of melamine formaldehyde (MF) usually blended with UF.

RESIN CONTENT. The amount of dry solids in the resin usually expressed as a percentage of the dry weight of the wood in a finished board. Any other relation used should be indicated.

RESIN TRANSFER. The transfer of resin from one particle to another in the blending operation.

RING FLAKER. A flaking machine in which flaking is accomplished by a ring of axially aligned knives mounted at the perimeter of a cylindrical casing in which a concentric rotary impeller forces wood chips, shavings, or other material against the knife edges for flaking.

SANDER DUST. Dust particles produced when a panel is surfaced in a sanding machine.

SAWDUST. Wood particles resulting from the cutting and breaking action of saw teeth.

SET. To convert an adhesive into a fixed or hardened state by chemical or physical action, such as condensation, polymerization, oxidation, vulcanization, gelation, hydration, or evaporation of volatile constituents. *See also* CURE.

SHAVING. *See* PARTICLE.

SIMULTANEOUS CLOSING. In a multi-opening press, the closing of the individual openings at the same time.

SINGLE-OPENING PRESS. A press that has only one opening between platens. It presses one board at a time.

SIZE. Alum, wax, petrolatum, asphalt, resin or other additive introduced to the stock for fiberboard or to the agglomerate for a particleboard prior to forming, primarily to improve water resistance. *See also* ADDITIVE and BINDER.

SLIVER. *See* PARTICLE.

SPRINGBACK. Tendency of a pressed particle mat to return to its original uncompressed state.

STORAGE LIFE. The period of time during which a packaged adhesive can be stored under specified temperature conditions and remain suitable for use. Sometimes called *Shelf Life*. *See also* WORKING LIFE.

STRAND. *See* PARTICLE.

STRIATED. A term used to describe a panel with a face which has been milled or embossed with a pattern or parallel grooves.

STRUCTURAL BOARD. A board of high strength and stiffness suitable for use in load-bearing structures.

SUBSTRATE. A material upon the surface of which an adhesive-containing substance is spread for any purpose, such as bonding or coating. A broader term than adherend.

SULFITE-BONDED BOARD. Particleboard made with sulfite spent liquor as a binder. Sulfite spent liquor contains the lignin removed from wood in the sulfite pulping process.

TACK. Stickiness of a resin in the uncured state. A high-tack resin gives useful integrity to the uncured mat in caulless systems.

TANNIN. Polyphenol wood or bark extractive that can be reacted with formaldehyde to form a binder.

TEMPERATURE, CURING. The temperature to which an adhesive or an assembly is subjected to cure the adhesive. *See also* TEMPERATURE, SETTING.

NOTE. The temperature attained by the adhesive in the process of curing it (adhesive curing temperature) may differ from the temperature of the atmosphere surrounding the assembly (assembly curing temperature).

TEMPERATURE, SETTING. The temperature to which an adhesive or an assembly is subjected to set the adhesive. *See also* TEMPERATURE, CURING.

NOTE. The temperature attained by the adhesive in the process of setting it (adhesive setting temperature) may differ from the temperature of the atmosphere surrounding the assembly (assembly setting temperature).

TEMPERED HARDBOARD. Hardboard which has been specially treated in manufacture to improve its physical properties considerably, usually by oil or heat treating with or without first applying a drying oil.

TEMPERING. The manufacturing process of adding to a fiber or particle panel material a siccative material such as drying oil blends of oxidizing resins which are stabilized by baking or other heating after introduction.

THERMAL CONDUCTIVITY. The quantity of heat which flows in unit time across unit area of a substance of unit thickness when the temperature of the faces differs by one degree.

THERMOPLASTIC, *n*. A material which will repeatedly soften when heated and harden when cooled.

THERMOPLASTIC, *adj*. Capable of being repeatedly softened by heat and hardened by cooling.

THERMOPLASTIC (RESIN). A synthetic resin or plastic that softens on heating.

THERMOSET, *n*. A material which will undergo or has undergone a chemical reaction by the action of heat, catalysts, ultraviolet light, etc., leading to a relatively infusible state.

THERMOSET, *adj*. Pertaining to the state of a resin in which it is relatively infusible.

THERMOSETTING. Having the property of undergoing a chemical reaction by the action of heat, catalysts, ultraviolet light, etc., leading to a relatively infusible state.

THERMOSETTING RESIN. A synthetic resin which, once hardened, does not soften or melt when heated.

THREE-LAYER PARTICLEBOARD. A particleboard in which the core particles and face particles are of different qualities.

VISCOSITY. The internal frictional resistance of a liquid.

WET FELTING. Forming of a fibrous-felted board mat from a water suspension of fibers and fiber bundles by means of a deckle box, fourdrinier, or cylinder board machine.

WOOD FLOUR. Very fine wood particles. Wood reduced by a ball or other mill until it resembles wheat or other flour in appearance.

WOOD WOOL. *See* PARTICLE.

WORKING LIFE. The period of time during which an adhesive, after mixing with catalyst, solvent, or other compounding ingredients, remains suitable for use. *See also* STORAGE LIFE.

REFERENCES

American Society for Testing Materials. Standard Definitions of Terms Relating to Wood-Base Fiber and Particle Panel Materials. ASTM Designation D 1554.

FAO. 1958. *Fibreboard and Particle Board*. Report of an International Consultation on Insulation Board, Hardboard and Particle Board. Geneva, Switzerland, January-February 1957.

———. 1966. Plywood and Other Wood-Based Panels. Report of an International Consultation on Plywood and other Wood-Based Panel Products. Rome, Italy, July 8-19, 1963.

Skeist, I. 1962. Handbook of Adhesives. New York: Reinhold Publishing Co.

U.S. Department of Commerce. 1966. Mat-Formed Wood Particleboard. Commercial Standard CS236-66. Washington, DC: Superintendent of Documents.

INDEX